Probability and its Applications

Series Editors

Thomas Liggett
Charles Newman
Loren Pitt

Bert Fristedt   Lawrence Gray

A Modern Approach
to Probability Theory

Birkhäuser
Boston • Basel • Berlin

Bert Fristedt
Lawrence Gray
School of Mathematics
University of Minnesota
Minneapolis, MN 55455
U.S.A.

Library of Congress Cataloging-in-Publication Data

Fristedt, Bert, 1937-
A modern approach to probability theory / Bert Fristedt,
Lawrence Gray.
p. cm. -- (Probability and its applications)
Includes bibliographical references (p.   -) and index.
ISBN 0-8176-3807-5 (hardcover : alk. paper)   ISBN 3-7643-3807-5
(hardcover: alk. paper)
1. Probabilities I. Gray, Lawrence F.   II. Title. III. Series.
QA273.F92 1996   96-5687
519.2--dc20   CIP

Printed on acid-free paper

© 1997 Birkhäuser Boston   *Birkhäuser* ®

ISBN 0-8176-3807-5
ISBN 3-7643-3807-5

Typeset by the authors in $\mathcal{AMS}$-LATEX.
Printed and bound by Quinn-Woodbine, Woodbine, NJ.
Printed in the U.S.A.
9 8 7 6 5 4 3 2 1

Contents

Part 2. Independence and Sums

Part 3. Convergence in Distribution

Part 5. Random Sequences

Part 7. Appendices

List of Tables

Preface

Overview

This book is intended as a textbook in probability for graduate students in mathematics and related areas such as statistics, economics, physics, and operations research. Probability theory is a 'difficult' but productive marriage of mathematical abstraction and everyday intuition, and we have attempted to exhibit this fact. Thus we may appear at times to be obsessively careful in our presentation of the material, but our experience has shown that many students find themselves quite handicapped because they have never properly come to grips with the subtleties of the definitions and mathematical structures that form the foundation of the field. Also, students may find many of the examples and problems to be computationally challenging, but it is our belief that one of the fascinating aspects of probability theory is its ability to say something concrete about the world around us, and we have done our best to coax the student into doing explicit calculations, often in the context of apparently elementary models.

The practical applications of probability theory to various scientific fields are far-reaching, and a specialized treatment would be required to do justice to the interrelations between probability and any one of these areas. However, to give the reader a taste of the possibilities, we have included some examples, particularly from the field of statistics, such as order statistics, Dirichlet distributions, and minimum variance unbiased estimation. We have also given several examples of random geometrical structures, involving exact computations where possible. And of course, a variety of models such as coin-tossing and urns appear repeatedly.

If little or no material is omitted, the book is suitable for a 3-semester sequence. We feel that an incoming graduate student in mathematics could begin her or his graduate studies with this book and by the middle of the second year be ready to do the reading necessary to begin research in probability theory. Later in this preface we give some suggestions for 2-semester sequences.

What is 'modern' about our approach? Three main features come to mind. First, there is our philosophy that random variables are functions that may take values in a variety of spaces, not just $\mathbb{R}$ or $\mathbb{R}^d$. Second, we have endeavored to employ the most up-to-date methodology in constructions and proofs. Third, we have included material that is of relatively recent vintage. Here are some of the consequences. (i) Random sequences, random functions, random sets, and random distributions are all presented within the same framework as $\mathbb{R}$-valued random variables. (ii) We accommodate the value ∞ wherever possible, such as in our treatment of nonnegative random variables. (iii) Except in our treatment of continuous-time stochastic processes, abstract topology plays little or no role in the book. (iv) We minimize the technicalities associated with continuous-time stochastic processes by using constructions based on almost sure convergence and convergence in probability. (v) Conditional expectations are defined as means of random distributions, and as such, naturally inherit properties of (unconditional) expectations. (vi) Relatively recent proofs are given of several standard results, including the Strong Law of Large Numbers, the Renewal Theorem, the De Finetti Theorem, and the representation theorem for Lévy processes. (vii) Poisson point processes, interacting particle systems, coupling methods, optimal strategies in gambling, and the martingale problem are all examples of important newer topics not typically presented in a general introductory graduate textbook.

Structure of book

The prerequisite knowledge needed is an understanding of elementary linear algebra and advanced calculus. Measure theory is introduced as needed mostly within the first nine chapters. One of the pleasant aspects of approaching measure theory in this manner is that probability nicely motivates many of its abstract ideas in a way that is not possible in a standard real analysis course.

At the beginning of the book, the pace is somewhat leisurely as we introduce probability spaces, random variables, and some important families of distributions. As the book progresses our expectations of the student gradually increase, as we endeavor to help her or him prepare to read various specialized books and papers.

There are more than 1200 problems. The asterisks (*) next to about 300 of them indicate that some sort of 'answers' are available at the Web site

http://www.birkhauser.com/books/ISBN/0-8176-3807-5

Such an 'answer' may be a complete solution, a solution to part of the problem, just a numerical answer, or merely a hint.

The problems are integrated with the text. Thus a problem coming immediately after a theorem and its proof is likely to be related to that theorem, but often in a way that forces the student to review earlier material. The asterisk is attached to a wide variety of problems: easy and hard, theoretical and calcula-

tional, specific and open-ended. Readers should choose to do all sorts, including
many for which an 'answer' is not given on the Web. One of the many skills that
the problems are designed to foster is that of learning to self-check solutions.

Despite its advantages, there is one significant drawback to interweaving
problems with the textual material itself. This format tends to make one feel
that one cannot continue until one has done all relevant problems, especially
the problems that request a proof of the preceding proposition or theorem. We
believe this approach to be self-defeating. Often by moving on, one gains a
perspective that enables one to return successfully to a difficult problem.

We urge the reader to become familiar with the appendices early so as to
be aware of what can be found there as needed.

The latter section or sections of some chapters are listed as optional. This
practice does not reflect our view of how important certain material is, but it
does indicate our view that a coherent well-structured body of knowledge is
contained in the nonoptional sections of any given chapter.

Organization of a sequence of courses

It is not required that chapters be studied in the order presented. In particular,
some may want to study Part 4 before Part 3. The dependency relations among
the chapters are shown in the chart below. There is one minor exception: the
proof of Lemma 20 in Chapter 21 relies on material from Section 18.7.

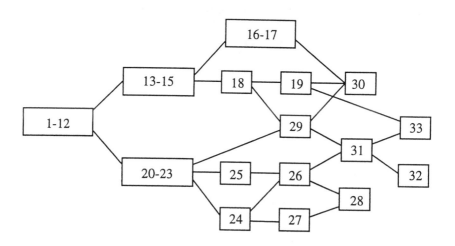

Those who have had a course in measure-theoretic analysis may find that only
a light reading of some chapters is necessary, primarily to become familiar with
terminology used in probability theory. These chapters are

- Chapter 2: Random Variables
- Chapter 4: Expectations: Theory
- Chapter 8: Integration Theory
- Chapter 20: Spaces of Random Variables

Also, light reading may be all that is necessary for Sections 2 and 3 of Chapter 7, although most who have learned an existence proof for measure have probably learned one that is different from the one in this book. Finally, Sections 2 and 3 of Chapter 9 are review for those who have studied product measure and the Fubini Theorem, and large portions of Chapter 13 may be review for those who have studied Fourier analysis.

In some cases, the proof of a result is best skipped on a first reading, even though an awareness of the result itself is important. Here is a list of proofs which might fall into this category:

- the proof of Theorem 14 of Chapter 5, which characterizes the functions that are probability generating functions;
- the proofs of the lemmas and theorems in Sections 2 and 3 of Chapter 7, which are used for proving the extension theorem for measure;
- the proof of Theorem 13 of Chapter 13, which identifies the class of characteristic functions of integer-valued random variables;
- the proof of Lemma 23 of Chapter 14, which is the cornerstone for proving that the limiting type of a normalized sequence is unique;
- the proofs of the lemma and theorem in Section 10 of Chapter 14, the latter of which asserts that certain positive definite functions are characteristic functions of real-valued random variables;
- the proof of Lemma 21 of Chapter 21, which says that the space of probability measures on a Borel space is a Borel space.

The optional sections at the end of some chapters are indicated by daggers (†) or double daggers (‡). The symbol ‡ is a warning that this optional section has as a prerequisite either a chapter that is not indicated by the dependency chart or another optional section. Here is a description of the additional prerequisite material needed for such optional sections:

- Section 10.4 requires Section 9.8
- Section 17.4 requires Section 17.3
- Section 19.7 requires Chapter 16
- Section 21.6 requires Chapter 13
- Section 27.5 requires Sections 10.4 and 27.4
- Section 28.7 requires Chapter 19
- Section 29.6 requires Section 29.5
- Section 29.7 requires Section 29.5
- Section 30.6 requires Chapter 25
- Section 30.7 requires Chapter 24
- Section 31.7 requires Section 30.6

Occasionally some optional section is relevant for a later problem in a nonoptional section. When this happens, the presentation of the problem is such as to provide a warning. For instance, when the zeta distribution, treated in an optional section, is relevant for a later problem, the term 'zeta distribution' appears in that problem. Similarly, the 'Wiener sausage' problem in Chapter 28 makes explicit reference to the chapter in which the standard Wiener process is introduced, since that chapter is not a prerequisite for Chapter 28.

Here are the chapter numbers for some possible 2-semester sequences:

- 1-15, 18-24
- 1-15, 20-25, 26 or 27
- 1-15, 18, 20-24, 29
- 1-17, 20-24

1 -12, 20-28

Unless the students entering the course are already comfortable with measure-theoretic analysis, one will probably have to omit some optional sections in order to complete any of these programs in two semesters. By omitting all optional sections of chosen chapters, one may be able to treat more chapters than are listed in the above schedules. We recommend against such a plan. We also do not recommend covering only a portion of the nonoptional sections of a chapter solely for the purpose of being able to touch on other chapters. We think that by moving rapidly but lightly one deprives the students of the in-depth training they will need in order to handle further reading and problem-solving on their own.

Acknowledgements

Preliminary versions of this text have been used for several years at the University of Minnesota and at some other schools. We thank many students and faculty for feedback. In particular, when Bisser Roussanov was a student at the University of Minnesota, he found many errors of the hard-to-spot variety in an early version.

In many cases, we have not identified a particular source in the literature for a theorem or proof, because in our judgment, the result may be viewed as somewhat widely known. However, Appendix G does cite sources for some proofs and theorems that we feel are not widely known. Some of these sources are colleagues at the University of Minnesota who have made conversational contributions to this book beyond the specific ones mentioned in Appendix G, and for these mathematical communications we are grateful. Finally we thank Chuck Newman, Loren Pitt, and Tom Liggett for looking at preliminary versions of the text and making several useful comments.

PART 1
Probability Spaces, Random Variables, and Expectations

A probability space is the mathematical abstraction of an experiment involving some randomness, a randomness that may only be in the eye of the observer as a consequence of a lack of full knowledge. The basic ingredients that constitute a probability space are discussed in Chapter 1.

'Random variables' represent observations or measurements based on the outcomes of experiments. Chapter 2 contains basic definitions and examples concerning random variables, while Chapter 3 introduces a fundamental concept that is used to both categorize and analyze random variables, namely the 'distribution' of a random variable.

The 'expectation' of a real-valued random variable denotes its *a priori* average value. The theory of expectations is developed in Chapter 4. An important theorem in probability theory, the Law of Large Numbers, shows that the expectation also represents the long term average of what is observed or measured when an experiment is repeated many times. Chapter 5 contains the statement and proof of the Law of Large Numbers, along with definitions and further results about expectations.

The theoretical foundations of probability theory are to be found in measure and integration theory. Quite a bit of measure theory is already contained in Chapters 1 through 4, but a more thorough treatment, including a detailed solution of certain significant technical problems involving existence and uniqueness, is reserved for Chapters 6 and 7. Integration theory, which appears in somewhat specialized and slightly disguised form in Chapter 4, is given a general treatment in Chapter 8.

CHAPTER 1
Probability Spaces

In modern probability theory, a fundamental building block is the 'probability space', a concept that is to be precisely defined in the latter portion of this chapter. We begin the chapter informally by giving some concrete examples of probability spaces. In particular, we model the experiment of tossing a fair coin infinitely many times.

1.1. Introductory examples

In each example given below, we will define three objects: (i) a set Ω, representing the possible outcomes of an experiment; (ii) a collection $\mathcal{F}$ of subsets of Ω; and (iii) a function $P\colon \mathcal{F} \to [0,1]$.

Example 1. To model the experiment of rolling an ordinary die one time, we let Ω denote the set of outcomes of this experiment. Ignoring such aspects of the experiment as the duration of time it takes for the die to stop rolling or the distance it travels, we focus on the six outcomes that correspond to the number of dots that show on the top face when the die comes to rest. So, for our model we set

$$\Omega = \{1, 2, 3, 4, 5, 6\}\,.$$

We have idealized somewhat by not including as possible outcomes such unusual occurrences as the die coming to rest on an edge.

In some circumstances the word 'outcome' might be used with a different meaning. For example, a person might say that he or she is hoping that the outcome is an even number. In so doing, this person is using 'outcome' to denote a subset of Ω rather than a member of Ω. With an appropriate description one might refer to any subset of Ω. We let $\mathcal{F}$ denote the collection of all subsets of Ω. Three terms are often used in lieu of the usual set-theoretic terminology: Ω is called the 'sample space', the members of Ω are called 'sample points', and the members of $\mathcal{F}$ are called 'events'. If A is an event, we say that A 'occurs' if the result of the experiment is a sample point that is a member of A. Thus, in the experiment under discussion, if $A = \{2, 4, 6\}$, then saying that "A occurs" is the same as saying that an "even number is rolled".

The probability of an event A, denoted by $P(A)$, indicates the likelihood that the event occurs. It is usual in the case of a balanced die to define

$$P(A) = \frac{\sharp A}{6}, \quad A \in \mathcal{F},$$

where $\sharp A$ indicates the number of members of A. In particular, $P(\Omega) = 1$ and $P(\emptyset) = 0$, which is consistent with the fact that in our idealized experiment, the outcome of the experiment is always a member of Ω and never a member of $\emptyset$. Another way of saying this is that Ω is an event which is certain to occur, while $\emptyset$ is an event which is certain to not occur. The function P is called a 'probability measure' and the number $P(A)$ is called the 'probability of A'. The particular P we have chosen reflects the geometrical symmetry of a die, but it should be emphasized that what makes a particular choice of P correct is not abstract reasoning but the degree to which P reflects the nature of the experiment. In fact, one of the main tasks for statisticians is to devise ways of choosing approximately correct probability measures P for various experiments. Probabilists, on the other hand, usually take P as given and obtain various consequences.

According to the principles of classical mechanics, one might argue that whenever a die is rolled, there is one sample point that has probability 1; that sample point could be determined via a complicated calculation involving the laws of physics and an exact knowledge of the initial conditions and the motion of the hand rolling the die. Typically, this knowledge and the required calculational ability is missing, and it is for this reason that a probabilistic model is appropriate.

Example 2. We wish to model the experiment of drawing an object at random from a container that contains several objects. We assume that each of the objects is equally likely to be drawn. In probability theory, it is customary to call such a container an *urn*. Suppose that the urn contains 3 green balls and 5 blue balls. If we have no reason to distinguish the balls any further than by color, we might choose

$$\Omega = \{\text{green, blue}\},$$

and let $\mathcal{F}$ be the collection of all subsets of Ω. There are only 4 events in $\mathcal{F}$: Ω, $\emptyset$, $\{\text{green}\}$, and $\{\text{blue}\}$. We assign these events probabilities 1, 0, 3/8 and 5/8 respectively. If we had wanted to consider like-colored balls to be different from one another, we could have instead chosen a sample space with 8 sample points, and then followed the pattern of Example 1 by defining the probability of an event A in terms of $\sharp A$.

Example 3. Consider the experiment consisting of n successive flips of a coin. Then

$$\Omega = \{(\omega_1, \ldots, \omega_n) \colon \omega_i \in \{\text{heads, tails}\}, i = 1, \ldots, n\}.$$

For convenience we may want codes for 'heads' and 'tails'—say, 1 for 'heads' and 0 for 'tails'. Thus,

$$\Omega = \{(\omega_1, \ldots, \omega_n) \colon \omega_i \in \{1, 0\}, i = 1, \ldots, n\}.$$

There are 2^n sample points. As in Example 1 we let $\mathcal{F}$ be the collection of all subsets of Ω. Set

$$P(A) = \frac{\sharp A}{2^n}, \quad A \in \mathcal{F}.$$

Suppose $0 \le k \le n$. In order that $\sum_{i=1}^{n} \omega_i = k$, it is necessary and sufficient that exactly k of the ω_i's equal 1. There are

$$\binom{n}{k} \overset{\text{def}}{=} \frac{n!}{k!(n-k)!}$$

sample points that have this property. Hence,

$$P\big(\{(\omega_1, \ldots, \omega_n) \colon \sum_{i=1}^{n} \omega_i = k\}\big) = \binom{n}{k} 2^{-n}, \quad 0 \le k \le n.$$

In coin-flipping terminology this is the probability of obtaining exactly k heads in n flips of a fair coin.

Problem 1. In Example 3 how many members does $\mathcal{F}$ have?

* **Problem 2.** For Example 3 calculate the probability of obtaining an even number of heads, as a function of n.

Problem 3. Example 3 really consists of infinitely many examples, one for each n. For each $n \ge 2$ it is meaningful to speak of the probability that the first two flips are both tails. Show that this probability is independent of n. Generalize.

The next example encompasses the infinitely many examples in Example 3 within one big example.

Example 4. Consider the experiment consisting of an infinite sequence of coin flips. Then

$$\Omega = \{(\omega_1, \omega_2, \ldots) \colon \omega_i \in \{1, 0\}, i = 1, 2, \ldots\}.$$

Since Ω has infinitely many members, the procedure used in Example 1 and Example 3 for defining P cannot be used here. However, we do know what we want $P(A)$ to equal for certain subsets of Ω, in order to have consistency with Example 3. For any set A_k of k-dimensional vectors of 0's and 1's, we want

$$(1.1) \qquad P(\{(\omega_1, \omega_2, \dots) \colon (\omega_1, \dots, \omega_k) \in A_k\}) = \frac{\sharp A_k}{2^k}.$$

We must wait for precise definitions before completing the description of P and considering $\mathcal{F}$.

* **Problem 4.** In Example 4 calculate, for $j = 1, 2, \dots$, the probability that the first head occurs on flip number j.

1.2. Ingredients of probability spaces

The first object needed for the definition of 'probability space' is a set, often denoted by Ω. The second object is found in the following definition.

Definition 1. A σ-*field* of subsets of a set Ω is a collection $\mathcal{F}$ of subsets of Ω that has $\emptyset$ as a member and is closed under complementation and countable unions. The sets that are members of the σ-field are said to be *measurable with respect to $\mathcal{F}$*, or $\mathcal{F}$-*measurable*, or just *measurable* if the σ-field is understood from the context.

We will only consider probability models in which the collection of events $\mathcal{F}$ is a σ-field. Here are some reasons for that restriction. When we construct a probability space, we would ideally want the collection $\mathcal{F}$ of events to contain all the subsets of Ω that might possibly be of interest to us. It seems reasonable that a theory that allows us to speak of an event occurring should also allow us to speak of it not occurring, so $\mathcal{F}$ should be closed under complementation. Also, we would like to speak of 'at least one' of several things occurring, provided that we can speak of them individually occurring, so $\mathcal{F}$ should be closed under unions. It is not so clear whether 'several' should mean 'finitely many', 'countably many', or even more. By requiring that $\mathcal{F}$ be a σ-field, we have chosen to have 'several' mean 'countably many'. It is to be understood that 'countably many' encompasses 'finitely many', and that 'countable' encompasses 'finite'. When we want to exclude finiteness, we will use a phrase like 'countably infinite'.

Problem 5. Show that if $\mathcal{F}$ is a σ-field of subsets of Ω, then Ω is a member of $\mathcal{F}$ and $\mathcal{F}$ is closed under countable intersections. Also show that $\mathcal{F}$ is closed under set differences: if A and B are in $\mathcal{F}$, then so is $A \setminus B$.

A sequence or collection $\mathcal{S}$ of sets is said to be *pairwise disjoint* if $A \cap B = \emptyset$ whenever A and B are distinct terms or members of $\mathcal{S}$. This terminology is used in the following definition, which introduces the third and final object needed in the definition of 'probability space'.

Definition 2. A *probability measure* P on a σ-field $\mathcal{F}$ of subsets of a set Ω is a function from $\mathcal{F}$ to the unit interval $[0, 1]$ such that $P(\Omega) = 1$ and

$$P\left(\bigcup_{m=1}^{\infty} A_m\right) = \sum_{m=1}^{\infty} P(A_m)$$

for each pairwise disjoint sequence $(A_m : m = 1, 2, 3, \ldots)$ of members of $\mathcal{F}$. Because P satisfies this summation condition it is said to be *countably additive*. For $A \in \mathcal{F}$, $P(A)$ is the *probability of A*.

Problem 6. Show that countable additivity encompasses *finite additivity:*

$$P\left(\bigcup_{m=1}^{n} A_m\right) = \sum_{m=1}^{n} P(A_m)$$

for each pairwise disjoint finite sequence $(A_1, \ldots, A_n)$ of members of $\mathcal{F}$.

Definition 3. A *probability space* is a triple $(\Omega, \mathcal{F}, P)$, where Ω is a set, $\mathcal{F}$ is a σ-field of subsets of Ω, and P is a probability measure on $\mathcal{F}$. The set Ω is called the *sample space* and its members are called *sample points*. The members of $\mathcal{F}$ are called *events*.

Given a probability space $(\Omega, \mathcal{F}, P)$ we sometimes say that P is a probability measure on Ω or on $(\Omega, \mathcal{F})$, rather than on $\mathcal{F}$.

Proposition 4. Let $A \subseteq B$ and A_n, $n = 1, 2, \ldots$, be events in a probability space $(\Omega, \mathcal{F}, P)$. Then $P(\emptyset) = 0$, $P(A) \leq P(B)$, $P(B \setminus A) = P(B) - P(A)$, $P(A^c) = 1 - P(A)$, and

$$P\left(\bigcup_{n=1}^{\infty} A_n\right) \leq \sum_{n=1}^{\infty} P(A_n).$$

Problem 7. Prove the preceding proposition.

Problem 8. Verify that probability spaces have been defined in Example 1, Example 2, and Example 3.

In most probability spaces, if ω is a sample point, the set $\{\omega\}$ is an event. When it is, we will often not distinguish between the sample point ω and the event $\{\omega\}$. Thus, for example, we talk of the probability of a sample point when we mean the probability of the event consisting of that one sample point.

For models with countably many sample points it is usual for $\mathcal{F}$ to be the σ-field of all subsets of Ω; indeed, this must be the case if each sample point constitutes an event.

Problem 9. Let A be an event which contains countably many sample points. Assume that each of the sample points in A is an event. Find a formula for the probability of A in terms of the probabilities of the sample points in A.

* **Problem 10.** Use the product of the sample space of Example 1 with itself to construct a probability space for the experiment of rolling two ordinary dice, one of them being red and the other green. Your model should contain 36 equally likely sample points.

* **Problem 11.** For the probability space of Example 4 show that each sample point is an event that has probability 0. What implicit role does this last fact play in the command: Flip a fair coin repeatedly until the first tails occurs?

Example 5. Suppose two identical dice are rolled. One way of constructing a sample space for this experiment is to use 21 sample points (ω_1, ω_2), $1 \leq \omega_1 \leq \omega_2 \leq 6$, with the smaller of the two numbers on the dice equaling ω_1 and the larger equaling ω_2. Consistency with the model constructed in Problem 10 requires that each of the 6 sample points of the form (ω_1, ω_1) have probability $1/36$, and that each of the remaining 15 sample points have probability $1/18$.

There is another way of making the natural connection between the experiment of Problem 10 and the example under discussion. For the experiment of throwing two identical dice, use the same 36 sample points that were used in Problem 10 but do not use the σ-field of all subsets of Ω. Rather, use the smallest σ-field containing all sets of the forms $\{(\omega_1, \omega_1)\}$ and $\{(\omega_1, \omega_2), (\omega_2, \omega_1)\}$.

1.3. σ-fields

It is often the case that in a probabilistic model, the events of interest form a collection $\mathcal{E}$ of subsets of Ω that is not a σ-field. Since the theory requires that we work with σ-fields, we replace $\mathcal{E}$ in such cases by a larger collection of subsets that does form a σ-field.

Definition and Proposition 5. Let $\mathcal{E}$ be a collection of subsets of a set Ω. Then there exists a unique σ-field $\mathcal{F} \supseteq \mathcal{E}$ such that if $\mathcal{G} \supseteq \mathcal{E}$ and $\mathcal{G}$ is a σ-field, then $\mathcal{G} \supseteq \mathcal{F}$; $\mathcal{F}$ is the smallest σ-field containing $\mathcal{E}$. It is called the σ-field *generated by* $\mathcal{E}$; we write $\mathcal{F} = \sigma(\mathcal{E})$.

PROOF. The uniqueness is clear. To prove the existence define

$$\mathcal{F} = \bigcap_{\substack{\mathcal{G} \supseteq \mathcal{E}: \\ \mathcal{G} \text{ is a } \sigma\text{-field}}} \mathcal{G}.$$

We finish the proof by showing that $\mathcal{F}$ is a σ-field. Clearly, $\emptyset \in \mathcal{F}$. Fix $A \in \mathcal{F}$ and let $\mathcal{G}$ be any σ-field that contains $\mathcal{E}$. Since $A \in \mathcal{G}$ and $\mathcal{G}$ is a σ-field, A^c is a member of $\mathcal{G}$. Hence $A^c \in \mathcal{F}$. Next let $(A_1, A_2, \ldots)$ be any sequence of members of $\mathcal{F}$, and again let $\mathcal{G}$ be any σ-field that contains $\mathcal{E}$. Then each A_n is a member of $\mathcal{G}$, and since $\mathcal{G}$ is a σ-field, $\bigcup_{n=1}^{\infty} A_n$ is a member of $\mathcal{G}$. It follows that $\bigcup_{n=1}^{\infty} A_n$ is a member of $\mathcal{F}$. We have shown that $\mathcal{F}$ is closed under complementation and countable unions, so $\mathcal{F}$ is a σ-field. $\square$

In the context of Example 4, let $\mathcal{E}$ be the collection of sets of the form

$$(1.2) \qquad \{(\omega_1, \omega_2, \dots) : (\omega_1, \dots, \omega_k) = (\varepsilon_1, \dots, \varepsilon_k)\}$$

for positive integers k and vectors $(\varepsilon_1, \dots, \varepsilon_k)$ of 0's and 1's. The formula in (1.1) defines probabilities $P(A)$ for all sets $A \in \mathcal{E}$. By using countable additivity in various ways, the reader should be able to determine probabilities for many events in $\sigma(\mathcal{E})$. It will be shown in Chapter 7 that probabilities can be determined in a unique way for all of the events in $\sigma(\mathcal{E})$.

* **Problem 12.** In each of the following, a subset of the probability space in Example 4 is described. In each case, show that this subset is an event (that is, a member of $\sigma(\mathcal{E})$), and calculate its probability.

    (i) The first head comes immediately after an even number of tails.

    (ii) At six flips, but no earlier, the number of heads equals the number of tails.

    (iii) The sequence 'heads, tails, heads' occurs before the sequence 'heads, heads, heads'.

    (iv) The sequence 'heads, heads, tails' occurs before the sequence 'tails, heads, tails'.

More generally, consider an arbitrary sample space Ω. One may have a family $\mathcal{E}$ of subsets that are particularly interesting or easy to describe. Then, when trying to build a probability space $(\Omega, \mathcal{F}, P)$, it is natural to take $\mathcal{F} = \sigma(\mathcal{E})$ and to begin constructing P by specifying its values on $\mathcal{E}$. Two questions immediately arise. Can the domain of P be extended to $\mathcal{F}$ so that it is a probability measure? If so is such an extension unique? Both questions will be treated in Chapter 7. Here is a counterexample that sheds some light on the uniqueness question.

(Counter)example 6. Let $\Omega = \{\alpha, \beta, \gamma, \delta\}$, let $\mathcal{F}$ denote the σ-field of all subsets of Ω, and let $\mathcal{E} = \{\{\alpha, \beta\}, \{\beta, \gamma\}\}$. Then $\sigma(\mathcal{E}) = \mathcal{F}$. We define two distinct probability measures P and Q on $\mathcal{F}$ whose restrictions to $\mathcal{E}$ are identical: define P by $P(\{\beta\}) = P(\{\delta\}) = 1/2$ and Q by $Q(\{\alpha\}) = Q(\{\gamma\}) = 1/2$.

1.4. Borel σ-fields

When constructing a probability space starting with a set Ω, it is often the case that Ω is a topological space (see Appendix C). When this is the case it is implicit, unless otherwise stated, that the σ-field of interest is the one generated by the collection of open sets. This σ-field is called the *Borel σ-field.* The members of the Borel σ-field are called *Borel sets,* or *Borel subsets* if the relation to the entire topological space is being emphasized. The symbols $\mathcal{A}$, $\mathcal{B}$, and $\mathcal{C}$ will be reserved for Borel σ-fields, but this does not mean that a σ-field carrying, say, the name $\mathcal{F}$ is not a Borel σ-field.

In any topological space, all closed sets, being the complements of open sets, are Borel sets. Typically there are Borel sets that are neither open nor closed; the intersection of countably many open sets is Borel but may be neither open nor closed.

The real line, which will be denoted by $\mathbb{R}$ throughout the book, is a topological space. In $\mathbb{R}$, the interval $[0, 1)$, for instance, is a Borel set because

$$[0,1) = \bigcap_{n=1}^{\infty} \left(-\tfrac{1}{n}, 1\right),$$

a countable intersection of open sets. Indeed all intervals are Borel sets. Singletons, being closed, are Borel. Thus countable sets, such as the set of rational numbers, are Borel.

Problem 13. Show that any of the following collections of subsets of $\mathbb{R}$ generate the Borel σ-field of subsets of $\mathbb{R}$: the collection of all closed sets, the collection of all open intervals, the collection of all bounded intervals open on the left and closed on the right, the collection of all open intervals having left endpoint $-\infty$, the collection of all closed intervals, the collection of all open intervals having rational endpoints, the collection of all intervals of the form $(-\infty, r]$ where r is rational, and the collection of intervals of the form $(j/2^n, (j+1)/2^n]$, where j is an integer and n is a nonnegative integer. Add a few more similar collections to the list.

Here is a list of several more topological spaces that are of interest in probability theory: (i) the set of nonnegative real numbers, denoted by $\mathbb{R}^+$; (ii) d-dimensional Euclidean space, denoted by $\mathbb{R}^d$; (iii) the infinite-dimensional space $\mathbb{R}^\infty$ (see Example 2 of Appendix C); (iv) the extended real line $\overline{\mathbb{R}} = \mathbb{R} \cup \{-\infty\} \cup \{\infty\}$; (v) the nonnegative members of $\overline{\mathbb{R}}$, denoted by $\overline{\mathbb{R}}^+$. The topology in (i) is the relative topology induced by the usual topology on $\mathbb{R}$. The topologies in (ii) and (iii) are product topologies. The spaces in (iv) and (v) are compactifications of $\mathbb{R}$ and $\mathbb{R}^+$, respectively, and they receive their topologies accordingly, as described in Example 1 of Appendix C.

* **Problem 14.** Show that the members of the Borel σ-field of $\mathbb{R}^+$ with the relative topology are those Borel subsets of $\mathbb{R}$ that contain only nonnegative numbers.

Problem 15. Show that the Borel σ-field of $[0, 1)$ with the relative topology induced by the usual topology on $\mathbb{R}$ is generated by the family of those subsets of $[0, 1)$ that are open in $\mathbb{R}$.

* **Problem 16.** Show that the collection of d-dimensional 'open boxes' generates the Borel σ-field in $\mathbb{R}^d$. (An *open box* is a set of the form $I_1 \times \cdots \times I_d$, where each I_j is an open interval in $\mathbb{R}$.)

Problem 17. Show that a subset of $\overline{\mathbb{R}}$ is Borel if and only if it is the union of a Borel subset of $\mathbb{R}$ and one of the four subsets of the set $\{\infty, -\infty\}$. What are the Borel subsets of $\overline{\mathbb{R}}^+$?

CHAPTER 2
Random Variables

This chapter treats certain functions having as their domain a probability space. Such functions, known as 'random variables', have the property that they transform one probability space into another. In applications, random variables often represent what is actually observed in an experiment. Thus, in specific examples, it may be more descriptive to call them by such names as 'random numbers', 'random sequences of heads and tails of a coin', and 'random chords of a circle'.

2.1. Definitions and basic results

Many of the functions studied in probability theory are $\mathbb{R}$-valued. But functions that take values in other spaces are also of interest, the only restriction being that these spaces satisfy the following definition.

Definition 1. A *measurable space* is a pair $(\Psi, \mathcal{G})$, where Ψ is a nonempty set and $\mathcal{G}$ is a σ-field of subsets of Ψ.

Note that if $(\Omega, \mathcal{F}, P)$ is a probability space, then $(\Omega, \mathcal{F})$ is a measurable space.

Definition 2. Let $(\Omega, \mathcal{F})$ and $(\Psi, \mathcal{G})$ be measurable spaces. A *measurable function* from $(\Omega, \mathcal{F})$ to $(\Psi, \mathcal{G})$ is a function $X \colon \Omega \to \Psi$ such that $X^{-1}(B) \in \mathcal{F}$ for every $B \in \mathcal{G}$. When a probability measure P is attached to the measurable space $(\Omega, \mathcal{F})$, so that $(\Omega, \mathcal{F}, P)$ is a probability space, X is also called a *random variable* from $(\Omega, \mathcal{F}, P)$ to $(\Psi, \mathcal{G})$.

The language of the preceding definition is often shortened. For instance, if the σ-fields $\mathcal{F}$ and $\mathcal{G}$ and the probability measure P are understood from context, we may say that X is a *random variable* from Ω to Ψ, or simply that X is a Ψ-valued *random variable*.

The next two propositions indicate useful methods for deciding whether a function from one measurable space to another is measurable.

Proposition 3. *Let X be a function from a measurable space $(\Omega, \mathcal{F})$ to a measurable space $(\Psi, \mathcal{G})$. Suppose $\mathcal{E}$ is a family of subsets of Ψ that generates $\mathcal{G}$ and that $X^{-1}(B) \in \mathcal{F}$ for every $B \in \mathcal{E}$. Then X is a measurable function.*

Problem 1. Prove the preceding proposition. *Hint:* Consider the collection $\mathcal{H}$ of all subsets B of Ψ for which $X^{-1}(B) \in \mathcal{F}$ and show that $\mathcal{H}$ is a σ-field.

Proposition 4. *Every continuous function from one topological space to another (or to itself) is measurable, where the relevant σ-fields are the Borel σ-fields.*

* **Problem 2.** Prove the preceding proposition.

We have already seen that the domain of a random variable X is a measurable space that has been fitted with a probability measure P. As the following result shows, X transfers P to a probability measure Q on the target of X in a natural way. This 'induced' probability measure Q is of central importance in probability theory.

Definition and Proposition 5. Let X be a random variable from a probability space $(\Omega, \mathcal{F}, P)$ to a measurable space $(\Psi, \mathcal{G})$. For $B \in \mathcal{G}$ let $Q(B) = P(X^{-1}(B))$. Then $(\Psi, \mathcal{G}, Q)$ is a probability space. The probability measure Q is called the *distribution* of the random variable X and is said to be *induced by* X (or by X *from* P).

PROOF. Since $X^{-1}(\Psi) = \Omega$, $Q(\Psi) = 1$. Let $B_1, B_2, \ldots$ be a pairwise disjoint sequence of members of $\mathcal{G}$. Then $X^{-1}(B_m) \cap X^{-1}(B_n) = \emptyset$ whenever $m \neq n$. Hence,

$$Q\left(\bigcup_{n=1}^{\infty} B_n\right) = P\left(X^{-1}\left(\bigcup_{n=1}^{\infty} B_n\right)\right)$$

$$= P\left(\bigcup_{n=1}^{\infty} X^{-1}(B_n)\right)$$

$$= \sum_{n=1}^{\infty} P(X^{-1}(B_n))$$

$$= \sum_{n=1}^{\infty} Q(B_n). \quad \square$$

Note that every distribution is a probability measure, and that every probability measure is a distribution (induced by the identity function), so 'distribution' and 'probability measure' are essentially synonymous terms that tend to be used in somewhat different contexts.

* **Problem 3.** Suppose that X and Y are random variables defined on the same probability space. Assume that the set

$$\{\omega : X(\omega) \neq Y(\omega)\}$$

is an event having probability 0. Prove that the distributions of X and Y are equal.

The preceding problem describes one of many situations in which an event having probability 0 can be ignored. Events having probability 0 are called *null events*. The union of countably many null events, being a null event itself, can also often be ignored. But a word of caution: One cannot ignore an uncountable union of null events—and the temptation to do so comes in many disguises. An uncountable union of null events may have positive probability, may indeed be the entire probability space, or may not even be an event.

Random variables X and Y for which the set $\{\omega: X(\omega) \neq Y(\omega)\}$ is a null event are said to be equal *almost surely*. The expression 'almost surely' is often abbreviated *a.s.* We say that two events A and B are equal a.s. if both of the events $A \setminus B$ and $B \setminus A$ are null events.

Problem 4. Prove that two events that are equal a.s. have the same probability.

Problem 5. Show that events A and B in a probability space $(\Omega, \mathcal{F}, P)$ are equal a.s. if and only if

$$P(A \cap B) = P(A) \vee P(B).$$

The property of being equal a.s. is an equivalence relation, whether applied to random variables or events. All random variables in an equivalence class have the same distribution and all events in an equivalence class have the same probability. (The converse is not true. Events can have the same probability even though they are not almost surely equal, and random variables can have the same distribution even though they are not almost surely equal.) It is sometimes more convenient to consider equivalence classes rather than individual random variables or events.

Consider a random variable X on a probability space $(\Omega, \mathcal{F}, P)$ with values in the measurable space $(\Psi, \mathcal{G})$. By Definition 5, X induces a probability measure Q on the measurable space $(\Psi, \mathcal{G})$ thereby transforming it into a probability space $(\Psi, \mathcal{G}, Q)$, a probability space on which one can contemplate defining random variables. These observations point the way to the next two propositions.

Proposition 6. *Let $(\Omega, \mathcal{F})$, $(\Psi, \mathcal{G})$, and $(\Theta, \mathcal{H})$ be measurable spaces, and let $X : \Omega \to \Psi$ and $Y : \Psi \to \Theta$ be measurable functions. Then $Y \circ X$ is a measurable function.*

Problem 6. Prove the preceding proposition.

Proposition 7. *Let X be a random variable from a probability space $(\Omega, \mathcal{F}, P)$ to a measurable space $(\Psi, \mathcal{G})$, and let Y be a measurable function from $(\Psi, \mathcal{G})$ to a measurable space $(\Theta, \mathcal{H})$. Let Q denote the probability measure on $(\Psi, \mathcal{G})$ induced by X. Then Y is a random variable on the probability space $(\Psi, \mathcal{G}, Q)$ with the same distribution as that of the random variable $Y \circ X$ defined on the probability space $(\Omega, \mathcal{F}, P)$.*

Problem 7. Prove the preceding proposition.

2.2. $\mathbb{R}^d$-valued random variables

Focusing first on $\mathbb{R}^1$-valued random variables, we start with some random variables that are easy to describe but, nevertheless, play an important role for probability calculations and theory.

Definition 8. Let $(\Omega, \mathcal{F})$ be a measurable space. For each subset A of Ω, define a function I_A on Ω by

$$I_A(\omega) = \begin{cases} 1 & \text{if } \omega \in A \\ 0 & \text{otherwise}. \end{cases}$$

The function I_A is called the *indicator function* of A. A finite linear combination of indicator functions is called a *simple function*.

Subsets C_j, $j \in J$, of a set Ψ form a *partition* of Ψ if: (i) $\bigcup_{j \in J} C_j = \Psi$, (ii) each $C_j \neq \emptyset$, and (iii) $C_j \cap C_k = \emptyset$ whenever $j \neq k$. A partition is *finite* if the corresponding set J is finite, and it is *countable* if J is countable.

Lemma 9. *Let $X \colon \Psi \to \mathbb{R}$ be a simple function:*

$$X = \sum_{i=1}^{m} a_i I_{A_i} \ .$$

Let $\mathcal{G}$ be a σ-field of subsets of Ψ such that $A_i \in \mathcal{G}$ for each i. Then there exists a unique positive integer n, a unique partition $\{C_j \colon 1 \leq j \leq n\}$ of Ψ, and unique real numbers c_j, $1 \leq j \leq n$, such that

$$(2.1) \qquad\qquad X = \sum_{j=1}^{n} c_j I_{C_j}$$

and $c_j \neq c_k$ whenever $j \neq k$. Moreover, each $C_j \in \mathcal{G}$.

PROOF. The image of X consists of finitely many values, the only possibilities being 0 and sums of one or more of the a_i's. Call the members of the image $c_1, c_2, \ldots, c_n$, and, for $j = 1, 2, \ldots, n$, set

$$(2.2) \qquad\qquad C_j = \{\psi \colon X(\psi) = c_j\} \ .$$

It is clear that (2.1) holds and that $\{C_j : 1 \leq j \leq n\}$ is a partition of Ψ. On the other hand, if (2.1) holds, then (2.2) holds because the numbers c_j are distinct. For each j,

$$C_j = \bigcup_{(i_1,\ldots,i_k):\, a_{i_1}+\cdots+a_{i_k}=c_j} \left[\left(\bigcap_{r=1}^{k} A_{i_r} \right) \cap \left(\bigcap_{\substack{i=1 \\ i \neq i_r,\, 1 \leq r \leq k}}^{n} A_i^c \right) \right].$$

Therefore, $C_j \in \mathcal{G}$. $\square$

Corollary 10. *Let $(\Psi, \mathcal{G})$ be a measurable space. If $f \colon \Psi \to \mathbb{R}$ is a simple function that is a finite linear combination of indicator functions of members of $\mathcal{G}$, then f is a measurable function into $(\mathbb{R}, \mathcal{H})$ for any choice of the σ-field $\mathcal{H}$.*

Problem 8. Prove the preceding corollary.

We may call the indicator function of an event an *indicator random variable*, and, in view of the preceding corollary, a finite linear combination of indicator random variables a *simple random variable*.

We now discuss various examples of random variables that are not simple.

Example 1. Let Ω be the sample space for an infinite sequence of coin flips, as described in Example 4 of Chapter 1. Let $\mathcal{F}$ be the σ-field generated by sets of the form (1.2) and let P denote the unique probability measure on $\mathcal{F}$ satisfying (1.1). (We will prove the existence and uniqueness of such a P in Chapter 7.) Let X denote the function from $(\Omega, \mathcal{F}, P)$ to $(\mathbb{R}, \mathcal{B})$ defined by

$$X((\omega_1, \omega_2, \ldots)) = 0.\omega_1 \omega_2 \ldots {}_{\text{two}} ,$$

where the symbol on the right represents a member of the interval $[0,1]$ written in binary notation. Let us show that X is a random variable and calculate its distribution.

We consider the set $X^{-1}((j2^{-n}, (j+1)2^{-n}])$, where j and n are nonnegative integers. If $j2^{-n} \geq 1$, then $X^{-1}((j2^{-n}, (j+1)2^{-n}]) = \emptyset$ and is thus a member of $\mathcal{F}$. In the remaining cases j satisfies $0 \leq j < 2^n$. Let $j = \varepsilon_1 \varepsilon_2 \ldots \varepsilon_n {}_{\text{two}}$ be the binary representation of j (written with possibly some superfluous leading zeroes in order to have n binary digits). Then

$$j2^{-n} = 0.\varepsilon_1 \ldots \varepsilon_n 000 \ldots {}_{\text{two}}$$

$$\text{and } (j+1)2^{-n} = 0.\varepsilon_1 \ldots \varepsilon_n 111 \ldots {}_{\text{two}} .$$

Therefore, it is almost true that

(2.3) $X^{-1}((j2^{-n}, (j+1)2^{-n}]) = \{(\omega_1, \omega_2, \ldots) : \omega_i = \varepsilon_i \text{ for } 1 \leq i \leq n\},$

which is a member of $\mathcal{F}$. The 'almost' arises for two reasons. First, the set on the right contains an extra point, namely $\omega = (\varepsilon_1, \ldots, \varepsilon_n, 0, 0, \ldots)$. And second, the set on the left contains an extra point because of the fact that there are two different binary representations of $(j+1)2^{-n}$. By Problem 11 of Chapter 1, individual sample points constitute measurable sets, so these differences between the two sets do not affect the measurability of the set on the left side. That X is a random variable now follows from Problem 13 of Chapter 1 and Proposition 3 of this chapter.

The preceding argument that X is a random variable also gives much information about its distribution Q. Since the P-probability of the event on the right side of (2.3) is 2^{-n} and the P-probability of the extra sample points on the two sides equals 0,

$$Q((j2^{-n}, (j+1)2^{-n}]) = 2^{-n} .$$

By taking finite disjoint unions and using the additivity of P, it is easily shown that

$$Q((i2^{-n}, j2^{-n}]) = (j-i)2^{-n}$$

for nonnegative integers i, j, n such that $i \leq j \leq 2^n$. Also, since individual sample points have probability 0, one may add or delete the endpoints of these intervals without changing their probabilities. Thus, for all intervals $I \subseteq [0,1]$ with binary rational endpoints, $Q(I)$ equals the length of I. It will develop, from the theory in Chapter 7, that these values of Q determine the function Q on the Borel sets $\mathcal{B}$ and that, in particular, the Q-probability of any subinterval of $[0,1]$ equals the length of that interval.

The probability measure Q is called *Lebesgue measure* on $[0,1]$ and can be regarded as a generalization of the notion of length to sets that are not intervals. In probabilistic language, Q is called the *uniform distribution* on $[0,1]$ and is likely the distribution someone has in mind when discussing, without clarification, the experiment of choosing a number at random from the unit interval. In fact, successive coin tosses constitute an effective but cumbersome way of approximately choosing such a random number. For instance, if the first four flips are 'heads, tails, tails, heads', then we conclude, successively, that the random number is a member of $[\frac{1}{2}, 1]$, of $[\frac{1}{2}, \frac{3}{4}]$, of $[\frac{1}{2}, \frac{5}{8}]$, and then of $[\frac{9}{16}, \frac{5}{8}]$.

The probability space $([0,1], \mathcal{B}, Q)$ arises often in probability theory. In this space, all countable sets are null events, so there is no essential change in the probability space if a countable subset of $[0,1]$ is ignored. Thus, one also uses phrases such as "the probability space consisting of Lebesgue measure on the interval $(0,1]$" and "the uniform distribution on $(0,1)$".

The term 'random number' was used in the preceding example to refer to an $\mathbb{R}$-valued random variable with a uniform distribution. In general, the term *random number* is synonymous with the term '$\mathbb{R}$-valued random variable', regardless of its distribution.

The first problem in the following set is useful for proving that certain $\mathbb{R}$-valued functions are random variables. The next five problems are relevant for $\mathbb{R}^2$-valued random variables, and the one following that gives an example of an

$\mathbb{R}^d$-valued random variable for arbitrary d. The final problem of the section indicates some methods of obtaining new random variables from old ones.

* **Problem 9.** Show that every increasing function from $\mathbb{R}$ to $\mathbb{R}$ is measurable.

Problem 10. Let $(\Omega, \mathcal{F}, P)$ denote the usual probability space for an infinite sequence of coin flips. Let X be the function from Ω to $(\mathbb{R}^2, \mathcal{B})$ defined by

$$X((\omega_1, \omega_2, \dots)) = (0.\omega_1\omega_3 \dots \text{two}, \ 0.\omega_2\omega_4 \dots \text{two}).$$

Do enough calculations and reasoning to become convinced that X is a random variable and that the distribution of X is a generalization of area, restricted to the unit square $[0, 1]^2$. This distribution is called *Lebesgue measure* or *uniform distribution* on $[0, 1]^2$. Some people would prefer to call the random variable X of this example a *random vector*.

Problem 11. Prove that the uniform distribution on the unit square assigns 0 probability to the boundary of the square. How may this fact be used when defining the uniform distribution on $[0, 1)^2$, $(0, 1]^2$, or $(0, 1)^2$?

* **Problem 12.** For the random variable $X = (X_1, X_2)$ of Problem 10, calculate the probability that $X_1 \vee X_2 > 2/3$; that is, calculate

$$P(\{\omega : X_1(\omega) \vee X_2(\omega) > 2/3\}).$$

Problem 13. Let $(\Omega, \mathcal{F}, P)$ be the probability space of Problem 10 of Chapter 1 for the rolling of two dice. Set

$$X(\omega_1, \omega_2) = (\omega_1 \wedge \omega_2, \omega_1 \vee \omega_2).$$

Relate this random vector to Example 5 of Chapter 1.

* **Problem 14.** Let Ω be the unit square, $\mathcal{B}$ the Borel σ-field of subsets of Ω, and P the uniform distribution on Ω. Let X be the random variable on $(\Omega, \mathcal{F}, P)$ defined for $\omega = (\omega_1, \omega_2)$ by

$$X(\omega) = (\omega_1 \wedge \omega_2, \omega_1 \vee \omega_2).$$

Prove that X is a random variable and describe its distribution.

Problem 15. Let Ω consist of all infinite sequences of 0's and 1's. Let $\mathcal{F}$ be the σ-field generated by sets of the form (1.2) and let P be the unique probability measure on $\mathcal{F}$ satisfying (1.1). Fix a positive integer d and let Ψ denote the set of all d-dimensional vectors of 0's and 1's. Let $\mathcal{G}$ denote the σ-field consisting of all subsets of Ψ. Define the function X from Ω to Ψ by the formula

$$X((\omega_1, \omega_2, \dots)) = (\omega_1, \dots, \omega_d).$$

Prove that X is a random variable and that its distribution is the probability measure introduced in Example 3 of Chapter 1.

Problem 16. Let $X_1, \ldots, X_d$ be $\mathbb{R}$-valued functions defined on the same space. Show that the vector $(X_1, \ldots, X_d)$ is an $\mathbb{R}^d$-valued measurable function if and only if each X_i is measurable, and deduce that the following are $\mathbb{R}$-valued measurable functions: the sum, product, max, and min of $X_1, \ldots, X_d$. *Hint:* Proposition 4 and Proposition 6 will be useful.

Notice that no knowledge of the underlying probability space is needed in order to do Problem 12. The only necessary information is the distribution of X. Since this situation is quite common, language like the following is often used: "Let X be a random variable uniformly distributed on $[0, 1]^2$." The existence of an appropriate underlying probability space is implicit in such a statement.

2.3. $\mathbb{R}^\infty$-valued random variables

The following problem, which involves the product topology of a countably infinite number of factors, is a continuation of the last problem of the preceding section.

Problem 17. Let $X_1, X_2, X_3, \ldots$ be $\mathbb{R}$-valued functions defined on the same space. Show that the $\mathbb{R}^\infty$-valued function $X = (X_1, X_2, X_3, \ldots)$ is measurable if and only if each X_i is measurable.

An $\mathbb{R}^\infty$-valued random variable $X = (X_1, X_2, X_3, \ldots)$ is also called a *random sequence*.

It is easy to see that one can replace $\mathbb{R}$ and $\mathbb{R}^\infty$ in Problem 17 by $\overline{\mathbb{R}}$ and $\overline{\mathbb{R}}^\infty$, respectively. That is, there is no essential difference between an infinite sequence of $\overline{\mathbb{R}}$- or $\overline{\mathbb{R}}^+$-valued measurable functions defined on the same measurable space and an $\overline{\mathbb{R}}^\infty$- or $(\overline{\mathbb{R}}^+)^\infty$-valued measurable function. The remainder of this section will be devoted to $\overline{\mathbb{R}}^\infty$- and $(\overline{\mathbb{R}}^+)^\infty$-valued measurable functions. Of course, our main concern is with measurable functions that also happen to be random variables, but for these particular results, the presence of a probability measure plays no essential role.

When we write

$$\lim_{n \to \infty} X_n = X$$

we mean

$$\lim_{n \to \infty} X_n(\omega) = X(\omega) \text{ for all } \omega.$$

Proposition 11. *Let $(\Omega, \mathcal{F})$ be a measurable space, and let $(X_1, X_2, \ldots)$ be an increasing sequence of measurable $\overline{\mathbb{R}}$-valued functions defined on $(\Omega, \mathcal{F})$. Then $X = \lim_{n \to \infty} X_n$ is measurable.*

PROOF. By Proposition 3, it is enough to show that $X^{-1}((a, \infty])$ is a measurable set for all real numbers a, since the collection of intervals of the form

$(a, \infty]$ generates the Borel field of $\overline{\mathbb{R}}$ (see Problem 13 of Chapter 1). Since X is the increasing pointwise limit of the sequence $(X_1, X_2, \dots)$,

$$X^{-1}((a, \infty]) = \bigcup_{n=1}^{\infty} X_n^{-1}((a, \infty]) \, .$$

Each of the sets in the union is measurable since each of the functions X_n is measurable. It follows that the union is measurable. $\square$

Corollary 12. *Let $(X_1, X_2, \dots)$ be a sequence of $\overline{\mathbb{R}}$-valued measurable functions defined on a measurable space $(\Omega, \mathcal{F})$. Then the following functions are measurable:*

$$\sup_n X_n, \quad \inf_n X_n, \quad \limsup_{n \to \infty} X_n, \quad \liminf_{n \to \infty} X_n \, .$$

Furthermore, if

$$\lim_{n \to \infty} X_n$$

exists, it is a measurable function.

Problem 18. Prove the preceding corollary. *Hint:* The supremum of any sequence of real numbers is the increasing pointwise limit of another appropriately defined sequence.

The following lemma is useful for proving facts about $\overline{\mathbb{R}}^+$-valued measurable functions, since it shows that all such functions can be seen as monotonically increasing limits of measurable simple functions. See Theorem 15 of Chapter 4 for an example of such an application.

Lemma 13. *An $\overline{\mathbb{R}}^+$-valued function X defined on a measurable space $(\Omega, \mathcal{F})$ is measurable if and only if there exists a sequence $(X_1, X_2, \dots)$ of measurable simple functions defined on $(\Omega, \mathcal{F})$ such that $0 \le X_1 \le X_2 \le \dots$, and*

$$\lim_{n \to \infty} X_n = X \, .$$

PROOF. The 'if' portion of the lemma is contained in Proposition 11. For the 'only if' part, suppose that X is a measurable $\overline{\mathbb{R}}^+$-valued function. For $k, n \ge 1$, let

$$A_{k,n} = \{\omega \colon (k-1)/2^n \le X(\omega) < k/2^n\} \text{ and } B_n = \{\omega \colon X(\omega) \ge n\}$$

and

$$X_n = nI_{B_n} + \frac{1}{2^n} \sum_{k=1}^{n2^n} (k-1) I_{A_{k,n}} \, .$$

Since X is a measurable function, all of the sets B_n and $A_{k,n}$ are measurable. It follows that each X_n is a nonnegative measurable simple function. It is easily checked that $X_n \nearrow X$ as $n \nearrow \infty$. $\square$

2.4. Further examples

In this section, we look at examples of random variables that are not $\mathbb{R}^d$- or $(\mathbb{R}^+)^d$-valued for any $d = 1, 2, \ldots, \infty$. We look first at examples of random variables whose values are continuous functions.

Let $\mathbf{C}[a, b]$ denote the space of continuous real-valued functions defined on the interval $[a, b]$. We regard $\mathbf{C}[a, b]$ as a metric space (see Appendix B), with the distance between two functions being the maximum of the absolute value of their difference. It therefore makes sense to talk about $\mathcal{B}$, the σ-field of Borel subsets of $\mathbf{C}[a, b]$.

Example 2. Let $(\Omega, \mathcal{F}, P)$ denote the usual probability space for an infinite sequence of coin flips. For each positive integer k we will define a random variable $X^{(k)} = (X_t^{(k)}: t \in \mathbb{R}^+)$ from $(\Omega, \mathcal{F}, P)$ to $(\mathbf{C}[0, 1], \mathcal{B})$. For each k and each $\omega = (\omega_1, \omega_2, \ldots)$ we must specify the values $X_t^{(k)}(\omega)$ of the continuous function $t \rightsquigarrow X_t^{(k)}(\omega)$ at each $t \in [0, 1]$. We first specify these values for t equal to a multiple of $1/k$: for $j = 0, 1, \ldots, k$, let

$$X_{j/k}^{(k)}(\omega) = \frac{1}{\sqrt{k}}\left(\sum_{i=1}^{j}(2\omega_i - 1)\right).$$

For all other values of t, the value of $X_t^{(k)}(\omega)$ is determined by linear interpolation. The graphs of $X^{(k)}(\omega)$ are shown in Figures 2.1, 2.2, and 2.3, for

$$\omega = (1, 1, 0, 1, 0, 0, 0, 1, 0, 0, 1, 1, 1, 1, 0, 1, \ldots)$$

and $k = 4$, 9, and 16.

For any subset B of $\mathbf{C}[0, 1]$, $[X^{(k)}]^{-1}(B)$ is a finite union of sets of the form (1.2). Hence $X^{(k)}$ is a random variable; but many would prefer to use one of the terms *random function* or *stochastic process* to emphasize that the target of $X^{(k)}$ is a set of functions. The term 'stochastic process' is especially prevalent when t represents time.

In the context of coin-tossing, the sum in the expression for $X^{(k)}(\omega)$ equals the difference between the number of heads and the number of tails after j tosses. The quantity $\sqrt{k}$ serves to give a common scale to the random functions $X^{(k)}$, $k = 1, 2, \ldots$. We can be more precise about this scaling factor after we introduce 'standard deviation' in Chapter 5 (see Problem 6 of that chapter).

Let Q_k denote the distribution induced on $(\mathbf{C}[0, 1], \mathcal{B})$ by $X^{(k)}$. For each k, $(\mathbf{C}[0, 1], \mathcal{B}, Q_k)$ is a probability space. The sample space and σ-field are the same for the various k but the probability measures are different. We can use the calculation done in Example 3 of Chapter 1 to conclude, for instance, that

$$Q_k(\{g: g(1) = 0\}) = \begin{cases} \binom{k}{k/2} 2^{-k} & \text{if } k \text{ is even} \\ 0 & \text{otherwise}. \end{cases}$$

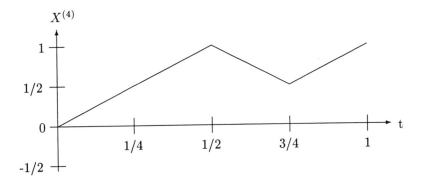

FIGURE 2.1. The graph of $X^{(4)}((1,1,0,1,\dots))$

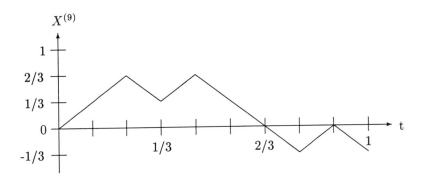

FIGURE 2.2. The graph of $X^{(9)}((1,1,0,1,0,0,0,1,0,\dots))$

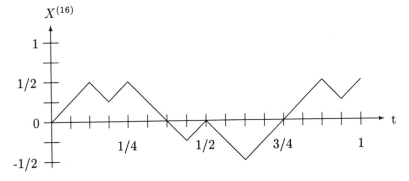

FIGURE 2.3. The graph of $X^{(16)}((1,1,0,1,0,0,0,1,0,0,1,1,1,1,0,1,\dots))$

* **Problem 19.** In the context of the preceding example, calculate

$$Q_k(\{g\colon g(1/2) = 0 \text{ and } g(1) = \sqrt{k}/2\})\,.$$

If one were interested, say, in the probability space $(\mathbf{C}[0,1], \mathcal{B}, Q_6)$, one could study it without ever introducing the random variables X_k or the underlying coin-flip probability space $(\Omega, \mathcal{F}, P)$. Indeed, before the 1930's, much probability and statistics were done without any systematic use of random variables. The preceding example does, however, indicate two good reasons for using random variables. First, it is easier to think in terms of sequences of coin flips for the construction of the probability space, rather than in terms of randomly choosing a function of a certain type. Second, defining several different random variables on the same probability space makes it possible to use one underlying experiment to study several probability measures simultaneously. For example, in Chapter 19, we will be interested in calculating the limit as $k \to \infty$ of Q_k, and only one underlying probability space, namely the space $(\Omega, \mathcal{F}, P)$ of Example 2, will be needed.

We have seen so far that the target of a random variable can consist of numbers, vectors, sequences, or functions, in which cases one may use terms that are more specific than 'random variable', such as 'random number', 'random vector', 'random sequence', and 'random function'. The forthcoming Example 3 treats a random compact set. In preparation we construct a metric space whose members are compact sets. Let $\mathcal{X}$ be a metric space with metric ρ, and let Ψ be the collection of compact subsets of $\mathcal{X}$. For members U and V of Ψ, set

$$\begin{aligned} d(U,V) = \tfrac{1}{2} \big[&\max\{\min\{\rho(u,v)\colon u \in U\}\colon v \in V\} \\ &+ \max\{\min\{\rho(u,v)\colon v \in V\}\colon u \in U\} \big]. \end{aligned}$$

Problem 20. Prove that the function d is a metric on Ψ. This metric is called the *Hausdorff metric* on Ψ.

* **Problem 21.** Calculate the three Hausdorff distances between the various pairs of the following subsets of the Euclidean plane:

$$\{(x_1, x_2)\colon -1 \le x_1 \le 0, |x_2| \le 1\}\,,$$
$$\{(x_1, x_2)\colon 0 \le x_1 \le 1, x_2 = 0\}\,,$$
$$\{(x_1, x_2)\colon x_1 = -1, |x_2| \le 1\}\,.$$

In the following example the random variables have values that are compact subsets of the Euclidean plane. In fact, the values are line segments, so we will work with a subspace of the metric space treated in the preceding problem. Of course, the metric for the subspace has the same formula as the metric for the whole space.

Example 3. Let us consider the experiment of choosing at random a chord of the unit circle in $\mathbb{R}^2$. Assume that the circle is given parametrically by $(\cos 2\pi t, \sin 2\pi t)$, $0 \leq t < 1$.

Here is one possible interpretation of 'at random'. Let Ω be the unit square $[0,1)^2$. Let $\mathcal{A}$ denote the Borel σ-field of subsets of Ω and let P be the uniform distribution on Ω (see Problem 11). Let Ψ be the space of chords of the given circle and let $\mathcal{B}$ denote the Borel σ-field of subsets of Ψ, with Ψ being regarded as a subspace of the metric space of all compact subsets of $\mathbb{R}^2$ (with the Hausdorff metric). For $\omega = (\alpha, \beta) \in \Omega$, let $X_1(\omega)$ denote the line segment having endpoints $(\cos 2\pi\alpha, \sin 2\pi\alpha)$ and $(\cos 2\pi\beta, \sin 2\pi\beta)$. Notice that $X_1(\omega)$ is a chord of the circle of interest for any choice of ω, provided we allow for the degenerate case of a chord being a single point.

Here are two more interpretations of 'at random', using the same spaces $(\Omega, \mathcal{A}, P)$ and $(\Psi, \mathcal{B})$. Let X_2 be the chord that both passes through the point whose polar coordinates are $(2\alpha - 1, \pi\beta)$ and is perpendicular to the line segment from the origin to that point. In case $\alpha = 1/2$, this line segment consists of a single point, but we can still define its direction to be $\pi\beta$. Let $X_3(\omega)$ be the chord which has $(\cos 2\pi\alpha, \sin 2\pi\alpha)$ for one endpoint and whose angle measured counterclockwise from the positive horizontal direction is $\pi\beta$.

A fourth interpretation can be obtained by letting Ω denote the interior of the circle in question. Let P denote the uniform distribution on Ω (meaning?) and let $\mathcal{A}$ be the Borel σ-field of subsets of Ω. Let $X_4(\omega)$ denote the chord whose midpoint is ω.

Example 4. Let us look more closely at the first interpretation in the preceding example of the experiment of choosing a random chord of the unit circle. We first ask whether the function X_1 defined in that example is a measurable function from $(\Omega, \mathcal{F}, P)$ to the space $(\Psi, \mathcal{B})$. We leave it to the reader to prove that a sequence of chords in Ψ converges in the Hausdorff metric if and only if the endpoints of these chords converge in $\mathbb{R}^2$. This fact makes it natural for us to include the degenerate case of chords of length 0 in order to make Ψ a complete metric space. (However, this is purely a matter of taste. There is nothing that compels us to complete Ψ.) Our description of convergence in Ψ should make it clear that X_1 is a continuous function from Ω to Ψ, so X_1 is measurable and may be legitimately called a 'random set'.

Let us calculate the probability of the event that X_1 intersects both the positive vertical axis and the negative horizontal axis. For definiteness, we exclude the origin from these two sets. As illustrated in Figure 2.4, this event equals the union of

$$A = \left\{ \omega = (\alpha, \beta) : \tfrac{1}{2} \leq \alpha < \tfrac{3}{4} \text{ and } \alpha - \tfrac{1}{2} < \beta \leq \tfrac{1}{4} \right\}$$

and the reflection of A about the line $\alpha = \beta$ in $\mathbb{R}^2$. These two sets are disjoint, and they each have area $1/32$, so the probability is $1/16$.

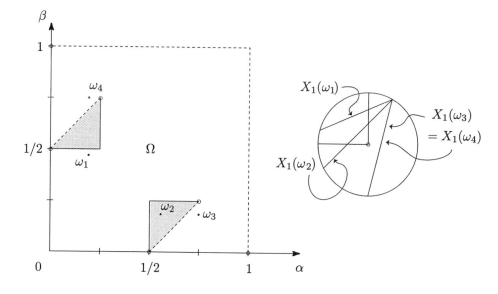

FIGURE 2.4. Random chord: four sample points

* **Problem 22.** Repeat the work done in the preceding example for each of the other three interpretations found in Example 3. That is, in each case, decide whether one-point subsets of the circle are to be considered chords, prove the necessary measurability, and compute the probability of the event that the random chord intersects the positive vertical and negative horizontal axes. Also comment on and fix an ambiguity in the definition of X_4.

CHAPTER 3
Distribution Functions

The main purpose of this chapter is to classify all probability measures on the measurable space $(\mathbb{R}, \mathcal{B})$. We will accomplish this task by establishing a one-to-one correspondence between such probability measures and a certain class of functions, known as 'distribution functions'. Many important probability measures and their corresponding distribution functions will be identified, including the binomial, normal, Poisson, gamma, and beta families of probability measures.

3.1. Basic theory

We introduce the class of functions to which the preceding paragraph refers.

Definition 1. A real-valued function F defined on $\mathbb{R}$ is called a *distribution function* for $\mathbb{R}$ if it is increasing and right-continuous and satisfies

$$\lim_{x \to -\infty} F(x) = 0 \quad \text{and} \quad \lim_{x \to \infty} F(x) = 1 \,.$$

Often the phrase 'for $\mathbb{R}$' will be omitted.

Let Q denote a probability measure on the measurable space $(\mathbb{R}, \mathcal{B})$. We want to show that the function $x \rightsquigarrow Q((-\infty, x])$ is a distribution function. For this purpose we need the following useful result.

Theorem 2. [Continuity of Measure] *Let $(\Omega, \mathcal{F}, P)$ be a probability space, and let $(A_1, A_2, \dots)$ be a sequence of events in $\mathcal{F}$. If $A_1 \subseteq A_2 \subseteq \dots$, then*

$$P\left(\bigcup_{n=1}^{\infty} A_n \right) = \lim_{n \to \infty} P(A_n) \,.$$

If $A_1 \supseteq A_2 \supseteq \dots$, then

$$P\left(\bigcap_{n=1}^{\infty} A_n \right) = \lim_{n \to \infty} P(A_n) \,.$$

PROOF. To prove the first assertion we suppose that $A_1 \subseteq A_2 \subseteq \ldots$. Let $B_1 = A_1$ and $B_m = A_m \setminus A_{m-1}$ for $m = 2, 3, \ldots$. Note that

$$A_n = \bigcup_{m=1}^{n} B_m \quad \text{and} \quad \bigcup_{m=1}^{\infty} A_m = \bigcup_{m=1}^{\infty} B_m.$$

Since the B_m's are disjoint, we can use the countable additivity of P to make the following calculation:

$$P\left(\bigcup_{m=1}^{\infty} A_m\right) = P\left(\bigcup_{m=1}^{\infty} B_m\right)$$
$$= \sum_{m=1}^{\infty} P(B_m) = \lim_{n \to \infty} \sum_{m=1}^{n} P(B_m)$$
$$= \lim_{n \to \infty} P\left(\bigcup_{m=1}^{n} B_m\right) = \lim_{n \to \infty} P(A_n).$$

This proves the first assertion of the theorem.

Now suppose that $A_1 \supseteq A_2 \supseteq \ldots$. Then $A_1^c \subseteq A_2^c \subseteq \ldots$. Therefore, we can apply the first part of the theorem to the sequence $(A_1^c, A_2^c, \ldots)$ to obtain the following:

$$P\left(\bigcap_{n=1}^{\infty} A_n\right) = 1 - P\left(\bigcup_{n=1}^{\infty} A_n^c\right)$$
$$= 1 - \lim_{n \to \infty} P(A_n^c)$$
$$= 1 - \lim_{n \to \infty} [1 - P(A_n)] = \lim_{n \to \infty} P(A_n). \quad \square$$

Proposition 3. *Let Q be a probability measure on $(\mathbb{R}, \mathcal{B})$. Then the function $F(x) = Q((-\infty, x])$ is a distribution function.*

PROOF. Note that

$$x \le y \implies (-\infty, x] \subseteq (-\infty, y] \implies Q((-\infty, x]) \le Q((-\infty, y]).$$

Hence, F is an increasing function. To prove right continuity, fix a real number x and let $(x_1, x_2, \ldots)$ be a decreasing sequence which converges to x. Since $(-\infty, x_1] \supseteq (-\infty, x_2] \supseteq \ldots$, Theorem 2 implies that

$$\lim_{n \to \infty} Q((-\infty, x_n]) = Q\left(\bigcap_{n=1}^{\infty} (-\infty, x_n]\right)$$
$$= Q((-\infty, x]).$$

Thus, F is right-continuous. The same reasoning, with $x_n \searrow -\infty$, shows that F has the desired behavior at $-\infty$. For the behavior at ∞, let $x_n \nearrow \infty$ and use the Continuity of Measure Theorem again:

$$Q((-\infty, x_n]) \to Q\left(\bigcup_{n=1}^{\infty} (-\infty, x_n]\right) = Q(\mathbb{R}) = 1 . \quad \square$$

If a distribution Q and a distribution function F are related as in the previous theorem, then we will call F the *distribution function* of Q. If X is a random variable with distribution Q, then we will also call F the *distribution function* of X.

Problem 1. Show that the distribution function of the uniform distribution on $[0, 1]$ is given by

$$F(x) = \begin{cases} 0 & \text{if } x < 0 \\ x & \text{if } 0 \leq x < 1 \\ 1 & \text{if } 1 \leq x . \end{cases}$$

What is the distribution function for the uniform distribution on $(0, 1)$?

Problem 2. Give a precise description of the probability measure Q on $(\mathbb{R}, \mathcal{B})$ that has distribution function F given by

$$F(x) = \begin{cases} 0 & \text{if } x < 0 \\ 1 & \text{if } x \geq 0 . \end{cases}$$

Also find a probability space $(\Omega, \mathcal{F}, P)$ and a real-valued random variable X defined on $(\Omega, \mathcal{F}, P)$ that has distribution function F. The probability measure Q is sometimes called the *unit point mass* or the *delta distribution* at 0.

The preceding exercise is an example of the converse of Proposition 3: For each distribution function F there exists a unique probability Q on $(\mathbb{R}, \mathcal{B})$ such that $Q((-\infty, x]) = F(x)$. This fact is included in the next theorem, which also provides a recipe for constructing a random variable with distribution function F. From a logical point of view, this result belongs in Chapter 7, since the necessary tools for proving the existence and uniqueness of probability measures are to be found there. However, if the reader is willing to accept uniqueness in general and the existence of the uniform distribution on $(0, 1)$ in particular, then the characterization of all other distributions on $(\mathbb{R}, \mathcal{B})$ can be derived using the concepts already introduced.

Proposition 4. *Let F be a distribution function. Then there exists a unique probability measure Q on $(\mathbb{R}, \mathcal{B})$ such that $Q((-\infty, x]) = F(x)$. Moreover, a random variable X with distribution function F can be constructed as follows: Let $\Omega = (0, 1)$, let P be the uniform distribution on Ω, and define*

$$(3.1) \qquad X(\omega) = \inf\{x \colon F(x) \geq \omega\}, \quad 0 < \omega < 1 .$$

PROOF. The uniqueness will follow from Theorem 3 of Chapter 7. We assume the existence of the uniform distribution P on $(0,1)$, which follows from Theorem 14 of Chapter 7. Further details about the existence of the uniform distribution are given in Example 1 of the same chapter.

By Problem 9 of Chapter 2, X is a random variable. Let Q be the distribution of X. We will complete the proof of this theorem by showing that F is the distribution function of Q, or in other words, that

$$F(y) = P(\{\omega \colon X(\omega) \leq y\})$$

for all $y \in \mathbb{R}$. Since X is an increasing function on $(0,1)$, the event $A = \{\omega \colon X(\omega) \leq y\})$ is an interval with endpoints 0 and $\sup A$. Under the uniform distribution, the probability of any sub-interval of $(0,1)$ is the length of that sub-interval, so we want to show that $F(y) = \sup A$.

The definition of X and the right continuity of F imply that $F(X(\omega)) \geq \omega$; so, if $\omega \in A$, then $F(y) \geq F(X(\omega)) \geq \omega$. Hence, $F(y)$ is an upper bound of A. On the other hand, $F(y) \in A$, because $X(F(y)) \leq y$. It follows that $F(y) = \sup A$. □

The relationship between F and X given in the preceding result is most easily understood when F is strictly increasing and continuous. In this case, X and F are inverse functions of each other and X is strictly increasing and continuous. In general X, defined by (3.1), is left-continuous. Jumps of F correspond to intervals of constancy of X, and bounded intervals of constancy of F correspond to jumps of X. Unbounded intervals of constancy of F correspond to finite limits of X at 0 and 1. It has become quite common to refer to X as the 'left-continuous inverse' of F and F as the 'right-continuous inverse' of X, even in the cases where there are jumps or intervals of constancy. Figure 3.1 shows an F that has both jumps and intervals of constancy. The corresponding X is shown below the graph of F, with its domain pictured vertically.

* **Problem 3.** Prove that X as defined in the preceding theorem is left-continuous and satsifies $X(\omega) = \sup\{x \colon F(x) < \omega\}$.

Problem 4. Discuss the options of making either X right-continuous or F left-continuous or both in Proposition 4.

Problem 5. Let X be an $\mathbb{R}$-valued random variable with distribution function F. Prove that for all real numbers $a < b$,

$$P(\{\omega \colon X(\omega) \in (a,b]\}) = F(b) - F(a).$$

Find analogous formulas involving intervals of the form (a,b), $[a,b)$, and $[a,b]$. For $x \in \mathbb{R}$, show that

$$P(\{\omega \colon X(\omega) = x\}) = F(x) - F(x-).$$

As a consequence, conclude that if F is continuous, then the events $\{\omega \colon X(\omega) = x\}$ are all null events.

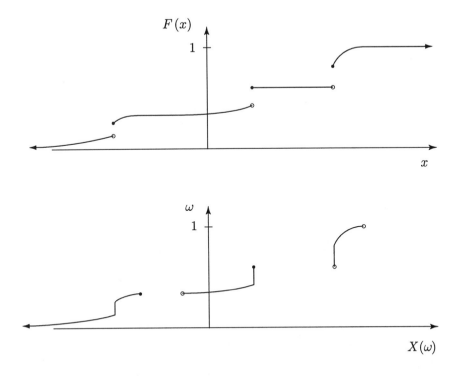

FIGURE 3.1. Distribution function and corresponding random variable

3.2. Examples of distributions

In the remainder of this chapter, we will illustrate Proposition 3 and Proposition 4 by introducing, through examples and exercises, some of the more important distributions on $(\mathbb{R}, \mathcal{B})$.

Problem 6. [Delta distributions] If X is equal to a constant a, show that the distribution function of X is given by

$$F(x) = \begin{cases} 0 & \text{if } x < a \\ 1 & \text{if } x \geq a. \end{cases}$$

The distribution of X is called the *delta distribution* or *unit point mass* at a. Often δ_a will be used for the delta distribution at a.

Problem 7. [Bernoulli distributions] Fix $p \in [0, 1]$, and let X be a random variable that equals 1 with probability p and equals 0 with probability $1 - p$. Calculate the distribution function F of X. Conversely, starting with F, construct a random variable whose distribution function is F. *Hint:* See Figure 3.2.

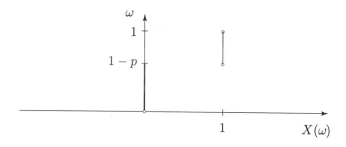

FIGURE 3.2. Bernoulli random variable defined on $(0, 1)$

Problem 8. [Cauchy distribution] For $x \in \mathbb{R}$ let

$$F(x) = \frac{1}{2} + \frac{\arctan x}{\pi},$$

and let Q be the corresponding distribution. It is easy to check that F is a continuous distribution function. Calculate a random variable X with distribution function F by using (3.1). For an arbitrary interval $[a, b)$ write $Q([a, b))$ in the form $\int_a^b f$ for an appropriate f. Do the same for intervals of the form $[a, b]$, (a, b), and $(a, b]$. In any of these, is it permissible to use $a = -\infty$ or $b = \infty$? At the end of Example 1 of Chapter 2, a method was described for approximately choosing a random number according to the uniform distribution by using a sequence of coin flips. Describe how to transform that method, using the construction in Proposition 4, to obtain a procedure for approximately choosing a Cauchy-distributed random number.

Problem 9. Define X on the probability space $((0, 1), \mathcal{B}, P)$, where P denotes Lebesgue measure, by

$$X(\omega) = \sqrt{5} \tan\left(\pi\omega - \tfrac{\pi}{2}\right).$$

Find the distribution function of X.

Problem 10. For the coin-flip probability space of Example 4 of Chapter 1, let

$$X((\omega_1, \omega_2, \dots)) = \inf\{i \colon \omega_i = 0\}$$

if $\omega_i = 0$ for some i, and let

$$X((1, 1, 1, 1, \dots)) = 23.$$

Prove that X is a random variable and calculate its distribution function.

Problem 11. [Geometric distributions] Fix $p \in [0, 1)$ and let X be a random variable which, for each nonnegative integer x, equals x with probability $(1 - p)p^x$, where 0^0 is understood to equal 1. Calculate the distribution function F of X. Conversely, starting with F, use the methods of this chapter to construct a random variable whose distribution function is F.

3.3. Some descriptive terminology

In this section we introduce some definitions that are useful in comparing and describing different distributions.

Definition 5. Let X and Y be $\mathbb{R}$-valued random variables. Then Y is *of the same type* as X if Y has the same distribution as $aX + b$ for some constants $a \in (0, \infty)$ and $b \in \mathbb{R}$. It is *of the same strict type* if b can be chosen equal to 0.

We apply the phrase *of the same type* to distributions and distribution functions of random variables as well as to the random variables themselves. Thus, two distributions are *of the same type* if they are the distributions of random variables of the same type.

$*$ **Problem 12.** Explain why two distribution functions F_1 and F_2 are of the same type if and only if

$$(3.2) \qquad\qquad F_2(x) = F_1((x - b)/a)$$

for some constants $a \in (0, \infty)$ and $b \in \mathbb{R}$ and all x. Also, explain why replacing b by 0 gives necessary and sufficient conditions for the distributions to be of the same strict type.

Problem 13. Prove that the relation of being of the same type is an equivalence relation.

Problem 14. Show that the distribution function that is the answer to Problem 10 is of the same type as a distribution function described in Problem 11.

It is clear that the delta distributions are all of the same type. These distributions are said to be *of degenerate type*. The delta distributions constitute three strict types: those at negative points, those at positive points, and the single delta distribution at 0.

Proposition 6. *Let Y be an $\mathbb{R}$-valued random variable of nondegenerate type, and let $a > 0$ and b be constants. If Y has the same distribution as $aY + b$, then $a = 1$ and $b = 0$.*

PROOF. Let F denote the distribution function of Y and let $(\Psi, \mathcal{G}, R)$ denote the probability space on which Y is defined. Define X, Ω, and P as in Proposition 4. By that proposition, the distribution function of X is F and thus X has the same distribution as Y. For any real number y,

$$R(\{\psi \in \Psi \colon aY(\psi) + b \leq y\}) = R\big(\{\psi \in \Psi \colon Y(\psi) \leq \frac{y - b}{a}\}\big)$$
$$= P\big(\{\omega \in \Omega \colon X(\omega) \leq \frac{y - b}{a}\}\big)$$
$$= P(\{\omega \in \Omega \colon aX(\omega) + b \leq y\}).$$

Thus $aX + b$ has the same distribution as $aY + b$, and, therefore, X and $aX + b$ have the same distribution.

Since X is increasing and left-continuous and since $a > 0$, $aX + b$ is also increasing and left-continuous. The construction in Proposition 4 shows that the distribution functions of X and $aX + b$ are their 'right-continuous inverses'. Since they both have the same distribution function, they have the same 'right-continuous inverse'. It follows that they are the same function. Since their common distribution is nondegenerate, they are not constant functions, so there exist two members ω_1 and ω_2 of $(0,1)$ such that $X(\omega_1) \neq X(\omega_2)$. The constants a and b must satisfy

$$aX(\omega_1) + b = X(\omega_1)$$
$$aX(\omega_2) + b = X(\omega_2).$$

This can be regarded as a system of two linear equations in the unknowns a and b. Since $X(\omega_1) \neq X(\omega_2)$, the determinant of the coefficient matrix is nonzero, so there exists a unique solution, which is obviously $a = 1$, $b = 0$. $\square$

Problem 15. Show that if Y is of degenerate type, then there exist constants $b \neq 0$ and $a > 0$ such that Y has the same distribution as $aY + b$.

Problem 16. Show by example that Proposition 6 is false if we allow $a < 0$.

Problem 17. Let Y be a $\mathbb{R}$-valued random variable whose distribution is not the delta distribution at 0. Add whatever is necessary to Proposition 6 to prove that if $a > 0$ and Y has the same distribution as aY, then $a = 1$.

Remark 1. We have introduced several named distributions or families of distributions: uniform, delta, Cauchy, Bernoulli, and geometric. We will be introducing other families of distributions later in this chapter and in other chapters. There are times when it is convenient to include in a family of distributions all those distributions that are of the same type as any member of the family. For example, the distribution of Problem 10 is often called a geometric distribution. This ambiguity will not usually cause any confusion. For example, when we use a phrase like "the family of Cauchy distributions" it is clear that we mean all distributions of the same type as the distribution introduced in Problem 8. If we want to be more precise, we can use language like "the standard geometric distribution with parameter p" when we wish to refer specifically to one of the distributions introduced in Problem 11. A phrase like "distributions of geometric type" can be used when we want to make it clear that we are speaking of all those distributions that are of the same type as the ones in Problem 11. We warn the reader, however, that the word 'standard' does not have a generally accepted precise meaning in this context, and we will not try to give it one here. The term 'family' is also imprecise. For instance, the geometric distributions of

Problem 11 could all be said to belong to the same family, but those for different p are not of the same type. On the other hand, the "family of uniform distribution functions" would usually denote the set of distribution functions that are of the same type as the distribution function $x \rightsquigarrow (x \vee 0) \wedge 1$.

Definition 7. Let $(\Omega, \mathcal{F}, P)$ be a probability space such that Ω is a topological space and $\mathcal{F}$ is the corresponding Borel σ-field. If there exists a closed set $C \subseteq \Omega$ such that

(i) $P(C) = 1$ and
(ii) $P(C') < 1$ for all closed proper subsets C' of C,

then C is called the *support* of P. If the support of P is contained in a set D, then we say that P is *supported by* D. In the case that $(\Omega, \mathcal{F}) = (\mathbb{R}, \mathcal{B})$, the support of P is also called the *support* of the distribution function corresponding to P.

It can be shown that the support of a probability measure P exists if Ω is a sufficiently nice topological space. The following result shows that if Ω is the real line, the support exists and can be identified explicitly in terms of the distribution function.

Proposition 8. *If P is a probability measure on $(\mathbb{R}, \mathcal{B})$ with distribution function F, then the support of P is the set*

$$C = \{x \in \mathbb{R} \colon F(x + \varepsilon) - F(x - \varepsilon) > 0 \text{ for all } \varepsilon > 0\}.$$

PROOF. We first show that C is a closed set. Choose a point $y \notin C$. By the definition of C, there exists an $\varepsilon > 0$ such that $F(y + \varepsilon) \leq F(y - \varepsilon)$. It follows from the fact that F is increasing that F is constant on the interval $[y - \varepsilon, y + \varepsilon]$. Thus, by the definition of C, $(y - \varepsilon, y + \varepsilon) \subseteq C^c$. We have shown that every point in C^c has an open neighborhood that also lies in C^c, so C is closed.

We have also shown that every point $y \notin C$ is the midpoint of an open interval on which F is constant. For each $y \notin C$, let J_y be the maximal such interval. (Clearly, J_y exists since arbitrary unions of open intervals centered at y are themselves open intervals centered at y.) By Problem 5, $P(J_y) = 0$ for all $y \notin C$. It is easily seen that

$$C^c = \bigcup_{y \notin C, y \text{ rational}} J_y,$$

so C^c is a countable union of null events. Thus $P(C) = 1$.

It remains to show that if C' is a closed proper subset of C, then $P(C') < 1$. Let C' be a closed proper subset of C, and let x be a point in $C \setminus C'$. Since C' is closed, there exists an $\varepsilon > 0$ such that $[x - \varepsilon, x + \varepsilon] \subseteq (C')^c$. By the definition of C, $F(x + \varepsilon) - F(x - \varepsilon) > 0$. By Problem 5, $P([x - \varepsilon, x + \varepsilon]) > 0$, from which it follows immediately that $P(C') < 1$. $\square$

Problem 18. Prove that if the support of a probability measure P exists, then it is unique.

Problem 19. Find the uniform distribution function whose support is $[c, d]$, where $-\infty < c < d < \infty$.

Problem 20. Show that all Cauchy distributions have the same support.

Problem 21. Let X be an $\mathbb{R}$-valued random variable, a a positive constant, and b a real constant. Set $Y = aX + b$. Let c and d denote the infimum and the supremum, respectively, of the support of the distribution of X. Prove that $ac + b$ and $ad + b$ are the infimum and supremum, respectively, of the support of the distribution of Y. Make sure the proof encompasses the possibilities $c = -\infty$ and $d = \infty$.

Problem 22. For any $\mathbb{R}$-valued random variable X describe the support of the distribution of X^2 in terms of the support of the distribution of X. *Hint:* Be careful.

* **Problem 23.** An $\mathbb{R}$-valued random variable X and its distribution are said to be *symmetric* about a point $b \in \mathbb{R}$ if $X - b$ and $b - X$ have the same distribution. The distribution of such a random variable X is also said to be *symmetric* about b. Reformulate this definition solely in terms of distribution functions. Also show that the standard Cauchy distribution is symmetric about 0. What other distributions introduced so far are symmetric about some point?

Problem 24. Let X be a real-valued random variable, and suppose that b is a real number such that both the quantities

$$P(\{\omega \colon X(\omega) < b\}) \text{ and } P(\{\omega \colon X(\omega) > b\})$$

are less than or equal to $1/2$. Then b is called a *median* of the distribution of X. Show that every distribution has a least one median, and find some distributions with more than one median, as well as distributions with exactly one median. Show that if a distribution is symmetric about b, then b is a median of that distribution. Is it possible for a distribution that is symmetric about b to have a median that is not equal to b?

Problem 25. Describe a graphical procedure for finding all the medians of a distribution, given the graph of its distribution function.

Problem 26. Show that if the support of a distribution is the entire real line, then the distribution has exactly one median.

Problem 27. Which of the families uniform, delta, Bernoulli, Cauchy, and geometric, contain distributions of different types? Which of them contain more than one member having the same support?

3.4. Distributions with densities

Some important distribution functions can be represented as integrals.

Definition 9. Let F be a distribution function that can be represented in the form

$$F(x) = \int_{-\infty}^{x} f(t)\, dt\,.$$

Then f is a *density* of F, and also of the corresponding probability measure of any corresponding random variable.

Whenever a density exists for a distribution function F, then F has infinitely many densities. For instance, the value of f can be changed at finitely many points without changing F. In spite of this fact, we will sometimes loosely speak of 'the' density of F. In many important cases, a distribution function F will have a density that is continuous on the support of F and vanishes elsewhere. There can be at most one such density, and when there is one, the phrase 'the density' usually refers to it. More will be said about the concept of density in Chapter 8.

* **Problem 28.** Find the densities of all Cauchy and uniform distributions.

Problem 29. Show that if f is a nonnegative function that is Riemann-integrable on every closed bounded interval in $\mathbb{R}$ and if

$$\int_{-\infty}^{\infty} f(t)\, dt = 1\,,$$

then f is the density of some distribution.

* **Problem 30.** [Exponential distributions] Let a be a positive real parameter. Show that

$$f(x) = \begin{cases} ae^{-ax} & \text{if } x \geq 0 \\ 0 & \text{if } x < 0 \end{cases}$$

is a density. Let X be a random variable with density f. Calculate the probability that X belongs to the interval $[2, 3]$. Also, calculate the median of the distribution of X.

Problem 31. Let X have the distribution of the preceding problem. For $n = 1, 2, \ldots$, let Y_n equal $\lfloor X \rfloor / n$. Prove that Y_n is a random variable, and calculate and identify its distribution function.

Problem 32. Let X be a random variable with density f. Let $a > 0$ and b be constants. Show that $x \rightsquigarrow \frac{1}{a} f((x - b)/a)$ is a density of the random variable $aX + b$. Formulate a theorem about densities of the same type.

* **Problem 33.** Let X be a random variable with density f. Show that X^2 has a density g, and find a formula for g in terms of f.

Example 1. [Normal or Gaussian distributions] We show that

$$f(x) = \tfrac{1}{\sqrt{2\pi}} e^{-x^2/2}$$

is a density. Obviously it is nonnegative. The square of its integral over $\mathbb{R}$ is

$$\frac{1}{2\pi} \int_{-\infty}^{\infty} e^{-x^2/2}\, dx \int_{-\infty}^{\infty} e^{-y^2/2}\, dy = \frac{1}{2\pi} \int_{-\infty}^{\infty} \int_{-\infty}^{\infty} e^{-(x^2+y^2)/2}\, dx\, dy$$

$$= \frac{1}{2\pi} \int_{0}^{\infty} \int_{0}^{2\pi} e^{-r^2/2} r\, dr = 1\,.$$

(Since the integrand is nonnegative and continuous, the replacement of an iterated integral by a double integral and the change of variables in the double integral are justified by results from advanced calculus. The validity of these steps is also a consequence of Theorem 10 and Theorem 15, both of Chapter 9.) Thus, f is a density. A simple change of variables shows that densities for other distributions of the same type can be written in the form

$$\tfrac{1}{\sqrt{2\pi a^2}} e^{-(x-b)^2/2a^2}\,.$$

Example 2. [Gamma distributions] For γ a positive real parameter, the *gamma function* Γ is defined by

$$\Gamma(\gamma) = \int_{0}^{\infty} u^{\gamma-1} e^{-u}\, du\,,$$

which is easily seen to be a convergent improper integral. It follows that the function

$$f(x) = \begin{cases} \dfrac{a^{\gamma} x^{\gamma-1} e^{-ax}}{\Gamma(\gamma)} & \text{if } x > 0 \\ 0 & \text{if } x \leq 0 \end{cases}$$

is a density for all $a > 0$. When the parameter γ equals 1, a gamma distribution is an exponential distribution.

* **Problem 34.** Prove the following four facts about the gamma function Γ:
   (i) $\Gamma(\gamma + 1) = \gamma \Gamma(\gamma)$ for $\gamma > 0$.
   (ii) $\Gamma(\gamma) = (\gamma - 1)!$ for $\gamma = 1, 2, \dots$.
   (iii)

$$\Gamma(\gamma) = \frac{\sqrt{\pi}(2\gamma - 1)!}{((2\gamma - 1)/2)!\, 2^{2\gamma-1}} \quad \text{for } \gamma = \frac{1}{2}, \frac{3}{2}, \frac{5}{2}, \dots.$$

   (iv)

$$\Gamma(\alpha)\Gamma(\beta) = \Gamma(\alpha + \beta) \int_{0}^{1} x^{\alpha-1}(1 - x)^{\beta-1}\, dx \quad \text{for } \alpha, \beta > 0\,.$$

Problem 35. Let Y denote a random variable having some gamma distribution as described in Example 2 with $a = 1$. Find the density of the random variable $\exp \circ (-Y)$.

Example 3. [Beta distributions] Fix parameters $\alpha, \beta > 0$. From the last fact in Problem 34 we see that

$$f(x) = \begin{cases} \dfrac{\Gamma(\alpha + \beta)x^{\alpha-1}(1 - x)^{\beta-1}}{\Gamma(\alpha)\Gamma(\beta)} & \text{if } 0 < x < 1 \\ 0 & \text{otherwise} \end{cases}$$

is a density, called the *beta density*. The beta distribution with parameters $\alpha = 1$ and $\beta = 1$ is a uniform distribution.

Problem 36. [Arcsin distribution] Calculate the distribution function of the beta distribution having parameters $\alpha = \beta = 1/2$. Sketch the graph of its density; the distribution function itself is shown in Figure 3.3.

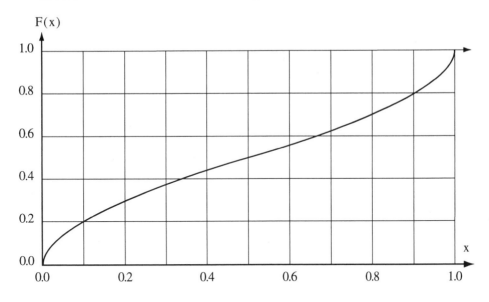

FIGURE 3.3. Arcsin distribution function: $F(x) = \frac{2}{\pi} \arcsin \sqrt{x}$

3.5. Further examples

A variety of distribution functions will be introduced in this section, including one (Problem 42) that does not have a density even though it is continuous. The distributions described in the next three problems play important roles in probability theory.

Problem 37. [Poisson distributions] Let $\lambda \in (0, \infty)$. Prove that there exists a distribution function that has a jump of size

$$\frac{\lambda^x e^{-\lambda}}{x!}$$

at each nonnegative integer x.

Problem 38. The distribution obtained in Example 3 of Chapter 1 of the number of heads in n flips of a fair coin is called a 'binomial distribution'. Sketch the distribution function for the case $n = 4$.

Problem 39. [Binomial distributions] Fix a positive integer n and a number $p \in (0, 1)$. Let $q = 1 - p$. Prove that there exists a distribution function that, for each integer x satisfying $0 \leq x \leq n$, has a jump at x of size

$$\binom{n}{x} p^x q^{n-x} .$$

* **Problem 40.** Calculate and name the distribution function of $-\log \circ [X/b]$, where X is a random number uniformly distributed on $(0, b]$.

Example 4. As in Example 3 of Chapter 2, let Ψ be the space of chords of the unit circle in $\mathbb{R}^2$, and let $\mathcal{B}$ denote the Borel field of subsets of Ψ. Let $Y : \Psi \to \mathbb{R}$ be the function that assigns to each chord in Ψ its length. The reader may check that Y is continuous, and hence measurable.

In Chapter 2 four different interpretations of the experiment of choosing a random chord were given by defining four Ψ-valued random variables $X_i, i = 1, 2, 3, 4$. Each of these random variables induces a distribution Q_i on $(\Psi, \mathcal{B})$. By Proposition 7 of Chapter 2, the distribution of Y as a random variable on the probability space $(\Psi, \mathcal{B}, Q_i)$ is the same as the distribution of $Y \circ X_i$ on the underlying probability space, the members of which are ordered pairs (α, β).

In this example, we compute the distribution function G_1 of $Y \circ X_1$. A little trigonometry gives

$$Y \circ X_1(\omega) = 2|\sin \pi(\beta - \alpha)| .$$

Clearly $G_1(y) = 0$ for $y < 0$ and $G_1(y) = 1$ for $y \geq 2$. For $y \in [0, 2)$, $\{\omega : Y \circ X_1(\omega) \leq y\}$ consists of a strip centered on the line of slope 1 through the origin, together with two right triangles having right angles at $(1, 0)$ and $(0, 1)$ and having heights (measured from the hypotenuse) equal to half the width of the strip, as illustrated by the shaded region in Figure 3.4. The area of this three-part region is $2^{1/2}$ multiplied by the width of the strip. The width of the strip is $2^{1/2}$ multiplied by the value of β at the intersection of the upper edge of the strip with the axis $\alpha = 0$. Setting $\alpha = 0$ and solving for β in the expression for $Y \circ X_1$, we find that this intersection occurs at $\beta = [\arcsin(y/2)]/\pi$. Thus,

$$G_1(y) = \frac{2}{\pi} \arcsin \frac{y}{2} \quad \text{for } 0 \leq y < 2 .$$

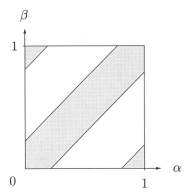

FIGURE 3.4. $\{\omega \colon Y \circ X_1(\omega) \le y\}$

* **Problem 41.** For Y as defined in the preceding example, calculate the distribution function of $Y \circ X_i$ for $i = 2, 3, 4$.

Problem 42. [Cantor distribution] Consider the coin-toss probability space of Example 4 of Chapter 1, and let

$$X((\omega_1, \omega_2, \dots)) = 0.\varepsilon_1\varepsilon_2\varepsilon_3 \cdots {}_{\text{three}} \,,$$

where each $\varepsilon_i = 2\omega_i$ and the subscript 'three' indicates that the expression is to be regarded as a base-three numeral. Sketch a graph of the distribution function F of X. Prove that F is continuous. Prove that there exists a Borel set $A \subseteq [0, 1]$ with Lebesgue measure equal to 1, such that for all $x \in A$, $F'(x) = 0$.

3.6. Distribution functions for the extended real line

The value 23 used in Problem 10 is artificial. A natural value to use for the first time that the coin comes up tails is ∞ in case all of the infinitely many flips come up heads. Indeed, it is usual to define the infimum of the empty set in $\mathbb{R}$ to equal ∞. Thus, in this example it is natural to consider a random variable that is $\overline{\mathbb{R}}$- or $\overline{\mathbb{R}}^+$-valued.

Definition 10. An $\mathbb{R}$-valued function F defined on $\mathbb{R}$ is a *distribution function for* $\overline{\mathbb{R}}$ if it is increasing and right-continuous and $0 \le F(x) \le 1$ for every $x \in \mathbb{R}$.

If Q is a distribution on $\overline{\mathbb{R}}$, then the *distribution function* of Q is the function $x \rightsquigarrow Q([-\infty, x])$. A theory for distribution functions for $\overline{\mathbb{R}}$ can be given that parallels that for distribution functions for $\mathbb{R}$. Since the similarities with the theory for $\mathbb{R}$ are so strong, we omit the details and only comment that if F is the distribution function of a distribution Q on $\overline{\mathbb{R}}$ and if $F(-\infty), F(\infty)$ are the limits, respectively, of F at $-\infty, \infty$, then $F(-\infty) = Q(\{-\infty\})$ and $1 - F(\infty) = Q(\{\infty\})$.

Example 5. For the fair-coin probability space of Example 4 of Chapter 1, let
$$N(\omega) = \inf\{n : \omega_n = \omega_{n+1} = \cdots = \omega_{2n} = 1\}.$$
That is, N is the first time at which there begins a sequence of heads longer than the number of flips up to and including the beginning of that sequence; $N = \infty$ if there is no such time. Let us show that $N = \infty$ with positive probability. Clearly,
$$P(\{\omega : N(\omega) = n\}) \le P(\{\omega : \omega_n = \omega_{n+1} = \cdots = \omega_{2n} = 1\}) = 2^{-(n+1)}.$$
Hence, with F denoting the distribution function of N,
$$F(\infty) = \sum_{n=1}^{\infty} P(\{\omega : N(\omega) = n\}) \le \sum_{n=1}^{\infty} 2^{-(n+1)} = 1/2.$$
By proving that $P(\{\omega : N(\omega) = \infty\}) > 0$, we have shown that ∞ is in the support of N. However, it should be noted that ∞ can be in the support of a random variable X even if $P(\{\omega : X(\omega) = \infty\}) = 0$. For an example, take any geometrically distributed random variable with parameter $p > 0$, regarded as an $\overline{\mathbb{R}}^+$-valued random variable.

CHAPTER 4

Expectations: Theory

The 'expectation' of an $\mathbb{R}$-valued random variable is a weighted average of the values taken by that random variable. It is a useful tool for the description and analysis of random variables and their distributions. Properties of expectations treated in this chapter include linearity and an important convergence theorem. The calculation of expectations is facilitated by establishing a connection with Riemann-Stieltjes integration.

4.1. Definitions

The expectation is first defined for simple random variables.

Definition 1. Let X be a random variable defined on a probability space $(\Omega, \mathcal{F}, P)$ and having the form

$$(4.1) \qquad X = \sum_{j=1}^{n} c_j I_{C_j}$$

for some distinct real constants c_j and events C_j that constitute a partition of Ω. Then the *expectation* of X equals

$$(4.2) \qquad \sum_{j=1}^{n} c_j P(C_j)$$

and is denoted by $E(X)$.

By definition, every random variable of the form (4.1) is simple, and by Lemma 9 of Chapter 2, every simple random variable can be written uniquely in the form (4.1).

Problem 1. Part of the preceding sentence assures us that Definition 1 is not ambiguous. Explain.

Problem 2. Prove that if X is a simple random variable, then $E(X)$ can be written in terms of the distribution function F of X:

$$E(X) = \sum_{x \in \mathbb{R}} x[F(x) - F(x-)],$$

the summation being meaningful since there are only finitely many nonzero terms.

The set of simple random variables on a probability space is a real vector space since it is closed under multiplication by a real number and under addition. We may think of E as a function defined on that vector space. Since the elements of the vector space are themselves functions, it is customary to call E an *operator* rather than a function. Thus, E may be called the *expectation operator*, with domain equal to the set of simple random variables defined on any particular probability space. We will see in Lemma 3 that E is *linear*. The following lemma is the first step towards the proof of Lemma 3.

Lemma 2. *Let* $X = \sum_{j=1}^{n} c_j I_{C_j}$ *be a simple random variable and suppose that* $(C_j : 1 \leq j \leq n)$ *is a finite sequence of pairwise disjoint events whose union is* Ω. *Then*

$$E(X) = \sum_{j=1}^{n} c_j P(C_j).$$

Problem 3. Prove the preceding lemma. (Notice that it is not assumed that the real constants c_j are distinct, nor is it assumed that the events C_j are nonempty.)

In some vector spaces there is a natural concept of positiveness. For instance, in $\mathbb{R}^2$, any ordered pair, other than $(0,0)$, that has two nonnegative coordinates might be called positive. It is required that the set of *positive* members of a vector space be closed under both multiplication by positive scalars and addition. For the vector space of simple random variables on a probability space, the positive members are those simple random variables, other than the zero function, whose values are nonnegative. An operator on a vector space is said to be *positive* if it maps all positive members of the vector space into $\mathbb{R}^+$.

Lemma 3. *The expectation operator E on the vector space of simple random variables on some probability space is both linear and positive.*

PROOF. Consider an arbitrary simple random variable X and a real number b. By Lemma 9 of Chapter 2, X can be represented in the form (4.1), and so

$$bX = \sum_{j=1}^{n} (bc_j) I_{C_j}.$$

By Lemma 2, $E(bX) = \sum_{j=1}^{n} (bc_j) P(C_j) = bE(X)$.

To complete the proof of linearity of E, consider two simple random variables

$$X = \sum_{j=1}^{n} c_j I_{C_j} \quad \text{and} \quad Y = \sum_{i=1}^{p} d_i I_{D_i},$$

where both $\{C_j : 1 \le j \le n\}$ and $\{D_k : 1 \le k \le p\}$ are partitions of the underlying probability space Ω. Then $(C_j \cap D_k : 1 \le j \le n, 1 \le k \le p)$ consists of pairwise disjoint events whose union is Ω and

$$X + Y = \sum_{j=1}^{n} \sum_{k=1}^{p} (c_j + d_k) I_{C_j \cap D_k}.$$

By Lemma 2

$$E(X + Y) = \sum_{j=1}^{n} \sum_{k=1}^{p} (c_j + d_k) P(C_j \cap D_k)$$

$$= \left(\sum_{j=1}^{n} \sum_{k=1}^{p} c_j P(C_j \cap D_k) \right) + \left(\sum_{k=1}^{p} \sum_{j=1}^{n} d_k P(C_j \cap D_k) \right)$$

$$= \left(\sum_{j=1}^{n} c_j P(C_j) \right) + \left(\sum_{k=1}^{p} d_k P(D_k) \right)$$

$$= E(X) + E(Y),$$

as desired.

Turning to positivity, we suppose that X is given by (4.1) and that its image consists only of nonnegative numbers. Then each c_j in (4.1) is nonnegative and, hence, by (4.2), $E(X) \ge 0$. Therefore, E is a positive operator. $\square$

Problem 4. Suppose that

$$X = \sum_{i=1}^{m} a_i I_{A_i}$$

for some integer m, some events A_i, and some real constants a_i. Prove that

$$E(X) = \sum_{i=1}^{m} a_i P(A_i).$$

(Notice that there is no assumption that $\{A_i : 1 \le i \le m\}$ is a partition of the underlying probability space.)

Lemma 4. *If X and Y are two simple $\mathbb{R}$-valued random variables for which $X \le Y$ a.s., then $E(Y) \le E(Z)$. If $X = Y$ a.s., then $E(X) = E(Y)$.*

Problem 5. Prove the preceding lemma.

Expectations are called by other names as well: *expected value* and *mean* are two common synonyms for 'expectation'.

Example 1. The mean of a simple binomially distributed random variable (defined in Problem 39 of Chapter 3) is, by Definition 1,

$$\sum_{k=0}^{n} k \frac{n!}{k!(n-k)!} p^k q^{n-k} = \sum_{k=1}^{n} \frac{n!}{(k-1)!(n-k)!} p^k q^{n-k}$$

$$= np \sum_{j=0}^{n-1} \frac{(n-1)!}{j!(n-1-j)!} p^j q^{n-1-j}$$

$$= np(p+q)^{n-1} = np.$$

Problem 6. Comment on the presence of the adjective 'simple' in the preceding example.

* **Problem 7.** Calculate $E(X^2)$ for a simple binomially distributed random variable X.

* **Problem 8.** Calculate the expected value of the outcome of a single roll of a fair six-sided die.

The next definition extends the concept of expectation to all $\overline{\mathbb{R}}^{+}$ random variables.

Definition 5. The *expectation* $E(X)$ of an $\overline{\mathbb{R}}^{+}$-valued random variable X defined on a probability space $(\Omega, \mathcal{F}, P)$ equals $\sup_Z E(Z)$ where the supremum is taken over all nonnegative simple random variables Z on $(\Omega, \mathcal{F}, P)$ that satisfy $Z \leq X$.

We notice several things about the preceding definition. It applies when X is a nonnegative simple random variable; it had better agree with the previously given definition in this case. It is conceivable that a random variable can have expectation equal to ∞. The zero random variable necessarily qualifies as one of the random variables Z over which the supremum is to be taken, so E as thus extended satisfies $E(X) \geq 0$ for nonnegative random variables X. Lemma 6 below also shows that the extended version of E is still linear.

* **Problem 9.** For X a nonnegative simple random variable prove that $E(X)$ as defined in Definition 5 equals $E(X)$ as defined in Definition 1.

* **Problem 10.** Let X be an $\overline{\mathbb{R}}^{+}$-valued random variable, and let $A = \{\omega: X(\omega) = \infty\}$. Show that if $P(A) > 0$, then $E(X) = \infty$. Find an example for which $P(A) = 0$, but the expectation of X is still infinite.

Lemma 6. *Let X and Y be $\overline{\mathbb{R}}^+$-valued random variables defined on a common probability space, and let a and b be constants belonging to $\overline{\mathbb{R}}^+$. Then,*

$$E(aX + bY) = aE(X) + bE(Y),$$

whether finite or infinite, provided that $0 \cdot \infty$ and $\infty \cdot 0$ are understood to equal 0.

* **Problem 11.** Prove the preceding lemma.

Lemma 7. *Let X and Y be $\overline{\mathbb{R}}^+$-valued random variables defined on a common probability space. If $X \leq Y$ a.s., then $E(X) \leq E(Y)$. If $X = Y$ a.s., then $E(X) = E(Y)$.*

Problem 12. Prove the preceding lemma, making sure to encompass the cases of infinite expectation. *Hint:* Use Lemma 4.

In preparation for extending the concept of expectation to an $\overline{\mathbb{R}}$-valued random variable X, we introduce two $\overline{\mathbb{R}}^+$-valued random variables related to X. For each ω, set

$$X^+(\omega) = 0 \vee X(\omega) \quad \text{and} \quad X^-(\omega) = 0 \vee [-X(\omega)].$$

The function X^+ is called the *positive part* of X, and X^- is called its *negative part*.

Problem 13. For X an $\overline{\mathbb{R}}$-valued random variable, prove that X^+ and X^- as just defined are $\overline{\mathbb{R}}^+$-valued random variables for which $X = X^+ - X^-$ and $|X| = X^+ + X^-$.

Since X^+ and X^- are both $\overline{\mathbb{R}}^+$-valued random variables, $E(X^+)$ and $E(X^-)$ are both defined. We will define the expectation of X to be the difference of these two values, when the difference is meaningful.

Definition 8. Let X be a random variable taking values in $\overline{\mathbb{R}}$. If $E(X^+)$ and $E(X^-)$ are not both infinite, the *expectation* of X is given by

$$E(X) = E(X^+) - E(X^-).$$

Otherwise, the expectation of X *does not exist.*

Since $X^- = 0$ for a $\overline{\mathbb{R}}^+$-valued random variable X, it is obvious that this new definition of $E(X)$ agrees with the previous one for such X. The following exercise asks the reader to prove that no ambiguity arises from the new definition of $E(X)$ when X is a simple (not necessarily $\mathbb{R}^+$-valued) random variable.

* **Problem 14.** For X a simple random variable, show that $E(X)$ as defined by Definition 1 equals $E(X)$ as defined by Definition 8.

A glance at Definition 1, which is needed to make sense of Definition 5 and Definition 8, shows that the definition of expectation involves the underlying probability measure P. Thus, we have been somewhat imprecise in speaking of 'the' expectation operator; changing the underlying probability space changes the expectation operator. This lack of precision is justified in those circumstances when the underlying probability space is clear from the context. And, as we will soon see, the expectation of a random variable is determined by its distribution, so when the distribution of a random variable X is known, knowledge of the underlying probability space is not necessary for the calculation of $E(X)$.

Nevertheless, it is sometimes necessary to distinguish expectation operators defined for different probability spaces. (See Theorem 15 for an example.) In such cases, the expectation operator on the space of $\overline{\mathbb{R}}$-valued random variables with underlying probability space $(\Omega, \mathcal{F}, P)$ is denoted by E_P and is called the *expectation operator with respect to P*. The quantity $E_P(X)$ is called the *expectation of X with respect to P*.

4.2. Linearity and positivity

In this section, we prove linearity and other basic properties of the expectation operator. Preliminary versions of some of these properties have appeared in Lemma 3, Lemma 4, Lemma 6, and Lemma 7.

The following convention, already used in Lemma 6, will simplify our discussion. We will adhere to it throughout the book.

CONVENTION. Unless explicitly stated otherwise, the products $0 \cdot \infty$ and $\infty \cdot 0$ will be interpreted to equal 0.

Theorem 9. *Let X and Y be $\overline{\mathbb{R}}$-valued random variables defined on a probability space $(\Omega, \mathcal{F}, P)$, and let a, b, and c be real constants.*

(i) If $aX(\omega) + bY(\omega)$ is defined for all ω, then $E(aX + bY) = aE(X) + bE(Y)$, provided the expression on the right is meaningful.

(ii) If $X = c$ a.s., then $E(X) = c$.

(iii) If $X = Y$ a.s., then either the expectations of X and Y both exist and are equal, or neither exists.

(iv) If $X \leq Y$ a.s. and either $E(X)$ exists and is different from $-\infty$ or $E(Y)$ exists and is different from ∞, then the other of $E(Y)$ and $E(X)$ exists and $E(X) \leq E(Y)$.

(v) If $E(X) = E(Y)$ is finite and $X \leq Y$ a.s., then $X = Y$ a.s.

(vi) If $E(X)$ exists, then $|E(X)| \leq E(|X|)$.

(vii) If $E(X)$ does not exist, then $E(|X|) = \infty$.

(viii) If $X(\omega) + Y(\omega)$ is defined for all ω, then $E(|X + Y|) \leq E(|X|) + E(|Y|)$.

PARTIAL PROOF. We first prove a special case of (i) —namely that

(4.3) $E(X + Y) = E(X) + E(Y)$

whenever the expression on the right is meaningful. We use the following identity:

$$(X + Y)^+ + Y^- + X^- = (X + Y)^- + Y^+ + X^+.$$

On each side we have a sum of three $\overline{\mathbb{R}}^+$-valued random variables. By Lemma 6

$$E((X + Y)^+) + E(Y^-) + E(X^-) = E((X + Y)^-) + E(Y^+) + E(X^+).$$

If all terms are finite, they can be rearranged to give the desired conclusion:

$$E((X + Y)^+) - E((X + Y)^-) = (E(X^+) - E(X^-)) + (E(Y^+) - E(Y^-)).$$

Consideration of the cases involving some infinite terms is left for the reader.

We also leave it to the reader to use the definition of $E(X)$ in conjunction with Lemma 6 to prove that $E(aX) = aE(X)$ for all real a, provided that $E(X)$ exists. Assertion (i) follows from this fact and (4.3).

We leave it to the reader to prove (iv). Then (iii) follows from two applications of (iv), one using (iv) as it stands and the other using (iv) with X and Y interchanged. Since the expected value of a constant random variable is that constant, (ii) as a special case of (iii). The contrapositive of (vii) is a consequence of (iv) and the inequality $X \leq |X|$. Thus, it remains for us to prove (v), (vi), and (viii).

Assertion (viii) follows from the inequality $|X + Y| \leq |X| + |Y|$, (iv), and the consequence $E(|X| + |Y|) = E(|X|) + E(|Y|)$ of (i).

Suppose that $E(X)$ exists. By (i), $E(-X) = -E(X)$ also exists. Then two applications of (iv)—one to X and $|X|$ and the other to $-X$ and $|X|$ —gives (vi).

To prove (v) we suppose that $E(X) = E(Y)$ is finite and $X \leq Y$ a.s.. The expected value of the nonnegative random variable $(Y - X)$ exists, so that (i) may be applied to the equality $Y = X + (Y - X)$ to give $E(Y) = E(X) + E(Y - X)$ from which it follows that $E(Y - X) = 0$. Thus, for any $\varepsilon > 0$, the simple function that equals 0 when $Y - X < \varepsilon$ and equals ε when $Y - X \geq \varepsilon$ has 0 expectation. Therefore,

$$P(\{\omega : Y(\omega) - X(\omega) \geq \varepsilon\}) = 0.$$

Let $\varepsilon \searrow 0$ through a sequence and use the Continuity of Measure Theorem to obtain $P(\{\omega : Y(\omega) - X(\omega) > 0\}) = 0$, as desired. □

Problem 15. Complete the proof of the preceding theorem.

For a random variable X, let $[X] = \{Y : Y = X \text{ a.s.}\}$. Thus, $[X]$ is the equivalence class of X mentioned in Chapter 2. In view of property (iii) in the preceding theorem, we may, with no ambiguity, define $E([X]) = E(X)$, with the understanding that the left side is defined if and only if the right side is defined. Therefore, E becomes an operator on equivalence classes.

From (v) of the preceding theorem we see that $E([Y]) > 0$ if $Y \geq 0$ and $[Y] \neq [0]$. A positive operator on a vector space (in which certain members have been identified as positive) is said to be *strictly positive* if the value it assigns to each positive member of the vector space is positive (not just nonnegative). In the following corollary we use the notation $\|[X]\|$; doing so is legitimate since $\|[X]\| = \|[Y]\|$ whenever $[X] = [Y]$.

Corollary 10. *The equivalence classes $[X]$ of random variables X on a probability space $(\Omega, \mathcal{F}, P)$ for which $E(\|[X]\|) < \infty$ constitute a vector space on which the expectation operator is linear and strictly positive.*

Problem 16. Prove the preceding corollary.

In practice one often does not use an equivalence class notation such as $[X]$, even when one is taking an equivalence-class point of view.

Suppose that X is a random variable defined on some sample space Ω, and that Y is a function that is defined at some but not necessarily all of the points in Ω. If the set of points where Y is undefined is contained in some null event A and if $Y(\omega) = X(\omega)$ for $\omega \in A^c$, then we write $X = Y$ a.s. and say that Y is an *a.s.-defined random variable*. It is sometimes convenient to include such functions Y in the equivalence class of X, and to define $E(Y)$ to equal $E(X)$ whenever the latter exists. Generally speaking, the theory of random variables and their expectations is easily adapted to accommodate a.s.-defined random variables.

Problem 17. Let X and Y be a.s.-defined $\overline{\mathbb{R}}$-valued random variables on the same probability space. Prove that if $E(X) + E(Y)$ is meaningful, then $X + Y$ is an a.s.-defined random variable. Use this fact to improve Theorem 9, especially parts (i) and (viii).

The following example illustrates an important way in which linearity is used to compute expectations.

Example 2. Consider a deck of n cards, labeled from 1 to n. Suppose they are shuffled so that each of the $n!$ possible arrangements is equally likely. Let X be the number of cards which occupy positions equal to their labels. We will calculate the expected value of X. Let I_m equal the indicator function of the event that the card with label m is in position m. Then $X = \sum_{m=1}^{n} I_m$, and

$$E(I_m) = P(\{\omega \colon I_m(\omega) = 1\}) = \frac{(n-1)!}{n!} = \frac{1}{n}.$$

Hence, by property (i) in Theorem 9,

$$E(X) = E\left(\sum_{m=1}^{n} I_m\right) = \sum_{m=1}^{n} E(I_m) = \sum_{m=1}^{n} \frac{1}{n} = 1.$$

The technique of using indicator random variables to switch from probabilities to expectations so that linearity of expectation can then be used is a commonly used method for solving various problems.

Problem 18. Consider a deck of n cards labeled 1 to n, arranged so that, for each m, the card labeled m is in position $n - m + 1$. Suppose that the deck is 'cut at random'. That is, one of the cards is chosen on an equiprobable basis, dividing the deck into two packets, one packet containing the chosen card and all cards below the chosen card, and the other (possibly empty) packet containing the remaining cards; the order of the cards within each packet is left unchanged, but the order of the packets within the deck is reversed. What is the expected number of cards which end up in positions equal to their labels?

4.3. Monotone convergence

The following important theorem is the first of several results concerning circumstances under which it is appropriate to interchange the taking of limits with the taking of expectations. Other such results will be found in Chapter 8.

Theorem 11. [Monotone Convergence] *Let* $(X_1, X_2, \ldots)$ *be an increasing sequence of* $\overline{\mathbb{R}}^+$*-valued random variables on a common probability space* $(\Omega, \mathcal{F}, P)$. *For each* $\omega \in \Omega$ *set*

$$X(\omega) = \lim_{n \to \infty} X_n(\omega).$$

Then X *is a random variable and, as* $n \nearrow \infty$, $E(X_n) \nearrow E(X)$ *(finite or infinite).*

PROOF. That X is a random variable follows from Proposition 11 of Chapter 2. That

$$E(X_1) \le E(X_2) \le \cdots \le E(X)$$

follows from Lemma 7. Thus, by Definition 5, we can finish the proof by showing $\sup_n E(X_n) \ge E(X)$. In view of the definition of $E(X)$ we only need prove

$$\sup_n E(X_n) \ge E(Z) - \varepsilon$$

for every $\varepsilon > 0$ and every simple random variable Z that satisfies $Z \le X$.

Fix such ε and Z, and, for each n, let

$$A_n = \{\omega \colon X_n(\omega) < Z - \tfrac{\varepsilon}{2}\}.$$

Using the linearity and positivity of E, we obtain

$$E(Z) = E(ZI_{A_n}) + E((Z - \tfrac{\varepsilon}{2})I_{A_n^c}) + E(\tfrac{\varepsilon}{2}I_{A_n^c})$$
$$\leq \max\{Z(\omega) : \omega \in \Omega\}P(A_n) + E(X_n) + \tfrac{\varepsilon}{2}.$$

We only need show this last expression to be no larger than $E(X_n) + \varepsilon$ for some n. We will perform this task by showing that $P(A_n) \to 0$ as $n \to \infty$.

For each ω, $X_n(\omega) \nearrow X(\omega) \geq Z(\omega)$ as $n \nearrow \infty$. Hence, $A_1 \supseteq A_2 \supseteq \dots$ and $\bigcap_{n=1}^{\infty} A_n = \emptyset$. By the Continuity of Measure Theorem, $P(A_n) \to 0$ as $n \to \infty$. $\square$

Corollary 12. *If* $(X_1, X_2, \dots)$ *is a sequence of* $\overline{\mathbb{R}}^+$*-valued random variables, then*

$$E\left(\sum_{n=1}^{\infty} X_n\right) = \sum_{n=1}^{\infty} E(X_n).$$

Problem 19. Prove the preceding corollary.

Problem 20. Let $(A_j : j = 1, 2, \dots)$ be a sequence of events in a probability space $(\Omega, \mathcal{F}, P)$, and let $(a_j : j = 1, 2, \dots)$ be a sequence of real numbers. Assume that

$$\sum_{j=1}^{\infty} |a_j| P(A_j) < \infty.$$

Prove that

$$X(\omega) = \sum_{j=1}^{\infty} a_j I_{A_j}(\omega)$$

is an almost surely defined random variable, and that

$$E(X) = \sum_{j=1}^{\infty} a_j P(A_j).$$

The next corollary generalizes the Monotone Convergence Theorem in a useful way.

Corollary 13. *Let* $(X_1, X_2, \dots)$ *be an increasing sequence of* $\overline{\mathbb{R}}$*-valued random variables on a common probability space* $(\Omega, \mathcal{F}, P)$. *For each* $\omega \in \Omega$ *set*

$$X(\omega) = \lim_{n \to \infty} X_n(\omega).$$

If $E(X_1) > -\infty$, *then* $E(X_n) \to E(X)$ *as* $n \to \infty$.

* **Problem 21.** Prove the preceding corollary.

* **Problem 22.** Calculate $E(X)$ and $E(X^2)$ for a random variable X having a geometric distribution by first writing expressions for $E(X \wedge n)$ and $E((X \wedge n)^2)$ and then applying the Monotone Convergence Theorem as $n \to \infty$.

* **Problem 23.** Calculate $E(X)$ and $E(X^2)$ for a random variable X having a Poisson distribution by first finding expressions for $E(X \wedge n)$ and $E((X \wedge n)^2)$ and then applying the Monotone Convergence Theorem as $n \to \infty$.

4.4. Expectation of compositions

Given a random variable X which takes values in some measurable space $(\Psi, \mathcal{G})$, it is often the case that we wish to find the expected value of a random variable of the form $\varphi \circ X$, where φ is a measurable function from Ψ to $\overline{\mathbb{R}}$. In making such a calculation, it is often convenient to be able to work directly with X and its distribution, rather than to try to apply the definitions to the random variable $\varphi \circ X$. The results of this section are designed for this purpose.

We begin with a result that, despite its simplicity, is very useful when working with random variables X that take on only countably many different values. In particular, it applies when X is $\overline{\mathbb{Z}}$-valued. For simplicity, the result is stated for $\overline{\mathbb{R}}^+$-valued functions φ. It is easily extended to $\overline{\mathbb{R}}$-valued functions φ by applying it to φ^+ and φ^-.

Proposition 14. *Let X be a random variable from a probability space $(\Omega, \mathcal{F}, P)$ into a measurable space $(\Psi, \mathcal{G})$. Assume that for each $x \in \Psi$, $\{x\}$ is a measurable set. Let Q be the distribution of X, and suppose that there exists a countable set $A \subseteq \Psi$ such that $Q(A) = 1$. Then for any measurable function $\varphi \colon \Psi \to \overline{\mathbb{R}}^+$,*

$$E(\varphi \circ X) = \sum_{x \in A} \varphi(x) Q(\{x\}).$$

(The summation on the right side may be taken in any order.)

PROOF. Let $(x_1, x_2, \dots)$ be an ordering of the members of A. (We will assume for simplicity that A is infinite. The modifications required for the finite case are obvious.) For $n = 1, 2, \dots$, let

$$\varphi_n(x) = \begin{cases} \varphi(x) & \text{if } x = x_1, \dots, x_n \\ 0 & \text{otherwise.} \end{cases}$$

Then $\varphi_n \circ X$ is a simple function. By Lemma 2,

$$E(\varphi_n \circ X) = \sum_{m=1}^{n} \varphi(x_m) Q(\{x_m\}).$$

To complete the proof, let $n \nearrow \infty$, note that $\varphi_n \circ X \nearrow \varphi \circ X$ a.s., and apply the Monotone Convergence Theorem in connection with (iii) of Theorem 9. $\square$

Problem 24. Discuss the similarities and differences between the proof of the preceding proposition and the method suggested in Problem 22 and Problem 23 for computing the expected values of certain $\mathbb{Z}$-valued random variables.

Note that the sum on the right side of the formula given in Proposition 14 involves only the distribution Q of X and the function φ. It does not require knowledge of the underlying probability space $(\Omega, \mathcal{F}, P)$. The following theorem shows that in general, the quantity $E(\varphi \circ X)$ depends only on Q and the function φ, and the corollary following the theorem shows that for $\overline{\mathbb{R}}$-valued random variables X, Q determines $E(X)$. (Incidentally, in view of property (iii) in Theorem 9, this corollary allows us to drop the adjective 'simple' from Example 1 and Problem 7.)

Theorem 15. *Let X be a random variable from a probability space $(\Omega, \mathcal{F}, P)$ into a measurable space $(\Psi, \mathcal{G})$, and let Q denote the distribution of X. Denote the expectation operators with respect to P and Q by E_P and E_Q, respectively. Then, for any measurable function φ on $(\overline{\mathbb{R}}, \mathcal{B})$,*

$$E_P(\varphi \circ X) = E_Q(\varphi)$$

in the sense that if either side exists, then so does the other and they are equal.

PROOF. Suppose, first, that φ is simple. Then $\varphi \circ X$ is simple. Thus, there exists a finite partition $\{C_j : 1 \leq j \leq n\}$ of Ω and distinct real constants c_j, $1 \leq j \leq n$ such that

$$\varphi \circ X = \sum_{j=1}^{n} c_j I_{C_j}.$$

By Definition 1,

(4.4) $$E_P(\varphi \circ X) = \sum_{j=1}^{n} c_j P(C_j).$$

For $1 \leq j \leq n$, set $B_j = \{x : \varphi(x) = c_j\}$. By definition of Q, $Q((\bigcup_{j=1}^{n} B_j)^c) = 0$. Hence,

$$\varphi = \sum_{j=1}^{n} c_j I_{B_j} \quad Q\text{-a.s.}$$

By Problem 4 and property (iii) in Theorem 9,

$$E_Q(\varphi) = \sum_{j=1}^{n} c_j Q(B_j),$$

which, as desired, equals the right side of (4.4).

Now let φ be an arbitrary $\overline{\mathbb{R}}^+$-valued random variable on $(\overline{\mathbb{R}}, \mathcal{B}, Q)$. By Lemma 13 of Chapter 2, it equals the limit of an increasing sequence $(\varphi_n : n = 1, 2, \dots)$ of nonnegative simple random variables. The sequence $(\varphi_n \circ X : n = 1, 2, \dots)$ is also an increasing sequence of nonnegative random variables—on the probability space $(\Omega, \mathcal{F}, P)$ rather than on $(\overline{\mathbb{R}}, \mathcal{B}, Q)$—and its limit is $\varphi \circ X$. Two applications of the Monotone Convergence Theorem give

$$E_Q(\varphi) = \lim_{n \to \infty} E_Q(\varphi_n)$$

and

$$E_P(\varphi \circ X) = \lim_{n \to \infty} E_P(\varphi_n \circ X).$$

By the preceding paragraph the right sides of these two equalities are equal term by term, so the left sides are equal also.

Clearly, $\varphi^+ \circ X = (\varphi \circ X)^+$ and $\varphi^- \circ X = (\varphi \circ X)^-$. Hence,

$$E_P((\varphi \circ X)^+) = E_P(\varphi^+ \circ X) = E_Q(\varphi^+)$$

and

$$E_P((\varphi \circ X)^-) = E_P(\varphi^- \circ X) = E_Q(\varphi^-).$$

The left sides of these equalities both equal ∞ if and only if the same is true of the right sides. If they do not both equal ∞, subtraction gives $E_P(\varphi \circ X) = E_Q(\varphi)$, as desired. $\square$

The following corollary shows that the expected value of a random variable depends only on its distribution.

Corollary 16. *Let X be an $\mathbb{R}$-valued random variable on a probability space $(\Omega, \mathcal{F}, P)$, and let Q be the distribution of X. Then*

$$E_P(X) = E_Q(x \rightsquigarrow x)$$

in the sense that if either side exists, then so does the other and they are equal.

PROOF. Apply the theorem with φ equal to the identity function $x \rightsquigarrow x$. $\square$

Problem 25. Discuss the difficulties one might face in trying to prove either Theorem 15 or Corollary 16 by approximating X (rather than φ) with simple measurable functions.

* **Problem 26.** Let X be an $\mathbb{R}$-valued random variable that is symmetric about a point b (see Problem 23 of Chapter 3). Show that if $E(X)$ exists, then $E(X) = b$.

4.5. The Riemann-Stieltjes integral and expectations

In Problem 2 we saw that the expected value of a simple random variable X on a probability space $(\Omega, \mathcal{F}, P)$ can be represented as a Riemann-Stieltjes integral with respect to its distribution function F, since the expression given in that problem for $E(X)$ equals $\int_{-\infty}^{\infty} x \, dF(x)$ (see Appendix D). Similarly, if X is an $\mathbb{R}$-valued random variable whose distribution Q is supported by a countable set A, and if $\varphi \colon \mathbb{R} \to \mathbb{R}$ is an appropriately nice function, then the formula in Proposition 14 for $E(\varphi \circ X)$ can also be written in terms of a Riemann-Stieltjes integral:

$$(4.5) \qquad E(\varphi \circ X) = \sum_{x \in A} \varphi(x)[F(x) - F(x-)] = \int_{-\infty}^{\infty} \varphi(x) \, dF(x),$$

since $Q(\{x\}) = F(x) - F(x-)$ for all $x \in \mathbb{R}$.

We will show in Theorem 17 that the right and left sides of (4.5) are equal for general $\mathbb{R}$-valued random variables X, as long as the Riemann-Stieltjes integrals of φ^+ and φ^- with respect to F exist and are not both infinite. Such an equality gives us a powerful calculational tool because it is often straightforward to calculate Riemann-Stieltjes integrals, especially when the integrator F has a continuous density f: $F(x) = \int_{-\infty}^{x} f(u)\,du$ (see Definition 9 of Chapter 3). In this case, the Riemann-Stieltjes integral can then be replaced by a Riemann integral to obtain

$$E(\varphi \circ X) = \int_{-\infty}^{\infty} \varphi(x) f(x)\,dx\,.$$

Once we have such formulas in place, the usual techniques of calculus are available. See, for example, Proposition 19, where integration by parts is used.

Let X be an $\mathbb{R}$-valued random variable with distribution Q and distribution function F. In order to compute $E_P(\varphi \circ X)$, it is sufficient by Theorem 15 to compute $E_Q(\varphi)$. We begin by considering a function $\varphi \colon \mathbb{R} \to \mathbb{R}^+$ for which there are real numbers $x_0 < x_1 < \cdots < x_m$ such that φ is constant on each of the half-open intervals $(x_{j-1}, x_j]$, $1 \le j \le m$, and is equal to 0 on the intervals $(-\infty, x_0]$ and (x_m, ∞). In this case, the Riemann-Stieltjes integral of φ with respect to F exists and we have

$$\int_{-\infty}^{\infty} \varphi(x)\,dF(x) = \int_{a}^{b} \varphi(x)\,dF(x)$$
$$= \sum_{j=1}^{m} \varphi(x_j)\,[F(x_j) - F(x_{j-1})]$$
$$= \sum_{j=1}^{m} \varphi(x_j)\,Q((x_{j-1}, x_j])\,,$$

which by Definition 1 equals $E_Q(\varphi)$.

Now drop the assumption that φ is constant on intervals, but keep the assumptions that φ is nonnegative and that $\varphi(x) = 0$ for x outside some bounded closed interval. Assume further that φ is bounded and Riemann-Stieltjes integrable with respect to F. Fix a and b to the left and to the right, respectively, of that closed interval. Choose a refining sequence $(\Pi_n : n = 1, 2, \dots)$ of point partitions of the interval $[a, b]$ such that the lower and upper Riemann-Stieltjes sums for $\int_{a}^{b} \varphi\,dF$ corresponding to Π_n converge to that integral as $n \to \infty$.

For each $x \in (r, s]$, where $r < s$ are two adjacent members of Π_n, set

$$\theta_n(x) = \inf\{\varphi(u) \colon r \le u \le s\}$$

and

$$\psi_n(x) = \sup\{\varphi(u) \colon r \le u \le s\}\,;$$

and, for $x \le a$ and for $x > b$, set $\theta_n(x) = \psi_n(x) = 0$.

Note that for each n, θ_n, and ψ_n are $\mathbb{R}^+$-valued functions that take only finitely many values and are constant on each of the intervals $(r, s]$ corresponding to adjacent members of the point partition Π_n. Thus, each θ_n and ψ_n is left-continuous and has finite right limits everywhere, so

$$E_Q(\theta_n) = \int_{-\infty}^{\infty} \theta_n(x)\, dF(x) \quad \text{and} \quad E_Q(\psi_n) = \int_{-\infty}^{\infty} \psi_n(x)\, dF(x).$$

It is easily checked that the quantities on the right sides of these two equations are respectively the lower and upper Riemann-Stieltjes sums corresponding to the point partition Π_n for the integral of φ with respect to F. Thus,

$$\lim_{n \to \infty} E_Q(\theta_n) = \lim_{n \to \infty} E_Q(\psi_n) = \int_{-\infty}^{\infty} \varphi(x)\, dF(x).$$

Since for each n, $0 \leq \theta_n \leq \varphi \leq \psi_n$, we have by part (iv) of Theorem 9 that $E_Q(\theta_n) \leq E_Q(\varphi) \leq E_Q(\psi_n)$, and the desired conclusion follows.

We have done most of the work involved in proving the following theorem.

Theorem 17. *Let X be an $\mathbb{R}$-valued random variable with distribution function F, and let φ be an $\mathbb{R}$-valued function that is Riemann-Stieltjes integrable with respect to F on every bounded interval. If $E(\varphi \circ X)$ exists, then*

$$E(\varphi \circ X) = \int_{-\infty}^{\infty} \varphi(x)\, dF(x).$$

Problem 27. Complete the proof of the preceding theorem. In particular, you will need to treat functions φ that are not necessarily 0 outside some bounded interval and functions φ that take both positive and negative values.

Corollary 18. *Let X be an $\mathbb{R}$-valued random variable with distribution function F. Then*

$$E(X) = \int_{-\infty}^{\infty} x\, dF(x),$$

in the sense that if one side exists, then both sides exist and are equal.

Problem 28. Except for the assertion that the existence of $\int_{-\infty}^{\infty} x\, dF(x)$ (possibly ∞ or $-\infty$) entails the existence of $E(X)$, the corollary is an immediate consequence of Theorem 17 and Theorem 15. Finish the proof of the corollary.

Example 3. Let X be a random variable with a gamma distribution function F (defined in Example 2 of Chapter 3). Fix $\varepsilon > 0$. By Corollary 18

$$E(X) = \int_{-\infty}^{\infty} x\, dF(x)$$

$$= \int_{-\infty}^{0} x\, dF(x) + \int_{0}^{\varepsilon} x\, dF(x) + \int_{\varepsilon}^{\infty} x\, dF(x)$$

$$= 0 + \int_{0}^{\varepsilon} x\, dF(x) + \int_{\varepsilon}^{\infty} \frac{(ax)^{\gamma} e^{-ax}}{\Gamma(\gamma)}\, dx .$$

The next to last integral is positive but no larger than $\varepsilon F(\varepsilon)$, which approaches 0 as $\varepsilon \searrow 0$. Therefore, as $\varepsilon \searrow 0$ the last integral approaches $E(X)$:

$$E(X) = \int_{0}^{\infty} \frac{(ax)^{\gamma} e^{-ax}}{\Gamma(\gamma)}\, dx = \frac{1}{a\Gamma(\gamma)} \int_{0}^{\infty} u^{\gamma} e^{-u}\, du = \frac{\Gamma(\gamma + 1)}{a\Gamma(\gamma)} = \frac{\gamma}{a} .$$

The reason for the introduction of ε into the discussion is that, for $0 < \gamma < 1$, X does not have a continuous bounded density, so the Riemann-Stieltjes integral cannot, in that case, be transformed directly into a Riemann integral. (In order to have emphasized the dependence on γ throughout we could have placed the subscript γ on F and on E.)

* **Problem 29.** Calculate the expected value of a normally distributed random variable.

* **Problem 30.** Calculate the mean of an arbitrary beta distribution.

It is important to note that Theorem 17 does not apply directly when $E(\varphi \circ X)$ does not exist. As the following counterexample shows, it is possible for the Riemann-Stieltjes integral to exist even when the expected value does not.

(Counter)example 4. Let us try to calculate $E(X^+ \sin X)$, where X is a Cauchy random variable (defined in Problem 8 of Chapter 3). Theorem 17 might lead us to believe that the answer is

$$\int_{0}^{\infty} \frac{x \sin x}{\pi(1 + x^2)}\, dx .$$

The alternating series test can be used to show that this improper integral converges to a finite value. However, comparison with $1/x$ shows that

$$\int_{0}^{\infty} \frac{x(\sin x)^+}{\pi(1 + x^2)}\, dx = \int_{0}^{\infty} \frac{x(\sin x)^-}{\pi(1 + x^2)}\, dx = \infty.$$

Hence, by Theorem 17,

$$E((X^+ \sin X)^+) = E((X^+ \sin X)^-) = \infty ,$$

so $E(X^+ \sin X)$ does not exist. Thus, it is possible for the Riemann-Stieltjes integral $\int_{-\infty}^{\infty} \varphi(x)\, dF(x)$ in Theorem 17 to exist even though $E_Q(\varphi)$ does not exist.

Besides pointing out the dangers of applying Theorem 17 without verifying its hypothesis, this example also illustrates another important point. By considering $\varphi^+ \circ X$ and $\varphi^- \circ X$ separately, Theorem 17 can be used to show that $E(\varphi \circ X)$ does not exist, even though the formula in the theorem does not apply directly to $\varphi \circ X$.

Of course, there are many situations in which $E(\varphi \circ X)$ exists, even as a finite number, and $\int_{-\infty}^{\infty} \varphi(x)\, dF(x)$ does not exist, because φ may not be Riemann-Stieltjes integrable with respect to the distribution function F. It is not wise to use $\int_{-\infty}^{\infty} \varphi(x)\, dF(x)$ as a synonym for $E(\varphi \circ X)$ in such cases because the symbolism $\int \varphi(x)\, dF(x)$ suggests that the usual integration techniques of calculus apply. In particular, the technique of integration by parts requires that the relevant integrals exist in the Riemann-Stieltjes sense.

We now use integration by parts to obtain a useful formula for calculations of expected values connected with nonnegative random variables.

Proposition 19. *Let F be the distribution function of a $\mathbb{R}^+$-valued random variable X, and let $\varphi \colon \mathbb{R} \to \mathbb{R}$ be monotonic and left-continuous. Then the expectation of $\varphi \circ X$ exists as a member of $\overline{\mathbb{R}}$, and*

$$(4.6) \qquad E(\varphi \circ X) = \varphi(0) + \int_0^{\infty} [1 - F(x)]\, d\varphi(x).$$

PROOF. We assume that φ is decreasing; the result for φ increasing then follows by considering $-\varphi$. For φ decreasing, it is easy to see that each of the left and right sides of (4.6) exists as a member of $[-\infty, \infty)$.

Let $\varepsilon > 0$. Since $X \geq 0$, $F(x) = 0$ for $x \leq -\varepsilon$. It follows from Theorem 17 and integration by parts that

$$
\begin{aligned}
E(\varphi \circ X) &= \int_{-\infty}^{\infty} \varphi(x)\, dF(x) \\
&= \lim_{M \to \infty} \left(\int_{-\infty}^{-\varepsilon} \varphi(x)\, dF(x) + \int_{-\varepsilon}^{M} \varphi(x)\, dF(x) \right) \\
&= \lim_{M \to \infty} \left(-\int_{-\varepsilon}^{M} \varphi(x)\, d[1 - F(x)] \right) \\
&= \lim_{M \to \infty} \left(-\varphi(M)[1 - F(M)] + \varphi(-\varepsilon) + \int_{-\varepsilon}^{M} [1 - F(x)]\, d\varphi(x) \right).
\end{aligned}
$$

(4.7)

If $E(\varphi \circ X) = -\infty$, then $\varphi(M) \to -\infty$ as $M \to \infty$, so $-\varphi(M)[1 - F(M)] \geq 0$ for all sufficiently large M. It follows from (4.7) that $\int_0^{\infty} [1 - F(x)]\, d\varphi(x) = -\infty$. Thus, (4.6) is correct when $E(\varphi \circ X) = -\infty$.

Now suppose that $E(\varphi \circ X) \in \mathbb{R}$. We plan to show that $\varphi(M)[1 - F(M)] \to 0$ as $M \to \infty$ in order to conclude from (4.7) that

$$E(\varphi \circ X) = \varphi(-\varepsilon) + \int_{-\varepsilon}^{\infty} [1 - F(x)] \, d\varphi(x) \,,$$

for then the desired conclusion follows by letting $\varepsilon \searrow 0$.

It is clear that $\varphi(M)[1 - F(M)] \to 0$ as $M \to \infty$ if $\lim_{x \to \infty} \varphi(x) > -\infty$. Thus, we suppose that $\varphi(x) \searrow -\infty$ as $x \nearrow \infty$. Then, for sufficiently large M,

$$0 \geq \varphi(M)[1 - F(M)] \geq E\big((\varphi \cdot I_{(M,\infty)}) \circ X\big) \,,$$

which, by Corollary 13 of the Monotone Convergence Theorem, approaches 0 as $M \to \infty$. $\square$

We highlight the most important special case of the preceding proposition as a corollary.

Corollary 20. *For an $\mathbb{R}^+$-valued random variable X,*

$$E(X) = \int_0^{\infty} [1 - F(x)] \, dx \,,$$

where F is the distribution function of X. If X is $\mathbb{Z}^+$-valued, then

$$E(X) = \sum_{k=1}^{\infty} P(\{\omega : X(\omega) \geq k\}) \,.$$

* **Problem 31.** Let $k > 0$. For an exponential random variable X with density $x \rightsquigarrow ke^{-kx}$, $x \in \mathbb{R}^+$, use the preceding corollary to calculate $E(X)$ and use Proposition 19 to calculate $E(\exp \circ X)$.

Problem 32. For a geometrically distributed random variable X, use Corollary 20 to calculate $E(X)$, previously calculated when solving Problem 22.

Problem 33. Use Proposition 19 to Calculate $E(X^r)$ for $r \geq 0$ and X having a standard uniform distribution in two different ways: by using Theorem 17 and by using Proposition 19.

Problem 34. For X a random variable having the Cantor distribution, calculate $E(X)$ and $E(X^2)$.

* **Problem 35.** Calculate the expected length of a random chord of a circle under various reasonable interpretations of 'random' (see Example 3 of Chapter 2 and Example 4 of Chapter 3). Compare your different answers and discuss the relations among them.

CHAPTER 5
Expectations: Applications

Expectations are amazingly useful in the study of random variables and their distributions. Some of the reasons for this statement are contained in this chapter. In the first section, we introduce the 'variance' of an $\mathbb{R}$-valued random variable. Variance is used to obtain one version of the Law of Large Numbers, also known informally as the Law of Averages. The 'covariance' of two random variables is also presented. In Section 2, variance and covariance are defined for $\mathbb{R}^d$-valued random variables, and Section 3 concerns the expectations of various functions of $\mathbb{R}$-valued random variables. The chapter concludes with a discussion of 'probability generating functions', used in the study of distributions on $\mathbb{Z}^+$. Several useful inequalities, including those of Chebyshev, Cauchy-Schwarz, and Jensen, are scattered throughout.

5.1. Variance and the Law of Large Numbers

Let X be a random variable with finite expected value μ, and consider the function $\varphi(x) = (x - \mu)^2$. The quantity $E(\varphi \circ X)$, denoted by $\mathrm{Var}(X)$, is called the *variance* of X. It is one measure of the degree to which X differs from its mean. A random variable has zero variance if and only if it is almost surely equal to a finite constant.

Proposition 1. *Let X be a random variable with finite mean μ. Then*

$$\mathrm{Var}(X) = E(X^2) - \mu^2 \,,$$

whether finite or infinite.

PROOF.

$$E((X - \mu)^2) = E(X^2 - 2\mu X + \mu^2) = E(X^2) - 2\mu E(X) + \mu^2$$
$$= E(X^2) - 2\mu^2 + \mu^2 = E(X^2) - \mu^2 \,. \quad \square$$

Example 1. Let us calculate the variance of the Poisson distribution. For a Poisson random variable X having the parameter λ, we obtain from Problem 23 of Chapter 4 that $E(X) = \lambda$ and $E(X^2) = \lambda^2 + \lambda$. In view of the preceding proposition, the variance equals λ.

Problem 1. Let X be a random variable with finite variance. Prove that for all real numbers a, $E((X - a)^2) \geq \mathrm{Var}(X)$.

Problem 2. Prove the following Monotone Convergence Theorem for variances: Let $(X_1, X_2, \dots)$ be an increasing sequence of $\overline{\mathbb{R}}^+$-valued random variables on a common probability space $(\Omega, \mathcal{F}, P)$. Set $X = \lim_n X_n$. If $E(X) < \infty$, then $\mathrm{Var}(X) = \lim_n \mathrm{Var}(X_n)$.

The nonnegative square root of the variance is called the *standard deviation*. It has the same physical units as the mean. If the variance is infinite, then we also say that the standard deviation is infinite. If the mean is infinite or undefined, we say that the standard deviation and variance are undefined.

Problem 3. Other than probability generating functions, which will be defined in the last section of this chapter, verify the entries in Table 5.1 that have not previously been calculated.

Distribution on $\mathbb{Z}^+$	probability assigned to $\{k\}$	mean	variance	prob. gen. function: $s \rightsquigarrow$
Bernoulli	$\begin{cases} p & \text{if } k = 1 \\ 1 - p & \text{if } k = 0 \end{cases}$	p	$p(1 - p)$	$ps + (1 - p)$
Binomial	$\binom{n}{k} p^k (1 - p)^{n-k}, \; k \leq n$	np	$np(1 - p)$	$[ps + (1 - p)]^n$
Poisson	$\dfrac{\lambda^k}{k!} e^{-\lambda}$	λ	λ	$e^{-\lambda(1-s)}$
Geometric	$(1 - p)p^k$	$\dfrac{p}{1 - p}$	$\dfrac{p}{(1 - p)^2}$	$\dfrac{1 - p}{1 - ps}$

TABLE 5.1. Basic facts about some distributions on $\mathbb{Z}^+$

Problem 4. Verify the entries in Table 5.2 that have not previously been calculated.

Distribution on $\mathbb{R}$	density: $x \rightsquigarrow$	mean	variance
Exponential	ae^{-ax}, $x \geq 0$	$\dfrac{1}{a}$	$\dfrac{1}{a^2}$
Gamma	$\dfrac{a^\gamma}{\Gamma(\gamma)} x^{\gamma-1} e^{-ax}$	$\dfrac{\gamma}{a}$	$\dfrac{\gamma}{a^2}$
Uniform	$\dfrac{1}{b-a}$, $a \leq x \leq b$	$\dfrac{a+b}{2}$	$\dfrac{(b-a)^2}{12}$
Beta	$\dfrac{\Gamma(\alpha+\beta)}{\Gamma(\alpha)\Gamma(\beta)} x^{\alpha-1}(1-x)^{\beta-1}$, $0 \leq x \leq 1$	$\dfrac{\alpha}{\alpha+\beta}$	$\dfrac{\alpha\beta}{(\alpha+\beta)^2(\alpha+\beta+1)}$
Gaussian	$\dfrac{1}{\sqrt{2\pi a^2}} e^{-(x-b)^2/2a^2}$	b	a^2
Normal	another name for Gaussian		

TABLE 5.2. Basic facts about some distributions with densities

Problem 5. Let X be a random variable with finite mean μ and standard deviation σ. Let $Y = aX + b$, where a and b are real constants. Prove that the mean of Y is $a\mu + b$ and that the standard deviation of Y is $|a|\sigma$. Notice, in particular, that if $a = 1/\sigma$ and $b = -aE(X)$, then Y is of the same type as X, the mean of Y is 0, and the variance and standard deviation of Y are both equal to 1.

Problem 6. Let X_k be the random function defined in Example 2 of Chapter 2, $k = 1, 2, \ldots$. Show that

$$\text{Var}(X_k(j/k)) = j/k, \quad \text{for } j = 1, \ldots, k.$$

Comment on the use of the term 'scaling factor' in Example 2 of Chapter 2.

* **Problem 7.** Calculate the variances of the various distributions arising in connection with Example 4 of Chapter 3, the random chord example. This problem is a continuation of Problem 35 of Chapter 4.

We turn to a very simple inequality that relates the variance of a random variable to the probability that the random variable deviates a given amount from its mean. The inequality is quite crude in the sense that for many distributions that arise in practice, the bound given can be considerably improved by other methods. Nevertheless, because of its great generality, it is an important theoretical tool that will appear in many proofs.

Proposition 2. [Chebyshev Inequality] *If X is an $\overline{\mathbb{R}}$-valued random variable with finite mean μ and standard deviation σ, then*

$$P(\{\omega: |X(\omega) - \mu| \geq z\}) \leq (\sigma/z)^2$$

for each $z > 0$.

PROOF. Let $A = \{\omega: |X(\omega) - \mu| \geq z\}$. Then

$$\sigma^2 = E((X - \mu)^2) \geq E(I_A(X - \mu)^2) \geq E(I_A z^2) = z^2 P(A).$$

Now divide by z^2 to obtain the desired result. $\square$

Problem 8. Give a sense in which the Chebyshev Inequality is best possible. Prove your assertions.

The next result is an easy generalization of the Chebyshev Inequality.

Proposition 3. [Markov Inequality] *Let Y be an $\mathbb{R}$-valued random variable, and let f be an $\mathbb{R}^+$-valued function which is increasing on some interval $J \subseteq \mathbb{R}$ containing the support of Y. Then, for all $z \in J$ such that $f(z) > 0$,*

$$P(\{\omega: Y(\omega) \geq z\}) \leq E(f \circ Y)/f(z).$$

Problem 9. Prove the preceding proposition. For what choice of Y and f is the Markov Inequality equivalent to the Chebyshev Inequality?

Problem 10. For $z > 0$, use the Markov Inequality to obtain one upper bound for $P(|X| \geq z)$ in terms of $E(|X|)$ and another in terms of $E(X^2)$, and also an upper bound for $P(X \geq z)$ in terms of $E(X^+)$.

Proposition 4. [Cauchy-Schwarz Inequality] *Let X and Y be two $\overline{\mathbb{R}}$-valued random variables defined on a probability space $(\Omega, \mathcal{F}, P)$. Assume that $E(X^2)$ and $E(Y^2)$ are both finite. Then $E(XY)$ exists and*

$$[E(XY)]^2 \leq E(X^2)E(Y^2).$$

Moreover, equality holds if and only if one of the random variables X or Y almost surely equals a constant multiple of the other.

PROOF. If either X or Y is almost surely a multiple of the other, the result is an immediate consequence of linearity. Suppose that neither X nor Y is almost surely a multiple of the other. For each real α,

$$E((\alpha X - Y)^2) - \alpha^2 E(X^2) - E(Y^2)$$

is meaningful and greater than $-\infty$. By linearity, this quantity equals $-2E(XY)$ when $\alpha = 1$ and $2E(XY)$ when $\alpha = -1$. We conclude that $E(XY)$ exists and is finite. Applying linearity again, we have

(5.1) $$E((\alpha X - Y)^2) = \alpha^2 E(X^2) - 2\alpha E(XY) + E(Y^2),$$

which is a quadratic function of α. The left side of (5.1) is positive for all α, so the discriminant of the quadratic, $4[E(XY)]^2 - 4E(X^2)E(Y^2)$, must be negative. The desired inequality follows. □

One consequence of the Cauchy-Schwarz Inequality is the following, which provides a useful complement to the Chebyshev Inequality because it gives an inequality in the opposite direction:

Corollary 5. *Let X be an $\mathbb{R}$-valued random variable such that $0 \leq E(X)$ and $0 < E(X^2) < \infty$, and let $\lambda \in [0, 1]$. Then*

$$P(\{\omega : X(\omega) > \lambda E(X)\}) \geq (1 - \lambda)^2 \frac{[E(X)]^2}{E(X^2)}.$$

PROOF. Let A be the event $\{\omega : X(\omega) > \lambda E(X)\}$. By the linearity and positivity properties of the expectation and by the Cauchy-Schwarz Inequality,

$$E(X) = E(XI_A) + E(XI_{A^c}) \leq [E(X^2)P(A)]^{1/2} + \lambda E(X),$$

from which the desired inequality follows. □

Problem 11. Prove the following: Let X be an $\mathbb{R}$-valued random variable with finite mean μ and standard deviation σ. Then for $0 \leq z \leq \sigma$,

$$P(\{\omega : |X - \mu| \geq z\}) \geq \left(1 - \frac{z^2}{\sigma^2}\right)^2 \frac{\sigma^4}{E(|X - \mu|^4)}.$$

For random variables X and Y defined on a common probability space and having finite means, we define their *covariance* by

$$\mathrm{Cov}(X, Y) = E[(X - E(X))(Y - E(Y))].$$

Note that $\mathrm{Cov}(X, X) = \mathrm{Var}(X)$. Of course, the covariance need not be finite, or even exist. However, it is finite if both random variables have finite variance.

Corollary 6. *If X and Y are random variables satisfying the conditions of the Cauchy-Schwarz Inequality, $\mathrm{Cov}(X, Y)$ is a well-defined finite number, and*

$$[\mathrm{Cov}(X, Y)]^2 \leq \mathrm{Var}(X)\,\mathrm{Var}(Y).$$

PROOF. Replace X by $X - E(X)$ and Y by $Y - E(Y)$ in the Cauchy-Schwarz Inequality. $\square$

Proposition 7. *Let X, Y, and Z be $\overline{\mathbb{R}}$-valued random variables defined on a probability space $(\Omega, \mathcal{F}, P)$. Assume that X, Y, and Z all have finite variances. Then*

(i) $\mathrm{Cov}(X, Y) = E(XY) - E(X)E(Y)$;
(ii) $\mathrm{Cov}(X, X) \geq 0$;
(iii) $\mathrm{Cov}(X, Y) = \mathrm{Cov}(Y, X)$;
(iv) $\mathrm{Cov}(aX + bY, Z) = a\,\mathrm{Cov}(X, Z) + b\,\mathrm{Cov}(Y, Z)$ *for all real a and b.*

Problem 12. Prove the preceding proposition.

* **Problem 13.** Find a formula for $\mathrm{Cov}(aX + b, cY + d)$ in terms of a, b, c, d, and $\mathrm{Cov}(X, Y)$.

Properties (iii) and (iv) of Proposition 7 have the following consequence.

Corollary 8. *Let $(X_1, \ldots, X_m)$ and $(Y_1, \ldots, Y_n)$ be sequences of random variables defined on a common probability space and having finite variances. Then*

$$\mathrm{Cov}\left(\sum_{j=1}^{m} X_j, \sum_{k=1}^{n} Y_k\right) = \sum_{j=1}^{m} \sum_{k=1}^{n} \mathrm{Cov}(X_j, Y_k).$$

In particular,

$$\mathrm{Var}\left(\sum_{j=1}^{m} X_j\right) = \sum_{j=1}^{m} \sum_{k=1}^{m} \mathrm{Cov}(X_j, X_k).$$

The *correlation* between two random variables X and Y is defined by

$$\mathrm{Corr}(X, Y) = \frac{\mathrm{Cov}(X, Y)}{\sqrt{\mathrm{Var}(X)\,\mathrm{Var}(Y)}}$$

whenever the covariance of X and Y is defined. (In this definition, $0/0 = 0$.) Random variables X and Y are said to be *positively correlated, negatively correlated,* or *uncorrelated* if their correlation is defined and is positive, negative, or zero, respectively.

* **Problem 14.** Let X be an $\mathbb{R}$-valued random variable and let $\varphi \colon \mathbb{R} \to \mathbb{R}$ be an increasing function. Assume that X and $\varphi \circ X$ both have finite variance. Show that $\mathrm{Cov}(X, \varphi \circ X) \geq 0$.

Problem 15. Prove that whenever the correlation between two random variables is defined, it lies in the interval $[-1, 1]$. Further prove that if the two random variables each have nonzero variance, then the correlation equals ± 1 if and only if one of the random variables is almost surely a linear function of the other. Also show that correlation is 'dimensionless' and translation invariant:

$$\text{Corr}(aX + b, Y) = \text{Corr}(X, Y)$$

for all real numbers a and b such that $a > 0$.

Problem 16. Let $X_1, \ldots, X_n$ be pairwise uncorrelated $\overline{\mathbb{R}}$-valued random variables. Prove that

$$\text{Var}\left(\sum_{i=1}^{n} X_i\right) = \sum_{i=1}^{n} \text{Var}(X_i).$$

* **Problem 17.** Calculate the variance of the number of cards whose positions equal their labels for each of the settings in Example 2 and Problem 18 both of Chapter 4.

We now have the tools necessary to prove a simple version of a Law of Large Numbers, also known as a 'Law of Averages'. The setting for this law is one in which an experiment is repeated a large number of times. The outcome of each trial (or of a measurement taken after each trial) is a real number. In other words, we consider a sequence $(X_1, X_2, \ldots)$ of random numbers. These random variables are all assumed to have the same distribution, which is a mathematical way of saying that the same experiment is performed at each trial. After performing the experiment n times, we take the average of the outcomes, obtaining the random variable $\sum_{1}^{n} X_k / n$. The Law of Large Numbers states that for large n, this average is likely to be close to the mean of the common distribution of the random variables X_k, provided that the random variables are either negatively correlated or uncorrelated. A stronger version of this result will be proved in Chapter 12.

Proposition 9. [Law of Large Numbers] *Let $(X_1, X_2, \ldots)$ be a sequence of identically distributed $\overline{\mathbb{R}}$-valued random variables defined on a common probability space $(\Omega, \mathcal{F}, P)$. Suppose that the common distribution of the random variables has finite mean μ and finite variance σ^2. Further assume that each pair of random variables in the sequence is either uncorrelated or negatively correlated. Then, for each $\varepsilon > 0$,*

$$\lim_{n \to \infty} P\left(\left\{\omega: \left|\frac{\sum_{k=1}^{n} X_k(\omega)}{n} - \mu\right| > \varepsilon\right\}\right) = 0.$$

PROOF. Let $S_n = \sum_{1}^{n} X_k$. Then $E(S_n/n) = \mu$ and $\text{Var}(S_n) \leq n\sigma^2$. (Why?) Thus $\text{Var}(S_n/n) \leq \sigma^2/n$, and so the Chebyshev Inequality implies that

$$P(\{\omega: |S_n(\omega)/n - \mu| > \varepsilon\}) \leq \sigma^2/n\varepsilon^2.$$

Since the right side goes to 0 as $n \to \infty$, the proof is complete. $\square$

Problem 18. Find an example to show that the correlation condition in the Law of Large Numbers cannot be eliminated.

Problem 19. The proof of Proposition 9 implies more than is stated in the proposition. State a more general proposition to which the proof of Proposition 9 applies.

Problem 20. Let $(A_1, A_2, \dots)$ be events in a probability space $(\Omega, \mathcal{F}, P)$. Assume that all of the events A_i have the same probability p. For $n = 1, 2, \dots$, let $M_n(\omega)$ be the number of events in the finite sequence $(A_1, \dots, A_n)$ which contain the sample point ω. Find conditions on the quantities $P(A_i \cap A_j)$ for $i \neq j$ that imply

$$\lim_{n \to \infty} P\left(\left\{\omega: \left|\frac{M_n(\omega)}{n} - p\right| > \varepsilon\right\}\right) = 0$$

for all $\varepsilon > 0$. *Hint:* Use indicator random variables.

5.2. Mean vectors and covariance matrices

Let $X = (X_1, X_2, \dots, X_d)$ be an $\mathbb{R}^d$-valued random variable. If the means of the coordinate random variables X_k exist, the vector $(E(X_1), E(X_2), \dots, E(X_d))$ is called the *mean vector*, of X, and, if all the variances are finite, the matrix with elements $\text{Cov}(X_i, X_j)$, $i, j = 1, 2, \dots, d$, is the *covariance matrix* of X. The *mean matrix* of a random matrix is the matrix of expectations of the individual entries. Thus, if m denotes the mean vector of X (regarded as a row vector), then the covariance matrix of X, sometimes written as $\text{Cov}(X, X)$, equals the mean of the random matrix $(X - m)^T(X - m)$, where 'T' denotes transpose; that is,

$$\text{Cov}(X, X) = E\big((X - m)^T(X - m)\big).$$

The terminology just described is also used for random infinite sequences $X = (X_1, X_2, \dots)$. For a random function X, such as defined in Example 2 of Chapter 2, it is often useful to consider the *mean function* $t \leadsto E[X_t]$ and the *covariance function:* $(s, t) \leadsto \text{Cov}(X_s, X_t)$.

Problem 21. Let $(\Omega, \mathcal{F}, P)$ denote the usual probability space for an infinite sequence of coin-flips. Define $X_i(\omega) = 2\omega_i - 1$, $i = 1, 2, \dots$. Show that the sequence $(X_1, X_2, \dots)$ is pairwise uncorrelated. Use the formula in the Problem 16 to repeat Problem 6. Then find the covariance function of each of the random functions X_k mentioned in Problem 6. What is the limit of the covariance function as $k \to \infty$?

Proposition 10. *Let X be an $\mathbb{R}^d$-valued random variable with mean equal to the zero vector $\mathbf{0}_d$ and finite covariance matrix Σ. Let Λ be some nonrandom $k \times d$ matrix for some k. Then the random vector*

$$X\Lambda^T = \left[\Lambda X^T\right]^T$$

has mean $\mathbf{0}_k$ and covariance matrix $\Lambda \Sigma \Lambda^T$.

PROOF. By linearity of expectation, coordinate by coordinate,

$$E(X\Lambda^T) = E(X)\Lambda^T = 0_d\Lambda^T = 0_k \,.$$

Again using linearity of expectation we obtain the covariance matrix of $X\Lambda^T$:

$$E([X\Lambda^T]^T X\Lambda^T) = E(\Lambda X^T X\Lambda^T) = \Lambda E(X^T X)\Lambda^T = \Lambda\Sigma\Lambda^T \,,$$

as desired. $\square$

Problem 22. Use the preceding proposition to prove that if X has a finite covariance matrix $\mathbf{X}$, then the variance of the random inner product $\langle z, X\rangle$ equals $z\Sigma z^T$, for any constant $z \in \mathbb{R}^d$.

Problem 23. Observe that covariance matrices are necessarily symmetric, and then use the preceding problem to prove that every eigenvalue of a covariance matrix is nonnegative.

A square matrix is said to be *positive definite* if every eigenvalue is nonnegative. It is *strictly positive definite* if every eigenvalue is positive. According to the preceding problem, every covariance matrix is symmetric and positive definite. This fact and its converse constitute the following theorem.

Theorem 11. *A $d \times d$ matrix is the covariance matrix of some $\mathbb{R}^d$-valued random variable if and only if it is symmetric and positive definite.*

PROOF. In view of the Problem 23 we only need prove the 'if' part. Let Υ be symmetric and positive definite. Denote the eigenvalues of Υ (counting multiplicity) by v_i^2, $1 \le i \le d$, where each $v_i \ge 0$. Since Υ is symmetric there exists a matrix $\mathbf{O}$ such that $\mathbf{O}^{-1} = \mathbf{O}^T$ and

$$\Upsilon = \mathbf{O}^T \begin{pmatrix} v_1^2 & 0 & \cdots & 0 \\ 0 & v_2^2 & \cdots & 0 \\ \vdots & \vdots & & \vdots \\ 0 & 0 & \cdots & v_d^2 \end{pmatrix} \mathbf{O} \,.$$

Set

$$\Lambda = \mathbf{O}^T \begin{pmatrix} v_1 & 0 & \cdots & 0 \\ 0 & v_2 & \cdots & 0 \\ \vdots & \vdots & & \vdots \\ 0 & 0 & \cdots & v_d \end{pmatrix} \mathbf{O} \,.$$

A straightforward calculation shows that $\Upsilon = \Lambda\Lambda^T$.

Let X_i, $1 \le i \le d$, be as in Problem 21 (ignoring $X_{d+1}, X_{d+2}, \ldots$ as defined there). By that problem and a simple calculation of variances, the covariance matrix of $X = (X_1, \ldots, X_d)$ is the identity matrix. Insert the identity matrix for Υ in Proposition 10 to obtain that the covariance matrix of $X\Lambda^T$ is $\Lambda\Lambda^T$, which equals Σ. $\square$

Problem 24. Decide which of the following matrices are covariance matrices. For those which are, find the matrix Λ described in the preceding proof.

$$\begin{pmatrix} 3 & 2 \\ 2 & 1 \end{pmatrix}, \quad \begin{pmatrix} 7 & -2\sqrt{3} \\ -2\sqrt{3} & 3 \end{pmatrix}, \quad \begin{pmatrix} 2 & -4 \\ -4 & 8 \end{pmatrix}$$

$$\begin{pmatrix} 1 & -2 & 4 \\ -2 & 4 & -8 \\ 4 & -8 & 16 \end{pmatrix}, \quad \begin{pmatrix} 0 & -2 & 4 \\ -2 & 3 & -8 \\ 4 & -8 & 15 \end{pmatrix}$$

The notation $\mathrm{Cov}(X, X)$ for the covariance matrix of the random vector X can be extended to accommodate two random vectors X and Y, not necessarily having the same number of coordinates. The entry in position (i, j) of $\mathrm{Cov}(X, Y)$ is, by definition, $\mathrm{Cov}(X_i, Y_j)$. Thus,

$$\mathrm{Cov}(X, Y) = E\big([X - E(X)]^T[Y - E(Y)]\big).$$

Similarly, one can speak of the covariance function of two random functions.

Problem 25. Let X_1, X_2, X_3, and X_4 be as in Problem 21 and set $Y_1 = X_1 + X_2$, $Y_2 = X_2 + X_3$, $Y_3 = X_3 + X_4$, $Y_4 = X_4 + X_1$, $Y_5 = X_1 + X_3$, $Y_6 = X_2 + X_4$, and $Y_7 = X_1 + X_2 + X_3 + X_4$. Find the 4×7 covariance matrix of X and Y.

5.3. Moments and the Jensen Inequality

For an integer $n \neq 0$, $E(X^n)$ is called the nth *moment* of the $\overline{\mathbb{R}}$-valued random variable X. (0^n is undefined for negative odd values of n, and is defined to be ∞ for negative even values of n in this definition.) If $E(X^n)$ fails to exist, then we say that the nth moment of X does not exist. For a real number $r \neq 0$, $E(|X|^r)$ is called the rth *absolute moment* of X. ($0^r = \infty$ for $r < 0$ in this definition.) If the mean μ of X is finite, the nth moment of $X - \mu$ is called the nth *moment of X about its mean*, and the rth moment of $|X - \mu|$ is called the rth *absolute moment* of X *about its mean*.

The next result can be used to obtain inequalities between different moments of a random variable. For it we need a definition: an $\mathbb{R}$-valued function φ defined on an interval $J \subseteq \mathbb{R}$ is *convex* if

$$(5.2) \qquad \varphi((1 - \lambda)x + \lambda z) \leq (1 - \lambda)\varphi(x) + \lambda\varphi(z)$$

whenever x and z belong to J and $0 \leq \lambda \leq 1$. Set $y = (1 - \lambda)x + \lambda z$ and observe that a geometrical interpretation of (5.2) is that the point $(y, \varphi(y))$ lies below or on the line segment with endpoints $(x, \varphi(x))$ and $(z, \varphi(z))$. The condition (5.2) can be rewritten as

$$(5.3) \qquad (1 - \lambda)[\varphi(y) - \varphi(x)] \leq \lambda[\varphi(z) - \varphi(y)].$$

Here x, y, and z are arbitrary members of J satisfying $x \leq y \leq z$, and λ is chosen so that $y = (1 - \lambda)x + \lambda z$. If we ignore the cases $y = x$ and $y = z$ for which (5.2) is trivially satisfied, we can divide (5.3) by $(1 - \lambda)(y - x) = \lambda(z - y)$ to conclude that φ is convex on J if and only if

$$(5.4) \qquad \frac{\varphi(y) - \varphi(x)}{y - x} \leq \frac{\varphi(z) - \varphi(y)}{z - y}$$

whenever $x < y < z$ belong to J. The function φ is said to be *concave* if its negative is convex.

Problem 26. Show that if φ is a $\mathbb{R}$-valued function which is convex or concave on an interval $J \subseteq \mathbb{R}$, then φ is measurable. *Hint:* First use (5.4) to show that φ must be continuous on the interior of J.

Proposition 12. [Jensen Inequality] *Let X be a random variable with finite mean, and let J be an interval that supports the distribution of X. Let φ be an $\mathbb{R}$-valued function which is convex on J. Then*

$$\varphi(E(X)) \leq E(\varphi \circ X).$$

PROOF. The result is obvious if X is almost surely equal to a constant. So, suppose that this is not the case. Then by property (v) in Theorem 9 of Chapter 4, $E(X)$ is not an endpoint of J. Let $y = E(X)$, and set

$$M = \sup \left\{ \frac{\varphi(y) - \varphi(u)}{y - u} : u < y \text{ and } u \in J \right\}.$$

The quantity M is finite since it is bounded above by the right side of (5.4) for some $z > y$, with $z \in J$. It follows from the definition of M and (5.4) that

$$M(v - E(X)) \leq \varphi(v) - \varphi(E(X))$$

for all $v \in J$. Since J supports the distribution of X, we conclude that

$$M(X - E(X)) \leq \varphi \circ X - \varphi(E(X)) \text{ a.s.}$$

The result of taking expected values is

$$0 \leq E(\varphi \circ X) - \varphi(E(X)). \quad \square$$

The reader may find it useful to think about the Jensen Inequality and its proof for the special case in which the support of X contains just two points, x_1 and x_2. Let $p = P(\{\omega : X(\omega) = x_1\})$. Then

$$E(X) = px_1 + (1 - p)x_2 \quad \text{and} \quad E(\varphi \circ X) = p\varphi(x_1) + (1 - p)\varphi(x_2),$$

and the Jensen Inequality follows immediately from the definition of convexity. See Figure 5.1, where this argument is illustrated graphically.

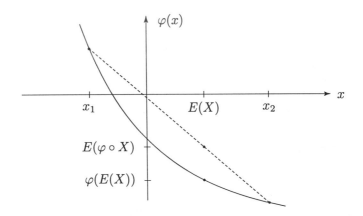

FIGURE 5.1. The Jensen Inequality for a random variable X with support $\{x_1, x_2\}$

Problem 27. Prove that if $\varphi \colon J \to \mathbb{R}$ is a convex function and $\lambda_1, \ldots, \lambda_n$ are nonnegative numbers that sum to 1, then for all $x_1, \ldots, x_n \in J$,

$$\varphi(\lambda_1 x_1 + \cdots + \lambda_n x_n) \leq \lambda_1 \varphi(x_1) + \cdots + \lambda_n \varphi(x_n).$$

Use this fact to extend the argument given in the preceding paragraph to prove the Jensen Inequality for random variables X with finite support.

Problem 28. Let X be an $\overline{\mathbb{R}}$-valued random variable. Use the Jensen Inequality to prove that the function $f(p) = [E(|X|^p)]^{1/p}$ is an increasing $\overline{\mathbb{R}}$-valued function for $0 < p < \infty$. Show that if X is not a constant a.s., then $\lim_{p \to \infty} f(p) = \sup\{|x| \colon 0 < F(x) < 1\}$, where F is the distribution function of X. Describe the situation in which X is a constant.

5.4. Probability generating functions

We conclude this chapter with an important application of expectations to the study of distributions of certain random variables.

CONVENTION. In the probability generating function setting, $0^0 = 1^\infty = 1$, derivatives of functions defined on $[0, 1]$ are understood to be right derivatives when taken at 0 and left derivatives when taken at 1, and if a derivative at 1 of order m equals ∞, then all derivatives at 1 of order $n \geq m$ also equal ∞.

Let X be a random variable whose distribution is supported by the set $\overline{\mathbb{Z}}^+ = \{0, 1, 2, \ldots, \infty\}$. The *probability generating function* of X (and also of the distribution of X) is defined by

$$\rho(s) = E(s^X) = \sum_{k=0}^{\infty} P(\{\omega \colon X(\omega) = k\})s^k + \begin{cases} 0 & \text{if } 0 \le s < 1 \\ P(\{\omega \colon X(\omega) = \infty\}) & \text{if } s = 1. \end{cases}$$

Theorem 13. *Let ρ be the probability generating function of a $\overline{\mathbb{Z}}^{+}$-valued random variable X, and let $\rho^{(k)}$ denote the derivative of ρ of order k. Then:*

$$P(\{\omega \colon X(\omega) = \infty\}) = 1 - \rho(1-) = \rho(1) - \rho(1-);$$

$$P(\{\omega \colon X(\omega) = k\}) = \frac{\rho^{(k)}(0)}{k!}, \quad k = 0, 1, 2, \ldots;$$

$$E\left(\prod_{m=0}^{n-1}(X - m)\right) = \rho^{(n)}(1), \quad n = 1, 2, \ldots.$$

PROOF. As $s \nearrow 1$ through any sequence, $s^X \nearrow I_{\{\omega \colon X(\omega) < \infty\}}$. The first formula in the theorem then follows by the Monotone Convergence Theorem. The second formula is the standard formula for the coefficients of a power series. The third formula is obvious if $\rho(1-) < 1$, since in this case, X is infinite with positive probability. Thus we restrict our attention to the case $\rho(1-) = 1$.

By definition

$$\rho'(1) = \lim_{s \nearrow 1} \frac{1 - \rho(s)}{1 - s} = \lim_{s \nearrow 1} E\left(\frac{1 - s^X}{1 - s}\right).$$

As s increases to 1 through an arbitrary sequence, the nonnegative random variables $(1 - s^X)/(1 - s)$ form an increasing sequence whose limit is X. It follows from the Monotone Convergence Theorem that $\rho'(1) = E(X)$ (whether finite or infinite) and, therefore, that the third formula holds for $n = 1$. To continue the proof by induction, assume the formula is true for n. Then, with $p_k = P(\{\omega \colon X(\omega) = k\})$,

$$\rho^{(n+1)}(1) = \lim_{s \nearrow 1} \frac{\rho^{(n)}(1) - \rho^{(n)}(s)}{1 - s}$$

$$= \lim_{s \nearrow 1} \frac{E\left(\prod_{m=0}^{n-1}(X - m)\right) - \sum_{k=n}^{\infty} \frac{k!}{(k-n)!} p_k s^{k-n}}{1 - s}$$

$$= \lim_{s \nearrow 1} \frac{\sum_{k=n}^{\infty} \frac{k!}{(k-n)!} p_k - \sum_{k=n}^{\infty} \frac{k!}{(k-n)!} p_k s^{k-n}}{1 - s}$$

$$= \lim_{s \nearrow 1} \sum_{k=n}^{\infty} \frac{(1 - s^{k-n})k!}{(1 - s)(k - n)!} p_k$$

$$= \lim_{s \nearrow 1} E\left(\frac{(1 - s^{X-n})}{(1 - s)} \prod_{m=0}^{n-1}(X - m)\right).$$

We have applied Proposition 14 of Chapter 4 twice in this series of equalities. Note that, because of the inductive hypothesis and the convention concerning infinite derivatives, these equalities are correct, even in the case that $\rho^{(n)}(1) = \infty$. To complete the proof, the Monotone Convergence Theorem can now be applied as above, this time using the fact that $0 \leq (1 - s^{X(\omega)-n})/(1 - s) \nearrow (X(\omega) - n)$ as $s \nearrow 1$ for those ω for which $\prod_{m=0}^{n-1}(X(\omega) - m) \neq 0$. $\square$

The quantity $E(\prod_{m=0}^{n-1}(X - m))$ is called the nth *factorial moment* of the random variable X. The preceding proposition and our interest in (ordinary) moments motivates us to consider the relation between the two sequences of polynomials:

$$\left(\prod_{m=0}^{n-1}(x - m): n = 0, 1, 2, \dots\right) \quad \text{and} \quad (x^n: n = 0, 1, 2, \dots),$$

where the empty product obtained for $n = 0$ in the left sequence is defined to equal the multiplicative identity 1. Consideration of the degrees of the various polynomials makes it clear that there exist two doubly indexed sequences of constants

$$(s(n, k): 0 \leq k \leq n, n \in \mathbb{Z}^+) \quad \text{and} \quad (S(n, k): 0 \leq k \leq n, n \in \mathbb{Z}^+)$$

such that

$$\prod_{m=0}^{n-1}(x - m) = \sum_{k=0}^{n} s(n, k)x^n \quad \text{and} \quad x^n = \sum_{k=0}^{n} S(n, k)\left(\prod_{m=0}^{k-1}(x - m)\right).$$

The numbers $s(n, k)$ and $S(n, k)$ are called *Stirling numbers of the first and second kinds,* respectively. In order to define 'infinite square matrices' of Stirling numbers one sets $s(n, k) = S(n, k) = 0$ for $k > n$. It is easy to show that the two matrices are inverses of each other.

* **Problem 29.** Calculate the Stirling numbers $s(n, k)$ and $S(n, k)$ for $n \leq 3$ and all k. Make sure that your answers are consistent with the assertion that the matrix of Stirling numbers of the first kind is the inverse of the matrix of Stirling numbers of the second kind. Also, use your calculations to express: the second factorial moment of a random variable in terms of its first and second (ordinary) moments; the second (ordinary) moment in terms of the first and second factorial moments; the third factorial moment in terms of the first, second, and third moments; and the third moment in terms of the first, second, and third factorial moments.

Problem 30. Verify the entries in the probability generating function column of 5.1 presented earlier. Then use the probability generating functions, Theorem 13, and Problem 29 to confirm the means and variances given in that table.

Problem 31. Prove that the function

$$\rho(s) = \frac{2}{(2 - s)(3 - s)}, \quad 0 \leq s \leq 1,$$

is a probability generating function of a distribution. Calculate this distribution and its mean, variance, and standard deviation. *Hint:* Use partial fractions.

* **Problem 32.** Replace the function in the preceding problem by

$$\rho(s) = \frac{8}{(2-s)^2(3-s)^3}, \quad 0 \le s \le 1.$$

* **Problem 33.** Fix $p \in [0,1]$. Prove that the function

$$\rho(s) = \begin{cases} 1 - \sqrt{1 - 4p(1-p)s^2} & \text{if } 0 \le s < 1 \\ 1 & \text{if } s = 1 \end{cases}$$

is a probability generating function of a distribution. Calculate the distribution and its mean and variance—writing the probabilities in the distribution in a form that makes it immediately apparent that they are nonnegative. (The number $\frac{1}{q+1}\binom{2q}{q}$ involved in the solution is called the qth *Catalan number.*)

5.5. Characterization of probability generating functions

The following result characterizes those functions which are probability generating functions of distributions supported by $\overline{\mathbb{Z}}^+$. The 'only if' aspect of the theorem is obvious and also the most used, often without comment.

Theorem 14. *An $\mathbb{R}$-valued function ρ defined on $[0,1]$ is a probability generating function of a $\overline{\mathbb{Z}}^+$-valued random variable if and only if $\rho(1) = 1$, $\rho(1-) \le 1$, $\rho(0) \ge 0$, and all derivatives of ρ are finite and nonnegative on $[0,1)$, in which case ρ equals its Maclaurin series on $[0,1)$.*

PROOF. We use $\rho^{(k)}$ to denote the derivative of order k of ρ; it equals ρ itself if $k = 0$. In view of the paragraph preceding the theorem, we can complete the proof by assuming that ρ has the properties given in the theorem and proving that

$$(5.5) \qquad \rho(s) = \sum_{k=0}^{\infty} \frac{\rho^{(k)}(0)}{k!} s^k$$

for $0 \le s < 1$, because one can then define a corresponding distribution by $Q(\{k\}) = \rho^{(k)}(0)/k!$ for $k \in \mathbb{Z}^+$ and $Q(\{\infty\}) = 1 - \rho(1-)$.

We begin by obtaining some consequences of the Taylor Formula with remainder. For $0 \le r < s < 1$,

$$\rho(s) = \sum_{k=0}^{n-1} \frac{\rho^{(k)}(r)}{k!}(s-r)^k + \frac{\rho^{(n)}(z_{n,r,s})}{n!}(s-r)^n$$

for some $z_{n,r,s}$ satisfying $r < z_{n,r,s} < s$. It follows that

$$(5.6) \qquad \sum_{k=0}^{\infty} \frac{\rho^{(k)}(r)}{k!}(s-r)^k \le \rho(s) \le \sum_{k=0}^{n-1} \frac{\rho^{(k)}(r)}{k!}(s-r)^k + \frac{\rho^{(n)}(s)}{n!}(s-r)^n$$

because the derivatives of ρ are nonnegative. Suppose that $2s - r < 1$ (in addition to $0 \le r < s < 1$). Setting $r = s$ and $s = 2s - r$ in the left inequality in (5.6), we conclude that

$$\sum_{k=0}^{\infty} \frac{\rho^{(k)}(s)}{k!} (s - r)^k \le \rho(2s - r) < \infty,$$

so that $\frac{\rho^{(n)}(s)}{n!}(s - r)^n \to 0$ as $n \to \infty$. We use this fact to conclude from (5.6) (as it stands) that

$$(5.7) \qquad \rho(s) = \sum_{k=0}^{\infty} \frac{\rho^{(k)}(r)}{k!}(s - r)^k \quad \text{for } 0 \le r < s < 2s - r < 1.$$

The special case $r = 0$ gives (5.5) for $s < \frac{1}{2}$.

For a proof by contradiction of the theorem, suppose that (5.5) fails for some $s \in [\frac{1}{2}, 1)$. Let t denote the greatest lower bound of the set of such s. Choose $r < t$ and $s > t$ so that $2s - r < 1$. Because ρ equals its MacLaurin series in the interval $[0, t)$, the derivatives of ρ in the interval $[0, t)$ can be calculated by term-by-term differentiation. Thus,

$$\rho^{(k)}(r) = \sum_{m=k}^{\infty} \frac{\rho^{(m)}(0)}{(m - k)!} r^{m-k}.$$

Insertion of this expression into (5.7) and an interchange of order of summation in the resulting double sum gives

$$\rho(s) = \sum_{m=0}^{\infty} \sum_{k=0}^{m} \binom{m}{k}(s - r)^k r^{m-k} \frac{\rho^{(m)}(0)}{m!} = \sum_{m=0}^{\infty} \frac{\rho^{(m)}(0)}{m!} s^m,$$

as desired. Since the summands are nonnegative, the interchange of order of summation is justified according to a result from advanced calculus, which is repeated in this book as Corollary 10 of Chapter 6. $\square$

Problem 34. Let

$$f(s) = \begin{cases} e^{-1/s} & \text{if } s > 0 \\ 0 & \text{if } s = 0. \end{cases}$$

Use the preceding theorem to prove that there does not exist $\varepsilon > 0$ such that $f^{(k)}(s) \ge 0$ for all $k \in \mathbb{Z}^+$ and all $s \in [0, \varepsilon)$.

Calculating Probabilities and Measures

In this chapter, after looking at several ways in which an event A can be defined in terms of other events $A_1, A_2, \ldots$, we will develop methods for calculating $P(A)$ in terms of the quantities $P(A_1), P(A_2), \ldots$. These methods include the Kochen-Stone and Borel-Cantelli Lemmas, the inclusion-exclusion formula, and some convergence theorems. Also, in the last section of this chapter, we will discuss measures other than probability measures.

6.1. Operations on events

In many cases, it is useful to describe operations on sets in terms of operations on the corresponding indicator functions. Here are indicator-function versions of some of the set-theoretic operations that we have already been using:

$$I_{A^c} = 1 - I_A \,;$$
$$I_{A \setminus B} = I_A(1 - I_B) \,;$$
$$I_{A \cap B} = I_A \wedge I_B = I_A I_B \,;$$
$$I_{A \cup B} = I_A \vee I_B = I_A + I_B - I_A I_B \,;$$
$$I_{\cap_n A_n} = \inf_n I_{A_n} = \prod_n I_{A_n} \,;$$
$$I_{\cup_n A_n} = \sup_n I_{A_n} \,.$$

The set $(A \setminus B) \cup (B \setminus A)$, denoted by $A \bigtriangleup B$ is called the *symmetric difference* of A and B. If A and B are events in some probability space, then so is $A \bigtriangleup B$. The symmetric difference of two sets consists of the points where the indicator functions of the two sets are different. In terms of indicator functions, we have

$$I_{A \bigtriangleup B} = I_A + I_B - 2I_A I_B \,.$$

Problem 1. Let A and B be two events in a probability space $(\Omega, \mathcal{F}, P)$. Show that $P(A \triangle B) = 0$ if and only if $I_A = I_B$ a.s., and that each of these conditions is equivalent to the condition that A and B be equal a.s.

If $I = \lim_{n \to \infty} I_{A_n}$ exists, then I is a function which takes only the values 0 and 1, so it is itself an indicator function of some set A contained in Ω. In this case, we say that the *limit* of the sequence $(A_1, A_2, \dots)$ exists and equals A, and we write

$$A = \lim_{n \to \infty} A_n .$$

Problem 2. A sequence of sets is said to be *increasing* or *decreasing*, respectively, if the corresponding sequence of indicator functions is increasing or decreasing. Prove that

$$\lim_{n \to \infty} A_n = \bigcup_{n=1}^{\infty} A_n \quad \text{if the sequence } (A_1, A_2, \dots) \text{ is increasing}$$

and

$$\lim_{n \to \infty} A_n = \bigcap_{n=1}^{\infty} A_n \quad \text{if the sequence } (A_1, A_2, \dots) \text{ is decreasing}.$$

In general, the limit of a sequence of sets may not exist. Let $J = \liminf I_{A_n}$ and $K = \limsup I_{A_n}$. Then J and K are indicator functions of sets B and C. We call B the *limit infimum* and C the *limit supremum* of the sequence $(A_1, A_2, \dots)$ and write

$$B = \liminf_{n \to \infty} A_n \quad \text{and} \quad C = \limsup_{n \to \infty} A_n .$$

From the definitions it follows that the limit of a sequence of sets exists if and only if the limit infimum and limit supremum are equal, in which case the limit equals their common value.

Proposition 1. *The limit infimum and limit supremum of a sequence of events is an event. If the limit of the sequence of events exists, it is an event.*

Problem 3. Prove the preceding proposition.

Problem 4. Let $(A_1, A_2, \dots)$ be a sequence of subsets of a space Ω. Prove that $\liminf A_n \subseteq \limsup A_n$, and that

$$\limsup_{n \to \infty} A_n = \bigcap_{n=1}^{\infty} \bigcup_{m=n}^{\infty} A_m \quad \text{and} \quad \liminf_{n \to \infty} A_n = \bigcup_{n=1}^{\infty} \bigcap_{m=n}^{\infty} A_m .$$

Problem 5. Show that the limit supremum of a sequence of sets is the set of points that are members of infinitely many of the sets in the sequence. Give a similar description of the limit infimum of the sequence. Also give a description in terms of set membership of what it means for the limit of a sequence of sets to exist.

* **Problem 6.** Show that $(\liminf_{n\to\infty} A_n)^c = \limsup_{n\to\infty} A_n^c$.

Problem 7. Suppose that $A_n \subseteq B_n \subseteq C_n$ for all sufficiently large $n \in \mathbb{Z}^+$ and that $\liminf_{n\to\infty} A_n \supseteq \limsup_{n\to\infty} C_n$. Prove that each of $\lim_{n\to\infty} A_n$, $\lim_{n\to\infty} B_n$, and $\lim_{n\to\infty} C_n$ exists and that all three limits are the same.

* **Problem 8.** Let $(A_1, A_2, \dots)$ and $(B_1, B_2, \dots)$ be sequences of subsets of a set Ω. Prove that each of the following equalities holds whenever the right side is meaningful:

$$\lim_{n\to\infty} (A_n \cup B_n) = (\lim_{n\to\infty} A_n) \cup (\lim_{n\to\infty} B_n);$$
$$\lim_{n\to\infty} (A_n \cap B_n) = (\lim_{n\to\infty} A_n) \cap (\lim_{n\to\infty} B_n);$$
$$\lim_{n\to\infty} (A_n \setminus B_n) = (\lim_{n\to\infty} A_n) \setminus (\lim_{n\to\infty} B_n);$$
$$\lim_{n\to\infty} (A_n \bigtriangleup B_n) = (\lim_{n\to\infty} A_n) \bigtriangleup (\lim_{n\to\infty} B_n);$$
$$\lim_{n\to\infty} A_n^c = (\lim_{n\to\infty} A_n)^c.$$

Also state and prove similar results involving limits suprema and limits infima, giving examples to show that equality is not valid in cases where your results are subset relations. Use indicator functions for some of the proofs and set-theoretic arguments for others.

The following exercise brings probability measures into the picture.

* **Problem 9.** Let $(A_1, A_2, \dots)$ be events in a probability space $(\Omega, \mathcal{F}, P)$. Show that

$$P(\limsup_{n\to\infty} A_n) \geq \limsup_{n\to\infty} P(A_n) \geq \liminf_{n\to\infty} P(A_n) \geq P(\liminf_{n\to\infty} A_n).$$

As an immediate consequence of the preceding exercise, we obtain an improved version of the Continuity of Measure Theorem, Theorem 2 of Chapter 3.

Theorem 2. [Continuity of Measure] *Let $(A_1, A_2, \dots)$ be a sequence of events in a probability space $(\Omega, \mathcal{F}, P)$. If $A = \lim_{n\to\infty} A_n$, then $P(A) = \lim_{n\to\infty} P(A_n)$.*

6.2. The Borel-Cantelli and Kochen-Stone Lemmas

This section is devoted chiefly to the computation of the probability of the limit supremum of a sequence of events. According to Problem 5, $\limsup_n A_n$ is the event consisting of those sample points that lie in infinitely many of the events A_n. For a simple example in which such an event is of interest, consider the coin-flip space, Example 4 of Chapter 1. Let A_n be the event that the nth flip is heads. Then $A = \limsup_n A_n$ is the event that there are infinitely many heads in the entire sequence of coin flips. The following three lemmas are often quite useful in calculating or estimating the probability of the limit supremum of a sequence.

Lemma 3. [Borel] *Let $(A_1, A_2, \dots)$ be a sequence of events in a common probability space $(\Omega, \mathcal{F}, P)$ and set $A = \limsup_{n \to \infty} A_n$. If $\sum_{n=1}^{\infty} P(A_n) < \infty$, then $P(A) = 0$.*

PROOF. By Problem 4,

$$A \subseteq \bigcup_{n=m}^{\infty} A_n$$

for each m. It follows from the monotonicity and subadditivity of probability measures that

$$P(A) \leq P\left(\bigcup_{n=m}^{\infty} A_n \right) \leq \sum_{n=m}^{\infty} P(A_n) .$$

Since $\sum_n P(A_n) < \infty$, the right side approaches 0 as $m \to \infty$; so $P(A) = 0$. $\square$

Problem 10. Let $(X_1, X_2, \dots)$ be a sequence of random variables defined on a common probability space. Assume that each random variable X_n is uniformly distributed on $[0, 1]$. Prove that for all $\alpha > 1$,

$$\lim_{n \to \infty} \frac{1}{n^{\alpha} X_n} = 0 \text{ a.s.}$$

Lemma 4. [Kochen-Stone] *Let $(A_1, A_2, \dots)$ be a sequence of events in a probability space $(\Omega, \mathcal{F}, P)$ and set $A = \limsup_{n \to \infty} A_n$. If $\sum_{n=1}^{\infty} P(A_n) = \infty$, then*

$$(6.1) \qquad P(A) \geq \limsup_{n \to \infty} \frac{\left[\sum_{m=1}^{n} P(A_m) \right]^2}{\sum_{k=1}^{n} \sum_{m=1}^{n} P(A_k \cap A_m)} .$$

PROOF. Let I_m denote the indicator function of A_m. For $n \geq 1$ let

$$J_n = \sum_{m=1}^{n} I_m .$$

Thus, $J_n(\omega)$ is the number of events $A_1, \dots, A_n$ that contain the point ω. Since

$$(6.2) \qquad E(J_n) = \sum_{m=1}^{n} P(A_n) ,$$

we have by hypothesis that $\lim_n E(J_n) = \infty$. By Problem 5,

$$A = \{\omega : \lim_{n \to \infty} J_n(\omega) = \infty\} .$$

It follows that for all $\lambda > 0$,

$$A \supseteq \limsup_{n \to \infty} B_{n, \lambda} ,$$

where

$$B_{n, \lambda} = \{\omega : J_n(\omega) > \lambda E(J_n)\} .$$

By Problem 9,

$$(6.3) \qquad P(A) \geq P(\limsup_{n\to\infty} B_{n,\lambda}) \geq \limsup_{n\to\infty} P(B_{n,\lambda}) \text{ for all } \lambda > 0.$$

By Corollary 5 of Chapter 5,

$$(6.4) \qquad P(B_{n,\lambda}) \geq \frac{(1-\lambda)^2 (E(J_n))^2}{E(J_n^2)}$$

for $0 < \lambda < 1$. By (6.3) and (6.4)

$$P(A) \geq \limsup_{n\to\infty} \frac{(E(J_n))^2}{E(J_n^2)}.$$

The proof is now completed by applying (6.2) and noting that

$$E(J_n^2) = \sum_{k=1}^{n}\sum_{m=1}^{n} E(I_{A_k} I_{A_m}) = \sum_{k=1}^{n}\sum_{m=1}^{n} P(A_k \cap A_m). \qquad \square$$

The following important special case of the Kochen-Stone Lemma is a partial converse of the Borel Lemma. For it, we need the following terminology: two events are *positively correlated*, *negatively correlated*, or *uncorrelated* according to what their indicator functions are.

Lemma 5. [Borel-Cantelli] *Let* $(A_1, A_2, \dots)$ *be a sequence of events in a probability space* $(\Omega, \mathcal{F}, P)$. *Assume that for each* $i \neq j$, *the events* A_i *and* A_j *are either negatively correlated or uncorrelated. Let* $A = \limsup_{n\to\infty} A_n$. *If* $\sum_{n=1}^{\infty} P(A_n) = \infty$, *then* $P(A) = 1$.

Problem 11. Prove the preceding lemma. *Hint:* The correlation condition is equivalent to

$$P(A_k \cap A_m) \leq P(A_k)P(A_m) \text{ if } k \neq m.$$

Thus it is enough to show that the summands with $k = m$ in the denominator of the formula in the Kochen-Stone Lemma can be ignored.

The preceding problem requests a proof that is rather straightforward only because it is based on a result whose proof is somewhat difficult. Under a somewhat more restrictive hypothesis there is a simple direct proof (requested in Problem 38 of Chapter 9).

Problem 12. In the context of the coin-flip space of Example 4 of Chapter 1, let A_n be the event that the nth flip is heads. Use the Borel-Cantelli Lemma to prove that $P(\limsup_{n\to\infty} A_n) = 1$.

* **Problem 13.** Let $(X_1, X_2, \ldots)$ be a sequence of pairwise negatively correlated or uncorrelated random variables, each having a standard Bernoulli distribution. Show that

$$\sum_{n=1}^{\infty} X_n \begin{cases} = \infty \text{ a.s.} \\ < \infty \text{ a.s.} \end{cases}$$

if and only if

$$E\left(\sum_{n=1}^{\infty} X_n\right) \begin{cases} = \infty \\ < \infty, \end{cases}$$

respectively. Construct an example to show that the correlation assumption cannot be dropped.

6.3. Inclusion-exclusion

We turn now to a formula that expresses the probability of a union of events in terms of sums and differences of probabilities of intersections. This formula generalizes the identity $P(A \cup B) = P(A) + P(B) - P(A \cap B)$. It is instructive to note that here, as in the proof of the Kochen-Stone Lemma, the use of indicator functions facilitates the argument.

Theorem 6. [Inclusion-Exclusion] *Let $C_1, \ldots, C_n$ be events in a probability space $(\Omega, \mathcal{F}, P)$. Then*

$$P\left(\bigcup_{m=1}^{n} C_m\right)$$

$$= \sum_{m=1}^{n} P(C_m) - \sum_{m=2}^{n} \sum_{l=1}^{m-1} P(C_l \cap C_m)$$

$$+ \sum_{m=3}^{n} \sum_{l=2}^{m-1} \sum_{k=1}^{l-1} P(C_k \cap C_l \cap C_m) - \cdots + (-1)^{n+1} P\left(\bigcap_{q=1}^{n} C_q\right).$$

PROOF. Let I be the indicator function of $\bigcup_{m=1}^{n} C_m$ and, for $m = 1, 2, \ldots, n$, let I_m be the indicator function of C_m. It suffices to prove that

(6.5)
$$I = \sum_{m=1}^{n} I_m - \sum_{m=2}^{n} \sum_{l=1}^{m-1} I_l I_m$$

$$+ \sum_{m=3}^{n} \sum_{l=2}^{m-1} \sum_{k=1}^{l-1} I_k I_l I_m - \cdots + (-1)^{n+1} \prod_{q=1}^{n} I_q,$$

for then the result follows by taking expectations. We will check that both sides of (6.5) have the same value for each $\omega \in \Omega$. Consider an arbitrary ω and let $\rho(\omega)$ equal the number of events C_n that contain ω as a member.

If $\rho(\omega) = 0$, then every term on both sides of (6.5) equals 0. Suppose that $\rho(\omega) > 0$. Then the left side of (6.5) equals 1. The value of the right side can be expressed in the form

(6.6)
$$\sum_{i=1}^{\rho(\omega)} (-1)^{i+1} S_i(\omega),$$

where $S_i(\omega)$ equals the value of the ith term in (6.5), a term that itself involves i summation symbols. Each summand of the ith term is the product of indicator functions and, thus, equals 1 or 0. Therefore, $S_i(\omega)$ equals the number of summands in the ith term that equal 1—that is, the number of ways of choosing i events from among all those events C_n that contain ω. It follows that

$$S_i(\omega) = \frac{\rho(\omega)!}{i!(\rho(\omega) - i)!}.$$

Insertion into (6.6) gives 1 minus the binomial series for $(1 - 1)^{\rho(\omega)}$, so the right side of (6.5) equals 1. $\square$

Problem 14. Suppose two dice are rolled. What is the probability that at least one five appears? Use two different methods for this problem. Follow the same instructions for three dice, then for n dice.

* **Problem 15.** In the context of Example 2 of Chapter 4, calculate the probability that at least one card occupies a position equal to its label. Discuss the limiting behavior as the number of cards approaches ∞.

Problem 16. In the context of Problem 18 of Chapter 4, calculate the probability that at least one card occupies a position equal to its label. Discuss the limiting behavior as the number of cards approaches ∞.

Problem 17. We know $P(\bigcup C_n) \le \sum_n P(C_n)$. It develops that other inequalities of a similar nature are true. They can be obtained by truncating the right side of the inclusion-exclusion equality and then proved by a modification of the proof of that equality. Carry out this program.

6.4. Finite and σ-finite measures

For the remainder of this chapter, we consider some of the consequences of dropping the assumption that $P(\Omega) = 1$. We are mainly concerned with definitions and examples here. Further developments will be found in Chapters 7 and 8.

Definition 7. Let $(\Omega, \mathcal{F})$ be a measurable space. A *measure* on $(\Omega, \mathcal{F})$ is a countably additive $\overline{\mathbb{R}}^+$-valued function μ defined on $\mathcal{F}$. For $A \in \mathcal{F}$, the quantity $\mu(A)$ is called the *measure* of A. If $\mu(\Omega) < \infty$, then μ is a *finite measure* on $(\Omega, \mathcal{F})$. Otherwise, μ is an *infinite measure*.

If $(\Omega, \mathcal{F}, \mu)$ is a finite measure space and if $\mu(\Omega) > 0$, then $(\Omega, \mathcal{F}, P)$ is a probability space, where $P(A) = \mu(A)/\mu(\Omega)$ for all $A \in \mathcal{F}$. Thus, the theory developed so far for probability spaces is easily generalized to finite measure spaces, and no further comment is required.

Before focusing on infinite measure spaces, we treat some needed results from advanced calculus.

Proposition 8. Let $\mathcal{S} = (s_{m,n} : m, n = 1, 2, \ldots)$ be a doubly indexed set of members of $\overline{\mathbb{R}}$. Suppose that $s_{m,n}$ is increasing in both m and n, in the sense that if $j \leq m$ and $k \leq n$, then $s_{j,k} \leq s_{m,n}$. Then

$$\lim_{m \to \infty} \lim_{n \to \infty} s_{m,n} = \lim_{n \to \infty} \lim_{m \to \infty} s_{m,n} = \sup \mathcal{S}.$$

PROOF. It is clear that $\sup \mathcal{S}$ is at least as large as either of the double limits. On the other hand, if $s < \sup \mathcal{S}$, then by definition of the supremum, there exist integers j, k such that $s_{j,k} > s$. It follows from the monotonicity assumption that $s_{m,n} > s$ for all $m \geq j$ and $n \geq k$. Thus, both double limits are greater than s. Since s is an arbitrary number less than $\sup \mathcal{S}$, both double limits are greater than or equal to $\sup \mathcal{S}$. Equality of the double limits and $\sup \mathcal{S}$ now follows. $\square$

Corollary 9. [Monotone Convergence for Sums] Let $(a_{m,n} : m, n = 1, 2, \ldots)$ be a doubly indexed set of members of $[0, \infty]$. Suppose that $a_{m,n}$ is increasing in n, in the sense that if $k \leq n$, then $a_{m,k} \leq a_{m,n}$ for all m. Then

$$\sum_{m=1}^{\infty} \lim_{n \to \infty} a_{m,n} = \lim_{n \to \infty} \sum_{m=1}^{\infty} a_{m,n}.$$

PROOF. Let

$$s_{m,n} = \sum_{j=1}^{m} a_{m,n}$$

and apply Proposition 8. $\square$

Corollary 10. [Fubini Theorem for Nonnegative Sums] Suppose that $\mathcal{A} = (a_{m,n} : m, n = 1, 2, \ldots)$ is a doubly indexed set of members of $[0, \infty]$. Then

$$\sum_{m=1}^{\infty} \sum_{n=1}^{\infty} a_{m,n} = \sum_{n=1}^{\infty} \sum_{m=1}^{\infty} a_{m,n} = \sup \mathcal{S},$$

where $\mathcal{S}$ is the set of all sums of finitely many members of $\mathcal{A}$.

PROOF. Set

$$s_{m,n} = \sum_{j=1}^{m} \sum_{k=1}^{n} a_{j,k}$$

and let $S = \sup\{s_{m,n} : m, n = 1, 2, \ldots\}$. By Proposition 8, the two double sums in the statement of the corollary are both equal to S. Clearly, $S \leq \sup \mathcal{S}$. On the other hand, for any $s \in \mathcal{S}$, there is a partial sum $s_{m,n}$ that is at least as large as s, so $S \geq \sup \mathcal{S}$. The desired equality follows. $\square$

Problem 18. Let $\mathcal{A} = (a_n : n = 1, 2, \ldots)$ be a sequence of members of $[0, \infty]$, and let π be any *permutation* of the natural numbers (in other words, π is a bijection from the natural numbers to the natural numbers). Prove that

$$\sum_{n=1}^{\infty} a_n = \sum_{n=1}^{\infty} a_{\pi(n)} = \sup \mathcal{S},$$

where $\mathcal{S}$ is the set of all sums of finitely many of the terms in $\mathcal{A}$.

With the preceding tools in hand we return to the study of measures.

Problem 19. Let μ_j, $j \in J$, be countably many measures on a measurable space $(\Omega, \mathcal{F})$. Set

$$\mu(A) = \sum_{j \in J} \mu_j(A)$$

for all $A \in \mathcal{F}$. Prove that μ is a measure.

Problem 18 ensures that the order of terms is irrelevant in the sum in Problem 19, and Corollary 10 implies that a true assertion would be obtained even if that sum were replaced by a double sum of doubly-index measures.

Problem 19 suggests: Write a measure that one does not understand as a sum of measures one does understand, then study the summands, and finally draw conclusions about the original measure. This plan of attack will work best when the summands do not 'overlap'.

Definition 11. Measures μ and ν on a measure space $(\Omega, \mathcal{F})$ are *mutually singular* if there exists a set $A \in \mathcal{F}$ such that $\mu(A) = \nu(A^c) = 0$.

Definition 12. A measure that can be written as a countable sum of pairwise mutually singular finite measures is said to be *σ-finite*.

Proposition 13. *Let $(\Omega, \mathcal{F}, \mu)$ be a measure space. Then μ is σ-finite if and only if there exists a pairwise disjoint sequence $(B_j : j = 1, 2, \ldots)$ of members of $\mathcal{F}$ such that $\mu(B_j) < \infty$ for every j and*

$$\mu(B) = \sum_{j=1}^{\infty} \mu(B \cap B_j)$$

for every $B \in \mathcal{F}$.

Problem 20. Prove the preceding proposition.

Example 1. [Lebesgue measure on $\mathbb{R}$] For each integer j, let λ_j be the uniform probability measure on the interval $[j, j+1)$, with each λ_j taken to be a probability distribution on $\mathbb{R}$. Thus, λ_j gives 0 measure to the set $[j, j+1)^c$, and measure 1 to the set $[j, j+1)$, so we have defined a pairwise mutually singular collection of finite measures on the Borel sets of $\mathbb{R}$. For any Borel set $B \subseteq \mathbb{R}$, let $\lambda(B) = \sum_{j \in \mathbb{Z}} \lambda_j(B)$. A straightforward calculation shows that $\lambda(I) =$ the length of I for any interval I. The σ-finite measure λ is called *Lebesgue measure on $\mathbb{R}$*, or *one-dimensional Lebesgue measure*.

If A is a Borel subset of $\mathbb{R}$, then the restriction of λ to Borel subsets of A is called *Lebesgue measure* on A. If $0 < \lambda(A) < \infty$, then we obtain the *uniform distribution* on A by dividing Lebesgue measure on A by $\lambda(A)$.

Example 2. [Counting measure] Let Ω be an arbitrary set and $\mathcal{F}$ the σ-field of all subsets of Ω. For $B \subseteq \Omega$, let $\mu(B)$ be the cardinality of B if B is finite; otherwise, let $\mu(B) = \infty$. It is easily seen that μ is a measure, and that μ is σ-finite if and only if Ω is countable. The measure μ is called *counting measure* on Ω. If Ω is finite and nonempty, then we obtain the *uniform distribution* on Ω by dividing the counting measure μ by $\mu(\Omega)$.

Although most of the concepts and theorems for probability spaces can also be extended to σ-finite measure spaces, and, in many cases, to general measure spaces, some cannot. Thus, care must be taken with infinite measure spaces, even σ-finite ones. For instance, let $A_n = (n, \infty), n = 1, 2, \ldots$, and let λ be Lebesgue measure on $(\mathbb{R}, \mathcal{B})$. Then $A_n \to \emptyset$ as $n \to \infty$, but, contrary to what one might expect based on the Continuity of Measure Theorem, $\lambda(A_n) = \infty$ for all n, and hence does not go to 0. We will have more to say about the similarities and differences between finite and infinite measure spaces in subsequent chapters, particularly in Chapter 8.

Problem 21. State and prove finite and σ-finite versions of the Inclusion-Exclusion Proposition.

Problem 22. Show that every translation invariant σ-finite measure on $\mathbb{Z}$ (with the σ-field of all subsets) is a multiple of counting measure.

CHAPTER 7

Measure Theory:
Existence and Uniqueness

In some of the examples of previous chapters, most notably the coin-flip space, we defined a sample space Ω and a σ-field $\mathcal{F}$, but we did not completely specify probabilities $P(A)$ for all $A \in \mathcal{F}$. Instead, we only gave the values of $P(A)$ for events A in a smaller collection $\mathcal{E}$ such that $\mathcal{F} = \sigma(\mathcal{E})$, and then we assumed without proof that P could be extended in a unique way to all of $\mathcal{F}$. In this chapter, we close this gap by showing that under certain natural assumptions, a function P defined on a collection $\mathcal{E}$ of subsets of a sample space Ω can be extended in a unique way to a probability measure on $\mathcal{F} = \sigma(\mathcal{E})$. Once probability measures are constructed, we can piece them together to form σ-finite measures. Interesting examples include Lebesgue measure in $\mathbb{R}^d$ and a certain measure on the space of all lines in $\mathbb{R}^2$, both of which are invariant under rigid motions. In preparation for this chapter, the reader may want to review (Counter)example 6 in Chapter 1 and to reread the paragraphs leading up to that example and to Definition 5 of the same chapter.

7.1. The Sierpiński Class Theorem and uniqueness

The uniqueness question can be resolved without much difficulty, so it is treated first. This question may be formulated as follows: Given a probability space $(\Omega, \sigma(\mathcal{E}), P)$, what conditions on $\mathcal{E}$ ensure that the values of $P(A)$ for $A \in \mathcal{E}$ uniquely determine the values of $P(A)$ for all $A \in \sigma(\mathcal{E})$? The following definition and theorem will set the stage for an answer to this question. The term *proper set difference,* referring to a set difference $B \setminus A$, entails the assumption that $A \subseteq B$. (Some use the notation $B - A$ for a proper set difference.)

Definition 1. A *Sierpiński class* of subsets of a set Ω is a class that is closed under limits of increasing sequences of sets and proper set differences.

Problem 1. Let $\mathcal{E}$ be a collection of subsets of a set Ω. Prove that there is a smallest Sierpiński class of subsets of Ω that contains $\mathcal{E}$.

Problem 2. Show that if a Sierpiński class of subsets of Ω is closed under pairwise intersections and contains Ω, then it is a σ-field.

Theorem 2. [Sierpiński Class] *Let $\mathcal{E}$ be a collection of subsets of a set Ω and suppose that $\mathcal{E}$ is closed under pairwise intersections and contains Ω. Then the smallest Sierpiński class of subsets of Ω that contains $\mathcal{E}$ equals $\sigma(\mathcal{E})$.*

PROOF. Let $\mathcal{D}$ denote the smallest Sierpiński class containing $\mathcal{E}$. Clearly, $\mathcal{D} \subseteq \sigma(\mathcal{E})$. Since $\Omega \in \mathcal{E}$ and $\mathcal{E} \subseteq \mathcal{D}$, it follows from Problem 2 that to show $\sigma(\mathcal{E}) \subseteq \mathcal{D}$, we need only prove that $\mathcal{D}$ is closed under pairwise intersections.

For $A \subseteq \Omega$, define

$$(7.1) \qquad \mathcal{N}_A = \{B : A \cap B \in \mathcal{D}\}.$$

It is easy to check that $\mathcal{N}_A$ is a Sierpiński class for any $A \in \mathcal{D}$. Consider $\mathcal{N}_A$ for $A \in \mathcal{E}$. Since $\mathcal{E}$ is closed under pairwise intersections, $\mathcal{N}_A \supseteq \mathcal{E}$. Therefore, if $A \in \mathcal{E}$, $\mathcal{N}_A$ is a Sierpiński class that contains $\mathcal{D}$.

Now consider $\mathcal{N}_A$ for $A \in \mathcal{D}$. By the preceding paragraph, the intersection of A with any member of $\mathcal{E}$ is a member of $\mathcal{D}$. Hence, $\mathcal{N}_A \supseteq \mathcal{E}$. We again conclude that $\mathcal{N}_A$ is a Sierpiński class that contains $\mathcal{D}$. Therefore, $\mathcal{D}$ is closed under pairwise intersections. $\square$

Theorem 3. [Uniqueness of Measure] *Let P and Q be probability measures on the measurable space $(\Omega, \sigma(\mathcal{E}))$, where $\mathcal{E}$ is a collection of sets closed under pairwise intersections. If $P(A) = Q(A)$ for every $A \in \mathcal{E}$, then $P = Q$.*

* **Problem 3.** Prove the preceding theorem.

Problem 4. Show how the Uniqueness of Measure Theorem can be used to prove the uniqueness assertion of Proposition 4 of Chapter 3. Incorporate some of Problem 13 of Chapter 1 into your discussion.

Problem 5. Which of the collections given in Problem 13 of Chapter 1 are closed under pairwise intersections?

Problem 6. Verify the assertions of uniqueness in Example 1 and Problem 15, both of Chapter 2.

The Uniqueness of Measure Theorem assures us that we can describe a probability measure in an unambiguous manner by giving its values for a relatively small collection of events, as long as this collection is closed under pairwise intersections. It does not, however, say that probabilities for sets in such a collection can be specified arbitrarily. For instance, consider the following collection of subsets of $\Omega = \{1, 2, 3\}$:

$$\{\emptyset, \{1\}, \{1, 2\}, \{1, 3\}, \Omega\}.$$

This collection is closed under intersections and generates the σ-field $\mathcal{F}$ of all subsets of Ω. Nevertheless, there is no way to extend the function R, defined by

$$R(\Omega) = R(\{1,2\}) = R(\{1,3\}) = 1, \quad R(\{1\}) = 1/2, \quad R(\emptyset) = 0,$$

to a probability measure on $\mathcal{F}$. It is easy to check in this example that the function R does not violate the conditions required of a probability measure on $\mathcal{E}$. The problem is that while $\mathcal{E}$ is large enough to ensure uniqueness, it is not large enough to ensure existence.

7.2. Finitely additive functions defined on fields

In general, there are two parts to the existence problem. The first is to find a tractable way to check that a function R satisfies all the properties of a probability measure on its domain of definition $\mathcal{E}$. The second part is to find a set of conditions on $\mathcal{E}$ that will guarantee that if R satisfies all the properties of a probability measure on $\mathcal{E}$, then R can be extended to a probability measure on $\sigma(\mathcal{E})$. Some of the work done in Chapter 6 concerning sequences of events will be useful here, although we will need to extend it to a more general setting than that of σ-fields.

Definition 4. A *field* of subsets of a set Ω is a collection of subsets of Ω that has $\emptyset$ as a member and is closed under complementation and pairwise unions.

Definition 5. A real-valued function R defined on a field $\mathcal{E}$ of subsets of a set Ω is said to be *finitely additive* if $R(A \cup B) = R(A) + R(B)$ for every disjoint pair A and B of members of $\mathcal{E}$. The function R is said to be *countably additive* if

$$R\left(\bigcup_{n=1}^{\infty} A_n\right) = \sum_{n=1}^{\infty} R(A_n)$$

whenever $(A_1, A_2, \dots)$ is a sequence of pairwise disjoint members of $\mathcal{E}$ whose union is also a member of $\mathcal{E}$.

Problem 7. Let R be a nonnegative finitely additive function defined on a field $\mathcal{E}$ of subsets of a space Ω, such that $R(\Omega) = 1$. Let A and B be members of $\mathcal{E}$ such that $A \subseteq B$. Prove that $R(\emptyset) = 0$, $R(A) \leq R(B)$, $R(B \setminus A) = R(B) - R(A)$, and $R(A^c) = 1 - R(A)$.

Problem 8. Show that if R is countably additive, then

$$R\left(\bigcup_{n=1}^{\infty} A_n\right) \leq \sum_{n=1}^{\infty} R(A_n)$$

whenever $(A_1, A_2, \dots)$ is a sequence of members of $\mathcal{E}$ whose union is also in $\mathcal{E}$.

The preceding problem shows that if R is countably additive, it possesses many of the properties associated with probability measures, even if its domain is only a field. It will be seen in the forthcoming lemma that such a function R automatically possesses another important property of probability measures, namely continuity of measure. Thus, this lemma is a strengthening of the Continuity of Measure Theorem of Chapter 3. It is recommended that the reader attempt the exercise preceding the lemma before reading the proof of the lemma.

Problem 9. Prove the following lemma under the additional assumption that the sequence $(A_1, A_2, \dots)$ is decreasing. *Hint:* Review the proof of the Continuity of Measure Theorem of Chapter 3.

Lemma 6. *Let $\mathcal{E}$ be a field of subsets of a set Ω, and let R be a nonnegative countably additive function defined on $\mathcal{E}$ such that $R(\Omega) = 1$. Let $(A_1, A_2, \dots)$ be a sequence in $\mathcal{E}$ with the property that $\lim_{n\to\infty} A_n = \emptyset$. Then $\lim_{n\to\infty} R(A_n) = 0$.*

PROOF. For $m = 1, 2, \dots$, the function

$$p \rightsquigarrow R\left(\bigcup_{n=m}^{p} A_n\right)$$

increases to a limit v_m as $p \to \infty$. Fix $\varepsilon > 0$. Choose integers $N(m)$, increasing to ∞ as m increases to ∞, such that

$$v_m - R(B_m) < \frac{\varepsilon}{2^m},$$

where for each m,

(7.2)
$$B_m = \bigcup_{n=m}^{N(m)} A_n.$$

Clearly, $B_m \to \emptyset$ as $m \to \infty$.

For $k = 1, 2, \dots$, let

(7.3)
$$C_k = \bigcap_{m=1}^{k} B_m.$$

Since $C_k \searrow \emptyset$ as $k \nearrow \infty$, it follows from Problem 9 that $R(C_k) \to 0$. By (7.2) and (7.3),

$$R(A_k \setminus C_k) \le R\left(\bigcup_{m=1}^{k} (B_k \setminus B_m)\right)$$

$$\le \sum_{m=1}^{k} R(B_k \setminus B_m)$$

$$\le \sum_{m=1}^{k} R\left(\left(\bigcup_{n=m}^{N(k)} A_n\right) \setminus B_m\right)$$

$$\le \sum_{m=1}^{k} (v_m - R(B_m)) \le \sum_{m=1}^{k} \frac{\varepsilon}{2^m} < \varepsilon.$$

Since $R(C_k) \to 0$ and $R(A_k) \le R(C_k) + R(A_k \setminus C_k)$, we conclude that

$$\limsup_{k \to \infty} R(A_k) \le \varepsilon.$$

Now let $\varepsilon \searrow 0$. $\quad\square$

Lemma 7. *Let $\mathcal{E}$ be a field of subsets of a set Ω and R a nonnegative count-ably additive function defined on $\mathcal{E}$ such that $R(\Omega) = 1$. Let $(A_1, A_2, \ldots)$ and $(B_1, B_2, \ldots)$ be sequences in $\mathcal{E}$ whose limits exist and are equal (the common limit need not be a member of $\mathcal{E}$). Then $\lim_{n \to \infty} R(A_n)$ and $\lim_{n \to \infty} R(B_n)$ exist and are equal.*

PROOF. It is sufficient to show that if $(m_1 < m_2 < \ldots)$ is any strictly in-creasing sequence of positive integers, then $R(A_n) - R(B_{m_n}) \to 0$ as $n \to \infty$. By Problem 7,

(7.4)
$$\begin{aligned}|R(A_n) - R(B_{m_n})| &\le R(A_n \cup B_{m_n}) - R(A_n \cap B_{m_n}) \\ &= R\big((A_n \cup B_{m_n}) \setminus (A_n \cap B_{m_n})\big).\end{aligned}$$

By Problem 8 of Chapter 6,

$$\lim_{n \to \infty} (A_n \cup B_{m_n}) \setminus (A_n \cap B_{m_n}) = \emptyset.$$

It follows from Lemma 6 that

$$\lim_{n \to \infty} R\big((A_n \cup B_{m_n}) \setminus (A_n \cap B_{m_n})\big) = 0.$$

Therefore, by (7.4), $R(A_n) - R(B_{m_n}) \to 0$, as desired. $\quad\square$

Corollary 8. *Let $\mathcal{E}$ and R be as in the preceding lemma. If $(A_1, A_2, \ldots)$ is a sequence in $\mathcal{E}$ that converges to $A \in \mathcal{E}$, then $R(A) = \lim_{n \to \infty} R(A_n)$.*

* **Problem 10.** Prove the preceding corollary.

We have seen a variety of conclusions that follow from countable additivity. The following proposition provides a useful condition under which finite additivity implies countable additivity.

Proposition 9. *Let R be a nonnegative finitely additive function defined on a field $\mathcal{E}$ of subsets of a set Ω, such that $R(\Omega) = 1$. Then R is countably additive if and only if $R(A_n) \to 0$ for every decreasing sequence $(A_1, A_2, \ldots)$ in $\mathcal{E}$ for which $\lim A_n = \emptyset$.*

PROOF. The 'only if' is a special case of Lemma 6 (see also Problem 9). To prove the 'if' part, let $(B_1, B_2, \ldots)$ be a sequence of pairwise disjoint members of $\mathcal{E}$ whose union B is also a member of $\mathcal{E}$. Let

$$A_n = B \setminus \bigcup_{m=1}^{n} B_m.$$

Then $(A_1, A_2, \ldots)$ is a decreasing sequence in $\mathcal{E}$ whose limit equals $\emptyset$. Hence $R(A_n) \to 0$. It follows that

$$R(B) - \sum_{m=1}^{\infty} R(B_m) = \lim_{n \to \infty} \left(R(B) - \sum_{m=1}^{n} R(B_m) \right)$$

$$= \lim_{n \to \infty} \left(R(B) - R\left(\bigcup_{m=1}^{n} B_m \right) \right)$$

$$= \lim_{n \to \infty} R(A_n)$$

$$= 0.$$

Therefore, R is countably additive. $\square$

7.3. Existence, extension, and completion of measures

Here is a typical situation in which the existence question arises. We start with a field $\mathcal{E}$ of subsets of a space Ω, and we are able to specify probabilities $R(A)$ for sets A that are members of $\mathcal{E}$. It usually happens naturally that R is nonnegative and finitely additive, and that $R(\Omega) = 1$. If we can verify the condition in Proposition 9, then we know that R is countably additive. We will now prove that if R is countably additive, it can be extended to a probability measure defined on $\sigma(\mathcal{E})$.

The proof requires several steps which we briefly describe here. First we let $\mathcal{E}_1$ be the collection of subsets of Ω which are limits of sequences of members of $\mathcal{E}$. The collection $\mathcal{E}_1$ is easily shown to be a field which contains $\mathcal{E}$. Next we use Lemma 7 and Proposition 9 to show that R can be extended to a nonnegative countably additive function defined on $\mathcal{E}_1$. The procedure is then repeated, with $\mathcal{E}_1$ in the place of $\mathcal{E}$. The result is that R is extended to a nonnegative countably additive function defined on a field $\mathcal{E}_2$ which contains all limits of sequences of members of $\mathcal{E}_1$. We could continue to repeat this procedure, successively extending R to fields $\mathcal{E}_3, \mathcal{E}_4, \ldots$. However, it develops that the collection $\mathcal{E}_2$

plays a special role. We introduce a collection $\mathcal{D}$ of subsets of Ω called the *completion* of $\mathcal{E}_2$. A set B is a member of $\mathcal{D}$ if there exist sets A and C in $\mathcal{E}_2$ such that $A \subseteq B \subseteq C$ and $R(A) = R(C)$. We show that $\mathcal{D}$ is a σ-field which contains $\sigma(\mathcal{E})$. Finally we extend the function R to $\mathcal{D}$ in a natural way and show that the extension is countably additive.

Lemma 10. *Let $\mathcal{E}$ be a field of subsets of a space Ω. Define $\mathcal{E}_1$ to be the collection of subsets of Ω which are limits of sequences of members of $\mathcal{E}$. Then $\mathcal{E}_1$ is a field which contains $\mathcal{E}$.*

*   **Problem 11.** Prove the preceding lemma.

Let $\mathcal{E}$ and $\mathcal{E}_1$ be as in Lemma 10, and let R be a nonnegative countably additive function defined on $\mathcal{E}$ such that $R(\Omega) = 1$. For $A \in \mathcal{E}_1$ and $(A_1, A_2, \dots)$ a sequence in $\mathcal{E}$ whose limit is A, define

$$R_1(A) = \lim_{n \to \infty} R(A_n).$$

It follows from Lemma 7 that this limit exists and does not depend on the choice of the sequence $(A_1, A_2, \dots)$. If $A \in \mathcal{E}$, it follows from Corollary 8 that $R_1(A) = R(A)$. Thus R_1 is an extension of R to $\mathcal{E}_1$. It is clear that R_1 is nonnegative and $R_1(\Omega) = 1$.

Problem 12. Show that R_1 is finitely additive. *Hint:* If A and B are disjoint members of $\mathcal{E}_1$, show that there are sequences $(A_1, A_2, \dots)$ and $(B_1, B_2, \dots)$ in $\mathcal{E}$, with limits A and B, respectively, such that for each $n = 1, 2, \dots, A_n$ and B_n are disjoint.

Lemma 11. *The extension R_1 is countably additive.*

PROOF. By the preceding problem, R_1 is finitely additive. Thus, by Proposition 9, it is enough to show that if $(A_1, A_2, \dots)$ is a decreasing sequence in $\mathcal{E}_1$ that converges to $\emptyset$, then $R_1(A_n) \to 0$ as $n \to \infty$. By the definition of $\mathcal{E}_1$, for each m there exists a sequence $(B_{m,n}, n = 1, 2, \dots)$ in $\mathcal{E}$ such that $A_m = \lim_{n \to \infty} B_{m,n}$. Let

$$C_{m,n} = \bigcap_{j=1}^{m} B_{j,n}.$$

Then since the sequence $(A_1, A_2, \dots)$ is decreasing,

$$\lim_{n \to \infty} C_{m,n} = \lim_{n \to \infty} \left(\bigcap_{j=1}^{m} B_{j,n} \right)$$
$$= \bigcap_{j=1}^{m} A_j = A_m.$$

By the definition of R_1, $R_1(A_m) = \lim_{n\to\infty} R(C_{m,n})$, so for $m = 1, 2, \ldots$ we can choose integers n_m increasing to ∞ as m increases to ∞, such that

$$(7.5) \qquad\qquad R_1(A_m) - R(C_{m,n_m}) \to 0 \text{ as } m \to \infty.$$

To finish the proof that $R_1(A_m) \to 0$ we will show that $R(C_{m,n_m}) \to 0$ by demonstrating that $C_{m,n_m} \to \emptyset$ and appealing to Lemma 6. Since for fixed n the sequence $(C_{k,n}, k = 1, 2, \ldots)$ is decreasing, we can conclude that for each fixed k,

$$\limsup_{m\to\infty} C_{m,n_m} \subseteq \limsup_{m\to\infty} C_{k,n_m} = A_k.$$

Now let $k \to \infty$ to obtain $\limsup C_{m,n_m} = \emptyset$, as desired. $\quad\square$

It is clear that the procedure of the preceding two lemmas can be repeated to produce an increasing sequence of fields $\mathcal{E}_n$, with corresponding nonnegative countably additive functions R_n defined on $\mathcal{E}_n$. The field $\mathcal{E}_{n+1}$ is obtained by taking all limits of sequences of members of $\mathcal{E}_n$, and R_{n+1} is an extension of R_n. Each of the fields $\mathcal{E}_n$ contains $\mathcal{E}$, and each R_n extends R to $\mathcal{E}_n$.

> **Problem 13.** Consider the collection of intervals in $\mathbb{R}$ of the following four types: $(a, b]$ for $a < b$, $(-\infty, b]$, (a, ∞), and $(-\infty, \infty)$. Let $\mathcal{E}$ be the collection consisting of all finite unions of such intervals. We include the empty set in $\mathcal{E}$ as the union of an empty collection of intervals. Show that $\mathcal{E}$ is a field. To which of the fields $\mathcal{E}_n$ described above do the finite subsets of $\mathbb{R}$ belong? What about the set of rational numbers? The open sets? The set $[0, 1)$?

It is possible for each field $\mathcal{E}_{n+1}$ to contain new sets that are not members of $\mathcal{E}_n$, so that in general, $\sigma(\mathcal{E})$ is strictly larger than any of the fields $\mathcal{E}_n$. In fact, in many important cases, $\sigma(\mathcal{E})$ is strictly larger than the union of all the fields $\mathcal{E}_n$. Fortunately, as the following lemma indicates, there is a sense in which nothing essentially new appears after the extension to $\mathcal{E}_2$.

Lemma 12. *Let $B \in \sigma(\mathcal{E})$. Then there exist $A, C \in \mathcal{E}_2$ such that $A \subseteq B \subseteq C$ and $R_2(A) = R_2(C)$.*

PROOF. Let $\mathcal{G}$ be the collection of all sets that are the limits of decreasing sequences of members of $\mathcal{E}$, and let $\mathcal{H}$ be the collection of all sets that are the limits of increasing sequences of members of $\mathcal{E}$. Note that $\mathcal{G}$ and $\mathcal{H}$ are contained in $\mathcal{E}_1$, so R_1 is defined on $\mathcal{G}$ and $\mathcal{H}$. Define

$$\mathcal{D} = \{B : \forall \varepsilon > 0, \exists G \in \mathcal{G}, H \in \mathcal{H}$$
$$\text{such that } G \subseteq B \subseteq H \text{ and } R_1(H) - R_1(G) < \varepsilon\}.$$

It is clear that the conclusion of the lemma holds for every $B \in \mathcal{D}$ and that $\mathcal{E} \subseteq \mathcal{D}$. Accordingly, we can finish the proof by showing that $\mathcal{D}$ is a Sierpiński class. It is easy to see that it is closed under proper set differences. All that remains is to show that it is closed under increasing limits.

Suppose that $B_m \nearrow B$ as $m \nearrow \infty$ and $B_m \in \mathcal{D}$ for each m. Let $\varepsilon > 0$. For each positive integer m, choose and $G_m \in \mathcal{G}$ and $H_m \in \mathcal{H}$ so that:

$$G_m \subseteq B_m \subseteq H_m \quad \text{and} \quad R_1(H_m) - R_1(G_m) < \frac{\varepsilon}{2^{m+1}}.$$

Let

$$H = \bigcup_{m=1}^{\infty} H_m.$$

Since a countable union of members of $\mathcal{H}$ is itself a limit of an increasing sequence of members of $\mathcal{E}$, the set H is a member of $\mathcal{H}$. By Lemma 7 applied to R_1,

$$R_1(H) = \lim_{n \to \infty} R_1\left(\bigcup_{m=1}^{n} H_m\right).$$

Choose n large enough so that

$$R_1(H) - R_1\left(\bigcup_{m=1}^{n} H_m\right) < \frac{\varepsilon}{2},$$

and let

$$G = \bigcup_{m=1}^{n} G_m.$$

Any finite union of members of $\mathcal{G}$ is also a limit of a decreasing sequence of members of $\mathcal{E}$, so the set G is a member of $\mathcal{G}$. Note that since the sequence $B_1, B_2, \ldots,$ is increasing,

$$G \subseteq B_n \subseteq B.$$

The definition of H ensures that $B \subseteq H$, so $G \subseteq B \subseteq H$. Also note that

$$G \subseteq \bigcup_{m=1}^{n} H_m \subseteq H,$$

so finite additivity implies that

$$R_1(H \setminus G) = R_1\left(H \setminus \bigcup_{m=1}^{n} H_m\right) + R_1\left(\bigcup_{m=1}^{n} H_m \setminus \bigcup_{m=1}^{n} G_m\right)$$

$$< \frac{\varepsilon}{2} + \sum_{m=1}^{n} R_1(H_m \setminus G_m) < \varepsilon.$$

It follows that B is a member of $\mathcal{D}$ as desired. $\quad \square$

Problem 14. Let Q be a probability measure defined on $(\mathbb{R}^d, \mathcal{B})$. A set B is called *regular* for Q if for each $\varepsilon > 0$, there exists a compact set K and an open set O such that $K \subseteq B \subseteq O$ and $Q(O \setminus K) < \varepsilon$. Show that every set in the completion of the Borel σ-field is regular. *Hint:* Mimic the proof of Lemma 12 to show that the collection of sets that are regular for Q is a Sierpiński class. Then show that every open rectangular box is regular for Q.

Lemma 12 motivates the following definition.

Definition 13. Let $\mathcal{E}$ be a field of subsets of a space Ω, and let R be a nonnegative countably additive function defined on $\mathcal{E}$ such that $R(\Omega) = 1$. The *completion* of $\mathcal{E}$ with respect to R is defined to be the collection of all sets $B \subseteq \Omega$ such that there exist sets A and C in $\mathcal{E}$ satisfying $A \subseteq B \subseteq C$ and $R(C \setminus A) = 0$.

Lemma 12 says that $\sigma(\mathcal{E})$ is contained in the completion of $\mathcal{E}_2$ with respect to R_2. We finally have all the ingredients that we need for the existence theorem, which may be more properly called the 'Extension Theorem'.

Theorem 14. [Extension] *Let $\mathcal{E}$ be a field of subsets of a space Ω and R a nonnegative countably additive function defined on $\mathcal{E}$ such that $R(\Omega) = 1$. Then there exists a unique probability measure P defined on $\sigma(\mathcal{E})$ such that $P(A) = R(A)$ for every $A \in \mathcal{E}$.*

PROOF. Since $\mathcal{E}$ is a field, it is closed under pairwise intersections. Thus the uniqueness assertion follows from the Uniqueness of Measure Theorem.

Let R_2 be the extension of R to $\mathcal{E}_2$, as in the paragraph preceding Problem 13. By Lemma 12, $\sigma(\mathcal{E})$ is in the completion of $\mathcal{E}_2$ with respect to R_2. Thus, for each set $B \in \sigma(\mathcal{E})$ we can choose sets A and C which are members of $\mathcal{E}_2$ such that $A \subseteq B \subseteq C$ and $R_2(C \setminus A) = 0$. Assuming such a choice has been made, define $P(B) = R_2(A)$. That $P(B)$ is well-defined follows from the fact that for any fixed choice of C, $R_2(A) = R_2(C)$ independently of the choice of A. It follows in particular that $P(B) = R(B)$ for $B \in \mathcal{E}$.

It remains to show that P is a probability measure. Clearly P is nonnegative and $P(\Omega) = 1$. We now show that P is finitely additive. Let B_1, B_2 be disjoint members of $\sigma(\mathcal{E})$. There exist sets $A_1, A_2, C_1, C_2 \in \mathcal{E}_2$ such that $A_i \subseteq B_i \subseteq C_i$ and $P(B_i) = R_2(A_i) = R_2(C_i)$ for $i = 1, 2$. Since A_1 and A_2 are clearly disjoint,

$$R_2(A_1 \cup A_2) = R_2(A_1) + R_2(A_2) = P(B_1) + P(B_2)$$

by the finite additivity of R_2. So to prove the finite additivity of P, it is enough to show that $R_2(A_1 \cup A_2) = P(B_1 \cup B_2)$. But this last equality follows from the definition of P and the fact that

$$R_2((C_1 \cup C_2) \setminus (A_1 \cup A_2)) \leq R_2((C_1 \setminus A_1) \cup (C_2 \setminus A_2))$$
$$\leq R_2(C_1 \setminus A_1) + R_2(C_2 \setminus A_2) = 0.$$

To show that P is a probability measure, it remains to show that P is countably additive. We will apply Proposition 9. Let $(B_1, B_2, \ldots)$ be a decreasing sequence in $\sigma(\mathcal{E})$ that converges to $\emptyset$. By the definition of P, there exists a sequence $(A_1, A_2, \ldots)$ in $\mathcal{E}_2$ such that $A_n \subseteq B_n$ and $P(B_n) = R_2(A_n)$ for each n. Clearly $A_n \to \emptyset$ as $n \to \infty$. So, by Lemma 6, $R_2(A_n) \to 0$. Thus, $P(B_n) \to 0$, and the countable additivity of P follows from Proposition 9. $\square$

Problem 15. Let $(\Omega, \mathcal{F}, P)$ be a probability space. Show that the completion of $\mathcal{F}$ with respect to P is a σ-field, and that P can be extended uniquely to this σ-field. (This extension is called the *completion* of P.)

7.4. Examples

The following example shows how to apply the Extension Theorem to a generalization of the coin-flip probability space, a generalization in which 'biased' coins are considered. For the special case of a fair coin, it establishes that there really is a probability space as described in Example 4 of Chapter 1 for the experiment of flipping a fair coin an infinite number of times. An approach that avoids topology is given in the proof of Theorem 16 in Chapter 9.

Example 1. Let Ω equal the set of infinite sequences $\omega = (\omega_1, \omega_2, \dots)$, where each $\omega_i \in \{0, 1\}$. Thus, Ω is the product of countably many copies of the set $\{0, 1\}$. We let Ω have the product topology, where each copy of $\{0, 1\}$ is given the discrete topology. Since the topological product of compact spaces is compact, Ω is a compact topological space. Let $\mathcal{E}$ be the collection of all sets of the form

$$(7.6) \qquad \{\omega : (\omega_1, \omega_2, \dots, \omega_n) \in C_n\} \quad \text{for } C_n \subseteq \{0, 1\}^n.$$

It is easily seen that $\mathcal{E}$ is a field, and that every set in $\mathcal{E}$ is a closed set. Since Ω is a compact set, it follows that every member of $\mathcal{E}$ is compact. By the finite intersection property of compact sets, if $A_1, A_2, \dots$ is any sequence of members of $\mathcal{E}$ that decreases to $\emptyset$, then $A_n = \emptyset$ for all sufficiently large n. It follows from Proposition 9 that if R is *any* nonnegative finitely additive function defined on $\mathcal{E}$, such that $R(\Omega) = 1$, then R is countably additive. By the Extension Theorem, any such R can be extended to a probability measure on the σ-field $\sigma(\mathcal{E})$. (It is not hard to prove that every open subset of Ω can be written as a countable union of members of $\mathcal{E}$, so in fact, $\sigma(\mathcal{E})$ is the σ-field of Borel subsets of Ω.)

We will construct a class of nonnegative finitely additive functions defined on $\mathcal{E}$, such that $R(\Omega) = 1$. The model for repeated flips of a fair coin is a special case. Let $p_1, p_2, \dots$ be a sequence of real numbers in the interval $[0, 1]$. For $C_n \subseteq \{0, 1\}^n$, set

$$R(\{\omega : (\omega_1, \dots, \omega_n) \in C_n\}) = \sum_{(\omega_1, \dots, \omega_n) \in C_n} \prod_{m=1}^{n} p_m^{\omega_m} (1 - p_m)^{1 - \omega_m}.$$

(In this formula, we understand 0^0 to equal 1.) There is an apparent ambiguity in this definition, which arises from the fact that any set which can be expressed in the form (7.6) can also be written in the form

$$\{\omega : (\omega_1, \dots, \omega_{n+1}) \in D_{n+1}\},$$

where D_{n+1} is the set of members of $\{0,1\}^{n+1}$ whose first n coordinates form a sequence of 0's and 1's which lies in C_n. That there is no real ambiguity is shown by a straightforward computation which is left to the reader.

If A and B are two disjoint members of $\mathcal{E}$, then they can be written in the form (7.6) using a common value of n. The summation in the definition of R now makes the finite additivity of R clear.

The particular case in which $p_n = 1/2$ for each n gives the coin-flip probability space of Example 4 of Chapter 1. For the general case, we can think of an experiment in which a different coin is tossed at each stage, with p_n being the probability that the result of the nth toss is heads. We will have more to say about the general case in Chapter 9, when the notion of 'independence' is introduced.

Problem 16. Suppose, in the preceding example, that each p_n equals a common value p. Let X_n denote the number of heads in the first n coin flips. Calculate and name the distribution of X_n.

* **Problem 17.** Suppose, in Example 1, that $p_n = n^{-\beta}$ for some $\beta > 0$. Calculate the probability that at least two of the first four flips are heads. For $n = 1, 2, \ldots$, let A_n be the event that the nth flip is heads. Describe in terms of coin flips the events $\liminf A_n$ and $\limsup A_n$, and calculate their probabilities. Let

$$Y(\omega) = \inf\{n \colon \omega \in A_n\} \quad \text{and} \quad Z(\omega) = \sup\{n \colon \omega \in A_n\}.$$

Calculate the distributions of Y and Z, and describe in terms of coin flips what these random variables represent.

In Example 1 of Chapter 2 we showed how to use the coin-flip model to construct a probability space $(\Omega, \mathcal{F}, P)$ in which Ω is the unit interval $[0,1]$ and P is Lebesgue measure. In Proposition 4 of Chapter 3 we used this probability space $(\Omega, \mathcal{F}, P)$ to construct all possible distributions on $\mathbb{R}$. Example 1 puts this work on a solid theoretical foundation by showing that the coin-flip probability space exists. Nevertheless, a more direct approach to constructing distributions on $\mathbb{R}$ may be instructive.

Example 2. Let F be an arbitrary distribution function for $\mathbb{R}$, and let $\mathcal{E}$ be the field of subsets of $\mathbb{R}$ described in Problem 13. For $-\infty < a < b < \infty$, set

$$R((a,b]) = F(b) - F(a) \qquad R((-\infty, b]) = F(b)$$
$$R((a,\infty)) = 1 - F(a) \quad \text{and} \quad R((-\infty, \infty)) = 1.$$

If $A_1, \ldots, A_n$ are disjoint intervals which are members of $\mathcal{E}$, then we define

$$R\left(\bigcup_{k=1}^{n} A_k\right) = \sum_{k=1}^{n} R(A_n).$$

The key to showing that this definition is unambiguous is the following calculation, together with others similar to it involving unbounded intervals:

$$R((a,b]) + R((b,c]) = (F(b) - F(a)) + (F(c) - F(b)) = F(c) - F(a) = R((a,c]).$$

Thus R is a well-defined nonnegative function defined on $\mathcal{E}$ such that $R(\Omega) = 1$. Finite additivity is immediate from the definition of R.

The field $\mathcal{E}$ generates the Borel σ-field $\mathcal{B}$ of subsets of $\mathbb{R}$. To show that R can be extended uniquely to a probability measure on $\mathcal{B}$, it is enough to show that R is countably additive on $\mathcal{E}$. By Proposition 9, this is equivalent to showing that $R(A_n) \to 0$ for each decreasing sequence $(A_1, A_2, \dots)$ of members of $\mathcal{E}$ such that $\lim A_n = \emptyset$. The reader is asked to prove this fact in the following exercise.

Problem 18. Let $\mathcal{E}$ and R be as in the preceding example. Prove that R is countably additive on $\mathcal{E}$. Hint: Fix $\varepsilon > 0$ and show that for each set A_k in the decreasing sequence $(A_1, A_2, \dots)$, there exists a compact set C_k and a set $B_k \in \mathcal{E}$ such that $B_k \subseteq C_k \subseteq A_k$ and $R(A_k \setminus B_k) < \varepsilon/2^k$. Since $A_k \to \emptyset$, the finite intersection property implies that

$$\bigcap_{k=1}^{n} B_k \subseteq \bigcap_{k=1}^{n} C_k = \emptyset$$

for all sufficiently large n. It follows for such n that

$$A_n = \bigcup_{k=1}^{n} (A_n \setminus B_k) \subseteq \bigcup_{k=1}^{n} (A_k \setminus B_k).$$

Conclude that $R(A_n) < \varepsilon$ for all sufficiently large n.

Example 2 shows that Lebesgue measure on $[0,1]$ exists, and Example 1 of Chapter 6 shows how to use that fact to construct Lebesgue measure on $\mathbb{R}$. The following example explores an important property of Lebesgue measure on $\mathbb{R}$.

Example 3. Let μ be Lebesgue measure on $\mathbb{R}$. For $x \in \mathbb{R}$ and $A \subseteq \mathbb{R}$, let $A + x$ denote the set $\{y + x : y \in A\}$. The continuity of addition and subtraction implies that $A \in \mathcal{B}$ if and only if $A + x \in \mathcal{B}$ for all $x \in \mathbb{R}$. We want to prove that $\mu(A) = \mu(A + x)$ for all $A \in \mathcal{B}$ and $x \in \mathbb{R}$. This property of μ is called *translation invariance*. Fix $x \in \mathbb{R}$, and let $\mathcal{E}$ be the class of all sets $A \in \mathcal{B}$ such that $\mu(A) = \mu(A + x)$. It is easy to check that $\mathcal{E}$ is a Sierpiński class. If I is an interval, $\mu(I) = \mu(I + x)$ since I and $I + x$ have the same length. Thus $\mathcal{E}$ contains the collection of all intervals. Since the collection of all intervals is closed under pairwise intersections, contains $\mathbb{R}$, and generates $\mathcal{B}$, it follows from the Sierpiński Class Theorem that $\mathcal{E} = \mathcal{B}$. Thus, one-dimensional Lebesgue measure is translation invariant.

Problem 19. Show that if μ is a translation-invariant measure on $(\mathbb{R}, \mathcal{B})$ such that $\mu([0,1)) = 1$, then μ is one-dimensional Lebesgue measure.

Problem 20. Show that if μ is a translation invariant measure on $(\mathbb{R}, \mathcal{B})$, then either $\mu(I) = \infty$ for all nonempty open intervals I, or μ is a finite multiple of one-dimensional Lebesgue measure.

Example 4. Let $\Omega = [0,1) \times [0,1)$. Consider the collection of rectangular boxes of the form $[a,b) \times [c,d)$, and let $\mathcal{E}$ be the collection of all finite unions of such boxes. For $A \in \mathcal{E}$, define $R(A)$ to be the area of A. We leave it as a problem to show that $\mathcal{E}$ is a field, and that R is a nonnegative countably additive function defined on $\mathcal{E}$, such that $R(\Omega) = 1$. Thus, R can be extended to a probability measure P defined on the Borel subsets of Ω. This probability measure was introduced in Problem 10 of Chapter 2 as the uniform distribution on $[0,1)^2$. An alternative construction of P will be given in Chapter 9.

Problem 21. For the preceding example, prove that $\mathcal{E}$ is a field, and that R is a nonnegative countably additive function such that $R(\Omega) = 1$. *Hint:* To prove that $\mathcal{E}$ is closed under complementation, first show that the complement of a rectangular box can be written as the union of at most four disjoint rectangular boxes. To prove countable additivity, follow the hint given in Problem 18.

Example 5. [Lebesgue measure on $\mathbb{R}^2$] Let $\Omega = \mathbb{R}^2$ and let $\mathcal{B}$ be the Borel σ-field of Ω. For integers m and n, let $A_{m,n} = [m, m+1) \times [n, n+1)$, and let $\mu_{m,n}$ be the uniform distribution on $A_{m,n}$, defined according to the pattern given in Example 4. Think of each measure $\mu_{m,n}$ as a probability measure on $(\mathbb{R}, \mathcal{B})$. For $B \in \mathcal{B}$, set

$$\mu(B) = \sum_{m=1}^{\infty} \sum_{n=1}^{\infty} \mu_{m,n}(B \cap A_{m,n}).$$

The σ-finite measure μ is called *two-dimensional Lebesgue measure*, or *Lebesgue measure on $\mathbb{R}^2$*.

Problem 22. Prove that 2-dimensional Lebesgue measure is translation invariant. Also prove that 2-dimensional Lebesgue measure is unique in the same sense in which 1-dimensional Lebesgue measure is unique (see Problem 20).

Problem 23. Prove that 2-dimensional Lebesgue measure is rotation invariant. In other words, if $T: \mathbb{R}^2 \to \mathbb{R}^2$ is any rotation of $\mathbb{R}^2$ and if B is any Borel subset of $\mathbb{R}^2$, show that $\mu(B) = \mu(T(B))$, where μ is 2-dimensional Lebesgue measure. *Hint:* In view of Problem 22, it suffices to consider rotations about the origin.

Example 6. Let $\mathcal{L}$ denote the set of all lines in $\mathbb{R}^2$. We establish a one-to-one correspondence between $\mathcal{L}$ and

$$S = \{(s, \phi): s \in \mathbb{R}, 0 \le \phi < \pi\}$$

as follows. The line corresponding to (s, ϕ) is that line whose distance from the origin equals $|s|$ and which meets a certain ray emanating from the origin at a right angle, that ray being the one with direction ϕ if $s \ge 0$ and the one with direction $\phi + \pi$ if $s < 0$. Figure 7.1 illustrates this correspondence for a certain infinite set of lines, namely those that intersect a given segment. The shaded region in the (s, ϕ)-plane corresponds to that infinite set of lines, and the points marked in the (s, ϕ)-plane correspond to the lines drawn in the right half of the figure.

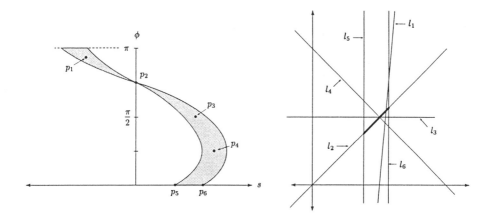

FIGURE 7.1. A subset of the space of lines

We give S the relative topology inherited from the usual topology on $\mathbb{R}^2$, and let $\mathcal{B}$ be the σ-field of Borel subsets of S. Let μ be the restriction of 2-dimensional Lebesgue measure to $\mathcal{B}$. The one-to-one correspondence between S and $\mathcal{L}$ just described serves to turn $\mathcal{L}$ into a topological space and produces a σ-finite measure space $(\mathcal{L}, \mathcal{A}, \nu)$, by transferring the corresponding structures from S. The following exercises concern properties of the measure space $(\mathcal{L}, \mathcal{A}, \nu)$.

* **Problem 24.** Let $(\mathcal{L}, \mathcal{A}, \nu)$ be the measure space of the preceding example. Show that ν is invariant under rigid motions in $\mathbb{R}^2$. You may use the fact that every rigid motion in $\mathbb{R}^2$ is a composition of rotations about the origin and translations.

* **Problem 25.** For the setting of the preceding problem, find the measure of the set of lines that intersect a fixed but arbitrary line segment. *Hint:* Use the preceding problem to show that it is sufficient to consider the case shown in Figure 7.1.

* **Problem 26.** Let $(\mathcal{L}, \mathcal{A}, \nu)$ be as in the preceding exercise. Let A be the set of lines that intersect a fixed convex polygon. Calculate $\nu(A)$. *Hint:* First prove that $\nu(A)$ is finite, either by a direct method or by using the preceding problem. Let $\Omega = A$ and let $\mathcal{B}$ be the σ-field of Borel subsets of A. For $B \in \mathcal{B}$, define $P(B) = \nu(B)/\nu(A)$, so that $(\Omega, \mathcal{B}, P)$ is a probability space. For each line $\omega \in \Omega$, let $X(\omega)$ equal the number of intersections of ω with the polygon. Calculate $E(X)$ in two ways, one using the answer to the preceding problem.

Problem 27. Let $(\mathcal{L}, \mathcal{A}, \nu)$ be as in the preceding exercise. Let A be the set of lines that intersect a fixed circle of radius r. Calculate $\nu(A)$. As in the preceding problem, turn A into a probability space, and for that space, calculate the probability that a line in A intersects a given second circle of radius $s < r$ inside the first circle.

Problem 28. In the preceding exercise, a random line intersecting a circle is defined. Such a random line determines a random chord of the circle. Is this random chord equivalent to any of the interpretations given in Example 3 of Chapter 2?

* **Problem 29.** Let C be a circle, and let D be a curve contained in the interior of C. Assume that D consists of straight line segments (possibly infinitely many), and that D does not intersect itself. Let ω be a random line intersecting C, as in the preceding two problems, and let $X(\omega)$ be the number of intersections of D and ω. Calculate $E(X)$ in terms of the radius r of the circle C and some constant associated with D.

CHAPTER 8
Integration Theory

In this chapter we have three main goals. The first is to extend the concept of expectation to general measure spaces. In this context, we do not use the phrase "expectation of a random variable", but instead we introduce the 'integral' of a measurable function. This new kind of integral is called the 'Lebesgue integral'. Our second goal is to introduce several tools, which along with the Monotone Convergence Theorem, are valuable for interchanging limit operations with expectation and integration. Our third goal is to explore some of the similarities and differences between Lebesgue integration and Riemann integration. As in the case of expectations, the Riemann-Stieltjes integral will be useful for computations of Lebesgue integrals. Also useful in such calculations is the 'Radon-Nikodym derivative', introduced near the end of the chapter.

8.1. Lebesgue integration

The definition of the Lebesgue integral in a general measure space is completely analogous to the definition of the expectation in a probability space. Nevertheless, we quickly repeat here the steps in that definition, both to establish some new notation, and also to provide a natural opportunity for review.

We start with measurable simple functions. However, since measurable sets can have infinite measure in a general measure space, we find it easiest to restrict our attention at first to nonnegative measurable simple functions.

Definition 1. Let f be a measurable function defined on a measure space $(\Omega, \mathcal{F}, \mu)$ and having the form

$$f = \sum_{j=1}^{n} c_j I_{C_j}$$

for some nonnegative constants c_j and measurable sets C_j. Then the *Lebesgue integral* of f equals

$$\sum_{j=1}^{n} c_j \mu(C_j)$$

and is denoted by

$$\int f \, d\mu \, .$$

The reader may note that we did not assume in this definition that the constants c_j are distinct, nor did we assume that the sets C_j form a partition of Ω, in contrast to Definition 1 of Chapter 4. The reason is that we already know from our work in Chapter 4 that no ambiguity can result from this relaxing of assumptions (see Problem 4 of Chapter 4).

Going from nonnegative measurable simple functions to arbitrary measurable functions is just as before. The fact that the following definition is unambiguous is proved just as in Chapter 4.

Definition 2. Let f be a measurable function defined on a measure space $(\Omega, \mathcal{F}, \mu)$. If f is $\overline{\mathbb{R}}^+$-valued, then

$$\int f \, d\mu = \sup_{h} \int h \, d\mu \, ,$$

where the supremum is taken over all nonnegative measurable simple functions h such that $h \leq f$. If f is $\overline{\mathbb{R}}$-valued, then

$$\int f \, d\mu = \int f^+ \, d\mu - \int f^- \, d\mu \, ,$$

provided the two integrals on the right side of this expression are not both infinite; otherwise, $\int f \, d\mu$ *does not exist*. When it exists, $\int f \, d\mu$ is called the *Lebesgue integral* of f with respect to μ.

It should be clear to the reader that the Lebesgue integral is indeed a generalization of expectation; that is, if μ happens to be a probability measure, then $\int f \, d\mu = E_\mu(f)$. In fact, as the following result shows, if μ can be written as a countable sum of finite measures, then $\int f \, d\mu$ can be expressed directly in terms of expectations.

Proposition 3. *Let $(\Omega, \mathcal{F}, \mu)$ be a measure space such that $\mu(\Omega) > 0$. Suppose there exist finite measures μ_j, $j = 1, 2, \ldots,$ on $(\Omega, \mathcal{F})$ such that $\mu(A) = \sum_{j=1}^{\infty} \mu_j(A)$ for all $A \in \mathcal{F}$. We may assume without loss of generality that $\mu_j(\Omega) > 0$ for each j, so that the formula*

$$P_j(A) = \mu_j(A)/\mu_j(\Omega) \text{ for } A \in \mathcal{F}$$

defines probability measures P_j on $(\Omega, \mathcal{F})$. Then for all measurable functions $f: \Omega \to \overline{\mathbb{R}}^+$,

$$(8.1) \qquad \int f \, d\mu = \sum_{j=1}^{\infty} \mu_j(\Omega) E_{P_j}(f) \,.$$

PROOF. We leave it to the reader to check that (8.1) holds for nonnegative measurable simple functions f. It follows immediately that the left side of (8.1) is less than or equal to the right side for all measurable $\overline{\mathbb{R}}^+$-valued functions f. On the other hand, for any integer n and nonnegative measurable simple functions $f_j \leq f, j = 1, \ldots, n$,

$$\sum_{j=1}^{n} \mu_j(\Omega) E_{P_j}(f_j) \leq \sum_{j=1}^{n} \mu_j(\Omega) E_{P_j}\left(\sup_{1 \leq j \leq n} f_j\right) \leq \int \sup_{1 \leq j \leq n} f_j \, d\mu \,,$$

since the supremum of finitely many nonnegative measurable simple functions is a nonnegative measurable simple function. By the definition of the integral, it follows that

$$\sum_{j=1}^{n} \mu_j(\Omega) E_{P_j}(f_j) \leq \int f \, d\mu \,.$$

Take the supremum over all nonnegative measurable simple functions $f_j \leq f$ to obtain

$$\sum_{j=1}^{n} \mu_j(\Omega) E_{P_j}(f) \leq \int f \, d\mu \,.$$

Now let n go to ∞ to see that the left side of (8.1) is greater than or equal to the right side. It follows that the two sides are equal. $\square$

The preceding result makes it easy to generalize many of the results about expectation to Lebesgue integration on σ-finite measure spaces. The following result often allows us to go beyond σ-finite measure spaces to general measure spaces.

Proposition 4. *Let $(\Omega, \mathcal{F}, \mu)$ be a measure space, and let f be an $\overline{\mathbb{R}}^+$-valued measurable function defined on Ω. Let $B = \{x \in \Omega : f(x) > 0\}$, and define $\nu(A) = \mu(A \cap B)$ for all $A \in \mathcal{F}$. Then*

$$\int f \, d\mu = \int f \, d\nu \,.$$

Moreover, if $\int f \, d\mu < \infty$, then $(\Omega, \mathcal{F}, \nu)$ is a σ-finite measure space.

Problem 1. Prove the preceding proposition. *Hint:* For $j = 1, 2, \ldots$, let $B_j = \{x \in \Omega : 1/(j-1) \geq f(x) > 1/j\}$ and define $\nu_j(A) = \mu(A \cap B_j)$.

The following result generalizes Theorem 9 of Chapter 4 to the Lebesgue integration setting. Each part is proved either exactly as in the proof of the corresponding part of Theorem 9 of Chapter 4, or by using Proposition 3 and Proposition 4 in a straightforward way to generalize the corresponding part of Theorem 9 of Chapter 4 to the Lebesgue integral.

In the statement of this theorem, we use the term 'μ-a.e.' ('μ-almost everywhere'). This term, which is analogous to 'a.s.', means that the statement to which it is attached is true except on a set of μ-measure 0. Often, when the measure μ is understood from the context, we use 'a.e.' in the place of 'μ-a.e.', and 'almost everywhere' in the place of 'μ-almost everywhere'.

Theorem 5. Let f and g be $\overline{\mathbb{R}}$-valued functions defined on a measure space $(\Omega, \mathcal{F}, \mu)$, and let a, b, and c be real constants.

(i) $\int (af + bg)\, d\mu = a \int f\, d\mu + b \int g\, d\mu$, provided the expression on the right is meaningful.

(ii) If $f = c$ a.e., then $\int f\, d\mu = c\mu(\Omega)$, where as usual, $0 \cdot \infty$ is understood to be 0.

(iii) If $f = g$ a.e., then either the integrals of f and g with respect to μ both exist and are equal, or neither exists.

(iv) If $f \leq g$ a.e. and either $\int f\, d\mu$ exists and is different from $-\infty$ or $\int g\, d\mu$ exists and is different from ∞, then both integrals exist and $\int f\, d\mu \leq \int g\, d\mu$.

(v) If $\int f\, d\mu = \int g\, d\mu$ is finite and $f \leq g$ a.e., then $f = g$ a.e.

(vi) If $\int f\, d\mu$ exists, then $|\int f\, d\mu| \leq \int |f|\, d\mu$.

(vii) If $\int f\, d\mu$ does not exist, then $\int |f|\, d\mu = \infty$.

(viii) $\int |f + g|\, d\mu \leq \int |f|\, d\mu + \int |g|\, d\mu$.

The following notation for the integral of the product of a measurable function f and a measurable indicator function I_B is useful:

$$\int_B f\, d\mu \overset{\text{def}}{=} \int f I_B\, d\mu.$$

One speaks of "integrating the function f with respect to μ over the set B". The corresponding notation in the context of expectations is

$$E(X \,;\, B) \overset{\text{def}}{=} E(X I_B).$$

Problem 2. Show that if $\int f\, d\mu$ exists, then $\int_B f\, d\mu$ exists for all measurable sets B.

Problem 3. Given a measure space $(\Omega, \mathcal{F}, \mu)$ and a measurable set $B \in \mathcal{F}$, define $\mu_B(A) = \mu(A \cap B)$ for all $A \in \mathcal{F}$. Show that

$$\int_B f\, d\mu = \int f\, d\mu_B$$

in the sense that if one side exists, then both sides exist and are equal.

8.2. Convergence theorems

We start by generalizing the Monotone Convergence Theorem to the Lebesgue integral.

Theorem 6. [Monotone Convergence] *Let* $0 \le f_1 \le f_2 \le \dots$ *be* $\overline{\mathbb{R}}^+$ *-valued measurable functions defined on a common measure space* $(\Omega, \mathcal{F}, \mu)$, *and let* $f = \lim_{n \to \infty} f_n$. *Then*

$$\int f \, d\mu = \lim_{n \to \infty} \int f_n \, d\mu \,.$$

PROOF. By Theorem 5,

$$\int f_1 \, d\mu \le \int f_2 \, d\mu \le \dots \le \int f \, d\mu \,.$$

Thus, if $\int f_n \, d\mu = \infty$ for any positive integer n, we are done.

Suppose, then, that $\int f_n \, d\mu < \infty$ for all n. Let B and ν be as in the statement of Proposition 4, let $B_n = \{x \in \Omega \colon f_n(x) > 0\}$, and define $\nu_n(A) = \mu(A \cap B_n)$ for $A \in \mathcal{F}$. By Proposition 4, each of the measures ν_n is σ-finite. Let $B_0 = \emptyset$ and define $\tilde{\nu}_n(A) = \nu_n(A \cap B_n \cap B_{n-1}^c)$ for $n = 1, 2, \dots$. It is easily checked that $(\tilde{\nu}_1, \tilde{\nu}_2, \dots)$ is a pairwise mutually singular sequence of σ-finite measures. By countable additivity,

$$\nu(A) = \sum_{n=1}^{\infty} \tilde{\nu}_n(A) \,,$$

so ν is σ-finite by Problem 19 of Chapter 6. By Proposition 4, $\int f \, d\nu = \int f \, d\mu$, and for similar reasons, $\int f_n \, d\nu = \int f_n \, d\mu$ for all n. Thus, it is enough to prove the theorem with μ replaced by the σ-finite measure ν.

Since ν is a countable sum of finite measures, it follows from Proposition 3 that there are probability measures $(P_j \colon j = 1, 2, \dots)$ on $(\Omega, \mathcal{F})$ and nonnegative constants $c_j, j = 1, 2, \dots$, such that

$$\int f \, d\nu = \sum_{j=1}^{\infty} c_j E_{P_j}(f) \,.$$

The same formula holds with f replaced by f_n. By first applying the Monotone Convergence Theorem for expectations, and then applying the Monotone Convergence Theorem for sums (Corollary 9 of Chapter 6), we have

$$\sum_{j=1}^{\infty} c_j E_{P_j}(f) = \sum_{j=1}^{\infty} \left(\lim_{n \to \infty} c_j E_{P_j}(f_n) \right)$$

$$= \lim_{n \to \infty} \sum_{j=1}^{\infty} c_j E_{P_j}(f_n) = \lim_{n \to \infty} \int f_n \, d\nu \,. \quad \square$$

Of course, Corollary 12 of the Monotone Convergence Theorem of Chapter 4 generalizes as well:

Corollary 7. *If $(f_1, f_2, \ldots)$ is a sequence of $\overline{\mathbb{R}}^+$-valued measurable functions defined on a measure space $(\Omega, \mathcal{F}, \mu)$, then*

$$\int \left(\sum_{n=1}^{\infty} f_n \right) d\mu = \sum_{n=1}^{\infty} \int f_n \, d\mu \,.$$

The following exercise is an application of the Monotone Convergence Theorem to obtain a limited extension of the Continuity of Measure Theorem to infinite measure spaces. See also Problem 5 and Problem 9 for further developments in this direction. (We have already seen in Chapter 6 that the full Continuity of Measure Theorem does not generalize to infinite measure spaces.)

Problem 4. [Monotone Continuity of Measure Theorem] Let $(A_n : n = 1, 2, \ldots)$ be an increasing sequence of measurable sets in a measure space $(\Omega, \mathcal{F}, \mu)$. Show that

$$\mu(A_n) \nearrow \mu(A) \text{ as } n \nearrow \infty \,,$$

where $A = \lim_{n \to \infty} A_n$.

The following useful result is an easy consequence of the Monotone Convergence Theorem.

Lemma 8. [Fatou] *Let $(f_1, f_2, \ldots)$ be a sequence of $\overline{\mathbb{R}}^+$-valued measurable functions defined on a measure space $(\Omega, \mathcal{F}, \mu)$. Then*

$$\int (\liminf_{n \to \infty} f_n) \, d\mu \leq \liminf_{n \to \infty} \int f_n \, d\mu \,.$$

PROOF. For $m = 1, 2, \ldots$, let $g_m = \inf\{f_n : n \geq m\}$. Then for each m, $g_m \leq f_m$, and

$$g_m \nearrow \liminf_{n \to \infty} f_n \quad \text{as} \quad m \nearrow \infty \,.$$

By the Monotone Convergence Theorem,

$$\int (\liminf_{n \to \infty} f_n) \, d\mu = \lim_{m \to \infty} \int g_m \, d\mu = \liminf_{m \to \infty} \int g_m \, d\mu \leq \liminf_{m \to \infty} \int f_m \, d\mu \,. \quad \square$$

Problem 5. Let $(A_n : n = 1, 2, \ldots)$ be a sequence of measurable sets in a measure space $(\Omega, \mathcal{F}, \mu)$. Show that

$$\liminf_{n \to \infty} \mu(A_n) \geq \mu(\liminf_{n \to \infty} A_n) \,.$$

Also show that the following inequality does *not* hold in the general measure space setting:

$$\limsup_{n \to \infty} \mu(A_n) \leq \mu(\limsup_{n \to \infty} A_n) \,.$$

(Compare Problem 9 of Chapter 6.)

Problem 6. For the coin-flip probability space, Example 4 of Chapter 1, let X_n denote the indicator function of the event that the nth flip is heads. Calculate $\liminf E(X_n)$ and $E(\liminf X_n)$.

Problem 7. Let $(X_1, X_2, \dots)$ be a sequence of random variables that converges almost surely to a random variable X. Show that if $\sup_n EX_n^2 < \infty$, then $EX^2 < \infty$.

The purpose of the next convergence theorem is similar to that of the Monotone Convergence Theorem. The hypothesis is not that the sequence $(f_1, f_2, \dots)$ be monotone but instead that it be dominated by a measurable function g having finite integral. The role of this 'dominating function' g should be compared with the role of X_1 in Corollary 13 of Chapter 4.

Theorem 9. [Dominated Convergence] *Let f_1, f_2, ... be $\overline{\mathbb{R}}$-valued measurable functions defined on a measure space $(\Omega, \mathcal{F}, \mu)$, and suppose that g is a nonnegative measurable function defined on $(\Omega, \mathcal{F}, \mu)$ such that, for each n, $|f_n| \le g$ a.e.. If $\int g \, d\mu < \infty$, then*

$$
(8.2) \quad
\begin{aligned}
-\infty < \int (\liminf_{n \to \infty} f_n) \, d\mu &\le \liminf_{n \to \infty} \int f_n \, d\mu \\
&\le \limsup_{n \to \infty} \int f_n \, d\mu \le \int (\limsup_{n \to \infty} f_n) \, d\mu < \infty.
\end{aligned}
$$

If, in addition, $f = \lim_{n \to \infty} f_n$ exists almost everywhere, then

$$
\int |f| \, d\mu < \infty, \quad \lim_{n \to \infty} \int f_n \, d\mu = \int f \, d\mu, \quad \text{and} \quad \lim_{n \to \infty} \int (|f - f_n|) \, d\mu = 0.
$$

* **Problem 8.** Prove the preceding theorem. *Hint:* For the first part, apply the Fatou Lemma to the functions $g - f_n$ and $g + f_n$.

Problem 9. [Dominated Continuity of Measure Theorem] Let $(A_n : n = 1, 2, \dots)$ be a sequence of measurable sets in a measure space $(\Omega, \mathcal{F}, \mu)$. Show that if $A = \lim_n A_n$ exists and if

$$
\mu\left(\bigcup_{n=1}^{\infty} A_n\right) < \infty,
$$

then

$$
\mu(A) = \lim_{n \to \infty} \mu(A_n).
$$

Our remaining convergence results apply only to finite measure spaces. In order to emphasize this restriction, we state them in the context of probability spaces.

Theorem 10. [Bounded Convergence] *Let* $(X_1, X_2, \ldots)$ *be a sequence of* $\overline{\mathbb{R}}$-*valued random variables on a probability space* $(\Omega, \mathcal{F}, P)$. *Assume that* $X = \lim_{n\to\infty} X_n$ *exists almost surely. Suppose that there exists a finite constant* M *such that for all* $n \geq 1$, $|X_n| \leq M$ *a.s.. Then*

$$E(|X|) \leq M, \quad \lim_{n\to\infty} E(X_n) = E(X), \quad and \quad \lim_{n\to\infty} E(|X - X_n|) = 0.$$

Problem 10. Prove the preceding theorem. Also provide an appropriate counterexample to show that it is false in any infinite measure space.

Problem 11. Let Y be an $\mathbb{R}$-valued random variable for which $E(|Y|) < \infty$. For $c \geq 1$, let I_c be the indicator function of the event $\{\omega : |Y(\omega)| \geq c\}$. Prove that

$$\lim_{c\to\infty} E(YI_c) = \lim_{c\to\infty} E(|Y|I_c) = 0.$$

For our last convergence result we need a definition. It is motivated in part by the preceding problem.

Definition 11. For $\overline{\mathbb{R}}$-valued random variables X_t, $t \in T$, defined on a probability space $(\Omega, \mathcal{F}, P)$, $(X_t : t \in T)$ is *uniformly integrable* if

$$\lim_{c\to\infty} \sup_{t\in T} E(|X_t|I_{t,c}) = 0,$$

where, for each $t \in T$ and $c > 0$, $I_{t,c}$ is the indicator function of the set $\{\omega \in \Omega : |X_t(\omega)| \geq c\}$.

Theorem 12. [Uniform Integrability Criterion] *Let* $(X_1, X_2, \ldots)$ *be a sequence of* $\overline{\mathbb{R}}$-*valued random variables on a probability space* $(\Omega, \mathcal{F}, P)$. *Assume that* $X = \lim_{n\to\infty} X_n$ *exists almost surely, and that* $E(|X_n|) < \infty$ *for all* n. *Then the following three statements are equivalent:*

(i) $(X_n : n = 1, 2, \ldots)$ *is uniformly integrable;*
(ii) $E(|X|) < \infty$ *and* $\lim_{n\to\infty} E(|X_n - X|) = 0$;
(iii) $\lim_{n\to\infty} E(|X_n|) = E(|X|) < \infty$.

Each of these three conditions implies

(iv) $\lim_{n\to\infty} E(X_n) = E(X)$.

PROOF. By using the basic properties of expectations found in Theorem 9 of Chapter 4, we obtain

$$\big|E(|X_n| - |X|)\big| \leq E\big(\big|(|X_n| - |X|)\big|\big) \leq E(|X_n - X|)$$

and

$$|E(X_n - X)| \leq E(|X_n - X|).$$

Thus, (ii) implies (iii) and (iv).

Next we suppose that (i) holds and prove (ii). For $c \in (0, \infty)$ and $n = 1, 2, \ldots$, set

$$Z_{n,c} = X_n \cdot \left(I_{\{x\,:\,|x|>c\}} \circ X_n\right) \quad \text{and} \quad Z_c = X \cdot \left(I_{\{x\,:\,|x|>c\}} \circ X\right).$$

Using uniform integrability, we choose c so that $E(|Z_{n,c}|) < \frac{\varepsilon}{3}$ for all n. Since $|Z_c| \leq \liminf_{n\to\infty} |Z_{n,c}|$, we obtain from the Fatou Lemma that $E(|Z_c|) \leq \frac{\varepsilon}{3}$. Hence,

$$E(|X|) = E(|Z_c|) + E(|X| - |Z_c|) \leq \frac{\varepsilon}{3} + c < \infty,$$

proving the first part of (ii). Also,

$$E(|X_n - X|) \leq E(|Z_{n,c}|) + E(|Z_c|) + E\left(|(X_n - Z_{n,c}) - (X - Z_c)|\right)$$
$$< \tfrac{2\varepsilon}{3} + E\left(|(X_n - Z_{n,c}) - (X - Z_c)|\right).$$

The Bounded Convergence Theorem implies that the last expectation on the right is less than $\frac{\varepsilon}{3}$ for all sufficiently large n. The rest of (ii) follows.

It remains to prove that (iii) implies (i). For this part of the proof, there is no loss of generality in assuming that each X_n is $\mathbb{R}^+$-valued. This assumption allows us to use (iv) as well as (iii).

For $c \in (0, \infty)$ and $n = 1, 2, \ldots$, set

$$Y_{n,c} = X_n \cdot \left(I_{[c,\infty)} \circ X_n\right) \quad \text{and} \quad Y_c = X \cdot \left(I_{[c,\infty)} \circ X\right).$$

Clearly, $X_n - Y_{n,c} \geq 0$ for each n and c. By the Fatou Lemma,

$$\liminf_{n\to\infty} E(X_n - Y_{n,c}) \geq E(\liminf_{n\to\infty}(X_n - Y_{n,c})) = E(X - \limsup_{n\to\infty} Y_{n,c}).$$

Condition (iv), linearity of expectation, and the preceding inequality yield

$$\begin{aligned}
E(X) - \limsup_{n\to\infty} E(Y_{n,c}) &= \lim_{n\to\infty} E(X_n) - \limsup_{n\to\infty} E(Y_{n,c}) \\
&= \liminf_{n\to\infty} [E(X_n) - E(Y_{n,c})] \\
&= \liminf_{n\to\infty} E(X_n - Y_{n,c}) \\
&\geq E(X) - E(\limsup_{n\to\infty} Y_{n,c}),
\end{aligned}$$

from which it follows that

$$(8.3) \qquad \limsup_{n\to\infty} E(Y_{n,c}) \leq E(\limsup_{n\to\infty} Y_{n,c}) \leq E(Y_c).$$

Let $\varepsilon > 0$. Choose c_0 so that $E(Y_{c_0}) < \frac{\varepsilon}{2}$ (see Problem 11). By (8.3) we may choose m so that $E(Y_{n,c_0}) < E(Y_{c_0}) + \frac{\varepsilon}{2}$ for $n > m$. Then choose c_n, $1 \leq n \leq m$, so that $E(Y_{n,c_n}) < \varepsilon$ for $1 \leq n \leq m$ (again, see Problem 11). Set $c^* = \max\{c_n : 0 \leq n \leq m\}$. Then $E(Y_{n,c^*}) < \varepsilon$ for all n. Therefore, (i) holds. $\square$

The following problem gives a useful condition for checking uniform integrability.

* **Problem 12.** Let X_t, $t \in T$, be $\overline{\mathbb{R}}$-valued random variables on a probability space $(\Omega, \mathcal{F}, P)$, and suppose that there exists $p > 1$ and $k < \infty$ such that $E(|X_t|^p) \leq k$ for all $t \in T$. Prove that $(X_t : t \in T)$ is uniformly integrable.

Problem 13. Let $(X_1, X_2, \dots)$ be a sequence of random variables, each with finite mean. Assume that $\lim_{n \to \infty} X_n = 0$ almost surely and that $\sup_n \text{Var}(X_n) < \infty$. Show that $\lim_{n \to \infty} E(|X_n|) = 0$. *Hint:* First use the Chebyshev Inequality to show that $\sup_n |E(X_n)| < \infty$.

8.3. Probability measures and infinite measures compared

In this section, we wish to summarize the most important similarities and differences between probability spaces and infinite measure spaces.

First, let us compare expectations and Lebesgue integrals. We have seen that the Monotone and Dominated Convergence Theorems and the Fatou Lemma are valid for both. We also obtained, in Theorem 5, generalizations of the eight properties from Theorem 9 of Chapter 4, with only property (ii) requiring any significant modification. It turns out that the Cauchy-Schwarz Inequality also holds in general; the proof in Chapter 4 applies to the general setting without change.

On the other hand, the Bounded Convergence Theorem and the Uniform Integrability Criterion fail in general. If the measure is infinite, boundedness of a function does not even ensure that its integral exists, and, as defined in Definition 11, 'uniform integrability' is useless.

The Jensen Inequality also fails in the infinite setting. For a simple counterexample, let μ be Lebesgue measure on the interval $[1, \infty)$, and let $f(x) = 1/x$ and $\varphi(x) = x^2$. Then

$$\varphi\left(\int f \, d\mu\right) = \infty \quad \text{and} \quad \int \varphi \circ f \, d\mu = 1.$$

Problem 14. What is the appropriate statement of the Jensen Inequality for a finite measure space?

Turning from results about integration with respect to infinite measures to results about infinite measures themselves, we note first that the Continuity of Measure Theorem does not generalize completely, but that restricted versions of it do (see Problem 4, Problem 5, and Problem 9). The Borel Lemma and its proof generalize without significant change (but this is not true of the Borel-Cantelli Lemma—see below). A great deal of the existence and uniqueness theory from Chapter 7 also carries over, but we do not concern ourselves with the details in this book, since Problem 19 of Chapter 6 provides us with a sufficiently powerful tool for constructing infinite measures from finite ones.

Finally, we note several concepts and results from probability theory that are either meaningless, useless, or obviously wrong in the infinite setting: variance, covariance, correlation, the Chebyshev and Markov Inequalities, and the Law of Large Numbers. The following example also shows that any attempt to generalize the Kochen-Stone and Borel-Cantelli Lemmas is doomed to failure: Let μ be Lebesgue measure on $(\mathbb{R}, \mathcal{B})$, and let $A_n = (n, n + 1)$.

8.4. Lebesgue integrals and Riemann-Stieltjes integrals

A measure μ on a Borel σ-field is called a *Radon measure* if $\mu(C) < \infty$ for every compact set C. In this section we consider Radon measures on the measurable space $(\mathbb{R}, \mathcal{B})$.

Problem 15. Prove that a Radon measure μ on $(\mathbb{R}, \mathcal{B})$ is necessarily σ-finite.

Definition 13. Let F be a function defined on $\mathbb{R}$, and let μ be a Radon measure on $\mathbb{R}$. We say that F is a *distribution function* for μ if

$$\mu((a, b]) = F(b) - F(a)$$

for all real $a < b$.

It is clear that two distribution functions correspond to the same Radon measure if and only if their difference is a constant. By methods used in the proofs of Proposition 3 and Proposition 4, both of Chapter 3, it can be shown that every distribution function is increasing and right-continuous, and that to a function F on $\mathbb{R}$ with these properties there corresponds a unique Radon measure μ such that F is a distribution function of μ. Given a Radon measure μ on $(\mathbb{R}, \mathcal{B})$, a corresponding distribution function F may be constructed by defining $F(x) = \mu([0, x])$ for $x \geq 0$, and $F(x) = -\mu((x, 0))$ for $x < 0$. For Lebesgue measure on $\mathbb{R}$, this construction gives us the distribution function $x \rightsquigarrow x$.

Often, integrals with respect to μ can be calculated as Riemann-Stieltjes integrals with respect to F. The following statement, a generalization of Theorem 17 of Chapter 4, is a straightforward consequence of that theorem.

Theorem 14. *Let μ be a Radon measure on $(\mathbb{R}, \mathcal{B})$ with distribution function F, and let φ be an $\mathbb{R}$-valued function that is Riemann-Stieltjes integrable with respect to F on every bounded interval. If $\int \varphi \, d\mu$ exists, then*

$$\int \varphi \, d\mu = \int_{-\infty}^{\infty} \varphi(x) \, dF(x) \, .$$

Remark 1. The two integrals in the preceding theorem are both limits of sums involving functions that approximate φ. The approximating functions for the integral on the left are simple functions, while the ones for the integral on the right are step functions. Using simple functions to approximate φ amounts to partitioning the target space of φ, while using step functions involves partitioning the domain.

Problem 16. Let μ be a Radon measure on $(\mathbb{R}, \mathcal{B})$ with distribution function F, and let φ be a monotone function with a continuous derivative defined on an interval $[a, b]$ contained in $\mathbb{R}$. Find a formula for

$$\int_{(a,b]} \varphi \, d\mu$$

which involves only a Riemann integral and the values of φ and F at a and b. *Hint:* If the domain of φ is extended to $(-\infty, \infty)$ by defining $\varphi(x) = 0$ for $x \notin (a, b]$, φ is Riemann-Stieltjes integrable with respect to every distribution function F.

Problem 17. What changes are necessary in the formula obtained in the preceding exercise if $(a, b]$ is replaced by: (i) $[a, b]$, (ii) $[a, b)$, and (iii) (a, b)? Prove the assertions you make.

Problem 18. Find an example of a bounded function whose improper Riemann integral on $(-\infty, \infty)$ exists but whose integral with respect to one-dimensional Lebesgue measure does not exist. Also find a function whose integral with respect to Lebesgue measure exists, but which is not Riemann integrable on some bounded interval. *Hint:* For the first part, look at (Counter)example 4 of Chapter 4.

In view of the Theorem 14, the Monotone and Dominated Convergence Theorems are available for the study of Riemann integrals. However, one must be cautious in using these theorems for the general theory, because the limit of a sequence of Riemann integrable functions may be a function without a Riemann integral, even if it is bounded by a constant and is equal to 0 outside some closed interval. The Dominated Convergence Theorem plays a role in the following sequence of exercises, designed to yield an asymptotic formula for the gamma function.

Problem 19. Use the substitution $u = \gamma + v\sqrt{\gamma}$, suggested by the fact that the mean and standard deviation of a gamma distribution are γ and $\sqrt{\gamma}$, respectively, to obtain the following formula for the gamma function Γ:

$$\Gamma(\gamma) = \left(\frac{\gamma}{e}\right)^\gamma \gamma^{-1/2} \int_{-\sqrt{\gamma}}^\infty (1 + v\gamma^{-1/2})^{\gamma-1} e^{-v\gamma^{1/2}} \, dv.$$

Problem 20. Use the preceding exercise to show that

$$\frac{\Gamma(\gamma)}{\gamma^{\gamma-(1/2)} e^{-\gamma}} = \int_{-\infty}^\infty \varphi_\gamma(v) \, dv + \int_{-\infty}^\infty \theta_\gamma(v) \, dv$$

where

$$\varphi_\gamma(v) = \begin{cases} (1 + v\gamma^{-1/2})^{\gamma-1} e^{-v\gamma^{1/2}} & \text{if } -\gamma^{1/2} \le v \le 0 \\ 0 & \text{otherwise} \end{cases}$$

and
$$\theta_\gamma(v) = \begin{cases} (1 + v\gamma^{-1/2})^{\gamma-1} e^{-v\gamma^{1/2}} & \text{if } v \geq 0 \\ 0 & \text{otherwise}. \end{cases}$$

Problem 21. Prove that the function
$$x \rightsquigarrow \log(1+x) - \frac{x(2+x)}{2(1+x)}, \quad x > -1,$$
is decreasing.

* **Problem 22.** Use the Dominated Convergence Theorem and Theorem 14 to prove that
$$\lim_{\gamma \to \infty} \int_{-\infty}^{\infty} \theta_\gamma(v)\, dv = \int_0^{\infty} e^{-v^2/2}\, dv,$$
with θ_γ as defined in Problem 20. *Hint:* Use the preceding problem at an appropriate point.

Problem 23. Use Theorem 14 to prove that
$$\lim_{\gamma \to \infty} \int_{-\infty}^{\infty} \varphi_\gamma(v)\, dv = \int_{-\infty}^0 e^{-v^2/2}\, dv,$$
with φ_γ as defined in Problem 20. *Hint:* For $\gamma \geq 2$, the Taylor Formula can be used to show that the Dominated Convergence Theorem is applicable.

The next result gives an asymptotic formula for the gamma function. We write '$f(\gamma) \sim g(\gamma)$ as $\gamma \to \infty$' to mean that $f(\gamma)/g(\gamma) \to 1$ as $\gamma \to \infty$. Table 8.1 shows this formula gives a surprisingly good approximation to the gamma function, even for small values of the argument.

Theorem 15. [Stirling Formula] *The gamma function Γ satisfies*
$$\Gamma(\gamma) \sim \sqrt{2\pi}\, \gamma^{\gamma-(1/2)} e^{-\gamma} \quad \text{as } \gamma \to \infty.$$

In particular,
$$n! \sim \sqrt{2\pi n}\, n^n e^{-n} \quad \text{as } n \to \infty.$$

Problem 24. Prove the preceding theorem. *Hint:* Use Problem 20, Problem 22, and Problem 23.

The following sequence of problems gives further practice in using the convergence theorems in combination with other techniques. In the process it will be shown that $\Gamma'(1)$ is the negative of 'Euler's constant'.

γ	$\frac{1}{2}$	$1\frac{1}{2}$	π	4	6	10
$\Gamma(\gamma)$	$\sqrt{\pi}$	$\dfrac{\sqrt{\pi}}{2}$	2.287	6	120	362880
$\sqrt{2\pi}\,\gamma^{\gamma-\frac{1}{2}}e^{-\gamma}$	1.520	.8390	2.228	5.877	118.3	359869.5
$\dfrac{\sqrt{2\pi}\,\gamma^{\gamma-\frac{1}{2}}e^{-\gamma}}{\Gamma(\gamma)} - 1$	-0.142	-0.053	-0.026	-0.021	-0.014	-0.008

TABLE 8.1. Gamma function, Stirling Formula, and relative error

Problem 25. Let Γ denote the gamma function. Prove that

$$\Gamma'(1) = \int_0^\infty (\log x)\, e^{-x}\, dx = \frac{1}{n!} \int_0^\infty (\log x)\, x^n\, e^{-x}\, dx - \sum_{k=1}^n \frac{1}{k}$$

for each positive integer n. *Hint:* Use mathematical induction and integration by parts.

* **Problem 26.** Prove that

$$\lim_{n\to\infty} \frac{1}{n!} \int_{n-\sqrt[3]{n^2}}^{n+\sqrt[3]{n^2}} \left(\log \frac{x}{n}\right) x^n\, e^{-x}\, dx = 0\,.$$

Problem 27. Prove that

$$\lim_{n\to\infty} \frac{1}{n!} \int_0^{n-\sqrt[3]{n^2}} \left(\log \frac{x}{n}\right) x^n\, e^{-x}\, dx = 0\,.$$

Hint: Show that integrand is negative and decreasing.

Problem 28. Prove that

$$\lim_{n\to\infty} \frac{1}{n!} \int_{n+\sqrt[3]{n^2}}^\infty \left(\log \frac{x}{n}\right) x^n\, e^{-x}\, dx = 0\,.$$

Problem 29. Use the preceding four exercises to prove that

$$\int_0^\infty (\log x) e^{-x}\, dx = -C\,,$$

where $C \approx 0.577$ is Euler's constant defined by

$$C = \lim_{n\to\infty} \left(\sum_{k=1}^n \frac{1}{k} - \log n \right).$$

Decide whether your proof establishes the existence of this last limit. If not, supply appropriate additional arguments.

8.5. Absolute continuity and densities

Riemann-Stieltjes integration is not always easy, so it is desirable to supplement Theorem 14 with further computational techniques. For this purpose, we use the density concept, which was introduced briefly in Chapter 3.

Definition 16. Let $(\Omega, \mathcal{F})$ be a measurable space, and let μ and ν be two measures on $(\Omega, \mathcal{F})$. We say that ν is *absolutely continuous* with respect to μ, written $\nu \ll \mu$, if $\nu(A) = 0$ for every $A \in \mathcal{F}$ for which $\mu(A) = 0$.

Proposition 17. *Let $(\Omega, \mathcal{F}, \mu)$ be a measure space and let f be an $\mathbb{R}^+$-valued measurable function. For $A \in \mathcal{F}$ define*

$$\nu(A) = \int_A f \, d\mu \, .$$

Then ν is a σ-finite measure on $(\Omega, \mathcal{F})$ satisfying $\nu \ll \mu$.

PROOF. Clearly ν is nonnegative. Countable additivity follows from the Monotone Convergence Theorem, as shown by the following computation:

$$\nu\left(\bigcup_{n=1}^{\infty} B_n\right) = \int f I_{\cup B_n} \, d\mu = \int f \sum_{n=1}^{\infty} I_{B_n} \, d\mu$$

$$= \sum_{n=1}^{\infty} \int f I_{B_n} \, d\mu = \sum_{n=1}^{\infty} \nu(B_n) \, ,$$

where we have assumed that the sets B_n are pairwise disjoint. Since $f I_A = 0$ μ-a.e. if $\mu(A) = 0$, the fact that ν is absolutely continuous with respect to μ follows from property (ii) of Theorem 5.

We will show that ν is σ-finite under the assumption that μ is finite. The extension to the case in which μ is σ-finite is straightforward. For each $n = 1, 2, \ldots$, let $B_n = \{x \colon n - 1 \le f(x) < n\}$. For $A \in \mathcal{F}$, define

$$\nu_n(A) = \int_A (f I_{B_n}) \, d\mu \quad n = 1, 2, \ldots \, .$$

The argument in the first paragraph of this proof shows that ν_n is a measure for each n. Since μ is assumed to be finite, and since f is bounded on each of the sets B_n, it is clear that each ν_n is finite. The Monotone Convergence Theorem implies that $\nu(A) = \sum_n \nu_n(A)$, so ν is σ-finite, as desired. $\square$

The preceding proposition has a converse which is important in measure theory but which plays a minor role in this book. For completeness, we state this converse here. The proof appears in Chapter 23 as an application of 'conditional expectation'.

Theorem 18. [Radon-Nikodym] *Let μ and ν be σ-finite measures defined on a common measurable space and satisfying $\nu \ll \mu$. Then there exists an $\mathbb{R}^+$-valued measurable function f such that*

$$\nu(A) = \int_A f \, d\mu$$

for all $A \in \mathcal{F}$.

The function f in the preceding theorem is called the *density* or the *Radon-Nikodym derivative* of ν with respect to μ. We write

$$f = \frac{d\nu}{d\mu} \, .$$

Problem 30. Justify the use of the word 'the' in the preceding statement by showing that if f and g are both densities of ν with respect to μ, then $f = g$ μ-a.e.

If ν has a density with respect to μ, we can use this density to compute the Lebesgue integral of a function with respect to ν in terms of an integral with respect to μ.

Proposition 19. *Let μ and ν denote σ-finite measures defined on a common measurable space, and suppose that $d\nu/d\mu$ exists. Then, for every measurable $\overline{\mathbb{R}}$-valued function g defined on $(\Omega, \mathcal{F})$,*

$$\int g \, d\nu = \int g \, \frac{d\nu}{d\mu} \, d\mu \, ,$$

where the product $g(x)(d\nu/d\mu)(x)$ is understood to equal 0 if either factor equals 0, even if the other factor equals ∞. The assertion is that if either side exists, then both sides exist and are equal.

Problem 31. Prove the preceding proposition. *Hint:* First consider simple functions g.

Proposition 20. [Chain Rule] *Let λ, μ, and ν be σ-finite measures defined on a common measurable space. Suppose that $d\nu/d\mu$ and $d\mu/d\lambda$ exist. Then*

$$\frac{d\nu}{d\lambda} = \frac{d\nu}{d\mu} \frac{d\mu}{d\lambda} \quad \lambda\text{- a.e.} \, ,$$

part of this conclusion being that $d\nu/d\lambda$ exists.

Problem 32. Prove the preceding proposition.

Problem 33. [Reciprocal Rule] Let μ, ν be σ-finite measures for which ν has a density φ with respect to μ. Prove that if φ is μ-a.e. nonzero, then $1/\varphi$ (defined to be some arbitrary constant for x such that $\varphi(x) = 0$) is the density of μ with respect to ν.

Proposition 19 is especially useful for calculations when μ is Lebesgue measure on $\mathbb{R}$. We have already seen this in the applications of Definition 9 of Chapter 3. The word 'density' there is consistent with our present usage. However, even in the special case of a probability space, our current definition is a strict generalization of Definition 9 of Chapter 3, since we now allow densities which are Lebesgue integrable even if they are not Riemann integrable.

Problem 34. Let X be a random variable with the gamma distribution, with parameters $a, \gamma > 0$. In Example 3 of Chapter 4, we showed that $E(X) = \gamma/a$. In the calculation, some effort was needed to accommodate $\gamma < 1$. Use aspects of this section to describe another way of accommodating $\gamma < 1$.

* **Problem 35.** Let μ be a Radon measure on $(\mathbb{R}, \mathcal{B})$ with distribution function F. Prove that if $F' = f$ exists and is Riemann integrable on every bounded interval, then μ is absolutely continuous with respect to Lebesgue measure on $\mathbb{R}$, and the density of μ is f. (The result is still true without the hypothesis that f be Riemann integrable on bounded intervals, but the proof is considerably more difficult.)

In the preceding section of this chapter, we studied the gamma function by a combination of Riemann integration theory and Lebesgue integration theory. In particular we used a change of variables in a Riemann integral. The next proposition, which does not require Riemann integrability, shows that we could have made the change of variables in the Lebesgue setting instead.

Proposition 21. [Change of Variables] *Let φ be a strictly increasing differentiable function from an interval J onto an interval K, and let λ be Lebesgue measure on $\mathbb{R}$. For f a measurable function from K to $\overline{\mathbb{R}}$,*

$$\int_K f \, d\lambda = \int_J (f \circ \varphi)\varphi' \, d\lambda$$

in the sense that if one side exists, then both exist and are equal.

Problem 36. Prove the preceding proposition. *Hint:* Show that φ' is the density with respect to Lebesgue measure of the measure induced by φ^{-1} on J.

Problem 37. Let f be an $\overline{\mathbb{R}}$-valued measurable function defined on $(\mathbb{R}, \mathcal{B})$. Let $g(x) = f(-x)$, and $h(x) = f(x + c)$, where c is a fixed constant. Show that

$$\int f \, d\lambda = \int g \, d\lambda = \int h \, d\lambda,$$

where λ is Lebesgue measure on $\mathbb{R}$.

Problem 38. Show that the Cantor distribution is not absolutely continuous with respect to Lebesgue measure on $\mathbb{R}$.

8.6. Integration with respect to counting measure

Lebesgue integration reduces to summation when the underlying measure is counting measure.

Proposition 22. *Let μ be counting measure on a measurable space $(\Omega, \mathcal{F})$, where $\mathcal{F}$ consists of all subsets of Ω, and let f be an $\overline{\mathbb{R}}$-valued function defined on Ω. Define*

$$\mathcal{S}^{\pm} = \{\sum_{x \in A} f^{\pm}(x) \colon A \text{ is a finite subset of } \Omega\}.$$

Then

$$\int f \, d\mu = \sup \mathcal{S}^+ - \sup \mathcal{S}^-,$$

in the sense that if either side exists, then so does the other and they are equal. In addition, if Ω is countable and $\int f \, d\mu$ exists, then

$$\int f \, d\mu = \sum_{x \in \Omega} f(x).$$

The order of the terms in this summation does not affect the value of the sum.

Problem 39. Prove the preceding result. *Hint:* See Problem 18 of Chapter 6.

Problem 40. Let μ be counting measure on a countable space Ω. Suppose that f is $\mathbb{R}$-valued and $\int f \, d\mu$ does not exist. Investigate the effect that the order of terms in the sum

$$\sum_{x \in \Omega} f(x)$$

has on its existence and possible value.

Problem 41. Let μ be counting measure on a space Ω. Show that if $\int f \, d\mu$ is finite, then $\{x \colon f(x) \neq 0\}$ is a countable set. *Hint:* Use Proposition 4.

Problem 42. State the Dominated Convergence Theorem, the Fatou Lemma, and the Cauchy-Schwarz Inequality for sums.

Problem 43. On $\mathbb{Z}$, let μ be counting measure and ν an arbitrary σ-finite measure. Find a formula in terms of ν and μ for the density g of ν with respect to μ. For functions $f \colon \mathbb{Z} \to \mathbb{R}$, find a summation formula for $\int f \, d\nu$ in terms of f and g.

Problem 44. Interpret certain quantities found in Table 5.1 of Chapter 5 as densities with respect to counting measure on $\mathbb{Z}$.

PART 2
Independence and Sums

In this part of the book, we introduce a very important concept: 'stochastic independence'. Roughly speaking, two random experiments are independent if knowledge of the outcome of one of the experiments does not affect one's assessment about the distribution of outcomes in the other experiment. This notion permeates probability theory, and in so doing distinguishes probability theory from general measure theory. Even random objects that are not independent are often analyzed in terms of related structures that are independent. An attractive feature of probability theory lies in its ability to model such an important heuristic notion with mathematical precision.

The mathematical definition of independence involves 'product measure', to be defined in Chapter 9. Calculations involving product measure are facilitated by the use of a famous theorem from integration theory, the Fubini Theorem. Other basic definitions and results and several examples are also found in Chapter 9.

Many applications of probability theory involve sums of independent random variables, and these form the subject matter of most of the rest of Part 2. General definitions and examples are found in Chapter 10. Random walks, which involve sums of identically distributed independent random variables, are studied in Chapter 11. Chapter 12 contains several convergence results related to sequences and sums of independent random variables, including the Strong Law of Large Numbers. Chapter 13 introduces an important tool known as the 'characteristic function'. This tool is useful in analyzing the distributions of sums of independent random variables. Some interesting applications of characteristic functions will be given in Chapter 13, but their real power will be revealed in Part 3.

It will sometimes be convenient to have an alternate notation for integrals, one which makes explicit the variable of integration. The new notation for $\int f \, d\mu$ is

$$\int f(x) \, \mu(dx) \, .$$

This notation will be particularly useful when we work with iterated integrals. See, for example, the statement of the Fubini Theorem in Chapter 9.

CHAPTER 9
Stochastic Independence

The first six sections of this chapter describe the measure-theoretic foundation for 'stochastic independence': products of probability spaces. After giving the basic definitions, we prove the existence of 'product measure' and also give an important result concerning integration with respect to product measure (the Fubini Theorem). Important relations among expectations, independence, and densities are described. The last three sections of the chapter do not depend on each other. The first treats the asymptotic behavior of sequences of independent identically distributed random variables. The second concerns 'order statistics' of finite sequences of such random variables. The last introduces some new distributions.

9.1. Definition and basic properties

We begin with a general example intended to establish the basic ideas behind our definitions.

Example 1. Consider two experiments, represented by probability spaces $(\Omega_1, \mathcal{F}_1, P_1)$ and $(\Omega_2, \mathcal{F}_2, P_2)$. The product space $\Omega = \Omega_1 \times \Omega_2$ is a natural sample space to use for a compound experiment in which both are performed. The coordinate maps $X_i \colon \Omega \to \Omega_i$, defined by

$$X_i(\omega_1, \omega_2) = \omega_i, \quad i = 1, 2,$$

link the compound experiment to the two original experiments.

For $A_1 \in \mathcal{F}_1$, let

$$Q_1(A_1 \times \Omega_2) = P_1(A_1).$$

This defines a probability measure on the measurable space $(\Omega, \mathcal{G}_1)$, where

$$\mathcal{G}_1 = \{A_1 \times \Omega_2 : A_1 \in \mathcal{F}_1\}.$$

The probability spaces $(\Omega, \mathcal{G}_1, Q_1)$ and $(\Omega_1, \mathcal{F}_1, P_1)$ are equivalent from a theoretical point of view, and hence either one can be used to model the first experiment. In analogous fashion, we can define a probability space $(\Omega, \mathcal{G}_2, Q_2)$, with $\mathcal{G}_2 = \{\Omega_1 \times A_2 \colon A_2 \in \mathcal{F}_2\}$ and $Q_2(\Omega_1 \times A_2) = P_2(A_2)$, which is equivalent to the probability space originally representing the second experiment.

Let
$$\mathcal{F} = \sigma(\mathcal{G}_1, \mathcal{G}_2),$$
the smallest σ-field containing both $\mathcal{G}_1$ and $\mathcal{G}_2$. We are interested in defining a probability measure P on $\mathcal{F}$ that agrees with Q_1 on $\mathcal{G}_1$ and with Q_2 on $\mathcal{G}_2$. There are typically many such measures P, each of which models a compound experiment involving the two original experiments. The choice of P reflects the relationship between the two original experiments.

In this chapter we will construct a particular P that models the situation in which the original two experiments have no influence on each other. Let $\mathcal{R}$ be the collection of all 'measurable rectangles', that is,
$$\mathcal{R} = \{A_1 \times A_2 \colon A_1 \in \mathcal{F}_1 \text{ and } A_2 \in \mathcal{F}_2\}.$$

Note that
$$\sigma(\mathcal{R}) = \mathcal{F} = \sigma(\mathcal{G}_1, \mathcal{G}_2),$$
since every measurable rectangle $A_1 \times A_2$ is the intersection of the set $A_1 \times \Omega_2 \in \mathcal{G}_1$ and the set $\Omega_1 \times A_2 \in \mathcal{G}_2$.

We define P for sets in $\mathcal{R}$ by
$$P(A_1 \times A_2) = P_1(A_1)P_2(A_2).$$

Note that this definition implies that

(9.1) $\qquad P(B_1 \cap B_2) = P(B_1)P(B_2) \quad$ for all $B_1 \in \mathcal{G}_1, B_2 \in \mathcal{G}_2,$

since we may write $B_1 = A_1 \times \Omega_2$ and $B_2 = \Omega_1 \times A_2$ and calculate as follows:
$$P(B_1 \cap B_2) = P(A_1 \times A_2) = P_1(A_1)P_2(A_2) = P(A_1 \times \Omega_2)P(\Omega_1 \times A_2).$$

The relationship between $\mathcal{G}_1$ and $\mathcal{G}_2$ that is expressed in (9.1) is known as 'stochastic independence'. This relationship has the interpretation that in the combined experiment, the two original experiments do not affect each other.

Of course, we have not yet completed the process of combining the two original experiments in this example, because we have not defined P on all of $\mathcal{F}$. We will prove in Theorem 7 that P can be extended to a probability measure on $\mathcal{F}$. Since $\mathcal{R}$ is closed under pairwise intersections, the Uniqueness of Measure Theorem implies that this extension is unique.

---

* **Problem 1.** Match the situation described in Problem 10 of Chapter 1 with the preceding example. For the situation in that problem decide how many members each of the following sets has: Ω_i, $\mathcal{F}_i$, $\mathcal{G}_i$, Ω, $\mathcal{F}$, and $\mathcal{R}$.

We are now ready to give a formal definition of stochastic independence. Notice that when this definition is applied to the pair $\mathcal{G}_1, \mathcal{G}_2$ in Example 1, it is equivalent to (9.1).

Definition 1. Let $(\Omega, \mathcal{F}, P)$ denote a probability space and let $\mathcal{F}_k$, $k \in K$, be sub-σ-fields of $\mathcal{F}$.

(i) If K is finite, $(\mathcal{F}_k : k \in K)$ is *stochastically independent* (or *independent* when there is no danger of confusion) if

$$(9.2) \qquad P\left(\bigcap_{k \in K} A_k\right) = \prod_{k \in K} P(A_k)$$

for all $A_k \in \mathcal{F}_k$.

(ii) If K is infinite, $(\mathcal{F}_k : k \in K)$ is *stochastically independent* if for all finite sets $J \subseteq K$, $(\mathcal{F}_j : j \in J)$ is stochastically independent.

We also want to define independence for random variables and events. To do so we speak of the σ-field *generated* by a $(\Psi, \mathcal{G})$-valued random variable X:

$$\sigma(X) = \{X^{-1}(B) : B \in \mathcal{G}\}.$$

Note that $\sigma(X)$ is the smallest σ-field with respect to which X is measurable. The σ-field *generated* by an event A in a probability space $(\Omega, \mathcal{F}, P)$ is the σ-field generated by the indicator function of A; it equals $\{\emptyset, A, A^c, \Omega\}$ and is denoted by $\sigma(A)$. Independence for random variables and events is defined in terms of independence of the σ-fields that they generate.

Definition 2. Let $(\Omega, \mathcal{F}, P)$ be a probability space, and let

$$\mathcal{M} = (\mathcal{E}_j, X_k, A_l : j \in J, k \in K, l \in L)$$

consist of collections $\mathcal{E}_j \subseteq \mathcal{F}$ of events, random variables X_k defined on $(\Omega, \mathcal{F}, P)$, and events $A_l \in \mathcal{F}$. Then $\mathcal{M}$ is said to be *stochastically independent* (or *independent* when there is no danger of confusion) if

$$(\sigma(\mathcal{E}_j), \sigma(X_k), \sigma(A_l) : j \in J, k \in K, l \in L)$$

is stochastically independent.

It is common to speak of two events A and B as being independent, or to say that A is independent of B, even though one really means that the pair (A, B) is independent; stochastic independence is not a property of the two events individually, but instead of the relationship between them. Such language is also often used for random variables and σ-fields. (A similar lack of precision occurs when the term 'linear independence' is used in linear algebra.)

Problem 2. For $k = 1, 2, \ldots$, let X_k be a $(\Psi_k, \mathcal{G}_k)$-valued random variable, and let φ_k be a measurable function from $(\Psi_k, \mathcal{G}_k)$ to $(\Theta_k, \mathcal{H}_k)$. Prove that if the sequence $(X_1, X_2, \ldots)$ is independent, then the sequence $(\varphi_1 \circ X_1, \varphi_2 \circ X_2, \ldots)$ is independent. *Hint:* Show that $\sigma(\varphi_i \circ X_i) \subseteq \sigma(X_i)$ for each i.

Problem 3. Show that a pair of events (A, B) is independent if and only if $P(A \cap B) = P(A)P(B)$. Find an example of a triple (A, B, C) of events that is not independent but still satisfies $P(A \cap B \cap C) = P(A)P(B)P(C)$.

Problem 4. Let $(A_1, A_2, \ldots)$ be an independent sequence of events. Prove that

$$P\left(\bigcap_{n=1}^{\infty} A_n\right) = \prod_{n=1}^{\infty} P(A_n).$$

Problem 5. Consider three events for the fair coin-flip probability space of Example 4 of Chapter 1:

$$A_1 = \{\omega : \omega_1 = 1\}, \quad A_2 = \{\omega : \omega_2 = 1\}, \quad A_3 = \{\omega : \omega_1 + \omega_2 = 1\}.$$

Show that any pair of them is independent, but that the triple (A_1, A_2, A_3) is not. Describe this phenomenon in an intuitive manner.

* **Problem 6.** Let $(X_1, X_2, \ldots)$ be an independent sequence of $\mathbb{R}$-valued random variables and $(F_1, F_2, \ldots)$ the corresponding sequence of distribution functions. Let $Y(\omega)$ equal the greatest lower bound and $Z(\omega)$ the least upper bound of the set $\{X_n(\omega) : n = 1, 2, \ldots\}$. Find a formula for the distribution functions of Y and Z in terms of the functions F_n. Comment on the situation if the random variables X_n are assumed to be $\overline{\mathbb{R}}$-valued.

* **Problem 7.** Let (X_1, X_2) be an independent pair of exponentially distributed random variables with means λ_1 and λ_2, respectively. Calculate and name the distribution of $X_1 \wedge X_2$.

Problem 8. Let (X_1, X_2) be an independent pair of random variables, each having the same distribution—beta with parameter $(\frac{1}{2}, 1)$. Calculate and name the distribution of $X_1 \vee X_2$.

Problem 9. Let $(X_1, X_2, \ldots)$ be an independent sequence of identically distributed $\mathbb{R}$-valued random variables. Let

$$U(\omega) = \liminf_{n \to \infty} X_n(\omega) \quad \text{and} \quad V(\omega) = \limsup_{n \to \infty} X_n(\omega).$$

Find the distribution of the ordered pair (U, V) in terms of the common distribution function F of the X_n. Is the pair (U, V) independent?

* **Problem 10.** Let $(\Omega, \mathcal{F}, P)$ be a probability space, and for each k in some finite index set K, let $\mathcal{E}_k$ be a countable partition of Ω. Prove that $(\mathcal{E}_k : k \in K)$ is independent if condition (9.2) holds for all events $A_k \in \mathcal{E}_k$, $k \in K$.

Problem 11. Show that if A is an event, then the pair (A, A) is independent if and only if $P(A) = 0$ or 1. Also show that if $\mathcal{G}$ is a σ-field, then $(\mathcal{G}, \mathcal{G})$ is independent if and only if $P(A) = 0$ or 1 for all $A \in \mathcal{G}$.

Problem 12. Show that if X is a random variable that takes values in $(\overline{\mathbb{R}}^d, \mathcal{B})$, then (X, X) is independent if and only if X is a constant a.s. Give an example to show that this statement fails to be true in general if X is merely assumed to take values in some measurable space $(\Psi, \mathcal{G})$. *Hint:* For the first part, use the previous exercise to show that if (X, X) is independent, the distribution function of each component of X can only take the values 0 and 1. For the example, take $\mathcal{G}$ to be the trivial σ-field $\{\Psi, \emptyset\}$.

Problem 13. Let Ω consist of the 36 ordered pairs $\omega = (\omega_1, \omega_2)$, where $1 \le \omega_i \le 6$. Let $\mathcal{F}$ denote the σ-field consisting of all 2^{36} subsets of Ω. Set $P(A) = \sharp A / 36$ for each $A \in \mathcal{F}$. This is the typical sample space used for the experiment of rolling two fair dice. Let

$$X_1(\omega) = \omega_1, \quad X_2(\omega) = \omega_2, \quad X_3(\omega) = (-1)^{\omega_1}, \quad X_4(\omega) = 5,$$
$$A_1 = \{\omega : \omega_1 \text{ is divisible by } 3\}, \quad A_2 = \{\omega : \omega_1 \text{ is divisible by } 2\},$$
$$A_3 = \{\omega : X_1(\omega) + X_2(\omega) = 7\}, \quad A_4 = \{\omega : X_1(\omega) + X_2(\omega) \text{ is odd}\},$$
$$\mathcal{E} = \{\{(\omega_1, \omega_2), (\omega_2, \omega_1)\} : \omega \in \Omega\}.$$

Consider the sequence

$$\mathcal{M} = (X_2, X_3, X_4, X_1 + X_2, (X_1, X_2), A_1, A_2, A_3, A_4, \Omega, \mathcal{E}).$$

Which subsequences are independent? *Hint:* Any further subsequence of an independent subsequence is independent, so one only needs to be concerned with independent subsequences that are maximal in length.

The next three propositions are useful for checking independence. The first one generalizes Problem 10.

Proposition 3. *Let* $(\Omega, \mathcal{F}, P)$ *denote a probability space, and let* $\mathcal{M} = (\mathcal{E}_l : l \in L)$ *consist of subcollections* $\mathcal{E}_l$ *of* $\mathcal{F}$, *each one of which is closed under pairwise intersections. Then* $\mathcal{M}$ *is independent if and only if (9.2) holds for all finite sets* $K \subseteq L$ *and events* $A_k \in \mathcal{E}_k, k \in K$.

* **Problem 14.** Prove the preceding proposition in the case that L has only 2 members. *Hint:* Apply the Sierpiński Class Theorem of Chapter 7 twice. (Generalizing the proof to treat arbitrary L is straightforward but notationally messy.)

* **Problem 15.** Use Proposition 3 to obtain a simple criterion that a sequence of events be independent.

When we are dealing with a collection or sequence of σ-fields, it is useful to have some streamlined notation for the smallest σ-field containing the collection or sequence. We write

$$\sigma(\mathcal{F}_k \colon k \in K) \quad \text{and} \quad \sigma(\mathcal{F}_1, \mathcal{F}_2, \dots)$$

for

$$\sigma\left(\bigcup_{k\in K} \mathcal{F}_k\right) \quad \text{and} \quad \sigma\left(\bigcup_{k=1}^{\infty} \mathcal{F}_k\right).$$

Similar notation is used with events, collections of events, and random variables. For example, we write $\sigma(X, Y)$ for $\sigma(\sigma(X) \cup \sigma(Y))$.

Proposition 4. *Let K and M denote sets and let $\{K_m, m \in M\}$ be a partition of K. Let $(\Omega, \mathcal{F}, P)$ be a probability space, and let $\mathcal{F}_k$, $k \in K$, be sub-σ-fields of $\mathcal{F}$. Suppose that $(\mathcal{F}_k \colon k \in K)$ is independent. Then $(\sigma(\mathcal{F}_k \colon k \in K_m) \colon m \in M)$ is independent.*

PROOF. By the definition of independence, it is enough to prove the result for finite sets M. For each $m \in M$, let $\mathcal{G}_m = \sigma(\mathcal{F}_k \colon k \in K_m)$ and let $\mathcal{E}_m$ be the collection of all finite intersections of members of

$$\bigcup_{k\in K_m} \mathcal{F}_k.$$

Each $\mathcal{E}_m$ is closed under pairwise intersections and $\mathcal{G}_m = \sigma(\mathcal{E}_m)$. Therefore, by Proposition 3, it is enough to check that

$$P\left(\bigcap_{m\in M} A_m\right) = \prod_{m\in M} P(A_m)$$

for all events $A_m \in \mathcal{E}_m, m \in M$. If A_m is an event in $\mathcal{E}_m$, then there exist events $B_k \in \mathcal{F}_k$ for $k \in K_m$ such that $B_k = \Omega$ for all but finitely many $k \in K_m$ and

$$A_m = \bigcap_{k\in K_m} B_k.$$

(Since the sets K_m are disjoint, there is no need to put additional subscripts on the events B_k to distinguish them for different values of m.) Since $(\mathcal{F}_k, k \in K)$ is independent and $B_k = \Omega$ for all but finitely many $k \in K$,

$$P\left(\bigcap_{m\in M} A_m\right) = P\left(\bigcap_{m\in M}\bigcap_{k\in K_m} B_k\right) = P\left(\bigcap_{k\in K} B_k\right) = \prod_{k\in K} P(B_k)$$

$$= \prod_{m\in M}\prod_{k\in K_m} P(B_k) = \prod_{m\in M} P\left(\bigcap_{k\in K_m} B_k\right) = \prod_{m\in M} P(A_m),$$

where we have used (9.2) twice and also have used the fact (or convention) that infinitely many (not necessarily countably many) factors equaling 1 in a product do not affect the value of that product. $\square$

Proposition 5. *Let $(\Omega, \mathcal{F}, P)$ be a probability space, and let $\mathcal{F}_k$, $k = 1, 2, \ldots$, be sub-σ-fields of $\mathcal{F}$. Then the sequence $(\mathcal{F}_k : k = 1, 2, \ldots)$ is independent if and only if each of the pairs $(\sigma(\mathcal{F}_1, \ldots, \mathcal{F}_n), \mathcal{F}_{n+1})$ is independent for $n = 1, 2, \ldots$.*

Problem 16. Prove Proposition 5.

Problem 17. Let $X_1, \ldots, X_n$ be $\overline{\mathbb{R}}$-valued random variables. Show that the sequence $(X_1, \ldots, X_n)$ is independent if and only if

$$P(\{\omega : X_1(\omega) \le a_1, \ldots, X_n(\omega) \le a_n\}) = \prod_{i=1}^{n} P(\{\omega : X_i(\omega) \le a_i\})$$

for all $a_1, \ldots, a_n \in \overline{\mathbb{R}}$. Also show that if the random variables X_i are $\overline{\mathbb{Z}}$-valued, then $(X_1, \ldots, X_n)$ is independent if and only if

$$P(\{\omega : X_1(\omega) = k_1, \ldots, X_n(\omega) = k_n\}) = \prod_{k=1}^{n} P(\{\omega : X_i(\omega) = k_i\})$$

for all $k_1, \ldots, k_n \in \overline{\mathbb{Z}}$.

Problem 18. Prove that if the triple (X_1, X_2, X_3) of $\mathbb{R}$-valued random variables is independent, then so is the pair $(X_1, X_2 + X_3)$. Also find an example for which $(X_1, X_2 + X_3)$ is independent, but (X_1, X_2, X_3) is not.

Example 2. For the coin-flip probability space of Example 1 of Chapter 7, let $X_n(\omega) = \omega_n$ and $\mathcal{F}_n = \sigma(X_n)$. It is straightforward to check that $(\mathcal{F}_1, \mathcal{F}_2, \ldots)$ is an independent sequence of σ-fields. Equivalently, the sequence $(X_1, X_2, \ldots)$ is independent, which is often expressed more informally by saying that the random variables X_n are independent. By Proposition 4 we can conclude such things as: the number of heads in the first ten flips is independent of the number of heads in the eleventh through eighteenth flips.

9.2. Product measure: finitely many factors

Example 1 and Example 2 indicate that product spaces can be used in a natural way in the construction of independent random variables. The key to the construction is to carry out the extension that was advertised in Example 1. In this section we will first do this for the product of a pair of probability spaces and then in a problem have the reader extend the construction to the product of a finite number of probability spaces. In a subsequent section we will treat countably many factors.

The first definition provides terminology for some of the objects that were already introduced in Example 1.

Definition 6. Let $(\Omega_1, \mathcal{F}_1)$ and $(\Omega_2, \mathcal{F}_2)$ be measurable spaces, and let $\Omega = \Omega_1 \times \Omega_2$. Let $\mathcal{R}$ be the collection of *measurable rectangles* in Ω:

$$\mathcal{R} = \{A_1 \times A_2 : A_1 \in \mathcal{F}_1 \text{ and } A_2 \in \mathcal{F}_2\}.$$

The σ-field

$$\mathcal{F} = \sigma(\mathcal{R}),$$

denoted by

$$\mathcal{F}_1 \times \mathcal{F}_2,$$

is called the *product σ-field* of $\mathcal{F}_1$ and $\mathcal{F}_2$. The measurable space $(\Omega, \mathcal{F})$, denoted by

$$(\Omega_1, \mathcal{F}_1) \times (\Omega_2, \mathcal{F}_2),$$

is called the *product* of $(\Omega_1, \mathcal{F}_1)$ and $(\Omega_2, \mathcal{F}_2)$.

Theorem 7. *Let $(\Omega_1, \mathcal{F}_1, \mu_1)$ and $(\Omega_2, \mathcal{F}_2, \mu_2)$ be σ-finite measure spaces. There exists a unique measure μ on the measurable space $(\Omega_1, \mathcal{F}_1) \times (\Omega_2, \mathcal{F}_2)$ such that*

$$(9.3) \qquad \mu(A_1 \times A_2) = \mu_1(A_1)\mu_2(A_2)$$

for all $A_1 \in \mathcal{F}_1$ and $A_2 \in \mathcal{F}_2$. The measure μ is σ-finite. Moreover, if μ_1 and μ_2 are probability measures, then so is μ.

PROOF. We give the proof in the case that μ_1 and μ_2 are finite. The extension to the σ-finite case is straightforward (see Problem 19 of Chapter 6 and the accompanying discussion), as is the proof that μ is σ-finite.

Since the collection of measurable rectangles is closed under pairwise intersections and generates $\mathcal{F}_1 \times \mathcal{F}_2$, the uniqueness follows from the Uniqueness of Measure Theorem, even though that theorem treats probability measures rather than arbitrary finite measures. The remainder of the proof will be devoted to the existence issue.

Consider the collection of $B \subseteq \Omega_1 \times \Omega_2$ for which the following function is defined and measurable:

$$\omega_1 \rightsquigarrow \int_{\Omega_2} I_B(\omega_1, \omega_2)\, \mu_2(d\omega_2).$$

Clearly, it contains all measurable rectangles. By linearity of integration, it is closed under proper differences; and by the Monotone Convergence Theorem, it is closed under increasing limits. Therefore, by the Sierpiński Class Theorem, it contains all members of $\mathcal{F}_1 \times \mathcal{F}_2$. Accordingly, we may define

$$(9.4) \qquad \mu(B) = \int_{\Omega_1} \left(\int_{\Omega_2} I_B(\omega_1, \omega_2)\, \mu_2(d\omega_2) \right) \mu_1(d\omega_1).$$

for $B \in \mathcal{F}_1 \times \mathcal{F}_2$.

That μ is countably additive follows from a corollary of the Monotone Convergence Theorem. For B of the form $A_1 \times A_2$ for some $A_1 \in \mathcal{F}_1$ and $A_2 \in \mathcal{F}_2$,

$I_B(\omega_1, \omega_2) = I_{A_1}(\omega_1)I_{A_2}(\omega_2)$. Insertion of this product into (9.4) gives (9.3). In particular, $\mu(\Omega_1 \times \Omega_2) = \mu_1(\Omega_1)\mu_2(\Omega_2) < \infty$. Therefore, μ is a finite measure, and furthermore, it is a probability measure if both μ_1 and μ_2 are probability measures. $\square$

The measure μ defined in the preceding theorem is called the *product measure* of μ_1 and μ_2. It is denoted by

$$\mu_1 \times \mu_2.$$

The measure space $(\Omega_1 \times \Omega_2, \mathcal{F}_1 \times \mathcal{F}_2, \mu_1 \times \mu_2)$ is called the *product space* of $(\Omega_1, \mathcal{F}_1, \mu_1)$ and $(\Omega_2, \mathcal{F}_2, \mu_2)$, and is denoted by

$$(\Omega_1, \mathcal{F}_1, \mu_1) \times (\Omega_2, \mathcal{F}_2, \mu_2).$$

In Example 1, we were given probability spaces $(\Omega_i, \mathcal{F}_i, P_i), i = 1, 2$, and then we constructed the measurable space $(\Omega, \mathcal{F})$ as the product of the spaces $(\Omega_i, \mathcal{F}_i)$. We also defined sub-$\sigma$-fields $\mathcal{G}_1, \mathcal{G}_2$ of $\mathcal{F}$ and a function P on the collection of measurable rectangles $\mathcal{R}$. We now see, according to Theorem 7, that the product measure $P_1 \times P_2$ is the desired extension of P to $\mathcal{F}$, and that the pair $(\mathcal{G}_1, \mathcal{G}_2)$ is independent.

Problem 19. Let X_1, X_2 be the two random variables defined in Example 1. Show that on the probability space described in the preceding paragraph, (X_1, X_2) is an independent pair.

Problem 20. Let $(\Psi_i, \mathcal{F}_i, \mu_i)$ be σ-finite measure spaces for $i = 1, 2, 3$. Show that

$$(\mathcal{F}_1 \times \mathcal{F}_2) \times \mathcal{F}_3 = \mathcal{F}_1 \times (\mathcal{F}_2 \times \mathcal{F}_3).$$

and

$$(\mu_1 \times \mu_2) \times \mu_3 = \mu_1 \times (\mu_2 \times \mu_3).$$

Extend to products of n measure spaces, and use the result to explain and justify the notation

$$\bigotimes_{i=1}^{n} (\Psi_i, \mathcal{F}_i, \mu_i).$$

Problem 21. Let $Q_1, Q_2, \ldots, Q_d$ denote the distributions of some random variables $X_1, X_2, \ldots, X_d$ defined on a common probability space. Prove that $X_i, 1 \le i \le d$, are independent random variables if and only if the distribution of the random d-tuple $(X_1, X_2, \ldots, X_d)$ equals $Q_1 \times Q_2 \times \cdots \times Q_d$.

Problem 22. Let λ be Lebesgue measure on $(\mathbb{R}, \mathcal{B})$, and let $(\mathbb{R}^d, \mathcal{B}^d, \lambda^d)$ be the d-fold product of $(\mathbb{R}, \mathcal{B}, \lambda)$ with itself. The measure λ^d is called d-*dimensional Lebesgue measure*. Explain why d-dimensional Lebesgue measure generalizes d-dimensional volume, and why integration with respect to d-dimensional Lebesgue measure generalizes Riemann integration in $\mathbb{R}^d$.

9.3. The Fubini Theorem

The following result is a preliminary step in obtaining an important tool for computing integrals on products of σ-finite measure spaces.

Proposition 8. *Let f be a measurable function from a product measurable space $(\Psi, \mathcal{G}) \times (\Theta, \mathcal{H})$ to a measurable space. Then $x \rightsquigarrow f(x, y)$ is measurable for each fixed $y \in \Theta$ and $y \rightsquigarrow f(x, y)$ is measurable for each fixed $x \in \Psi$.*

* **Problem 23.** Prove the preceding proposition. *Hint:* It is enough to prove that $x \rightsquigarrow (I_B \circ f)(x, y)$ is measurable for fixed y and measurable B.

Here is a partial converse of Proposition 8; the full converse is not true.

Proposition 9. *Let $(\Theta, \mathcal{H})$ be a measurable space, and let A be a Borel subset of $\mathbb{R}^d$ (with the usual topology). Suppose that $f: A \times \Theta \to \mathbb{R}$ is such that: (i) $x \rightsquigarrow f(x, y)$ is continuous for each $y \in \Theta$ and (ii) $y \rightsquigarrow f(x, y)$ is measurable for each $x \in A$. Then f is measurable.*

PROOF. Let $(a_1, a_2, \dots)$ be a countable dense sequence in A. For each positive integer n, define f_n by

$$f_n(x, y) = f(a_{j_{x,n}}, y),$$

where

$$j_{x,n} = \min\{i : |x - a_i| \le |x - x_k|, 1 \le k \le n\}.$$

In other words, $a_{j_{x,n}}$ is the element in the ordered n-tuple $(a_1, \dots, a_n)$ closest to x, with ties being broken by the ordering. For any Borel subset B of $\mathbb{R}$,

$$f_n^{-1}(B) = \bigcup_{i=1}^{n} \left(A_{i,n} \times \{y : f(a_i, y) \in B\} \right),$$

where $A_i = \{x : j_{x,n} = i\}$. Hypothesis (ii) shows that $f_n^{-1}(B)$ is a finite union of measurable rectangles and, hence, that f_n is measurable. Hypothesis (i) shows that, for each (x, y), $f_n(x, y) \to f(x, y)$ as $n \to \infty$. Therefore, f is measurable. $\square$

Theorem 10. [Fubini] *Let $(\Psi, \mathcal{G}, \mu)$ and $(\Theta, \mathcal{H}, \nu)$ be two σ-finite measure spaces, and let φ be an $\overline{\mathbb{R}}$-valued measurable function defined on the product measure space $(\Psi, \mathcal{G}, \mu) \times (\Theta, \mathcal{H}, \nu)$. If*

$$\int_{\Psi \times \Theta} \varphi \, d(\mu \times \nu)$$

exists, then

(9.5) $$x \rightsquigarrow \int_{\Theta} \varphi(x, y) \, \nu(dy)$$

is a μ-almost everywhere defined measurable function from $(\Psi, \mathcal{G})$ to $\overline{\mathbb{R}}$, and

(9.6) $$\int_{\Psi \times \Theta} \varphi \, d(\mu \times \nu) = \int_{\Psi} \left(\int_{\Theta} \varphi(x, y) \, \nu(dy) \right) \mu(dx).$$

PROOF. We give the proof for the case in which μ is a finite measure. The extension to σ-finite measures is straightforward, using either the Fubini Theorem for sums (Corollary 10 of Chapter 6) or the Monotone Convergence Theorem.

The theorem obviously holds when f is the indicator function of a measurable rectangle. The linearity of integration and the Monotone Convergence Theorem allow us to use the Sierpiński Class Theorem to extend the result to all indicator functions of measurable sets in $\mathcal{G} \times \mathcal{H}$. We then can extend it to simple functions f by again using the linearity of the integral, and to arbitrary nonnegative measurable functions by using Lemma 13 of Chapter 2 and the Monotone Convergence Theorem.

Now consider an arbitrary measurable $\overline{\mathbb{R}}$-valued function φ. Write $\varphi = \varphi^+ - \varphi^-$. Since the result has been proved for nonnegative measurable functions,

$$\int_{\Psi \times \Theta} \varphi^+ \, d(\mu \times \nu) = \int_{\Psi} \left(\int_{\Theta} \varphi^+(x, y) \, \nu(dy) \right) \mu(dx)$$

and

$$\int_{\Psi \times \Theta} \varphi^- \, d(\mu \times \nu) = \int_{\Psi} \left(\int_{\Theta} \varphi^-(x, y) \, \nu(dy) \right) \mu(dx).$$

Subtraction of the left sides gives the integral of φ. Since this integral is assumed to exist, at least one of the two iterated integrals on the right must be finite. It follows that at least one of the two inside integrals on the right must be μ-almost everywhere finite. Both of the inside integrals are measurable functions of x. Therefore, their difference is a μ-almost everywhere defined measurable function of x. The desired conclusion now follows by subtracting the right sides. $\square$

Remark 1. It is important to note carefully the hypotheses of the Fubini Theorem concerning the integrand: the function φ must be measurable and its integral with respect to $\mu \times \nu$ must exist. In particular, one should avoid the temptation to use the iterated integral on the right side of (9.6) directly to determine whether $\int \varphi \, d(\mu \times \nu)$ exists (see Example 3 in the next section). However, if φ is measurable, the Fubini Theorem does apply to the functions φ^+ and φ^-, since the integrals of $\overline{\mathbb{R}}^+$-valued measurable functions always exist. Thus, (9.6) can be used to determine the finiteness of the integrals of φ^+ and φ^-, from which the existence of the integral of φ can be determined.

Remark 2. It is not asserted that the integral in (9.5) is defined for all x. An exceptional set of measure 0 is included in the statement because both positive and negative parts might have infinite integrals.

Remark 3. The Fubini Theorem does not generalize to arbitrary measure spaces. To construct a simple counterexample, let $\Omega_1 = \Omega_2 = [0,1]$ with the corresponding Borel σ-fields, and let μ_1 be Lebesgue measure and μ_2 counting measure. If f is the indicator function of the diagonal of the square $\Omega_1 \times \Omega_2$, then one of the iterated integrals in Fubini Theorem is 0 and the other is 1. This example also indicates that there are difficulties with our construction of product measure in the general setting, since we used the iterated integral in that construction.

Problem 24. Compute the following iterated integral by interchanging the order of integration. Be sure to justify all the steps.

$$\int_0^\infty \int_y^\infty \frac{y \sin(2\pi y/x)}{(x^3 + 1)^2} \, dx \, dy \, .$$

Problem 25. Give an alternative proof of Corollary 20 of Chapter 4 using the Fubini Theorem.

Problem 26. State the Fubini Theorem for the product of three σ-finite measure spaces. Describe how the theorem you have stated follows from Problem 20, Proposition 8, and the Fubini theorem for the product of two σ-finite measure spaces.

The following two examples serve as warnings against trying to apply the Fubini Theorem without carefully checking that the hypotheses are satisfied.

Example 3. Let $\Psi = \{1, 2, 3, \ldots\}$, $\mathcal{G} = $ all subsets of Ψ, and $\mu(\{i\}) = 2^{-i}$. Define a random variable X on $(\Psi, \mathcal{G}, \mu)^2$ by $X(i,j) = a_{ij}$, where (a_{ij}) is the following infinite matrix:

$$\begin{pmatrix} 4 & -4 & 0 & 0 & 0 & \ldots \\ -4 & 16 & -16 & 0 & 0 & \ldots \\ 0 & -16 & 64 & -64 & 0 & \ldots \\ 0 & 0 & -64 & 256 & -256 & \ldots \\ \vdots & \vdots & \vdots & \vdots & \vdots & \ddots \end{pmatrix}$$

To compute $E(X)$, we try to sum the entries in the matrix $(a_{ij}\mu(\{i\})\mu(\{j\}))$, which looks like this:

$$\begin{pmatrix} 1 & -\frac{1}{2} & 0 & 0 & 0 & \ldots \\ -\frac{1}{2} & 1 & -\frac{1}{2} & 0 & 0 & \ldots \\ 0 & -\frac{1}{2} & 1 & -\frac{1}{2} & 0 & \ldots \\ 0 & 0 & -\frac{1}{2} & 1 & -\frac{1}{2} & \ldots \\ \vdots & \vdots & \vdots & \vdots & \vdots & \ddots \end{pmatrix}$$

Note that the sum of the positive terms in this second matrix is ∞, and the sum of the negative terms is $-\infty$. As a consequence, the expectation of X does not exist. On the other hand, calculating the two iterated integrals in the Fubini Theorem is equivalent to summing the row sums and summing the column sums of this matrix. In both cases, the result is $1/2$.

Example 4. For the probability space $(\Psi, \mathcal{G}, \mu)^2$ of the preceding example, define a random variable Y by $Y(i,j) = b_{ij}$, where (b_{ij}) is the infinite matrix

$$
\begin{pmatrix}
4 & -8 & 0 & 0 & 0 & \cdots \\
0 & 16 & -32 & 0 & 0 & \cdots \\
0 & 0 & 64 & -128 & 0 & \cdots \\
0 & 0 & 0 & 256 & -512 & \cdots \\
\vdots & \vdots & \vdots & \vdots & \vdots & \ddots
\end{pmatrix}
$$

For computing $E(Y)$, the relevant matrix is

$$
\begin{pmatrix}
1 & -1 & 0 & 0 & 0 & \cdots \\
0 & 1 & -1 & 0 & 0 & \cdots \\
0 & 0 & 1 & -1 & 0 & \cdots \\
0 & 0 & 0 & 1 & -1 & \cdots \\
\vdots & \vdots & \vdots & \vdots & \vdots & \ddots
\end{pmatrix}
$$

The two iterated integrals in the Fubini Theorem both exist but are unequal; one of them equals 1 and the other equals 0. It follows from the Fubini Theorem that the expected value of X does not exist.

9.4. Expectations and independence

Here is a version of the Fubini Theorem for a probabilistic setting.

Proposition 11. *Let (X, Y) be an independent pair of $\mathbb{R}$-valued random variables with finite expectation. Then $E(XY) = E(X)E(Y)$, or equivalently, $\mathrm{Cov}(X, Y) = 0$.*

* **Problem 27.** Prove the preceding proposition and then deduce the following corollary.

Corollary 12. *Let $(X_1, \ldots, X_n)$ be an independent sequence of random variables. For each i, let X_i take values in $(\Psi_i, \mathcal{G}_i)$, let φ_i be a measurable function from $(\Psi_i, \mathcal{G}_i)$ to $\mathbb{R}$, and suppose that $E(|\varphi_i \circ X_i|) < \infty$. Then*

$$
E\left(\prod_{i=1}^{n} \varphi_i \circ X_i\right) = \prod_{i=1}^{n} E(\varphi_i \circ X_i).
$$

From Proposition 11 above and Corollary 8 of Chapter 5 we conclude that if $(X_1, \ldots, X_n)$ is independent and if each X_k has finite mean, then

$$\mathrm{Var}\Big(\sum_{k=1}^{n} X_k\Big) = \sum_{k=1}^{n} \mathrm{Var}(X_k).$$

Moreover, the assumption of independence can be replaced by that of pairwise independence.

It is possible for $\mathrm{Cov}(X,Y)$, and therefore $\mathrm{Corr}(X,Y)$, to equal 0 even if (X,Y) is not independent. Nevertheless, $\mathrm{Corr}(X,Y)$ is often used as a rough measure of the dependence between two random variables. If Y is a constant multiple of X, then the correlation of X and Y is 1 or -1 according as the constant multiple is positive or negative. On the other hand, $|\,\mathrm{Corr}(X,Y)|$ may be less than 1 even if $Y = \varphi \circ X$ for some (nonrandom) function φ.

Problem 28. Provide an example of two dependent random variables X and Y whose correlation is 0.

* **Problem 29.** Let (X,Y) be an independent pair of $\mathbb{R}$-valued random variables. Show that $E(X+Y) = E(X) + E(Y)$ if *either* side of the equation makes sense. Give an example to show that this statement is not true if $\mathbb{R}$ is replaced by $\overline{\mathbb{R}}$.

Problem 30. Let (X,Y) be an independent pair of $\mathbb{R}$-valued random variables. Suppose that $E(X^4), E(Y^4) < \infty$. Show that

$$E([X+Y-E(X+Y)]^4) = E([X-E(X)]^4) + E([Y-E(Y)]^4) + 6\,\mathrm{Var}(X)\,\mathrm{Var}(Y).$$

9.5. Densities and independence

The following proposition can be especially useful in computations with random vectors having independent components in case the components have densities with respect to 1-dimensional Lebesgue measure or counting measure.

Proposition 13. *Let X_i, $1 \le i \le d$, denote random variables on a common probability space, and suppose that for each i, the distribution of X_i has density f_i with respect to a σ-finite measure μ_i. Then the random variables $X_1, \ldots, X_d$ are independent if and only if the random vector $(X_1, \ldots, X_d)$ has density*

$$(x_1, \ldots, x_d) \rightsquigarrow f_1(x_1) \cdot f_2(x_2) \cdot \ldots \cdot f_d(x_d)$$

with respect to $\mu_1 \times \cdots \times \mu_d$.

PARTIAL PROOF. We will only consider the case $d = 2$. For the 'only if' part we assume that X_1 and X_2 are independent and intend to show that

(9.7)

$$P(\{\omega \colon (X_1(\omega), X_2(\omega)) \in A\}) = \int_A f_1(x_1) f_2(x_2)\, (\mu_1 \times \mu_2)(d(x_1, x_2))$$

for every set $A \in \mathcal{G}_1 \times \mathcal{G}_2$.

We will prove (9.7) for $A = A_1 \times A_2$, with $A_1 \in \mathcal{G}_1$ and $A_2 \in \mathcal{G}_2$. Once we have done this, the rest is a straightforward application of the Sierpiński Class Theorem which is left to the reader. Since each f_i is a density, it is measurable and nonnegative. Thus the product $f_1 f_2$ is nonnegative and measurable with respect to $\mathcal{G}_1 \times \mathcal{G}_2$ (proof?). By the Fubini Theorem, the right side of (9.7) equals

$$\int_{A_1} \left(\int_{A_2} f_1(x_1) f_2(x_2)\, \mu_2(dx_2) \right) \mu_1(dx_1).$$

The quantity $f_1(x_1)$ does not depend on x_2 and may therefore be taken out of the inside integral. The inside integral then no longer depends on x_1, so it may be taken outside of the outer integral, leaving the product

$$\int_{A_2} f_2(x_2)\, \mu_2(dx_2) \int_{A_1} f_1(x_1)\, \mu_1(dx_1).$$

By the definition of density, this expression equals

$$P(\{\omega : X_2(\omega) \in A_2\})\, P(\{\omega : X_1(\omega) \in A_1\}),$$

which equals the left side of (9.7) by the definition of independence.

We leave the proof of the 'if' part to the reader. $\square$

Problem 31. Complete the preceding proof for the case $d = 2$ by doing the two things mentioned as being left for the reader.

Problem 32. Let (X_1, X_2) be an independent pair of exponentially distributed random variables, with parameters a_1, a_2, respectively. Use Proposition 13 to compute $E|X_1 - X_2|$. Then use this answer, the answer to Problem 6, and the fact that $|x_1 - x_2| = x_1 \vee x_2 - x_1 \wedge x_2$ to calculate $E(X_1 \vee X_2)$. Finally, check this answer via a direct calculation.

* **Problem 33.** Let (X_1, X_2) be an independent pair of random variables each having a beta distribution with parameter pair $(\frac{1}{2}, 1)$. Find $P(\{\omega : X_1(\omega) > [X_2(\omega)]^2\})$.

The following result illustrates the usefulness of the Fubini Theorem in probability theory, even in the absence of independence.

Proposition 14. *Let $X = (X_1, X_2)$ be a random vector that takes values in a product space*

$$(\Psi, \mathcal{G}, \mu) = (\Psi_1, \mathcal{G}_1, \mu_1) \times (\Psi_2, \mathcal{G}_2, \mu_2),$$

where μ_1 and μ_2 are σ-finite measures. Suppose that the distribution of X has density f with respect to μ. Then the distributions of X_1 and X_2 have densities

$$f_1(x_1) = \int_{\Psi_2} f(x_1, x_2)\, \mu_2(dx_2) \quad and \quad f_2(x_2) = \int_{\Psi_1} f(x_1, x_2)\, \mu_1(dx_1)$$

with respect to μ_1 and μ_2, respectively.

PROOF. Let A be a member of $\mathcal{G}_1$. Then

$$P(\{\omega\colon X_1(\omega) \in A\}) = P(\{\omega\colon Z(\omega) \in A \times \Psi_2\}) = \int_{A \times \Psi_2} f(x_1, x_2)\, \mu(d(x_1, x_2))\,.$$

By the Fubini Theorem, the integral equals

$$\int_A \left(\int_{\Psi_2} f(x_1, x_2)\, \mu_2(dx_2) \right) \mu_1(dx_1)\,.$$

Thus, the density of X_1 is as claimed. The proof for X_2 is similar. $\square$

The densities f_1 and f_2 are called *marginal densities*, although the term is misleading. A 'marginal density' is just a 'density'; the adjective 'marginal' reflects the fact that the density in question has arisen from integrating the density of a random vector. When densities under consideration are with respect to Lebesgue measure in $\mathbb{R}^d$, the following result is useful—for instance, for calculating the probability that a random vector with a known distribution belongs to some particular set.

Theorem 15. [Change of Variables] *Let φ be an $\mathbb{R}^d$-valued invertible continuously differentiable function defined on an open set $U \subseteq \mathbb{R}^d$. Let B be a Borel subset of U and let $A = \varphi^{-1}(B)$. If f is a measurable function from $\mathbb{R}^d$ to $\overline{\mathbb{R}}$, then*

$$\int_B f\, d\lambda^d = \int_A (f \circ \varphi)|J|\, d\lambda^d\,,$$

where J is the Jacobian determinant of the transformation φ and λ^d is d-dimensional Lebesgue measure. The two integrals are equal in the sense that if one exists, then both exist and are equal.

Problem 34. Prove the preceding proposition. *Hint:* Use a theorem from advanced calculus to show that the proposition is true for continuous functions f that are 0 off a bounded set. Then use the Monotone Convergence Theorem to show it is true for indicators of open rectangles. Extend to indicators of measurable rectangles by using the Sierpiński Class Theorem.

9.6. Product probability measure: infinitely many factors

Many theorems in probability begin with a statement such as "Let $(X_n\colon n = 1, 2, \dots)$ be an infinite sequence of independent identically distributed random variables having common distribution Q." Such a theorem would be vacuous for any Q for which there were no such sequence. More generally, one might want to drop the assumption of identical distributions while keeping the independence and specifying the distribution of each term in the sequence. The highpoint of this section, Corollary 17, says that such a specification is possible.

Let $(\Omega_n, \mathcal{F}_n)$, $n = 1, 2, \ldots$, be an infinite sequence of measurable spaces, and let

$$\Omega = \bigotimes_{n=1}^{\infty} \Omega_n.$$

A *measurable rectangle* in Ω is a set of the form

$$\bigotimes_{n=1}^{\infty} A_n,$$

where for each n, $A_n \in \mathcal{F}_n$. Let $\mathcal{R}$ be the collection of measurable rectangles in Ω, and let

$$\mathcal{F} = \sigma(\mathcal{R}).$$

The σ-field $\mathcal{F}$ is called the *infinite product* of $\mathcal{F}_1, \mathcal{F}_2, \ldots$ and is denoted by

$$\bigotimes_{n=1}^{\infty} \mathcal{F}_n.$$

Theorem 16. *Let $((\Omega_n, \mathcal{F}_n, P_n), n = 1, 2, \ldots)$ be a sequence of probability spaces. Let Ω be the infinite product of the spaces Ω_n, and let $\mathcal{F}$ be the infinite product of the σ-fields $\mathcal{F}_n$. Then there exists a unique probability measure P on the measurable space $(\Omega, \mathcal{F})$ such that*

$$(9.8) \qquad P\left(\bigotimes_{n=1}^{\infty} A_n\right) = \prod_{n=1}^{\infty} P_n(A_n)$$

for events $A_n \in \mathcal{F}_n$, $n = 1, 2, \ldots$.

PROOF. For $n = 1, 2, \ldots$ let $\mathcal{H}_n = \sigma(\mathcal{R}_n)$, where $\mathcal{R}_n$ is the collection of all measurable rectangles of the form

$$\left(\bigotimes_{k=1}^{n} A_k\right) \times \left(\bigotimes_{k=n+1}^{\infty} \Omega_k\right).$$

Thus, any member of $\mathcal{H}_n$ can be written in the form

$$(9.9) \qquad B \times \Omega_{n+1} \times \Omega_{n+2} \times \cdots$$

for some unique set

$$B \in \mathcal{G}_n = \bigotimes_{k=1}^{n} \mathcal{F}_k.$$

Let

$$\mathcal{E} = \bigcup_{n=1}^{\infty} \mathcal{H}_n.$$

Note that $\mathcal{E}$ is a field, and that $\mathcal{F} = \sigma(\mathcal{E})$. (The members of $\mathcal{E}$ are called *cylinder sets*, and $\mathcal{E}$ itself is the *field of cylinder sets*.)

We define P on $\mathcal{E}$ by defining P on each $\mathcal{H}_n$. For $A \in \mathcal{H}_n$, written in the form (9.9), define

$$P(A) = (P_1 \times \cdots \times P_n)\,(B)\,.$$

A set A which is a member of $\mathcal{H}_n$ is also a member of $\mathcal{H}_{n+1}$ with $B \times \Omega_{n+1}$ taking the place of B. Thus we need to check that the definition of P is consistent. We apply the associative law of Problem 20:

$$
\begin{aligned}
(P_1 \times \cdots \times P_{n+1})(B \times \Omega_{n+1}) &= ((P_1 \times \cdots \times P_n) \times P_{n+1})(B \times \Omega_{n+1}) \\
&= (P_1 \times \cdots \times P_n)(B) P_{n+1}(\Omega_{n+1}) \\
&= (P_1 \times \cdots \times P_n)(B)\,.
\end{aligned}
$$

Thus, P has been consistently defined on the field $\mathcal{E}$ of cylinder sets.

It is clear from the definition of P that (9.8) is satisfied for measurable rectangles in each of the collections $\mathcal{R}_n$. We have not yet defined P on all of $\mathcal{R}$. However, any member of $\mathcal{R}$ can be written as the limit of members of $\cup_n \mathcal{R}_n$, since for any sequence $(A_1, A_2, \dots)$ such that $A_n \in \mathcal{F}_n, n = 1, 2, \dots$,

$$
\bigotimes_{n=1}^{\infty} A_n = \lim_{m \to \infty} \left[\left(\bigotimes_{n=1}^{m} A_n \right) \times \left(\bigotimes_{n=m+1}^{\infty} \Omega_n \right) \right].
$$

Thus, the Continuity of Measure Theorem implies that any extension of P to a probability measure on $\mathcal{F}$ will satisfy (9.8) on all of $\mathcal{R}$.

We turn our attention to the task of extending P to a probability measure on $\mathcal{F}$. Since $\mathcal{E}$ is a field which generates $\mathcal{F}$, it suffices, by the Extension Theorem and the companion Proposition 9 of Chapter 7, to show that the limit of a decreasing sequence $(C_1, C_2, \dots)$ of members of $\mathcal{E}$ is nonempty if $\lim P(C_n) \neq 0$. So, suppose that the limit equals $\varepsilon > 0$. By inserting initial terms and repetitions into the sequence if necessary, we may assume that $C_n \in \mathcal{H}_n$ for each n. We will complete the proof by identifying a point ψ in $\lim C_n$.

Let $Y_{n,n}(\omega_1, \dots, \omega_n) = I_{C_n}(\omega_1, \omega_2, \dots)$; and for $0 \le m < n$, let

$$
\begin{aligned}
Y_{m,n}(\omega_1, \dots, \omega_m) = \\
\int_{\Omega_{m+1} \times \cdots \times \Omega_n} Y_{n,n}(\omega_1, \dots, \omega_n)\,(P_{m+1} \times \cdots \times P_n)(d(\omega_{m+1}, \dots, \omega_n))\,,
\end{aligned}
$$

so that, in particular, $Y_{0,n} = P(C_n)$. By the Fubini Theorem, each $Y_{m,n}$ is $\mathcal{G}_m$-measurable, and

$$(9.10) \quad Y_{m-1,n}(\omega_1, \dots, \omega_{m-1}) = \int_{\Omega_m} Y_{m,n}(\omega_1, \dots, \omega_{m-1}, \omega_m)\,P_m(d\omega_m)$$

for $0 < m \le n$.

For fixed m, the sequence $(Y_{m,n}, n = m, m+1, m+2, \dots)$ decreases to a nonnegative $\mathcal{G}_m$-measurable random variable Y_m, since the sequence $(C_1, C_2, \dots)$

is decreasing. The random variables $Y_{m,n}$ are all bounded by 1, so it follows from the Bounded Convergence Theorem and (9.10) that

$$Y_{m-1}(\omega_1, \ldots, \omega_{m-1}) = \int_{\Omega_m} Y_m(\omega_1, \ldots, \omega_{m-1}, \omega_m) \, P_m(d\omega_m).$$

By Theorem 9 (v) of Chapter 4, for $m \geq 1$ and each $(\omega_1, \ldots, \omega_{m-1})$, there exists an ω_m such that

$$Y_m(\omega_1, \omega_2, \ldots, \omega_{m-1}, \omega_m) \geq Y_{m-1}(\omega_1, \omega_2, \ldots, \omega_{m-1}).$$

Thus, we may find a sequence $\psi_1, \psi_2, \ldots$, such that for all $m \geq 1$,

$$Y_m(\psi_1, \psi_2, \ldots, \psi_{m-1}, \psi_m) \geq Y_{m-1}(\psi_1, \psi_2, \ldots, \psi_{m-1}).$$

Let $\psi = (\psi_1, \psi_2, \ldots)$.

We now have that

$$
\begin{aligned}
I_{C_n}(\psi) = Y_{n,n}(\psi_1, \ldots, \psi_n) \\
\geq Y_n(\psi_1, \ldots, \psi_n) \\
\geq Y_0 = \lim_{n \to \infty} P(C_n) \geq \varepsilon > 0
\end{aligned}
$$

for all $n > 0$. Since indicator functions can only take the values 0 and 1, it follows that $\psi \in C_n$ for each n. Thus, $\lim C_n$ is nonempty as desired. □

The probability measure P constructed in Theorem 16 is called the *infinite product* of $P_1, P_2, \ldots$ and is denoted by

$$\bigotimes_{n=1}^{\infty} P_n.$$

The probability space $(\Omega, \mathcal{F}, P)$ is the *infinite product* of the probability spaces $(\Omega_n, \mathcal{F}_n, P_n)$. The construction of the product of an infinite sequence of probability measures does not generalize to infinite measures, or even to finite measures which are not probability measures.

Corollary 17. *Let $(\Omega_n, \mathcal{F}_n, P_n)$, $n = 1, 2, \ldots$, be as in Theorem 16. There exists an independent sequence $(X_1, X_2, \ldots)$ such that for each n, X_n has distribution P_n.*

PROOF. Let $(\Omega, \mathcal{F}, P)$ be the infinite product of the spaces $(\Omega_n, \mathcal{F}_n, P_n)$, $n = 1, 2, \ldots$. For $n = 1, 2, \ldots$, and $\omega = (\omega_1, \omega_2, \ldots) \in \Omega$, define

$$X_n(\omega) = \omega_n.$$

A little thought convinces one that the sequence $(X_1, X_2, \ldots)$ has the desired properties. □

The random sequence $(X_1, X_2, \ldots)$ described in the preceding proof is merely the identity function on an appropriate probability space, and so the target and the domain of this random sequence are the same. It should be emphasized that there are many natural situations in which the target and the domain are very different and yet the sequence is independent.

Problem 35. Where in the proof of Theorem 16 did we make use of the hypothesis that the measures P_n are probability measures? What happens when you try to construct $(J, \mathcal{B}, \lambda)^\infty$ where J is an interval in $\mathbb{R}$ and λ is 1-dimensional Lebesgue measure?

The following problem connects intuitive ideas about independence with the definitions we have given.

Problem 36. Let (X, Y) be an independent pair of random variables, and let

$$\big((X_1, Y_1), (X_2, Y_2), \ldots\big)$$

be an independent sequence of random vectors, each of which has the same distribution as (X, Y). (Such a sequence exists by Theorem 16.) Let A and B be Borel subsets of $\mathbb{R}$ and assume that $P(\{\omega : X(\omega) \in A\}) > 0$. Define

$$S_n(\omega) = \sum_{i=1}^n I_{A \times B}(X_i(\omega), Y_i(\omega)) \quad \text{and} \quad M_n(\omega) = \sum_{i=1}^n I_A(X_i(\omega)).$$

Prove that for all $\varepsilon > 0$,

$$(9.11) \qquad \lim_{n \to \infty} P\big(\big\{\omega : \big|\frac{S_n(\omega)}{M_n(\omega)} - P(\{\psi : Y(\psi) \in B\})\big| > \varepsilon\big\}\big) = 0.$$

Think of the sequence $(X_1, Y_1), (X_2, Y_2), \ldots$ as representing a sequence of pairs of measurements which are done on the outcomes of independent repetitions of an experiment, and give an interpretation of (9.11) in terms of your intuitive ideas about the meaning of independence.

From (9.8) we see that for any $\varepsilon > 0$, there exists m such that

$$P\left(\left[\bigotimes_{n=1}^\infty A_n\right] \triangle \left[\bigotimes_{n=1}^m A_n \times \bigotimes_{n=m+1}^\infty \Omega_n\right]\right) < \varepsilon.$$

Thus all measurable rectangles C in the product σ-field can be approximated by cylinder sets D in the sense that $P(C \triangle D)$ can be made arbitrarily small by appropriate choices of D. More is said in the following lemma.

Lemma 18. *Let*

$$(\Omega, \mathcal{F}, P) = \bigotimes_{n=1}^\infty (\Omega_n, \mathcal{F}_n, P_n),$$

where each $(\Omega_n, \mathcal{F}_n, P_n)$ is a probability space. Then for any $C \in \mathcal{F}$ and any $\varepsilon > 0$, there exists a cylinder set D for which $P(C \triangle D) < \varepsilon$.

Problem 37. Prove the preceding lemma. *Hint:* Use the Sierpiński Class Theorem.

9.7. The Borel-Cantelli Lemma and independent sequences

As we have already seen, an independent sequence of events is pairwise uncorrelated, so the Borel-Cantelli Lemma applies.

Problem 38. State the Borel-Cantelli Lemma for independent sequences of events. Provide a new proof based on the formula in Problem 4.

The next example illustrates an important way in which the Borel-Cantelli Lemma can be used to analyze the asymptotic behavior of independent sequences of random variables.

Example 5. Let $(X_1, X_2, \dots)$ be an independent sequence of random variables having the standard exponential distribution with mean λ. We will describe rather precisely the large values of X_n as $n \to \infty$ by proving that, with probability 1,

$$\limsup_{n \to \infty} \frac{X_n}{\lambda \log n} = 1.$$

The two steps are to first show that the limit supremum is greater than or equal to 1 a.s., and then to prove that the limit supremum is a.s. less than or equal to $1 + \delta$ for all $\delta > 0$.

For each n, let $A_n = \{\omega: X_n(\omega) \geq \lambda \log n\}$. In order to apply the Borel-Cantelli Lemma, we calculate

$$\sum P(A_n) = \sum \exp(-\log n) = \sum (1/n) = \infty.$$

Therefore, $P(\limsup A_n) = 1$ and thus

$$P\left(\{\omega: \limsup \frac{X_n(\omega)}{\lambda \log n} \geq 1\}\right) = 1.$$

Now fix $\delta > 0$. For each n, let $B_n = \{\omega: X_n(\omega) > (1 + \delta)\lambda \log n\}$. Now we obtain a convergent series:

$$\sum P(B_n) = \sum \exp(-(1 + \delta) \log n) = \sum (1/n^{1+\delta}) < \infty.$$

By the Borel Lemma, $P(\limsup B_n) = 0$ and thus

$$P\left(\{\omega: \limsup \frac{X_n(\omega)}{\lambda \log n} \leq 1 + \delta\}\right) = 1.$$

Now $\limsup X_n/(\lambda \log n) \leq 1$ a.s. follows by letting $\delta \to 0$ along a countable sequence.

Problem 39. Let X_n be as in the preceding example, and let

$$Y_n = \max\{X_1, X_2, \ldots, X_n\}$$

for each n. Prove without using probability theory that

$$\limsup_{n\to\infty} \frac{Y_n}{\lambda \log n} = \limsup_{n\to\infty} \frac{X_n}{\lambda \log n}.$$

Example 6. This is a continuation of the preceding example. We will prove that

$$\lim_{n\to\infty} \frac{Y_n}{\lambda \log n} = 1 \text{ a.s.}$$

By the preceding problem, it is enough to prove that for each $\delta > 0$,

$$\liminf \frac{Y_n}{\lambda \log n} > 1 - \delta \text{ a.s.}$$

Let $C_n = \{\omega : Y_n(\omega) \leq (1-\delta)\lambda \log n\}$. Then

$$\sum P(C_n) = \sum (1 - \exp(-(1-\delta)\log n))^n$$
$$= \sum (1 - n^{-(1-\delta)})^n$$
$$\leq \sum \left(\exp(-n^{-(1-\delta)})\right)^n = \sum e^{-n^\delta} < \infty.$$

It follows from the Borel Lemma that $P(\limsup C_n) = 0$, and the desired result follows as in the previous example by letting δ go to zero along a countable sequence.

Problem 40. For the sequence $(X_1, X_2, \ldots)$ of Example 5, prove that for each decreasing sequence $(b_1, b_2, \ldots)$ of positive numbers, either

$$\liminf_{n\to\infty} \frac{X_n}{b_n} = \infty \text{ a.s.} \quad \text{or} \quad \liminf_{n\to\infty} \frac{X_n}{b_n} = 0 \text{ a.s.}$$

* **Problem 41.** Let X be a normally distributed random variable having mean 0 and standard deviation σ. Prove that, as $x \to \infty$,

$$(9.12) \qquad P(\{\omega : X(\omega) > x\}) \sim \frac{\sigma}{x\sqrt{2\pi}} \exp(-x^2/2\sigma^2).$$

* **Problem 42.** Let $(X_1, X_2, \ldots)$ be an independent sequence of normally distributed random variables with mean 0 and standard deviation σ. Find an increasing sequence $(a_1, a_2, \ldots)$ such that $\limsup X_n/a_n = 1$ a.s. Let

$$Y_n = \max_{1 \leq k \leq n} X_k.$$

Is it true that $\lim Y_n/a_n = 1$ a.s.? *Hint:* Use the preceding exercise.

Problem 43. For $x \to \infty$, find a simple asymptotic formula for $P(\{\omega: X(\omega) > x\})$, where X has a gamma distribution.

Problem 44. Let $(X_1, X_2, \dots)$ be an independent sequence of identically distributed gamma random variables. Construct an increasing sequence $(a_1, a_2, \dots)$ such that $\limsup X_n/a_n = 1$ a.s. Let $Y_n = \max\{X_1, \dots, X_n\}$. Is it true that $\lim Y_n/a_n = 1$ a.s.? *Hint:* Use the preceding exercise.

* **Problem 45.** Let $(X_1, X_2, \dots)$ be an independent sequence of random variables, where for each n, X_n is uniformly distributed on $[0, n]$. Calculate $P(\{\omega: X_n(\omega) \to \infty \text{ as } n \to \infty\})$.

Problem 46. Let $(X_1, X_2, \dots)$ be an independent sequence of $\mathbb{R}^2$-valued random variables, where for each n, X_n is uniformly distributed on $[-n, n] \times [-n, n]$. Calculate $P(\{\omega: \|X_n(\omega)\| \to \infty \text{ as } n \to \infty\})$, where $\|\cdot\|$ denotes the usual Euclidean norm in $\mathbb{R}^2$.

9.8. † Order statistics

Suppose that an independent sequence of five random numbers—say, rather untypically, $1/3, 1/\sqrt{2}, 4/5, 1/\pi, 3/7$—are drawn according to the uniform distribution on $(0, 1)$. They can be arranged in increasing order: $1/\pi, 1/3, 3/7, 1/\sqrt{2}, 4/5$. When so arranged these are, according to a definition to be given shortly, called the first through fifth 'order statistics'. Of course, if someone gives you the order statistics only, you have not obtained the full information about the result of the experiment; because the same order statistics typically come from $5! = 120$ different members of $\mathbb{R}^5$. However, for many purposes the information contained in the order statistics is the important information. The following exercise will help one develop facility with some of the technicalities involved in the definition of order statistics.

* **Problem 47.** For $x \in \mathbb{R}^d$, use the notation x_i for the ith coordinate of x. Let $\chi^{(d)}$ be the function from $\mathbb{R}^d$ to $\mathbb{R}^d$ defined by

$$(9.13) \qquad [\chi^{(d)}(x)]_j = \min\{v \in \mathbb{R}: \#\{i: x_i \le v\} \ge j\}.$$

Discuss the appearance of minimum rather than just infimum in this definition. Say why

$$(9.14) \qquad [\chi^{(d)}(x)]_i \le [\chi^{(d)}(x)]_j \quad \text{for } i \le j$$

and

$$(9.15) \qquad \#\{j: [\chi^{(d)}(x)]_j = v\} = \#\{i: x_i = v\}$$

for each $x \in \mathbb{R}^d$ and $v \in \mathbb{R}$. Describe (and, in case $d = 2$, draw) the image of $\chi^{(d)}$. Prove that $\chi^{(d)}$ is continuous, hence measurable. Calculate the cardinality of $(\chi^{(d)})^{-1}(\{y\})$ for all $y \in \mathbb{R}^d$.

Definition 19. Let $X = (X_1, \ldots, X_d)$ be a random vector consisting of d $\mathbb{R}$-valued components, and let $Y = \chi^{(d)}(X)$. Then for $j = 1, \ldots, d$, Y_j is called the jth *order statistic* of X. If the components of X form an independent sequence and are identically distributed according to a distribution Q, then we call Y_j the jth order statistic *based on d observations of Q*.

The following proposition shows that order statistics based on observations of the uniform distribution can be used to analyze order statistics based on observations of arbitrary distributions on $\mathbb{R}$.

Proposition 20. *Let G_j be the distribution function of the jth order statistic based on d observations of the uniform distribution on $[0, 1]$. Let Q denote a distribution on $\mathbb{R}$ with corresponding distribution function F. Then $G_j \circ F$ is the distribution function of the jth order statistic based on d observations of Q.*

Problem 48. Prove the preceding proposition. Generalize to obtain a result for the distribution of the random vector whose components are the order statistics based on d observations of Q.

* **Problem 49.** Let $Y_1 \leq Y_2 \leq \cdots \leq Y_d$ be order statistics based on d observations of the uniform distribution on $[0, 1]$. Calculate the density of the vector $(Y_1, \ldots, Y_d)$. (One assumes the problem is requesting the density with respect to d-dimensional Lebesgue measure, since there is no indication of some other measure.)

Problem 50. Let $Y_1 \leq Y_2 \leq \cdots \leq Y_d$ be order statistics based on d observations of a distribution Q on $\mathbb{R}$. Assume that Q has a continuous distribution function. Prove that the distribution of the vector $Y = (Y_1, \ldots, Y_d)$ on $\mathbb{R}^d$ is the restriction of

$$d! \bigotimes_{i=1}^{d} Q$$

to the set $\{y \in \mathbb{R}^d : y_1 < y_2 < \cdots < y_d\}$. Conclude that the distribution of Y has a density with respect to d-dimensional Lebesgue measure if Q has a density with respect to 1-dimensional Lebesgue measure.

* **Problem 51.** Let $(X_1, X_2, \ldots)$ be an independent sequence of random variables, each of which is uniformly distributed on the interval $[0, 1]$. Let

$$N = \sup\{k \geq 1 : X_k \leq X_{k-1} \leq \cdots \leq X_1\}.$$

Find the distribution and mean of N. Let $Z = X_N$. Find the density of Z.

The closed interval with endpoints equal to the first and the nth order statistics of a sequence of n $\mathbb{R}$-valued random variables is the smallest closed interval that contains all n values. We will extend this idea to $\mathbb{R}^d$-valued random variables.

A set $C \subseteq \mathbb{R}^d$ is *convex* if the line segment containing each pair of points in A is a subset of A. The intersection of any collection of closed convex sets is easily seen to be closed and convex. Therefore, there is for each $B \subseteq \mathbb{R}^d$, a smallest closed convex set containing B. It is called the *convex hull* of B and is denoted by conv(B). For $(X_1, X_2, \ldots)$, an independent sequence of identically distributed random vectors in $\mathbb{R}^d$, one studies the sequence conv$(\{X_1\})$, conv$(\{X_1, X_2\})$, conv$(\{X_1, X_2, X_3\})$, ... of random sets in $\mathbb{R}^d$. In $\mathbb{R}$ this amounts to studying the sequence of intervals with endpoints equal to the largest and smallest order statistics.

* **Problem 52.** Let Ψ be the union of the x- and y-axes in $\mathbb{R}^2$, and let $\mathcal{G}$ be the σ-field of sets of the form $A \cup B$, where A and B are Borel subsets of the x- and y-axes, respectively. For such sets $A \cup B$, define $\mu(A \cup B) = \lambda(A) + \lambda(B)$, where λ denotes 1-dimensional Lebesgue measure. Define a probability measure Q by $\frac{dQ}{d\mu}(u, v) = 1/2$ if either $v = 0$ and $0 < u < 1$ or $u = 0$ and $0 < v < 1$. Let (X_1, X_2, X_3) be independent random variables, each having the distribution Q. Calculate the mean area of conv$(\{X_1, X_2, X_3\})$. *Hint:* It is not necessary to calculate the distribution function of the area.

9.9. † Some new distributions involving independence

The *Riemann zeta function* is relevant to the next exercise: for $z > 1$, let

$$\zeta(z) = \sum_{n=1}^{\infty} \frac{1}{n^z}.$$

This definition is meaningful for complex z whose real part is greater than 1, but we will be primarily interested in real $z > 1$.

* **Problem 53.** [Zeta Distributions] Let

$$\Omega = \{1, 2, 3, \ldots\}$$

and let $\mathcal{F}$ be the σ-field consisting of all subsets of Ω. For each real $z > 1$, P_z, defined by

$$P_z(\{\omega\}) = \frac{1}{\zeta(z)\omega^z}, \quad \omega \in \Omega,$$

is a probability measure on the measurable space $(\Omega, \mathcal{F})$. We can also think of P_z as being the distribution of the random variable X defined by $X(\omega) = \omega$. Evaluate its mean and variance in terms of the Riemann zeta function. For each m calculate the probability of the event

$$\{\omega : \omega/m \text{ is an integer}\}.$$

Calculate the limit of this probability as $z \searrow 1$.

Problem 54. This exercise is a continuation of the previous exercise. For each positive integer ω, write

$$\omega = \prod_{p \text{ prime}} p^{X_p(\omega)},$$

where $X_p(\omega) \in \{0, 1, 2, \dots\}$ denotes the power of the prime p in the prime factorization of ω. For the probability space $(\Omega, \mathcal{F}, P_z)$, prove that the sequence $(X_p : p \text{ prime})$ is independent. Also calculate and name the distribution of each X_p. Notice that an independent sequence of random variables has been defined in a natural manner on a probability space that is not a product space. Discuss the pros and cons of constructing, on a computer say, a random natural number according to a zeta distribution by constructing the powers of the primes in its prime factorization.

Problem 55. Let $X = (X_1, X_2)$ be an independent pair of normally distributed random variables having mean 0 and standard deviation σ. Calculate the distributions of the Euclidean norm $\|X\|$ and the polar coordinate angle $\arg X$. Show that this pair of random variables is independent.

Problem 56. Let X be an $\mathbb{R}$-valued random variable with a distribution that has a density and is symmetric about 0. Suppose that the distribution of X^2 is gamma with parameters γ, a. Find the density of X. In particular, find the value of γ for which X is normally distributed.

* **Problem 57.** Let $U = (X, Y)$ be an independent pair of identically symmetrically distributed random variables, each of whose squares has the same gamma distribution. Calculate the distributions of $\|U\|$ and $\arg U$. Show that this pair of random variables is independent.

CHAPTER 10

Sums of Independent Random Variables

In Example 3 of Chapter 1 we calculated the probability that exactly k heads appear in n flips of a fair coin. In view of the construction of that example and the definition of independence given in the preceding chapter, we see that what we calculated is the distribution of the sum of n independent random variables, each of which has the Bernoulli distribution with parameter $p = 1/2$. Sums of independent random variables constitute a major theme in probability theory.

From a theoretical point of view, this chapter contains essentially only two main results: some formulas for the distribution and distribution function of the sum of two independent $\mathbb{R}^d$-valued or $\overline{\mathbb{R}}^+$-valued random variables (Proposition 3 and Proposition 4), and a formula for the probability generating function of the sum of two independent $\overline{\mathbb{Z}}^+$-valued random variables (Theorem 5). While the proofs of these results are relatively simple, some practice is required in their application. The rest of the chapter is devoted to examples intended to provide such practice. These examples include some new families of distributions (multinomial, negative binomial, Dirichlet). We also introduce a new concept ('infinite divisibility') that will be central to much of the theory in Part 3. At the end of the chapter, we explore sums of independent random variables in settings other than $\mathbb{R}^d$ and $\overline{\mathbb{R}}^+$.

10.1. Convolutions of distributions

Let (X, Y) be an independent pair of $\mathbb{R}^d$- or $\overline{\mathbb{R}}^+$-valued random variables with distributions Q and R, respectively. Then the distribution of the random variable (X, Y) is $Q \times R$, and therefore the distribution of the sum $X + Y$ is given by

$$P(\{\omega \colon X(\omega) + Y(\omega) \in B\}) = (Q \times R)(\{(x, y) \colon x + y \in B\}).$$

Let us introduce symbolism to describe the distribution of $X + Y$ in terms of the distributions of X and Y.

Definition 1. Let Q and R be two distributions on $\mathbb{R}^d$ or $\overline{\mathbb{R}}^+$. The *convolution* of Q and R, written $Q * R$, is the distribution on $(\mathbb{R}, \mathcal{B})$ defined by

$$(Q * R)(B) = (Q \times R)(\{(x,y): x + y \in B\}).$$

The shaded region in Figure 10.1 indicates the region C in $\mathbb{R}^2$ corresponding to the event $\{\omega: X(\omega) + Y(\omega) \in [a,b]\}$. That is, $(Q * R)([a,b]) = (Q \times R)(C)$. It should be emphasized that $Q * R$ is defined whether or not (X, Y) is an independent pair but the conclusion that it is the distribution of $X + Y$ is based upon the assumption of independence.

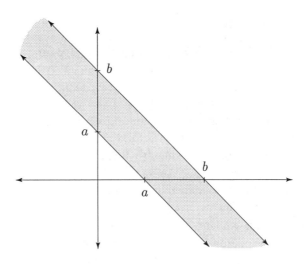

FIGURE 10.1. The region in $\mathbb{R}^2$ corresponding to $\{\omega: X(\omega) + Y(\omega) \in [a,b]\}$

Proposition 2. *Convolution is commutative and associative. The delta distribution at 0 is the unique identity for convolution.*

PROOF. The commutativity and associativity of convolution follow immediately from the corresponding properties for addition of random variables. The random variable which is identically 0 is an identity for the addition of random variables, so its distribution δ_0 is an identity for convolution. If R_0 is any identity for convolution, then

$$\delta_0 = R_0 * \delta_0 = R_0,$$

so the identity is unique. $\square$

Problem 1. Why do you think the uniqueness of the additive identity for $\mathbb{R}$ was not used in the preceding proof to show uniqueness of the identity for convolution? Why is there no mention of an inverse for convolution?

In view of Proposition 2 we can speak of the convolution of a finite number of distributions without regard to order or grouping. The convolution of n copies of Q is denoted by Q^{*n}, where, of course, Q^{*0} denotes the identity for convolution—namely, the delta distribution at 0. The convolution terminology and notation is also used for distribution functions. Thus, if F_1 and F_2 are the distribution functions corresponding to Q_1 and Q_2 respectively, we write $F_1 * F_2$ for the distribution function corresponding to $Q_1 * Q_2$. And the notation F^{*n} is also used where appropriate.

If Q and P are distributions such that $P = Q^{*n}$ for some positive integer n, then Q is called an nth *convolution root* of P, and we write

$$Q = P^{*\frac{1}{n}} .$$

It will be seen in Part 3 that an important class of distributions consists of those that have an nth convolution root for every positive integer n. Such distributions are called *infinitely divisible*.

Example 1. Let Q_1, Q_2 be Poisson distributions with means λ_1, λ_2. We wish to calculate $Q_1 * Q_2$. Since random variables with a Poisson distribution are nonnegative integer-valued, so are their sums. Thus, it is enough to compute $(Q_1 * Q_2)(\{x\})$ for nonnegative integers x.

$$(Q_1 * Q_2)(\{x\}) = (Q_1 \times Q_2)(\{(x_1, x_2) : x_1 + x_2 = x\})$$

$$= \sum_{x_1=0}^{x} Q_2(\{x - x_1\}) Q_1(\{x_1\})$$

$$= \sum_{x_1=0}^{x} \frac{\lambda_2^{x-x_1} e^{-\lambda_2}}{(x - x_1)!} \frac{\lambda_1^{x_1} e^{-\lambda_1}}{x_1!} .$$

Combine the exponentials, multiply and divide by $x!$, and apply the Binomial Theorem in this last expression to obtain

$$(Q_1 * Q_2)(\{x\}) = \frac{(\lambda_1 + \lambda_2)^x e^{-(\lambda_1 + \lambda_2)}}{x!} ,$$

thus identifying $Q_1 * Q_2$ as Poisson with mean $\lambda_1 + \lambda_2$. A simple inductive argument implies that the convolution of n Poisson distributions with respective means $\lambda_1, \ldots, \lambda_n$ is Poisson with mean $\lambda_1 + \cdots + \lambda_n$. It follows that for any $\lambda \geq 0$, the Poisson distribution with mean λ/n is an nth convolution root of the Poisson distribution with mean λ, so every Poisson distribution is infinitely divisible. We will see in Section 3 that each Poisson distribution has only one n^{th} convolution root.

Problem 2. Calculate and identify by name Q^{*n} where Q is Bernoulli with parameter p.

The notation $B - x$ for the set $\{y - x : y \in B\}$ will be used in the following consequence of the Fubini Theorem, a consequence that does not apply to $\overline{\mathbb{R}}^{+}$ where subtraction is not universally defined. The roles of Q_1 and Q_2 in the statement may be reversed because, according to Proposition 2, convolution is a commutative operation.

Proposition 3. *Let Q_1, Q_2 be distributions on $(\mathbb{R}^d, \mathcal{B})$. Then, for $B \in \mathcal{B}$,*

$$(10.1) \quad (Q_1 * Q_2)(B) = \int Q_2(B - y)\, Q_1(dy) = \int Q_1(B - y)\, Q_2(dy)\,.$$

Suppose further that $dQ_2/d\lambda^d = f_2$, where λ^d is Lebesgue measure on $\mathbb{R}^d$. Then $d(Q_1 * Q_2)/d\lambda^d$ exists and

$$(10.2) \qquad \frac{d(Q_1 * Q_2)}{d\lambda^d}(x) = \int f_2(x - y)\, Q_1(dy)$$

for λ^d-almost every x. If also $dQ_1/d\lambda^d = f_1$, then

$$(10.3) \qquad \frac{d(Q_1 * Q_2)}{d\lambda^d}(x) = \int f_2(x - y) f_1(y)\, \lambda^d(dy)$$

for λ^d-almost every x.

PROOF. Let $A = \{(x_1, x_2) : x_1 + x_2 \in B\}$. Then A is measurable since the function $(x_1, x_2) \rightsquigarrow x_1 + x_2$ is continuous. By the definition of convolution and the Fubini Theorem,

$$
\begin{aligned}
(Q_1 * Q_2)(B) &= \int I_A \, d(Q_1 \times Q_2) \\
&= \int \left(\int I_A(x_1, x_2)\, Q_2(dx_2) \right) Q_1(dx_1) \\
&= \int Q_2(B - x_1)\, Q_1(dx_1)\,,
\end{aligned}
$$

giving the first equality in (10.1).

Suppose $dQ_2/d\lambda = f_2$ and set

$$f(x) = \int f_2(x - y) Q_1(dy)\,.$$

The function $(x, y) \rightsquigarrow f_2(x - y)$, being the composition of the measurable functions $(x, y) \rightsquigarrow x - y$ and f_2, is measurable. So by the Fubini and Change of

Variables Theorems,

$$\int_B f \, d\lambda^d = \int_B \left(\int f_2(x - y) \, Q_1(dy) \right) \lambda^d(dx)$$

$$= \int \left(\int_B f_2(x - y) \, \lambda^d(dx) \right) Q_1(dy)$$

$$= \int \left(\int_{B-y} f_2(u) \, \lambda^d(du) \right) Q_1(dy)$$

$$= \int Q_2(B - y) \, Q_1(dy) = (Q_1 * Q_2)(B)$$

for every $B \in \mathcal{B}$. Hence, $f = d(Q_1 * Q_2)/d\lambda^d$.

In case the other density $dQ_1/d\lambda^d$ also exists, (10.3) follows from (10.2) in combination with Proposition 19 of Chapter 8. $\square$

Problem 3. Where does the proof of Proposition 3 break down if λ^d is replaced by an arbitrary Radon measure on $\mathbb{R}^d$?

In case, Q_1 and Q_2 have densities with respect to Lebesgue measure and thus (10.3) holds, the notation $f_1 * f_2$ is often used for $d(Q_1 * Q_2)/d\lambda^d$, although, in case $d = 1$, this notation is somewhat at variance with the notation $F_1 * F_2$ for the distribution function corresponding to $Q_1 * Q_2$, where F_1 and F_2 are the distribution functions of Q_1 and Q_2. But since a distribution function can never be a probability density function, there is no real danger of confusion.

In the 1-dimensional case (10.1) can be written in terms of F_1 and F_2 in a manner that is valid even in the $\overline{\mathbb{R}}^+$ setting.

Proposition 4. Let F_1, F_2 be distribution functions for $\mathbb{R}$ or for $\overline{\mathbb{R}}^+$. Then, for $x \in \mathbb{R}$,

$$(F_1 * F_2)(x) = \int_{-\infty}^{\infty} F_2(x - y) \, dF_1(y) = \int_{-\infty}^{\infty} F_1(x - y) \, dF_2(y) \, .$$

Problem 4. Prove the preceding proposition.

* **Problem 5.** Find the density of the sum of two independent normally distributed random variables having means μ_1 and μ_2 and standard deviations σ_1 and σ_2 respectively.

Problem 6. Use the preceding problem to show that every normal distribution is infinitely divisible.

* **Problem 7.** Find the continuous density of the sum of two independent random variables each of which is uniformly distributed on the interval $(0, 1)$.

Problem 8. Show that the convolution of two not necessarily identical densities of Cauchy type is of Cauchy type.

Problem 9. Suppose that (X_1, X_2) is an independent pair of random variables distributed according to gamma distributions with parameters (a, γ_1) and (a, γ_2). Show that $X_1 + X_2$ is gamma distributed with parameter $(a, \gamma_1 + \gamma_2)$.

Problem 10. Let $F(x) = 1 - e^{-x}$ for $x \geq 0$ and $F(x) = 0$ otherwise. Represent $F^{*n}(x)$ for n a positive integer and $x \geq 0$ as a single definite integral on an appropriate interval. Relate your answer to the preceding exercise.

Remark 1. Convolutions can be defined more generally than has been done in this section. Except for the places where λ^d is involved, the space $\mathbb{R}^d$ can be replaced in this section by any space on which an operation analogous to sums of vectors can be defined, an operation that is commutative and associative and has the properties that an identity exists and every element has an inverse. Such a structure is a *commutative group*. To rigorously speak of sums of random variables having values in a commutative group, a measurable structure consistent with the group operations should be imposed on the group. For the next two problems the usual Borel structure will suffice even though the operation is not standard addition, but rather an operation appropriate for viewing $(-\pi, \pi]$ as representing the set of all rotations in $\mathbb{R}^2$ about the origin.

* **Problem 11.** Consider the binary operation $\oplus$ on $(-\pi, \pi]$ defined by

$$a \oplus b = a + b - 2\pi \lceil (a + b - \pi)/(2\pi) \rceil.$$

In other words, $a \oplus b$ is obtained by adding a and b and then adding or subtracting the appropriate multiple of 2π so that the result lies in $(-\pi, \pi]$. Corresponding to this operation calculate $Q * R$, where Q assigns probability $\frac{1}{4}$ to each member of $\{-\frac{\pi}{2}, 0, \frac{\pi}{2}, \pi\}$ and R assigns probability $\frac{1}{3}$ to each member of $\{-\frac{2\pi}{3}, 0, \frac{2\pi}{3}\}$.

Problem 12. For the structure of the preceding problem calculate $Q * Q$, where $Q = \frac{1}{3}\delta_0 + \frac{1}{3}\delta_\pi + \frac{1}{6\pi}\lambda$ with δ_a denoting the delta distribution at a and λ denoting Lebesgue measure on $(-\pi, \pi]$. More generally, calculate Q^{*n} for $n \in \mathbb{Z}^+$.

10.2. Multinomial distributions

Let $\mathbf{e}_i$, $i = 1, \ldots, d$, be the standard basis vectors in $\mathbb{R}^d$. A distribution on $\mathbb{R}^d$ that is supported by the set $\{\mathbf{e}_1, \ldots, \mathbf{e}_d\}$ is called a *multivariate Bernoulli* distribution. The same adjective is used for any random vector having such a distribution.

Problem 13. [Multinomial Distributions] Let Q be a multivariate Bernoulli distribution on $\mathbb{R}^d$, with $p_i = Q(\{e_i\})$, $i = 1, \ldots, d$. If $(X_1, X_2, \ldots, X_n)$ is an independent sequence of random vectors, each having distribution Q, then the distribution R of $X_1 + \cdots + X_n$ is called the *multinomial distribution* with parameters $p_1, \ldots, p_d$, and n. Show, for nonnegative integers $x_1, \ldots, x_d$, that

$$R(\{(x_1, \ldots, x_d)\}) = \begin{cases} n! \prod_{i=1}^d \dfrac{p_i^{x_i}}{x_i!} & \text{if } \sum_{i=1}^d x_i = n \\ 0 & \text{otherwise}. \end{cases}$$

Problem 14. Describe the connection between multivariate Bernoulli distributions with $d = 2$ and Bernoulli distributions, and also the relationship between multinomial distributions with $d = 2$ and binomial distributions.

Problem 15. Describe the relationship between the multinomial distribution and the experiment of placing n balls into d urns according to certain probabilities.

Problem 16. Describe the relationship between the multinomial distribution and the experiment of throwing a possibly unbalanced d-faced die n times.

* **Problem 17.** Fix $p \in [0, 1]$ and let $(X_1, X_2, \ldots)$ be a (not necessarily independent) sequence of random variables, with X_n being binomially distributed with parameters p and n. Define $X_0 = 0$. Let N be Poisson with mean λ and assume that N is independent of the sequence $(X_1, X_2, \ldots)$. Show that the distribution of the random variable

$$\omega \rightsquigarrow X_{N(\omega)}(\omega)$$

is Poisson with mean $p\lambda$.

Problem 18. Fix $p_1, \ldots, p_d$ nonnegative with sum 1. Let $Z = (Z_1, Z_1, Z_3, \ldots)$ be an independent sequence of $\mathbb{R}^d$-valued random vectors, with Z_n being multinomially distributed with parameters $p_1, \ldots, p_d$, and n. Let N be Poisson with mean λ and assume that N is independent of the sequence Z. Show that the random vector

$$\omega \rightsquigarrow Z_{N(\omega)}(\omega)$$

has independent coordinates, and that the ith coordinate is Poisson with mean λp_i. Use the result of Example 1 as a partial check of this conclusion.

10.3. Probability generating functions and sums in $\overline{\mathbb{Z}}^+$

Convolutions of a large number of distributions, or even of just two, are in many cases difficult to calculate. The following result says that the probability generating function of a convolution may be easy to calculate.

Theorem 5. *For $i = 1, 2$, let ρ_i be the probability generating function of a distribution Q_i supported by $\overline{\mathbb{Z}}^+$. Then the probability generating function of $Q_1 * Q_2$ is $\rho_1 \rho_2$.*

PROOF. Let X_1 and X_2 be independent random variables having distributions Q_1 and Q_2. In view of Proposition 11 of Chapter 9, the probability generating function of $Q_1 * Q_2$ is given by

$$E\left(s^{X_1+X_2}\right) = E\left(s^{X_1}s^{X_2}\right) = E\left(s^{X_1}\right)E\left(s^{X_2}\right) = \rho_1(s)\rho_2(s). \qquad \square$$

By Problem 30 of Chapter 5, the probability generating function of a Poisson distribution with mean λ is $\rho_\lambda(s) = e^{-\lambda(1-s)}$. We see that

$$\rho_{\lambda_1+\lambda_2} = \rho_{\lambda_1}\rho_{\lambda_2}$$

and

$$\rho_\lambda = \rho_{\lambda/n}^n$$

for each positive integer n, two facts that were already discussed in another form in Example 1.

In general, probability generating functions can be useful for determining whether or not a distribution on $\overline{\mathbb{Z}}^+$ has an nth convolution root. The requirement that the power series of a probability generating function have nonnegative coefficients implies that only a positive root of a probability generating function can itself be a probability generating function. It follows that a distribution on $\overline{\mathbb{Z}}^+$ can have at most one n^{th} convolution root supported by $\overline{\mathbb{Z}}^+$. It turns out that it can have at most one n^{th} convolution root of any sort, and, in certain cases, there is such a convolution root that is not supported by $\overline{\mathbb{Z}}^+$. The full story is worked out in the following problem.

Problem 19. Let P be a distribution supported by $\overline{\mathbb{Z}}^+$ and suppose that $P = Q^{*n}$. Prove that Q is uniquely determined by P and is supported by $\{0 + k/n, 1 + k/n, \ldots, \infty\}$, where k is the smallest integer such that $P(\{k\}) > 0$. *Hint:* First consider the case $k = 0$. For that case show that Q is supported by $\overline{\mathbb{R}}^+$, then show that $Q(\{0\}) > 0$, and finally show that Q is supported by $\overline{\mathbb{Z}}^+$.

Problem 20. Let $(X_1, \ldots, X_n)$ and $(Y_1, \ldots, Y_n)$ each be independent sequences of $\overline{\mathbb{Z}}^+$-valued random variables. Suppose that the random variables $X_1, \ldots, X_n$ all have distribution Q and that the random variables $Y_1, \ldots, Y_n$ all have distribution R. If

$$X_1 + \cdots + X_n \quad \text{and} \quad Y_1 + \cdots + Y_n$$

have the same distribution, what is the relationship between Q and R?

* **Problem 21.** Let P be the distribution defined by $P(\{0\}) = P(\{2\}) = 1/4$ and $P(\{1\}) = 1/2$. Show that P has a second convolution root but no third convolution root.

We conclude this section by introducing a new family of distributions on $\mathbb{Z}^+$. We will see that the distributions in this family are infinitely divisible.

It will be convenient to define some new notation. The product $x(x - 1)(x - 2)\ldots(x - k + 1)$ arises sufficiently often to deserve a name. It is called a *falling factorial*, and we denote it by $(x)_k^\downarrow$. Here x is any real number and k is a nonnegative integer; $(x)_0^\downarrow = 1$, consistent with the convention that empty products equal 1. (A variety of notations are used by other authors for falling factorials.) It is also convenient to define the *rising factorial*: $(x)_k^\uparrow = x(x+1)\ldots(x+k-1)$. Note that

$$(x)_k^\uparrow = (x + k - 1)_k^\downarrow$$

for all real numbers x and nonnegative integers k.

Problem 22. Let $0 \leq p < 1$. Use the Binomial Theorem to show that

$$\sum_{k=0}^{\infty}(1 - p)^r \frac{(-r)_k^\downarrow}{k!}(-p)^k = 1$$

for every real number r. Also show that the summands in the above expression are equal to

$$(1 - p)^r \frac{(r)_k^\uparrow}{k!}p^k ,$$

$k = 0, 1, 2, \ldots$.

Problem 23. [Negative Binomial Distributions] Let $0 \leq p < 1$ and $r > 0$. Use the preceding exercise to show that

$$k \rightsquigarrow (1 - p)^r \frac{(r)_k^\uparrow}{k!}p^k , \quad k \in \mathbb{Z}^+,$$

is the density with respect to counting measure of a probability distribution; it is called the *negative binomial distribution* with parameters p and r. Show that the probability generating function of this distribution is $s \rightsquigarrow [(1 - p)/(1 - ps)]^r$. Also calculate the mean and variance of each negative binomial distribution. For which r is a negative binomial distribution a geometric distribution?

Problem 24. Use probability generating functions to show that the sum of n independent geometrically distributed random variables with the same mean has a negative binomial distribution, and describe the relation of the parameters of the negative binomial distribution to n and the parameter of the geometric distribution. Calculate the distribution of the sum of two independent random variables having negative binomial distributions with the same p but possibly different r's, where p and r have the same meaning as in the preceding problem. Is it meaningful to consider $r = 0$, and, if so, do your calculations encompass that case?

Problem 25. Prove that geometric distributions and, more generally, negative binomial distributions are infinitely divisible.

Problem 26. Show that binomial distributions are not infinitely divisible.

10.4. ‡ Dirichlet distributions

In this section we discuss a family of distributions that are important in statistics. We warm up with an exercise.

Problem 27. Let X_1 and X_2 be as in Problem 9. Show that $X_1/(X_1 + X_2)$ has a beta distribution. Calculate its parameters. Prove that the distribution of the random vector $(X_1, X_2)/(X_1 + X_2)$ is absolutely continuous with respect to 1-dimensional Lebesgue measure on the line $x_1 + x_2 = 1$ and calculate its derivative with respect to λ. (See also the following example.)

Example 2. Let d be an integer greater than 1. Let $X_1, \ldots, X_d$ be random variables distributed according to gamma distributions with corresponding parameters (a, γ_i), and suppose that the sequence $(X_1, \ldots, X_d)$ is independent. Let us calculate the distribution of the random vector

$$(Y_1, \ldots, Y_d) = \frac{1}{X_1 + \cdots + X_d}(X_1, \ldots, X_d).$$

It is clear that its distribution is supported by the intersection T of the $(d-1)$-dimensional hyperplane $y_1 + \cdots + y_d = 1$ and the orthant $\{(y_1, \ldots, y_d) : y_i \geq 0$ for each $i\}$.

For A a Borel subset of T, let

$$B = \{(x_1, \ldots, x_d) : \frac{1}{x_1 + \cdots + x_d}(x_1, \ldots, x_d) \in A\}.$$

By Proposition 13 of Chapter 9 we can write

$$P\left(\left\{\omega : \frac{(X_1(\omega), \ldots, X_d(\omega))}{X_1(\omega) + \cdots + X_d(\omega)} \in A\right\}\right) = \int_B \prod_{i=1}^d \frac{a^{\gamma_i} x_i^{\gamma_i - 1} e^{-ax_i}}{\Gamma(\gamma_i)} \, d(x_1, \ldots, x_d),$$

where $d(x_1, \ldots, x_d)$ indicates that the integration is with respect to Lebesgue measure in $\mathbb{R}^d$. We make a change of variables:

$$s = x_1 + \cdots + x_d \quad \text{and} \quad y_i = \frac{x_i}{s}, \ 1 \leq i \leq d - 1.$$

The Jacobian of $(x_1, \ldots, x_d)$ with respect to $(y_1, \ldots, y_{d-1}, s)$ is s^{d-1}. Thus, with $\gamma = \gamma_1 + \cdots + \gamma_d$ and

$$C = \{(y_1, \ldots, y_{d-1}) : (y_1, \ldots, y_{d-1}, 1 - y_1 - \cdots - y_{d-1}) \in A\},$$

the above integral equals

$$\int_{C\times(0,\infty)} a(as)^{\gamma-1}e^{-as}\frac{(1-y_1-\cdots-y_{d-1})^{\gamma_d-1}}{\Gamma(\gamma_d)}\prod_{i=1}^{d-1}\frac{y_i^{\gamma_i-1}}{\Gamma(\gamma_i)}\,d(y_1,\ldots,y_{d-1},s)$$

$$=\int_C\int_{(0,\infty)} u^{\gamma-1}e^{-u}\frac{(1-y_1-\cdots-y_{d-1})^{\gamma_d-1}}{\Gamma(\gamma_d)}\prod_{i=1}^{d-1}\frac{y_i^{\gamma_i-1}}{\Gamma(\gamma_i)}\,du\,d(y_1,\ldots,y_{d-1})$$

$$=\Gamma(\gamma)\int_C\frac{(1-y_1-\cdots-y_{d-1})^{\gamma_d-1}}{\Gamma(\gamma_d)}\prod_{i=1}^{d-1}\frac{y_i^{\gamma_i-1}}{\Gamma(\gamma_i)}\,d(y_1,\ldots,y_{d-1}).$$

Thus we have a density for $(X_1,\ldots,X_{d-1})/(X_1+\cdots+X_d)$ with respect to Lebesgue measure on

$$S=\{(y_1,\ldots,y_{d-1})\colon y_i\geq 0\text{ for each }i\text{ and }y_1+\cdots+y_{d-1}\leq 1\}.$$

The function $y_d=1-y_1-\cdots-y_{d-1}$ gives a natural one-to-one correspondence between S and T. This correspondence maps C onto A. Also, the Lebesgue measure of a subset of S equals $1/\sqrt{d}$ times the Lebesgue measure of its image in T under this map (proof?). Therefore, with respect to Lebesgue measure on T, the random vector $(Y_1,\ldots,Y_d)$ has the density

(10.4)
$$\frac{1}{\sqrt{d}}\Gamma(\gamma)\prod_{i=1}^{d}\frac{y_i^{\gamma_i-1}}{\Gamma(\gamma_i)}$$

for $y\in T$, where $\gamma=\gamma_1+\cdots+\gamma_d$.

Problem 28. Explain how calculations in the preceding example show that $X_1+\cdots+X_d$ is independent of the random vector $(Y_1,\ldots,Y_d)$.

Problem 29. It is apparent that the expression in (10.4) factors as a product of terms $f_i(y_i)$, $i=1,\ldots,d$. Can we conclude from Proposition 13 of Chapter 9 that the sequence $(Y_1,\ldots,Y_d)$ is independent? Why or why not?

* **Problem 30.** [Dirichlet Distributions] Let T be the subset of $\mathbb{R}^d$ defined in Example 2, and let $Y=(Y_1,\ldots,Y_d)$ be a T-valued random vector with the density given in (10.4). We say that Y has a *Dirichlet distribution* with parameters $\gamma_1,\ldots,\gamma_d$. Compute the mean vector and covariance matrix of Y. Compute the variance of $Y_1+\cdots+Y_d$ and use your answer to show that the determinant of the covariance matrix of Y is zero.

Problem 31. Let $Y=(Y_1,\ldots,Y_d)$ be as in the preceding exercise. Choose a partition $\{K_1,\ldots,K_m\}$, $m>1$ of the set $\{1,\ldots,d\}$, and let Z be the $\mathbb{R}^m$-valued random vector with coordinates

$$Z_i=\sum_{j\in K_i}Y_j.$$

Show that Z has a Dirichlet distribution and find the corresponding parameters.

Problem 32. Let $Y = (Y_1, \ldots, Y_d)$ be as in the preceding exercise. Choose integers $1 \leq j_1 < j_2 < \cdots < j_k \leq d$, and let $\widetilde{Y}$ be the $\mathbb{R}^k$-valued random variable with components $\widetilde{Y}_i = Y_{j_i}, i = 1, \ldots, k$. Define Z to be the $\mathbb{R}^k$-valued random variable obtained by dividing $\widetilde{Y}$ by the sum of its components. Show that Z has a Dirichlet distribution and find the corresponding parameters.

The following theorem describes how a Dirichlet distribution arises in connection with order statistics.

Theorem 6. *Let* $Y_1 \leq Y_2 \leq \cdots \leq Y_d$ *be order statistics based on* d *observations of the uniform distribution on the unit interval* $[0, 1]$. *Set* $Y_0 = 0$ *and* $Y_{d+1} = 1$. *Then the distribution of the random vector*

$$((Y_1 - Y_0), (Y_2 - Y_1), \ldots, (Y_d - Y_{d-1}), (Y_{d+1} - Y_d))$$

is the Dirichlet distribution having $d + 1$ *parameters each equal to 1.*

PROOF. According to Problem 50 of Chapter 9, the distribution of the vector $(Y_1, \ldots, Y_d)$ is the uniform distribution on

$$\{(y_1, y_2, \ldots, y_d) : 0 \leq y_1 \leq y_2 \leq \cdots \leq y_d \leq 1\}.$$

Consider the linear transformation from this region to the region

(10.5) $\{(z_1, z_2, \ldots, z_d) : 0 \leq z_i$ for all i and $z_1 + z_2 + \cdots + z_d \leq 1\}$

defined by

$$z_1 = y_1 \quad \text{and} \quad z_i = y_i - y_{i-1} \text{ for } 2 \leq i \leq d.$$

Its determinant is easily shown to equal 1. Any linear transformation with nonzero determinant takes a uniform distribution on any set to a uniform distribution on the image of that set, because it preserves ratios of volumes. Therefore,

(10.6) $((Y_1 - Y_0), (Y_2 - Y_1), \ldots, (Y_d - Y_{d-1}))$

is uniformly distributed on the region (10.5). As the discussion in the paragraph preceding Problem 28 indicates, the uniform distribution for the random vector (10.6) corresponds to the Dirichlet distribution with all parameters equal to 1 for the random vector

$$((Y_1 - Y_0), (Y_2 - Y_1), \ldots, (Y_d - Y_{d-1}), (Y_{d+1} - Y_d)). \quad \square$$

* **Problem 33.** Let $d = 2$ in the notation of the preceding theorem. Calculate the distribution function of

$$Y_1 \wedge (Y_2 - Y_1) \wedge (1 - Y_2).$$

Problem 34. Let $d = 2$ in the notation of Theorem 6. Calculate the distribution function of

$$Y_1 \vee (Y_2 - Y_1) \vee (1 - Y_2).$$

Problem 35. Let $d = 2$ in the notation of Theorem 6. Calculate the probability that at one of the three random variables Y_1, $(Y_2 - Y_1)$, and $1 - Y_2$ is greater than $\frac{1}{2}$ and at least one other of them is less than $\frac{1}{6}$.

* **Problem 36.** In the notation of Theorem 6 find the distribution of $Y_d - Y_1$.

* **Problem 37.** Find the distribution function of the random area described in Problem 52 of Chapter 9.

10.5. † Random sums in various settings

In Remark 1 we noted that much of the theory of sums of independent $\mathbb{R}^d$-valued random variables carries over to the sums of independent random variables taking values in a measurable commutative group. Here is another example illustrating this fact.

Example 3. The *Galois field* $GF(2)$ has two members, 0 and 1. Addition and multiplication are defined as in the real numbers, with the exception that $1 + 1 = 0$. Thus, 0 is the identity for the binary operation of addition, and 1 is the identity for the binary operation of multiplication. Let $\mathcal{G}$ be the σ-field of all four subsets of $GF(2)$. Notice that $GF(2)$ is a commutative group with the operation of addition.

The distribution of a random variable from a probability space $(\Omega, \mathcal{F}, P)$ into $(GF(2), \mathcal{G})$ is characterized by a number $p \in [0, 1]$, defined to be the probability that the random variable equals 1. For this setting, we can use the notation

$$p_1 * \cdots * p_n = P(\{\omega \colon \sum_1^n X_i(\omega) = 1\}),$$

where $(X_1, \ldots, X_n)$ is an independent sequence of $GF(2)$-valued random variables and $p_i = P(\{\omega \colon X_i(\omega) = 1\})$, $1 \le i \le n$. If the random variables X_i are identically distributed with $p_i = p$, we can use the notation p^{*n}. We might imagine the following experiment. On day 0, an 'on-off' switch is placed in the 'off' position. On each succeeding day, the position of the two-way switch is changed from its position of the previous day with probability p, independent of past history. On day n, p^{*n} is the probability that the switch is in the 'on' position.

Problem 38. This problem concerns the preceding example. Find a formula for $p_1 * p_2$. Suppose that $0 < p < 1$. Prove that $p^{*n} \to 1/2$ as $n \to \infty$. *Hint:* Obtain a recursive formula for p^{*n}.

Problem 39. Make appropriate adjustments in Example 3 to focus on the operation of multiplication in $GF(2)$ instead of addition.

Remark 2. The existence of inverses, required in a group, is not used in the definition of convolution, although it is used in Proposition 3. Thus, one may speak of sums of independent random variables in structures having an operation that is commutative and associative and for which there is an identity. Such a structure is a *commutative semigroup*. One example of a semigroup that is not a group is $\mathbb{R}^+$. Another is found in Problem 39. For semigroups to be turned into measurable spaces it is required that the semigroup operation be measurable, but we will not explicitly treat measurability issues in the examples that we give.

We conclude with some examples involving 'sums' of random sets. We first need some terminology and notation. For two compact convex sets of A and B contained in $\mathbb{R}^2$ we let $A \vee B$ denote the convex hull of $A \cup B$; that is, $A \vee B$ is the intersection of all compact convex sets containing $A \cup B$. The next problem treats the only nonobvious aspect of showing that the set of compact convex sets with the operation $\vee$ is a semigroup.

* **Problem 40.** Prove that the operation $\vee$ on convex compact subsets of $\mathbb{R}^2$ is associative.

Problem 41. Sketch a picture of

$$\{x \in \mathbb{R}^2 : x_1 = 0, |x_2| \le 1\} \vee \{x \in \mathbb{R}^2 : x_2 = 0, |x_1| \le 1\}.$$

Let A be a compact convex subset of $\mathbb{R}^2$. For $\varphi \in [0, 2\pi)$, let

$$h(\varphi) = \sup\{x_1 \cos \varphi + x_2 \sin \varphi : (x_1, x_2) \in A\}.$$

The function h, illustrated in Figure 10.2, is called the *support function* of A.

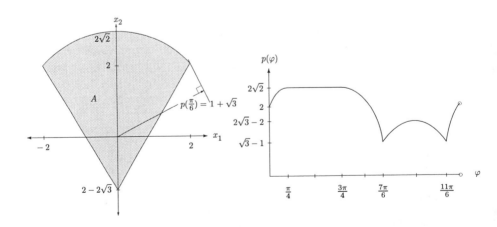

FIGURE 10.2. The support function h of a set A

Problem 42. Calculate the support function of an arbitrary one-point set in $\mathbb{R}^2$.

* **Problem 43.** Calculate the support functions of the three compact convex sets involved in Problem 41.

Problem 44. Let A and B be two convex compact subsets of $\mathbb{R}^2$ having respective support functions h_A and h_B. Prove that the support function of $A \vee B$ is the function $h_A \vee h_B$.

Problem 45. Prove that distinct compact convex subsets of $\mathbb{R}^2$ have distinct support functions.

The collection of compact convex subsets of $\mathbb{R}^2$ is easily shown to be a closed subcollection in the space of all compact subsets of $\mathbb{R}^2$. Therefore, it is a metric space with the Hausdorff metric, and as such it is a measurable space with the Borel σ-field. In view of the preceding problem there is a one-to-one correspondence between the space of compact convex subsets of $\mathbb{R}^2$ and the set of support functions. This correspondence can be used to define a σ-field of subsets of the set of support functions.

Problem 46. Let $h \rightsquigarrow h(\varphi)$ be the function which assigns to each support function its value at φ for some fixed φ. Prove that this function is measurable, when the measurable sets of support functions are taken to be those described in the last sentence of the preceding paragraph.

* **Problem 47.** Let (X_1, X_2, X_3) be an independent triple of $\mathbb{R}^2$-valued random variables, each of which is uniformly distributed on the unit disk $\{x \colon \|x\| \leq 1\}$. For each ω in the underlying probability space, let $\varphi \rightsquigarrow H_\varphi(\omega)$ denote the support function of $\{X_1(\omega)\} \vee \{X_2(\omega)\} \vee \{X_3(\omega)\}$. For each φ calculate the distribution of the $\mathbb{R}$-valued random variable H_φ.

For two compact convex subsets A and B of $\mathbb{R}^2$ let

$$A + B = \{x + y \colon x \in A, y \in B\}.$$

The set $A + B$, which is necessarily compact and convex, is called the *Minkowski sum* of A and B. As an illustration, Figure 10.3 shows how adding a small disc to a rectangular region rounds the corners of the rectangular region.

* **Problem 48.** Prove that the Minkowski sum of two compact convex subsets of $\mathbb{R}^2$ is compact and convex.

Problem 49. Let A and B be two compact convex subsets of $\mathbb{R}^2$ having respective support functions h_A and h_B. Prove that the support function of $A + B$ is $h_A + h_B$.

Problem 50. Fix $\alpha \in [0, 2\pi)$. Calculate the support function of the compact convex set, the boundary of which is the square with vertices at $(\cos\alpha, \sin\alpha)$, $(-\sin\alpha, \cos\alpha)$, $(-\cos\alpha, -\sin\alpha)$, $(\sin\alpha, -\cos\alpha)$.

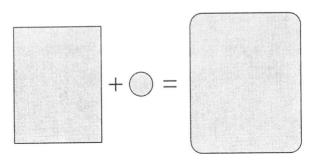

FIGURE 10.3. The Minkowski sum of a rectangular region and a disc

Problem 51. By randomly choosing α in the preceding exercise according to the uniform distribution on $[0, 2\pi)$, a random compact convex set, and thus a random support function, is obtained. For each fixed $\varphi \in [0, 2\pi)$ calculate the distribution of the value of the random support function at φ.

* **Problem 52.** Consider an independent pair of random square regions, each having the distribution described in the preceding exercise. Let

$$\omega \rightsquigarrow [\varphi \rightsquigarrow H_\varphi(\omega)]$$

denote the random support function of their Minkowski sum. For each $\varphi \in [0, 2\pi)$ calculate the mean and variance of the $\mathbb{R}$-valued random variable H_φ.

CHAPTER 11
Random Walk

In this chapter, we will study certain sequences of random variables, known as 'random walks'. These are defined in terms of sums of independent identically distributed random variables. Important in the study of random walks (and of more general random sequences) are 'filtrations' and 'stopping times'. A filtration is a sequence of σ-fields representing the information available at various stages of an experiment. A stopping time is a $\overline{\mathbb{Z}}^+$-valued random variable whose value may be regarded as the time at which an experiment is to be terminated. In applications, such as gambling theory, important stopping times are the time at which a random walk reaches a certain goal and the time at which it returns to its original position. These will be treated in the latter part of the chapter for several special random walks.

11.1. Random sequences

In our analysis of a sequence of random variables, it will often be useful to think of the sequence itself as a single random object. In Chapter 2, we briefly discussed this point of view in the $\mathbb{R}$-valued setting by stating that an infinite sequence of $\mathbb{R}$-valued random variables is also a single $\mathbb{R}^\infty$-valued random variable.

Problem 1. Let $(\Psi, \mathcal{G})$ be a measurable space, and let $Y = (Y_0, Y_1, \dots)$ be a sequence of $(\Psi, \mathcal{G})$-valued functions, all defined on the same probability space. Prove that Y is a $(\Psi, \mathcal{G})^\infty$-valued random variable if and only if Y_n is a $(\Psi, \mathcal{G})$-valued random variable for each n, and that in this case, $\sigma(Y) = \sigma(Y_0, Y_1, \dots)$. Also prove that two $(\Psi, \mathcal{G})^\infty$-valued random variables $Y = (Y_0, Y_1, \dots)$ and $Z = (Z_0, Z_1, \dots)$ have the same distribution if and only if the random vectors $(Y_0, \dots, Y_n)$ and $(Z_0, \dots, Z_n)$ have the same distribution for all n.

According to the preceding exercise, given a sequence $Y = (Y_0, Y_1, \dots)$ of random variables, we may view Y_n as the nth component of the $(\Psi, \mathcal{G})^\infty$-valued random variable Y. One could use the notation $(Y(\omega))_n$ instead of $Y_n(\omega)$ in order to emphasize this point of view, although we will not use such notation even though it reflects the point of view that we often take. Thus, when we write $Y_n(\omega)$, we may be thinking of first fixing n and then looking at the random variable Y_n, or we may have in mind first fixing ω, and then looking at the nth term in the sequence $Y(\omega)$. The random object Y is called a *random sequence*. The coordinate index n is often called the *(discrete) time parameter* or, more briefly *(discrete) time*.

> **Problem 2.** Let Y be a random sequence of $(\Psi, \mathcal{G})$-valued random variables defined on a probability space $(\Omega, \mathcal{F}, P)$. Let $\mathcal{K}$ be the collection of all subsets of $\mathbb{Z}^+$. Prove that $(n, \omega) \rightsquigarrow Y_n(\omega)$ is a measurable function from $(\mathbb{Z}^+, \mathcal{K}) \times (\Omega, \mathcal{F})$ to $(\Psi, \mathcal{G})$.

The preceding exercise introduces yet another point of view, namely that a random sequence Y is a measurable function of $(n, \omega) \in \mathbb{Z}^+ \times \Omega$. To emphasize the fact that this type of measurability refers simultaneously to n and ω, we say that Y is *jointly measurable*. As can be seen from working the exercises, joint measurability is a straightforward consequence of the measurability of each of the individual random variables Y_n, so that it is a general property of random sequences. In Part 6 we will encounter 'random functions', which arise when the time parameter set is $\mathbb{R}^+$ instead of $\mathbb{Z}^+$. In that setting joint measurability is not automatic, and some additional assumptions are needed.

The following proposition asserts that evaluating a random sequence at a random (time) coordinate gives a random variable. It is an easy consequence of joint measurability. (We have already used this result implicitly; see Problem 51 of Chapter 9.)

> **Proposition 1.** *Let N be a $\mathbb{Z}^+$-valued random variable and let Y be a random sequence of $(\Psi, \mathcal{G})$-valued random variables, all defined on a probability space $(\Omega, \mathcal{F}, P)$. Then*
>
> $$\omega \rightsquigarrow Y_{N(\omega)}(\omega)$$
>
> *is a $(\Psi, \mathcal{G})$-valued random variable.*

> **Problem 3.** Use Problem 2 to prove the preceding proposition.

If a sequence $Y = (Y_0, Y_1, \dots)$ is independent, and if all of the components Y_n have the same distribution, then we call Y an *iid random sequence*, or simply an *iid sequence*. (The letters 'iid' stand for "independent and identically distributed".) Such sequences are quite important in probability theory, and in particular, they constitute the basic ingredient in the definition of random walks.

11.2. Definition and examples

We introduce a famous type of random sequence.

Definition 2. Let $X = (X_1, X_2, \dots)$ be an iid sequence of random variables each taking its values in $\mathbb{R}^d$ or $\overline{\mathbb{R}}^+$. The sequence $S = (S_0, S_1, S_2, \dots)$, where $S_0(\omega) = 0$ for all ω and

$$(11.1) \qquad S_n = \sum_{k=1}^{n} X_k,$$

is a *random walk* with *steps* $X_1, X_2, \dots$. The common distribution of the steps is the *step distribution*.

We see that a random walk is the sequence of partial sums of an iid sequence. We will typically take the point of view that S_0 is the empty sum, so that its equaling 0 need not be mentioned explicitly. We may then use (11.1) to define S_n for all $n \geq 0$.

The space $\mathbb{R}^d$ or $\overline{\mathbb{R}}^+$ in which the various S_n take their values is called the *state space* of S, and, for instance, the phrase "S is a random walk in $\mathbb{R}^2$" indicates that the state space is $\mathbb{R}^2$. For random walks in $\mathbb{R}^d$ whose steps have integral coordinates it is natural to call $\mathbb{Z}^d$ the *state space*, and similarly for $\overline{\mathbb{R}}^+$ and $\overline{\mathbb{Z}}^+$.

Let $X = (X_1, X_2, \dots)$ be the sequence of steps of a random walk S with step distribution Q. The random sequence X has distribution Q^∞. Since S is the composition of a measurable function and X (can you prove this?), the distribution of S is determined by Q. Thus, we may speak of 'the' random walk with step distribution Q, even though there are many possible underlying probability spaces $(\Omega, \mathcal{F}, P)$. If, say, the state space is $\mathbb{R}$, a canonical choice for Ω is $\mathbb{R}^\infty$. Then we may take X to be the identity function on $\Omega = \mathbb{R}^\infty$ and S a certain measurable function from $\mathbb{R}^\infty$ to $\mathbb{R}^\infty$.

Problem 4. Write down the measurable function to which the preceding sentence refers.

Problem 5. Let S be a random walk with step sequence X. For integers $0 \leq i < j$, define

$$S_{i,j} = X_{i+1} + \dots + X_j.$$

Thus, $S_n = S_{0,n}$. Prove that for all positive integers k, the doubly indexed sequences

$$(S_{i,j} : 0 \leq i < j) \quad \text{and} \quad (S_{i+k, j+k} : 0 \leq i < j)$$

have the same distribution. (The random variables $S_{i,j}$ are called the *increments* of S, and the result just stated is often expressed by saying that S has *stationary increments*.) *Hint:* Express the first doubly indexed sequence as a certain function φ of the step sequence X. The second doubly indexed sequence is $\varphi \circ Y$, where $Y = (X_k, X_{k+1}, \dots)$. Use the fact that X and Y have the same distribution.

Problem 6. This is a continuation of the preceding problem. Show that for integers

$$0 \le i_1 < j_1 \le i_2 < j_2 \le \cdots \le i_n < j_n ,$$

the collection of increments

$$(S_{i_k, j_k} : k = 1, \ldots, n)$$

is independent. (This property is often expressed by saying that S has *independent increments*.)

Example 1. [Simple Random Walks in $\mathbb{Z}^d$] A *simple random walk* in $\mathbb{Z}^d$ is a random walk with state space $\mathbb{Z}^d$ (or $\mathbb{R}^d$) whose step distribution Q is supported by the set $\mathcal{N} = \{\pm \mathbf{e}_i : i = 1, \ldots, d\}$, where $\mathbf{e}_i$, $i = 1, \ldots, d$, are the standard basis vectors in $\mathbb{R}^d$. A simple random walk on $\mathbb{Z}^d$ is also called a *nearest neighbor random walk* on $\mathbb{Z}^d$, because successive states $S_n(\omega)$ and $S_{n+1}(\omega)$ are always one unit apart ('nearest neighbors'). The first eleven states of a simple random walk in $\mathbb{Z}^2$ are illustrated in Figure 11.1. If Q is the distribution which assigns equal probability to each of the $2d$ points in $\mathcal{N}$, then the random walk is called a simple *symmetric* random walk on $\mathbb{Z}^d$. In particular, the simple symmetric random walk on $\mathbb{Z}$ makes steps of size 1 in either the positive or the negative direction, each with probability $1/2$.

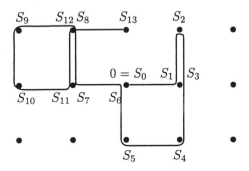

FIGURE 11.1. A simple random walk in $\mathbb{Z}^2$

Problem 7. Describe the relation between Example 2 of Chapter 2 and the simple symmetric random walk in $\mathbb{Z}$.

Problem 8. Let S be a simple random walk in $\mathbb{Z}$, with step distribution Q. Show that for each n, the distribution of S_n is of the same type as the binomial distribution with parameters n and p, where $p = Q(\{1\})$.

Remark 1. In view of Remark 2 of Chapter 10, random walks can be defined in arbitrary commutative semigroups. Although our main interest is in the two state spaces $\mathbb{R}^d$ and $\mathbb{R}^+$, most assertions in this chapter and later, such as Theorem 12, are for state spaces that are arbitrary commutative semigroups. The semigroups $\overline{\mathbb{R}}^+$ and $\overline{Z}^+$ are special because they arise naturally as state spaces in the study of random walks in $\mathbb{R}$.

There are some general results for random walks that require the state space to be a group, not just a semigroup (for example, see Problem 15). For simplicity, we will state such results in the $\mathbb{R}^d$-setting only, even though they hold for arbitrary commutative groups.

The next example treats a random walk in a group different from $\mathbb{R}^d$, the group $(-\pi, \pi]$ of rotations that was introduced in Problem 11 of Chapter 10.

Example 2. In the set $(-\pi, \pi]$ with addition defined as in Problem 11 of Chapter 10 consider the random walk S with step distribution Q given by

$$Q(\{\tfrac{2\pi}{5}\}) = Q(\{-\tfrac{2\pi}{5}\}) = \tfrac{1}{2}.$$

Let us calculate the distribution of S_n for each n. It is clear from symmetry that, for each n, there exist nonnegative numbers p_n, q_n, r_n such that $p_n + 2q_n + 2r_n = 1$ and

$$
\begin{aligned}
p_n &= P(\{\omega\colon S_n(\omega) = 0\})\,; \\
q_n &= P(\{\omega\colon S_n(\omega) = \tfrac{2\pi}{5}\}) = P(\{\omega\colon S_n(\omega) = -\tfrac{2\pi}{5}\})\,; \\
r_n &= P(\{\omega\colon S_n(\omega) = \tfrac{4\pi}{5}\}) = P(\{\omega\colon S_n(\omega) = -\tfrac{4\pi}{5}\})\,.
\end{aligned}
$$

Clearly, $p_0 = 1$ and $q_0 = r_0 = 0$. For $n \geq 1$ the distribution of S_n is the convolution of the distributions of X_n and S_{n-1}, where X_n denotes the nth step. Thus, we have

$$
\begin{aligned}
r_n &= P(\{\omega\colon X_n(\omega) = 0\})\,P(\{\omega\colon S_{n-1}(\omega) = \tfrac{4\pi}{5}\}) \\
&\quad + P(\{\omega\colon X_n(\omega) = \tfrac{2\pi}{5}\})\,P(\{\omega\colon S_{n-1}(\omega) = \tfrac{2\pi}{5}\}) \\
&\quad + P(\{\omega\colon X_n(\omega) = \tfrac{4\pi}{5}\})\,P(\{\omega\colon S_{n-1}(\omega) = 0\}) \\
&\quad + P(\{\omega\colon X_n(\omega) = -\tfrac{4\pi}{5}\})\,P(\{\omega\colon S_{n-1}(\omega) = -\tfrac{2\pi}{5}\}) \\
&\quad + P(\{\omega\colon X_n(\omega) = -\tfrac{2\pi}{5}\})\,P(\{\omega\colon S_{n-1}(\omega) = -\tfrac{4\pi}{5}\}) \\
&= 0 + \tfrac{1}{2}q_{n-1} + 0 + 0 + \tfrac{1}{2}r_{n-1}\,.
\end{aligned}
$$

Similar calculations give expressions for p_n and q_n. The result, for $n \geq 1$, is, in matrix notation,

$$
\begin{pmatrix} p_n \\ q_n \\ r_n \end{pmatrix} =
\begin{pmatrix} 0 & 1 & 0 \\ \tfrac{1}{2} & 0 & \tfrac{1}{2} \\ 0 & \tfrac{1}{2} & \tfrac{1}{2} \end{pmatrix}
\begin{pmatrix} p_{n-1} \\ q_{n-1} \\ r_{n-1} \end{pmatrix} =
\begin{pmatrix} 0 & 1 & 0 \\ \tfrac{1}{2} & 0 & \tfrac{1}{2} \\ 0 & \tfrac{1}{2} & \tfrac{1}{2} \end{pmatrix}^n
\begin{pmatrix} 1 \\ 0 \\ 0 \end{pmatrix}.
$$

We compute the nth power of the matrix by diagonalizing:

$$\begin{pmatrix} 0 & 1 & 0 \\ \frac{1}{2} & 0 & \frac{1}{2} \\ 0 & \frac{1}{2} & \frac{1}{2} \end{pmatrix}^n$$

$$= \frac{1}{20} \begin{pmatrix} 4 & 4 & 4 \\ 4 & -1+\sqrt{5} & -1-\sqrt{5} \\ 4 & -1-\sqrt{5} & -1+\sqrt{5} \end{pmatrix}$$

$$\cdot \begin{pmatrix} 1 & 0 & 0 \\ 0 & \frac{-1+\sqrt{5}}{4} & 0 \\ 0 & 0 & \frac{-1-\sqrt{5}}{4} \end{pmatrix}^n \begin{pmatrix} 1 & 2 & 2 \\ 2 & -1+\sqrt{5} & -1-\sqrt{5} \\ 2 & -1-\sqrt{5} & -1+\sqrt{5} \end{pmatrix}.$$

The appropriate matrix arithmetic gives

$$p_n = \frac{1}{5} + \frac{2}{5}\left(\frac{-1+\sqrt{5}}{4}\right)^n + \frac{2}{5}\left(\frac{-1-\sqrt{5}}{4}\right)^n ;$$

$$q_n = \frac{1}{5} + \frac{2}{5}\left(\frac{-1+\sqrt{5}}{4}\right)^{n+1} + \frac{2}{5}\left(\frac{-1-\sqrt{5}}{4}\right)^{n+1} ;$$

$$r_n = \frac{1}{5} - \frac{1}{10}\left(\frac{-1+\sqrt{5}}{4}\right)^{n-1} - \frac{1}{10}\left(\frac{-1-\sqrt{5}}{4}\right)^{n-1} .$$

It is interesting to note that p_n, q_n, r_n all approach $\frac{1}{5}$ as $n \to \infty$.

Problem 9. Use the result of the preceding example to obtain some formulas for certain sums of binomial coefficients.

Problem 10. Modify Example 2 and Problem 9 by defining Q by

$$Q(\{\tfrac{2\pi}{3}\}) = Q(\{-\tfrac{2\pi}{3}\}) = \tfrac{1}{2}.$$

Problem 11. Modify Example 2 and Problem 9 by defining Q by

$$Q(\{\tfrac{\pi}{2}\}) = Q(\{-\tfrac{\pi}{2}\}) = \tfrac{1}{2}.$$

Find a qualitative difference between this random walk and the random walks described in Example 2 and Problem 10.

* **Problem 12.** In the set $(-\pi, \pi]$ with addition defined as in Problem 11 of Chapter 10 consider the random walk S with step distribution Q given by

$$Q(\{0\}) = Q(\{\pi\}) = \tfrac{1}{3}$$

and for all Borel sets $B \subseteq (-\pi, \pi] \setminus \{0, \pi\}$,

$$Q(B) = \tfrac{1}{6\pi}\lambda(B),$$

where λ denotes Lebesgue measure. Calculate the distribution of S_n for $n \geq 1$.

* **Problem 13.** Let p, q, r denote positive numbers whose sum equals 1. Let Q be the distribution on $\overline{\mathbb{Z}}^+$ given by $Q(\{1\}) = p$, $Q(\{2\}) = q$ and $Q(\{\infty\}) = r$. Consider a random walk S in $\overline{\mathbb{Z}}^+$ (or $\overline{\mathbb{R}}^+$) whose step distribution equals Q, and let $N(\omega) = \inf\{n \colon S_n(\omega) = \infty\}$. Prove that $N < \infty$ a.s., so that S_{N-1} is a.s.-defined. Calculate the distribution of the random vector $(N - 1, S_{N-1})$ and the expected value of S_{N-1}.

* **Problem 14.** Let Q be a distribution on $\overline{\mathbb{R}}^+$ for which $0 < Q(\{\infty\}) < 1$. Let S be the random walk with step distribution Q and define N as in the preceding exercise. Follow the instructions in the last two sentences of that exercise, except that here one cannot expect such an explicit formula for the distribution of $(N - 1, S_{N-1})$ as in that very specific situation; convolutions may appear in the answer.

Example 3. Consider a random walk S in $\overline{\mathbb{R}}^+$ having exponential step distribution with mean 1. By Problem 9 of Chapter 10, for $k = 1, 2, \ldots$, S_k is gamma distributed with parameters $1, k$.

Fix $\beta \in (0, \infty)$ and let

$$N(\omega) = \#\{n \colon 0 < S_n(\omega) \leq \beta\}.$$

Since each step is a.s. positive, N a.s. equals the number of steps taken by the random walk before it reaches the interval (β, ∞).

For $k \in \mathbb{Z}^+$,

$$
\begin{aligned}
P(\{\omega \colon N(\omega) = k\}) &= P(\{\omega \colon N(\omega) \geq k\}) - P(\{\omega \colon N(\omega) \geq k + 1\}) \\
&= P(\{\omega \colon S_k(\omega) \leq \beta\}) - P(\{\omega \colon S_{k+1}(\omega) \leq \beta\}) \\
&= \int_0^\beta \left(\frac{x^{k-1} e^{-x}}{\Gamma(k)} - \frac{x^k e^{-x}}{\Gamma(k+1)} \right) dx \\
&= \frac{x^k e^{-x}}{\Gamma(k+1)} \bigg|_0^\beta = \frac{\beta^k e^{-\beta}}{k!}.
\end{aligned}
$$

Therefore, the random variable N is Poisson distributed with mean β. This fact will play a role in the study of 'Poisson point processes' in Chapter 29.

If a random sequence has the same distribution as a random walk, then it deserves the name 'random walk', even if its description or construction does not explicitly involve iid steps. In the $\mathbb{R}^d$-setting, the steps can be obtained from the original sequence by subtraction, so a random sequence $(S_0 = 0, S_1, S_2, \ldots)$ of $\mathbb{R}^d$-valued random variables has the same distribution as a random walk if and only if the random sequence $((S_n - S_{n-1}) \colon n = 1, 2, \ldots)$ is iid. Since subtraction is not universally defined in $\overline{\mathbb{R}}^+$, the story there is somewhat subtle.

Proposition 3. *Let $T = (T_0 = 0, T_1, T_2, \ldots)$ be a sequence of $\overline{\mathbb{R}}^+$-valued random variables. Then T has the same distribution as a random walk in $\overline{\mathbb{R}}^+$ if* *and only if*

$$P(\{\omega: T_k(\omega) - T_{k-1}(\omega) \in B_k \text{ for } 1 \le k \le n\}) = \prod_{k=1}^{n} P(\{\omega: T_1(\omega) \in B_k\})$$

for every choice of the positive integer n and Borel subsets B_k of $\mathbb{R}^+$, with the understanding that the undefined expression $\infty - \infty$ is not a member of $\mathbb{R}^+$. In this case, the step distribution of the random walk is the distribution of T_1.

PROOF. The 'only if' assertion is obvious. For the 'if' assertion, let $(\Omega, \mathcal{F}, P_1)$ denote the probability space on which T is defined. Let $Y = (Y_j: j \ge 1)$ be an iid sequence defined on another probability space $(\Psi, \mathcal{G}, P_2)$, with the common distribution of the Y_j equal to the distribution of T_1. Set

$$(\Theta, \mathcal{H}, P) = (\Omega, \mathcal{F}, P_1) \times (\Psi, \mathcal{G}, P_2).$$

We may regard each Y_j as defined on Θ as follows:

$$Y_j((\omega, \psi)) = Y_j(\psi).$$

Similarly we may regard T as defined on Θ via its values at first coordinates of members of Θ. We have thus arranged for Y to be independent of T.

For $k \ge 1$ define random variables Z_k on $(\Theta, \mathcal{H}, P)$ via

$$Z_k(\theta) = \begin{cases} T_k(\theta) - T_{k-1}(\theta) & \text{if } T_{k-1}(\theta) < \infty \\ Y_k(\theta) & \text{if } T_{k-1}(\theta) = \infty. \end{cases}$$

Clearly $T_n(\theta) = \sum_{k=1}^{n} Z_k(\theta)$ for all n and θ, and the distributions of Z_1, Y_1, and T_1 are identical. Thus we only need prove that $(Z_k: k \ge 1)$ is an iid sequence.

In view of Proposition 5 of Chapter 9 we only need prove that Z_n is independent of $(Z_1, \ldots, Z_{n-1})$ and has the same distribution as Z_1 for $n = 2, 3, \ldots$. Hence we may finish the proof by showing

$$(11.2) \quad \begin{aligned} & P(\{\theta: Z_n(\theta) \in B_n, (Z_1(\theta), \ldots, Z_{n-1}\theta)) \in C_{n-1}\}) \\ & = P(\{\theta: Z_1(\theta) = B_n\}) P(\{\theta: (Z_1(\theta), \ldots, Z_{n-1}(\theta)) \in C_{n-1}\}) \end{aligned}$$

for every choices of $n \ge 2$, Borel subset B_n of $\overline{\mathbb{R}}^+$, and Borel subset C_{n-1} of $(\mathbb{R}^+)^{n-1}$.

For fixed n and B_n the collection of C_{n-1} for which (11.2) holds is a Sierpinski class, by additivity of probability measures and continuity of measure. Thus, we only need consider sets C_{n-1} of the form

$$C_{n-1} = B_1 \times B_2 \times \cdots \times B_{n-1},$$

where each B_k is either the one-point set $\{\infty\}$ or a Borel subset of $\mathbb{R}^+$. In case $B_k \subseteq \mathbb{R}^+$ for $k \le n$, (11.2) is an immediate consequence of the hypothesis in the theorem, so we assume that there exists $q \le n$ such that $B_q = \{\infty\}$ but $B_k \subseteq \mathbb{R}^+$ for $k < q$. If $q < n$, then Z_n can be replaced by Y_n on the left side of (11.2), and then (11.2) follows from the fact that Y_n has the same distribution as Z_1 and is independent of $(T, Y_1, \ldots, Y_{n-1})$. Therefore we only need consider $q = n$. In this case the left side of (11.2) equals

$$P\big(\{\theta : T_k(\theta) - T_{k-1}(\theta) \in B_k \text{ for } k < n\}\big)$$
$$- P\big(\{\theta : T_k(\theta) - T_{k-1}(\theta) \in B_k \text{ for } k < n \text{ and } T_n(\theta) - T_{n-1}(\theta) < \infty\}\big),$$

which by the hypothesis of the theorem, equals

$$P\big(\{\theta : T_k(\theta) - T_{k-1}(\theta) \in B_k \text{ for } k < n\}\big) \big[1 - P\big(\{T_n(\theta) - T_{n-1}(\theta) < \infty\}\big)\big]$$
$$= P\big(\{\theta : T_k(\theta) - T_{k-1}(\theta) \in B_k \text{ for } k < n\}\big) P\big(\{\theta : T_n(\theta) - T_{n-1}(\theta) = \infty\}\big),$$

as desired. $\square$

11.3. Filtrations and stopping times

One might observe a random sequence S over time, with S_n being the observation made at time n. The σ-field $\sigma(S_0, S_1, \ldots, S_n)$ represents the information accumulated up to and including time n by making such observations. The following definition introduces a sequence of σ-fields designed to represent information accumulated over time.

Definition 4. Let $(\Omega, \mathcal{F})$ be a measurable space. A sequence $(\mathcal{F}_0, \mathcal{F}_1, \mathcal{F}_2, \ldots)$ of sub-σ-fields of $\mathcal{F}$ is a *filtration* in $(\Omega, \mathcal{F})$ if it is increasing, that is, if

$$\mathcal{F}_0 \subseteq \mathcal{F}_1 \subseteq \mathcal{F}_2 \subseteq \ldots .$$

A sequence $Y = (Y_0, Y_1, \ldots)$ of measurable functions defined on $(\Omega, \mathcal{F})$ is said to be *adapted* to a filtration $(\mathcal{F}_n : n \geq 0)$ if, for each n, Y_n is measurable with respect to $\mathcal{F}_n$. Corresponding to a filtration $(\mathcal{F}_n : n \geq 0)$, we use the notation

$$\mathcal{F}_\infty = \sigma(\mathcal{F}_0, \mathcal{F}_1, \ldots) .$$

The *minimal filtration* $(\mathcal{F}_0, \mathcal{F}_1, \ldots)$ of a random sequence $Y = (Y_0, Y_1, \ldots)$ is given by $\mathcal{F}_n = \sigma(Y_0, \ldots, Y_n)$. The random sequence Y is clearly adapted to its minimal filtration. The minimal filtration is the one that is implicit in any discussion for which a filtration is required and none has been explicitly mentioned.

Problem 15. Let S be a random walk in $\mathbb{R}^d$ and denote the corresponding sequence of steps by X. Prove that

$$\sigma(X_1, \ldots, X_n) = \sigma(S_0, S_1, \ldots, S_n)$$

for each positive integer n. Is the same conclusion true for $n = 0$ if, in that case, the σ-field on the left is interpreted to be the smallest of all σ-fields? Give an example to show that $\sigma(X_1, \ldots, X_n)$ and $\sigma(S_0, S_1, \ldots, S_n)$ need not be equal for random walks in $\overline{\mathbb{R}}^+$.

Problem 16. Prove that if a random sequence Y is adapted to a filtration $(\mathcal{F}_n : n \geq 0)$, then $\sigma(Y) \subseteq \mathcal{F}_\infty$, with equality in case $(\mathcal{F}_n : n \geq 0)$ is the minimal filtration of Y.

The next definition introduces a class of random variables that is fundamental to the study of random sequences.

Definition 5. Let $(\Omega, \mathcal{F})$ be a measurable space, and let $(\mathcal{F}_n : n \geq 0)$ be a filtration of sub-σ-fields of $\mathcal{F}$. A $\overline{\mathbb{Z}}^+$-valued measurable function N defined on $(\Omega, \mathcal{F})$ is a *stopping time* with respect to this filtration if

$$\{\omega : N(\omega) \leq n\} \in \mathcal{F}_n$$

for all $n \in \overline{\mathbb{Z}}^+$.

* **Problem 17.** Show that N is a stopping time with respect to a filtration $(\mathcal{F}_n : n \geq 0)$ if and only if $\{\omega : N(\omega) = n\} \in \mathcal{F}_n$ for all $n \in \overline{\mathbb{Z}}^+$.

It is important to develop intuition for filtrations and stopping times. A filtration represents the information obtained by observing an experiment up to time n. One may think of a stopping time N as the time at which the observations of the experiment are to be stopped. The definition of stopping time requires that the decision to stop observing at a certain time n be based on the information available up to time n. Of course, the definition does not require that there actually be an observer who stops observing at time n; the term 'stopping time' is used regardless of whether such an interpretation is intended.

Problem 18. In this problem the infimum of the empty set is defined to be $+\infty$ as usual, and the supremum of the empty set is defined, somewhat unconventionally, to be 0. Let Y be a sequence of $\mathbb{R}$-valued random variables. Show that for all Borel sets $B \subseteq \mathbb{R}$,

$$N(\omega) = \inf\{n \geq 0 : Y_n(\omega) \in B\}$$

is a stopping time with respect to the minimal filtration of Y, but that

$$M(\omega) = \sup\{n : Y_n(\omega) \in B\}$$

is in general not, even if it is assumed that, for every ω, $\{n : Y_n(\omega) \in B\} \neq \emptyset$. *Hint:* One approach for the first part is to use Problem 17.

Problem 19. Let S and N be as in Example 3. Is N a stopping time with respect to the minimal filtration of S?

The random variable N in the preceding exercise is called the *hitting time* of the set B. Many (but certainly not all) stopping times are hitting times.

Proposition 6. *Let $N_1, N_2, \ldots$ be stopping times with respect to a filtration. Then $N_1 \wedge N_2$, $N_1 \vee N_2$, $N_1 + N_2$, $\limsup_{k \to \infty} N_k$, and $\liminf_{k \to \infty} N_k$ are also stopping times with respect to that filtration, as is any constant $n \in \overline{\mathbb{Z}}^+$, regarded as a constant random variable.*

Problem 20. Prove the preceding proposition. Also provide an example of two stopping times $N_1 \leq N_2$ with respect to some filtration, such that $N_2 - N_1$ is not a stopping time with respect to that filtration.

When one observes a random sequence Y up to a random time N, one obtains a certain amount of information. If Y is adapted to a filtration $(\mathcal{F}_n \colon n \geq 0)$ and N is a stopping time with respect to that filtration, then, according to the following definition, this information is contained in a σ-field that we call $\mathcal{F}_N$.

Definition and Proposition 7. Let N be a stopping time with respect to a filtration $(\mathcal{F}_n \colon n \geq 0)$. The collection of events A such that

$$A \cap \{\omega \colon N(\omega) \leq n\} \in \mathcal{F}_n$$

for all $n \in \overline{\mathbb{Z}}^+$ is a σ-field. It is denoted by $\mathcal{F}_N$.

Problem 21. Prove the preceding proposition.

Problem 22. What is wrong with defining

$$\mathcal{F}_N = \sigma(\mathcal{F}_n \colon n \leq N)?$$

Proposition 8. *A stopping time N is measurable with respect to the σ-field $\mathcal{F}_N$.*

Problem 23. Prove the preceding proposition.

* **Problem 24.** Prove that if $M \leq N$ are two stopping times with respect to a common filtration $(\mathcal{F}_n \colon n \geq 0)$, then $\mathcal{F}_M \subseteq \mathcal{F}_N$.

We have already seen in Proposition 1 that if Y is a sequence of random variables and N is a $\mathbb{Z}^+$-valued random variable, then Y_N is a random variable. In the case of stopping times, we can say more, provided we introduce an appropriate convention for Y_∞.

Proposition 9. *Let Y be a random sequence adapted to a filtration $(\mathcal{F}_n \colon n \geq 0)$, and let N be a stopping time with respect to the same filtration. Let Y_∞ denote any random variable that is measurable with respect to $\mathcal{F}_\infty$. Then Y_N is measurable with respect to $\mathcal{F}_N$.*

Problem 25. Prove the preceding proposition. *Hint:* Break up events involving Y_N into disjoint pieces, according to the value of N.

11.4. Stopping times and random walks

The main reason that stopping times are useful in the study of random walks is contained in the following result, which says that the iid property persists after a stopping time.

Proposition 10. *Let* $X = (X_n : n \geq 1)$ *be an iid sequence of* $(\Psi, \mathcal{G})$*-valued random variables, where* $(\Psi, \mathcal{G})$ *is a measurable space. Let* $(\mathcal{F}_n : n \geq 1)$ *be the corresponding minimal filtration, and let* N *be an a.s. finite stopping time with respect to this filtration. For* $n \geq 1$*, define*

$$Y_n(\omega) = X_{N(\omega)+n}(\omega).$$

Then the sequence $Y = (Y_1, Y_2, \dots)$ *has the same distribution as* X*, and is independent of* $\mathcal{F}_N$*.*

PROOF. It follows from Proposition 1 that Y is a $(\Psi, \mathcal{G})^\infty$-valued random sequence. Fix $A \in \mathcal{G}^\infty$ and $B \in \mathcal{F}_N$. Then

$$P(B \cap \{\omega : Y(\omega) \in A\})$$

$$= \sum_{m=0}^{\infty} P(B \cap \{\omega : N(\omega) = m\} \cap \{\omega : X^{(m)}(\omega) \in A\}),$$

where $X^{(m)}$ is the random sequence $(X_{m+1}, X_{m+2}, \dots)$. By Problem 17,

$$\{\omega : N(\omega) = m\} \in \mathcal{F}_m,$$

and, by the definition of $\mathcal{F}_N$,

$$B \cap \{\omega : N(\omega) \leq m\} \in \mathcal{F}_m.$$

Therefore, the intersection of these two events, $B \cap \{\omega : N(\omega) = m\}$, is a member of $\mathcal{F}_m$. The random sequence $X^{(m)}$ is independent of $\mathcal{F}_m$ and has the same distribution as X, so

$$P\big(B \cap \{\omega : N(\omega) = m\} \cap \{\omega : X^{(m)}(\omega) \in A\}\big)$$

$$= P(B \cap \{\omega : N(\omega) = m\}) \, P(\{\omega : X^{(m)}(\omega) \in A\})$$

$$= P(B \cap \{\omega : N(\omega) = m\}) \, P(\{\omega : X(\omega) \in A\}).$$

Now sum on m to obtain

$$P(B \cap \{\omega : Y(\omega) \in A\}) = P(B) \, P(\{\omega : X(\omega) \in A\}) = P(B) \, Q^\infty(A),$$

where Q is the common distribution of the X_i's. By letting $B = \Omega$, we find that Y has the same distribution as X, namely Q^∞. It then follows that B and Y

are independent. Since B is an arbitrary member of $\mathcal{F}_N$, Y is independent of $\mathcal{F}_N$. $\square$

By Proposition 8 and Proposition 9, N and X_N are measurable with respect to $\mathcal{F}_N$, and so, by the preceding proposition, the random vector (N, X_N) is independent of Y.

Corollary 11. *Let* $(\mathcal{F}_n : n \geq 0$ *be the minimal filtration of a random walk* S, *and let* N *be an a.s. finite stopping time with respect to this filtration. Denote by* X *the step sequence of* S. *Then*

$$\left(X_{N+1}, \, [X_{N+1} + X_{N+2}], \, [X_{N+1} + X_{N+2} + X_{N+3}], \, \dots \right)$$

is a random walk that has the same distribution as S *and is independent of* $\mathcal{F}_N$.

The next problem, which should be compared with Problem 6 where it was shown that a random walk has independent increments, transforms the preceding corollary into statement about the increments of a random walk.

Problem 26. Let $N_1 \leq N_2 \leq \dots$ be a sequence of a.s. finite stopping times with respect to the minimal filtration of a random walk S in $\mathbb{R}^d$. Prove that the random variables

$$S_{N_1}, \, S_{N_2} - S_{N_1}, \, S_{N_3} - S_{N_2}, \, \dots$$

are independent.

11.5. A hitting-time example

This section treats hitting times of simple random walks in $\mathbb{Z}$.

Example 4. Consider a simple random walk S in $\mathbb{Z}$ and set

$$p = P(\{\omega : S_1(\omega) = 1\}) = 1 - P(\{\omega : S_1(\omega) = -1\}) .$$

To avoid trivialities we assume that $p \in (0, 1)$. Fix a positive integer c and for $b = 0, 1, \dots, c$ let N_b denote the hitting time of the two-point set $\{b - c, b\}$. For $c = 6$ the random variable N_2 is illustrated in Figure 11.2. We set the goal of calculating the probability generating function ρ_b of N_b.

Since $N_0 = N_c = 0$, $\rho_0(s) = \rho_c(s) = 1$ for $0 \leq s \leq 1$. Since $N_b > 0$ for $0 < b < c$, $\rho_b(0) = 0$.

Denote the sequence of steps by $X = (X_1, X_2, \dots)$, and let $\widetilde{S}$ be the random walk obtained by deleting the first step of S; thus, $\widetilde{S}_0 = 0, \widetilde{S}_1 = X_2, \widetilde{S}_2 = X_2 + X_3, \dots$. Let $\widetilde{N}_b$ equal the hitting time of the set $\{b - c, b\}$ by $\widetilde{S}$. The random walks S and $\widetilde{S}$ are identically distributed, as are the hitting times N_b and $\widetilde{N}_b$.

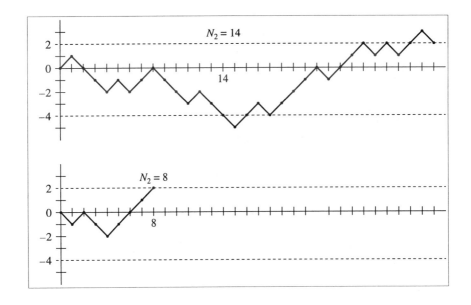

FIGURE 11.2. The hitting time of $\{-4, 2\}$

For $0 < b < c$ and $n > 0$,

$$
\begin{aligned}
P(\{\omega\colon & N_b(\omega) = n\}) \\
&= P(\{\omega\colon S_1(\omega) = 1, \widetilde{N}_{b-1}(\omega) = n - 1\}) \\
&\quad + P(\{\omega\colon S_1(\omega) = -1, \widetilde{N}_{b+1}(\omega) = n - 1\}) \\
&= P(\{\omega\colon S_1(\omega) = 1\})\, P(\{\omega\colon \widetilde{N}_{b-1}(\omega) = n - 1\}) \\
&\quad + P(\{\omega\colon S_1(\omega) = -1\})\, P(\{\omega\colon \widetilde{N}_{b+1}(\omega) = n - 1\}) \\
&= pP(\{\omega\colon N_{b-1}(\omega) = n - 1\}) + (1 - p)P(\{\omega\colon N_{b+1}(\omega) = n - 1\})\,.
\end{aligned}
$$

(11.3)

We have used the fact that S_1 and $\widetilde{S}$ are independent. Now multiply by $s^n, 0 < s < 1$, sum from $n = 1$ to ∞, and use the fact that $P(\{\omega\colon N_b(\omega) = 0\}) = 0$ to obtain

(11.4) $$\rho_b(s) = ps\rho_{b-1}(s) + (1 - p)s\rho_{b+1}(s)\,.$$

For each fixed s this is a linear homogeneous second-order difference equation. As such, its solution set consists of all linear combinations of any two linearly independent solutions. We look for solutions of the form $\rho_b(s) = [\lambda(s)]^b$. Inserting this expression into (11.4) we obtain a quadratic equation for $\lambda(s)$:

$$(1 - p)s[\lambda(s)]^2 - \lambda(s) + ps = 0.$$

We solve for the two solutions and conclude that, for appropriate $\alpha(s)$ and $\beta(s)$,

$$\rho_b(s) = \alpha(s)\left(\frac{1 + \sqrt{1 - 4p(1-p)s^2}}{2(1-p)s}\right)^b + \beta(s)\left(\frac{1 - \sqrt{1 - 4p(1-p)s^2}}{2(1-p)s}\right)^b.$$

The conditions $\rho_0(s) = \rho_c(s) = 1$ determine the functions α and β. The result, after some algebraic simplification, is:

$$\rho_b(s) = (2ps)^b \frac{\left(1 + \sqrt{1 - 4p(1-p)s^2}\right)^{c-b} - \left(1 - \sqrt{1 - 4p(1-p)s^2}\right)^{c-b}}{\left(1 + \sqrt{1 - 4p(1-p)s^2}\right)^c - \left(1 - \sqrt{1 - 4p(1-p)s^2}\right)^c}$$

$$+ (2(1-p)s)^{c-b} \frac{\left(1 + \sqrt{1 - 4p(1-p)s^2}\right)^b - \left(1 - \sqrt{1 - 4p(1-p)s^2}\right)^b}{\left(1 + \sqrt{1 - 4p(1-p)s^2}\right)^c - \left(1 - \sqrt{1 - 4p(1-p)s^2}\right)^c}.$$

Even though we have assumed $s > 0$ in the calculation, the preceding formula is also correct for $s = 0$, by continuity.

The expected value of the hitting time can be calculated by differentiating ρ_b. However, we find it here by multiplying (11.3) by n and summing from 1 to ∞. The result is the following inhomogeneous linear second-order difference equation:

$$(11.5) \qquad E(N_b) = pE(N_{b-1}) + (1-p)E(N_{b+1}) + 1, \quad 0 < b < c.$$

We begin by solving this difference equation under the assumption that all quantities are finite, first for the case $p \neq \frac{1}{2}$. The general solution of the homogeneous equation

$$E(N_b) = pE(N_{b-1}) + (1-p)E(N_{b+1})$$

can be obtained by the same method that was used for (11.4). It is

$$\gamma + \delta\left(\frac{p}{1-p}\right)^b.$$

The function $b/(2p-1)$ is easily seen to be a particular solution of (11.5), so that the general solution is

$$\gamma + \delta\left(\frac{p}{1-p}\right)^b + \frac{b}{2p-1}.$$

The conditions $E(N_0) = E(N_c) = 0$ determine γ and δ. The result is

$$E(N_b) = \frac{b[1 - (p/(1-p))^c] - c[1 - (p/(1-p))^b]}{(2p-1)[1 - (p/(1-p))^c]}.$$

When $p = 1/2$, the general solution to (11.5) is of the form $\gamma + \delta b - b^2$. Solving for γ and δ as above gives

$$E(N_b) = b(c - b).$$

To confirm that our formulas are really correct, that is, that both sides of (11.5) are finite, we note first that, for each b, $N_b < Mc$, where

$$M(\omega) = \inf\{n : \omega \in A_n\}$$

and

$$A_n = \{\omega\colon X_{nc}(\omega) = X_{nc-1}(\omega) = \cdots = X_{nc-(c-1)}(\omega) = 1\}.$$

The events A_n are independent and the probability of each equals p^c. Therefore,

$$P(\{\omega\colon M(\omega) > m\}) = (1 - p^c)^m.$$

Thus, M is geometrically distributed and, hence, has finite mean. So, each N_b also has finite mean.

Problem 27. Show that $\rho_b(1-) = 1$, where ρ_b is as in the last example. What conclusion can you draw? Does that conclusion seem intuitively reasonable? Also, calculate the expected value of the hitting time of the set $\{-(c-b), b\}$ by using the derivative of ρ_b. *Hint:* You can simplify your computations by noting that when $s = 1$, the expression under the radical is a perfect square.

* **Problem 28.** For the random walks of Example 4, let σ_b be the probability generating function of the hitting time of $\{b\}$. Prove that $\sigma_b = (\sigma_1)^b$ for $b > 0$ and obtain a similar formula in case $b < 0$. By an argument similar to that used in Example 4, find another relation between σ_1 and σ_2 and then use the two relations between σ_1 and σ_2 to evaluate σ_1. Alternatively, evaluate σ_1 by letting c go to ∞ in Example 4. Calculate the distribution of the hitting time of $\{1\}$ (illustrated in Figure 11.3); in particular evaluate the probability that the random walk never hits $\{1\}$. Finally, calculate the distribution of the global maximum (or supremum in case the global maximum does not exist) of the random walk. *Hint:* After obtaining a formula for σ_1 rationalize the denominator if necessary. Also, see Problem 33 in Chapter 5.

The following exercise shows how a hitting time may be used to simplify a formula that seems difficult to treat by direct algebraic techniques.

Problem 29. Evaluate the following sum for $0 < p < 1$:

$$\sum_{n=k}^{\infty} \binom{n}{k} p^k (1 - p)^{n-k}.$$

Hint: Consider a random walk in $\mathbb{Z}$ whose step distribution is supported by $\{0, 1\}$.

11.6. Returns to 0

The hitting time of $\{0\}$ for a random walk $(S_0, S_1, \dots)$ adapted to a filtration $(\mathcal{F}_n\colon n \geq 0)$ equals 0 (with probability 1). Of more interest is the *first return time to 0*,

$$T_1(\omega) = \inf\{n > 0\colon S_n(\omega) = 0\},$$

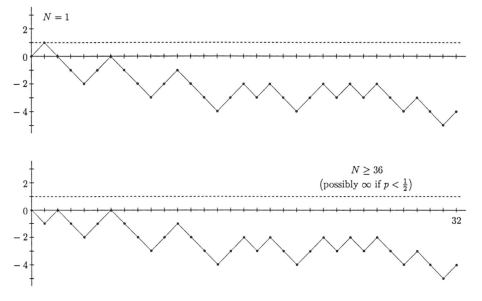

FIGURE 11.3. The hitting time of $\{1\}$

with, as is usual, $\inf \emptyset$ defined to equal ∞. Since

$$\{\omega\colon T_1(\omega) \le n\} = \bigcup_{m=1}^{n} \{\omega\colon S_m(\omega) = 0\} \in \mathcal{F}_n,$$

T_1 is a stopping time.

Example 5. Let us calculate the distribution of the first return time T_1 to 0 for the simple random walk S in $\mathbb{Z}$ with step distribution Q given by

$$Q(\{1\}) = p = 1 - Q(\{-1\}).$$

We take $0 < p < 1$.

As in Example 4 we let $\widetilde{S}$ denote the random walk obtained by using the steps of S beginning with the second step. Thus, $\widetilde{S}$ is independent of S_1 and, for $n = 0, 1, 2, \ldots$, $\widetilde{S}_n = S_{n+1} - S_1$. Let $\widetilde{M}_b$ denote the hitting time of $\{b\}$ by the process $\widetilde{S}$. Then, for $n > 0$,

$$\begin{aligned}
P(\{\omega\colon T_1(\omega) = n\}) \\
= P(\{\omega\colon S_1(\omega) = 1, \widetilde{M}_{-1}(\omega) = n - 1\}) \\
\quad + P(\{\omega\colon S_1(\omega) = -1, \widetilde{M}_1(\omega) = n - 1\}) \\
= P(\{\omega\colon S_1(\omega) = 1\}) \, P(\{\omega\colon \widetilde{M}_{-1}(\omega) = n - 1\}) \\
\quad + P(\{\omega\colon S_1(\omega) = -1\}) \, P(\{\omega\colon \widetilde{M}_1(\omega) = n - 1\}) \\
= p \, P(\{\omega\colon \widetilde{M}_{-1}(\omega) = n - 1\}) + (1 - p) \, P(\{\omega\colon \widetilde{M}_1(\omega) = n - 1\}).
\end{aligned}$$

Using the formula obtained for the distribution of $\widetilde{M}_1$ in Problem 28 and a similar formula for $\widetilde{M}_{-1}$, we obtain, for positive even n,

$$
\begin{aligned}
&P(\{\omega\colon T_1(\omega)=n\}) \\
&= p\,\frac{[p(1-p)]^{n/2}}{p}\,\frac{(n-2)!}{(\frac{n}{2})!(\frac{n}{2}-1)!} + (1-p)\,\frac{[p(1-p)]^{n/2}}{1-p}\,\frac{(n-2)!}{(\frac{n}{2})!(\frac{n}{2}-1)!} \\
&= 2[p(1-p)]^{n/2}\,\frac{(n-2)!}{(\frac{n}{2})!(\frac{n}{2}-1)!} \\
&= \frac{2}{\frac{n-2}{2}+1}\binom{n-2}{\frac{n-2}{2}}[p(1-p)]^{n/2}\,;
\end{aligned}
$$

and

$$
P(\{\omega\colon T_1(\omega)=\infty\}) = |2p-1|\,.
$$

We notice that: (i) twice the Catalan numbers (see Problem 33 of Chapter 5) appear as coefficients of $[p(1-p)]^{n/2}$ and (ii) $T_1 < \infty$ a.s. if and only if $p = \frac{1}{2}$. For the case $p = \frac{1}{2}$, the distribution function of T_1 is shown via the solid graph in Figure 11.4, with the jumps being filled in as a visual aid. (The dashed graph is related to the Glivenko-Cantelli Theorem described in the next chapter.)

* **Problem 30.** For T_1 defined as in the preceding example show that $E(T_1) = \infty$, even if $p = 1/2$. For the case $p = 1/2$, decide which moments of T_1 are finite.

Let T_1 be the first return time to 0 of a random walk S. For $j > 1$ recursively define the jth *return time to 0* of S by

$$
T_j(\omega) = \inf\{n > T_{j-1}(\omega)\colon S_n(\omega) = 0\}
$$

where, as usual, $\inf\emptyset = \infty$. The proof that each T_j is a stopping time with respect to the minimal filtration of S is similar to that for the first return time to 0.

Theorem 12. *For $j = 1, 2, \ldots$, let T_j denote the jth return time of some random walk to 0. Let R be the distribution of T_1. Then the distribution of $(0, T_1, T_2, \ldots)$ is that of a random walk in $\overline{\mathbb{Z}}^+$ with step distribution R.*

Problem 31. Prove the preceding theorem. *Hint:* Use Proposition 3. Show that one may take each B_k in that theorem to be a one-point set in the current situation.

Corollary 13. *Let Q denote the step distribution of a random walk S, and let*

$$
V(\omega) = \inf\{j\colon T_j(\omega)=\infty\} = \sharp\{n \geq 0\colon S_n(\omega)=0\}\,.
$$

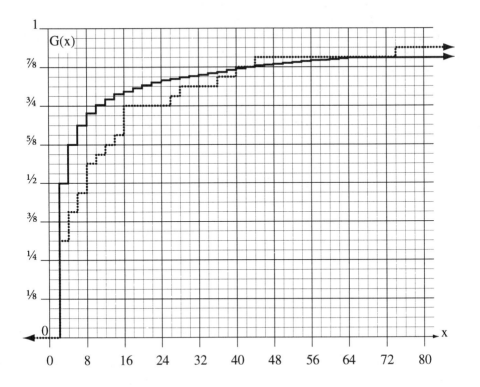

FIGURE 11.4. Distribution function G of first return time to 0, and empirical distribution function

Then either

$$\sum_{n=0}^{\infty} Q^{*n}(\{0\}) = \infty,$$

in which case $V = \infty$ a.s., or

$$\sum_{n=0}^{\infty} Q^{*n}(\{0\}) < \infty,$$

in which case the distribution of V is of geometric type, supported by $\{1, 2, \ldots\}$, and

(11.6) $P(\{\omega \colon V(\omega) = 1\}) = P(\{\omega \colon T_1(\omega) = \infty\}) = \dfrac{1}{\sum_{n=0}^{\infty} Q^{*n}(\{0\})}.$

(The possibility that $P(\{\omega \colon V(\omega) = 1\}) = 1$ is not excluded by the phrase "of geometric type".)

---

*   **Problem 32.** Prove the preceding corollary. *Hint:* Once it is known that V has geometric type, a relation between $E(V)$ and $P(\{\omega \colon V(\omega) = 1\})$ can be obtained.

The random variable V in the preceding result counts the number of times (including the time $n = 0$) at which the random walk is at 0. By Problem 5 of Chapter 6, if we define $A_n = \{\omega \colon S_n(\omega) = 0\}$, then $V(\omega) = \infty$ if and only if $\omega \in \limsup_n A_n$. Thus, part of Corollary 13 may be rephrased in a way that is reminiscent of the Borel and Borel-Cantelli Lemmas:

$$(11.7) \qquad P(\limsup_{n \to \infty} A_n) = \begin{cases} 1 & \text{if } \sum_{n=0}^{\infty} P(A_n) = \infty \\ 0 & \text{if } \sum_{n=0}^{\infty} P(A_n) < \infty. \end{cases}$$

When $P(\limsup_n A_n) = 1$, the random walk S is called *recurrent*. Otherwise, S is called *transient*. Note that the Borel Lemma can be used to get part of (11.7), but that the Borel-Cantelli Lemma cannot be used to get the other part, since typically the events A_n are neither uncorrelated nor negatively correlated.

Problem 33. Apply Corollary 13 to the random walk of Problem 12.

Problem 34. Let S be simple symmetric random walk in $\mathbb{Z}$. Show that

$$\lim_{n \to \infty} \sqrt{\pi n}\, P(\{\omega \colon S_{2n}(\omega) = 0\}) = 1.$$

Conclude that S is recurrent. *Hint:* Use the Stirling Formula.

Problem 35. Let $S^{(i)}$, $i = 1, \ldots, d$, be independent simple symmetric random walks in $\mathbb{Z}$, and let $\widetilde{S}$ be the sequence of $\mathbb{Z}^d$-valued random variables defined by

$$\widetilde{S}_n = (S_n^{(1)}, \ldots, S_n^{(d)}), \quad n = 0, 1, 2, \ldots.$$

Show that $\widetilde{S}$ is a random walk on $\mathbb{Z}^d$ and describe its step distribution. Use Problem 34 to show that $\widetilde{S}$ is recurrent if $d = 2$ and transient if $d \geq 3$.

Problem 36. Show that simple symmetric random walk in $\mathbb{Z}^2$ is recurrent. *Hint:* Find a linear transformation L from $\mathbb{R}^2$ onto $\mathbb{R}^2$ such that S has the same distribution as the sequence $(L(\widetilde{S}_0), L(\widetilde{S}_1), L(\widetilde{S}_2), \ldots)$, where $\widetilde{S}$ is the random walk from the preceding exercise with $d = 2$.

Problem 37. Let S be simple symmetric random walk in $\mathbb{Z}^3$. Show that

$$P(\{\omega \colon S_{2n}(\omega) = 0\}) = \frac{(2n)!}{6^{2n} n! n!} \sum_{j=0}^{n} \sum_{k=0}^{n-j} \left(\frac{n!}{j! k! (n-j-k)!} \right)^2$$

for $n = 0, 1, 2, \ldots$. Conclude that S is transient. *Hint:* Use the fact that for nonnegative $a_1, \ldots, a_m$,

$$\sum_{\ell=1}^{m} a_\ell^2 \leq \left(\max_{1 \leq \ell \leq m} a_\ell \right) \sum_{\ell=1}^{m} a_\ell.$$

Problem 38. Show that simple symmetric random walk in $\mathbb{Z}^d$ is transient for $d \geq$ 3. (This problem can be done in a straightforward manner by generalizing the calculation done in the preceding problem. But there is an interesting alternative method: Project the d-dimensional random walk onto $\mathbb{Z}^3$, and then use the result of the preceding problem to show that this projection is transient. Some care is required to carry out this second method rigorously.)

In Chapters 12, 13, and 25 we will develop further tools for analyzing the returns to 0 of random walks. In particular, we will prove that if S is a random walk on $\mathbb{Z}$ such that $E(S_1)$ exists, then the probability that S returns to 0 is less than 1 if $E(S_1) \neq 0$ (Chapter 12) and equal to 1 if $E(S_1) = 0$ (Chapter 13).

11.7. † Random walks in various settings

As indicated earlier, one can define random walks in semigroups other than $\mathbb{R}^d$, $\mathbb{Z}^d$, $\mathbb{R}^+$ and $\overline{\mathbb{Z}}^+$. Some random walks in various semigroups are described in the following problems.

Problem 39. Consider the collection Ψ of all subsets of the finite set $\{1, \ldots, m\}$. Thus Ψ has 2^m members. Prove that $(\Psi, \mathcal{G})$ is a commutative group under the symmetric difference operation $(\psi, \chi) \rightsquigarrow \psi \triangle \chi$. In particular, identify the identity in Ψ as well as the inverse of each member of the group.

* **Problem 40.** Let S denote the random walk in the group Ψ of the preceding problem whose step distribution Q assigns probability $1/m$ to each singleton. Calculate

$$P\big(\{\omega \colon S_m(\omega) = \{1, \ldots, m\}\}\big).$$

* **Problem 41.** Consider the random walk of the preceding problem for the case $m = 2$. Calculate the distribution of S_n for each n. Do the same for the case $m = 3$.

* **Problem 42.** Consider the semigroup Ψ of all 2^m subsets of $\{1, \ldots, m\}$ under the operation of intersection. Let S denote the random walk having the step distribution that assigns equal probability to each member of Ψ. For $j = 1, \ldots, m$, let N_j be the hitting time of the collection of sets not containing j. Prove that the random variables $N_j, 1 \leq j \leq m$, are independent and geometrically distributed. Calculate the distribution of the hitting time of the one-point set $\{\emptyset\}$. For $1 \leq k < m$ find a formula for the distribution of the hitting time of the one-point set $\{\{1, 2, \ldots, k\}\}$. Calculate explicitly the probability that this hitting time equals ∞ for the case $k = m - 1$.

Problem 43. Let Ψ be the semigroup of all infinite sequences of 1's and 0's under the operation of term-by-term multiplication. Consider the random walk whose step distribution assigns probability 2^{-k} to the singleton $\{(1, 1, \ldots, 1, 0, 1, 1, \ldots)\}$, where the only 0 is in the kth position. For $i = 1, 2, 3$, calculate the distribution of the hitting time of the set $\{\psi \colon \psi_1 = \cdots = \psi_i = 0\}$.

Problem 44. Fix a positive integer m and consider the collection K of all size-two subsets of $\{1, \ldots, m\}$. Let Ψ denote the family of all subcollections of K. How many members does Ψ have? Notice that a member ψ of Ψ can be interpreted as a graph with m vertices: the graph corresponding to ψ contains an edge connecting the vertices i and j if and only the pair $\{i, j\}$ is a member of ψ. Regard Ψ as a semigroup under the operation of union. Consider the random walk whose step distribution assigns probability $\frac{2}{m(m-1)}$ to each singleton of Ψ. For $m = 2, 3, 4$, calculate the distribution of the hitting time of the collection of connected graphs. For general m calculate the distribution of the hitting time of the one-point set $\{K\}$. *Hint:* Use inclusion-exclusion for the last part.

* **Problem 45.** Consider a random walk S in a finite commutative group whose step distribution assigns equal probability to each member of the group. Calculate the distribution of each S_n, and of the sequence S. Calculate the distribution of the first return time to 0.

Problem 46. Show that any random walk in a finite commutative group is recurrent.

CHAPTER 12
Theorems of A.S. Convergence

In this chapter we study the convergence of certain sequences defined in terms of sums of independent random variables. We will chiefly be interested in almost sure convergence, but it will be useful to also consider another weaker type of convergence, called 'convergence in probability'. The two major results concerning almost sure convergence are the Strong Law of Large Numbers and the Kolmogorov Three-Series Theorem. Other results included in this chapter are three important tools: the Kolmogorov 0-1 Law, the Hewitt-Savage 0-1 Law, and an important inequality, known as the Etemadi Lemma. As an application of the ideas contained in the proof of the Strong Law of Large Numbers, we also determine the asymptotic behavior of the size of the image of a random walk.

12.1. Convergence in probability

The following definition gives a name to a type of convergence that we have already encountered in the Weak Law of Large Numbers.

Definition 1. Let $Y_n, n = 1, 2, \ldots$, and Y be $\overline{\mathbb{R}}$-valued random variables defined on a common probability space $(\Omega, \mathcal{F}, P)$. For each $\varepsilon > 0$, let

$$
\begin{aligned}
A_n^\varepsilon &= \{\omega : |Y(\omega)| < \infty \text{ and } |Y(\omega) - Y_n(\omega)| > \varepsilon\}, \\
B_n^\varepsilon &= \{\omega : Y(\omega) = \infty \text{ and } Y_n(\omega) < 1/\varepsilon\}, \\
C_n^\varepsilon &= \{\omega : Y(\omega) = -\infty \text{ and } Y_n(\omega) > -1/\varepsilon\}.
\end{aligned}
$$

Then $Y_n \to Y$ *in probability* as $n \to \infty$ if, for each $\varepsilon > 0$,

$$
\lim_{n \to \infty} P(A_n^\varepsilon \cup B_n^\varepsilon \cup C_n^\varepsilon) = 0.
$$

The phrase 'in probability' will often be abbreviated 'i.p.', and if $Y_n \to Y$ i.p. as $n \to \infty$, we may also write

$$
\lim_{n \to \infty} Y_n = Y \text{ i.p.}
$$

The Law of Large Numbers in Chapter 5 implies that if S is the sequence of partial sums of an iid sequence of $\mathbb{R}$-valued random variables having finite second moments, then $S_n/n \to E(S_1)$ i.p. The Strong Law of Large Numbers, to be stated and proved later in the present chapter, implies that $S_n/n \to E(S_1)$ a.s. (In addition, we will find that the second moment hypothesis can be replaced by the weaker assumption that $E(S_1)$ exists.) The term 'strong' is often used to describe a result that asserts almost sure convergence.

The next problem and the theorem and example following the problem give important information about the relationship between i.p. convergence and a.s. convergence.

Problem 1. Let Y, $A_n^\varepsilon, B_n^\varepsilon, C_n^\varepsilon$ be as in Definition 1. Prove that $\lim_n Y_n = Y$ a.s. if and only if

$$P\left(\limsup_n [A_n^\varepsilon \cup B_n^\varepsilon \cup C_n^\varepsilon]\right) = 0$$

for all ε.

Theorem 2. Let $Y_n, n = 1, 2, \ldots,$ and Y be $\overline{\mathbb{R}}$-valued random variables defined on a probability space $(\Omega, \mathcal{F}, P)$. If $Y_n \to Y$ a.s. as $n \to \infty$, then $Y_n \to Y$ i.p. as $n \to \infty$.

Problem 2. Prove the preceding theorem. *Hint:* Use Problem 1 in conjunction with Problem 9 of Chapter 6.

(Counter)example 1. Let P denote Lebesgue measure on $\Omega = [0, 1)$. For $m = 0, 1, 2, \ldots$ and $2^m \leq n < 2^{m+1}$, let Y_n denote the indicator function of the interval

$$[n2^{-m} - 1, (n + 1)2^{-m} - 1).$$

It is clear that, for each ω, $\limsup Y_n(\omega) = 1$ and $\liminf Y_n(\omega) = 0$ and, hence, that, for every ω, the sequence $(Y_n(\omega) \colon n = 1, 2, \ldots)$ fails to converge. Since $P(\{\omega \colon Y_n(\omega) \neq 0\}) \to 0$ we see that $Y_n \to 0$ i.p. Thus, a.s. convergence is strictly stronger than i.p. convergence.

In the setting of a general measure space $(\Omega, \mathcal{F}, \mu)$, 'convergence in probability' is replaced by *convergence in measure*. That is, we require that

$$\lim_{n \to \infty} \mu(A_n^\varepsilon \cup B_n^\varepsilon \cup C_n^\varepsilon) = 0$$

for all $\varepsilon > 0$, where $A_n^\varepsilon, B_n^\varepsilon, C_n^\varepsilon$ are defined as in Definition 1. It is important to remember that in an infinite measure space, convergence almost everywhere does not imply convergence in measure.

Problem 3. Provide an example in which there is almost everywhere convergence but not convergence in measure.

The following lemma enables us to replace the hypothesis of convergence almost everywhere in some theorems of Chapter 8 by the hypothesis of convergence in measure.

Lemma 3. *Let $f_n, n = 1, 2, \ldots$, and f be $\overline{\mathbb{R}}$-valued measurable functions defined on a common measure space. If $f_n \to f$ in measure as $n \to \infty$, then there exists a subsequence of (f_n) that converges to f almost everywhere.*

Problem 4. Prove the preceding lemma. *Hint:* Recall that the Borel Lemma is true for general measure spaces, and apply the general measure space version of Problem 1.

Each of the next three results is proved in the following manner: Use convergence in measure to find a subsequence that converges almost everywhere, apply the corresponding result from Chapter 8 to the subsequence, and then use Proposition 4 of Appendix B, if necessary, to get back to the original sequence.

Lemma 4. [Fatou] *Let f_n, $n = 1, 2, \ldots$, and f be $\mathbb{R}^+$-valued measurable functions defined on a measure space $(\Omega, \mathcal{F}, \mu)$. Suppose that $f_n \to f$ in measure as $n \to \infty$. Then*

$$\int f \, d\mu \leq \liminf_{n \to \infty} \int f_n \, d\mu \,.$$

Theorem 5. [Dominated Convergence] *Let f, g, and f_n, $n = 1, 2, \ldots$, be $\mathbb{R}$-valued measurable functions defined on a measure space $(\Omega, \mathcal{F}, \mu)$. Suppose that $|f_n(\omega)| \leq g(\omega)$ for almost every ω and every n, that $\int g \, d\mu < \infty$, and that $f_n \to f$ in measure as $n \to \infty$. Then*

$$\int |f| \, d\mu < \infty, \quad \lim_{n \to \infty} \int f_n \, d\mu = \int f \, d\mu, \quad \text{and} \quad \lim_{n \to \infty} \int |f - f_n| \, d\mu = 0 \,.$$

Theorem 6. [Uniform Integrability Criterion] *Let $(X_1, X_2, \ldots)$ be a sequence of $\mathbb{R}$-valued random variables whose limit X exists in probability. Assume that $E(|X_n|) < \infty$ for all n. Then, the following three statements are equivalent:*

    (i) the family $\{X_n : n = 1, 2, \ldots\}$ is uniformly integrable;

    (ii) $E(|X|) < \infty$ and $\lim_{n \to \infty} E(|X_n - X|) = 0$;

    (iii) $\lim_{n \to \infty} E(|X_n|) = E(|X|) < \infty$.

Each of these three conditions implies

    (iv) $\lim_{n \to \infty} E(X_n) = E(X)$.

Problem 5. Prove the preceding three theorems. *Hint:* For a proof by contradiction that (iii) $\implies$ (i) in the Uniform Integrability Criterion, find a subsequence (X_{n_k}) such that no further subsequence is uniformly integrable.

12.2. Laws of Large Numbers

Let $S = (S_n \colon n = 1, 2, \dots)$ be the sequence of partial sums of an iid sequence $(X_n \colon n = 1, 2, \dots)$ of $\mathbb{R}$-valued random variables. We will study the behavior of S_n/n as $n \to \infty$ by a sequence of lemmas, leading up to the main result, the Strong Law of Large Numbers, which is Theorem 14. These lemmas are interesting in their own right, because their proofs illustrate commonly used techniques. In some cases, the lemmas are given in greater generality than that needed for their application in the proof of Theorem 14. For example, our first four lemmas make no mention of independence. In fact, it will be seen that even the final result only requires pairwise-independence.

Lemma 7. *Suppose that a sequence* $(Y_n \colon n = 1, 2, \dots)$ *of* $\mathbb{R}$-*valued random variables satisfies* $E(Y_n) \to c$ *for some real constant* c *and* $\sum_{n=1}^{\infty} \mathrm{Var}(Y_n) < \infty$. *Then* $Y_n \to c$ *a.s. as* $n \to \infty$.

PROOF. Let $Z_n = Y_n - E(Y_n)$ for each n, and

$$W = \sum_{n=1}^{\infty} Z_n^2 \,,$$

an $\overline{\mathbb{R}}^+$-valued random variable. By the Monotone Convergence Theorem,

$$E(W) = \sum_{n=1}^{\infty} E(Z_n^2) = \sum_{n=1}^{\infty} \mathrm{Var}(Y_n) < \infty \,.$$

Therefore, $W < \infty$ a.s. and, hence, with probability one $Z_n \to 0$ as $n \to \infty$. $\square$

Problem 6. Give an alternative proof of the preceding lemma using the Chebyshev Inequality and the Borel Lemma.

Problem 7. Show that if the hypothesis $\sum_n \mathrm{Var}(Y_n) < \infty$ in Lemma 7 is replaced by the weaker hypothesis $\lim_n \mathrm{Var}(Y_n) = 0$, then we may conclude that $\lim_n Y_n = Y$ i.p.

Problem 8. Let $(S_n \colon n = 1, 2, \dots)$ be the sequence of partial sums of an iid sequence of $\mathbb{R}$-valued random variables having finite second moment. Show that if $(n_k \colon k = 1, 2, \dots)$ is an increasing sequence of positive integers such that for some $p > 1$, $n_k \geq k^p, k = 1, 2, \dots$, then

$$\lim_{k \to \infty} \frac{S_{n_k}}{n_k} = E(S_1) \text{ a.s.}$$

The preceding problem shows that Lemma 7 can be used to get the almost sure convergence of S_n/n along a subsequence, provided the summands are assumed to have finite second moments. But we will not be making such an assumption in the Strong Law of Large Numbers. In order to be able to apply results like Lemma 7 when second moments are not finite, we will need to work with summands that have been truncated. The following two lemmas introduce the appropriate truncation and show that when the summands have finite means, this truncation makes no essential difference in the limiting behavior of S_n/n.

Lemma 8. *Let $(X_n : n = 1, 2, \ldots)$ be a sequence of identically distributed random variables having finite mean. Then*

$$P(\limsup_{n \to \infty} \{\omega : |X_n(\omega)| > n\}) = 0.$$

PROOF. Let F denote the common distribution function of the random variables $|X_n|$. Using the expression for the mean of a nonnegative random variable given in Corollary 20 of Chapter 4, we obtain

$$\infty > \int_0^\infty [1 - F(x)]\, dx \geq \sum_{n=1}^\infty [1 - F(n)] = \sum_{n=1}^\infty P(\{\omega : |X_n(\omega)| > n\}).$$

An appeal to the Borel Lemma completes the proof. □

Problem 9. Generalize the conclusion of the preceding lemma by replacing " $> n$" by " $> cn$", where c is an arbitrary positive constant. Prove this generalization and then use it to conclude that

$$\lim_{n \to \infty} \frac{X_n}{n} = 0 \text{ a.s.}$$

for any sequence (X_n) of identically distributed $\mathbb{R}$-valued random variables having finite mean.

Lemma 9. *Let $(X_n : n = 1, 2, \ldots)$ be a sequence of identically distributed random variables having finite mean. For $n = 1, 2, \ldots$ let $Y_n = X_n I_{\{\omega : |X_n(\omega)| \leq n\}}$. Let $(S_n : n = 1, 2, \ldots)$ and $(T_n : n = 1, 2, \ldots)$ denote the sequences of partial sums of the sequences (X_n) and (Y_n), respectively. Then*

$$\lim_{n \to \infty} \left(\frac{S_n}{n} - \frac{T_n}{n} \right) = 0 \text{ a.s.}$$

and

$$\lim_{n \to \infty} \frac{E(T_n)}{n} = E(X_1).$$

PROOF. The first conclusion is left as an exercise. To prove the second conclusion, note that by an application of the Dominated Convergence Theorem, requested below in Problem 10, $E(Y_n) \to E(X_1)$. It is a standard exercise to show that if a sequence (a_n) converges, then the sequence $(a_1, (a_1 + a_2)/2, (a_1 + a_2 + a_3)/3, \ldots)$ converges to the same limit. Taking $a_n = E(Y_n)$, we see that $E(T_n)/n \to E(X_1)$ as $n \to \infty$. □

* **Problem 10.** Complete the proof of the preceding lemma, by doing the following: (i) use Lemma 8 to show that $(S_n/n - T_n/n) \to 0$ a.s. as $n \to \infty$; (ii) show how the Dominated Convergence Theorem applies to give $E(Y_n) \to E(X_1)$ as $n \to \infty$; (iii) prove the fact stated in the proof about convergent sequences (a_n).

Lemma 9 says that we can analyze the convergence of S_n/n by studying that of T_n/n. It is a standard technique to use the Borel Lemma to replace random variables of interest by truncated random variables, as in Lemma 8 and Lemma 9.

The next lemma gives us control over the second moments of the truncated random variables.

Lemma 10. *Let* X_n, *and* Y_n *be as in* Lemma 9. *Then*

$$\sum_{j=1}^{\infty} \frac{\mathrm{Var}(Y_j)}{j^2} < \infty \,.$$

PROOF. The function $(j, x) \leadsto x^2 I_{[-j,j]}(x)/j^2$ is a nonnegative measurable function on the product space $\{1, 2, \dots\} \times (-\infty, \infty)$. By the Fubini Theorem, its integral with respect to the product of counting measure and the common distribution Q of the X_n equals two iterated integrals. One of these is

$$\int_{(-\infty,\infty)} x^2 \left(\sum_{j \geq |x|} \frac{1}{j^2} \right) Q(dx)$$

$$\leq \int_{(-\infty,\infty)} x^2 \left(\frac{1}{x^2} + \int_{|x|}^{\infty} \frac{1}{j^2} \, dj \right) Q(dx)$$

$$= 1 + E(|X_1|) < \infty \,.$$

The other is

$$\sum_{j=1}^{\infty} \frac{E(Y_j^2)}{j^2} \,,$$

which must, therefore, also be finite. Since variances are no larger than second moments, the proof is complete. $\square$

Notice that the conclusion that $\sum \mathrm{Var}(Y_j)/j^2 < \infty$ in the preceding lemma is easy to prove if $E(X_1^2) < \infty$. In the current development we only want to assume the means to be finite. The history of probability contains many instances of theorems with superfluous assumptions on moments that have later been removed by better and often simpler proofs.

In order to proceed further, we need to assume some independence.

Lemma 11. *Let* $(X_n : n = 1, 2, \dots)$ *be a sequence of identically distributed pairwise independent* $\mathbb{R}$-*valued random variables with finite mean, and let* T_n *be defined as in* Lemma 9. *Fix* $c > 1$ *and, for* $m = 0, 1, 2, \dots$, *let* $b_m = \lceil c^m \rceil$. *Then* $\lim_{m \to \infty} T_{b_m}/b_m = E(X_1)$ *a.s.*

PROOF. By Lemma 9, $E(T_{b_m})/b_m \to E(X_1)$ as $m \to \infty$. In order to apply Lemma 7 we calculate

$$\sum_{m=0}^{\infty} \mathrm{Var}\left(\frac{T_{b_m}}{b_m}\right) \leq \sum_{m=0}^{\infty} \sum_{j=1}^{b_m} c^{-2m} \mathrm{Var}(Y_j)$$

$$= \frac{\mathrm{Var}\, Y_1}{1 - c^{-2}} + \sum_{j=2}^{\infty} \sum_{m > \frac{\log(j-1)}{\log c}} c^{-2m} \mathrm{Var}(Y_j)$$

$$\leq \frac{1}{1 - c^{-2}}\left(\mathrm{Var}(Y_1) + \sum_{j=2}^{\infty} \frac{\mathrm{Var}(Y_j)}{(j-1)^2}\right),$$

which is finite by Lemma 10. An appeal to Lemma 7 completes the proof. $\square$

Problem 11. In the setting of the preceding lemma, show that $T_n/n \to E(X_1)$ i.p. as $n \to \infty$. *Hint:* Use Lemma 10 to show that $\mathrm{Var}(T_n/n) \to 0$ as $n \to \infty$, then apply Problem 7.

The preceding problem, in conjunction with Lemma 3, shows that (T_n/n) (and hence (S_n/n)) converges almost surely to the desired limit along some subsequence. But we need more than that, and Lemma 11 gives it to us: It shows that this convergence occurs along certain specific subsequences (b_m). As the next lemma shows, when the random variables X_n are nonnegative, convergence along the subsequences (b_m) easily implies convergence along the full sequence.

Lemma 12. *Let T_n and X_n be as in the preceding lemma and suppose that each X_n is nonnegative. Then $T_n/n \to E(X_1)$ a.s. as $n \to \infty$.*

PROOF. Let $c > 1$ and define b_m as in Lemma 11. For $n = 1, 2, 3, \ldots$ let $M(n)$ equal the smallest integer m for which $n \leq c^m$. Then, by Lemma 11, for almost every ω,

$$\limsup_{n \to \infty} \frac{T_n(\omega)}{n} \leq \limsup_{n \to \infty} \frac{T_{b_{M(n)}}(\omega)}{c^{-1} b_{M(n)}} = cE(X_1).$$

For an inequality in the other direction let $L(n)$ equal the largest integer m for which $c^m \leq n$. Then, for almost every ω,

$$\liminf_{n \to \infty} \frac{T_n(\omega)}{n} \geq \liminf_{n \to \infty} \frac{T_{b_{L(n)}}(\omega)}{c\, b_{L(n)}} = c^{-1} E(X_1).$$

Let c decrease to 1 through a countable sequence to complete the proof. $\square$

Problem 12. What purpose, if any, is served by the phrase "countable sequence" in the last sentence of the preceding proof?

Problem 13. If $\mathrm{Var}(X_1) < \infty$, the preceding development can be simplified. In particular, truncation is not necessary. Carry out such a simplification, and, as you do so, weaken the assumption of identical distributions to one of identical means and variances. Also weaken the independence assumption to one concerning correlations.

In our final lemma before the main result of this section, we drop the assumption that the means be finite.

Lemma 13. *Let $(X_n: n = 1, 2, \ldots)$ be a sequence of nonnegative identically distributed pairwise-independent $\mathbb{R}$-valued random variables, and denote by $(S_n: n = 1, 2, \ldots)$ the corresponding sequence of partial sums. Then, as $n \to \infty$, $S_n/n \to E(X_1)$ a.s. Moreover, if $E(X_1) < \infty$, then $E(|S_n/n - E(X_1)|) \to 0$ as $n \to \infty$.*

PROOF. First assume that $E(X_1) < \infty$. Then, the conclusion $S_n/n \to E(X_1)$ a.s. follows immediately from Lemma 12 and Lemma 9. Since

$$E\left(\left|\frac{S_n}{n}\right|\right) = E\left(\frac{S_n}{n}\right) = E(X_1),$$

condition (iii) in the Uniform Integrability Criterion is satisfied, and, thus, $E(|S_n/n - E(X_1)|) \to 0$.

In case $E(X_1) = \infty$ we use a truncation argument and the Monotone Convergence Theorem:

$$\liminf_{n \to \infty} \frac{S_n(\omega)}{n} \geq \lim_{k \to \infty} \liminf_{n \to \infty} \frac{(k \wedge X_1(\omega)) + \cdots + (k \wedge X_n(\omega))}{n}$$
$$= \lim_{k \to \infty} E(k \wedge X_1) = E(X_1) = \infty \text{ a.s.} \quad \square$$

It remains to drop the assumption that the random variables X_n be nonnegative. We will assume that the means of the random variables X_n exist (not necessarily finite), and apply the preceding lemma separately to the positive and negative parts.

Theorem 14. [Strong Law of Large Numbers] *Let $(X_n: n = 1, 2, \ldots)$ be a sequence of identically distributed pairwise-independent $\mathbb{R}$-valued random variables for which $E(X_1)$ exists, and let $(S_n: n = 1, 2, \ldots)$ denote the corresponding sequence of partial sums. Then, as $n \to \infty$,*

$$\frac{S_n}{n} \to E(X_1) \text{ a.s.}$$

Moreover, if $|E(X_1)| < \infty$, then

$$E\left(\left|\frac{S_n}{n} - E(X_1)\right|\right) \to 0$$

as $n \to \infty$.

PROOF. Let $(U_n\colon n = 1, 2, \ldots)$ and $(V_n\colon n = 1, 2, \ldots)$ denote, respectively, the sequences of partial sums of the sequences (X_n^+) and (X_n^-). By Problem 2 of Chapter 9, each of these sequences is pairwise independent. By the preceding lemma,

$$\frac{S_n}{n} = \frac{U_n}{n} - \frac{V_n}{n} \to E(X_1^+) - E(X_1^-) = E(X_1) \text{ a.s.}$$

If, in addition, $|E(X_1)| < \infty$, then, from the triangle inequality, the positivity of the expectation operator, and the preceding lemma, we obtain

$$E\left(\left|\frac{S_n}{n} - E(X_1)\right|\right)$$

$$\leq E\left(\left|\frac{U_n}{n} - E(X_1^+)\right|\right) + E\left(\left|\frac{V_n}{n} - E(X_1^-)\right|\right) \to 0. \quad \square$$

Since almost sure convergence implies convergence in probability, we have also obtained $\lim_n S_n/n = E(X_1)$ i.p. under the assumption that $E(X_1)$ exists.

Problem 14. Compare the conclusion described in the preceding sentence with the Law of Large Numbers in Chapter 5.

Problem 15. Let $(S_0 = 0, S_1, S_2, \ldots)$ be the sequence of partial sums of an iid sequence having positive (possibly infinite) mean. Prove that $S_n \to \infty$ a.s. as $n \to \infty$.

* **Problem 16.** Let $(X_n\colon n = 1, 2, \ldots)$ be a sequence of identically distributed pairwise independent $\mathbb{R}$-valued random variables, and let $(S_n\colon n = 1, 2, \ldots)$ denote the corresponding sequence of partial sums. Assume that $E(|X_1|) = \infty$. Prove that for all $c > 0$,

(12.1) $P(\{\omega\colon |X_n(\omega)| > cn \text{ for infinitely many even } n\}) = 1$.

Conclude that

(12.2) $P\left(\left\{\omega\colon \dfrac{S_n(\omega)}{n} \text{ converges in } \mathbb{R} \text{ as } n \to \infty\right\}\right) = 0$.

We will discover in Chapter 15 that it is possible for the hypothesis $E(|X_1|) = \infty$ in the preceding exercise to hold and, nevertheless, for there to exist a finite number μ such that $\lim_n S_n/n = \mu$ i.p.

12.3. Applications

The following exercise shows that the expectation of a random variable need not be representative of typical values of that random variable.

* **Problem 17.** [Stick-breaking random walk] Let $(X_n\colon n = 1, 2, \ldots)$ be an iid sequence of random variables uniformly distributed on $(0, 1]$. For $n = 0, 1, 2, \ldots$, set $S_n = \prod_{m=1}^{n} X_k$. Calculate $E(S_n)$. Compare S_n and $E(S_n)$ for large n. Give an intuitive explanation of the result of your comparison. Describe the sequence (S_n) in terms of the successive lengths of the remaining part of a stick whose original length was one. (The sequence (S_n) is a random walk on the state space $(0, 1]$, which is a semigroup under the operation of ordinary multiplication. See Chapter 11.) *Hint:* In order to get an idea of how S_n behaves for large n, apply the Strong Law of Large Numbers to $\log(S_n)$.

For the next example, we extend the notion of mutual singularity introduced in Chapter 6. A family $(\mu_\alpha \colon \alpha \in \mathcal{A})$ of σ-finite measures on a measurable space $(\Omega, \mathcal{F})$ is *mutually singular* if, for each $\alpha \in \mathcal{A}$, there exists a measurable set $B_\alpha \subseteq \Omega$ such that $\{B_\alpha \colon \alpha \in \mathcal{A}\}$ is a family of pairwise disjoint sets and, for each $\alpha \in \mathcal{A}$, $\mu_\alpha(B_\alpha^c) = 0$.

Example 2. Fix a real number $p \in [0, 1]$, and let $X = (X_1, X_2, \ldots)$ be an iid sequence of Bernoulli random variables, each with parameter p. As in Example 1 of Chapter 2, for each ω, identify $X(\omega)$ with a point $Y(\omega)$ in $[0, 1]$ by way of binary expansions:

$$Y(\omega) = .X_1(\omega)X_2(\omega)X_3(\omega)\ldots_{\text{two}}.$$

Let Q_p be the distribution of Y.

Thus we have defined a family $(Q_p \colon p \in [0, 1])$ of probability measures on the Borel sets of $[0, 1]$. We will use the Strong Law of Large Numbers to show that this family is mutually singular. For each $p \in [0, 1]$, let

$$B_p = \left\{ y = .x_1 x_2 x_3 \ldots_{\text{two}} \colon \frac{1}{n} \sum_{i=1}^n x_i \to p \text{ as } n \to \infty \right\}.$$

By the Strong Law of Large Numbers, $Q_p(B_p^c) = 0$ for all $p \in [0, 1]$. Since the sets B_p are pairwise disjoint, the family $(Q_p \colon p \in [0, 1])$ is mutually singular, as desired.

The mutual singularity just demonstrated implies that no Q_p has a density respect to Q_q if $p \neq q$. In particular, since $Q_{1/2}$ is Lebesgue measure on $[0, 1]$, no Q_p, $p \neq 1/2$, has a density with respect to Lebesgue measure.

Problem 18. For $0 < p < 1$ and Q_p as defined in the preceding example, let F_p be the corresponding distribution function. Show that F_p is continuous and strictly increasing on $[0, 1]$, $F_p(0) = 0$, $F_p(1) = 1$, and

$$F_p(x) = \begin{cases} (1 - p)F_p(2x) & \text{if } 0 \leq x \leq 1/2 \\ 1 - p + pF_p(2x - 1) & \text{if } 1/2 \leq x \leq 1. \end{cases}$$

* **Problem 19.** Use the preceding exercise to calculate $F_p(3/4)$, $F_p(3/8)$, $F_p(1/3)$, and $F_p(7/10)$. Check each of the first three answers by interpreting it as the probability of the union of certain disjoint events each of whose probability is easily calculated.

Problem 20. Let Y be as defined in Example 2. Write Y as an infinite sum of independent random variables, and use this representation to calculate the mean and variance of Y under the various probability measures P_p. Also calculate the mean and variance of Y directly using the distribution function F_p.

Consider a product space $(\Omega, \mathcal{F}, P) = (\Psi, \mathcal{G}, Q)^\infty$ and write a member ω of Ω as $\omega = (\psi_1, \psi_2, \dots)$. Fix $B \in \mathcal{G}$ and define $X_n(\omega) = I_B(\psi_n)$ for $n = 1, 2, \dots$. We can think of the sequence $(\psi_n : n = 1, 2, \dots)$ as representing the outcomes of repeated trials in an experiment, with the random variable $S_n = \sum_{k=1}^n X_k$ counting the number of times in the first n trials that an outcome is in the set B, or stated more informally, "the number of times that B occurs by time n". Applying the Strong Law of Large Numbers, we conclude that the proportion of times that B occurs by time n converges a.s. to $Q(B)$ as $n \to \infty$.

The following result is both an application and an extension of the ideas introduced in the preceding paragraph. The function F_n, defined in the theorem in terms of an iid sequence, is often called the *empirical distribution function* of the sequence.

Theorem 15. [Glivenko-Cantelli] Let $(Y_n : n = 1, 2, \dots)$ be an iid sequence of $\mathbb{R}$-valued random variables defined on a probability space $(\Omega, \mathcal{F}, P)$ and having a common distribution function F. For each $x \in \mathbb{R}$, let I_x be the indicator function of the interval $(-\infty, x]$, and define random variables

$$F_n(x) = \frac{1}{n} \sum_{k=1}^n I_x(X_k)$$

for $\omega \in \Omega$ and $n = 1, 2, \dots$. Then

$$\lim_{n \to \infty} \sup_{x \in \mathbb{R}} |F_n(x) - F(x)| = 0 \text{ a.s.}$$

Problem 21. Prove the preceding theorem. *Hint:* The discussion in the paragraph preceding the statement of the theorem shows that for each fixed x, $F_n(x) \to F(x)$ a.s. as $n \to \infty$. Take the intersection of the relevant events for x rational, and then use the monotonicity of F and of each $F_n(\cdot, \omega)$ as functions of x to get the desired result.

Problem 22. Describe the changes, if any, needed in Theorem 15 to accommodate $\overline{\mathbb{R}}$-valued random variables.

In Figure 11.4 of Chapter 11 the distribution function of the time of first return to 0 of a simple symmetric random walk is shown. Also shown, in dashes, is the empirical distribution function obtained by having 32 people each flip a coin until the number of heads thus far obtained was equal to the number of tails thus far obtained or until the coin had been flipped 80 times whichever came first. To aid in visualization, the jumps in both graphs have been filled in.

12.4. 0-1 laws

In the Strong Law of Large Numbers we are interested in the event

$$A = \{\omega\colon \frac{S_n(\omega)}{n} \to E(X_1) \text{ as } n \to \infty\},$$

the occurrence of which is determined by an independent sequence $X = (X_n\colon n \geq 1)$ of random variables. Note that the occurrence of E does not depend on the values along any given *finite* subsequence of X. We are about to see that such events always have probability one or zero. We first need a definition.

Definition 16. Let $(\mathcal{G}_m\colon m = 1, 2, \ldots)$ be a sequence of sub-σ-fields of a σ-field $\mathcal{F}$. For each positive integer n let

$$\mathcal{H}_n = \sigma(\mathcal{G}_m\colon m \geq n).$$

The σ-field

$$\mathcal{T} = \bigcap_{n=1}^{\infty} \mathcal{H}_n$$

is called the *tail σ-field* of the sequence $(\mathcal{G}_m)$, and the members of $\mathcal{T}$ are *tail events*.

The following theorem reflects the fact that the tail σ-field of an infinite independent sequence of σ-fields is independent of itself.

Theorem 17. [Kolmogorov 0-1 Law] *Each tail event of an infinite independent sequence of σ-fields has probability 0 or 1.*

PROOF. Let $(\mathcal{G}_m\colon m = 1, 2, \ldots)$ denote an infinite independent sequence of σ-fields, and let $\mathcal{H}_n$ and $\mathcal{T}$ be defined as in Definition 16. By Proposition 4 of Chapter 9, $(\mathcal{H}_{m+1}, \mathcal{G}_1, \ldots, \mathcal{G}_m)$ is an independent finite sequence for each m. Since $\mathcal{T} \subseteq \mathcal{H}_{m+1}$, we see that $(\mathcal{T}, \mathcal{G}_1, \ldots, \mathcal{G}_m)$ is independent for each m. Hence the infinite sequence $(\mathcal{T}, \mathcal{G}_1, \mathcal{G}_2, \ldots)$ is independent. Applying Proposition 4 of Chapter 9 again, we conclude that $(\mathcal{T}, \mathcal{H}_1)$ is an independent pair. Since $\mathcal{T} \subseteq \mathcal{H}_1$, the pair $(\mathcal{T}, \mathcal{T})$ is independent. The result now follows from Problem 11 of Chapter 9. $\square$

Corollary 18. *Any random variable measurable with respect to the tail σ-field of an infinite independent sequence of σ-fields is equal to some constant a.s.*

Problem 23. Let $(S_n\colon n = 0, 1, 2, \ldots)$ be the sequence of partial sums of an independent (not necessarily identically distributed) sequence of $\mathbb{R}$-valued random variables. Prove that the $\overline{\mathbb{R}}$-valued random variables $\limsup(S_n/n)$ and $\liminf(S_n/n)$ each equal constants a.s.

Problem 24. Continue with the notation of the preceding exercise and, in addition, assume that the summands are identically distributed and do not have a mean (finite or infinite). Use Problem 16 to prove that $\limsup(S_n/n)$ and $\liminf(S_n/n)$ cannot both be finite. In addition, prove that $\limsup(S_n/n) = -\liminf(S_n/n) = \infty$ if the distribution of the steps is symmetric with respect to some real number.

Problem 25. Find an independent sequence $(X_1, X_2, \dots)$ of random variables such that

$$P(\limsup_{n \to \infty}\{\omega\colon X_n(\omega) = X_1(\omega)\}) = \frac{1}{2}.$$

Problem 26. Let $(S_0 = 0, S_1, S_2, \dots)$ be a random walk in $\mathbb{R}$ and let $(a_n > 0\colon n = 0, 1, 2, \dots)$ be an increasing sequence with limit ∞. Use the Kolmogorov 0-1 Law to show that

$$\limsup_{n \to \infty} \frac{S_n}{a_n}$$

is an almost surely constant $\overline{\mathbb{R}}$-valued random variable. Let c denote its value. Describe the difficulty in trying to use the Kolmogorov 0-1 Law to prove that the event

$$\limsup_{n \to \infty}[S_n > ca_n]$$

has probability 0 or 1. Then use the forthcoming Theorem 19 to prove that it does have probability 0 or 1.

We say that a σ-field of events is *0-1 trivial* if every event in it has probability 0 or 1. Of course, the trivial σ-field $\{\emptyset, \Omega\}$ is also 0-1 trivial. The Kolmogorov 0-1 Law says that the tail σ-field of an infinite independent sequence of σ-fields is 0-1 trivial. The focus of the remainder of this section is on another law that describes a 0-1 trivial σ-field.

Let Ψ denote any set. A permutation π on $\mathbb{Z}^+ \setminus \{0\}$ induces a permutation $\hat{\pi}$ of $\bigotimes_{n=1}^{\infty} \Psi$ via the formula

$$\hat{\pi}(\psi_1, \psi_2, \dots) = (\psi_{\pi^{-1}(1)}, \psi_{\pi^{-1}(2)}, \dots).$$

A subset A of $\bigotimes_{n=1}^{\infty} \Psi$ is *exchangeable* if $\hat{\pi}(A) = A$ for all permutations π of $\mathbb{Z}^+ \setminus \{0\}$ such that $\pi(k) = k$ for all but finitely many k. Let $\mathcal{G}$ denote a σ-field of subsets of Ψ. It is easy to see that the collection of all exchangeable members of $\bigotimes_{n=1}^{\infty} \mathcal{G}$ is a σ-field; it is called the *exchangeable σ-field* in $\bigotimes_{n=1}^{\infty}(\Psi, \mathcal{G})$.

Theorem 19. [Hewitt-Savage 0-1 Law] *Let*

$$(\Omega, \mathcal{F}, P) = \prod_{n=1}^{\infty}(\Psi, \mathcal{G}, Q),$$

where $(\Psi, \mathcal{G}, Q)$ is a probability space. Then the exchangeable σ-field is 0-1 trivial.

* **Problem 27.** Prove the preceding theorem. *Hint:* Use Lemma 18 of Chapter 9.

Problem 28. Let $X_n \colon (\Omega, \mathcal{F}, P) \to (\Psi, \mathcal{G})$ be independent identically distributed random variables and set $X = (X_1, X_2, \dots)$. On what basis do we know that

$$\left\{ X^{-1}(A) \colon A \text{ exchangeable in } \bigotimes_{n=1}^{\infty} \mathcal{G} \right\}$$

is a sub-σ-field of $\mathcal{F}$? Prove that it, known as the *exchangeable σ-field* in $\mathcal{F}$ induced by X, is 0-1 trivial (with respect to the probability measure P).

Problem 29. Let $X_n \colon (\Omega, \mathcal{F}, P) \to (\Psi_n, \mathcal{G}_n)$ be independent random variables and set $X = (X_1, X_2, \dots)$. On what basis do we know that

$$\left\{ X^{-1}(B) \colon B \text{ tail in } \bigotimes_{n=1}^{\infty} \mathcal{G}_n \right\}$$

is a sub-σ-field of $\mathcal{F}$. Prove that it, known as the *tail σ-field* in $\mathcal{F}$ induced by X, is 0-1 trivial (with respect to the probability measure P).

In view of Problem 28, we can, when discussing random walks, speak of *exchangeable events* and the *exchangeable σ-field* whether or not the underlying probability space is a product space. Similarly, when treating sequences of independent random variables, *tail events* may play a role even if the underlying probability space is not a product space.

* **Problem 30.** Let $(S_0 = 0, S_1, S_2, \dots)$ be a random walk the support of whose step distribution is $\mathbb{Z}$. With respect to the sequence of steps, decide, in each case, whether the given event is a tail event and whether it is an exchangeable event:

   (i) $\limsup_{n \to \infty} [S_n > 0]$;

   (ii) $\liminf_{n \to \infty} [S_n = S_{2n}]$;

   (iii) $\limsup_{n \to \infty} [S_n = S_2]$;

12.5. Random infinite series

This section and the two that follow it are concerned with the a.s. convergence of infinite series of independent $\mathbb{R}$-valued random variables. When we speak of convergence we will mean convergence within $\mathbb{R}$ (as opposed to convergence within $\overline{\mathbb{R}}$). The problem is rather uninteresting if the random variables are identically distributed, since in that case the sequence of partial sums will, with probability 1, fail to approach a finite limit, unless the summands equal 0 a.s.

Problem 31. Prove the assertion made in the preceding sentence.

Problem 32. Let $(X_1, X_2, \dots)$ be an independent sequence of random variables with

$$P(\{\omega \colon X_n(\omega) = 2^{-n}\}) = P(\{\omega \colon X_n(\omega) = 0\}) = 1/2.$$

Prove that $\sum X_n$ converges a.s. and calculate the distribution function of the limit.

Problem 33. Let $(X_1, X_2, \dots)$ be an independent sequence of random variables with

$$P(\{\omega \colon X_n(\omega) = 1\}) = 1 - P(\{\omega \colon X_n(\omega) = 0\}) = 1/n.$$

Prove that $\sum X_n$ diverges a.s.

Problem 34. Let $(X_1, X_2, \dots)$ be an independent sequence of random variables with

$$P(\{\omega \colon X_n(\omega) = 1\}) = 1 - P(\{\omega \colon X_n(\omega) = 0\}) = 2^{-n}.$$

Prove that $\sum X_n$ converges a.s. and that this infinite series equals 0 with probability

$$\prod_{n=1}^{\infty} (1 - 2^{-n}) > 0,$$

and equals 1 with probability

$$\left(\sum_{n=1}^{\infty} \frac{2^{-n}}{1 - 2^{-n}} \right) \prod_{n=1}^{\infty} (1 - 2^{-n}).$$

* **Problem 35.** Let $(X_1, X_2, \dots)$ be an independent sequence of $\mathbb{R}$-valued random variables, and suppose that

$$P(\{\omega \colon |X_n(\omega)| \le 1/n^2\}) \ge 1 - (1/n^2)$$

for all n. Prove that $\sum X_n$ converges a.s.

Problem 36. Find an independent sequence $(X_1, X_2, \dots)$ of nonnegative random variables such that for each n,

$$E(X_n) < \infty, \quad \sum_{n=1}^{\infty} X_n < \infty \text{ a.s.}, \quad \text{and} \quad E\left(\sum_{n=1}^{\infty} X_n \right) = \infty.$$

The following result says that it is not by chance that in each of the preceding exercises, there is either almost sure convergence or almost sure divergence of the infinite series.

Proposition 20. *A series $\sum X_n$ of independent $\mathbb{R}$-valued random variables either converges (in $\mathbb{R}$) a.s. or diverges a.s.*

Problem 37. Prove the preceding proposition by using the Kolmogorov 0-1 Law.

Problem 38. Show that the assumption of independence may not be deleted from the preceding proposition.

12.6. The Etemadi Lemma

In view of the preceding proposition it is natural to search for conditions that distinguish between the two cases—a.s. convergence and a.s. divergence. In the section following this section, we will obtain necessary and sufficient conditions on the distributions of independent $\mathbb{R}$-valued random variables X_n for the series $\sum X_n$ to converge a.s. In the present section, we lay some groundwork by proving that a.s. convergence is equivalent to i.p. convergence for such a series. This equivalence is a consequence of the following important lemma which is useful for many purposes besides that of obtaining conditions for almost sure convergence.

Lemma 21. [Etemadi] *Let $(X_1, X_2, \ldots)$ be an independent sequence of $\mathbb{R}$-valued random variables, and let $S_m = \sum_{i=1}^{m} X_i$ for $m > 0$. Then, for each $\varepsilon > 0$,*

$$(12.3) \qquad P(\{\omega: \sup_m |S_m(\omega)| > 4\varepsilon\}) \le 4 \sup_m P(\{\omega: |S_m(\omega)| > \varepsilon\}),$$

and for each positive integer n,

$$(12.4) \quad P(\{\omega: \max_{0 \le m \le n} |S_m(\omega)| > 4\varepsilon\}) \le 4 \max_{0 \le m \le n} P(\{\omega: |S_m(\omega)| > \varepsilon\}).$$

PROOF. It is enough to prove (12.4), since (12.3) follows from (12.4) and the Continuity of Measure Theorem.

Let $S_0 = 0$, and for $m = 1, 2, \ldots, n$, let

$$A_m = \{\omega: \max_{j < m} |S_j(\omega)| \le 4\varepsilon < |S_m(\omega)|\}.$$

The left side of (12.4) equals

$$\sum_{m=1}^{n} P(A_m)$$

$$= \sum_{m=1}^{n} P(A_m \cap \{\omega: |S_n(\omega)| > 2\varepsilon\}) + \sum_{m=1}^{n-1} P(A_m \cap \{\omega: |S_n(\omega)| \le 2\varepsilon\})$$

$$\le P(\{\omega: |S_n(\omega)| > 2\varepsilon\}) + \sum_{m=1}^{n-1} P(A_m \cap \{\omega: |S_n(\omega) - S_m(\omega)| > 2\varepsilon\})$$

$$= P(\{\omega: |S_n(\omega)| > 2\varepsilon\}) + \sum_{m=1}^{n-1} P(A_m) P(\{\omega: |S_n(\omega) - S_m(\omega)| > 2\varepsilon\})$$

$$\le P(\{\omega: |S_n(\omega)| > 2\varepsilon\}) + \max_{1 \le m \le n-1} P(\{\omega: |S_n(\omega) - S_m(\omega)| > 2\varepsilon\})$$

$$\le 2 \max_{0 \le m \le n-1} P(\{\omega: |S_n(\omega) - S_m(\omega)| > 2\varepsilon\})$$

$$\le 2 \max_{0 \le m \le n-1} [P(\{\omega: |S_m(\omega)| > \varepsilon\}) + P(\{\omega: |S_n(\omega)| > \varepsilon\})]$$

$$\le 4 \max_{0 \le m \le n} P(\{\omega: |S_m(\omega)| > \varepsilon\}). \quad \square$$

Theorem 22. *A series* $\sum X_n$ *of independent* $\mathbb{R}$-*valued random variables converges a.s. to an* $\mathbb{R}$-*valued random variable* Z *if and only if* $\sum X_n$ *converges i.p. to* Z.

PROOF. The 'only if' assertion follows from Theorem 2. To prove the 'if' assertion, fix $\varepsilon > 0$, let $S_n = \sum_{m=1}^{n} X_m$ for $n > 0$, and suppose that $S_n \to Z$ in probability as $n \to \infty$. By the Continuity of Measure Theorem,

$$(12.5) \qquad P(\{\omega \colon \limsup_{n \to \infty} |S_n(\omega) - Z(\omega)| > 8\varepsilon\})$$

$$= \lim_{m \to \infty} P(\{\omega \colon \sup_{n \geq m} |S_n(\omega) - Z(\omega)| > 8\varepsilon\})$$

$$(12.6) \qquad \leq \limsup_{m \to \infty} P(\{\omega \colon |S_m(\omega) - Z(\omega)| > 4\varepsilon\})$$

$$(12.7) \qquad + \limsup_{m \to \infty} P(\{\omega \colon \sup_{n \geq m} |S_n(\omega) - S_m(\omega)| > 4\varepsilon\}) .$$

Since S_n converges to Z in probability, (12.6) equals 0. We obtain an upper bound for (12.7) by applying (12.3) of the Etemadi Lemma to the sequence $(X_{m+1}, X_{m+2}, \dots)$; the bound is

$$4 \limsup_{m \to \infty} \sup_{n \geq m} P(\{\omega \colon |S_n(\omega) - S_m(\omega)| > \varepsilon\})$$

$$\leq 4 \limsup_{m \to \infty} P(\{\omega \colon |S_m(\omega) - Z(\omega)| > \varepsilon/2\})$$

$$+ 4 \limsup_{m \to \infty} \sup_{n \geq m} P(\{\omega \colon |S_n(\omega) - Z(\omega)| > \varepsilon/2\}) ,$$

which equals 0 as a consequence of the convergence of S_n to Z in probability. Thus (12.5) equals 0 for every ε. Now apply Problem 1. $\square$

In order to apply the preceding result to relate the two types of convergence in the present setting, we need to develop a criterion for convergence in probability that does not require us to have a candidate for the limiting random variable.

Definition 23. A sequence $(W_n \colon n = 1, 2, \dots)$ of $\mathbb{R}$-valued random variables defined on a probability space $(\Omega, \mathcal{F}, P)$ is *Cauchy in probability* if for every $\varepsilon > 0$ there exists an integer l such that

$$P(\{\omega \colon |W_n(\omega) - W_m(\omega)| > \varepsilon\}) < \varepsilon$$

whenever $n, m > l$.

Lemma 24. *A sequence* $(W_n \colon n = 1, 2, \dots)$ *of* $\mathbb{R}$-*valued random variables defined on a common probability space converges in probability to an* $\mathbb{R}$-*valued random variable if and only if it is Cauchy in probability.*

PARTIAL PROOF. Suppose that the sequence (W_n) is Cauchy in probability. By definition we may choose a subsequence $(W_{n_k} \colon n_1 < n_2 < \dots)$ such that

$$P(\{\omega \colon |W_{n_{k+1}}(\omega) - W_{n_k}(\omega)| > 2^{-k}\}) < 2^{-k}.$$

The sum of these probabilities is finite, so by the Borel Lemma,

$$|W_{n_{k+1}}(\omega) - W_{n_k}(\omega)| \leq 2^{-k}$$

for almost every ω and all sufficiently large k depending on ω. For such ω, repeated applications of the triangle inequality imply that

$$|W_{n_{k+m}}(\omega) - W_{n_k}(\omega)| \leq 2^{-k+1}$$

for all positive m and sufficiently large k. Thus, the sequence $(W_{n_k}(\omega): k = 1, 2, \dots)$ is a Cauchy sequence of real numbers and has a finite limit $W(\omega)$. W is defined a.s. Set $W(\omega)$ equal 0 for any ω for which $W(\omega)$ has not been defined by the preceding discussion.

We have a subsequence of $(W_n: n = 1, 2, \dots)$ that converges almost surely to W and thus in probability to W. To see that the full sequence converges in probability to W we note that, for $n > n_k$,

$$P(\{\omega: |W_n(\omega) - W(\omega)| > \varepsilon\})$$
$$\leq P(\{\omega: |W_n(\omega) - W_{n_k}(\omega)| > \tfrac{\varepsilon}{2}\}) + P(\{\omega: |W_{n_k}(\omega) - W(\omega)| > \tfrac{\varepsilon}{2}\})$$
$$< \tfrac{\varepsilon}{2} + \tfrac{\varepsilon}{2} = \varepsilon$$

for sufficiently large k.

The proof that convergence in probability implies Cauchy in probability is left for the next problem. $\square$

Problem 39. Complete the proof of the preceding lemma by showing that a sequence that converges in probability to a $\mathbb{R}$-valued limit is Cauchy in probability.

12.7. † The Kolmogorov Three-Series Theorem

We now search for conditions on the distributions of the individual summands X_n that will be necessary and sufficient for the almost sure convergence of $\sum_n X_n$, or, equivalently, its convergence in probability. Fix $b > 0$, and, for $n \geq 1$, define

$$Y_n(\omega) = \begin{cases} X_n(\omega) & \text{if } |X_n(\omega)| \leq b \\ 0 & \text{otherwise}. \end{cases}$$

If

$$\sum P(\{\omega: X_n(\omega) \neq Y_n(\omega)\}) = \sum P(\{\omega: |X_n(\omega)| > b\}) = \infty,$$

then it follows from the Borel-Cantelli Lemma that with probability 1, infinitely many of the X_n's will be larger than b in absolute value, and the series $\sum_n X_n$ will diverge a.s. On the other hand, if

$$\sum P(\{\omega: X_n(\omega) \neq Y_n(\omega)\}) < \infty,$$

then it follows from the Borel Lemma that with probability 1, only finitely many of the summands in the series $\sum X_n$ differ from the corresponding summands in the series $\sum Y_n$. Therefore, in this case, $\sum_n X_n$ converges a.s. if and only if $\sum_n Y_n$ converges a.s. In this way, we reduce the study of arbitrary series of independent $\mathbb{R}$-valued random variables to the study of series with finite variances.

Proposition 25. *Let* $(Z_1, Z_2, \ldots)$ *be an independent sequence of* $\mathbb{R}$-*valued random variables with finite variance. If* $\sum_n \mathrm{Var}(Z_n) < \infty$, *then* $\sum_n (Z_n - E(Z_n))$ *converges a.s. in* $\mathbb{R}$.

PROOF. We may assume without loss of generality that $E(Z_n) = 0$ for all n. For each n, let $S_n = \sum_{m=1}^n Z_m$. Let $\varepsilon > 0$. Choose an integer l such that, for $n > m \geq l$,

$$\mathrm{Var}(S_n - S_m) = \sum_{k=m+1}^n \mathrm{Var}(Z_k) < \varepsilon^3 \, .$$

By the Chebyshev Inequality,

$$P(\{\omega \colon |S_n(\omega) - S_m(\omega)| > \varepsilon\}) \leq \frac{\mathrm{Var}(S_n - S_m)}{\varepsilon^2} < \varepsilon \, .$$

Thus, the sequence $(S_n \colon n = 1, 2, \ldots)$ is Cauchy in probability. By Lemma 24 it converges in probability. By Theorem 22 it converges almost surely. $\square$

Corollary 26. *Let* $(Z_1, Z_2, \ldots)$ *be an independent sequence of* $\mathbb{R}$-*valued random variables with finite variance. If* $\sum_n \mathrm{Var}(Z_n) < \infty$, *then* $\sum_n Z_n$ *converges or diverges a.s. according as* $\sum_n E(Z_n)$ *converges or diverges.*

Corollary 26 and the discussion preceding Proposition 25 constitute a proof of the first 'if' assertion of the following result.

Theorem 27. [Kolmogorov Three-Series] *Let* $(X_1, X_2, \ldots)$ *be an independent sequence of* $\mathbb{R}$-*valued random variables, and let* b *be a positive real number. Define*

$$Y_n(\omega) = \begin{cases} X_n(\omega) & \text{if } |X_n(\omega)| \leq b \\ 0 & \text{otherwise} \, . \end{cases}$$

Then $\sum_n X_n$ *converges a.s. if and only if the following three series converge:*

(12.8)
$$\sum_n P(\{\omega \colon X_n(\omega) \neq Y_n(\omega)\}) \, ,$$

(12.9)
$$\sum_n E(Y_n) \, ,$$

(12.10)
$$\sum_n \mathrm{Var}(Y_n) \, .$$

If one of these three series diverges, then $\sum_n X_n$ *diverges a.s.*

PROOF. As stated above, the first 'if' assertion has already been proved. Once the 'only if' assertion is proved, the second 'if' assertion is an immediate consequence of Proposition 20. Thus we will focus our attention on the proof of the 'only if' assertion.

Suppose that $\sum_n X_n$ converges a.s. Then $X_n \to 0$ a.s.; so, for almost every ω, $X_n(\omega) = Y_n(\omega)$ for all but finitely many n. Therefore, $\sum_n Y_n$ converges a.s., and furthermore, as previously observed, (12.8) must converge as a consequence of the Borel-Cantelli Lemma. It follows from Corollary 26 that, to finish the proof, it is enough to show that (12.10) converges.

After possibly enlarging the original probability space, we may assume that there is a sequence $(Y_1', Y_2', \ldots)$ that has the same distribution as and is independent of the sequence $(Y_1, Y_2, \ldots)$. Since the two sequences have the same distribution, $\sum_n Y_n'$ converges a.s., and hence so does the series $\sum_n (Y_n - Y_n')$. Let $Z_n = Y_n - Y_n'$. For each n, $\mathrm{Var}(Z_n) = 2\,\mathrm{Var}(Y_n)$, since Y_n' has the same distribution and is independent of Y_n. Thus, $\sum_n \mathrm{Var}(Y_n)$ converges if and only if $\sum_n \mathrm{Var}(Z_n)$ converges. Note that the random variables Z_n are independent, bounded by $2b$, and have mean 0. Therefore, we have constructed an independent sequence $(Z_1, Z_2, \ldots)$ of uniformly bounded random variables with mean 0, such that $\sum_n Z_n$ converges a.s., and $\sum_n \mathrm{Var}(Y_n)$ is finite whenever $\sum_n \mathrm{Var}(Z_n)$ is finite. Thus, in order to complete the proof, it is enough to show that the a.s. convergence of the series $\sum_n Z_n$ implies that the series $\sum_n \mathrm{Var}(Z_n)$ has a finite sum.

Let $S_n = \sum_{i=1}^n Z_i$. Because each of the random variables Z_n has mean 0, $E(S_n^2) = \mathrm{Var}(S_n) = \sum_{i=1}^n \mathrm{Var}(Z_n)$. By removing all Z_i's that equal 0 a.s., we may further assume that $\mathrm{Var}(S_1) > 0$.

We wish to prove that the increasing sequence $(\mathrm{Var}(S_1), \mathrm{Var}(S_2), \ldots)$ has a finite limit. (For a nice alternative approach to this part of the proof, using an important tool that generalizes probability generating functions, see Problem 21 of Chapter 13.) In order to do so, we apply the inequality in Corollary 5 of Chapter 5, with $\lambda = 1/2$ and $X = S_n^2$, and obtain

$$(12.11) \qquad P(\{\omega \colon S_n^2(\omega) > \tfrac{1}{2}\,\mathrm{Var}(S_n)\}) \geq \frac{(\mathrm{Var}(S_n))^2}{4E(S_n^4)}.$$

We will prove that the right side of (12.11) does not converge to 0 as $n \to \infty$. Since the convergence in $\mathbb{R}$ of S_n implies the convergence in $\mathbb{R}$ of S_n^2, we can then immediately conclude that the sequence $(\mathrm{Var}(S_n))$ has a finite limit. In the following calculations, we will use the two facts that $E(Z_i Z_j^3) = E(Z_i)E(Z_j^3) = 0$ and $E(Z_i^2 Z_j^2) = E(Z_i^2)E(Z_j^2)$ for $i \neq j$, both of which follow from the independence of the Z_i's:

$$
\begin{aligned}
E(S_n^4) &= \sum_{i=1}^n E(Z_i^4) + 6 \sum_{1 \le i < j \le n} E(Z_i^2 Z_j^2) \\
&= \sum_{i=1}^n E(Z_i^4) + 6 \sum_{1 \le i < j \le n} E(Z_i^2) E(Z_j^2) \\
&\le \sum_{i=1}^n 4b^2 E(Z_i^2) + 3 \sum_{i=1}^n \sum_{j=1}^n E(Z_i^2) E(Z_j^2) \\
&= 4b^2 \sum_{i=1}^n E(Z_i^2) + 3 \left(\sum_{i=1}^n E(Z_i^2) \right)^2 \\
&= 4b^2 \operatorname{Var}(S_n) + 3(\operatorname{Var}(S_n))^2 \\
&\le [(4b^2 / \operatorname{Var}(S_1)) + 3](\operatorname{Var}(S_n))^2 .
\end{aligned}
$$

It follows that $E(S_n^4)$ is bounded above by a constant multiple of $(\operatorname{Var}(S_n))^2$, uniformly in n. Thus, the right side of (12.11) stays bounded away from 0 as $n \to \infty$, as desired. $\square$

The solution of the next problem indicates that when Corollary 26 applies, it is often easier to use than is the Kolmogorov Three-Series Theorem.

* **Problem 40.** Let $(X_1, X_2, \dots)$ be an independent sequence of nonnegative geometrically distributed random variables and suppose that $E(X_n) = 1/n^2$. Using three different methods, prove that $\sum_n X_n$ converges a.s.: first by using the Three-Series Theorem, second by using Corollary 26, and third by using the Monotone Convergence Theorem.

* **Problem 41.** [Random signs] Let $(X_1, X_2, \dots)$ be an iid sequence of random variables with mean 0 and support equal to the 2-point set $\{-1, 1\}$. Let $(c_1, c_2, \dots)$ be a sequence of positive constants, and define $Y_n = c_n X_n$. Find necessary and sufficient conditions on the sequence (c_n) for the a.s. convergence of $\sum_n Y_n$.

Problem 42. Find an independent sequence $(X_1, X_2, \dots)$ of $\mathbb{R}$-valued random variables and a positive constant b for which (12.8) diverges and both (12.9) and (12.10) converge.

Problem 43. Find an independent sequence $(X_1, X_2, \dots)$ of $\mathbb{R}$-valued random variables and a positive constant b for which (12.9) diverges and both (12.8) and (12.10) converge.

Problem 44. Find an independent sequence of random variables, each with finite variance and with distribution symmetric about 0, for which (12.8), (12.9), and (12.10) all converge, but $\sum_n \operatorname{Var}(X_n) = \infty$. (This shows that Proposition 25 has no converse.)

12.8. † The image of a random walk

We turn to a problem that has a treatment similar to the one we used for the Strong Law of Large Numbers. Let $(S_0 = 0, S_1, S_2, \dots)$ be a random walk. For $n = 0, 1, 2, \dots$ let $R_n = \sharp\{S_m : 0 \le m \le n\}$; that is, R_n equals the cardinality of the image of the random walk through time n. Our goal is to study the behavior of R_n/n as $n \to \infty$. We will prove that R_n/n converges a.s. to the probability that the random walks never returns to 0. For this purpose we introduce a sequence $(I_n : n = 0, 1, \dots)$ of indicator random variables:

$$(12.12) \qquad I_n(\omega) = \begin{cases} 1 & \text{if } S_n(\omega) \ne S_m(\omega) \text{ for } m < n \\ 0 & \text{otherwise}. \end{cases}$$

Clearly,

$$(12.13) \qquad R_n = \sum_{m=0}^{n} I_m .$$

Theorem 28. *Let* $(S_0 = 0, S_1, S_2, \dots)$ *be a random walk in* $\mathbb{R}^d$, *and, for* $n = 0, 1, 2, \dots$, *set* $R_n = \sharp\{S_m : 0 \le m \le n\}$. *Then, as* $n \to \infty$,

$$\frac{R_n}{n} \to P(\{\omega : S_m(\omega) \ne 0 \text{ for } m > 0\}) \text{ a.s.}$$

PROOF. Let I_n be defined by (12.12) and $(X_n : n = 1, 2, \dots)$ denote the step sequence of the random walk. Since the X_n are iid, the distribution of a random vector of the form $(X_{n_1}, \dots, X_{n_d})$ depends only on d and not on the choice of the subscripts $n_1, \dots, n_d$, provided only that these subscripts are distinct. This fact is used for the third equality below:

$$E(I_n) = P\big(\{\omega : \sum_{k=1}^{n} X_k(\omega) \ne \sum_{k=1}^{m} X_k(\omega) \text{ for } 0 \le m < n\}\big)$$

$$= P\big(\{\omega : \sum_{k=m+1}^{n} X_k(\omega) \ne 0 \text{ for } 0 \le m < n\}\big)$$

$$= P\big(\{\omega : \sum_{k=1}^{n-m} X_k(\omega) \ne 0 \text{ for } 0 \le m < n\}\big)$$

$$= P(\{\omega : S_{n-m} \ne 0 \text{ for } 0 \le m < n\})$$

$$= P(\{\omega : S_m \ne 0 \text{ for } 0 < m \le n\})$$

$$\searrow P(\{\omega : S_m \ne 0 \text{ for } 0 < m\})$$

as $n \nearrow \infty$. In view of this calculation, (12.13), and the fact proved in part (iii) of Problem 10, we conclude that

$$E\big(\frac{R_n}{n}\big) \to P(\{\omega : S_m(\omega) \ne 0 \text{ for } m > 0\}) .$$

We next wish to apply Lemma 7. We first obtain a bound on $\mathrm{Cov}(I_n, I_p)$ for $n < p$:

$$E(I_n I_p)$$
$$= P(\{\omega : S_n(\omega) \neq S_m(\omega) \text{ for } m < n, S_p(\omega) \neq S_q(\omega) \text{ for } q < p\})$$
$$\leq P(\{\omega : S_n(\omega) \neq S_m(\omega) \text{ for } m < n, S_p(\omega) \neq S_q(\omega) \text{ for } q \in [n, p)\})$$
$$= P(\{\omega : S_n(\omega) \neq S_m(\omega) \text{ for } m < n\})$$
$$\cdot P\left(\left\{\omega : \sum_{k=n+1}^{p} X_k(\omega) \neq \sum_{k=n+1}^{q} X_k(\omega) \text{ for } q \in [n, p)\right\}\right)$$
$$= P(\{\omega : S_n(\omega) \neq S_m(\omega) \text{ for } m < n\})$$
$$\cdot P(\{\omega : S_{p-n}(\omega) \neq S_r(\omega) \text{ for } r < p - n\})$$
$$= E(I_n) E(I_{p-n}) .$$

Let $b_m = \lceil c^m \rceil$ for fixed $c > 1$. Then, using (12.13) and the inequality just proved, along with the elementary facts that $\mathrm{Var}(I_k) \leq 1$ for all k and $\mathrm{Cov}(I_0 I_k) = 0$ for all k, we have

$$\sum_{m=0}^{\infty} \mathrm{Var}\left(\frac{R_{b_m}}{b_m}\right) = \sum_{m=0}^{\infty} b_m^{-2} \sum_{k=0}^{b_m} \mathrm{Var}(I_k) + 2 \sum_{m=0}^{\infty} b_m^{-2} \sum_{k=1}^{b_m} \sum_{j=0}^{k-1} \mathrm{Cov}(I_j I_k)$$
$$\leq \sum_{m=0}^{\infty} b_m^{-1} + 2 \sum_{m=1}^{\infty} \sum_{k=2}^{b_m} \sum_{j=1}^{k-1} b_m^{-2} E(I_j) [E(I_{k-j}) - E(I_k)] .$$

The first term is finite. We showed earlier in the proof that $n \rightsquigarrow E(I_n)$ is a decreasing function, so we may apply the Fubini Theorem to bound the second term by

$$2 \sum_{k=2}^{\infty} \sum_{j=1}^{k-1} \sum_{\{m : b_m \geq k\}} b_m^{-2} \sum_{i=k-j}^{k-1} [E(I_i) - E(I_{i+1})]$$
$$\leq \frac{2}{1 - c^{-2}} \sum_{k=2}^{\infty} \sum_{j=1}^{k-1} \sum_{i=k-j}^{k-1} k^{-2} [E(I_i) - E(I_{i+1})]$$
$$= \frac{2}{1 - c^{-2}} \sum_{i=1}^{\infty} \sum_{k=i+1}^{\infty} \sum_{j=k-i}^{k-1} k^{-2} [E(I_i) - E(I_{i+1})]$$
$$\leq \frac{2}{1 - c^{-2}} \sum_{i=1}^{\infty} \left(\int_i^{\infty} k^{-2} \, dk \right) i [E(I_i) - E(I_{i+1})]$$
$$= \frac{2}{1 - c^{-2}} \sum_{i=1}^{\infty} [E(I_i) - E(I_{i+1})]$$
$$\leq \frac{2 E(I_1)}{1 - c^{-2}} < \infty .$$

An appeal to Lemma 7 gives the desired limit for R_{b_m}/b_m. An argument that mimics the one used in the proof of Lemma 12 completes the proof. $\square$

Remark 1. The preceding theorem and proof are valid for random walks in groups.

* **Problem 45.** Find the place or places where the preceding proof breaks down if S is assumed to be a random walk in $\overline{\mathbb{R}}^+$ rather than in $\mathbb{R}^d$. Then decide whether the theorem itself is true in the $\overline{\mathbb{R}}^+$-setting.

Problem 46. For an arbitrary simple random walk in $\mathbb{Z}$ use the Strong Law of Large Numbers to calculate the almost sure limit of (R_n/n). Then use Theorem 28 to calculate the probability that the random walk returns to 0 at least once.

Problem 47. Modify the conclusions for the preceding exercise to encompass random walks in $\mathbb{Z}$ whose steps have absolute value 1 or 0.

CHAPTER 13
Characteristic Functions

'Characteristic functions' and 'moment generating functions' correspond to distributions on $\mathbb{R}$ and $\overline{\mathbb{R}}^+$, respectively, in a manner analogous to the correspondence between probability generating functions and distributions on $\overline{\mathbb{Z}}^+$. It will be seen here and in succeeding chapters that these tools are quite powerful, particularly when independence is involved. After completing our coverage of the theory for the real line, we generalize characteristic functions to $\mathbb{R}^d$ and discuss normal distributions in that setting. At the end of the chapter, we apply the 1-dimensional theory to random walk on $\mathbb{Z}$.

Remark 1. Unfortunately, terminology is not consistent in the literature. Many books on real analysis use the term 'Fourier transform' or 'Fourier-Stieltjes transform' for what we call a 'characteristic function', and use the term 'characteristic function' for what we have called an 'indicator function'. And 'moment generating functions' are known elsewhere as 'Laplace transforms' or 'Laplace-Stieltjes transforms'. Others use the term 'moment generating function' but with a sign difference from what is used here.

13.1. Definition and basic examples

We begin with

Definition 1. The *characteristic function* of an $\mathbb{R}$-valued random variable X is the $\mathbb{C}$-valued function

$$v \rightsquigarrow E(e^{ivX}) = E(\cos(vX)) + iE(\sin(vX)), \quad v \in \mathbb{R},$$

where i denotes a complex number whose square is -1.

Notice that for $v \in \mathbb{R}$, the function $x \rightsquigarrow e^{ivx}$ is bounded and continuous (and thus measurable), so $E(e^{ivX})$ exists and is finite. The characteristic function of a random variable is also called the *characteristic function* of its distribution and also of its distribution function.

Problem 1. Calculate the characteristic functions of the Bernoulli, binomial, geometric, and Poisson distributions.

Problem 2. Calculate the characteristic function of the constant random variable equal to c.

Example 1. Let
$$F(x) = \int_{-\infty}^{x} \frac{1}{\sqrt{2\pi}} e^{-u^2/2} \, du \,.$$
We will sketch a calculation of the characteristic function β of this normal distribution function F; some details will be omitted. For $v \in \mathbb{R}$,
$$\beta(v) = \frac{1}{\sqrt{2\pi}} \int_{-\infty}^{\infty} e^{ivx} e^{-x^2/2} \, dx \,.$$
We use the Dominated Convergence Theorem to differentiate with respect to v and then integration by parts to obtain a differential equation for β:
$$\beta'(v) = \frac{1}{\sqrt{2\pi}} \int_{-\infty}^{\infty} ixe^{ivx} e^{-x^2/2} \, dx = -\frac{1}{\sqrt{2\pi}} \int_{-\infty}^{\infty} ie^{ivx} \, de^{-x^2/2}$$
$$= -\frac{1}{\sqrt{2\pi}} \int_{-\infty}^{\infty} e^{-x^2/2} ve^{ivx} \, dx = -v\beta(v) \,.$$
So, β is the unique solution of the differential equation $\beta'(v) = -v\beta(v)$ satisfying $\beta(0) = 1$; thus, $\beta(v) = \exp(-v^2/2)$.

Problem 3. Supply the details for the preceding example by showing that $\beta'(v)$ does equal the expression obtained by differentiation inside the integral and also by commenting on the correctness of the manipulations involving $\mathbb{C}$-valued functions.

Problem 4. Calculate the characteristic function of a normally distributed random variable having arbitrary mean and variance.

Problem 5. Calculate the characteristic function of the exponential distribution with support $\mathbb{R}^+$ and mean $1/\lambda$, $\lambda > 0$.

Problem 6. [Bilateral exponential distributions] Let $\lambda > 0$ and let Q denote the probability measure whose density with respect to Lebesgue measure is the function $x \rightsquigarrow (\lambda/2)e^{-\lambda|x|}$. Show that the characteristic function of Q is the function
$$v \rightsquigarrow \frac{\lambda^2}{\lambda^2 + v^2} \,.$$

Problem 7. Let β denote the characteristic function of an $\mathbb{R}$-valued random variable X, and let a and b be two real constants. Show that the characteristic function of the random variable $aX + b$ is the function
$$v \rightsquigarrow e^{ivb} \beta(av) \,.$$

Problem 8. Calculate the characteristic function of the uniform distribution on the interval $[-1, 1]$. As a check confirm directly that your answer is continuous. Then use the preceding proposition to calculate the characteristic function of a uniform distribution on an arbitrary interval. Write the characteristic function in a form that displays the fact that it is $\mathbb{R}$-valued if and only if the support of the uniform distribution is $[-a, a]$ for some $a > 0$.

13.2. The Parseval Relation and uniqueness

The following lemma is a useful tool. In this section, we will use it to prove that distinct distributions have distinct characteristic functions.

Lemma 2. [Parseval Relation] *Let Q and R be two probability measures on $\mathbb{R}$, and denote their characteristic functions by β and γ, respectively. Then*

$$\int \gamma(x - v)\, Q(dx) = \int e^{-ivy}\, \beta(y)\, R(dy)$$

for each $v \in \mathbb{R}$.

PROOF. The function $(x, y) \rightsquigarrow e^{i(x-v)y}$ is bounded and continuous (and thus measurable). So its integral with respect to the product measure $Q \times R$ exists. By the Fubini Theorem this integral equals two different iterated integrals. These two iterated integrals are the integrals in the relation being proved. $\square$

There is an important special case of the Parseval Relation. Let R be the normal distribution with mean 0 and variance σ^2 and use the formula found in Problem 4 for the characteristic function of the normal distribution to obtain:

$$(13.1) \qquad \int e^{-\sigma^2(x-v)^2/2}\, Q(dx) = \frac{1}{\sqrt{2\pi\sigma^2}} \int_{-\infty}^{\infty} e^{-ivy}\, e^{-y^2/2\sigma^2}\, \beta(y)\, dy\,.$$

Theorem 3. *If Q_1 and Q_2 are probability measures on $\mathbb{R}$ with the same characteristic function, then $Q_1 = Q_2$.*

PROOF. Since Q_1 and Q_2 have the same characteristic function, (13.1) implies that

$$\int e^{-\sigma^2(x-v)^2/2}\, Q_1(dx) = \int e^{-\sigma^2(x-v)^2/2}\, Q_2(dx)$$

for $-\infty < v < \infty$ and $\sigma > 0$. Multiply both sides by $\sigma/\sqrt{2\pi}$, integrate over the interval $(-\infty, a)$ with respect to Lebesgue measure, and apply the Fubini Theorem to obtain

$$\int \int_{(-\infty, a)} \frac{\sigma e^{-\sigma^2(v-x)^2/2}}{\sqrt{2\pi}}\, dv\, Q_1(dx) = \int \int_{(-\infty, a)} \frac{\sigma e^{-\sigma^2(v-x)^2/2}}{\sqrt{2\pi}}\, dv\, Q_2(dx)$$

for all $a \in \mathbb{R}$ and $\sigma > 0$. On both sides of this equation, the integrand of the inside integral is the density of the normal distribution with mean x and variance $1/\sigma^2$, so the inside integral equals the probability that a random variable with such a distribution lies in the interval $(-\infty, a)$. As $\sigma \to \infty$, this probability converges to 1 if $x < a$, and to 0 if $x > a$. It converges to $1/2$ if $x = a$. It follows from the Bounded Convergence Theorem that

$$Q_1((-\infty, a)) + \frac{1}{2}Q_1(\{a\}) = Q_2((-\infty, a)) + \frac{1}{2}Q_2(\{a\})$$

for all $a \in \mathbb{R}$. Since there can be at most countably many values of a such that either $Q_1(\{a\})$ or $Q_2(\{a\})$ is nonzero, we may conclude that there is a dense set of real numbers a such that $Q_1(-\infty, a)) = Q_2((-\infty, a))$. A straightforward application of the Uniqueness of Measure Theorem now implies that $Q_1 = Q_2$. $\square$

Distribution	support	density: $x \rightsquigarrow$	characteristic function: $v \rightsquigarrow$
Gaussian	$\mathbb{R}$	$\frac{1}{\sqrt{2\pi}}e^{-x^2/2}$	$e^{-v^2/2}$
Exponential	$\mathbb{R}^+$	e^{-x}	$\dfrac{1}{1 - iv}$
Bilateral Exp.	$\mathbb{R}$	$\frac{1}{2}e^{-|x|}$	$\dfrac{1}{1 + v^2}$
Cauchy	$\mathbb{R}$	$\dfrac{1}{\pi(1 + x^2)}$	$e^{-|v|}$
Gamma	$\mathbb{R}^+$	$\frac{1}{\Gamma(\gamma)}x^{\gamma-1}e^{-x}$	$(1 - iv)^{-\gamma}$
Uniform	$[-1, 1]$	$\frac{1}{2}$	$\begin{cases} \frac{\sin v}{v} & \text{if } v \neq 0 \\ 1 & \text{if } v = 0 \end{cases}$
Triangular	$[-2, 2]$	$\frac{1}{2} - \frac{1}{4}|x|$	$\begin{cases} (\frac{\sin v}{v})^2 & \text{if } v \neq 0 \\ 1 & \text{if } v = 0 \end{cases}$

TABLE 13.1. Characteristic functions of some continuous distributions

The characteristic functions of some distributions are given in Tables 13.1 and 13.2. These are all worked out in examples and problems contained in this

Distribution	support	density: $x \rightsquigarrow$	characteristic function: $v \rightsquigarrow$
Delta	$\{0\}$	1	1
Bernoulli	$\{0,1\}$	$\begin{cases} p & \text{if } x = 1 \\ 1-p & \text{if } x = 0 \end{cases}$	$1 - p(1 - e^{iv})$
Binomial	$\{0,1,2,\ldots,n\}$	$\binom{n}{x}p^x(1-p)^{n-x}$	$[1 - p(1 - e^{iv})]^n$
Poisson	$\mathbb{Z}^+$	$\dfrac{\lambda^x}{x!}e^{-\lambda}$	$\exp(-\lambda(1 - e^{iv}))$
Geometric	$\mathbb{Z}^+$	$(1-p)p^x$	$\dfrac{1-p}{1 - pe^{iv}}$
Neg. Binomial	$\mathbb{Z}^+$	$(1-p)^r\dfrac{(r)_x^\uparrow}{x!}p^x$	$\left(\dfrac{1-p}{1 - pe^{iv}}\right)^r$

TABLE 13.2. Characteristic functions of some discrete distributions

chapter. In view of Problem 7 only one representative distribution of any type is included.

Example 2. We show here how to compute the characteristic function of a Cauchy distribution. Our method uses the Parseval Relation in a manner that is similar to that found in the proof of Theorem 3. Let Q be the Cauchy distribution with density $a/\pi(a^2 + x^2)$, where a is a positive parameter, and let β be its characteristic function. Also, let R be the normal distribution with mean c and variance σ^2. The density of R is

$$y \rightsquigarrow \frac{e^{-(y-c)^2/2\sigma^2}}{\sqrt{2\pi\sigma^2}}.$$

By Example 1 and Problem 7, the characteristic function of R is

$$\gamma(x) = e^{ixc}e^{-\sigma^2 x^2/2}.$$

By the Parseval Relation, with $v = 0$,

(13.2) $$\int e^{ixc} e^{-\sigma^2 x^2/2} \frac{a\,dx}{\pi(a^2 + x^2)} = \frac{1}{\sqrt{2\pi\sigma^2}} \int \beta(y) e^{-(y-c)^2/2\sigma^2}\,dy\,.$$

Now make the change of variables $y = -x$ in the left side of (13.2). After some minor rearrangement of constants, the left side of (13.2) becomes

(13.3) $$\sqrt{\frac{2}{\pi}} \int e^{-icy} \frac{a}{a^2 + y^2} \sqrt{\frac{\sigma^2}{2\pi\sigma^2}} e^{-\sigma^2 y^2/2}\,dy\,.$$

We wish to apply the Parseval Relation again, this time with (13.3) playing the role of the right side of the Parseval Relation. The function $a^2/(a^2+y^2)$ is the characteristic function of the bilateral exponential distribution with parameter $\lambda = a$ (see Problem 6), and we let this function play the role of β in the right side of the Parseval Relation. The function

$$\sqrt{\frac{\sigma^2}{2\pi}} e^{-\sigma^2 y^2/2}$$

is the density of the normal distribution with mean 0 and variance $1/\sigma^2$. This distribution plays the role of R as we apply the Parseval Relation to (13.3) and find that it equals

$$\sqrt{\frac{2}{\pi\sigma^2}} \int e^{-(x-c)^2/2\sigma^2} \frac{ae^{-a|x|}}{2}\,dx\,.$$

Substitute this expression into the left side of (13.2) and rearrange constants to obtain

$$\int e^{-a|x|} \frac{e^{-(x-c)^2/2\sigma^2}}{\sqrt{2\pi\sigma^2}}\,dx = \int \beta(y) \frac{e^{-(y-c)^2/2\sigma^2}}{\sqrt{2\pi\sigma^2}}\,dy$$

for all $c \in \mathbb{R}$ and $\sigma > 0$. In a manner similar to the end of the proof of Theorem 3, we may now conclude that $\beta(x) = e^{-a|x|}$ for all $x \in \mathbb{R}$.

Problem 9. Complete the details left at the end of the preceding example. *Hint:* Integrate in c from 0 to t, let σ^2 go to 0, and apply the Fundamental Theorem of Calculus.

Problem 10. (for those who know some complex variable theory) Use contour integration to calculate the characteristic function of a Cauchy distribution.

13.3. Characteristic functions of convolutions

Now that we have a one-to-one correspondence between the set of probability measures on $\mathbb{R}$ and the set of characteristic functions, it is natural to inquire about the details of this correspondence with respect to operations that are important for distributions. The subject of the next theorem is the operation of convolution.

Theorem 4. *Let Q and R denote two probability measures on $\mathbb{R}$. Then the characteristic function of $Q * R$ equals $\beta\gamma$, where β and γ are the characteristic functions of Q and R.*

PROOF. Let X and Y be independent random variables whose distributions are Q and R, respectively. Then the characteristic function α of $Q * R$ satisfies

$$\alpha(v) = E(e^{iv(X+Y)}) = E(e^{ivX} e^{ivY}) = E(e^{ivX}) E(e^{ivY}) = \beta(v)\gamma(v). \quad \square$$

Example 3. Our goal is to calculate the characteristic function $\beta_{p,r}$ of the negative binomial distribution $Q_{p,r}$ with parameters p and r. By using probability generating functions (see Theorem 5 and Problem 23, both of Chapter 10), we can easily show that

$$Q_{p,s} * Q_{p,t} = Q_{p,s+t}$$

for all $p \in (0,1)$ and all $s, t > 0$. It follows from Theorem 4 that, for all such p, s, t,

$$(13.4) \qquad \beta_{p,s}\beta_{p,t} = \beta_{p,s+t}.$$

When $r = 1$, $Q_{p,r}$ is the geometric distribution with parameter p. The characteristic function for the geometric distribution is easily calculated directly from the definition:

$$\beta_{p,1}(v) = \frac{1-p}{1-pe^{iv}}.$$

Repeated applications of (13.4) would therefore seem to show that

$$(13.5) \qquad \beta_{p,(m/n)} = (\beta_{p,m})^{1/n} = \left[(\beta_{p,1})^m\right]^{1/n} = \left(\frac{1-p}{1-pe^{iv}}\right)^{m/n}$$

for all positive integers m, n.

This formula is indeed correct, provided we properly interpret the $(m/n)^{\text{th}}$ power of the $\mathbb{C}$-valued function $\beta_{p,1}$. This is most efficiently done in terms of the complex logarithm. As indicated in Problem 7 of Appendix E, for any continuous function $f: \mathbb{R} \to \mathbb{C} \setminus \{0\}$ with $f(0) = 1$, there is a unique continuous function $\lambda: \mathbb{R} \to \mathbb{C}$ such that $\lambda(0) = 0$ and $e^{\lambda(v)} = f(v)$ for all real v. (We think of λ as the complex logarithm of f, but we cannot, at first, simply define $\lambda = \log \circ f$ because the complex logarithm is multivalued.) Applying this result to the functions $\beta_{p,r}$, we let $\psi_{p,r}$ be the unique continuous $\mathbb{C}$-valued function defined on $\mathbb{R}$ such that $\psi_{p,r}(0) = 0$ and

$$\beta_{p,r}(v) = e^{\psi_{p,r}(v)}$$

for all real v. It follows from the uniqueness of $\psi_{p,r}$ and (13.4) that

$$\psi_{p,s} + \psi_{p,t} = \psi_{p,s+t}$$

for all relevant p, s, t. In particular,

$$\psi_{p,(m/n)} = \frac{m}{n}\psi_{p,1},$$

so that

$$\beta_{p,(m/n)} = e^{(m/n)\psi_{p,1}} .$$

Note that this function is continuous, takes the value 1 at 0, and satisfies the relationship

$$(\beta_{p,(m/n)})^n = (\beta_{p,1})^m .$$

That no other function can have these properties follows from the facts that we have mentioned concerning the uniqueness of the continuous complex logarithm.

We have now given a precise and correct interpretation of (13.5), thereby providing a formula for $\beta_{p,r}$ for all positive rational r:

(13.6) $\beta_{p,r}(v) = e^{r\psi_{p,1}(v)} .$

We claim that this formula is correct for all $r > 0$. The expression on the right is certainly meaningful for all such r, and since the exponential function is continuous, it is also continuous as a function of r, for fixed p and v. By the Dominated Convergence Theorem and the definition of the characteristic function, the function $r \rightsquigarrow \beta_{p,r}(v)$ is also continuous for fixed p and v. It follows that (13.6) is correct for all $r > 0$, as claimed. For simplicity, the function on the right of (13.6) is usually written in the less precise form found in Table 13.2.

Problem 11. Give a rigorous interpretation of the formula in Table 13.1 for the characteristic function of the gamma distribution. Use arguments similar to those given in the preceding example to verify that your interpretation is correct.

Problem 12. Use characteristic functions to prove that the sum of independent Gaussian random variables is Gaussian with mean and variance equal to the respective sums of the means and variances of the summands.

Problem 13. Prove the following identity:

$$\frac{\sin u}{u} = \prod_{k=1}^{\infty} \cos\left(\frac{u}{2^k}\right) .$$

Hint: The right side is the characteristic function of a convergent series of independent random variables.

Problem 14. [Triangular distributions] For real constants $a > 0$ and b, let

$$f(x) = \begin{cases} a - a^2|x - b| & \text{if } |x - b| \le 1/a \\ 0 & \text{if } |x - b| > 1/a . \end{cases}$$

Show that f is the density with respect to Lebesgue measure of a probability measure. For the case $a = b = 1$ calculate the characteristic function of this measure in two ways: directly and by using Problem 7 of Chapter 10 in conjunction with Problem 8 and Theorem 4 of this chapter. Then use Problem 7 to calculate the characteristic function for arbitrary a and b, writing the characteristic function in a form that displays the fact that it is $\mathbb{R}$-valued if and only if $b = 0$.

* **Problem 15.** Use characteristic functions to decide when the sum of two independent random variables of uniform type is triangularly distributed (as described in the preceding problem). For instance, is it necessary that the two independent random variables have the same distribution? Is it sufficient?

13.4. Symmetrization

A common tool in probability theory is that of 'symmetrization'. If R is the distribution of an $\mathbb{R}$-valued random variable X and Q is the distribution of $-X$, then the *symmetrization* of R is $R * Q$. Thus the symmetrization of R is the distribution of the difference of two independent random variables, each of which has distribution R. The following exercises introduce some basic facts about symmetrization.

Problem 16. Let β denote the characteristic function of a random variable X. Show that the characteristic function of $-X$ is $\bar{\beta}$, the complex conjugate of β.

Problem 17. Let X and Y be independent random variables with common characteristic function β. Show that the characteristic function of $X - Y$ is $|\beta|^2$.

Problem 18. Use characteristic functions to show that the bilateral exponential with parameter λ is the symmetrization of the exponential with the same parameter.

* **Problem 19.** [Bilateral geometric distributions] Calculate the density (with respect to counting measure) and the characteristic function of the symmetrization of a geometric distribution.

Problem 20. Prove that a random variable X is symmetric about 0 if and only if its characteristic function β is $\mathbb{R}$-valued, in which case $\beta(v) = E[\cos(vX)], v \in \mathbb{R}$.

Problem 21. $(Z_1, Z_2, \dots)$ be an independent sequence of $\mathbb{R}$-valued random variables, each of which is symmetric about 0. Suppose there is a real constant c such that $|Z_n| \le c$ for all n. The Kolmogorov Three-Series Theorem implies that $\sum_n Z_n$ converges a.s. in $\mathbb{R}$ if and only if $\sum_n E(Z_n^2) < \infty$. Use characteristic functions to prove the 'only if' part of this result. (This approach is an alternative to the use of Corollary 5 of Chapter 5 in the proof of the Kolmogorov Three-Series Theorem.) *Hint:* For all v with $|v|$ sufficiently small (depending on c), $\cos(vZ_n) \le (1 - \frac{1}{4}v^2 Z_n^2)$. Thus, for such v, the characteristic function of $\sum_n Z_n$ is bounded above by $\prod_n [1 - \frac{1}{4}v^2 E(Z_n^2)]$. Use Problem 1 of Appendix E.

Problem 22. Find a distribution on $\mathbb{Z}$ that is symmetric about 0, but which is not the symmetrization of any distribution.

13.5. Moment generating functions

Probability generating functions and characteristic functions of distributions are called *transforms* of those distributions. We now introduce a third transform, defined for $\overline{\mathbb{R}}^+$-valued random variables X: $\varphi(u) = E(e^{-uX})$ for $0 \leq u < \infty$. It is called the *moment generating function* of X or of its distribution or of its distribution function. Just as we defined $1^\infty = 1$ in the definition of probability generating function, we define $e^{-0\cdot\infty} = 1$ in the definition of moment generating function.

The relationship between the probability generating function and the moment generating function is contained in the substitution $s = e^{-u}$. Note also that since the power series that defines a probability generating function converges for any complex number s such that $|s| \leq 1$, the substitution $s = e^{iv}$ determines a similar relationship between probability generating functions and characteristic functions. This fact implies that there is also a substitution that relates moment generating functions to characteristic functions (see Problem 24).

Any of the three transforms—characteristic functions, moment generating functions, probability generating functions—can be used for a random variable supported by $\mathbb{Z}^+$. Often a good strategy is to use probability generating functions for distributions supported by $\mathbb{Z}^+$ (or $\overline{\mathbb{Z}}^+$), moment generating functions for distributions supported by $\mathbb{R}^+$ (or $\overline{\mathbb{R}}^+$) but not by $\overline{\mathbb{Z}}^+$, and characteristic functions for other distributions on $\mathbb{R}$.

Problem 23. Prove that the moment generating function of an $\mathbb{R}^+$-valued random variable is continuous on $[0, \infty)$ and that the moment generating function of an $\overline{\mathbb{R}}^+$-valued random variable is continuous on $(0, \infty)$.

Problem 24. Find a change of variables that relates characteristic functions to moment generating functions of nonnegative finite random variables. Include appropriate comments about extending the domains of definition of these two types of transforms.

Theorem 5. *Let Q_1 and Q_2 be two probability measures on $\overline{\mathbb{R}}^+$. Then the moment generating function of $Q_1 * Q_2$ equals $\varphi_1\varphi_2$, where φ_1 and φ_2 are the moment generating functions of Q_1 and Q_2, respectively.*

Problem 25. Prove the preceding theorem.

Theorem 6. *Different probability measures on $[0, \infty]$ have different moment generating functions.*

PROOF. Let $f: \mathbb{R}^+ \to (0, 1]$ be the function $x \rightsquigarrow e^{-x}$. This function sets up a one-to-one correspondence between $\mathbb{R}^+$ and $(0, 1]$ which is continuous, and hence measurable, in both directions. Let Q_1 and Q_2 be two probability measures on $\mathbb{R}^+$, and let $R_i, i = 1, 2$, be the corresponding distributions induced by f on $(0, 1]$. Clearly, $Q_1 = Q_2$ if and only if $R_1 = R_2$. We will show that if Q_1 and Q_2 have the same moment generating function φ, then R_1 and R_2 have the same characteristic function. The result then follows immediately from Theorem 3.

By Theorem 15 of Chapter 4 (with the role of X in that result being played by $x \rightsquigarrow e^{-x}$),

$$\int_{[0,\infty)} (e^{-x})^n \, Q_1(dx) = \int_{(0,1]} x^n \, R_1(dx), \quad n = 0, 1, 2, \ldots .$$

The left side of this equation equals $\varphi(n)$ and the right side is the nth moment of R_1.

By expanding the function e^{ivx} in a power series and applying the Bounded Convergence Theorem, we can express the characteristic function of R_1 in terms of its moments, and hence in terms of φ:

$$\int_{(0,1]} e^{ivx} \, R_1(dx) = \sum_{n=0}^{\infty} \frac{(iv)^n}{n!} \int_{(0,1]} x^n \, R_1(dx) = \sum_{n=0}^{\infty} \frac{(iv)^n \varphi(n)}{n!} .$$

The characteristic function of R_2 is similarly determined by φ, so that R_1 and R_2 have the same characteristic function, as desired. $\square$

The preceding proof actually shows more than is asserted in the theorem.

Corollary 7. *Let Q_1, Q_2 be probability measures on $\mathbb{R}^+$ with moment generating functions φ_1, φ_2. If $\varphi_1(n) = \varphi_2(n)$ for all $n = 1, 2, \ldots$, then $Q_1 = Q_2$.*

Remark 2. There are two alternative proofs of Theorem 6 that may be of interest.

The first relies on complex variable theory to show that the moment generating function uniquely determines the characteristic function. Let Q be a distribution on $\mathbb{R}^+$ with moment generating function φ and characteristic function β. A standard technique ('analytic continuation') can be used to extend φ uniquely to a holomorphic function in the interior of the right half of the complex plane, and then to a continuous function on the entire right half of the complex plane, including the imaginary axis. The theory implies that when this is done, $\varphi(-iv) = \beta(v)$ for all $v \in \mathbb{R}$, as might be expected from the formulas for φ and β. Tables 13.1, 13.2, and 13.3 bear out this relationship.

The second alternative proof avoids the use of complex numbers and characteristic functions. Let f be any bounded continuous function with domain $[0, \infty)$. The Stone-Weierstrass Theorem implies that there is a uniformly bounded sequence $(f_n : n = 1, 2, \ldots)$ of functions of the form $f_n(x) = a_1 e^{-u_1 x} + \cdots + a_n e^{-u_n x}$, $a_1, \ldots, a_n \in \mathbb{R}$ and $u_1, \ldots, u_n \in \mathbb{R}^+$, such that f_n converges to f uniformly on compact sets as $n \to \infty$. Since $\int f_n(x) \, Q(dx) = a_1 \varphi(u_1) + \cdots + a_n \varphi(u_n)$, the Bounded Convergence Theorem implies that $\int f(x) \, Q(dx)$ is determined by φ. Since f is an arbitrary bounded continuous function on $[0, \infty)$, it is not hard to conclude that Q itself is determined by φ. (There is a similar proof of Theorem 3, based on (13.1).)

Problem 26. Verify the entries in the last column of Table 13.3.

Problem 27. Use part of Table 13.3 to give an alternative solution of Problem 9 of Chapter 10.

Problem 28. [The Moment Problem] Let R and Q be probability distributions on $\mathbb{R}$, each with bounded support. Show that if R and Q have the same n^{th} moments for $n = 0, 1, 2, \ldots$, then $R = Q$. *Hint:* See the proof of Theorem 6. (Note: The hypothesis of bounded support can be weakened but not wholly eliminated.)

Problem 29. [Yule-Furry distributions] For $k = 1, 2, \ldots, m$ and $x > 0$, let $F_k'(x) = ke^{-kx}$. Use moment generating functions to show that the density on $[0, \infty)$ of $F_1 * F_2 * \cdots * F_m$ is

$$x \rightsquigarrow me^{-x}(1 - e^{-x})^{m-1} \,.$$

Hint: Use partial fractions.

* **Problem 30.** Use the preceding problem to calculate the mean and variance of the Yule-Furry distributions.

The following proposition gives a formula for the moment generating function of the absolute value of an $\mathbb{R}$-valued random variable X in terms of the characteristic function of X.

Proposition 8. *Let Q be a probability distribution on $(\mathbb{R}, \mathcal{B})$, and let β denote its characteristic function. Then, for $u > 0$,*

$$\int e^{-u|x|} Q(dx) = \int_{-\infty}^{\infty} \frac{u}{\pi(u^2 + y^2)} \beta(y) \, dy \,.$$

Problem 31. Prove the preceding proposition. *Hint:* Apply the Parseval Relation, with $v = 0$ and R equaling a Cauchy distribution.

The standard normal distribution function arises so often that it has been studied as a function in much the same way that, say, the sine and exponential functions have been studied. In particular, tables of approximate values for it have been constructed. And some calculators give approximate values for it just as they give approximate values for the exponential function. Actually the values given may be for the distribution function of the absolute value of a normally distributed random variable having mean 0 and variance 1. This function, called the *error function*, is denoted by

$$\text{erf}(x) = \sqrt{\frac{2}{\pi}} \int_0^x e^{-w^2/2} \, dw \,, \quad x \in \mathbb{R}^+ \,.$$

Distribution	support	density: $x \rightsquigarrow$	moment generating function: $u \rightsquigarrow$
Exponential	$\mathbb{R}^+$	e^{-x}	$\dfrac{1}{1+u}$
Gamma	$\mathbb{R}^+$	$\frac{1}{\Gamma(\gamma)}x^{\gamma-1}e^{-x}$	$(1+u)^{-\gamma}$
Uniform	$[0,1]$	1	$\dfrac{e^{-u}-1}{u}$
Triangular	$[0,2]$	$1-|x-1|$	$\left(\dfrac{e^{-u}-1}{u}\right)^2$
Delta	$\{0\}$	1	1
Bernoulli	$\{0,1\}$	$\begin{cases} p & \text{if } x=1 \\ 1-p & \text{if } x=0 \end{cases}$	$1-p(1-e^{-u})$
Binomial	$\{0,1,2,\ldots,n\}$	$\binom{n}{x}p^x(1-p)^{n-x}$	$[1-p(1-e^{-u})]^n$
Poisson	$\mathbb{Z}^+$	$\frac{\lambda^x}{x!}e^{-\lambda}$	$\exp(-\lambda(1-e^{-u}))$
Geometric	$\mathbb{Z}^+$	$(1-p)p^x$	$\dfrac{1-p}{1-pe^{-u}}$
Neg. Binomial	$\mathbb{Z}^+$	$(1-p)^r\frac{(r)_x^\uparrow}{x!}p^x$	$\left(\dfrac{1-p}{1-pe^{-u}}\right)^r$

TABLE 13.3. Moment generating functions of some distributions

Problem 32. Prove that

$$\int_{-\infty}^{\infty} \frac{1}{\beta + y^2} e^{-\alpha y^2} \, dy = \frac{\pi}{\sqrt{\beta}} e^{\alpha \beta} \left[1 - \text{erf}\left(\sqrt{2\alpha\beta} \right) \right]$$

for $\alpha, \beta > 0$.

Problem 33. Use Proposition 8 and the preceding problem to show that the moment generating function of the absolute value of a normally distributed random variable X with mean 0 and variance σ^2 is the function

$$u \rightsquigarrow e^{(\sigma u)^2/2} [1 - \text{erf}(\sigma u)] \, .$$

Show that the derivative from the right at 0 of this function equals $-\sigma\sqrt{2/\pi}$.

It is useful to become familiar with the functions

$$\text{si}(x) = -\int_x^{\infty} \frac{\sin w}{w} \, dw \, , \quad x \geq 0 \, ,$$

and

$$\text{ci}(x) = -\int_x^{\infty} \frac{\cos w}{w} \, dw \, , \quad x > 0 \, ,$$

known as the *sine integral* and *cosine integral*, respectively; these integrals are improper Riemann integrals. As measure-theoretic integrals they do not exist.

* **Problem 34.** Show that

$$\int_0^{\infty} \frac{1}{u^2 + y^2} e^{-\alpha y} \, dy = \frac{1}{u} \Big(\text{ci}(\alpha u) \sin(\alpha u) - \text{si}(\alpha u) \cos(\alpha u) \Big)$$

for $\alpha, u > 0$.

Problem 35. Let $t > 0$. Use the preceding problem to show that the moment generating function of the absolute value of a Cauchy random variable X with density

$$x \rightsquigarrow \frac{t}{\pi(t^2 + x^2)}$$

is the function

(13.7) $u \rightsquigarrow \begin{cases} 1 & \text{if } u = 0 \\ \frac{2}{\pi} \Big(\text{ci}(tu) \sin(tu) - \text{si}(tu) \cos(tu) \Big) & \text{if } u > 0 \, . \end{cases}$

Problem 36. Use the preceding problem and Problem 23 to obtain the following formula for an important improper Riemann integral:

$$\int_0^{\infty} \frac{\sin x}{x} \, dx = \frac{\pi}{2} \, .$$

Problem 37. Show that the derivative from the right at 0 of the function (13.7) equals $-\infty$.

13.6. Moment theorems

In Chapter 5, we found a relationship between the moments of a distribution Q on $\mathbb{Z}^+$ and the derivatives of the probability generating function of Q. There are analogous relationships involving moment generating functions and characteristic functions.

Let Q be a distribution on $\mathbb{R}$, and assume that the nth moment of Q exists and is finite for some fixed positive integer n. For each $k = 0, \ldots, n$, consider the function $f_k \colon \mathbb{R} \to \mathbb{C}$ defined by

$$v \rightsquigarrow \int x^k e^{ivx}\, Q(dx)\,.$$

An easy argument based on the Dominated Convergence Theorem shows that each f_k is continuous on $\mathbb{R}$. We wish to calculate $f_k'(v)$ for $k \le n - 1$:

(13.8)
$$\lim_{h \to 0} \frac{f_k(v + h) - f_k(v)}{h} = \lim_{h \to 0} \int x^k e^{ivx} \frac{e^{ihx} - 1}{h}\, Q(dx)$$
$$= \int x^k e^{ivx} \lim_{h \to 0} \frac{e^{ihx} - 1}{h}\, Q(dx) = i f_{k+1}(v)\,.$$

Moving the limit inside the integral is justified by the Dominated Convergence Theorem, the assumption that Q has finite n^{th} moment, and the fact that for all $h \ne 0$,

$$\left| x^k e^{ivx} \frac{e^{ihx} - 1}{h} \right| = \left| x^k \frac{(\cos(hx) - 1) + i \sin(hx)}{h} \right| \le 2|x^{k+1}|\,.$$

Let β be the characteristic function of Q. Note that $\beta = f_0$. A simple inductive argument based on (13.8) shows that for $k = 0, \ldots, n$,

$$\beta^{(k)}(v) = i^k f_k(v)$$

so the derivatives of β exist and are finite up to order n. Setting $v = 0$ and $k = n$ in this equation gives the following result:

Proposition 9. *Let X be an $\mathbb{R}$-valued random variable with characteristic function β. If the nth moment of X exists and is finite for some nonnegative integer n, then the nth derivative of β exists, and*

$$E(X^n) = (-i)^n \beta^{(n)}(0)\,.$$

The following problem contains the analogous result for moment generating functions:

Problem 38. Let X be an $\mathbb{R}^+$-valued random variable with moment generating function φ. Prove: For all n, the nth derivative of φ exists on $(0, \infty)$, and

$$\varphi^{(n)}(u) = (-1)^n E(X^n e^{-uX})\,, \quad u > 0\,.$$

And if $E(X^n)$ is finite, then the nth derivative of φ exists at 0 (taken from the right), and

$$E(X^n) = (-1)^n \varphi^{(n)}(0).$$

The results we have obtained so far for the nth moment of a random variable are not as useful as we would like, since they require knowing in advance that the nth moment is finite. For the case of moment generating functions, this deficiency is remedied by the following theorem:

Theorem 10. *Let X be an $\mathbb{R}^+$-valued random variable with moment generating function φ. For each positive integer n, $E(X^n) < \infty$ if and only if $\varphi^{(n)}(0)$ exists as a finite number, in which case*

$$E(X^n) = (-1)^n \varphi^{(n)}(0),$$

where the derivatives at 0 are taken from the right.

PROOF. The 'only if' part of the result is contained in Problem 38. We also know from Problem 38 that

(13.9) $$\varphi^{(n)}(u) = (-1)^n E(X^n e^{-uX})$$

for $u > 0$. To prove the 'if' part, we need to extend this formula to allow $u = 0$ when $\varphi^{(n)}(0)$ exists and is finite (taken from the right).

Let us use induction on n. The assertion in the theorem makes sense and is obviously true when $n = 0$, so we can start our induction there. Now assume that the result is true when $n = k$ for some integer $k \geq 0$. This assumption together with (13.9) implies that

$$(-1)^{k+1} \varphi^{(k+1)}(0) = (-1)^k \lim_{u \searrow 0} \frac{\varphi^{(k)}(0) - \varphi^{(k)}(u)}{u}$$

$$= \lim_{u \searrow 0} E\left(X^k \frac{1 - e^{-uX}}{u} \right).$$

By the l'Hospital Rule, $[1 - e^{-ux}]/u \to x$ as $u \searrow 0$. Thus, we can complete the proof by an appeal to the Monotone Convergence Theorem provided that we can show $[1 - e^{-ux}]/u$ increases as $u \searrow 0$. This desired monotonicity is a consequence of the fact that the derivative of $[1 - e^{-ux}]/u$ with respect to u, namely

$$u \rightsquigarrow \frac{e^{-ux}[1 + ux] - 1}{u^2},$$

is negative for $x \geq 0$, a fact that can be seen in either of two ways. One way is to observe that, by the Taylor Formula, $1 + a < e^a$ for $a > 0$. The other is to note that the function $a \rightsquigarrow e^{-a}(1 + a) - 1$ is 0 at 0 and has negative derivative for $a > 0$. $\square$

Problem 39. Use the preceding theorem together with appropriate calculations as a check on the correctness of $-\sigma\sqrt{2/\pi}$ in Problem 33 and $-\infty$ in Problem 37.

Problem 40. Let $X = (X_n : n = 0, 1, 2, \dots)$ be an iid sequence of $\overline{\mathbb{R}}^+$-valued random variables with common moment generating function φ, and let N be an $\overline{\mathbb{Z}}^+$-valued random variable with probability generating function ρ. Assume that N is independent of X. Show that the moment generating function of the random variable $S = X_1 + \cdots + X_N$ is $\rho \circ \varphi$, and use this fact to find formulas for $E(S)$ and $\operatorname{Var}(S)$ in terms of the means and variances (or second moments) of X_1 and N.

Unfortunately, the story for characteristic functions is more complicated than the one for moment generating functions. For example, it is possible that the first derivative at 0 of the characteristic function be finite even though the first moment does not exist (see Problem 42).

Theorem 11. *Let β be the characteristic function of an $\mathbb{R}$-valued random variable X, and let n denote a positive integer. If $E(X^n)$ exists as a finite number, then $\beta^{(n)}$ exists as a $\mathbb{C}$-valued function on $\mathbb{R}$ and*

$$E(X^n) = (-i)^n \beta^{(n)}(0).$$

If n is even and $\beta^{(n)}(0)$ exists as a finite number, then $E(X^n)$ exists as a finite number.

PROOF. The first part of this result is contained in Proposition 9. It remains to show that if n is even and $\beta^{(n)}(0)$ is finite, then $E(X^n)$ is finite.

We use induction on n, beginning, as in the proof of Theorem 10, at $n = 0$, for which the assertion is obviously true. Suppose the result is true for $n = k - 2$ for some even integer $k \geq 2$, and suppose that $\beta^{(k)}(0)$ exists as a finite number. The inductive hypothesis implies that $E(X^{k-2})$ is finite. We want to show that $E(X^k)$ is also finite.

By the l'Hospital Rule

$$\lim_{v \to 0} \frac{2 - 2\cos vx}{v^2} \to x^2.$$

Thus, the Fatou Lemma applies to give

$$E(X^k) \leq \liminf_{v \to 0} E\left(X^{k-2}\frac{2 - 2\cos vX}{v^2}\right)$$

$$= \liminf_{v \to 0} \left(-E\left(X^{k-2}\frac{e^{ivX} - 2 + e^{-ivX}}{v^2}\right)\right)$$

(13.10) $$= \liminf_{v \to 0} \left(-\frac{f_{k-2}(v) - 2f_{k-2}(0) + f_{k-2}(-v)}{v^2}\right),$$

where $f_0, f_1, f_2, \dots$, are the functions defined in the discussion leading up to Proposition 9. Since we are given that $E(X^{k-2})$ is finite, we know from that discussion that $\beta^{(k-2)}(v) = i^{k-2}f_{k-2}(v)$ for all real v. Our assumption that

$\beta^{(k)}(0)$ exists and is finite thus implies that $f'_{k-2}(v)$ exists for v in a neighborhood of the origin, and that $f''_{k-2}(0)$ also exists and is finite.

These facts allow us to apply the l'Hospital Rule to (13.10), showing in the process that the limit infimum there is actually a limit:

$$
\begin{aligned}
\liminf_{v \to 0} &\left(-\frac{f_{k-2}(v) - 2f_{k-2}(0) + f_{k-2}(-v)}{v^2} \right) \\
&= -\lim_{v \to 0} \frac{f'_{k-2}(v) - f'_{k-2}(-v)}{2v} \\
&= \lim_{v \to 0} \frac{[f'_{k-2}(v) - f'_{k-2}(0)] + [f'_{k-2}(0) - f'_{k-2}(-v)]}{2v} \\
&= f''_{k-2}(0),
\end{aligned}
$$

the last step following from the definition of $f''_{k-2}(0)$. After substituting this expression into (13.10), we may conclude that $E(X^k)$ is finite, as desired. $\square$

Problem 41. Near the end of the preceding proof, why was the application of the l'Hospital Rule followed by a reference to the definition of the derivative, rather than by a second application of the l'Hospital Rule?

Problem 42. Let g denote a decreasing function on the interval $[1, \infty)$ such that $g(x) \to 0$ as $x \to \infty$. Let Q denote the probability measure whose density with respect to Lebesgue measure is the function

$$
x \rightsquigarrow \begin{cases} cx^{-2}g(|x|) & \text{if } |x| \geq 1 \\ 0 & \text{if } |x| < 1, \end{cases}
$$

where c is the appropriate constant. (i) Show that $\beta'(0) = 0$, where β denotes the characteristic function of Q. (ii) Give an example of a g for which the distribution Q does not have a mean. *Hint:* For part (i) make a change of variables before using a convergence theorem.

Problem 43. Let $X = (X_n : n = 0, 1, 2, \dots)$ be an iid sequence of $\mathbb{R}$-valued random variables, with common characteristic function β. Let N be a $\mathbb{Z}^+$-valued random variable with probability generating function ρ. Assume that N is independent of X. Show that $X_1 + \cdots + X_N$ has characteristic function $\rho \circ \beta$. Can you justify formulas for $E(S)$ and $\mathrm{Var}(S)$ that are similar to those obtained in Problem 40?

13.7. Inversion theorems

Let us start with a simple general result about characteristic functions:

Proposition 12. *Let β be the characteristic function of a distribution on $\mathbb{R}$. Then (i) $\beta(0) = 1$; (ii) β is continuous; and (iii) for all $n = 1, 2, \dots$, and all* *complex n-tuples* $(z_1, \ldots, z_n)$ *and real n-tuples* $(v_1, \ldots, v_n)$,

(13.11)
$$\sum_{k=1}^{n} \sum_{j=1}^{n} \beta(v_k - v_j) z_j \bar{z}_k \geq 0.$$

PROOF. Property (i) is obvious, and Property (ii) follows easily from the definition of β and the Bounded Convergence Theorem. We now prove Property (iii). Let Q be the distribution corresponding to β. By the definition of β and the linearity of the integral, the left side of (13.11) equals

$$\int \left(\sum_{k=1}^{n} \sum_{j=1}^{n} e^{i(v_k - v_j)x} z_j \bar{z}_k \right) Q(dx) = \int \left| \sum_{j=1}^{n} e^{-iv_j x} z_j \right|^2 Q(dx) \geq 0.$$

(We have used some familiar facts from complex number theory, namely that $|z|^2 = z\bar{z}$ for any complex number z, and that $e^{-ix} = \overline{e^{ix}}$ for all real x.) $\quad\square$

Any $\mathbb{C}$-valued function β on $\mathbb{R}$ that satisfies Property (iii) is *positive definite*. If the inequality at (13.11) is strict whenever $(z_1, \ldots, z_n)$ is not the zero vector, β is *strictly positive definite*.

It turns out that the converse of Proposition 12 is also true: If $\beta : \mathbb{R} \to \mathbb{C}$ is continuous and positive definite, and if $\beta(0) = 1$, then β is the characteristic function of some distribution on $\mathbb{R}$. This converse will be proved in Chapter 14. In the present section, we will prove two important special cases of the converse, and for those special cases, provide formulas for calculating the distribution corresponding to β. In particular, we will treat all distributions that have densities with respect to counting measure on $\mathbb{Z}$, and some of the distributions that have densities with respect to Lebesgue measure on $\mathbb{R}$.

Let X be a $\mathbb{Z}$-valued random variable, with characteristic function β. Then

$$\beta(v + 2\pi) = E(e^{i(v+2\pi)X}) = E(e^{ivX} e^{2\pi i x}) = E(e^{ivX}) = \beta(v).$$

Thus, the characteristic function of a distribution on $\mathbb{Z}$ is periodic with period 2π, in addition to having the three properties in Proposition 12. The converse of this result is also true, as stated in the following theorem, which also contains the first of our 'inversion formulas'.

Theorem 13. *A function* $\beta : \mathbb{R} \to \mathbb{C}$ *is the characteristic function of a* $\mathbb{Z}$-*valued random variable if and only if it satisfies Properties (i)-(iii) of Proposition 12 and, in addition, is periodic with period* 2π. *Furthermore, if* β *is the characteristic function of a* $\mathbb{Z}$-*valued random variable* X, *then*

(13.12)
$$P(\{\omega : X(\omega) = x\}) = \frac{1}{2\pi} \int_{-\pi}^{\pi} e^{-ixv} \beta(v) \, dv$$

for each integer x.

PROOF. We have already seen that any characteristic function of a distribution on $\mathbb{Z}$ satisfies Properties (i)-(iii) and is periodic with period 2π. So now we consider a function $\beta\colon \mathbb{R} \to \mathbb{C}$ which has these four properties. We wish to show that β is the characteristic function of the distribution on $\mathbb{Z}$ whose density with respect to counting measure on $\mathbb{Z}$ is given by

(13.13) $$p_x = \frac{1}{2\pi} \int_{-\pi}^{\pi} e^{-ixv} \beta(v)\,dv\,, \quad x \in \mathbb{Z}\,.$$

We must show three things: that each p_x, defined by (13.13), is nonnegative; that $\sum_{x \in \mathbb{Z}} p_x = 1$; and that the characteristic function of the distribution on $\mathbb{Z}$ thus defined is the function β.

To prove that each p_x is nonnegative, we first use periodicity to conclude that, for each u:

$$p_x = \frac{1}{2\pi} \int_{-\pi}^{\pi} e^{-ix(v-u)} \beta(v-u)\,dv\,.$$

Thus,

$$p_x = \frac{1}{2\pi} \int_{-\pi}^{\pi} p_x\,du = \frac{1}{(2\pi)^2} \int_{-\pi}^{\pi} \left[\int_{-\pi}^{\pi} e^{-ix(v-u)} \beta(v-u)\,dv \right] du\,.$$

This iterated integral can be written as the limit of Riemann sums, each of which takes the form appearing in (13.11) (with the z_j equal to e^{ixv} for various values of v in $[0, 2\pi]$). Since β is positive definite, each of the Riemann sums is nonnegative. We conclude that each p_x is nonnegative.

We now prove, simultaneously, that the quantities p_x sum to 1 and that the corresponding distribution has characteristic function β. In order to do this, we consider, for each positive integer m and each $u \in \mathbb{R}$, the quantity

(13.14) $$b_m(u) = \sum_{x=-m}^{m} \left(1 - \frac{|x|}{m} \right) e^{iux} p_x\,.$$

Suppose we can show that

(13.15) $$b_m(u) \to \beta(u)\,, \quad u \in \mathbb{R}\,.$$

Then it would follow from the Monotone Convergence Theorem that

$$\lim_{m \to \infty} b_m(0) = \sum_{-\infty}^{\infty} p_x\,.$$

Since $\beta(0) = 1$, we would have that $\sum p_x = 1$. The finiteness of $\sum p_x$ would then allow us to use the Dominated Convergence Theorem to justify taking the limit as $m \to \infty$ inside the summation in (13.14) to obtain

$$\lim_{m \to \infty} b_m(u) = \sum_{-\infty}^{\infty} e^{iux} p_x$$

for all $u \in \mathbb{R}$. The desired conclusion that β is the characteristic function of the distribution on $\mathbb{Z}$ corresponding to the quantities p_x would then follow.

Thus it remains to prove (13.15). Substitute the expression (13.13) into the definition of $b_m(u)$ and then change variables to obtain

$$b_m(u) = \frac{1}{2\pi} \int_{-\pi}^{\pi} \left[\sum_{x=-m}^{m} \left(1 - \frac{|x|}{m}\right) e^{-ix(v-u)} \right] \beta(v) \, dv$$

$$= \frac{1}{2\pi} \int_{-\pi}^{\pi} \sum_{x=-m}^{m} (e^{-iv})^x \beta(v+u) \, dv - \frac{1}{2\pi} \int_{-\pi}^{\pi} \sum_{x=-m}^{m} \frac{|x|}{m} (e^{-iv})^x \beta(v+u) \, dv \, .$$

We used the periodicity of both β and the function $v \rightsquigarrow e^{-ivx}$ to adjust the limits of integration after the change of variables. The first sum on the right of the preceding chain of equalities can be treated as a pair of finite geometric series, and the second term can be treated as the derivative of a pair of geometric series. Thus, the two sums can be calculated explicitly, and the result is

(13.16)
$$\frac{1}{2\pi} \int_{-\pi}^{\pi} \frac{1 - \cos[(m+1)v]}{2m \sin^2(v/2)} \beta(v+u) \, dv - \frac{1}{2\pi} \int_{-\pi}^{\pi} \frac{\sin[(m+\frac{1}{2})v]}{m \sin(v/2)} \beta(v+u) \, dv \, .$$

The integrands in (13.16) are to be understood as being defined at $v = 0$ so as to be continuous there. (Why?)

As $m \to \infty$, the second integral in the right side of (13.16) approaches 0 because its integrand approaches 0 and the Bounded Convergence Theorem applies by virtue of the inequalities

(13.17)
$$\left| m \sin \frac{v}{2} \right| \ge m \left(\left| \frac{v}{2} \right| - \left| \frac{v^3}{48} \right| \right) \ge m \left| \frac{v}{2} \right| \left(1 - \frac{\pi^2}{24} \right),$$

and

$$|\sin(m+\tfrac{1}{2})v| \le (m+\tfrac{1}{2})|v| \, ,$$

both of which are consequences of the Taylor Formula with remainder.

The integrand in the first integral in the right side of (13.16) also approaches 0, but this time neither the Bounded nor Dominated Convergence Theorems is applicable; and as we will see, the limit of the integral is not 0. We make the substitution $v = w/(m+1)$ in the first integral in the right side of (13.16), obtaining

(13.18)
$$\frac{1}{4\pi} \int_{-\pi(m+1)}^{\pi(m+1)} \frac{(1 - \cos w)\, \beta([w/(m+1)]+u)}{m(m+1)\sin^2(w/2(m+1))} \, dw \, .$$

Since β is continuous, we see that the integrand approaches $4\beta(u)(1 - \cos u)/u^2$ as $m \to \infty$, so that, were we able to move the limit inside the integral sign and into the endpoints of integration, we would obtain

$$b_m(u) \to \beta(u) \frac{1}{\pi} \int_{-\infty}^{\infty} \frac{1 - \cos w}{w^2} \, dw \, ,$$

which, by the forthcoming Problem 44, equals $\beta(u)$, as desired. To handle the endpoints of integration, we simply let the interval of integration be $(-\infty, \infty)$

and insert the indicator function of the interval $[-\pi(m+1), \pi(m+1)]$ into the integrand. We now wish to apply the Dominated Convergence Theorem. The function β is continuous and periodic, so it is bounded. The indicator of the interval $[-\pi(m+1), \pi(m+1)]$ is also bounded, so it is sufficient to find a function with finite integral that dominates

$$\frac{(1 - \cos w)}{m(m+1)\sin^2(w/2(m+1))}.$$

An argument similar to that at (13.17) shows that $m(m+1)\sin^2(w/2(m+1))$ is bounded below by a constant multiple of w^2, so some constant multiple of $(1 - \cos w)/w^2$ will serve as our dominating function, since we have already noted that this function has finite integral with respect to Lebesgue measure on $\mathbb{R}$. $\square$

Remark 3. In applications of Theorem 13, the most difficult hypothesis to check is often the positive definiteness. But the proof shows that this condition is only needed to prove that the quantities p_x are nonnegative. In many cases of interest, it is possible to compute these quantities explicitly from (13.13), thus making the verification of positive definiteness unnecessary.

Problem 44. Prove the following equality, which was used in the preceding proof:

$$\int_{-\infty}^{\infty} \frac{1 - \cos u}{u^2}\, du = 2 \int_0^{\infty} \frac{1 - \cos u}{u^2}\, du = \pi.$$

Hint: For $\alpha \geq 0$ let

$$f(\alpha) = \int_0^{\infty} \frac{e^{-\alpha u}(1 - \cos u)}{u^2}\, du$$

and use a convergence theorem to prove that f is continuous. Also, use a convergence theorem to prove that, for $\alpha > 0$,

$$f''(\alpha) = \int_0^{\infty} e^{-\alpha u}\, (1 - \cos u)\, du = \frac{1}{\alpha(1 + \alpha^2)}.$$

Show that $f(\alpha) \to 0$ and $f'(\alpha) \to 0$ as $\alpha \to \infty$ and use these facts in conjunction with two successive calculations of antiderivatives to obtain a formula for $f(\alpha)$, $\alpha > 0$. Then let $\alpha \searrow 0$ to calculate $f(0)$.

Problem 45. Suppose, in the proof of Theorem 13, that m had not been introduced, but that instead, the integral formula for p_x had been inserted immediately with the idea of then, at the very next step, using the Fubini Theorem to interchange the order of summation and integration. Would that procedure work?

Problem 46. In the proof of Theorem 13 it would seem more natural to introduce the indicator function of the set of integers having absolute value no larger than m rather than the factor $(1 - |x|/m) \vee 0$. However, some difficulties arise that did not arise in the proof as given. Explore this issue.

Problem 47. Use Theorem 13 to show that for all integers n, the function $v \rightsquigarrow \cos nv$ is the characteristic function of a distribution on $\mathbb{Z}$, and use the inversion formula contained there to calculate the corresponding density. *Hint:* See the remark following the proof of the theorem.

* **Problem 48.** Use Theorem 13 and the symmetrization of a geometric distribution (see Problem 19) to obtain a formula for the integral

$$\int_{-\pi}^{\pi} \frac{\cos nv}{a - b \cos v}\, dv\,,$$

where n is an integer and $0 < b < a$.

Problem 49. State and prove the analogue of Theorem 13 for distributions supported by the set $\{ax: x \in \mathbb{Z}\}$, where a is a fixed positive number.

Problem 50. Let β be the characteristic function of an $\mathbb{R}$-valued random variable X. Show that if $\beta(2\pi a) = 1$ for some $a \neq 0$, then aX is almost surely $\mathbb{Z}$-valued. Also show that if $|\beta(2\pi a)| = 1$ for some $a \neq 0$, then there exists a real constant b such that $aX + b$ is $\mathbb{Z}$-valued. *Hint:* If $\beta(2\pi a) = 1$, then $E(\cos(2\pi aX)) = 1$. If $|\beta(2\pi a)| = 1$, then $e^{ic}\beta(2\pi a) = 1$ for some real number c.

Problem 51. Let β be the characteristic function of an $\mathbb{R}$-valued random variable X. Show that if $|\beta(v)| = 1$ for all v in some nonempty open interval, then X is almost surely constant. *Hint:* See the previous problem.

We wish also to obtain an analogue of Theorem 13 for distributions that have densities with respect to Lebesgue measure. It turns out that there are no simple necessary and sufficient conditions that characterize all such distributions. The following result provides a set of sufficient conditions. It also contains a useful inversion formula.

Theorem 14. *Let* $\beta\colon \mathbb{R} \to \mathbb{C}$ *be a function that satisfies Properties (i)-(iii) of Proposition 12. Suppose, in addition, that*

$$\int |\beta(v)|\, dv < \infty\,.$$

Then β *is the characteristic function of a distribution* Q *on* $\mathbb{R}$ *that is absolutely continuous with respect to Lebesgue measure. Furthermore,* Q *has a continuous density* f, *given by the formula*

$$(13.19) \qquad f(x) = \frac{1}{2\pi} \int_{-\infty}^{\infty} e^{-ivx} \beta(v)\, dv\,.$$

PARTIAL PROOF. Let f be the function defined in (13.19). Because of our assumption about β, f is clearly well-defined and finite. The Dominated Convergence Theorem (with $|\beta|$ as the dominating function) implies that f is continuous. Thus, it remains to prove three facts: that f is nonnegative, that the integral of f with respect to Lebesgue measure equals 1, and that the distribution with density f has characteristic function β. The proof of the second and third facts is similar to (but simpler than) the analogous part of the proof of Theorem 13 and is left as an exercise for the reader. We prove here that f is nonnegative.

First note that a simple change of variables implies that for all $u \in \mathbb{R}$,

$$\int_{-\infty}^{\infty} e^{-ivx} \beta(v)\, dv = \int_{-\infty}^{\infty} e^{-i(v-u)x} \beta(v-u)\, dv.$$

Thus, for all positive numbers A,

$$2\pi f(x) = \frac{1}{2A} \int_{-A}^{A} \int_{-\infty}^{\infty} e^{-i(v-u)x} \beta(v-u)\, dv\, du$$

$$= \frac{1}{2A} \int_{-A}^{A} \int_{-A}^{A} e^{-i(v-u)x} \beta(v-u)\, dv\, du$$

$$+ \frac{1}{2A} \int_{-A}^{A} \int_{[-A,A]^c} e^{-i(v-u)x} \beta(v-u)\, dv\, du.$$

The first iterated integral on the right side of this expression is nonnegative for all A by the nonnegative definiteness of β (see the corresponding part of the argument in the proof of Theorem 13). To complete the proof, it is sufficient to show that the second iterated integral converges to 0 as $A \to \infty$.

This second iterated integral breaks naturally into two pieces, one of which is

$$\frac{1}{2A} \int_{-A}^{A} \int_{A}^{\infty} e^{-i(v-u)x} \beta(v-u)\, dv\, du.$$

We will show that this piece converges to 0 as $A \to \infty$. The proof for the other piece is similar.

$$\left| \frac{1}{2A} \int_{-A}^{A} \int_{A}^{\infty} e^{-i(v-u)x} \beta(v-u)\, dv\, du \right| \leq \frac{1}{2A} \int_{-A}^{A} \int_{A}^{\infty} |\beta(v-u)|\, dv\, du$$

$$= \frac{1}{2A} \int_{-A}^{A} \int_{A-u}^{\infty} |\beta(w)|\, dw\, du$$

$$= \frac{1}{2A} \int_{0}^{\infty} \int_{(A-w)\vee(-A)}^{A} |\beta(w)|\, du\, dw$$

$$= \frac{1}{2A} \int_{0}^{\infty} (w \wedge (2A)) |\beta(w)|\, dw$$

$$= \int_{0}^{\infty} \left(\frac{w}{2A} \wedge 1 \right) |\beta(w)|\, dw.$$

The integrand in the last expression converges to 0 as $A \to \infty$, and it is dominated by $|\beta|$, so the integral converges to 0 as $A \to \infty$ by the Dominated Convergence Theorem. $\square$

Problem 52. Complete the proof of the preceding theorem by showing that the integral of f with respect to Lebesgue measure equals 1 and that the distribution with density f has characteristic function β. *Hint:* Instead of summing from $-m$ to m, as was done in the proof of Theorem 13, you should integrate. You will obtain an expression similar to but simpler than the one in (13.16).

Problem 53. The inversion formula in Theorem 14 is valid for computing some, but not all, of the densities in Table 13.1. Which ones are to be excluded? What do these densities have in common? Does the beta distribution fit into the category of distributions covered by Theorem 14? Why or why not?

Problem 54. [Normal approximation to binomial] Let $X = (X_1, X_2, \dots)$ be an iid sequence of Bernoulli random variables with parameter $p = 1/2$, and let Z be a normally distribution random variable with mean 0 and variance 1. Assume that Z is independent of the sequence X. Fix $\delta > 0$, and let β_n be the characteristic function of

$$\delta Z + \frac{2(X_1 + \cdots + X_n) - n}{\sqrt{n}}.$$

Calculate β_n explicitly, and prove that the corresponding distribution has a density with respect to Lebesgue measure on $\mathbb{R}$. Then show that for all $v \in \mathbb{R}$,

$$\lim_{n \to \infty} \beta_n(v) = e^{-v^2(1+\delta^2)/2}.$$

Use this fact in conjunction with the inversion formula in Theorem 14 to show that for all real numbers $a \le b$,

$$\lim_{n \to \infty} P\Big(\big\{\omega: a \le \delta Z(\omega) + \frac{2(X_1(\omega) + \cdots + X_n(\omega)) - n}{\sqrt{n}} \le b\big\}\Big)$$
$$= \frac{1}{\sqrt{2\pi(1+\delta^2)}} \int_a^b e^{-x^2/2(1+\delta^2)}\, dx.$$

Conclude that

$$\lim_{n \to \infty} P\Big(\big\{\omega: \frac{n + a\sqrt{n}}{2} \le \sum_{k=1}^n X_k(\omega) \le \frac{n + b\sqrt{n}}{2}\big\}\Big) = \frac{1}{\sqrt{2\pi}} \int_a^b e^{-x^2/2}\, dx.$$

Problem 55. [Normal approximation to gamma] Let Q_γ be the gamma distribution on $[0, \infty)$ with parameters $1, \gamma$. Show that for all real $a \le b$,

$$\lim_{\gamma \to \infty} Q_\gamma([\gamma + \sqrt{\gamma}a, \gamma + \sqrt{\gamma}b]) = \frac{1}{\sqrt{2\pi}} \int_a^b e^{-x^2/2}\, dx.$$

Hint: See the previous exercise.

Problem 56. [Parseval Formula] Let β be a function satisfying the conditions of Theorem 14, and let f be a density of the corresponding distribution. Prove that

$$\int f^2(x)\, dx = \frac{1}{2\pi} \int |\beta(v)|^2\, dv < \infty.$$

Hint: Use the inversion formula in Theorem 14 for one of the factors of $f(x)$ in $\int f^2$, switch the order of integration, and make the change of variables $x \rightsquigarrow -x$. Be sure to justify your use of the Fubini Theorem.

Problem 57. Let β be a function that satisfies the conditions of Theorem 14, and let f be the continuous density of the corresponding distribution. Suppose that $\beta(v)$ is nonnegative for all $v \in \mathbb{R}$. Show that the function $v \rightsquigarrow f(v)/f(0)$ is the characteristic function of the distribution whose density with respect to Lebesgue measure on $\mathbb{R}$ is $\beta/(\int \beta(v)\,dv)$. What examples of this result can you find in Table 13.1?

13.8. Characteristic functions in $\mathbb{R}^d$

The spaces $\mathbb{R}^d$, $d < \infty$, have enough structure so that characteristic functions can be defined. For an $\mathbb{R}^d$-valued random variable X with distribution Q, the *characteristic function* β is given by

$$\beta(w) = E(e^{i\langle w, X\rangle}) = \int e^{i\langle w, x\rangle}\,Q(dx)\,, \quad w \in \mathbb{R}^d\,,$$

where $\langle w, x\rangle$ denotes the Euclidean inner product of w and x. Note that characteristic functions of distributions on $\mathbb{R}^d$ are bounded continuous functions from $\mathbb{R}^d$ to the complex numbers (the continuity following as in the one-dimensional case from the Bounded Convergence Theorem).

Problem 58. Let $w \to \beta(w)$ be the characteristic function of an $\mathbb{R}^d$-valued random variable X. Show that

$$w_j \rightsquigarrow \beta(0, \ldots, 0, w_j, 0 \ldots, 0)$$

is the characteristic function of X_j, the jth coordinate of X.

Problem 59. Calculate the characteristic function of an $\mathbb{R}^d$-valued random variable uniformly distributed on $[-1, 1]^d$.

Problem 60. Let $Z = (X + 2, Y - 2)$, where X and Y, respectively, denote the smaller and larger of the two components of a random vector uniformly distributed on the square region $[-6, 6]^2$ in $\mathbb{R}^2$. Show that the characteristic function of Z is the function on $\mathbb{R}^2$ given by

$$(u, v) \rightsquigarrow \frac{(u + v)e^{4i(v-u)} - ue^{-4i(u+2v)} - ve^{4i(2u+v)}}{72\,uv(u + v)}\,.$$

Problem 61. Show that the gradient of the characteristic function in Problem 60 is zero at the origin. Give a probabilistic interpretation of this result.

Problem 62. Let $X = (X_1, X_2, \ldots, X_d)$ be a random vector having independent coordinates each of which is normally distributed with mean 0. Denote the standard deviation of X_j by σ_j. Show that the characteristic function of X is

$$w \rightsquigarrow \exp\left(-\frac{1}{2}\sum_{j=1}^{d} w_j^2 \sigma_j^2\right) = \exp\left(-\frac{1}{2}w\Sigma w^T\right),$$

where the diagonal matrix Σ is the covariance matrix of X and w is viewed as a row matrix in the matrix product.

Problem 63. Let X and Y be independent $\mathbb{R}$-valued normally distributed random variables having mean 0 and variance 1. Find the characteristic functions of the $\mathbb{R}^2$-valued random variable $(X, X + Y)$ and the $\mathbb{R}^3$-valued random variable $(X - Y, X, X + Y)$.

As a general rule, everything about characteristic functions of probability measures on $\mathbb{R}$ which one could reasonably expect to carry over to the $\mathbb{R}^d$-setting does so. Here is one instance.

Lemma 15. [Parseval Relation (for $\mathbb{R}^d$)] *Let Q and R be two probability measures on $\mathbb{R}^d$, and denote their characteristic functions by β and γ, respectively. Then*

$$\int \gamma(x - v)\, Q(dx) = \int e^{-i\langle v,y\rangle}\, \beta(y)\, R(dy)$$

for each $v \in \mathbb{R}$.

Problem 64. Prove the preceding lemma.

We specialize Lemma 15 using Problem 62:

$$\int e^{-\sigma^2|x-v|^2}\, Q(dx)$$
$$= \frac{1}{(2\pi)^{d/2}\sigma^d} \int_{-\infty}^{\infty} \cdots \int_{-\infty}^{\infty} e^{-i\langle v,y\rangle}\, e^{(y_1^2+\cdots+y_d^2)/2\sigma^2}\, \beta(y)\, dy_1 \ldots dy_d\,.$$

Thus, if Q_1 and Q_2 have the same characteristic function β, then

(13.20) $$\int e^{-\sigma^2|x-v|^2}\, Q_1(dx) = \int e^{-\sigma^2|x-v|^2}\, Q_2(dx)$$

for every $\sigma > 0$ and $v \in \mathbb{R}^d$. Let $a_1, \ldots, a_d$ be numbers such that

$$Q_1(\{v\colon v_j = a_j \text{ for some } j\}) = 0 = Q_2(\{v\colon v_j = a_j \text{ for some } j\})\,.$$

Multiply both sides of (13.20) by $\sigma^d/(2\pi)^{d/2}$, integrate with respect to v over the region

(13.21) $$\{v\colon v_1 \le a_1, \ldots, v_d \le a_d\}\,,$$

and let $\sigma \to \infty$ to obtain (see the proof of Theorem 3)

$$Q_1(\{v\colon v_1 \le a_1, \ldots, v_d \le a_d\}) = Q_2(\{v\colon v_1 \le a_1, \ldots, v_d \le a_d\})\,.$$

The collection of sets of the form (13.21) is closed under pairwise intersections and generates the Borel σ-field. Thus, $Q_1 = Q_2$. We have proved the following important theorem.

Theorem 16. *Distinct probability measures on $\mathbb{R}^d$, $d < \infty$, have distinct characteristic functions.*

Corollary 17. *Suppose that X and Y are $\mathbb{R}^d$-valued random variables such that, for every $w \in \mathbb{R}^d$, $\langle w, X \rangle$ and $\langle w, Y \rangle$ have the same distribution. Then X and Y have the same distribution.*

Problem 65. Prove the preceding corollary.

Corollary 18. *Let $d < \infty$. The coordinates of an $\mathbb{R}^d$-valued random variable X are independent if and only if the characteristic function of X has the form*

$$(13.22) \qquad\qquad w \rightsquigarrow \prod_{j=1}^{d} \beta_j(w_j)$$

for some continuous functions β_j each of whose values at 0 is 1, in which case β_j is the characteristic function of X_j, the jth coordinate of X.

PROOF. Suppose the coordinates of X are independent. Then its characteristic function equals

$$w \rightsquigarrow E(e^{i\langle w, X \rangle}) = E\left(\prod_{j=1}^{d} e^{iw_j X_j}\right) = \prod_{j=1}^{d} E(e^{iw_j X_j}),$$

which has the form (13.22), with β_j equal to the characteristic function of X_j.

For the converse suppose that (13.22) holds and that $\beta_j(0) = 1$ for each j. By Problem 58, β_j is the characteristic function of X_j. Let Y be a vector with independent coordinates, the distribution of each Y_j being the same as the distribution of the corresponding coordinate X_j. By the part of the corollary already proved, the characteristic function of the random vector Y is the function (13.22)—that is, the characteristic function of the random vector X. By Theorem 16, X and Y have the same distribution and thus X, like Y, has independent coordinates. $\square$

Problem 66. Show that the sum and difference of two iid normally distributed $\mathbb{R}$-valued random variables are independent.

Problem 67. Suppose that X is an $\mathbb{R}^2$-valued random variable and that its characteristic function has the form $w \rightsquigarrow \beta_1(w_1)\beta_2(w_2)$, with $\beta_1(0) = 4 + 3i$. Can any interesting conclusion be drawn? If so, what?

The following proposition gives a formula for the moment generating function of the Euclidean norm of a random vector in terms of the characteristic function of the random vector itself. Thus, it generalizes Proposition 8.

Proposition 19. *Let β denote the characteristic function of a distribution Q on $\mathbb{R}^d$. Then, for $u > 0$,*

$$\int e^{-u\|x\|} Q(dx) = \frac{u\Gamma(\frac{d+1}{2})}{\pi^{(d+1)/2}} \int \frac{1}{\pi(|u|^2 + |y|^2)^{(d+1)/2}} \beta(y)\, \lambda^d(dy),$$

where λ^d denotes Lebesgue measure on $\mathbb{R}^d$.

Problem 68. Prove the preceding proposition.

13.9. Normal distributions on d-dimensional space

An $\mathbb{R}^d$-valued random vector X is said to be *normally distributed* if each inner product of X with a member of $\mathbb{R}^d$ is an $\mathbb{R}$-valued normally distributed random variable. The distribution of such a normally distributed random vector is called a *normal distribution* or *Gaussian distribution* on $\mathbb{R}^d$. The next result characterizes all normal distributions on $\mathbb{R}^d$.

Theorem 20. *There is a one-to-one correspondence between the family of normal distributions on $\mathbb{R}^d$ and the set of ordered pairs (c, Σ), where $c \in \mathbb{R}^d$ and Σ is a symmetric positive definite $d \times d$ matrix. The distribution corresponding to (c, Σ) has mean vector c, covariance matrix Σ, and characteristic function*

$$w \rightsquigarrow e^{-\frac{1}{2}w\Sigma w^T + i\langle w, c\rangle}.$$

PARTIAL PROOF. Let Q be a normal distribution on $\mathbb{R}^d$, and let X be a random vector having distribution Q. For each $z \in \mathbb{R}$, the inner product $\langle z, X \rangle$ is an $\mathbb{R}$-valued normally distributed random variable. By setting z equal to unit vectors in each coordinate direction we see that each coordinate X_i of X is normally distributed. In particular, $\text{Var}(X_i) < \infty$ for each i and, hence $|\text{Cov}(X_i, X_j)| < \infty$ for each i and j. Denote the mean vector and covariance matrix of X by c and Σ, respectively. By Theorem 11 of Chapter 5, Σ is symmetric and positive definite. $\square$

Problem 69. Complete the proof of the preceding theorem. *Hint:* For each $w \in \mathbb{R}^d$, find the mean and variance of the normally distributed random variable $\langle w, X \rangle$ in terms of w, c, and Σ.

Theorem 21. *Let Q be a normal distribution on $\mathbb{R}^d$ and denote its mean vector and covariance matrix by c and Σ, respectively. Then the support of Q is the subspace of $\mathbb{R}^d$ perpendicular to $\{x \in \mathbb{R}^d : \Sigma x^T = 0\}$. The support equals $\mathbb{R}^d$*

if and only if Σ is strictly positive definite in which case Q has a density with respect to d-dimensional Lebesgue measure given by

$$x \rightsquigarrow \frac{1}{\sqrt{(2\pi)^d \det(\Sigma)}} \, e^{-\frac{1}{2}(x-c)\Sigma^{-1}(x-c)^T}.$$

Problem 70. Prove the preceding theorem. *Hint:* Use the Change of Variables Theorem.

13.10. † An application to random walks on $\mathbb{Z}$

We will use characteristic functions to show that a $\mathbb{Z}$-valued random walk whose steps have mean 0 returns to its starting position with probability 1.

Theorem 22. *Let $S = (S_0 = 0, S_1, \dots)$ be a $\mathbb{Z}$-valued random walk for which $E(S_1) = 0$. Then,*

$$P(\{\omega \colon S_n(\omega) = 0 \text{ for some } n > 0\}) = 1.$$

PROOF. Denote the step distribution by Q and its characteristic function by β. In view of Corollary 13 of Chapter 11, we only need show that

$$\sum_{n=0}^{\infty} Q^{*n}(\{0\}) = \infty.$$

By the Monotone Convergence Theorem, this is equivalent to

(13.23) $$\sum_{n=0}^{\infty} s^n Q^{*n}(\{0\}) \to \infty$$

as $s \nearrow 1$.

Since β^n is the characteristic function of Q^{*n}, we conclude from (13.12) that

$$s^n Q^{*n}(\{0\}) = \frac{1}{2\pi} \int_{-\pi}^{\pi} (s\beta(v))^n \, dv.$$

For $0 \le s < 1$ we use the Fubini Theorem applied to the product of counting measure on the nonnegative integers and Lebesgue measure on $[-\pi, \pi]$ to obtain

$$\sum_{n=0}^{\infty} s^n Q^{*n}(\{0\}) = \frac{1}{2\pi} \int_{-\pi}^{\pi} \sum_{n=0}^{\infty} (s\beta(v))^n \, dv = \frac{1}{2\pi} \int_{-\pi}^{\pi} \frac{1}{1 - s\beta(v)} \, dv$$

for $s < 1$. Since the left side is real, we may replace the integrand by its real part. Thus, for $s < 1$,

(13.24) $$\sum_{n=0}^{\infty} s^n Q^{*n}(\{0\}) = \frac{1}{2\pi} \int_{-\pi}^{\pi} \frac{1 - s\Re(\beta(v))}{[1 - s\Re(\beta(v))]^2 + [s\Im(\beta(v))]^2} \, dv,$$

where $\Re$ and $\Im$ indicate the real and imaginary parts, respectively.

Let $\varepsilon > 0$. Since $\beta'(0) = 0$, there exists a positive v_ε such that $1 - \mathfrak{R}(\beta(v)) \leq \varepsilon|v|$ and $|\mathfrak{I}(\beta(v))| \leq \varepsilon|v|$ for $|v| \leq v_\varepsilon$. Using the inequality $2ab \leq a^2 + b^2$ for real numbers a and b, we obtain

$$[1 - s\mathfrak{R}(\beta(v))]^2 + [s\mathfrak{I}(\beta(v))]^2$$
$$= [1 - s]^2 + 2[1 - s]s[1 - \mathfrak{R}(\beta(v))] + s^2[1 - \mathfrak{R}(\beta(v))]^2 + s^2[\mathfrak{I}(\beta(v))]^2$$
$$\leq 2[1 - s]^2 + 2s^2[1 - \mathfrak{R}(\beta(v))]^2 + s^2[\mathfrak{I}(\beta(v))]^2$$
$$\leq 2[1 - s]^2 + 3\varepsilon^2 v^2$$

for $|v| \leq v_\varepsilon$. Combining this inequality with the obvious inequality

$$1 - s\mathfrak{R}(\beta(v)) \geq 1 - s,$$

we obtain

$$\sum_{n=0}^{\infty} s^n Q^{*n}(\{0\}) \geq \frac{1}{2\pi} \int_0^{v_\varepsilon} \frac{1 - s}{2[1 - s]^2 + 3\varepsilon^2 v^2}\, dv$$

$$= \frac{1}{2\pi\varepsilon\sqrt{6}} \arctan \frac{\varepsilon v_\varepsilon \sqrt{3}}{(1 - s)\sqrt{2}} \to \frac{1}{4\varepsilon\sqrt{6}} \text{ as } s \nearrow 1.$$

Since ε is arbitrary we conclude that $\sum s^n Q^{*n}(\{0\}) \to \infty$ as $s \nearrow 1$, as desired. $\square$

Problem 71. What step or steps in the preceding proof would break down were the factors s^n not introduced?

* **Problem 72.** In Theorem 22, can one replace the hypothesis $E(S_1) = 0$ by the weaker hypothesis that $\beta'(0) = 0$, where β denotes the characteristic function of S_1?

Problem 73. The Fatou Lemma, the Dominated Convergence Theorem, the Monotone Convergence Theorem, and the Fubini Theorem have been important in this chapter. Write a short essay describing some of these techniques, lifting, as appropriate, specific examples from this chapter.

13.11. † An application to the calculation of a sum

The sum $\sum_{k=1}^{\infty} k^{-2}$ arises fairly often in probabilistic settings. One of the many ways of showing that it equals $\pi^2/6$ is described in the next problem.

Problem 74. Complete the missing steps in the calculation outlined below. For $\varepsilon \geq 0$, let Q_ε be the probability measure on $\mathbb{Z}$ whose density with respect to counting measure is

$$k \rightsquigarrow \begin{cases} 0 & \text{if } k = 0 \\ \dfrac{g(\varepsilon)e^{-\varepsilon|k|}}{k^2} & \text{if } k \neq 0, \end{cases}$$

where $g(\varepsilon) > 0$ is determined by the condition that Q_ε be a probability measure. Our goal is to show that $g(0) = 3/\pi^2$. For each ε, let φ_ε denote the characteristic function of Q_ε.

Fix $\varepsilon > 0$. We have

$$\varphi_\varepsilon''(v) = -g(\varepsilon) \sum_{k=1}^{\infty} (e^{ivk} + e^{-ivk}) e^{-\varepsilon k} = 2g(\varepsilon) \frac{1 - e^\varepsilon \cos v}{1 - 2e^\varepsilon \cos v + e^{2\varepsilon}}.$$

Since $\varphi_\varepsilon'(0) = 0$, we can integrate φ_ε'' to obtain

$$\varphi_\varepsilon'(v) = g(\varepsilon)v - 2g(\varepsilon) \arctan\left[\left(\frac{e^\varepsilon + 1}{e^\varepsilon - 1}\right) \tan \frac{v}{2}\right], \quad |v| \le \pi,$$

with appropriate interpretations for $v = \pm\pi$. Hence, for $|v| \le \pi$,

$$\varphi_\varepsilon(v) = 1 + g(\varepsilon) \int_0^v \left(u - 2\arctan\left[\left(\frac{e^\varepsilon + 1}{e^\varepsilon - 1}\right) \tan \frac{u}{2}\right]\right) du.$$

Let $\varepsilon \searrow 0$ to obtain

$$\varphi_0(v) = 1 + g(0) \int_0^v [u - \pi \operatorname{sgn}(u)] \, du = 1 + g(0)\left(\frac{v^2}{2} - \pi|v|\right).$$

Since $\int_{-\pi}^{\pi} \varphi_0(v) \, dv = Q_0\{0\} = 0$,

$$0 = 2\pi + g(0)\left(\frac{-2\pi^3}{3}\right),$$

from which follows $g(0) = 3/\pi^2$, as desired. Rewriting this conclusion we have

$$\sum_{k=1}^{\infty} \frac{1}{k^2} = \frac{\pi^2}{6}.$$

PART 3
Convergence in Distribution

Two random variables having a common distribution can, for some purposes, be treated as the same. It follows that in some settings one should introduce a convergence concept that focuses on distributions of random variables rather than the random variables themselves. Such a convergence concept is the focus of this part of the book.

For distributions on $\mathbb{R}$ and $\overline{\mathbb{R}}$, the basic definitions, results, and examples are found in Chapter 14, along with several applications. Two of the best known results of classical probability theory are found in Chapter 15: the Classical Central Limit Theorem and the general Weak Law of Large Numbers. That chapter also contains a brief introduction to the theory of 'large deviations', and a 'local' version of the central limit theorem. Chapter 16 gives a characterization of infinitely divisible distributions, and uses that characterization to give a complete solution to the 'central limit problem' for 'triangular arrays', which concerns the limiting behavior of sequences of sums of independent random variables. The full story is somewhat long, but we have organized Chapter 16 in such a way that useful parts of the story can be learned without covering the entire chapter. Chapter 17 treats 'stable' distributions and their relationship to the limiting behavior of sums of iid random variables.

In Chapter 18, we leave the $\mathbb{R}$- and $\overline{\mathbb{R}}$-settings and turn our attention to the convergence of distributions on 'Polish spaces' (complete separable metric spaces). With appropriate care, most of the theory from the $\mathbb{R}$-setting can be adapted to Polish spaces. A major application of this generalization can be found in Chapter 19, where we show how to construct 'Brownian motion' and prove the Invariance Principle, which says that the distribution of Brownian motion is the 'scaling limit' of the distribution of every random walk in $\mathbb{R}$ whose steps have mean 0 and finite second moment.

Certain notational shortcuts have shown themselves to be useful in probability. Their frequent use will begin in this part. In expressions such as

$$\{\omega \colon X(\omega) \in A\},$$

the 'ω' will often be suppressed, and we will instead write

$$[X \in A].$$

This abbreviated notation is often combined with an omission of parentheses as illustrated by

$$P[X \in A] = P([X \in A]) = P(\{\omega \colon X(\omega) \in A\}).$$

Parentheses may also be omitted in other contexts. For example, if Q is a probability measure on $(\mathbb{R}, \mathcal{B})$, we may write $Q(a, b]$ for $Q((a, b])$ and $Q\{x\}$ for $Q(\{x\})$. And if f is a function with domain $\mathbb{R}^2$, we will often adopt the common convention of writing $f(x, y)$ for $f((x, y))$.

CHAPTER 14

Convergence in Distribution on the Real Line

In this chapter, we introduce a concept of convergence for sequences of distributions on $\mathbb{R}$ and $\overline{\mathbb{R}}$. This 'convergence in distribution' gives us a rigorous way to express the idea that two distributions are close to each other. For instance, we show in an example that a Poisson distribution can be approximated arbitrarily closely by binomial distributions. An important result, the Continuity Theorem, gives a criterion for the convergence of a sequence of distributions in terms of the corresponding sequence of characteristic functions. There are several other useful criteria as well, which are collected together in a result known as the Portmanteau Theorem. Although the most important applications of convergence in distribution will be found in later chapters, some are included here, including an introduction to the theory of 'extreme values', a discussion of the effects that 'scaling' and 'centering' have on sequences of distributions, and characterizations of moment generating functions and characteristic functions.

14.1. Definitions and examples

The following three examples serve as motivation for our main definition.

Example 1. [Exponential distribution as a limit] Let Q be the standard exponential distribution with mean 1. Thus, Q has a density f with respect to Lebesgue measure given by

$$f(x) = \begin{cases} e^{-x} & \text{if } x \geq 0 \\ 0 & \text{if } x < 0. \end{cases}$$

Let Q_n denote the distribution of geometric type that has mean 1 and whose support consists of all nonnegative integral multiples of $1/n$. So, for x a nonnegative integral multiple of $1/n$,

$$Q_n\{x\} = \frac{1}{n+1}\left(1 + \frac{1}{n}\right)^{-xn}.$$

Let F and F_n denote the distribution functions corresponding to Q and Q_n, respectively. The functions $F(x)$ and $F_n(x)$ are 0 for $x < 0$, and, for $x \geq 0$, $F(x) = 1 - e^{-x}$ and

$$F_n(x) = 1 - \left(1 + \frac{1}{n}\right)^{-k-1} \quad \text{for} \quad \frac{k}{n} \leq x < \frac{k+1}{n}, \quad k = 0, 1, \ldots .$$

Thus, $F_n(x) \to F(x)$ as $n \to \infty$ for all x. On the other hand, it is not true that $Q_n(B) \to Q(B)$ for all Borel subsets B of $\mathbb{R}$; for instance, if B is the set of rational numbers, then $Q_n(B) = 1$ for each n, but $Q(B) = 0$. Is it more natural to say that the sequence $(Q_1, Q_2, \ldots)$ converges to Q or to say that it does not?

Example 2. Let X be an $\mathbb{R}$-valued random variable whose distribution is Q. Suppose that $Q(B) = 0$ for the set B of rational numbers. For $n = 1, 2, \ldots$ let

$$X_n(\omega) = \frac{1}{n} \lfloor nX(\omega) \rfloor .$$

Notice that $Q_n(B) = 1$, where Q_n denotes the distribution of X_n. It is easy to check that $X_n(\omega) \to X(\omega)$ for each ω. It does not bother us that a sequence of random variables each of which is rational with probability 1 can converge almost surely to a random variable that is irrational with probability 1. So, it is natural to look for a definition of convergence of a sequence of probability measures that will entail $Q_n \to Q$ in this case, despite the fact that $Q_n(B) \not\to Q(B)$.

Example 3. Let Ω consist of the two points -1 and 1, each of which has probability $\frac{1}{2}$. Set $X_n(\omega) = \omega/n$ and $X(\omega) = 0$. Clearly, $X_n(\omega) \to X(\omega)$ as $n \to \infty$ for each of the two values of ω. Let F_n and F denote the distribution functions of X_n and X, respectively. Notice that $F_n(x) \to F(x)$ for $x \neq 0$, but

$$\tfrac{1}{2} = F_n(0) \not\to 1 = F(0) .$$

Since $X_n \to X$, we want a definition of convergence of sequences of distribution functions that will entail $F_n \to F$ in this case, even though pointwise convergence to F fails for the sequence (F_n).

Definition 1. Let F and F_n, $n = 1, 2, \ldots$, be distribution functions for $\mathbb{R}$. The sequence $(F_n : n = 1, 2, \ldots)$ *converges* to F if $F_n(x) \to F(x)$ as $n \to \infty$ for every x at which F is continuous. Let Q_n and Q be the probability measures corresponding to F_n and F, respectively. Then $(Q_n : n = 1, 2, \ldots)$ *converges* to Q if $(F_n : n = 1, 2, \ldots)$ converges to F.

The convergence just defined is denoted by

$$F_n \to F \text{ as } n \to \infty$$

for distribution functions, and by

$$Q_n \to Q \text{ as } n \to \infty$$

for distributions.

Proposition 2. *Let F and F_n, $n = 1, 2, \ldots$, be distribution functions for $\mathbb{R}$. Then $F_n \to F$ as $n \to \infty$ if and only if there is a dense subset D of $\mathbb{R}$ such that, for every $x \in D$, $F_n(x) \to F(x)$ as $n \to \infty$.*

Problem 1. Prove the preceding proposition.

* **Problem 2.** Let F_n, $n = 1, 2, \ldots$, F, and G be distribution functions for $\mathbb{R}$. Suppose that $F_n \to F$ and $F_n \to G$ as $n \to \infty$. Prove that $F = G$.

In view of the preceding problem we call F the *limit* of the sequence $(F_n : n = 1, 2, \ldots)$ if $F_n \to F$ as $n \to \infty$, and we write

$$\lim_{n \to \infty} F_n = F.$$

Similarly, if $Q_n \to Q$ as $n \to \infty$, we write

$$\lim_{n \to \infty} Q_n = Q.$$

Problem 3. Show that a sequence $(\delta_{x_n} : n = 1, 2, \ldots)$ of delta distributions converges if and only if the corresponding sequence $(x_n : n = 1, 2, \ldots)$ converges in $\mathbb{R}$.

* **Problem 4.** [Poisson limit of binomials] Fix $\lambda > 0$. For integers $n > \lambda$, let Q_n denote the binomial probability distribution:

$$Q_n\{x\} = \binom{n}{x} \left(\frac{\lambda}{n}\right)^x \left(1 - \frac{\lambda}{n}\right)^{n-x}, \quad 0 \le x \le n, \ x \in \mathbb{Z}.$$

Prove that $Q_n \to Q$, where Q is the Poisson distribution with mean λ.

Problem 5. [Binomial continuity] Fix a positive integer n. For $0 \le p \le 1$, let Q_p denote the binomial distribution given by

$$Q_p\{x\} = \binom{n}{x} p^x (1-p)^{n-x}, \quad 0 \le x \le n, \ x \in \mathbb{Z},$$

where the convention $0^0 = 1$ applies in case $p = 0$ or $p = 1$. Prove that the function $p \rightsquigarrow Q_p$ is continuous in the sense that if $p_k \to p$ as $k \to \infty$, then $Q_{p_k} \to Q_p$ as $k \to \infty$.

* **Problem 6.** Mimic the preceding exercise for other one-parameter families of distributions: geometric, exponential, standard gamma (that is, with $a = 1$), and Poisson. In cases where the parameter ranges over an interval, discuss the limiting behavior as the parameter approaches any finite endpoint.

Problem 7. [Beta continuity in two parameters] For $\alpha > 0$ and $\beta > 0$, let $Q_{\alpha,\beta}$ denote the beta distribution described in Example 3 of Chapter 3. Prove that the function $(\alpha, \beta) \rightsquigarrow Q_{\alpha,\beta}$ is a continuous function. Decide which of the following limits exist and evaluate those that do:

$$\lim_{\alpha \searrow 0} Q_{\alpha,\beta} ; \qquad \lim_{\beta \searrow 0} Q_{\alpha,\beta} ;$$

$$\lim_{\alpha \to \infty} Q_{\alpha,\beta} ; \qquad \lim_{\beta \to \infty} Q_{\alpha,\beta} ;$$

$$\lim_{\alpha \searrow 0} Q_{\alpha,\gamma\alpha} ; \qquad \lim_{\alpha \to \infty} Q_{\alpha,\gamma\alpha} ;$$

$$\lim_{(\alpha,\beta) \to (0,0)} Q_{\alpha,\beta} ; \qquad \lim_{(\alpha,\beta) \to (\infty,\infty)} Q_{\alpha,\beta} ;$$

$$\lim_{(\alpha,\beta) \to (0,\infty)} Q_{\alpha,\beta} ; \qquad \lim_{(\alpha,\beta) \to (\infty,0)} Q_{\alpha,\beta} .$$

Problem 8. [Continuity of the normal family] Prove that the family of distributions on $\mathbb{R}$ of normal type is continuous as a function of mean and standard deviation. Include the cases of zero standard deviation in your considerations.

Problem 9. [Negative binomial continuity] Let $Q_{p,r}, 0 \le p < 1$ and $r \ge 0$, denote negative binomial distributions as defined by

$$k \rightsquigarrow (1-p)^r \frac{(r)_k^\uparrow}{k!} p^k , \quad k \in \mathbb{Z}^+ .$$

(See Problem 23 of Chapter 10.) Investigate the continuity of the function $(p, r) \rightsquigarrow Q_{p,r}$.

* **Problem 10.** [Gamma distribution as a limit] Let $Q_{p,r}$ be as in the preceding exercise, fix $r > 0$, and let R_m denote the distribution of the random variable $m^{-1} X_m$, where the distribution of X_m is $Q_{(1-1/m),r}$. Prove that, as $m \to \infty$, $R_m \to R$, where R is the gamma distribution defined in Example 2 of Chapter 3, with parameters $a = 1, \gamma = r$. What is the situation for $r = 0$?

Problem 11. Modify the preceding exercise by letting the distribution of X_m be $Q_{(1-\lambda/m),r}$ for some fixed $\lambda > 0$.

14.2. Limit distributions for extreme values

In order to further illustrate of the concept of convergence for sequences of distributions and also to give a brief introduction to an important topic that will not be treated elsewhere in this book, we will examine distributions that arise naturally when studying the maximum of a large number of iid random variables.

Problem 12. Let $(X_1, X_2, \dots)$ be an iid sequence of $\mathbb{R}$-valued random variables having common distribution function F. Prove that the distribution function of $\max\{X_k : 1 \le k \le n\}$ is F^n.

Problem 13. [Gumbel distribution] Let

$$F(x) = e^{-e^{-x}}, \quad x \in \mathbb{R}.$$

Show that F is a distribution function that, for each positive integer n, is of the same type as F^n. (The density of F is shown in Figure 14.1.)

* **Problem 14.** Prove that if F is a distribution function of Gumbel type, then

$$F(x) = e^{-ce^{-ax}}$$

for some constants $a, c > 0$.

Problem 15. Let $(X_n : n = 1, 2, \dots)$ be an iid sequence of standard exponentially distributed random variables having mean 1. For $n = 1, 2, \dots$, let $M_n = \max\{X_k : 1 \le k \le n\}$ and let G_n denote the distribution function of $(M_n - \log n)$. Show that $G_n \to F$ as $n \to \infty$, where F is a standard Gumbel distribution function (defined in Problem 13).

* **Problem 16.** Decide whether the sequence $(M_n - \log n)$ of the preceding problem converges almost surely as $n \to \infty$. If not, does it converge in probability? Comment on any connections that this and the preceding problem have with Example 6 of Chapter 9.

Problem 17. Let X be distributed according to the Gumbel distribution of Problem 13. Show that

$$E(X) = -\Gamma'(1) = \int_0^\infty (-\log x) e^{-x} \, dx,$$

where Γ denotes the gamma function. (By Problem 29 of Chapter 8 this constant is Euler's constant.)

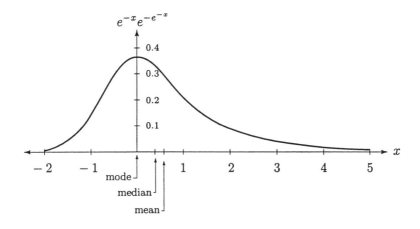

FIGURE 14.1. The density of the standard Gumbel distribution

Problem 18. [Weibull distributions] Let $\alpha > 0$. Show that the function

$$x \rightsquigarrow \begin{cases} e^{-|x|^\alpha} & \text{if } x < 0 \\ 1 & \text{if } x \geq 0 \end{cases}$$

is a distribution function that, for every n, is of the same strict type as the nth power of itself. (The parameter α is called the *index* of the Weibull distribution.)

Problem 19. Let X be a random variable having a Weibull distribution of index α. Calculate the distribution function of $-X$.

Remark 1. The term 'Weibull distribution' is used more often for the distribution of $-X$ obtained in the preceding problem than it is for the distribution in Problem 18. The reason is that the Weibull distribution arises in applications involving the minimum of a large number of nonnegative random variables. Note that X and $-X$ are not of the same type.

Example 4. Let $(X_1, X_2, \dots)$ be an iid sequence of random variables uniformly distributed on $[0, 1]$ and set $M_n = \max\{X_k : 1 \leq k \leq n\}$. It is easy to see that $M_n \to 1$ a.s.; so we consider random variables of the form $(M_n - 1)/a_n$ with the goal of obtaining a nondegenerate limit. To choose a_n we calculate:

$$P\left[\frac{M_n - 1}{a_n} \leq x\right] = P[M_n \leq 1 + x a_n] = \prod_{k=1}^{n} P[X_k \leq 1 + x a_n] = (1 + x a_n)^n,$$

valid for all $x \leq 0$ provided that a_n is sufficiently small (depending on x). It is clear that we can obtain the nondegenerate limit $e^x \wedge 1$ by taking $a_n = 1/n$. Letting G_n denote the distribution function of $n(M_n - 1)$, we conclude that the sequence $(G_n : n = 1, 2, \dots)$ converges to the distribution function of the negative of an exponentially distributed random variable having mean 1—that is, a Weibull distribution of index 1.

Problem 20. [Fréchet distributions] Let $\alpha > 0$. Show that the function

$$x \rightsquigarrow \begin{cases} 0 & \text{if } x \leq 0 \\ e^{-x^{-\alpha}} & \text{if } x > 0 \end{cases}$$

is a distribution function that, for every n, is of the same strict type as the nth power of itself. (The parameter α is called the *index* of the Fréchet distribution.)

Problem 21. Let $(X_1, X_2, \dots)$ be an iid sequence of random variables having a standard Cauchy distribution. For each n, let $M_n = \max\{X_k : 1 \leq k \leq n\}$. Find constants $a_n > 0$ and b_n such that the distribution of $\frac{M_n - b_n}{a_n}$ converges as $n \to \infty$ to a Fréchet distribution of index 1. *Hint:* You may want to prove and then use the fact that $u[\frac{\pi}{2} - \arctan u] \to 1$ as $u \to \infty$.

* **Problem 22.** Calculate the mean and variance of the Weibull and Fréchet distributions defined in Problem 18 and Problem 20, writing all finite answers in terms of the gamma function.

Problem 23. Show that all Fréchet distributions have continuous densities and calculate them. Show more: that each Fréchet distribution function has infinitely many continuous derivatives and that the value of each of the derivatives at the lower endpoint of the support of the distribution equals 0. *Hint:* This problem has some features in common with Problem 34 of Chapter 5.

Remark 2. The problems and examples of this section illustrate the main results of the theory of extreme values. It can be shown that, in general, if M_n denotes the maximum (or minimum) of the first n terms of an iid sequence and if the distribution of $\left(\frac{M_n - b_n}{a_n}\right)$ converges as $n \to \infty$ to a nondegenerate limit for some constants $a_n > 0$ and $b_n \in \mathbb{R}$, then the limiting distribution is of Gumbel, Weibull, or Fréchet type.

14.3. Relationships to other types of convergence

The concept of convergence of distribution functions defined in the first section has a nice relationship with almost sure convergence and convergence in probability.

Proposition 3. *Let X and X_n, $n = 1, 2, \ldots$, be $\mathbb{R}$-valued random variables on a common probability space and suppose that X_n converges to X either i.p. or a.s. as $n \to \infty$. Then $Q_n \to Q$ as $n \to \infty$, where Q_n and Q denote the distributions of X_n and X, respectively.*

PROOF. Since a.s. convergence implies convergence i.p. we will assume convergence i.p. throughout. Let F_n and F denote the distribution functions of X_n and X, x a point of continuity of F, and $\varepsilon > 0$. Choose $\delta > 0$ so that $F(x+\delta) - F(x-\delta) < \varepsilon/2$, and choose l so that $P[|X_n - X| > \delta] < \varepsilon/2$ whenever $n \geq l$. Then, for $n \geq l$,

$$
\begin{aligned}
F_n(x) &= P[X_n \leq x] \\
&\leq P[X \leq x + \delta] + P[|X_n - X| > \delta] \\
&< \left(F(x) + \tfrac{\varepsilon}{2}\right) + \tfrac{\varepsilon}{2} \\
&= F(x) + \varepsilon,
\end{aligned}
$$

and

$$F(x) = P[X \le x]$$
$$\le P[X \le x - \delta] + \tfrac{\varepsilon}{2}$$
$$\le P[X_n \le x] + P[|X_n - X| > \delta] + \tfrac{\varepsilon}{2}$$
$$< F_n(x) + \tfrac{\varepsilon}{2} + \tfrac{\varepsilon}{2}$$
$$= F_n(x) + \varepsilon. \quad \square$$

Let $(X_n : n = 1, 2, \dots)$ be a sequence of $\mathbb{R}$-valued random variables, and suppose that there is a random variable X such that $Q_n \to Q$, where Q_n and Q are the distributions of X_n and X, respectively. Then we say that the sequence (X_n) *converges* to X *in distribution* and write

$$X_n \xrightarrow{\,D\,} X.$$

Since it is possible to have convergence in distribution even if all the random variables involved are defined on different probability spaces, care must be taken when using the phrase "converges in distribution". When one has convergence in distribution, it is only the distribution of the limiting random variable that is uniquely determined by the sequence and not the limiting random variable itself. Thus, for instance, an iid sequence of random variables converges in distribution to any one of the random variables in the sequence. In particular, the converse of Proposition 3 is not true. Nevertheless, there are two important results that are in the converse direction.

Proposition 4. *Let $(X_n : n = 1, 2, \dots)$ be a sequence of $\mathbb{R}$-valued random variables on a common probability space. Then, for any $c \in \mathbb{R}$, $X_n \to c$ in probability as $n \to \infty$ if and only if $X_n \xrightarrow{\,D\,} c$ as $n \to \infty$.*

Problem 24. Prove the preceding proposition.

Proposition 5. *Let Q and Q_n, $n = 1, 2, \dots$, be probability measures on $\mathbb{R}$ and suppose that $Q_n \to Q$ as $n \to \infty$. Then there exists a probability space $(\Omega, \mathcal{F}, P)$ and $\mathbb{R}$-valued random variables X and X_n, $n = 1, 2, \dots$, defined on Ω, such that the distributions of X and X_n are Q and Q_n, respectively, and $X_n \to X$ a.s. as $n \to \infty$.*

PROOF. Let F and F_n, $n = 1, 2, \dots$, be the distribution functions corresponding to Q and Q_n, respectively. Let Ω equal the interval $(0, 1)$, $\mathcal{F}$ its Borel field, and P Lebesgue measure on $(\Omega, \mathcal{F})$. For $\omega \in \Omega$, set

$$X(\omega) = \inf\{x : F(x) \ge \omega\},$$

as in Proposition 4 of Chapter 3, and

$$X_n(\omega) = \inf\{x : F_n(x) \ge \omega\}.$$

(Recall that in a certain sense, X and F are inverses of each other, and similarly for X_n and F_n. Intervals of constancy of F correspond to jumps of X, and jumps of F correspond to intervals of constancy of X.)

Fix ω and then fix $\delta > 0$ such that $X(\omega) - \delta$ is a point at which F is continuous. Clearly

$$F(X(\omega) - \delta) < \omega,$$

and, hence,

$$F_n(X(\omega) - \delta) < \omega$$

for all sufficiently large n. For such n, $X_n(\omega) \geq X(\omega) - \delta$. Thus,

$$\liminf_{n \to \infty} X_n(\omega) \geq X(\omega) - \delta.$$

Now let $\delta \searrow 0$ to conclude that

$$\liminf_{n \to \infty} X_n(\omega) \geq X(\omega).$$

Since X is monotone, it has at most countably many points of discontinuity. Thus, for almost every ω, X is continuous at ω. For such an ω fix $\varepsilon > 0$ and then fix $\delta > 0$ such that $X(\omega + \varepsilon) + \delta$ is a point at which F is continuous. Clearly

$$F(X(\omega + \varepsilon) + \delta) \geq \omega + \varepsilon.$$

So for all n sufficiently large

$$F_n(X(\omega + \varepsilon) + \delta) \geq \omega,$$

and, hence,

$$X_n(\omega) \leq X(\omega + \varepsilon) + \delta.$$

Thus,

$$\limsup_{n \to \infty} X_n(\omega) \leq X(\omega + \varepsilon) + \delta.$$

Now let $\delta \searrow 0$ and then $\varepsilon \searrow 0$ to conclude that

$$\limsup_{n \to \infty} X_n(\omega) \leq X(\omega).$$

The preceding two paragraphs show that $X_n(\omega) \to X(\omega)$ as $n \to \infty$ for almost every ω, in particular those at which X is continuous. $\square$

Proposition 5 is very useful for the study of sequences of probability measures, for it enables one to use results about almost sure convergence. This feature is illustrated in Problem 25, Problem 26, Proposition 6, and Proposition 7.

Problem 25. [Bounded Convergence Theorem for Distributions] Let Q and $Q_n, n = 1, 2, \ldots$, be probability measures on $\mathbb{R}$, and suppose that $Q_n \to Q$ as $n \to \infty$ and that there exists a single bounded set that supports every Q_n. Prove that

$$\int x \, Q_n(dx) \to \int x \, Q(dx).$$

Problem 26. [Fatou Lemma for Distributions] Let Q and Q_n, $n = 1, 2, \ldots$, be probability measures, each of which is supported by $[0, \infty)$, and suppose that $Q_n \to Q$ as $n \to \infty$. Prove that

$$\int x \, Q(dx) \leq \liminf_{n \to \infty} \int x \, Q_n(dx).$$

Problem 27. [Uniform Integrability Criterion for Distributions] Prove that most of the Uniform Integrability Criterion holds with almost sure convergence replaced by convergence in distribution. Explain why one should not expect one aspect of the criterion to be valid for convergence in distribution.

Proposition 6. *Let Q, R, and Q_n and R_n, $n = 1, 2, \ldots$, be probability measures on $\mathbb{R}$, and suppose that $Q_n \to Q$ and $R_n \to R$ as $n \to \infty$. Then $Q_n * R_n \to Q * R$ as $n \to \infty$.*

PROOF. By Proposition 5 there exist random variables X, Y, and X_n and Y_n, $n = 1, 2, \ldots$, with respective distributions Q, R, and Q_n and R_n, such that $X_n \to X$ a.s. and $Y_n \to Y$ a.s. as $n \to \infty$. By using a product space, as in Theorem 7 of Chapter 9, we may arrange for the collection of random variables $\{X, X_1, X_2, \ldots\}$ to be independent of the collection $\{Y, Y_1, Y_2, \ldots\}$. So the distribution of $X_n + Y_n$ is $Q_n * R_n$, the distribution of $X + Y$ is $Q * R$, and $X_n + Y_n \to X + Y$ a.s. as $n \to \infty$. The desired conclusion follows from Proposition 3. □

Problem 28. Explore the possibility of proving the preceding proposition without using random variables. Either decide that there is such a proof that is straightforward and write it or identify specific difficulties that lead you to appreciate the value of Proposition 5.

14.4. Convergence conditions for sequences of distributions

The next proposition gives equivalent conditions for the convergence of a sequence of probability measures. These conditions are meaningful for probability distributions on spaces other than $\mathbb{R}$ and thus will be used for generalization in Chapter 18, where the corresponding result is known as the Portmanteau Theorem. In the proposition the notation ∂B is used for the boundary of a set B, the set obtained by removing the interior of B from the closure of B.

Proposition 7. *Let Q and $Q_n, n = 1, 2, \ldots$, be probability measures on $\mathbb{R}$. Then the following conditions are equivalent:*

(i) $Q_n \to Q$ as $n \to \infty$;
(ii) $\int g \, dQ_n \to \int g \, dQ$ as $n \to \infty$ for each bounded continuous function g on $\mathbb{R}$;

(iii) $\limsup_{n \to \infty} Q_n(C) \leq Q(C)$ *for each closed subset C of $\mathbb{R}$;*
(iv) $\liminf_{n \to \infty} Q_n(O) \geq Q(O)$ *for each open subset O of $\mathbb{R}$;*
(v) $\lim_{n \to \infty} Q_n(B) = Q(B)$ *for each Borel subset B of $\mathbb{R}$ for which* $Q(\partial B) = 0$.

PROOF. We will prove the chain of implications (v) $\Longrightarrow$ (i) $\Longrightarrow$ (ii) $\Longrightarrow$ (iii) as well as the implications (iii) $\Longleftrightarrow$ (iv) and {(iii), (iv)} $\Longrightarrow$ (v).

(v) $\Longrightarrow$ (i). For x a real number for which $Q\{x\} = 0$, let $B = (-\infty, x]$. By (v) we conclude that $Q_n(-\infty, x] \to Q(-\infty, x]$. Therefore, $Q_n \to Q$.

(i) $\Longrightarrow$ (ii). Let X and X_n be as in Proposition 5, and let g be a continuous bounded function on $\mathbb{R}$. Then $g \circ X_n \to g \circ X$ a.s. By the Bounded Convergence Theorem, $E(g \circ X_n) \to E(g \circ X)$; that is, $\int g \, dQ_n \to \int g \, dQ$.

(ii) $\Longrightarrow$ (iii). Let C be a closed subset of $\mathbb{R}$. In order to use (ii) we introduce continuous functions whose integrals with respect to Q approximate $Q(C)$ by virtue of the fact that they equal 1 on C and equal 0 on the complement of a neighborhood of C: for $k = 1, 2, \dots$ and $x \in \mathbb{R}$, define

$$g_k(x) = (1 - k \inf\{|y - x| : y \in C\}) \vee 0.$$

Thus, each g_k is continuous (by the triangle inequality) and $[0, 1]$-valued, has the value 1 on C, and has the value 0 at points whose distance from C is greater than $1/k$. So, for each $x \in \mathbb{R}$, $g_k(x) \searrow I_C(x)$ as $k \to \infty$. By the Bounded Convergence Theorem and (ii),

$$Q(C) = \lim_{k \to \infty} \int g_k \, dQ$$
$$= \lim_{k \to \infty} \lim_{n \to \infty} \int g_k \, dQ_n$$
$$\geq \limsup_{k \to \infty} \limsup_{n \to \infty} Q_n(C)$$
$$= \limsup_{n \to \infty} Q_n(C).$$

(iii) $\Longleftrightarrow$ (iv). For an arbitrary closed set C and its complement, an open set O, we have

$$Q(C) - \limsup_{n \to \infty} Q_n(C)$$
$$= 1 - Q(O) - \limsup_{n \to \infty}[1 - Q_n(O)]$$
$$= \liminf_{n \to \infty} Q_n(O) - Q(O).$$

If either the left or right side in this equality is nonnegative, then so is the other.

{(iii), (iv)} $\Longrightarrow$ (v). Let B be a Borel set for which $Q(\partial B) = 0$. Let C and O denote the closure and interior of B, respectively. Then,

$$\limsup Q_n(B) \leq \limsup Q_n(C) \leq Q(C)$$
$$= Q(B) = Q(O) \leq \liminf Q_n(O) \leq \liminf Q_n(B);$$

so, equalities hold throughout. □

Problem 29. Let $(x_n : n = 1, 2, \dots)$ denote a convergent sequence of real numbers with limit x. For $n = 1, 2, \dots$, let Q_n denote the delta distribution at x_n; also, let Q denote the delta distribution at x. Show directly that each of (i)-(v) of the preceding proposition holds. Also, show that strict inequality is possible in (iii) and (iv).

Problem 30. Suppose that $Q_n \to Q$ and that every Q_n is supported by a common closed set C. Use the preceding proposition to prove that Q is supported by C.

The following result concerns the convergence of probability measures with densities.

Proposition 8. *Suppose that probability measures Q and Q_n, $n = 1, 2, \dots$, on $\mathbb{R}$ are all absolutely continuous with respect to a common σ-finite measure μ, and denote their respective densities by f and f_n. If $f_n \to f$ μ-almost everywhere, then $Q_n \to Q$.*

Problem 31. Prove the preceding proposition.

Problem 32. Give an example that contradicts the converse of the last sentence in the preceding proposition.

Problem 33. Give an example of a sequence $(Q_n : n = 1, 2, \dots)$ of probability distributions on $\mathbb{R}$ which satisfies: (i) there is a σ-finite measure μ such that, for each n, Q_n is absolutely continuous with respect to μ and, moreover, its density f_n is continuous; (ii) as $n \to \infty$, the sequence (f_n) converges uniformly; (iii) the sequence (Q_n) does not converge.

14.5. Sequences of distributions on $\overline{\mathbb{R}}$

Let us turn our attention to probability measures on the extended real line $\overline{\mathbb{R}} = [-\infty, \infty]$. Definition 1 carries over directly to the $\overline{\mathbb{R}}$-setting: simply replace "$\mathbb{R}$" by "$\overline{\mathbb{R}}$" in the first sentence of the definition.

Some books and articles use a term different from 'convergence in distribution' when speaking of $\overline{\mathbb{R}}$. We will not adopt this practice and will pay the price of having to state explicitly whether we are working in $\mathbb{R}$ or $\overline{\mathbb{R}}$ whenever there is a chance of ambiguity.

Example 5. Let

$$F_n(x) = \begin{cases} 0 & \text{if } x < -n \\ 1/2 & \text{if } -n \le x < n \\ 1 & \text{if } n \le x. \end{cases}$$

Each F_n is a distribution function for $\mathbb{R}$, and in the $\mathbb{R}$-setting the sequence $(F_n: n = 1, 2, \ldots)$ fails to converge. However, each F_n is also a distribution function for $\overline{\mathbb{R}}$, and in the $\overline{\mathbb{R}}$-setting the sequence does converge—to the distribution function that is identically equal to $1/2$. In this case, the limiting distribution is $\frac{1}{2}(\delta_{-\infty} + \delta_{\infty})$.

Problem 34. For the one-parameter families of distributions—geometric, exponential, gamma, and Poisson—discuss, for the $\overline{\mathbb{R}}$-setting, the limiting behavior as the parameter approaches the endpoints, finite or infinite, of the interval in which it takes its values.

* **Problem 35.** Find a sequence $(Q_n: n = 1, 2, \ldots)$ of probability measures on $\overline{\mathbb{R}}$ such that $Q_n(\{\infty\} \cup [-\infty, 0)) = 0$ for each n, $\int x \, Q_n(dx) < \infty$ for each n, $\int x \, Q_n(dx) \to \infty$, and $Q_n \to Q$ for some Q for which $Q\{\infty\} = 0$ and $\int x \, Q(dx) < \infty$.

Problem 36. Adapt Problem 2, Proposition 3, Proposition 5, Proposition 7, and Problem 30 to the $\overline{\mathbb{R}}$-setting.

The next two problems involve the zeta distribution, which was introduced in an optional section of Chapter 9. We use the notation bB to denote the set $\{bx: x \in B\}$ for $b \in \mathbb{R}$ and $B \subseteq \mathbb{R}$.

* **Problem 37.** Let P_z denote the zeta distribution. That is, P_z is the distribution supported by $\mathbb{Z}^+ \setminus \{0\}$ and satisfying $P_z(\{x\}) = \frac{1}{\zeta(z)x^z}$, where ζ is the Riemann zeta function. Fix a number $c > 1$, and for $z > 1$ and each Borel set B, set $Q_z(B) = P_z(c^{1/(z-1)}B)$. For the $\overline{\mathbb{R}}$-setting show that, as $z \searrow 1$, $Q_z \to Q$, where $Q\{\infty\} = 1/c$ and $Q\{0\} = (c-1)/c$. (Of course, were we in the $\mathbb{R}$-setting we would say that $\lim_{z \searrow 1} Q_z$ does not exist.)

Problem 38. Let P_z be as in the preceding problem, and let X_z be a random variable with distribution P_z. Let Y be a random variable with the standard exponential distribution. Show that $(z - 1) \log X_z \xrightarrow{\mathcal{D}} Y$ as $z \searrow 1$.

14.6. Relative sequential compactness

In attempting to prove that a sequence of probability measures converges, often the first step is to show that it has a convergent subsequence. In this section we develop criteria for determining whether a sequence of probability measures on $\mathbb{R}$ or $\overline{\mathbb{R}}$ has a convergent subsequence.

Definition 9. For either the $\mathbb{R}$- or the $\overline{\mathbb{R}}$-setting, a family $\mathcal{Q}$ of probability distributions is *relatively sequentially compact* if every sequence $(Q_n: n = 1, 2, \ldots)$ of members of $\mathcal{Q}$ has a convergent subsequence.

The word 'relatively' is used in the preceding definition because it is not required that the limit belong to Q, but only that it be a probability distribution on $\mathbb{R}$ or $\overline{\mathbb{R}}$, whichever is relevant. Now we come to a fact that makes the study of convergence in distribution more pleasant in the $\overline{\mathbb{R}}$-setting than in the $\mathbb{R}$-setting.

Theorem 10. *Every set of probability distributions on $\overline{\mathbb{R}}$ is relatively sequentially compact; that is, every sequence of probability distributions on $\overline{\mathbb{R}}$ has a convergent subsequence.*

PROOF. Let $S = (F_n: n = 1, 2, \ldots)$ be a sequence of distribution functions for $\overline{\mathbb{R}}$. Since the values of distribution functions lie in the bounded interval $[0, 1]$, the sequence $(F_n(x): n = 1, 2, \ldots)$ has a convergent subsequence for any $x \in \overline{\mathbb{R}}$. By using the Cantor diagonalization procedure, we can find a single subsequence $(F_{n_k}: k = 1, 2, \ldots)$ such that $(F_{n_k}(r))$ converges to a limit $G(r)$ for every rational number r. Since each F_n is increasing, so is G. Hence, for $x \in \mathbb{R}$, we may define

$$F(x) = \lim_{\substack{r \searrow x \\ r \text{ rational}}} G(r).$$

We will finish the proof by showing that F is a distribution function for $\overline{\mathbb{R}}$ and that $F_{n_k} \to F$. That F is increasing follows from the fact that G is. Since all values of each F_n lie in $[0, 1]$, so do all values of G and, hence, all values of F. To show that F is right-continuous at x, let $\varepsilon > 0$ and choose a rational $r > x$ such that $G(r) < F(x) + \varepsilon$. Then, for any $y \in (x, r)$, $F(y) < F(x) + \varepsilon$. It follows that F is right-continuous, and so a distribution function for $\overline{\mathbb{R}}$.

Let x be a point of continuity of F and let $\gamma > 0$. We want to show that both

(14.1) $$F_{n_k}(x) < F(x) + \gamma$$

and

(14.2) $$F_{n_k}(x) > F(x) - \gamma$$

for all sufficiently large k. To show (14.1) choose a rational $r > x$ so that $G(r) < F(x) + \gamma/2$, and choose l_1 so that, for all $k \geq l_1$, $F_{n_k}(r) < G(r) + \gamma/2$. Then, for $k \geq l_1$,

$$F_{n_k}(x) \leq F_{n_k}(r) < G(r) + \gamma/2 < F(x) + \gamma,$$

as desired. To prove (14.2) choose a $z < x$ such that $F(z) > F(x) - \gamma/2$. Next fix a rational $s \in (z, x]$ and choose an integer l_2 so that, for all $k \geq l_2$, $F_{n_k}(s) > G(s) - \gamma/2$. Then, for $k \geq l_2$,

$$F_{n_k}(x) \geq F_{n_k}(s) > G(s) - \gamma/2 \geq F(z) - \gamma/2 > F(x) - \gamma.$$

Therefore, for $k \geq l_1 \vee l_2$, both (14.1) and (14.2) hold. $\square$

Problem 39. Create an example that shows the possibility of there existing a rational number r for which $F(r) \neq G(r)$, where F and G are as defined in the preceding proof.

Problem 40. Create an example that shows the possibility of $F(r) \neq G(r)$ for all rational r, where F and G are as defined in the preceding proof.

For the $\mathbb{R}$-setting, the set of all distribution functions is not relatively sequentially compact, since a sequence of distributions on $\mathbb{R}$ can converge to a distribution that gives positive probability to $\pm\infty$ (see, for instance, Problem 37). The phrase "mass can escape to infinity" is sometimes used as an informal description of this possibility. Any characterization of relative sequential compactness of a collection $\mathcal{Q}$ of probability measures on $\mathbb{R}$ will be equivalent to the following condition, which says in a formal way that mass does not escape to infinity:

$$(14.3) \qquad\qquad \lim_{b \to \infty} \sup_{Q \in \mathcal{Q}} Q[-b, b]^c = 0$$

(see Theorem 13).

It will become apparent in the remainder of this chapter and also in Chapters 15 and 16 that characteristic functions are important in the study of convergence in distribution. The next lemma will enable us to obtain a condition on the behavior of characteristic functions near 0 that is equivalent to (14.3).

Lemma 11. Let Q be a probability measure on $\mathbb{R}$. For all $b > 0$,

$$(1 - \sin 1)Q[-b, b]^c \leq b \int_0^{1/b} [1 - \Re(\beta(v))]\, dv \leq 2Q[-\sqrt{b}, \sqrt{b}]^c + \frac{1}{2b},$$

where β denotes the characteristic function of Q and $\Re(\beta(v))$ denotes the real part of $\beta(v)$.

PROOF. By the Fubini Theorem

$$b \int_0^{1/b} [1 - \Re(\beta(v))]\, dv = b \int \int_0^{1/b} (1 - \cos(vx))\, dv\, Q(dx)$$

$$= \int \left(1 - \frac{\sin(x/b)}{x/b}\right) Q(dx) \geq \int_{[-b,b]^c} \left(1 - \frac{\sin(x/b)}{x/b}\right) Q(dx)$$

$$\geq \int_{[-b,b]^c} (1 - \sin 1)\, Q(dx) = (1 - \sin 1)Q[-b, b]^c,$$

proving the first inequality. We have used the fact, easily verified by elementary calculus, that the function $y \rightsquigarrow 1 - \sin(y)/y$ takes its minimum value in the interval $[1, \infty)$ at $y = 1$.

The second inequality follows from the following calculation, which holds for all $v \in \mathbb{R}$ and $b > 0$:

$$1 - \Re(\beta(v)) = \int (1 - \cos(vx)) \, Q(dx)$$

$$= \int_{[-\sqrt{b}, \sqrt{b}]^c} (1 - \cos(vx)) \, Q(dx) + \int_{[-\sqrt{b}, \sqrt{b}]} (1 - \cos(vx)) \, Q(dx)$$

$$\leq 2Q[-\sqrt{b}, \sqrt{b}]^c + \int_{[-\sqrt{b}, \sqrt{b}]} \frac{v^2 x^2}{2} \, Q(dx) \leq 2Q[-\sqrt{b}, \sqrt{b}]^c + \frac{v^2 b}{2}. \quad \square$$

We need one more definition before stating a theorem which gives several equivalent criteria for relative sequential compactness.

Definition 12. A set $\mathcal{U}$ of $\mathbb{R}$-valued functions defined in a neighborhood of a point x_0 is *equicontinuous at x_0* if for every $\varepsilon > 0$ there is a neighborhood $\mathcal{N}$ of x_0 such that $|w(x) - w(x_0)| < \varepsilon$ whenever $x \in \mathcal{N}$ and $w \in \mathcal{U}$.

Note that condition (ii) in the following theorem is obviously equivalent to (14.3).

Theorem 13. *Let $\mathcal{Q}$ be a set of probability measures on $\mathbb{R}$, and $\mathcal{U}$ the set of their characteristic functions. Then the following four statements are equivalent:*

(i) $\mathcal{Q}$ is relatively sequentially compact;
(ii) for every $\varepsilon > 0$ there exists a bounded subset B of $\mathbb{R}$ such that $Q(B) > 1 - \varepsilon$ for all $Q \in \mathcal{Q}$;
(iii) $\mathcal{U}$ is equicontinuous at 0;
(iv) for every $\varepsilon > 0$ there exists a $b > 0$ such that

$$b \int_0^{1/b} [1 - \Re(\beta(v))] \, dv < \varepsilon$$

for all but finitely many $\beta \in \mathcal{U}$.

PROOF. We first show that (i) implies (ii). For a proof by contradiction, suppose that (i) holds but that (ii) does not hold. Since (ii) does not hold, there exists an $\varepsilon > 0$ and a sequence $(Q_n : n = 1, 2 \ldots)$ of measures in $\mathcal{Q}$ such that for every n, $Q_n[-n, n]^c > \varepsilon$. Thus, for all $b > 0$,

(14.4) $$\limsup_{n \to \infty} Q_n(-b, b) \leq 1 - \varepsilon.$$

By (i), (Q_n) has a convergent subsequence; call its limit Q. By (14.4) and Proposition 7, $Q(-b, b) \leq 1 - \varepsilon$ for all $b > 0$. This is a contradiction, since no probability measure on $\mathbb{R}$ has this property.

Next we show that (ii) implies (i). We may regard the members of $\mathcal{Q}$ as distributions on $\overline{\mathbb{R}}$. By Theorem 10, any sequence $(Q_n : n = 1, 2, \ldots)$ in $\mathcal{Q}$ has a convergent subsequence $(Q_{n_k} : k = 1, 2, \ldots)$ whose limit Q is a distribution on $\overline{\mathbb{R}}$. By the extension of Proposition 7 to the $\overline{\mathbb{R}}$ setting (see Problem 36), $Q[-b, b] \geq$

$\limsup_k Q_{n_k}[-b, b]$ for all $b > 0$. It follows from (ii) that $\lim_{b \to \infty} Q[-b, b] = 1$, so Q is a distribution on $\mathbb{R}$. Thus (i) holds.

So far we have shown that (i) and (ii) are equivalent. Since, as mentioned earlier, (ii) is equivalent to (14.3), the second inequality in Lemma 11 makes it clear that (ii) implies (iv). It is also easy to see that if $\mathcal{U}$ is equicontinuous at 0, then (iv) holds. So (iii) implies (iv).

Now we show that (iv) implies (ii). The first inequality in Lemma 11 shows that if (iv) holds, then for any $\varepsilon > 0$, there exists an $a > 0$ such that $Q[-a, a]^c < \varepsilon$ for all $Q \in \mathcal{Q}$ except possibly for Q in some finite set $\mathcal{Q}' \subseteq \mathcal{Q}$. For each $Q \in \mathcal{Q}'$, the Continuity of Measure Theorem implies that $\lim_{a \to \infty} Q[-a, a]^c = 0$. Since $\mathcal{Q}'$ is finite, there exists an $a' > 0$ such that $Q[-a', a']^c < \varepsilon$ for all $Q \in \mathcal{Q}'$. Now (ii) holds with $b = \max\{a, a'\}$.

Finally, to complete the proof, we show that (ii) implies (iii). Let Q be a measure in $\mathcal{Q}$, with characteristic function β. The following calculation is very similar to one made in greater detail in the proof of Lemma 11, except that this time we use the inequality $|1 - e^{ivx}| \leq 2|vx|$ (see E.9 of Appendix E):

$$|1 - \beta(v)| \leq \int |1 - e^{ivx}| \, Q(dx)$$
$$\leq 2Q[-b, b]^c + \int_{[-b,b]} |1 - e^{ivx}| \, Q(dx)$$
$$\leq 2\big(Q[-b, b]^c + |vb|\big).$$

Combining this inequality with (ii), we see that for any $\varepsilon > 0$, we may find a $b > 0$ such that

$$|1 - \beta(v)| \leq \varepsilon/2 + 2|vb|$$

for $\beta \in \mathcal{U}$. Let $\mathcal{N} = (-\varepsilon/2b, \varepsilon/2b)$. Then $|1 - \beta(v)| < \varepsilon$ for all $v \in \mathcal{N}$ and $\beta \in \mathcal{U}$. Thus $\mathcal{U}$ is equicontinuous at 0. $\square$

Problem 41. Show that the phrase "but finitely many" can be dropped from (iv) of Theorem 13, and that it can be added to (ii). (The phrase is included in the theorem to make it easier to use (iv) as a criterion for relative sequential compactness. See, for example, the proof of Theorem 15.)

We conclude this section with a result which is quite useful for showing that a sequence of probability measures has a limit. Its proof is very similar to the analogous result concerning sequences of real numbers, Proposition 4 of Appendix B.

Theorem 14. *In either the $\mathbb{R}$- or $\overline{\mathbb{R}}$-setting, suppose that $(Q_n : n = 1, 2, \dots)$ is a sequence of probability distributions that has the property that, for some probability distribution Q, every subsequence has a further subsequence that converges to Q. Then $Q_n \to Q$ as $n \to \infty$.*

Problem 42. Prove the preceding theorem.

14.7. The Continuity Theorem

From part (ii) of Proposition 7 we see that if a sequence of distributions converges, then the corresponding sequence of characteristic functions converges. The following theorem both strengthens this assertion and provides a converse. It is called the Continuity Theorem because it says that, in a certain sense, the function that maps every probability measure on $\mathbb{R}$ to its characteristic function is continuous and has a continuous inverse.

Theorem 15. [Continuity (for Characteristic Functions)] *A sequence of probability distributions on $\mathbb{R}$ converges to a probability distribution Q if and only if the sequence of corresponding characteristic functions converges pointwise to a function γ which is continuous at 0, in which case the convergence to γ is uniform on $[-u, u]$ for every $u \in \mathbb{R}^+$, and γ is the characteristic function of Q.*

PROOF. Suppose first that $Q_n \to Q$ as $n \to \infty$ and denote the characteristic functions of Q_n and Q by β_n and β. All characteristic functions are continuous, so, in particular, β is continuous at 0. We already know from the Bounded Convergence Theorem that $\beta_n(v) \to \beta(v)$ as $n \to \infty$ for all $v \in \mathbb{R}$. It is the uniform convergence on bounded intervals that needs to be proved.

By Proposition 5 there exists a probability space and random variables X and $X_n, n = 1, 2, \ldots$, defined on that probability space such that Q is the distribution of X, Q_n is the distribution of X_n for $n = 1, 2, \ldots$, and $X_n \to X$ a.s.

Fix $u \in \mathbb{R}^+$. By E.9 of Appendix E,

$$|e^{ivX_n} - e^{ivX}| \leq 2|v|\,|X_n - X|.$$

Hence

$$\sup_{|v| \leq u} |e^{ivX_n} - e^{ivX}| \leq 2u|X_n - X| \to 0 \text{ a.s. as } n \to \infty.$$

Then

$$
\begin{aligned}
\sup_{|v| \leq u} |\beta_n(v) - \beta(v)| &= \sup_{|v| \leq u} |E(e^{ivX_n}) - E(e^{ivX})| \\
&\leq \sup_{|v| \leq u} E(|e^{ivX_n} - e^{ivX}|) \\
&\leq E(\sup_{|v| \leq u} |e^{ivX_n} - e^{ivX}|)
\end{aligned}
$$

which, by the Bounded Convergence Theorem, approaches 0 as $n \to \infty$, giving the desired uniform convergence.

For the converse, let $\mathcal{Q} = (Q_n \colon n = 1, 2, \ldots)$ be a sequence of probability measures on $\mathbb{R}$ with corresponding characteristic functions β_n, and suppose that there is a function γ, continuous at 0, such that $\beta_n(v) \to \gamma(v)$ as $n \to \infty$ for each $v \in \mathbb{R}$.

Since γ is continuous at 0, for every $\varepsilon > 0$ there exists a $b > 0$ such that

$$b \int_0^{1/b} [1 - \Re(\gamma(v))]\, \lambda(dv) < \varepsilon .$$

By the Bounded Convergence Theorem,

$$b \int_0^{1/b} [1 - \Re(\beta_n(v))]\, dv < \varepsilon$$

for all but finitely many n. We have thus verified condition (iv) of Theorem 13, so the sequence $\mathcal{Q}$ is relatively sequentially compact. In particular, it has a convergent subsequence.

By the first paragraph of this proof, the characteristic function of the limit of any convergent subsequence of $\mathcal{Q}$ must be γ. So, all convergent subsequences have the same limit, namely the distribution Q whose characteristic function is γ. By Theorem 14 we conclude that $Q_n \to Q$ as $n \to \infty$. $\square$

The preceding proof illustrates a useful technique. When proving an implication and its converse, prove the easier of the two implications first and then use that implication, if possible, in the proof of the converse.

Problem 43. Use the preceding theorem to redo Problem 8 and Problem 10.

A consequence of Proposition 4 and Theorem 15 is that a sequence of random variables converges in probability to the zero random variable if and only if the corresponding sequence of characteristic functions converges to the function $v \rightsquigarrow 1$. The following useful fact strengthens the 'if part' of the last sentence.

Lemma 16. *Let $(\beta_k \colon k = 1, 2, \dots)$ be a sequence of characteristic functions such that $\beta_k(u) \to 1$ as $k \to \infty$ for every u in some open interval containing 0. Then, for every $u \in \mathbb{R}$, $\beta_k(u) \to 1$ as $k \to \infty$.*

* **Problem 44.** Prove the preceding lemma. *Hint:* Use positive definiteness.

Problem 45. Let $(X_n \colon n = 1, 2, \dots)$ be a sequence of $\mathbb{R}$-valued random variables defined on a common probability space and $(\beta_n \colon n = 1, 2, \dots)$ the corresponding sequence of characteristic functions. Prove that $X_n \to c$ in probability as $n \to \infty$ if and only if, for every u in some open interval containing 0, $\beta_n(u) \to e^{iuc}$ as $n \to \infty$.

The stories related to convergence of sequences of moment generating functions and probability generating functions are similar to that just described for characteristic functions. The monotonicity of the two types of generating functions makes it easy to prove the following six relevant results for these settings.

Theorem 17. *In the $\mathbb{R}^{+}$-setting, a set of probability distributions is relatively sequentially compact if and only if the set of corresponding moment generating functions is equicontinuous at 0.*

Theorem 18. *In the $\mathbb{Z}^{+}$-setting, a set of probability distributions is relatively sequentially compact if and only if the set of corresponding probability generating functions is equicontinuous at 1.*

Theorem 19. [Continuity (for Moment Generating Functions)] *A sequence of distributions on $\mathbb{R}^{+}$ converges to a distribution Q on $\mathbb{R}^{+}$ if and only if the sequence of corresponding moment generating functions converges pointwise to a function φ that is continuous at 0, in which case φ is the moment generating function of Q.*

Theorem 20. [Continuity (for Probability Generating Functions)] *A sequence of distributions on $\mathbb{Z}^{+}$ converges to a distribution Q on $\mathbb{Z}^{+}$ if and only if the sequence of corresponding probability generating functions converges pointwise to a function ρ for which $\rho(1-) = 1$, in which case ρ is the probability generating function of Q.*

Theorem 21. *A sequence of distributions on $\overline{\mathbb{R}}^{+}$ converges to a distribution Q if and only if the sequence of corresponding moment generating functions converges pointwise to a function φ, in which case φ is the moment generating function of Q.*

Theorem 22. *A sequence of distributions on $\overline{\mathbb{Z}}^{+}$ converges to a distribution Q if and only if the sequence of corresponding probability generating functions converges pointwise to a function ρ, in which case ρ is the probability generating function of Q.*

Problem 46. Prove a representative subset of the preceding six theorems.

Problem 47. Use probability generating functions to redo Problem 4 and Problem 5.

* **Problem 48.** Use probability generating functions to redo Problem 9.

* **Problem 49.** Use moment generating functions to redo Example 1 and Problem 10.

Problem 50. Do Problem 34 again, this time by using Theorem 21 and Theorem 22.

14.8. Scaling and centering of sequences of distributions

Let $(Q_n: n = 1, 2, \ldots)$ be a sequence of probability measures on $\mathbb{R}$. Suppose that there is some pattern in the definitions of the Q_n, so that one is motivated to describe the behavior of the sequence (Q_n) for large n. Often, the best way to try to describe such behavior is in terms of appropriately 'centered' and 'scaled' versions of the distributions Q_n. More precisely, as in Problem 21, one looks for a sequence $(a_n: n = 1, 2, \ldots)$ of positive constants (the 'scaling' constants) and another sequence $(b_n: n = 1, 2, \ldots)$ of real constants (the 'centering' constants), such that the sequence $(R_n: n = 1, 2, \ldots)$ of distributions converges, where, for each n,

$$R_n(B) = Q_n(a_n B + b_n), \quad B \in \mathcal{B}.$$

Notice that if Q_n is the distribution of an $\mathbb{R}$-valued random variable V_n, then R_n is the distribution of the random variable $(V_n - b_n)/a_n$.

Problem 51. Let $(Q_n: n = 1, 2, \ldots)$ be any sequence of distributions on $\mathbb{R}$. Show that for any $c \in \mathbb{R}$, scaling and centering constants can be chosen so that $R_n \to \delta_c$ as $n \to \infty$, where R_n is defined as in the preceding paragraph. Show that it is possible to choose the centering constants equal to 0 if $c = 0$ but that such a choice for the centering constants is not in general possible if $c \neq 0$.

In view of Problem 51, we do not accomplish much if we choose scaling and centering constants in such a way that the limit of the scaled and centered sequence is of degenerate type. It is thus natural to ask: (i) Can (a_n) and (b_n) be chosen so that (R_n) converges to a distribution on $\mathbb{R}$ that is not degenerate? (ii) If so, how should (a_n) and (b_n) be chosen and for such a choice, what is the limit? (iii) In what sense, if any, is $\lim_n R_n$ unique when it exists and is nondegenerate? The first two questions will be addressed in Chapters 15 and 17 as well as later in this chapter. The answer to the third question is a consequence of the following preliminary result:

Lemma 23. Let $(V_n: n = 1, 2, \ldots)$ be a sequence of $\mathbb{R}$-valued random variables that converges in distribution to an $\mathbb{R}$-valued random variable V that is not of degenerate type. Let $(a_n: n = 1, 2, \ldots)$ and $(b_n: n = 1, 2, \ldots)$ be sequences in $(0, \infty)$ and $\mathbb{R}$, respectively. Then the sequence $((V_n - b_n)/a_n: n = 1, 2, \ldots)$ converges in distribution to an $\mathbb{R}$-valued random variable that is not of degenerate type if and only if the sequence (a_n) converges to a member a of $(0, \infty)$ and the sequence (b_n) converges to a member b of $\mathbb{R}$, in which case

$$\frac{V_n - b_n}{a_n} \xrightarrow{\mathcal{D}} \frac{V - b}{a}.$$

PROOF. For each n, let β_n be the characteristic function of V_n, and let β be the characteristic function of V. By Problem 7 of Chapter 13, the characteristic function of $(V_n - b_n)/a_n$ is the function γ_n defined by

$$\gamma_n(v) = e^{-ivb_n/a_n}\beta_n(v/a_n).$$

Clearly, if $a_n \to a \in (0, \infty)$ and $b_n \to b \in \mathbb{R}$ as $n \to \infty$, then $\gamma_n(v) \to e^{-ivb/a}\beta(v/a)$, which is the characteristic function of $(V - b)/a$. So

$$\frac{V_n - b_n}{a_n} \xrightarrow{\mathcal{D}} \frac{V - b}{a}$$

as desired. Since V is not of degenerate type, neither is $(V - b)/a$.

It remains to prove the 'only if' part. Let β_n, γ_n, and β be as in the preceding paragraph, and let γ be the limit as $n \to \infty$ of γ_n. We assume that neither β nor γ is a characteristic function of a degenerate distribution. We must show that there exist constants $a > 0$ and $b \in \mathbb{R}$ such that $a = \lim_n a_n$ and $b = \lim_n b_n$.

Choose an increasing sequence $(n_k : k = 1, 2, \dots)$ of positive integers such that $a_{n_k} \to a$ and $b_{n_k} \to b$ as $k \to \infty$ for some constants $a \in [0, \infty]$ and $b \in [-\infty, \infty]$. We will show that a cannot be 0 or ∞, and that b cannot be $\pm\infty$. Suppose first that $a = \infty$. Since $|\gamma_n(v)| = |\beta_n(v/a_n)|$ for all $v \in \mathbb{R}$,

$$|\gamma(v)| = \lim_{k \to \infty} |\gamma_{n_k}(v)| = \lim_{k \to \infty} |\beta_{n_k}(v/a_{n_k})| = 1.$$

Let R be the symmetrization of the distribution corresponding to γ. The characteristic function of R is $|\gamma|^2$, which is identically equal to 1, so $R = \delta_0$. It follows that the distribution corresponding to γ is degenerate. Since we are assuming that this distribution is not degenerate, our argument shows that $a = \infty$ is impossible. The proof that $a \neq 0$ is carried out in a similar manner, using the fact that $|\beta_n(v)| = |\gamma_n(a_n v)|$ for $v \in \mathbb{R}$.

Thus we have shown that a cannot be 0 or ∞. Choose a bounded neighborhood $\mathcal{N}$ of the origin so that $\gamma(v) \neq 0$ and $\beta(v/a) \neq 0$ for $v \in \mathcal{N}$. Then

$$e^{-ivb_{n_k}/a_{n_k}} = \frac{\gamma_{n_k}(v)}{\beta_{n_k}(v/a_{n_k})} \to \frac{\gamma(v)}{\beta(v/a)} \quad \text{as} \quad k \to \infty$$

for $v \in \mathcal{N}$. The left side is continuous and never equal to 0, so it has a unique continuous logarithm that takes the value 0 at $v = 0$ (see Problem 7 of Appendix E). This logarithm is obviously equal to $-ivb_{n_k}/a_{n_k}$. The limit on the right side is also continuous and nonzero for $v \in \mathcal{N}$, so it also has a unique continuous logarithm that takes the value 0 at $v = 0$. The logarithm of the right side equals the limit of the logarithm of the left side as $k \to \infty$, which is $-ivb/a$. It follows that b is finite. Furthermore, the first paragraph of this proof shows that

$$\frac{V_{n_k} - b_{n_k}}{a_{n_k}} \xrightarrow{\mathcal{D}} \frac{V - b}{a} \quad \text{as } k \to \infty.$$

By hypothesis, the distribution of the random variable on the right has characteristic function γ.

It remains to show that the original sequences (a_n) and (b_n) have the limits a and b, respectively. The argument in the preceding two paragraphs shows that every subsequence of $((a_n, b_n) : n \geq 1)$ has a further subsequence that converges to a member of $(0, \infty) \times \mathbb{R}$ and that for any subsequential limit (a', b'), γ is the characteristic function of $(V - b')/a'$. Thus $(V - b')/a'$ has the same distribution as $(V - b)/a$. It follows from Proposition 6 of Chapter 3 that $(a', b') = (a, b)$. We have verified the conditions of Proposition 4 of Appendix B for the sequence $((a_n, b_n): n \geq 1)$. The conclusion is that $a_n \to a$ and $b_n \to b$ as $n \to \infty$. $\square$

* **Problem 52.** Give an example that shows that the hypothesis in the preceding proposition that V not be of degenerate type may neither be removed nor even be replaced by the hypothesis that V not be a.s. equal to 0.

Theorem 24. [Convergence of Types] *Let $(V_n : n = 1, 2, \ldots)$ be a sequence of $\mathbb{R}$-valued random variables, and suppose, for some $a_n \in (0, \infty)$ and $b_n \in \mathbb{R}$, $n = 1, 2, \ldots$, that the sequence $((V_n - b_n)/a_n : n = 1, 2, \ldots)$ converges in distribution to an $\mathbb{R}$-valued random variable Y that is not of degenerate type. Let $(a'_n : n = 1, 2, \ldots)$ and $(b'_n : n = 1, 2, \ldots)$ be sequences in $(0, \infty)$ and $\mathbb{R}$, respectively. Then the sequence $((V_n - b'_n)/a'_n : n = 1, 2, \ldots)$ converges in distribution to an $\mathbb{R}$-valued random variable Y' that is not of degenerate type if and only if a'_n/a_n converges to a constant $a' \in (0, \infty)$ and $(b'_n - b_n)/a_n$ converges to a constant $b' \in \mathbb{R}$ as $n \to \infty$. In this case, Y' has the same distribution as $(Y - b')/a'$.*

Problem 53. Use Lemma 23 to prove the preceding theorem.

The preceding theorem makes it natural to speak of the *limit type* (if it exists) of a given sequence, implicitly excluding the degenerate type. To avoid any ambiguity one sometimes uses the term *nondegenerate limit type*. If it happens that with a certain scaling and centering one gets convergence to a constant random variable, one tries to find another scaling and centering that gives a nondegenerate limit type. The term *normalization* is often used to describe the process of scaling and centering.

Sometimes one is willing to allow scaling but not centering. Then, in addition to nondegenerate limit types, degenerate limits are also of interest, with the exception of the delta distribution at 0. In this setting, one speaks of the *strict limit type* of a sequence (if it exists). The following result justifies this terminology.

Theorem 25. [Convergence of Strict Types] *Let $(V_n : n = 1, 2, \ldots)$ be a sequence of $\mathbb{R}$-valued random variables, and suppose, for some $a_n \in (0, \infty)$, $n = 1, 2, \ldots$, that the sequence $(V_n/a_n : n = 1, 2, \ldots)$ converges in distribution to an $\mathbb{R}$-valued random variable Y that is not a.s. equal to 0. Let $(a'_n : n = 1, 2, \ldots)$ be a sequence of positive numbers. Then the sequence $(V_n/a'_n : n = 1, 2, \ldots)$ converges in distribution to an $\mathbb{R}$-valued random variable Y' that is not a.s.*

equal to 0 if and only if a'_n/a_n converges to a constant $a' \in (0,\infty)$ as $n \to \infty$. In this case, Y' has the same distribution as Y/a'.

PROOF. The proof of the 'if' portion and the last sentence of the theorem is just as in the Convergence of Types Theorem. That theorem also implies the 'only if' part if either Y or Y' has a nondegenerate distribution. It remains to prove the 'only if' part under the assumption that both Y and Y' are almost surely equal to a nonzero constant. For this part of the proof, we revisit the proof of the 'only if' portion of Lemma 23.

Let β_n be the characteristic function of V_n/a_n, and let β and γ be the characteristic functions of Y and Y', respectively. Then by Problem 7 of Chapter 13, the characteristic function of V_n/a'_n is the function $v \rightsquigarrow \beta_n(a_n v/a'_n)$. By hypothesis,

$$\lim_n \beta_n(v) = \beta(v) \quad \text{and} \quad \lim_n \beta_n(a_n v/a'_n) = \gamma(v)$$

for all $v \in \mathbb{R}$. Now the same argument used in the proof of the 'only if' portion of Lemma 23 applies to show that the sequence $(a_n/a'_n : n = 1, 2, \dots)$ has a limit $a' \in (0,\infty)$. □

Problem 54. Let $(X_n : n = 1, 2, \dots)$ be an iid sequence of $\mathbb{R}$-valued random variables with nonzero finite mean. For every $c \neq 0$, characterize all sequences $(a_n : n = 1, 2, \dots)$ of positive constants such that

$$\frac{X_1 + \cdots + X_n}{a_n} \xrightarrow{\mathcal{D}} \delta_c \quad \text{as } n \to \infty.$$

Note: When the mean is 0 and the variance is finite, no such sequences exist. If the variance is infinite, such sequences may or may not exist. Further information on the case with mean 0 may be found in Chapters 15 and 16.

14.9. Characterization of moment generating functions

Theorem 14 of Chapter 5 identifies those functions that are probability generating functions of $\overline{\mathbb{Z}}^+$-valued random variables. We will now use that theorem in conjunction with Theorem 19 of this chapter to identify those functions that are moment generating functions.

Let φ be the moment generating function of a probability measure Q on $\mathbb{R}^+$:

$$\varphi(u) = \int_{\mathbb{R}^+} e^{-ux} Q(dx), \quad 0 \le u < \infty.$$

By Problem 38 of Chapter 13, φ is continuous, and, for $u > 0$, the kth derivative of φ at u is given by

$$\varphi^{(k)}(u) = (-1)^k \int_{\mathbb{R}^+} x^k e^{-ux} Q(dx).$$

An $\mathbb{R}$-valued function defined on an interval is said to be *completely monotone* if it and all its even-order derivatives (including the function itself as its own derivative of order 0) are nonnegative on the interval and all its odd-order derivatives are nonpositive on the interval. From the preceding discussion we see that moment generating functions of probability measures on $\mathbb{R}^+$ are completely monotone on $(0, \infty)$ and continuous at 0 with the value 1 there. This fact and its converse constitute a portion of the next theorem.

Theorem 26. *A function $\varphi \colon \mathbb{R} \to [0, \infty)$ is the moment generating function of a probability measure on $\mathbb{R}^+$ if and only if φ is completely monotone on $(0, \infty)$, continuous at 0, and $\varphi(0) = 1$, in which case the corresponding probability measure is the limit of the sequence $(Q_n \colon n = 1, 2, \dots)$, where Q_n is the probability measure supported by $\{\frac{k}{n} \colon k \in \mathbb{Z}^+\}$ and given by*

$$(14.5) \qquad Q_n\{k/n\} = \frac{(-n)^k \varphi^{(k)}(n)}{k!}, \quad k \in \mathbb{Z}^+,$$

where $\varphi^{(k)}$ denotes the kth derivative of φ when $k = 1, 2, \dots$, and φ itself when $k = 0$.

PROOF. The discussion preceding the theorem proves the 'only if' assertion. To complete the proof assume that φ is completely monotone on $(0, \infty)$ and continuous at 0 with the value 1 there. It then follows that (14.5) defines a σ-finite measure on $\{\frac{k}{n} \colon k \in \mathbb{Z}^+\}$.

Let

$$\rho_n(s) = \varphi(n - sn), \quad 0 \leq s \leq 1.$$

From the continuity and complete monotonicity of φ, we conclude that ρ_n is continuous with nonnegative derivatives on $(0, 1)$. By Theorem 14 of Chapter 5 and the continuity of ρ_n at 1, we conclude that

$$\varphi(n - sn) = \sum_{k=0}^{\infty} \frac{\rho_n^{(k)}(0) s^k}{k!} = \sum_{k=0}^{\infty} \frac{(-n)^k \varphi^{(k)}(n) s^k}{k!}.$$

Since the left side equals 1 when $s = 1$, so does the right side. Thus, (14.5) defines a probability measure, the moment generating function of which can be obtained by inserting $e^{-u/n}$ for s:

$$u \rightsquigarrow \varphi(n(1 - e^{-u/n})).$$

As $n \to \infty$, this function converges to φ. An appeal to Theorem 19 completes the proof. $\square$

Problem 55. State and prove an analogue of the preceding theorem for probability measures on $\overline{\mathbb{R}}^+$ and their moment generating functions.

Problem 56. Prove that a bounded completely monotone function on $(0, \infty)$ is either constant or else has the property that neither it nor any of its derivatives is 0 in the interval. Can the hypothesis of boundedness be removed? *Hint:* Use the preceding theorem.

Problem 57. Let φ and ψ be two continuous functions on $[0, \infty)$ that are completely monotone on $(0, \infty)$. Prove twice that $\varphi\psi$ is completely monotone on $(0, \infty)$ using different methods.

Problem 58. Show that the moment generating function of any gamma distribution on $\mathbb{R}^+$ has the form

$$u \rightsquigarrow e^{-\int_0^u \psi(t)\, dt}, \quad 0 \le u < \infty,$$

for some completely monotone function ψ on $(0, \infty)$.

The next theorem gives a general framework into which the preceding problem fits.

Theorem 27. *There is a one-to-one correspondence between the set of infinitely divisible distributions on $(0, \infty)$ and the set of completely monotone functions ψ on $(0, \infty)$ having finite improper Riemann integral on $(0, 1]$. The distribution corresponding to ψ is the one whose moment generating function is*

(14.6)
$$u \rightsquigarrow e^{-\int_0^u \psi(t)\, dt}, \quad 0 \le u < \infty.$$

(The integral is improper because $\psi(0)$ is not necessarily defined; it may or may not be the case that $\psi(0+) < \infty$ in which case $\psi(0)$ could be defined in order to change the improper integral into a proper integral.)

PROOF. Let ψ be a completely monotone function on $(0, \infty)$ whose integral near 0 is finite, and denote by φ the function (14.6). It is clear that $\varphi(0) = 1$ and φ is continuous at 0. We will prove by mathematical induction that the mth derivative of φ is the product of a completely monotone function and $(-1)^m \varphi$. This is obvious for $m = 0$ and we assume it be true for $m = k$. Thus, $\varphi^{(k)} = \eta_k(-1)^k \varphi$, where η_k is completely monotone on $(0, \infty)$. Then

$$\varphi^{(k+1)} = \eta_k \psi (-1)^{k+1} \varphi + [-\eta_k'](-1)^{k+1} \varphi.$$

By Problem 57, $\eta_k \psi$ is completely monotone. Clearly, $-\eta_k'$ is completely monotone. The observation that the sum of completely monotone functions is completely monotone completes the induction proof. In particular, $\varphi^{(m)}$ is nonnegative if m is even and nonpositive if m is odd. Theorem 26 thus applies to show that φ is the moment generating function of a distribution on $\mathbb{R}^+$.

The nth root of φ is

$$u \rightsquigarrow e^{-\int_0^u \frac{\psi(t)}{n}\, dt}, \quad 0 \le u < \infty,$$

which is also the moment generating function of a distribution on $\mathbb{R}^+$, because $\frac{\psi}{n}$ is completely monotone. Therefore φ is infinitely divisible.

For the converse assume that φ is the moment generating function of an infinitely divisible distribution on $\mathbb{R}$. Consider the functions

(14.7)
$$u \rightsquigarrow n\big([\varphi(u)]^{1/n} - 1\big)$$

for various positive integers n. By Theorem 26 the function $u \rightsquigarrow [\varphi(u)]^{1/n}$ is completely monotone for each n. So, for $m \geq 1$, the mth derivative of (14.7) is nonnegative or nonpositive according as m is even or odd. The limit as $n \to \infty$ equals $-\log \circ \varphi$, so we hope that for $m \geq 1$, the mth derivative of $-\log \circ \varphi$ is nonnegative or nonpositive according as m is even or odd. Were this the case, we could complete the proof by setting $\psi = [\log \circ \varphi]'$. Even though derivatives of limits do not in general equal limits of derivatives, an induction argument based on the fact that the various derivatives of the approximating functions are each of constant sign can be used to show that the above-mentioned hope is fulfilled in the situation at hand. $\square$

14.10. Characterization of characteristic functions

Theorem 13 of Chapter 13 identifies those functions that are characteristic functions of $\mathbb{Z}$-valued random variables, and Theorem 14 of the same chapter identifies those functions with finite Lebesgue integral that are characteristic functions of $\mathbb{R}$-valued random variables. In order to identify all functions that are characteristic functions of $\mathbb{R}$-valued random variables, we need a fact about positive definite functions.

Lemma 28. *Every positive definite function is bounded.*

PROOF. Let β be positive definite. By positive definiteness, $\beta(0 - 0)z\bar{z} \geq 0$ for all $z \in \mathbb{C}$. Hence $\beta(0) \geq 0$.

Again using the definition of positive definiteness, we obtain

$$(14.8) \quad \beta(0 - 0)z_1\bar{z}_1 + \beta(0 - v)z_2\bar{z}_1 + \beta(v - 0)z_1\bar{z}_2 + \beta(v - v)z_2\bar{z}_2 \geq 0 .$$

By setting $z_1 = 1$ and $z_2 = 1 + ci$, $c \in \mathbb{R}$, and using the fact that $\beta(0)$ is real, we conclude that $(1 + ci)\beta(-v) + (1 - ci)\beta(v)$ is real. By considering $c = 0$ and then $c = 1$, we see that $\beta(-v) = \overline{\beta(v)}$.

Now we use (14.8) four times with $z_1 = 1$: once with $z_2 = 1$; then with $z_2 = -1$; third with $z_2 = i$; and last with $z_2 = -i$. We deduce that the following four numbers are less than $2\beta(0)$: twice the real part of $-\beta(v)$; twice the real part of $\beta(v)$; twice the imaginary part of $\beta(v)$; and twice the imaginary part of $-\beta(v)$. We conclude that $|\beta(v)| \leq \beta(0)\sqrt{2}$ for all v and that, therefore, β is a bounded function. $\square$

Theorem 29. *A function $\beta \colon \mathbb{R} \to \mathbb{C}$ is the characteristic function of some $\mathbb{R}$-valued random variable if and only if it is continuous and positive definite, and satisfies $\beta(0) = 1$.*

PROOF. The content of Proposition 12 of Chapter 13 is that a characteristic function is continuous, positive definite, and has the value 1 at 0. For the converse, we assume that β is continuous and positive definite and that $\beta(0) = 1$.

For $b > 0$, let

$$\gamma_b(v) = e^{-b|v|}\beta(v) .$$

By Lemma 28, β is a bounded function and, therefore, each γ_b has finite integral. Recalling that $v \rightsquigarrow e^{-b|v|}$ is the characteristic function of a Cauchy distribution, we write

$$\gamma_b(v) = \int_{-\infty}^{\infty} \frac{b}{\pi(b^2 + x^2)} e^{ivx} \beta(v) \, dx \, .$$

Hence, for any choice of n and $v_1, \ldots, v_n \in \mathbb{R}$ and $z_1, \ldots, z_n \in \mathbb{C}$,

$$\sum_{k=1}^{n} \sum_{j=1}^{n} \gamma_b(v_k - v_j) z_j \bar{z}_k$$

$$= \int_{-\infty}^{\infty} \frac{b}{\pi(b^2 + x^2)} \left(\sum_{k=1}^{n} \sum_{j=1}^{n} \beta_b(v_k - v_j)(z_j e^{-iv_j x}) \overline{(z_k e^{-iv_k x})} \right) dx \, .$$

The integrand is nonnegative because β is positive definite. Hence, the integral is nonnegative and, therefore, γ_b is positive definite. Clearly, γ_b inherits the properties of being continuous and having the value 1 at 0 from β.

On the basis of the preceding paragraph we may apply Theorem 14 of Chapter 13 to conclude that each γ_b is a characteristic function. Let $b \searrow 0$ and apply Theorem 15 of this chapter to conclude that β is a characteristic function. $\square$

Let β be some continuous positive definite function for which $\beta(0) = 1$. By the preceding theorem it is the characteristic function of some distribution Q on $\mathbb{R}$. The inversion formulas given in Theorem 13 and Theorem 14 of Chapter 13 conveniently describe Q in terms of β in case β is periodic or has finite integral. The above proof yields a description in general: $Q = \lim_{b \searrow 0} Q_b$, where Q_b is the distribution with density

$$x \rightsquigarrow \frac{1}{2\pi} \int_{-\infty}^{\infty} e^{-ivx} \gamma_b(v) \, dv \, .$$

We could have used characteristic functions (with finite integral) other than $v \rightsquigarrow e^{-b|v|}$ in the proof of the preceding formula, and so the preceding sentence remains valid for a wide variety of definitions of γ_b.

Problem 59. Find a distribution whose characteristic function is neither periodic nor has finite integral. Then check your conclusion by actually calculating the characteristic function.

Problem 60. Let $(\Psi, \mathcal{G}, R)$ be a probability space. For each $\psi \in \Psi$, let β_ψ be a characteristic function and assume, for each v, that $\psi \rightsquigarrow \beta_\psi(v)$ is measurable. For each $v \in \mathbb{R}$ set

$$\gamma(v) = \int_\Psi \beta_\psi(v) \, R(d\psi) \, .$$

Prove that γ is a characteristic function. Describe a two-step experiment that is related to this problem.

CHAPTER 15
Distributional Limit Theorems
for Partial Sums

In this chapter we study convergence in distribution in settings involving sequences $(S_n: n = 1, 2, \ldots)$, where for each n, $S_n = X_1 + \cdots + X_n$ is the nth partial sum of a series of independent random variables. Our first result is that convergence in distribution of (S_n) is equivalent to a.s. convergence. Thereafter, we specialize to the case in which $(X_1, X_2, \ldots)$ is an iid sequence. Further limit theorems involving more general sums of independent random variables will be found in Chapter 16.

For the case of iid summands, we first look at the convergence in distribution of S_n/n, and obtain necessary and sufficient conditions for a Law of Large Numbers: for all $\varepsilon > 0$,

$$(15.1) \qquad \lim_{n \to \infty} P\left[\left|\frac{S_n}{n} - c\right| > \varepsilon\right] = 0,$$

where c is some constant. Unlike the Strong Law of Large Numbers, this result applies in some cases in which the mean does not exist. In general, it is difficult to obtain good estimates for the rate of convergence in (15.1). But in certain special cases, we can derive very useful and precise information about this rate, the so-called 'large deviations estimates' treated later in the chapter.

In Section 3, we examine the issue of convergence in distribution of $(S_n - b_n)/a_n$ for constants $a_n > 0$ and b_n. If the summands X_k have finite mean and variance the constants can be chosen to give the standard normal distribution as a limit, whatever the distribution of the summands. This result is known as the Classical Central Limit Theorem. Some related 'local limit theorems' are proved in the last section of the chapter.

When the mean or variance is not finite, other limiting distributions known as 'stable' distributions are possible limits for normalized sums of the form $(S_n - b_n)/a_n$. Some basic results concerning convergence to stable distributions are contained in this chapter. Chapter 17 contains deeper results along these lines.

15.1. Infinite series of independent random variables

Let $(X_k\colon k = 1, 2, 3, \ldots)$ be a sequence of independent $\mathbb{R}$-valued random variables and denote the characteristic function of X_k by β_k. When we say that the infinite series $\sum_{k=1}^{\infty} X_k$ converges in distribution, we mean that sequence of partial sums $(S_n\colon n = 1, 2, \ldots)$ converges in distribution as $n \to \infty$. By the independence of the summands and the Continuity Theorem of Chapter 14, convergence of this sequence is equivalent to the convergence of the infinite product of characteristic functions to an appropriate limit:

$$(15.2) \qquad \prod_{k=1}^{\infty} \beta_k(v) = \gamma(v), \quad v \in \mathbb{R},$$

for some function γ that is continuous at 0. In case γ is such a function, γ is the characteristic function of the limiting distribution. We will prove that in this case, the series $\sum_{k=1}^{\infty} X_k$ actually converges almost surely to a random variable S. Thus, for infinite series of independent random variables, a.s. convergence, convergence in probability, and convergence in distribution are equivalent.

* **Problem 1.** Suppose that $(X_1, X_2, \ldots)$ is an independent sequence for which

$$P[X_k = -m6^{-k}] = P[X_k = m6^{-k}] = \tfrac{1}{6} \quad \text{for } m = 1, 3, 5\,.$$

Decide if $\sum_{k=1}^{\infty} X_k$ converges in distribution, and if so, calculate its characteristic function and its distribution. Also, decide if $\sum X_k$ is almost surely convergent. Which of your conclusions remain valid if the independence assumption is dropped?

Problem 2. Suppose that $P[X_k = -k^{-1/2}] = P[X_k = k^{-1/2}] = \tfrac{1}{2}$. Decide if $\sum X_k$ converges in distribution.

Theorem 1. *Let $(X_k\colon k = 1, 2, \ldots)$ be a sequence of independent $\mathbb{R}$-valued random variables. Then*

$$\sum_{k=1}^{\infty} X_k$$

either converges in distribution, in probability, and almost surely, or else it diverges in all three senses.

PROOF. It is true in general that a.s. convergence implies convergence in probability and therefore convergence in distribution. By Theorem 22 of Chapter 12 convergence in probability of an infinite series of independent random variables implies its almost sure convergence. Thus it remains to prove that convergence in distribution implies convergence in probability, or equivalently, that convergence in distribution implies Cauchy in probability. For the sequence of partial sums S_n, Cauchy in probability means that for any sequence of positive integers $(m_n\colon n = 1, 2, \ldots)$,

$$(15.3) \qquad \lim_{n \to \infty} \left(X_{n+1} + \cdots + X_{n+m_n} \right) = 0 \text{ i.p.}$$

We will show that convergence in distribution of the partial sums S_n as $n \to \infty$ implies (15.3).

As indicated earlier, convergence in distribution implies that (15.2) holds for some characteristic function γ, where β_k denotes the characteristic function of X_k. Thus, if v is a real number such that $\gamma(v) \neq 0$, then $\beta_k(v) \neq 0$ for all k. Therefore, since $\gamma(0) = 1$ and γ is continuous, there exists an open interval B containing 0 such that neither γ nor any β_k is 0 at any point in B. Hence

$$\lim_{n \to \infty} \prod_{k=n+1}^{n+m_n} \beta_k(v) = \frac{\gamma(v)}{\lim_{n\to\infty} \prod_{k=1}^{n} \beta_k(v)} \cdot \frac{1}{\lim_{n\to\infty} \prod_{k=n+m_n+1}^{\infty} \beta_k(v)} = 1$$

for $v \in B$.

The left side of this last expression is the characteristic function of $\sum_{k=n+1}^{n+m_n} X_k$, which we denote by α_n. We have shown that $\alpha_n(v) \to 1$ as $n \to \infty$ for every $v \in B$. By Lemma 16 of Chapter 14, $\alpha_n(v) \to 1$ as $n \to \infty$ for every $v \in \mathbb{R}$. Hence, $\sum_{k=n+1}^{n+m_n} X_k \xrightarrow{D} 0$. So (15.3) now follows from Proposition 4 of Chapter 14. $\square$

15.2. The Law of Large Numbers revisited

For the remainder of the chapter, we focus our attention on normalized partial sums of iid sequences $(X_n: n = 1, 2, \dots)$. In this section, we concern ourselves with the convergence of $\frac{1}{n} \sum_{k=1}^{n} X_k$ as $n \to \infty$. Recall from Chapter 12 that there is almost sure convergence to a finite limit if and only if $E(|X_1|) < \infty$. Problem 42 of Chapter 13 and the following theorem show that this condition is not necessary for convergence in probability.

Theorem 2. [Law of Large Numbers] *Let β denote the common characteristic function of the terms of an iid sequence $(X_n: n = 1, 2, \dots)$ of $\mathbb{R}$-valued random variables. Then*

$$(15.4) \qquad \lim_{n \to \infty} \frac{\sum_{k=1}^{n} X_k}{n} = c \ \ i.p.$$

for some constant $c \in \mathbb{R}$ if and only if $\beta'(0)$ exists, in which case $c = -i\beta'(0)$.

PROOF. The characteristic function of $\frac{1}{n} \sum_{k=1}^{n} X_k$ is the function

$$v \rightsquigarrow \beta^n(v/n).$$

In view of the Continuity Theorem and Proposition 4 of Chapter 14, the assertion of the theorem is the same as the following statement : $\beta'(0)$ exists if and only if there exists a real constant c such that

$$\beta^n(\tfrac{v}{n}) \to e^{icv} \text{ as } n \to \infty,$$

in which case $\beta'(0) = ic$.

By the continuity of β, there exists some positive number b such that $\beta(w) \neq 0$ for $|w| \leq b$. For $n \geq |v|/b$ we may write

$$\beta^n(\tfrac{v}{n}) = \exp\big(n(\log \circ \beta)(\tfrac{v}{n})\big).$$

(See Problem 7 of Appendix E for the definition of $\log \circ \beta$.) It will be useful to make the substitution $w = \tfrac{v}{n}$. Thus,

$$(15.5) \qquad \beta^n(\tfrac{v}{n}) = \exp\big(v\tfrac{1}{w}(\log \circ \beta)(w)\big),$$

provided that $|w| \leq b$. We also need the consequence

$$\lim_{w \to 0} \frac{(\log \circ \beta)(w)}{\beta(w) - 1} = 1$$

of Problem 13 of Appendix E and the comment following that problem. We rewrite the preceding equality as follows:

$$(15.6) \qquad \lim_{w \to 0} \frac{\frac{(\log \circ \beta)(w)}{w}}{\frac{\beta(w)-1}{w}} = 1.$$

It follows from (15.6) that if $\beta'(0)$ exists, then the limit as $n \to \infty$ of (15.5) equals $e^{v\beta'(0)}$, because for each fixed v, $|w| = \tfrac{|v|}{n} \leq b$ for all sufficiently large n. Thus, the 'if' part of the proof is completed, as is the assertion that $ic = \beta'(0)$.

To complete the 'only if' portion of the proof, we assume that $\beta^n(\tfrac{v}{n}) \to e^{icv}$ for all $v \in \mathbb{R}$ as $n \to \infty$, and show that $\beta'(0) = ic$. Since e^{icv} is a characteristic function, the Continuity Theorem implies that the convergence of $\beta^n(\tfrac{v}{n})$ is uniform for $v \in [-1, 1]$. Keeping in mind that $\beta(\tfrac{v}{n}) \neq 0$ for $v \in [-1, 1]$ and sufficiently large n, we may take logarithms and conclude that

$$(15.7) \qquad n \log \circ \beta(\tfrac{v}{n}) \to icv \text{ as } n \to \infty, \text{ uniformly for } v \in [-1, 1].$$

For each $w \in [-1, 1]$ there is a v_n satisfying $1 - \tfrac{1}{n} \leq |v_n| \leq 1$ and a positive integer n such that $w = \tfrac{v_n}{n}$, and no matter how such choices are made, $w \to 0$ implies that $n \to \infty$. By the uniform convergence in (15.7), $\tfrac{1}{w} \log \circ \beta(w) \to ic$ as $w \to 0$. It now follows from (15.6) that $\beta'(0) = ic$. $\square$

The Law of Large Numbers concerns the convergence of (S_n/n) to a constant. As we have already seen, for such a limit it does not matter whether we talk about convergence in distribution or convergence in probability. By the Kolmogorov 0-1 Law, if we also insist on almost sure convergence, then convergence to a constant is the only possibility. But as the following exercise shows, in the case of convergence in distribution, it is possible for (S_n/n) to converge to a nonconstant $\mathbb{R}$-valued random variable.

Problem 3. In the setting of the preceding paragraph, suppose that the distribution of X_1 is Cauchy with characteristic function $v \rightsquigarrow \exp(-|v|)$. Show that S_n/n has the same distribution as X_1, and so,

$$\frac{S_n}{n} \xrightarrow{\mathcal{D}} X_1 .$$

15.3. The Classical Central Limit Theorem

In the Law of Large Numbers, convergence in distribution is obtained by dividing the nth partial sum S_n by n. We turn more generally to the issue of convergence in distribution of sequences of the form

$$\left(\frac{S_n - b_n}{a_n} : n = 1, 2, \ldots \right),$$

where a_n and b_n are constants. Special cases have already been treated in Problem 54 and Problem 55 of Chapter 13. The following theorem is the most famous general result along these lines.

Theorem 3. [Classical Central Limit] *Let* $(X_n : n \geq 1)$ *be an iid sequence of* $\mathbb{R}$-*valued random variables having finite mean and finite nonzero variance. For each* n, *let*

$$Z_n = \frac{\sum_{k=1}^{n} X_k - nE(X_1)}{\sqrt{n \operatorname{Var}(X_1)}} .$$

Then, for each $x \in \mathbb{R}$,

$$\lim_{n \to \infty} P[Z_n \leq x] = \frac{1}{\sqrt{2\pi}} \int_{-\infty}^{x} e^{-y^2/2} \, dy ;$$

that is, as $n \to \infty$,

$$Z_n \xrightarrow{\mathcal{D}} Z ,$$

where Z *is a normally distributed random variable with mean* 0 *and variance* 1.

PROOF. Let β denote the characteristic function of X_1. Since X_1 has finite mean and variance, both $\beta'(0)$ and $\beta''(0)$ exist:

$$\beta'(0) = iE(X_1),$$
$$\beta''(0) = -E(X^2).$$

The characteristic function of $\sum_{k=1}^{n} X_k$ is β^n. Since $E(X_1) = -i\beta'(0)$, we conclude from Problem 7 of Chapter 13 that the characteristic function of Z_n is the function

$$v \rightsquigarrow \beta^n \left(\frac{v}{\sqrt{n \operatorname{Var}(X_1)}} \right) e^{-v\sqrt{n}\,\beta'(0)/\sqrt{\operatorname{Var}(X_1)}} .$$

We will complete the proof by showing pointwise convergence of this sequence of functions to the characteristic function $v \rightsquigarrow e^{-v^2/2}$ of the standard normal distribution as $n \to \infty$.

We use Problem 7 of Appendix E as we work with logarithms of the relevant quantities, noting that for fixed v, these quantities are nonzero for sufficiently large n. As $n \to \infty$, it follows from Proposition 3 of Appendix E that

$$n(\log \circ \beta)\left(\frac{v}{\sqrt{n \operatorname{Var}(X_1)}}\right) - \frac{v\sqrt{n}\,\beta'(0)}{\sqrt{\operatorname{Var}(X_1)}}$$

(15.8)

$$= -n\left(1 - \beta\left(\frac{v}{\sqrt{n \operatorname{Var}(X_1)}}\right)\right) - \frac{n}{2}\left(1 - \beta\left(\frac{v}{\sqrt{n \operatorname{Var}(X_1)}}\right)\right)^2$$

$$+ n\,o(n^{-1}) - \frac{v\sqrt{n}\,\beta'(0)}{\sqrt{\operatorname{Var}(X_1)}}.$$

We first compute the limit of the sum of the first and last terms, using Proposition 3 of Appendix E:

$$-n\left(1 - \beta\left(\frac{v}{\sqrt{n \operatorname{Var}(X_1)}}\right)\right) - \frac{v\sqrt{n}\,\beta'(0)}{\sqrt{\operatorname{Var}(X_1)}}$$

$$= -n\left(-\frac{v}{\sqrt{n \operatorname{Var}(X_1)}}\beta'(0) - \frac{v^2}{2n \operatorname{Var}(X_1)}\beta''(0) + o(n^{-1})\right) - \frac{v\sqrt{n}\,\beta'(0)}{\sqrt{\operatorname{Var}(X_1)}}$$

$$\to \frac{v^2\,\beta''(0)}{2\operatorname{Var}(X_1)} \quad \text{as } n \to \infty.$$

For the limit of the second term in (15.8), we multiply and divide by $v^2/\operatorname{Var}(X_1)$ and use the definition of $\beta'(0)$:

$$\lim_{n\to\infty} -\frac{n}{2}\left(1 - \beta\left(\frac{v}{\sqrt{n \operatorname{Var}(X_1)}}\right)\right)^2 = -\frac{(v\beta'(0))^2}{2\operatorname{Var}(X_1)}.$$

Thus, the entire expression in (15.8) converges to

$$-\frac{v^2[-\beta''(0) + (\beta'(0))^2]}{2\operatorname{Var}(X_1)} = -\frac{v^2}{2},$$

as desired. $\square$

Problem 4. Give two variations of the preceding proof: (i) by initially showing that there is no loss of generality in assuming that the summands have mean 0 and (ii) by using Proposition 2 of Appendix E and then applying Proposition 3 of that appendix to $\log \circ \beta$, rather than to $\log$ and β separately.

Problem 5. Use the Classical Central Limit Theorem and facts about Poisson random variables to formulate and prove a convergence in distribution statement involving Poisson random variables with mean n. Then prove the same statement without using the Classical Central Limit Theorem. For each proof, decide whether the assumption that the means are $\mathbb{Z}$-valued is relevant. *Hint:* Figure 15.1 shows,

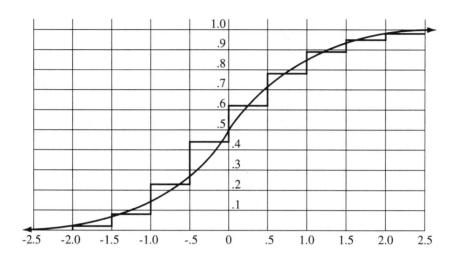

FIGURE 15.1. Normal and Poisson distribution functions

with jumps filled in as a visual aid, the normal distribution function and a normalized Poisson distribution function both having mean 0 and and variance 1. The mean of the unnormalized Poisson distribution function (not shown) is 4.

* **Problem 6.** Use tables of the normal distribution, possibly within your calculator, to approximate the probability that there are at least 520 heads in 1000 flips of a fair coin.

Problem 7. Let Z be a normally distributed random variable with mean 0 and variance 1. Let $(X_k : k = 1, 2, \dots)$ be an iid sequence of $\mathbb{R}$-valued random variables and set $S_n = \sum_{k=1}^{n} X_k$ for each n. Suppose that

$$\frac{S_n - b_n}{\sqrt{nc}} \xrightarrow{\mathcal{D}} Z$$

for some positive constant c and some sequence $(b_1, b_2, \dots)$ of constants. Prove that X_1 has finite variance.

15.4. The general setting for iid sequences

As in previous sections, we consider the partial sums $S_n = X_1 + \cdots + X_n$ of an iid sequence $(X_n : n = 1, 2, \dots)$. Denote the distribution of X_1 by Q. A general question in the spirit of Theorem 3 is: Do there exist positive constants a_n and real constants b_n such that $((S_n - b_n)/a_n : n = 1, 2, \dots)$ converges in distribution to a nondegenerate limit? An alternative version of the same question is: Do there exist positive constants a_n and real constants b_n such that $(Q_n : n =$

$1, 2, \dots$) converges to a nondegenerate limit, where Q_n is defined by

$$Q_n(B) = Q^{*n}(a_n B + b_n), \quad B \text{ Borel}.$$

Here, 'degenerate' signifies a delta distribution when applied to distributions and an almost surely constant random variable when applied to random variables. By Theorem 24 of Chapter 14 we know that if the answer to the above question is 'yes', then the type of limit is uniquely determined. In that case we say that Q is in the *domain of attraction* of that limiting type. For instance, the Classical Central Limit Theorem says that all distributions having finite nonzero variance are in the domain of attraction of the normal type. It develops that there are distributions having infinite variance that are in the domain of attraction of the normal type, although Problem 7 indicates that scaling proportional to $\sqrt{n}$ is not appropriate for such a distribution. Also, there are distributions that belong to no domain of attraction. Variations of terminology are used: the domain of attraction of a particular distribution really means the domain of attraction of the type to which that distribution belongs; and the terminology is also carried over to characteristic functions and random variables.

Similar questions arise when no centering is permitted. Chapter 14 applies to show that a limit different from the delta distribution at 0 is unique up to strict type, and thus, apart from the strict type consisting of the delta distribution at 0, each strict type has a well-defined *domain of strict attraction,* possibly empty. For instance, the Classical Central Limit Theorem says that all distributions having 0 mean and finite nonzero variance are in the domain of strict attraction of the normal distribution with mean 0 and variance 1.

The following definition is relevant to the preceding paragraphs.

Definition 4. Let Q be a distribution on $\mathbb{R}$. The distribution Q is *stable* if Q^{*n} is of the same type as Q for every n. It is *strictly stable* if Q^{*n} is of the same strict type as Q for every n.

The adjectives 'stable' and 'strictly stable' are used for random variables as well as for their distributions—and also for their distribution functions and characteristic functions (and moment generating functions when appropriate).

It is clear that every distribution of the same type as a stable distribution is also stable, and that every distribution of the same strict type as a strictly stable distribution is also strictly stable. Accordingly, we may speak of a type being or not being stable and of a strict type being or not being strictly stable.

Problem 8. Prove that every stable distribution on $\mathbb{R}$ is infinitely divisible.

* **Problem 9.** Decide if the Poisson distributions are strictly stable.

Problem 10. Decide if the gamma distributions are stable.

Problem 11. Decide if the Cauchy distributions are stable. Which if any of them are strictly stable?

Proposition 5. *The distributions on $\mathbb{R}$ that have nonempty domains of attraction are the stable distributions. Those that have nonempty domains of strict attraction are the strictly stable distributions.*

PARTIAL PROOF. We treat only domains of strict attraction in this proof. The proof for domains of attraction is requested in Problem 12. Suppose that a distribution Q is in the domain of strict attraction of a distribution R. Let $(X_k : k = 1, 2, \ldots)$ be an iid sequence with common distribution Q, Y a random variable with distribution R, and $(a_n : n = 1, 2, \ldots)$ a sequence of positive constants such that

$$\frac{\sum_{k=1}^n X_k}{a_n} \xrightarrow{\mathcal{D}} Y.$$

Fix a positive integer m and let $(Y_1, Y_2, \ldots, Y_m)$ be a sequence of independent random variables each having distribution R.

On the one hand

$$\frac{\sum_{k=1}^{mn} X_k}{a_{mn}} \xrightarrow{\mathcal{D}} Y$$

as $n \to \infty$, and on the other hand, by Proposition 6 of Chapter 14,

$$\frac{\sum_{k=1}^{mn} X_k}{a_n} = \sum_{k=1}^m \frac{\sum_{j=1+(k-1)n}^{kn} X_j}{a_n} \xrightarrow{\mathcal{D}} \sum_{k=1}^m Y_k$$

as $n \to \infty$. By Theorem 25 of Chapter 14, Y and $\sum_{k=1}^m Y_k$ are of the same strict type, and so, according to Definition 4, R is strictly stable.

Now suppose that Y is strictly stable. Let $(Y_1, Y_2, \ldots)$ be an iid sequence of random variables each having the same distribution as Y. By the definition of strict stability, there exists, for each n, a positive constant a_n such that $a_n^{-1} \sum_{k=1}^n Y_k$ has the same distribution as Y. Hence,

$$\frac{\sum_{k=1}^n Y_k}{a_n} \xrightarrow{\mathcal{D}} Y.$$

Therefore, Y is in its own domain of strict attraction. □

Problem 12. Complete the proof of the preceding proposition.

In Chapter 17, the issue of characterizing stable distributions and identifying their domains of attraction will be treated. Here, as an introduction to this topic, are exercises concerning some stable distributions.

Problem 13. Let α be a real number. Show that

$$u \rightsquigarrow e^{-u^\alpha}, \quad u \in [0, \infty),$$

is the moment generating function of a probability distribution on $\mathbb{R}^+$ if and only if $0 < \alpha \leq 1$, in which case the corresponding probability distribution is strictly stable. *Hint:* Use Theorem 27 of Chapter 14.

* **Problem 14.** Identify the strictly stable type obtained by setting $\alpha = 1$ in the preceding problem.

Problem 15. Use the Law of Large Numbers to identify many distributions in the domain of strict attraction of the type identified in the preceding problem.

* **Problem 16.** [Strictly stable distribution on $\mathbb{R}^+$ of index $\frac{1}{2}$] Show that the strictly stable distribution on $\mathbb{R}^+$ obtained by setting $\alpha = \frac{1}{2}$ in Problem 13 has a density with respect to Lebesgue measure of the form

$$x \rightsquigarrow \frac{a}{\sqrt{2\pi x^3}} e^{-\frac{a^2}{2x}}$$

for some positive constant a. Also, find a.

Problem 17. Calculate the moments of the distribution in Problem 16.

Problem 18. Which of the distributions of Problem 13 have finite expectations?

Problem 19. Let $\alpha \in (0,1]$ and let $(X_1, X_2, \dots)$ be an iid sequence of strictly stable $\mathbb{R}^+$-valued random variables having the moment generating function given in Problem 13. Find a_n for each positive n so that $\frac{1}{a_n}\sum_{k=1}^{n} X_k$ has the same distribution as X_1.

15.5. † Large deviations

The Law of Large Numbers implies that if $(X_n: n = 1, 2, \dots)$ is an iid sequence of $\mathbb{R}$-valued random variables with finite mean, then for all $\varepsilon > 0$,

$$P\left[\sum_{k=1}^{n} X_k > n[E(X_1) + \varepsilon]\right] \to 0 \text{ as } n \to \infty.$$

The event in this expression could be viewed as a 'large deviation' by the partial sum $S_n = X_1 + \cdots + X_n$ above its mean, which is $nE(X_1)$. A similar statement holds for the probability of large deviations below $E(S_n)$, that is, for the event that S_n is less than $n(E(X_1) - \varepsilon)$. Our original proof of the Law of Large Numbers in Chapter 5, which used the Chebyshev Inequality, provided us with an upper bound for the probability of large deviations, but this bound is not very good in many cases. Our more recent proof, using characteristic functions, was designed to include some cases in which $E(X_1)$ does not exist. In terms of providing useful upper bounds for large deviations probabilities, this second method is even worse than the first.

It turns out that under an additional assumption on the distribution of X_1, we can use the Markov Inequality to obtain vastly improved large deviations estimates. The extra assumption needed is that there exists a constant $a > 0$ such that

(15.9) $E\left(e^{a|X_1|}\right) < \infty.$

Throughout this section, we will let

$$\varphi(b) = E\!\left(e^{bX_1}\right).$$

Note that if $a > 0$ is such that (15.9) is satisfied, then $\varphi(b)$ exists and is finite for all $b \in [0, a]$.

Theorem 6. [Large Deviations] *Let $(X_n : n = 1, 2, \ldots)$ be an iid sequence of $\mathbb{R}$-valued random variables satisfying (15.9) for some constant $a > 0$. Then for all $b \in [0, a]$,*

$$(15.10) \qquad P\!\left[\sum_{k=1}^{n} X_k > n(E(X_1) + \varepsilon)\right] \le \left[e^{-b(E(X_1)+\varepsilon)}\varphi(b)\right]^n.$$

Furthermore, $b \in [0, a]$ may be chosen so that

$$(15.11) \qquad e^{-b(E(X_1)+\varepsilon)}\varphi(b) < 1.$$

PROOF. We rewrite the left side of (15.10) as

$$P\!\left[\sum_{k=1}^{n}[X_k - E(X_k)] > n\varepsilon\right],$$

and then apply the Markov Inequality (with $f(z) = e^{bz}$) to obtain the upper bound

$$e^{-bn\varepsilon}E\!\left(e^{b\sum_{k=1}^{n}[X_k-E(X_k)]}\right).$$

Since the random variables X_k are independent, the expectation in this expression factors by Corollary 12 of Chapter 9, and (15.10) follows.

To prove the rest of the theorem, we calculate the right-hand derivative of the left side of (15.11) with respect to b at $b = 0$. The result is $-(E(X_1) + \varepsilon) + E(X_1) = -\varepsilon$. (The reader may check that the differentiation inside the expectation is justified by splitting X_1 into positive and negative parts, and then applying the Monotone and Dominated Convergence Theorems. The argument is similar to that used to differentiate moment generating functions in Chapter 13.) Since this derivative is negative, and since the left side of (15.11) equals 1 when $b = 0$, there exists a $b \in (0, a]$ such that (15.11) is satisfied. $\square$

* **Problem 20.** Find the minimum value of the left side of (15.11) in each of the following cases: (i) X_1 has a Bernoulli distribution with parameter p; (ii) X_1 has a standard exponential distribution; (iii) X_1 has a standard normal distribution.

Problem 21. Use Theorem 6 to give an alternate proof of the Strong Law of Large Numbers for an iid sequence that satisfies (15.9) for some $a > 0$.

It turns out that the minima found in Problem 20 provide, in a certain sense, the sharpest possible large deviations estimates for the distributions treated in that problem. We will not pursue this matter in this book.

15.6. † Local limit theorems

The Classical Central Limit Theorem may be restated briefly as follows: if F_n is the distribution function of $[S_n - nE(X_1)]/\sqrt{n \operatorname{Var}(X_1)}$, where S_n is the nth partial sum of an iid sequence (X_k), then (F_n) converges pointwise as $n \to \infty$ to the standard normal distribution function. Thus, it is a result about the convergence of sequences of distribution functions. The main results of this section concern the pointwise convergence of sequences of density functions. Such results are known as 'local limit theorems'.

We will state and prove local limit theorems for two important cases, essentially corresponding to the two cases for which we obtained inversion formulas in Chapter 13. In each case, our proof will involve showing that we can move the relevant limit outside the integral sign in the inversion formula.

We begin with the case in which the distributions under consideration have a density with respect to Lebesgue measure on $\mathbb{R}$. For this case, we need two preliminary results concerning the characteristic function of a distribution that has a density.

Lemma 7. [Riemann-Lebesgue] *Let Q be a probability measure on $\mathbb{R}$ with characteristic function β. Suppose that Q has a density f with respect to Lebesgue measure on $\mathbb{R}$. Then $\lim_{v \to \infty} \beta(v) = 0$.*

PROOF. We will actually show that if $f \colon \mathbb{R} \to \mathbb{R}$ is a nonnegative Borel measurable function such that $\int f(x)\,dx < \infty$, then

$$(15.12) \qquad \lim_{v \to \infty} \int f(x)e^{ivx}\,dx = 0 .$$

It is easily checked using elementary calculus that (15.12) holds when f is the indicator function of a bounded open interval. Since every open subset of $\mathbb{R}$ is a countable union of pairwise disjoint open intervals, it is also easy to prove (15.12) when f is the indicator of a bounded open subset of $\mathbb{R}$.

Now suppose that $f = I_A$, where A is a bounded Borel subset of $\mathbb{R}$. Since Lebesgue measure is regular (see Problem 14 of Chapter 7), for each $\varepsilon > 0$, there exists an open set $B \supseteq A$ such that $\lambda(B \setminus A) < \varepsilon$, where λ is Lebesgue measure on $\mathbb{R}$. It follows that

$$\limsup_{v \to \infty}\left|\int f(x)e^{ivx}\,dx\right| \leq \lim_{v \to \infty}\left|\int I_B(x)e^{ivx}\,dx\right| + \lambda(B \setminus A) < \varepsilon .$$

Thus (15.12) holds when f is the indicator function of a bounded Borel subset of $\mathbb{R}$. The extension to bounded Borel measurable functions f that vanish off a bounded set is now routine.

To complete the proof, let f be an arbitrary nonnegative Borel measurable function such that $\int f(x)\,dx < \infty$. By the Monotone Convergence Theorem,

$\int f(x)\, dx = \lim_n \int f_n(x)\, dx$, where $f_n(x) = [f(x) \wedge n]I_{[-n,n]}$. Thus, for $\varepsilon > 0$, there exists an integer n such that $\int |f(x) - f_n(x)|\, dx < \varepsilon$. For such n,

$$\limsup_{v \to \infty} \left| \int f(x) e^{ivx}\, dx \right| \leq \lim_{v \to \infty} \left| \int f_n(x) e^{ivx}\, dx \right| + \int |f(x) - f_n(x)|\, dx < \varepsilon,$$

and (15.12) follows. $\square$

The second preliminary result is an improvement on Problem 56 of Chapter 13:

Lemma 8. [Parseval Formula] *Let Q be a probability measure on $\mathbb{R}$ with characteristic function β. Suppose that Q has a density f with respect to Lebesgue measure on $\mathbb{R}$, and that $\int f^2(x)\, dx < \infty$. Then*

$$\int f^2(x)\, dx = \frac{1}{2\pi} \int |\beta(v)|^2\, dv.$$

PROOF. Let

$$g(x) = \int f(y) f(x + y)\, dy = (f * \tilde{f})(-x) = (f * \tilde{f})(x),$$

where $\tilde{f}$ is the function $x \rightsquigarrow f(-x)$. Note that g is the density of the symmetrization of Q, and that the characteristic function of g is $|\beta|^2$ (see Problem 17 of Chapter 13). Also note that

$$g(0) = \int f^2(x)\, dx.$$

By the Cauchy-Schwarz Inequality and the definition of g,

$$g^2(x) \leq \left[\int f^2(y)\, dy \right] \left[\int f^2(x + y)\, dy \right] = \left[\int f^2(x)\, dx \right]^2,$$

so g is bounded.

We wish to show that g is continuous. Again applying the Cauchy-Schwarz Inequality, we have

$$(g(x) - g(z))^2 \leq \left[\int f^2(y)\, dy \right] \left[\int (f(x + y) - f(z + y))^2\, dy \right].$$

After making the change of variables $y \rightsquigarrow y - z$, we see that in order to prove the continuity of g, it is enough to prove that

$$(15.13) \qquad \lim_{x \to 0} \int |f(x + y) - f(y)|^2\, dy = 0.$$

The proof of (15.13) is similar to the proof of (15.12). It is easy to see that (15.13) holds if f is the indicator function of a bounded open set, and the extension to indicator functions of bounded Borel sets is also straightforward. In order to treat simple functions f that vanish off a bounded set, expand the integrand as a finite sum of terms of the form $a_{ij}[f_i(x+y) - f_i(x)][f_j(x+y) - f_j(x)]$, where each a_{ij} is a constant, and each f_i is the indicator function of a bounded Borel set. For the terms with $i = j$, apply the result of the previous step. For the terms with $i \neq j$, use the result of the previous step together with the Cauchy-Schwarz Inequality. The extension to bounded Borel measurable functions f that vanish off a bounded set is done by approximating such functions uniformly with measurable simple functions. And finally, the extension to nonnegative Borel functions f such that $\int f^2(x)\,dx < \infty$ is straightforward.

We have shown that g is a continuous bounded function whose value at 0 is $\int f^2(x)\,dx$. For $\delta > 0$, let h_δ be the continuous density function of the normal distribution with mean 0 and variance δ, and let φ_δ be the corresponding characteristic function. Then $g * h_\delta$ has characteristic function $|\beta|^2 \varphi_\delta$. This characteristic function has finite integral with respect to Lebesgue measure on $\mathbb{R}$, so the Inversion Formula, Theorem 14 of Chapter 13, applies to give

$$(g * h_\delta)(0) = \frac{1}{2\pi} \int |\beta(v)|^2 \varphi_\delta(v)\,dv .$$

As δ goes to 0, the right side of this equation goes to $\frac{1}{2\pi} \int |\beta(v)|^2\,dv$ by the Monotone Convergence Theorem. Since g is bounded and continuous, the left side is easily seen to converge to $g(0)$ as δ goes to 0. Since $g(0) = \int f^2(x)\,dx$, the proof is complete. $\square$

We are now ready to state and prove our first local limit theorem:

Theorem 9. [Local Limit—Continuous Case] *Let $(X_n : n = 1, 2, \dots)$ be an iid sequence of $\mathbb{R}$-valued random variables with finite mean μ and finite variance $\sigma^2 > 0$. Suppose that the distribution of X_1 has density f with respect to Lebesgue measure, and further suppose there exists a positive integer k such that*

$$\int \left[f^{*k}(x) \right]^2 dx < \infty .$$

Let p_n be the density with respect to Lebesgue measure of

$$\frac{X_1 + \dots X_n - n\mu}{\sqrt{n\sigma^2}} ,$$

and let p be the density with respect to Lebesgue measure of the standard normal distribution. Then

$$\lim_{n \to \infty} \sup_{x \in \mathbb{R}} |p_n(x) - p(x)| = 0 .$$

PROOF. For simplicity, we assume that $\mu = 0$ and $\sigma^2 = 1$. The generalization to arbitrary finite μ, σ^2 is routine. Let β be the characteristic function of X_1. Thus, for $n = 1, 2, \dots$, the characteristic function corresponding to f^{*n} is β^n. The assumption that $\int [f^{*k}(x)]^2 dx < \infty$ and Lemma 8 imply that $\int |\beta(v)|^{2k} dv < \infty$. Since $|\beta| \leq 1$, it follows that $\int |\beta(v)|^n dv < \infty$ for all $n \geq 2k$.

The characteristic function corresponding to p_n is $v \rightsquigarrow \beta^n(v/\sqrt{n})$. By the argument in the preceding paragraph, this function has finite integral with respect to Lebesgue measure on $\mathbb{R}$ when $n \geq 2k$. For such n, the Inversion Formula, Theorem 14 of Chapter 13, gives

$$p^n(x) = \frac{1}{2\pi} \int \beta^n(v/\sqrt{n}) e^{-ivx} \, dv \,, \quad x \in \mathbb{R}.$$

The same Inversion Formula gives

$$p(x) = \frac{1}{2\pi} \int \varphi(v) e^{-ivx} \, dv \,, \quad x \in \mathbb{R},$$

where φ is the characteristic function of the standard normal distribution. Thus, it is enough to show that

$$(15.14) \qquad \lim_{n \to \infty} \int |\beta^n(v/\sqrt{n}) - \varphi(v)| \, dv = 0 \,.$$

We will break the integral in (15.14) into three pieces:

$$(15.15) \qquad \int_{|v| < A_n} |\beta^n(v/\sqrt{n}) - \varphi(v)| \, dv \,;$$

$$(15.16) \qquad \int_{A_n \leq |v| \leq B\sqrt{n}} |\beta^n(v/\sqrt{n}) - \varphi(v)| \, dv \,;$$

$$(15.17) \qquad \int_{|v| > B\sqrt{n}} |\beta^n(v/\sqrt{n}) - \varphi(v)| \, dv \,.$$

The constants $A_n, n = 1, 2, \ldots$, and B are to be chosen later.

By the Classical Central Limit Theorem and the Continuity Theorem, the integrand in (15.15) goes to 0 uniformly on bounded intervals as $n \to \infty$. It follows that if the sequence of constants $(A_n : n \to \infty)$ increases to ∞ slowly enough, the integral in (15.15) converges to 0 as $n \to \infty$. Choose $(A_n : n = 1, 2, \ldots)$ to be any sequence with this property. To avoid notational technicalities with the limits of integration, we may assume that $A_n/\sqrt{n} \to 0$ as $n \to \infty$, so that no matter how $B > 0$ is chosen, we have $A_n < B\sqrt{n}$ for large n.

The integral in (15.17) is bounded above by

$$(15.18) \qquad \int_{|v| > B} \sqrt{n} |\beta(v)|^n \, dv + \int_{|v| > \sqrt{n}B} |\varphi(v)| \, dv \,.$$

The second term in this expression goes to 0 as $n \to \infty$ by the Dominated Convergence Theorem. By Problem 50 of Chapter 13, $|\beta(v)| < 1$ for $v \neq 0$, so since β is continuous, the Riemann-Lebesgue Lemma implies that $\sup_{|v| > B} |\beta(v)| < 1$ for any $B > 0$. Therefore, for any fixed B there exists an integer l such that the sequence $(\sqrt{n} |\beta(v)|^n : n \geq l)$ decreases for $|v| > B$. Since $\int |\beta|^n \, dv < \infty$ for all $n \geq 2k$, the Dominated Convergence Theorem implies that the first integral in (15.18) goes to 0 as $n \to \infty$ for any $B > 0$.

It remains to show that the integral in (15.16) goes to 0 as $n \to \infty$ for A_n as chosen earlier and for some $B > 0$. We apply the inequality in (E.10) of Appendix E to $\beta(v) = E(e^{ivX_1})$:

$$\left|\beta(v) - 1 + \frac{v^2}{2}\right| \le E\left[\frac{|vX_1|^3}{3} \wedge X_1^2 v^2\right] = v^2 h(v),$$

where $h(v)$ is a nonnegative function that goes to 0 as $v \to 0$ by the Dominated Convergence Theorem. Choose B small enough so that if $|v| \le B$, then $h(v) \le 1/4$ and $v^2 \le 2$. For such v we then have

$$|\beta(v)| \le 1 - \frac{v^2}{2} + \frac{v^2}{4} = 1 - \frac{v^2}{4} \le e^{-v^2/4}.$$

It follows that for $|v| \le B\sqrt{n}$, $|\beta(v/\sqrt{n})|^n \le e^{-\sigma^2 v^2/4}$, so (15.16) is bounded above by

$$2 \int_{A_n \le |v| \le B\sqrt{n}} e^{-v^2/4} \, dv,$$

which is easily seen to go to 0 as $n \to \infty$. $\square$

Problem 22. Show that the hypothesis $\int [f^{*k}(x)]^2 \, dx < \infty$ is satisfied for some positive integer k by every density f introduced so far in this book (see the index for a list). For which of these distributions does the conclusion of the Local Limit Theorem hold?

Problem 23. Let

$$f(x) = \begin{cases} \dfrac{1}{2|x|\log^2 |x|} & \text{if } 0 < |x| < \dfrac{1}{e} \\ 0 & \text{otherwise}. \end{cases}$$

Show that f is a density with respect to Lebesgue measure of some distribution Q on $\mathbb{R}$ with finite mean and variance, but that the conclusion of the Local Limit Theorem does not hold for Q.

We now turn our attention to a local limit theorem for the case in which the iid random variables $X_1, X_2, \ldots$, are $\mathbb{Z}$-valued. Since we will be scaling and centering, it turns out to be convenient to work in a more general setting.

Definition 10. A *lattice* in $\mathbb{R}$ is a set L of the form

$$L = \{ax + b \colon x \in \mathbb{Z}\},$$

where $a > 0$ and b are real constants, known respectively as the *span* and *shift* of the lattice L. If $b = 0$, L is a *centered lattice*. A distribution Q on $\mathbb{R}$ is a lattice distribution with *span* a if the support of Q is contained in a lattice with span a and is not contained in a lattice with larger span. A lattice distribution Q is a *centered lattice distribution* if 0 is a member of the support of Q.

Note that since we insist that the span of a lattice be real (∞ is not allowed), a single point is not a lattice, and a delta distribution is not a lattice distribution. In other words, lattice distributions are nondegenerate. Also note that the span of a distribution on $\mathbb{Z}$ is not necessarily 1. For example, the distribution $\frac{1}{2}(\delta_{-1} + \delta_1)$ has span 2.

Problem 24. Let Q be a distribution on $\mathbb{R}$ with characteristic function β. Show that Q is a lattice distribution if and only if the quantity

$$\lambda = \inf\{v > 0 : |\beta(v)| = 1\}$$

is positive and finite, in which case $2\pi/\lambda$ is the span of Q. Also show that if λ is positive and finite and $\beta(\lambda) = 1$, then Q is a centered lattice distribution and β is periodic with period λ.

Problem 25. Show that if Q is a lattice distribution with span a, then so is Q^{*n} for $n = 1, 2, \ldots$. *Hint:* The statement is easy to prove if "lattice" is replaced by "centered lattice".

Lattice distributions do not have densities with respect to Lebesgue measure, so a local limit theorem for lattice distributions will necessarily be worded somewhat differently than one for distributions with densities. Let $(X_n : n = 1, 2, \ldots)$ be an iid sequence of $\mathbb{R}$-valued random variables with finite mean μ, finite variance $\sigma^2 > 0$, and distribution Q. Assume that Q is a lattice distribution with span a. Thus, it is supported by $L = \{ax + b : x \in \mathbb{Z}\}$ for some real constant b. For each n, let

$$Z_n = \frac{X_1 + \cdots + X_n - n\mu}{\sqrt{n\sigma^2}}.$$

Then the support of Z_n is contained in the lattice

$$L_n = \left\{ \frac{a}{\sqrt{n\sigma^2}} x + \frac{n(b - \mu)}{\sqrt{n\sigma^2}} : x \in \mathbb{Z} \right\}.$$

Let m_n denote counting measure on L_n, and set

$$\lambda_n = \frac{a}{\sqrt{n\sigma^2}} m_n.$$

Since λ_n approximates Lebesgue measure in a certain sense (see Problem 27), we might hope that the density of Z_n with respect to λ_n will approximate the standard normal density in a certain sense, and the following result justifies this hope:

Theorem 11. [Local Limit—Lattice Case] *For $n = 1, 2, \ldots$, let Z_n, L_n, and λ_n be as in the preceding paragraph, and let $p_n : L_n \to \mathbb{R}^+$ be the density of the distribution of Z_n with respect to λ_n. Then*

$$\lim_{n \to \infty} \sup_{x \in L_n} |p_n(x) - p(x)| = 0,$$

where p is the density with respect to Lebesgue measure of the standard normal distribution.

Problem 26. Prove the preceding theorem. *Hint:* Assume $\mu = 0, \sigma^2 = 1$ and let β be the characteristic function of X_1. Use the Inversion Theorem, Theorem 13 of Chapter 13, appropriately modified for the lattice case, to show that

$$p_n(x) = \frac{1}{2\pi} \int_{-\pi\sqrt{n}/a}^{\pi\sqrt{n}/a} e^{-ivx} \beta^n(v/\sqrt{n})\, dv \,.$$

Then follow the proof of Theorem 9.

Problem 27. State and prove a precise statement about the sense in which λ_n approximates Lebesgue measure on $\mathbb{R}$.

* **Problem 28.** Let S_n be a simple random walk on $\mathbb{Z}$ with expected step size $2p - 1$. Restate Theorem 11 as a statement about the asymptotic behavior of the quantities $P[S_n = x]$, $x \in \mathbb{Z}$, as $n \to \infty$.

Problem 29. Describe the relation between the sizes of the jumps of the Poisson distribution function in Figure 15.1 and the values of the Poisson density shown in Figure 15.2. In this figure, the densities of Theorem 11 are shown for $n = 4$ and X_j Poisson distributed with mean 1.

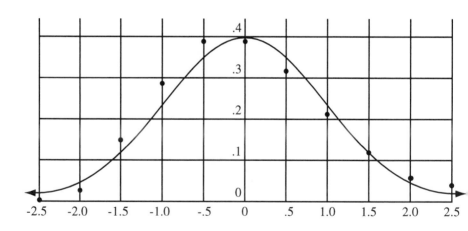

FIGURE 15.2. Normal and Poisson densities

CHAPTER 16

Infinitely Divisible Distributions as Limits

Recall that an infinitely divisible distribution is one that, for each n, is equal to Q_n^{*n} for some distribution Q_n. In preceding chapters several infinitely divisible distributions have appeared. In particular all the stable distributions are infinitely divisible, as is easily seen by comparing the definitions of these two concepts. In this chapter we characterize all infinitely divisible characteristic functions. This characterization is based on the family of 'compound Poisson distributions', to be introduced in the first section.

Infinitely divisible distributions arise naturally as limits of sequences of distributions of sums of independent random variables. The relevant story is told in this chapter with some repetition, starting with some important special cases and then moving on to more general cases that encompass the earlier special cases. The repetition is warranted since in the most general case the story is quite intricate.

In the $\mathbb{R}^+$-setting, there are fewer complications than in the $\mathbb{R}$-setting, especially if moment generating functions are used rather than characteristic functions. This issue is treated and the important extension to the $\overline{\mathbb{R}}^+$-setting is also examined.

16.1. Compound Poisson distributions

When speaking of a Poisson random variable we typically mean a standard Poisson random variable, a $\mathbb{Z}^+$-valued random variable that has positive probability of equaling 1. Accordingly, we introduce the terminology *generalized Poisson* for random variables that are nonzero (possibly negative) multiples of standard Poisson random variables. And, of course, we use the same adjective for corresponding distributions and characteristic functions.

Let ν be a finite measure with finite support $\{x_j : j = 1, 2, \ldots, m\} \subset \mathbb{R} \setminus \{0\}$. For each fixed j, the function $u \rightsquigarrow \exp[-(1 - e^{iux_j})\nu(\{x_j\})]$ is a generalized Poisson characteristic function. The corresponding distribution has support equal to the set of nonnegative integral multiples of x_j. It follows that the product over

$j = 1, 2, \ldots, m$ of these characteristic functions, which may be written as

$$(16.1) \qquad u \rightsquigarrow \exp\left(-\int_{\mathbb{R}\setminus\{0\}} (1 - e^{iuy})\, \nu(dy)\right),$$

is the characteristic function of a sum of m independent generalized Poisson random variables. By an easy limiting argument, using the Continuity Theorem, we see that we still have a characteristic function even if we drop the requirement that the finite measure ν have finite support. A distribution is *compound Poisson* if its characteristic function has the form (16.1) for some finite measure ν (possibly the zero measure) on $\mathbb{R} \setminus \{0\}$.

Replace ν by $\frac{1}{n}\nu$ in (16.1) to obtain a compound Poisson characteristic function whose n^{th} power is the characteristic function (16.1) as it stands. Therefore, all compound Poisson distributions are infinitely divisible.

Another feature that makes compound Poisson distributions central for this chapter is the way they arise as limits. We have already seen in Problem 4 of Chapter 14 that a sum of a large number of independent Bernoulli random variables, each having small probability of equaling 1, can have a distribution that is close to Poisson. In the following example we generalize this fact.

Example 1. Fix a distribution R. Let constants

$$p_{k,n} \in [0, 1], \quad 1 \le k \le n, n = 1, 2, \ldots,$$

satisfy

$$\lim_{n\to\infty} \left(\max_{1\le k\le n} p_{k,n}\right) = 0 \quad \text{and} \quad \lim_{n\to\infty} \sum_{k=1}^{n} p_{k,n} = \lambda,$$

for some constant $\lambda \in (0, \infty)$. For $1 \le k \le n$, $n = 1, 2, \ldots$, we let

$$Q_{k,n} = (1 - p_{k,n})\delta_0 + p_{k,n} R$$

and set the goal of deciding if

$$(16.2) \qquad \lim_{n\to\infty} (Q_{1,n} * Q_{2,n} * \cdots * Q_{n,n})$$

exists, and if so of evaluating it.

In terms of characteristic functions the limit of interest is

$$\lim_{n\to\infty} \prod_{k=1}^{n} \left((1 - p_{k,n}) + p_{k,n}\gamma(v)\right),$$

where γ is the characteristic function of R. We introduce well-defined logarithms as described in Problem 7 of Appendix E, and consider

$$\lim_{n\to\infty} \sum_{k=1}^{n} \log\left(1 - p_{k,n}(1 - \gamma(v))\right).$$

Since $|1 - \gamma(v)| \leq 2$ we see that the error in using only the first nonzero term of the power series for the logarithms is no more than

$$O\left(\sum_{k=1}^n p_{k,n}^2\right) \leq O\left(\left(\max_{1 \leq k \leq n} p_{k,n}\right)\left(\sum_{k=1}^n p_{k,n}\right)\right) \to 0 \quad \text{as } n \to \infty.$$

Hence, the issue has become: Does

$$\lim_{n \to \infty} \sum_{k=1}^n -p_{k,n}(1 - \gamma(v))$$

exist, and if so, what is its limit? It is clear from our assumptions that the limit does exist and is equal to

$$-\lambda(1 - \gamma(v)) = -\int_{\mathbb{R}\setminus\{0\}} (1 - e^{iuy})\,(\lambda R)(dy).$$

Therefore, the limit (16.2) exists and equals the compound Poisson distribution whose characteristic function is

$$v \rightsquigarrow \exp\left(-\int_{\mathbb{R}\setminus\{0\}} (1 - e^{iuy})\,(\lambda R)(dy)\right).$$

* **Problem 1.** [Law of Rare Events] Let $A_{k,n}$, $1 \leq k \leq n$, $n \geq 1$, be events in a probability space $(\Omega, \mathcal{F}, P)$ and suppose for each n, that $(A_{k,n} : 1 \leq k \leq n)$ is an independent n-tuple. Also, suppose that

$$\lim_{n \to \infty} \left(\max_{1 \leq k \leq n} P(A_{k,n})\right) = 0 \quad \text{and} \quad \lim_{n \to \infty} \sum_{k=1}^n P(A_{k,n}) = \lambda \in (0, \infty).$$

Let $Y_n = \sum_{k=1}^n I_{A_{k,n}}$. Prove that $(Y_n : n \geq 1)$ converges in distribution to a standard Poisson random variable with mean λ.

Problem 2. Explain how Problem 4 of Chapter 14 can be regarded as a special case of the preceding problem.

Problem 3. In Example 1 we assumed $\lambda \in (0, \infty)$. What new issues, if any, arise if $\lambda = 0$ is permitted.

Example 2. Let Y_k, $k = 1, 2, \ldots$, and M be independent random variables and suppose that the distribution of M is standard Poisson with mean λ and that the Y_k have a common distribution R. Set $S_0 = 0$ and, for $m = 1, 2, \ldots$, set $S_m = \sum_{k=1}^{m} Y_k$. Let us calculate the distribution Q of the random variable $\omega \rightsquigarrow S_{M(\omega)}(\omega)$. For any Borel set B we have

$$Q(B) = \sum_{m=0}^{\infty} P[M = m, S_m \in B] = \sum_{m=0}^{\infty} P[M = m]\, P[S_m \in B]$$

$$= \sum_{m=0}^{\infty} \frac{e^{-\lambda}\lambda^m}{m!}\, R^{*m}(B)\,;$$

that is,

(16.3) $$Q = \sum_{m=0}^{\infty} \frac{e^{-\lambda}\lambda^m}{m!}\, R^{*m}\,.$$

Using the Fubini Theorem, the calculation of the characteristic function β of Q in terms of the characteristic function γ of R, and thus in terms of R itself, is easy:

$$\beta(u) = \sum_{m=0}^{\infty} \frac{e^{-\lambda}\lambda^m}{m!}\gamma^m(u) = e^{-\lambda(1-\gamma(u))} = \exp\left(-\int_{\mathbb{R}\setminus\{0\}} (1 - e^{iuy})\,(\lambda R)(dy)\right).$$

We see that Q has a compound Poisson distribution. In this example, unlike Example 1, we have obtained a formula for Q itself, not just a formula for its characteristic function. Of course, this formula is also valid for Example 1.

We have seen that compound Poisson distributions can be represented as a limits of sequences of distributions of sums of independent generalized Poisson distributions, and also as the sum of a Poisson number of iid random variables. We have used characteristic functions to show the equivalence of these two types of representations. The following problem asks for a more direct approach.

Problem 4. Use Problem 18 of Chapter 10, rather than characteristic functions, to relate the two types of representations just described.

Problem 5. [Two-sided Poisson distributions] Let X and Y be iid standard Poisson random variables with mean ρ. Show that: (i) the distribution of $X - Y$ is given by

$$P[X - Y = k] = e^{-2\rho} I_{|k|}(2\rho)\,, \quad k \in \mathbb{Z}\,,$$

where

(16.4) $$I_r(x) = \sum_{s=0}^{\infty} \frac{1}{s!\,\Gamma(s+r+1)}\left(\frac{x}{2}\right)^{r+2s} \quad r \neq -1, -2, -3, \ldots,$$

with Γ denoting the gamma function; (ii) the characteristic function of the standard *two-sided Poisson* random variable $X - Y$ is the function $u \rightsquigarrow e^{-2\rho(1-\cos xu)}$; and (iii) $X - Y$ is compound Poisson. Also, identify the measure ν, using the notation of (16.1). The functions I_r are called *modified Bessel functions of the first kind*.

* **Problem 6.** For $\rho = 1$ in Problem 5 use a table of Bessel functions to approximate to three decimal places the density of $X - Y$ with respect to counting measure.

Problem 7. Find a formula for the distribution of the difference of two independent standard Poisson random variables. Express the answer in terms of modified Bessel functions of the first kind and the means, not necessarily identical, of the two Poisson random variables. Also show that the difference is compound Poisson and, in the notation of (16.1), find the measure ν.

16.2. Infinitely divisible distributions on $\mathbb{R}$

We begin with a few examples involving stability and infinite divisibility. The major issue treated in this section is that of identifying in an explicit manner all infinitely divisible distributions on $\mathbb{R}$.

Example 3. The characteristic functions of the Cauchy distributions symmetric about 0 are the functions $u \rightsquigarrow e^{-a|u|}$ for positive a. Not only is each nth root of such a characteristic function another such characteristic function, thus establishing the infinite divisibility of the Cauchy distributions, but all these characteristic functions are of the same strict type, thus establishing the strict stability of the Cauchy distributions symmetric about 0.

Problem 8. Show that every distribution of the same type as those in the preceding example is strictly stable. (Comment: The reason the phrase 'Cauchy type' is not used in this problem is that 'Cauchy' is often also used for certain asymmetric distributions.)

Problem 9. Which normal distributions are strictly stable, which are stable, and which are infinitely divisible?

In order not to give a misleading picture we turn to some infinitely divisible distributions that are not stable.

Example 4. The moment generating functions of the standard gamma distributions are the functions $v \rightsquigarrow (1 + v)^{-\gamma}$. We see that, for every n, the nth root of such a function is another function of the same form. This observation establishes the infinite divisibility of the gamma distributions.

Problem 10. Show that the gamma distributions are not stable.

The next two propositions indicate ways that 'new' infinitely divisible distributions can be obtained from 'old' infinitely divisible distributions. A similar approach was used in the first section to obtain compound Poisson distributions.

Proposition 1. *Linear combinations of independent infinitely divisible random variables are infinitely divisible.*

Problem 11. Prove the preceding proposition.

Proposition 2. *The limit of a sequence of infinitely divisible distributions is an infinitely divisible distribution.*

* **Problem 12.** Prove the preceding proposition.

When working with the characteristic function of an infinitely divisible distribution, it is often convenient to take the logarithm, which according to Problem 7 of Appendix E can be defined uniquely in any interval about 0 in which the characteristic function does not have the value 0. The following lemma says that this interval can be taken to be $\mathbb{R}$ when the characteristic function is infinitely divisible.

Proposition 3. *The complex number* 0 *does not belong to the image of any infinitely divisible characteristic function.*

PROOF. Let β denote an infinitely divisible characteristic function. For $n = 1, 2, \ldots$, let γ_n be a characteristic function having the property that $\gamma_n^n = \beta$. In an interval J about 0 in which β and, hence, γ_n are different from 0, we may take logarithms and divide by n:

$$(16.5) \qquad \log\big(\gamma_n(u)\big) = \frac{1}{n} \log\big(\beta(u)\big), \quad u \in J.$$

Let $n \to \infty$ to see that $\gamma_n(u) \to 1$ as $n \to \infty$ for $u \in J$. By Lemma 16 of Chapter 14, $\gamma_n(u) \to 1$ for every $u \in \mathbb{R}$. In particular, for every $u \in \mathbb{R}$ there exists an n such that $\gamma_n(u) \neq 0$. It follows that $\beta(u) \neq 0$ for every $u \in \mathbb{R}$. $\square$

We now know that J can be taken to equal $\mathbb{R}$ in (16.5), and thus we obtain the following corollary.

Corollary 4. *For any infinitely divisible characteristic function* β *and any positive integer* n, *there is exactly one characteristic function* γ_n *having the property that* $\gamma_n^n = \beta$.

In view of the preceding corollary we may, without ambiguity, write $\beta^{1/n}$ for the characteristic function called γ_n in the corollary, provided that β is an infinitely divisible characteristic function. If Q is an infinitely divisible distribution, there is, for each n, a unique distribution Q_n such that $Q = Q_n^{*n}$; we call Q_n the nth *convolution root* of Q. Proposition 3 and Corollary 4 have the following easy consequence.

Corollary 5. *As $n \to \infty$, the nth convolution root of an infinitely divisible distribution converges to the delta distribution at 0.*

The following proposition shows that compound Poisson distributions are more than just examples of infinitely divisible distributions.

Proposition 6. *Every infinitely divisible distribution is the limit of some sequence of compound Poisson distributions.*

PROOF. Let β be an infinitely divisible characteristic function and let Q_n denote the distribution corresponding to $\beta^{1/n}$. By Corollary 5, $\beta^{1/n}(u) \to 1$ as $n \to \infty$ for all $u \in \mathbb{R}$, so

$$\lim_{n \to \infty} \frac{\log \beta^{1/n}(u)}{-(1 - \beta^{1/n}(u))} = 1, \quad u \in \mathbb{R}.$$

Since $\log \circ \beta = n \log \circ \beta^{1/n}$ for all n, we see that

$$\begin{aligned}
\log(\beta(u)) &= \lim_{n \to \infty} n \log(\beta^{1/n}(u)) \\
&= -\lim_{n \to \infty} n(1 - \beta^{1/n}(u)) \\
&= -\lim_{n \to \infty} \int (1 - e^{iuy}) \, nQ_n(dy)
\end{aligned}$$

for all $u \in \mathbb{R}$. This last expression is the limit of a sequence of logarithms of compound Poisson characteristic functions, with nQ_n in the nth term of the sequence playing the role that ν plays in (16.1). $\square$

16.3. Lévy-Khinchin representations

As we work our way towards a characterization of infinitely divisible distributions, the function χ illustrated in Figure 16.1 will play a special role. Some flexibility in the definition of this function is possible, but once a choice has been made, it is best to stay with that choice. A formula for the choice we have made is $\chi(y) = (y \wedge 1) \vee (-1)$.

* **Problem 13.** Let $\eta \in \mathbb{R}$, $\sigma \in \mathbb{R}^+$, ν be a finite measure on $\mathbb{R} \setminus \{0\}$, and χ be as shown in Figure 16.1. Define $\psi \colon \mathbb{R} \to \mathbb{C}$ by

$$\psi(u) = -i\eta u + \frac{\sigma^2 u^2}{2} + \int_{\mathbb{R} \setminus \{0\}} \left(1 - e^{iuy} + iu\chi(y)\right) \nu(dy).$$

Explain why the function $\exp \circ (-\psi)$ is the characteristic function of an infinitely divisible distribution and why every compound Poisson characteristic function has this form.

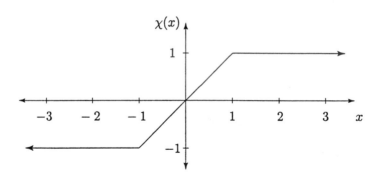

FIGURE 16.1. The function χ

We will eventually generalize the preceding problem to encompass all infinitely divisible distributions. The first step in that direction is a definition. A measure ν on $\mathbb{R} \setminus \{0\}$ is called a *Lévy measure* if

$$(16.6) \qquad \int_{\mathbb{R} \setminus \{0\}} (y^2 \wedge 1)\, \nu(dy) < \infty.$$

The next lemma establishes a useful connection, via Radon-Nikodym derivatives, between Lévy measures and general finite measures on $\mathbb{R}$.

Lemma 7. *The relation*

$$(16.7) \qquad \frac{d\zeta}{d(\sigma^2 \delta_0 + \nu)}(y) = \begin{cases} 1 & \text{if } y = 0 \\ 6\left(1 - \frac{\sin y}{y}\right) & \text{if } y \neq 0, \end{cases}$$

where δ_0 denotes the delta distribution at 0, establishes a one-to-one correspondence between the set of all finite measures ζ on $\mathbb{R}$ and the set of all pairs (σ, ν) where σ is a nonnegative real number and ν is a Lévy measure.

PROOF. Let φ be the function on the right of (16.7). Since φ is nonnegative and finite everywhere, (16.7) defines a σ-finite measure ζ whenever ν is a σ-finite measure and $\sigma \geq 0$, by Proposition 17 of Chapter 8. Thus ζ is determined by the pair (σ, ν).

For the other direction, we use the Reciprocal Rule for densities (Problem 33 of Chapter 8) to solve (16.7) for the density of $\sigma^2 \delta_0 + \nu$ with respect to ζ:

$$\frac{d(\sigma^2 \delta_0 + \nu)}{d\zeta}(y) = \begin{cases} 1 & \text{if } y = 0 \\ \frac{1}{\varphi(y)} & \text{if } y \neq 0. \end{cases}$$

Thus, the measure $\sigma^2 \delta_0 + \nu$ is determined by ζ. Since δ_0 and ν are mutually singular, ζ determines σ and ν. Since the density is everywhere finite, ν is σ-finite if ζ is σ-finite.

It remains to show that if ζ is finite, then the measure ν determined by ζ is a Lévy measure, and if ν is a Lévy measure and $\sigma \geq 0$, then the measure ζ determined by (σ, ν) is finite. Both of these statements follow immediately from the definition of Lévy measure and the fact that, by the l'Hospital Rule, the quotient

$$y \rightsquigarrow \frac{y^2 \wedge 1}{6\left(1 - \frac{\sin y}{y}\right)}$$

is bounded away from 0 and from ∞. $\square$

Lemma 8. *Let η be a real constant, ζ a finite measure on $\mathbb{R}$, and χ the function in Figure 16.1. Define the function ψ by*

$$(16.8) \qquad \psi(u) = -i\eta u + \int \left\{ \begin{array}{ll} u^2/2 & \text{if } y = 0 \\ \dfrac{1 - e^{iuy} + iu\chi(y)}{6(1 - \frac{\sin y}{y})} & \text{if } y \neq 0 \end{array} \right\} \zeta(dy) .$$

Then $\exp \circ (-\psi)$ is an infinitely divisible characteristic function.

PROOF. The result is obvious if ζ is the zero measure. For other ζ, we define finite measures ζ_k, $k = 1, 2, \ldots$, on $\mathbb{R}$ by

$$\zeta_k(B) = \zeta\left(B \cap \{y : y = 0 \text{ or } |y| \geq \tfrac{1}{k}\}\right) ,$$

noting that, for sufficiently large k, ζ_k is not the zero measure. Define ψ_k by (16.8) with ζ_k in place of ζ. A Dominated Convergence argument shows that ψ is continuous and Lemma 7 and Problem 13 imply that $\exp \circ (-\psi_k)$ is an infinitely divisible characteristic function for each k. We will finish the proof by showing that, for each u, $\psi_k(u) \to \psi(u)$ as $k \to \infty$, for then it will follow by the Continuity Theorem and Proposition 2 that $\exp \circ (-\psi)$ is an infinitely divisible characteristic function.

As $k \to \infty$, the sequence $(\zeta_k(\mathbb{R}))$ of numbers converges to the number $\zeta(\mathbb{R})$ by Continuity of Measure, and the sequence $(\frac{1}{\zeta_k(\mathbb{R})} \zeta_k)$ of probability measures converges to the probability measure $\frac{1}{\zeta(\mathbb{R})} \zeta$ by the Portmanteau Theorem. It follows that $\int g \, d\zeta_k \to \int g \, d\zeta$ as $k \to \infty$ for any bounded continuous g. For fixed u the integrand in (16.8) is a bounded continuous function. $\square$

The next lemma is called a lemma rather than a theorem only because it is incorporated in a later theorem.

Lemma 9. *Let η be a real number, σ a nonnegative real number, ν a Lévy measure, and χ the function shown in Figure 16.1. Define ψ by*

$$(16.9) \qquad \psi(u) = -i\eta u + \frac{\sigma^2 u^2}{2} + \int_{\mathbb{R} \setminus \{0\}} \left(1 - e^{iuy} + iu\chi(y)\right) \nu(dy) .$$

Then $\exp \circ (-\psi)$ is an infinitely divisible characteristic function.

PROOF. Apply Lemma 7 and Lemma 8. $\square$

* **Problem 14.** Let X be a random variable having the characteristic function given in the preceding lemma. Prove that

$$E(X) = \eta + \int_{\mathbb{R}\setminus\{0\}} [y - \chi(y)]\, \nu(dy)$$

in the sense that if either side exists, then so does the other and they are equal. Show also that if $E(|X|) < \infty$, then

$$\mathrm{Var}(X) = \sigma^2 + \int_{\mathbb{R}\setminus\{0\}} y^2\, \nu(dy)\,.$$

Lemma 9 identifies a certain class of infinitely divisible distributions. The goal of the next two lemmas is a proof that such an infinitely divisible distribution can arise from only one triple (η, σ, ν). It will be seen near the end of this section that this class includes all infinitely divisible distributions.

Lemma 10. *Let ψ be related to a and ζ as in Lemma 8. Then*

$$(16.10) \qquad \zeta(\mathbb{R}) = 3 \int_0^1 [\psi(v) + \psi(-v)]\, dv$$

and, if $\zeta(\mathbb{R}) \neq 0$,

$$(16.11) \quad \int e^{iuy} \frac{1}{\zeta(\mathbb{R})}\, \zeta(dy) = \frac{3}{\zeta(\mathbb{R})} \int_0^1 [\psi(u+v) - 2\psi(u) + \psi(u-v)]\, dv\,.$$

PROOF. From (16.9) we obtain

$$\psi(u+v) - 2\psi(u) + \psi(u-v) = \sigma^2 v^2 + 2 \int_{\mathbb{R}\setminus\{0\}} e^{iuy}(1 - \cos vy)\, \nu(dy)\,.$$

Multiplication by 3 and integration on v from 0 to 1 using the Fubini Theorem gives

$$3 \int_0^1 [\psi(u+v) - 2\psi(u) + \psi(u-v)]\, dv$$
$$= \sigma^2 + \int_{\mathbb{R}\setminus\{0\}} e^{iuy}\, 6(1 - \tfrac{\sin y}{y})\, \nu(dy) = \int e^{iuy}\, \zeta(dy)\,,$$

from which (16.10) and (16.11) follow. $\square$

Lemma 11. *The mapping $(\eta, \sigma, \nu) \rightsquigarrow \psi$ defined by (16.9), where $\eta \in \mathbb{R}$, $\sigma \in \mathbb{R}^+$, and ν is a Lévy measure, is one-to-one.*

PROOF. Let ζ correspond to (σ, ν), as in Lemma 7, in which case that lemma and the Reciprocal Rule for densities implies that ψ is given by (16.8). Thus, if we can show that ψ determines ζ, it will follow from (16.8) that it determines η and from Lemma 7 that it determines σ and ν, thus establishing one-to-oneness.

That the real number $\zeta(\mathbb{R})$ is determined by ψ is an immediate consequence of (16.10). If $\zeta(\mathbb{R}) \neq 0$, the probability measure $\frac{1}{\zeta(\mathbb{R})}\zeta$ is determined by its characteristic function, which by (16.11) is determined by ψ. Hence, ζ itself is determined by ψ. □

By Proposition 2 limits of sequences of infinitely divisible distributions are infinitely divisible. In view of the yet to be proven fact that every infinitely divisible distribution corresponds to a triple (η, σ, ν) via the relations in Lemma 9, it is natural to look for convergence conditions in terms of such triples.

Theorem 12. *Let $((\eta_n, \sigma_n, \nu_n),\ n = 1, 2, \ldots)$ be a sequence of triples such that, for each n, $\eta_n \in \mathbb{R}$, $\sigma_n \in \mathbb{R}^+$, and ν_n is a Lévy measure. For each n, let Q_n be the infinitely divisible distribution corresponding to $(\eta_n, \sigma_n, \nu_n)$ via Lemma 9. Then the sequence $(Q_n \colon n = 1, 2, \ldots)$ converges if and only if there exist $\eta \in \mathbb{R}$, $\sigma \in \mathbb{R}^+$, and a Lévy measure ν for which the following three conditions all hold:*

$$\nu(B) = \lim_{n\to\infty} \nu_n(B) \quad \text{for closed intervals } B \text{ such that } 0 \notin B \text{ and } \nu(\partial B) = 0\,;$$

$$\sigma^2 = \lim_{\varepsilon \searrow 0} \limsup_{n\to\infty} \left(\sigma_n^2 + \int_{[-\varepsilon,\varepsilon]\setminus\{0\}} y^2\, \nu_n(dy) \right)$$

$$= \lim_{\varepsilon \searrow 0} \liminf_{n\to\infty} \left(\sigma_n^2 + \int_{[-\varepsilon,\varepsilon]\setminus\{0\}} y^2\, \nu_n(dy) \right)\,;$$

$$\eta = \lim_{n\to\infty} \eta_n\,.$$

In case these conditions are satisfied the limit of the sequence $(Q_n \colon n \geq 1)$ is the infinitely divisible distribution corresponding to (η, σ, ν) via Lemma 9.

Problem 15. Prove the preceding theorem. *Hint:* Keep in mind that the assertion that when the limit exists, it has the form described in Lemma 9, is part of the conclusion, not one of the hypotheses.

Problem 16. Let X_λ denote a standard Poisson random variable having mean λ, and let Y be a normal random variable having mean 0 and variance 1. Use the preceding theorem to prove that

$$\frac{X_\lambda - \lambda}{\sqrt{\lambda}} \xrightarrow{\mathcal{D}} Y$$

as $\lambda \to \infty$.

Here is the theorem, promised earlier, that characterizes infinitely divisible distributions.

Theorem 13. [Lévy-Khinchin Representation] *In the $\mathbb{R}$-setting there is a one-to-one correspondence between the set of triples (η, σ, ν), where $\eta \in \mathbb{R}$, $\sigma \in \mathbb{R}^+$, and ν is a Lévy measure, and the set of infinitely divisible characteristic functions, with the characteristic function corresponding to the triple (η, σ, ν)*

being given by $\exp \circ (-\psi)$, *where*

$$\psi(u) = -i\eta u + \frac{\sigma^2 u^2}{2} + \int_{\mathbb{R}\setminus\{0\}} \left(1 - e^{iuy} + iu\chi(y)\right) \nu(dy)$$

and χ is the function in Figure 16.1.

PROOF. That every (η, σ, ν) leads to an infinitely divisible characteristic function is the content of Lemma 9. That different triples give different characteristic functions is a consequence of Lemma 11, since Proposition 3 implies that an infinitely divisible characteristic function has a unique continuous logarithm whose value at 0 is 0. To see that every infinitely divisible distribution has the given form, note first that, by Problem 13, every compound Poisson characteristic function has the form described. Then, by Theorem 12, the limit of any sequence of compound Poisson characteristic functions has this form. By Proposition 6 every infinitely divisible characteristic function is such a limit. $\square$

The function ψ is called the *characteristic exponent* of the corresponding infinitely divisible distribution. The formula for characteristic exponents of infinitely divisible distributions simplifies nicely for those distributions that are also symmetric about 0.

Proposition 14. *An infinitely divisible distribution characterized by (η, σ, ν), via its Lévy-Khinchin representation, is symmetric about 0 if and only if $\eta = 0$ and ν is symmetric about 0, in which case its characteristic exponent is*

$$u \rightsquigarrow \frac{\sigma^2 u^2}{2} + 2 \int_{(0,\infty)} (1 - \cos uy) \, \nu(dy).$$

* **Problem 17.** Prove the preceding proposition.

16.4. Infinitely divisible distributions on $\mathbb{R}^+$

Nonnegative infinitely divisible distributions are particularly important and the relevant theory is somewhat simpler than the general theory for the $\mathbb{R}$-setting. The following problem is a natural place to begin.

Problem 18. [Compound Poisson Distributions on $\mathbb{R}^+$] Prove directly, without using a change of variables involving complex numbers, that if the measure ν in (16.1) is supported by $(0, \infty)$, then the corresponding compound Poisson distribution is supported by $\mathbb{R}^+$ and has moment generating function

(16.12) $$v \rightsquigarrow \exp\left(-\int_{(0,\infty)} (1 - e^{-vy}) \, \nu(dy)\right).$$

Problem 19. Let X have the distribution described in the preceding problem. Verify that $P[X = 0] = e^{-\nu(0,\infty)}$.

* **Problem 20.** Prove that every compound Poisson distribution supported by $[0, \infty)$ has the form (16.12); that is, show that the Lévy measure of a compound Poisson distribution supported by $[0, \infty)$ is supported by $(0, \infty)$.

The point of the next result, in comparison with Proposition 6, is that here the compound Poisson distributions may be chosen to have support $\mathbb{R}^+$.

Proposition 15. *Every infinitely divisible distribution supported by $\mathbb{R}^+$ is the limit of some sequence of compound Poisson distributions supported by $\mathbb{R}^+$.*

Problem 21. Prove the preceding proposition. *Hint:* See the proof of Proposition 6.

A measure ν on $(0, \infty)$ is a *Lévy measure* for $\mathbb{R}^+$ if

$$(16.13) \qquad \int_{(0,\infty)} (y \wedge 1) \, \nu(dy) < \infty.$$

By analogy with this terminology, the phrase 'for $\mathbb{R}$' may, for clarity, be adjoined to the term 'Lévy measure' when the discussion concerns infinitely divisible distributions on $\mathbb{R}$.

Corresponding to a Lévy measure ν for $\mathbb{R}^+$ and a constant $\xi \in \mathbb{R}^+$, not both zero, it is convenient to introduce a finite measure μ on $\mathbb{R}^+$ defined by

$$(16.14) \qquad \frac{d\mu}{d(\xi \delta_0 + \nu)} = \begin{cases} 1 & \text{if } y = 0 \\ 1 - e^{-y} & \text{if } y \in (0, \infty). \end{cases}$$

Problem 22. Mimic the development from Lemma 7 through Lemma 11 to show that, for each pair (ξ, ν), where $\xi \in \mathbb{R}^+$ and ν is a Lévy measure for $\mathbb{R}^+$, there exists a unique infinitely divisible distribution on $\mathbb{R}^+$, the moment generating function of which equals $\exp \circ (-\theta)$, where

$$\theta(v) = \xi v + \int_{(0,\infty)} (1 - e^{-vy}) \, \nu(dy).$$

The constant ξ is sometimes called the *shift* of the corresponding infinitely divisible distribution.

Problem 23. How much of the preceding problem could have been done easily using Theorem 27 of Chapter 14?

Problem 24. Let X be a finite infinitely divisible random variable corresponding to (ξ, ν) as in Problem 22. Show that

$$E(X) = \xi + \int_{(0,\infty)} y \, \nu(dy).$$

* **Problem 25.** By Problem 58 of Chapter 14 (or Example 4 of this chapter), gamma distributions are infinitely divisible. Find the shifts and Lévy measures of all gamma distributions with support equal to $\mathbb{R}^+$.

Here are the analogues for the $\mathbb{R}^+$-setting of Theorem 12 and Theorem 13.

Theorem 16. *Let* $((\xi_n, \nu_n),\ n = 1, 2, \ldots)$ *be a sequence of pairs such that, for each* n, $\xi_n \in \mathbb{R}^+$ *and* ν_n *is a Lévy measure for* $\mathbb{R}^+$. *For each* n, *let* Q_n *be the infinitely divisible distribution on* $\mathbb{R}^+$ *corresponding to* (ξ_n, ν_n) *via Problem 22. Then the sequence* $(Q_n : n = 1, 2, \ldots)$ *converges to a distribution on* $\mathbb{R}^+$ *if and only if there exist* $\xi \in \mathbb{R}^+$ *and a Lévy measure* ν *for* $\mathbb{R}^+$ *for which the following two conditions both hold:*

$$\nu[x, \infty) = \lim_{n \to \infty} \nu_n[x, \infty) \quad \text{if } 0 < x \text{ and } \nu\{x\} = 0;$$

$$\xi = \lim_{\varepsilon \searrow 0} \limsup_{n \to \infty} \left(\xi_n + \int_{(0,\varepsilon]} y\, \nu_n(dy) \right)$$
$$= \lim_{\varepsilon \searrow 0} \liminf_{n \to \infty} \left(\xi_n + \int_{(0,\varepsilon]} y\, \nu_n(dy) \right).$$

In case these conditions are satisfied, the limit of the sequence $(Q_n : n \geq 1)$ *is the infinitely divisible distribution corresponding to* (ξ, ν) *via Problem 22.*

Theorem 17. [Lévy-Khinchin Representation for $\mathbb{R}^+$] *In the* $\mathbb{R}^+$-*setting there is a one-to-one correspondence between the set of pairs* (ξ, ν), *where* $\xi \in \mathbb{R}^+$ *and* ν *is a Lévy measure for* $\mathbb{R}^+$, *and the set of infinitely divisible moment generating functions for* $\mathbb{R}^+$, *with the moment generating function corresponding to the pair* (ξ, ν) *being given by* $\exp \circ (-\theta)$, *where*

$$\theta(v) = \xi v + \int_{(0,\infty)} (1 - e^{-vy})\, \nu(dy).$$

Problem 26. Prove the preceding two theorems.

Problem 27. Prove that every infinitely divisible moment generating function for $\mathbb{R}^+$ has the form described in Theorem 27 of Chapter 14.

Problem 28. Discuss why it is that certain Lévy measures ν for $\mathbb{R}$ for which $\nu(-\infty, 0) = 0$ have not been designated as Lévy measures for $\mathbb{R}^+$.

16.5. Extension to $\overline{\mathbb{R}}^+$

In the $\mathbb{R}$-setting, the sequence of nth convolution roots of an infinitely divisible distribution converges to the delta distribution at 0 (see Corollary 5). In the $\overline{\mathbb{R}}^+$-setting we incorporate this property into the definition.

Definition 18. A distribution Q on $\overline{\mathbb{R}}^+$ is *infinitely divisible* if there exists a sequence $(Q_n : n = 1, 2, \ldots)$ of distributions on $\overline{\mathbb{R}}^+$ such that $Q = Q_n^{*n}$ for each n and $\lim_{n \to \infty} Q_n = \delta_0$, where δ_0 denotes the delta distribution at 0.

Problem 29. Prove that Q_n in the preceding definition is necessarily unique. *Hint:* Use moment generating functions.

Problem 30. Prove that there is only one distribution Q on $\overline{\mathbb{R}}^+$, namely the delta distribution at ∞, that satisfies the first condition in Definition 18 (that $Q_n^{*n} = Q$ for some distributions Q_n on $\overline{\mathbb{R}}^+$) but not the second condition (that $\lim_{n \to \infty} Q_n = \delta_0$).

Problem 31. Show that a generalized Poisson distribution with support equal to $\{0, \infty\}$ is infinitely divisible by finding an explicit formula for its convolution roots.

A measure ν on $(0, \infty]$ is a *Lévy measure* for $\overline{\mathbb{R}}^+$ if

$$(16.15) \qquad \int_{(0,\infty]} (y \wedge 1)\, \nu(dy) < \infty.$$

Problem 32. Carry out for the $\overline{\mathbb{R}}^+$-setting the analogue of the program requested in Problem 22 for the $\mathbb{R}^+$-setting.

It is possible for a sequence of infinitely divisible distributions on $\overline{\mathbb{R}}^+$ to converge to the delta distribution at ∞, which according to Definition 18 is not infinitely divisible. This fact makes the wording of an analogue of Theorem 16 slightly more complicated than that of Theorem 16 itself.

* **Problem 33.** State and prove an analogue of Theorem 16 for the $\overline{\mathbb{R}}^+$-setting.

Problem 34. State and prove an analogue of Theorem 17 for the $\overline{\mathbb{R}}^+$-setting.

16.6. The triangular array problem: introduction

The remainder of this chapter is concerned with the limiting behavior of sequences of distributions of sums of independent random variables.

Example 1 describes a situation in which a compound Poisson distribution arises naturally as a limit connected with sums of independent random variables. The result in that example differs from the Weak Law of Large Numbers or the Classical Central Limit Theorem, because it is not centering or scaling that enables one to obtain a limit, but rather a more complicated changing of the probability structure. A general framework designed to accommodate both types of settings is the focus of this section.

Let $X_{k,n}$, $1 \le k \le n$, $1 \le n < \infty$, be $\mathbb{R}$-valued random variables. Assume, for each n, that $(X_{k,n} \colon k = 1, \ldots, n)$ is an independent n-tuple and let $S_n = \sum_{k=1}^{n} X_{k,n}$. Our goal is to give a criterion for the sequence $(S_n \colon n = 1, 2, \ldots)$ to converge in distribution, and, in the case of convergence, to identify the limit. It is common to depict $(X_{k,n} \colon 1 \le k \le n\, , n = 1, 2, \ldots)$ as follows:

$$
\begin{array}{llll}
X_{1,1} & & & \\
X_{1,2} & X_{2,2} & & \\
X_{1,3} & X_{2,3} & X_{3,3} & \\
\vdots & \vdots & \vdots & \ddots
\end{array}
$$

For this reason, it is called a *triangular array*, the independence described above is called *row-wise independence*, and the sums S_n are called *row sums*. Since the issue is that of convergence in distribution of the sequence of row sums, the joint distributions of random variables in different rows are irrelevant. In particular, the random variables in different rows need not be defined on the same probability space.

Sometimes one wants to consider a generalization in which the requirement that there be exactly n entries in the nth row of the array be dropped. One can easily turn such an array of random variables into a triangular array by first duplicating some rows to obtain an array in which none of the rows are too long, and then by inserting 0's into rows that are too short.

The *triangular array problem* is this: given a row-wise independent triangular array, characterize the limiting behavior of the sequence of distributions of the row sums in terms of data about the distributions of the row summands. In the succeeding sections of this chapter, we give a fairly detailed solution to this problem. In the remainder of this section, we introduce some simple concepts which will play an important role in this solution.

It turns out that the best way to approach the triangular array problem is via characteristic functions (or moment generating functions in the $\mathbb{R}^+$-setting). The characteristic function of the row sum is the product of the characteristic functions of the row summands, and as we have already seen, it is useful to take logarithms when dealing with such products. Of course, the logarithm of a characteristic function must be carefully defined because of the singularity of the logarithm at 0. This difficulty does not occur when dealing with infinitely divisible distributions, as we have already seen when we defined characteristic exponents. In the following paragraphs, we generalize the definitions of characteristic exponent.

Let Q be a distribution on $\mathbb{R}$ with characteristic function β, and let J_β be the largest open interval containing 0 on which β is never 0. The *characteristic exponent* of Q (and of β) is the function

$$
u \rightsquigarrow -\log \circ \beta(u)\,, \quad u \in J_\beta\,,
$$

where $\log \circ \beta$ denotes the unique continuous logarithm of β on J_β whose value at 0 is 0 (see Problem 7 of Appendix E). The following result is an easy consequence of the Continuity Theorem of Chapter 14 and Problem 8 of Appendix E:

Proposition 19. *Let $(Q_n \colon n = 1, 2, \dots)$ be a sequence of distributions on $\mathbb{R}$, and let $(\psi_n \colon n = 1, 2, \dots)$ be the corresponding sequence of characteristic exponents. If $Q_n \to Q$ as $n \to \infty$, where Q is a distribution on $\mathbb{R}$ with characteristic exponent ψ, then $\psi_n(u) \to \psi(u)$ for all u in the domain of ψ. Moreover, the convergence is uniform on compact subsets of the domain of ψ.*

We have one other issue to discuss now. It turns out that this issue is not relevant for triangular arrays in which the summands within each given row are identically distributed, so the reader who is only interested in that special case may skip ahead to the next section.

For the general triangular array problem, we will not be concerned with cases in which a relatively small number of summands in a row significantly affect the distribution of the row sum, so we wish to introduce a condition designed to eliminate such examples. We have already encountered a special case of this condition in Example 1, where it was assumed that $\max\{p_{k,n} \colon 1 \le k \le n\} \to 0$. For an arbitrary triangular array $(X_{k,n} \colon k \le n = 1, 2, \dots)$ the condition is

$$\lim_{n\to\infty} \sup_k P[|X_{k,n}| > \varepsilon] = 0 \quad \text{for all } \varepsilon > 0 \,.$$

A triangular array satisfying this assumption is *uniformly asymptotically negligible*, abbreviated *uan*. Typically, we use adjectives introduced for random variables also for entities that are associated with the random variables; for example, we speak of a 'uan triangular array of characteristic functions'.

If the uan condition is not assumed in a general triangular array, any limiting distribution R is possible, since one summand in each row could have the distribution R, and the other summands could all equal 0. When the uan condition is in force, a fixed finite number of summands in each row cannot influence the limiting distribution of the row sums. We will find that, under the uan condition, only infinitely divisible distributions are possible as limits.

As mentioned earlier, characteristic exponents will play an important role in our solution of the triangular array problem. The following lemma shows that the uan condition is just right for ensuring that the characteristic exponents of the row summands and row sums are defined on arbitrarily large domains as $n \to \infty$.

Lemma 20. *Let $(\beta_{k,n} \colon k \le n = 1, 2, \dots)$ be a triangular array of characteristic functions. Then the following three statements are equivalent:*

(i) *the triangular array is uan;*
(ii) $\lim_{n\to\infty} \left(\max_{1 \le k \le n} |1 - \beta_{k,n}(u)| \right) = 0$ *for each u;*
(iii) $\lim_{n\to\infty} \left(\max_{1 \le k \le n} \max\{|1 - \beta_{k,n}(u)| \colon |u| \le w\} \right) = 0$
$\qquad$ *for each $w > 0$;*

(iv) $\lim_{n\to\infty}\left(\max_{1\leq k\leq n}\max\{|\log\circ\beta_{k,n}(u)|:|u|\leq w\}\right)=0$
for each $w>0$.

Problem 35. Prove the preceding lemma. *Hint:* Use Lemma 11 of Chapter 14.

In addition to helping us deal with the domains of characteristic exponents, Lemma 20 is also quite useful for approximating these functions by using Taylor series. Here is analogue of Lemma 20 for moment generating functions.

Lemma 21. *Let $(\varphi_{k,n}:k\leq n=1,2,\ldots)$ be a triangular array of moment generating functions for either the $\mathbb{R}^+$- or $\overline{\mathbb{R}}^+$-setting. Then the following three statements are equivalent:*

(i) *the triangular array is uan;*
(ii) $\lim_{n\to\infty}\left(\max_{1\leq k\leq n}(1-\varphi_{k,n}(v))\right)=0$ *for each v;*
(iii) $\lim_{n\to\infty}\left(\max_{1\leq k\leq n}\max\{(1-\varphi_{k,n}(v)):v\leq w\}\right)=0$
for each $w>0$;
(iv) $\lim_{n\to\infty}\left(\max_{1\leq k\leq n}\max\{-\log\circ\varphi_{k,n}(u):|u|\leq w\}\right)=0$
for each $w>0$.

Problem 36. Prove the preceding lemma.

In the next section, we will solve the triangular array problem in the special case that the summands within each row are identically distributed. In the section following that, two more special cases are considered: summands that are symmetrically distributed about 0 and nonnegative summands. In the final section of the chapter, the general triangular array problem is solved.

16.7. Iid triangular arrays

Let $(Q_n:n=1,2,\ldots)$ be a sequence of distributions on $\mathbb{R}$, and consider a row-wise independent triangular array in which each summand in the nth row has distribution Q_n. Such a triangular array is an *iid triangular array*. The distribution of the nth row sum is Q_n^{*n}, and the triangular array problem in the iid case is to determine the possible limits of the sequence $(Q_n^{*n}:n=1,2,\ldots)$. The following theorem solves this problem.

Theorem 22. [iid case] *Let $(Q_n:n\geq 1)$ be a sequence of distributions on $\mathbb{R}$. In order that the sequence $(Q_n^{*n}:n\geq 1)$ converge it is necessary and sufficient that there exist a triple (η,σ,ν), where $\eta\in\mathbb{R}$, $\sigma\in\mathbb{R}^+$, and ν is a Lévy measure,* *satisfying the following three conditions:*

(16.16)
$$\nu(B) = \lim_{n\to\infty} (nQ_n)(B) \quad \text{for closed intervals } B$$
$$\text{such that } 0 \notin B \text{ and } \nu(\partial B) = 0\,;$$

(16.17)
$$\sigma^2 = \lim_{\varepsilon\searrow 0} \limsup_{n\to\infty} \int_{[-\varepsilon,\varepsilon]} x^2\,(nQ_n)(dx)$$
$$= \lim_{\varepsilon\searrow 0} \liminf_{n\to\infty} \int_{[-\varepsilon,\varepsilon]} x^2\,(nQ_n)(dx)\,;$$

(16.18)
$$\eta = \lim_{n\to\infty} \int \chi(x)\,(nQ_n)(dx)\,.$$

If these conditions are satisfied, the limit of the sequence $(Q_n^{*n} : n \geq 1)$ is the infinitely divisible distribution corresponding to (η, σ, ν) via the Lévy-Khinchin Representation Theorem, and $(Q_n : n \geq 1)$ converges to the delta distribution at 0.

PROOF. Let ψ_n denote the characteristic exponent of Q_n. Suppose that the sequence $(Q_n^{*n} : n \geq 1)$ converges to some limit Q with characteristic exponent ψ. By Proposition 19, $\psi(u) = \lim n\psi_n(u)$ for each u in the domain of ψ. Since $n \to \infty$, it must be that $\psi_n(u) \to 0$ for such u and hence that $(Q_n : n \geq 1)$ converges to the delta distribution at 0 (Lemma 16 of Chapter 14). In terms of Q_n we have

$$\psi(u) = \lim_{n\to\infty} -n \log\left(\int e^{iux} Q_n(dx)\right)$$
$$= \lim_{n\to\infty} -n \log\left(1 - \int (1 - e^{iux}) Q_n(dx)\right).$$

The logarithm must be defined for all sufficiently large n (possibly depending on u) and approach 0 as $n \to \infty$. Therefore, the limit is the same as that obtained by replacing each logarithm by the first term of its Taylor series:

(16.19)
$$\psi(u) = \lim_{n\to\infty} \int (1 - e^{iux})\,(nQ_n)(dx)\,.$$

Let J be the domain of ψ. We wish to show that $J = \mathbb{R}$. We will do so by showing that J cannot have any finite endpoints. By (16.19),

$$\mathfrak{R}(\psi(u)) = \lim_{n\to\infty} \int (1 - \cos(ux))(nQ_n)(dx)\,.$$

Since $1 - \cos 2\theta \leq 4(1 - \cos\theta)$ for all $\theta \in \mathbb{R}$ (see Problem 37 below), it follows that $\mathfrak{R}(\psi(2u)) \leq 4\mathfrak{R}(\psi(u))$ for all $u \in \mathbb{R}$. Since ψ is continuous on the interior of J, this last inequality implies that ψ cannot become unbounded near any finite endpoint of J. But the definition of J implies that ψ must become unbounded near any finite endpoint of J. Therefore, J cannot have any finite endpoints.

We now rewrite (16.19) using the function χ. Let $\eta_n = \int \chi(x)\,(nQ_n)(dx)$. Then

$$(16.20) \quad \psi(u) = \lim_{n \to \infty} \left(-i\eta_n u + \int_{\mathbb{R}\setminus\{0\}} \left(1 - e^{iux} + iu\chi(x) \right)(nQ_n)(dx) \right)$$

for all $u \in \mathbb{R}$. Let ν_n be the restriction of nQ_n to $\mathbb{R}\setminus\{0\}$. Then (16.20) expresses ψ as the limit of a sequence of characteristic exponents of infinitely divisible distributions, the nth one of which is characterized by the triple $(\eta_n, 0, \nu_n)$. An application of Theorem 12 shows that Q is infinitely divisible and characterized by the triple (η, σ, ν), where ν, σ, and η are defined by (16.16), (16.17), and (16.18).

For the converse suppose that (16.16), (16.17), and (16.18) hold for some $\eta \in \mathbb{R}$, $\sigma \in \mathbb{R}^+$, and Lévy measure ν, and let ψ denote the characteristic exponent corresponding to (η, σ, ν) via the Lévy-Khinchin Representation Theorem. As above, let ν_n be the restriction to $\mathbb{R}\setminus\{0\}$ of nQ_n. By Theorem 12, ψ is the pointwise limit of the sequence of characteristic exponents corresponding to the triples $(\eta_n, 0, \nu_n)$, where $\eta_n = \int \chi(x)\,\nu_n(dx)$. That is,

$$\psi(u) = \lim_{n \to \infty} \left(-i\eta_n u + \int_{\mathbb{R}\setminus\{0\}} \left(1 - e^{iux} + iu\chi(x) \right) \nu_n(dx) \right)$$

$$(16.21) \qquad = \lim_{n \to \infty} n\left(1 - \int e^{iux} Q_n(dx) \right);$$

moreover the convergence is uniform for u in any bounded interval, as it always is for convergence of sequences of characteristic functions to a characteristic function. In any such interval, the quantity within parentheses in (16.21) must be uniformly close to 0 and thus, for n sufficiently large, the last integral must have positive real part. For such n, the logarithm of the integral must have imaginary part between $-\frac{\pi}{2}$ and $\frac{\pi}{2}$ and thus be close to 0 when the quantity within parentheses is close to 0. Accordingly,

$$\psi(u) = \lim_{n \to \infty} -n \log\left(\int e^{iux} Q_n(dx) \right) = \lim_{n \to \infty} n\psi_n(u).$$

Therefore, $Q_n^{*n} \to Q$, as desired.

(Comment: The reason for the fuss in the latter portion of this proof is the fact that it is not generally true that characteristic exponents are close to 0 when characteristic functions are close to 1. That a characteristic function is close to 1 at some point only forces the corresponding characteristic exponent to be close to an integral multiple of $2\pi i$.) $\square$

Problem 37. Prove the inequality $1 - \cos 2\theta \leq 4(1 - \cos\theta)$, $\theta \in \mathbb{R}$. *Hint:* For $\theta \in [0, \pi]$, compare derivatives.

Corollary 23. [iid case] *In order that the sequence* $(Q_n^{*n}: n \geq 1)$ *converge it is necessary and sufficient that the following three conditions all hold:*

$$(16.22) \qquad \lim_{y \to \infty} \limsup_{n \to \infty} n Q_n(\{x: |x| > y\}) = 0;$$

$$(16.23) \qquad \eta = \lim_{n \to \infty} n \int \chi(x) \, Q_n(dx)$$

for some $\eta \in \mathbb{R}$;

$$(16.24) \qquad \sigma^2 \delta_0[x,y] + \int_{[x,y] \setminus \{0\}} t^2 \, \nu(dt) = \lim_{n \to \infty} n \int_{[x,y]} t^2 \, Q_n(dt)$$
$$\text{if } x, y \neq 0 \text{ and } \nu\{x\} = \nu\{y\} = 0$$

for some $\sigma \in \mathbb{R}^+$ *and Lévy measure* ν, *where* δ_0 *denotes the delta distribution at* 0. *If these conditions are satisfied, the limit of the sequence* $(Q_n^{*n}: n \geq 1)$ *is the infinitely divisible distribution corresponding to* (η, σ, ν).

Problem 38. Prove the preceding corollary.

Problem 39. Use the preceding theorem or corollary to redo Example 1.

Problem 40. Study the limiting behavior of the sequence of row sums of the row-wise independent triangular array of random variables $X_{k,n}$ having distributions $Q_{k,n}$ given by $Q_{k,n}\{1\} = \frac{1}{n}$ and $Q_{k,n}\{\frac{1}{n}\} = \frac{n-1}{n}$. Approach the problem by using Theorem 22 and then again by using moment generating functions.

* **Problem 41.** Define Q_n by $Q_n\{\frac{1}{n}\} = Q_n\{-\frac{1}{n}\} = \frac{n-1}{2n}$, $Q_n\{c\} = Q_n\{-c\} = \frac{1}{2n}$, for some constant $c > 1$. Use Theorem 22 to decide if the sequence $(Q_n^{*n}: n \geq 1)$ converges, and if so, identify the limit.

* **Problem 42.** Define Q_n by $Q_n\{1\} = Q_n\{-1\} = \frac{1}{2n}$, $Q_n\{n^{-1/2}\} = Q_n\{-n^{-1/2}\} = \frac{n-1}{2n}$. Decide if the sequence $(Q_n^{*n}: n \geq 1)$ converges, and if so, find the Lévy-Khinchin representation of the limit.

Problem 43. Define Q_n by $Q_n\{0\} = \frac{1}{2}$ and

$$Q_n\left\{\frac{1}{(2 + (-1)^n)n^2}\right\} = \frac{1}{2}.$$

Decide if $\lim Q_n^{*n}$ exists, and if so, find its Lévy-Khinchin representation.

Problem 44. Define Q_n by $Q_n\{0\} = \frac{1}{2}$ and

$$Q_n\left\{\frac{1}{(2 + (-1)^n)n}\right\} = \frac{1}{2}.$$

Decide if $\lim Q_n^{*n}$ exists, and if so, find its Lévy-Khinchin representation.

Problem 45. Define Q_n by $Q_n\{0\} = \frac{1}{2}$ and

$$Q_n\left\{\frac{1}{(2 + (-1)^n)n}\right\} = Q_n\left\{-\frac{1}{(2 + (-1)^n)n}\right\} = \frac{1}{4}.$$

Decide if $\lim Q_n^{*n}$ exists, and if so, find its Lévy-Khinchin representation.

Problem 46. Let Q_n have density with respect to Lebesgue measure λ given by

$$\frac{dQ_n}{d\lambda}(x) = \frac{1}{n}\left((1 - |x|) \vee 0\right) + \frac{n-1}{\sqrt{n}}e^{-2\sqrt{n}\,|x|}.$$

Decide if $\lim Q_n^{*n}$ exists, and if so, find it characteristic exponent.

Problem 47. Apply Theorem 22 to the sequence (Q_n) defined by

$$Q_n\{2^{-m} + (-8)^{-n}\} = \frac{1}{n}, \quad 1 \leq m \leq n.$$

If $\lim Q_n^{*n}$ exists, find its characteristic exponent.

Problem 48. Let $(Q_n : n \geq 1)$ be a sequence of distributions on $\mathbb{R}$ such that $\lim_{n\to\infty} Q_n^{*n}$ exists. Denote the Lévy measure of the limit by ν. Let c be a positive constant for which $\nu\{-c, c\} = 0$. Let Y_n denote the number of random variables in the nth row of a corresponding triangular array which have absolute value larger than c. Prove that $(Y_n : n \geq 1)$ converges in distribution to a standard Poisson random variable with mean $\nu\{y : |y| > c\}$.

Problem 49. Investigate the question: Can the hypothesis $\nu\{-c, c\} = 0$ be dropped from the preceding problem?

16.8. Symmetric and nonnegative triangular arrays

In this section we drop the assumption that all the distributions in each row of a triangular array are the same, but we specialize in other ways that make the triangular array problem quite simple to treat. Two cases are considered: all distributions symmetric about 0, and all distributions supported by $\mathbb{R}^+$. In each case the relevant theorem can be proved along the lines of the proof of Theorem 22. First, the special nature of each case and Lemma 20 or Lemma 21 justify the use of first-term Taylor approximations, as in the argument leading to (16.19). Then it can be seen (as in (16.20)) that such approximations are actually the negative logarithms of infinitely divisible characteristic functions or moment generating functions, to which either Theorem 12 or Theorem 16 can be applied.

It may be helpful to again read Proposition 14 about infinitely divisible distributions that are symmetric about 0 before examining the next theorem.

Theorem 24. [symmetric case] *Let* $(Q_{k,n} : 1 \leq k \leq n, n = 1, 2, \ldots)$ *be a uan triangular array of distributions on* $\mathbb{R}$, *each of which is symmetric about* 0. *For each* n, *let*

$$Q_n = Q_{1,n} * Q_{2,n} * \cdots * Q_{n,n}.$$

In order that the sequence $(Q_n: n = 1, 2, \ldots)$ *converge it is necessary and sufficient that there exist a nonnegative number* σ *and a Lévy measure* ν, *symmetric about 0, satisfying the following two conditions:*

$$(16.25) \qquad \nu[x, \infty) = \lim_{n \to \infty} \sum_{k=1}^{n} Q_{k,n}[x, \infty) \quad \text{if } 0 < x \text{ and } \nu\{x\} = 0;$$

$$(16.26) \qquad \begin{aligned} \sigma^2 &= \lim_{\varepsilon \searrow 0} \limsup_{n \to \infty} \sum_{k=1}^{n} \int_{[-\varepsilon, \varepsilon]} x^2 \, Q_{k,n}(dx) \\ &= \lim_{\varepsilon \searrow 0} \liminf_{n \to \infty} \sum_{k=1}^{n} \int_{[-\varepsilon, \varepsilon]} x^2 \, Q_{k,n}(dx). \end{aligned}$$

In case these conditions are satisfied, the sequence $(Q_n: n = 1, 2, \ldots)$ *converges to the infinitely divisible distribution corresponding to* $(0, \sigma, \nu)$ *via the Lévy-Khinchin Representation Theorem.*

* Problem 50. Prove the preceding theorem.

Theorem 25. [nonnegative case] *Let* $(Q_{k,n}: 1 \le k \le n, n = 1, 2, \ldots)$ *be a uan triangular array of distributions on* $\mathbb{R}^+$. *For each* n, *let*

$$Q_n = Q_{1,n} * Q_{2,n} * \cdots * Q_{n,n}.$$

In order that the sequence $(Q_n: n = 1, 2, \ldots)$ *converge to a distribution on* $\mathbb{R}^+$ *it is necessary and sufficient that there exist a nonnegative number* ξ *and a Lévy measure* ν *for* $\mathbb{R}^+$ *satisfying the following two conditions:*

$$(16.27) \qquad \nu[x, \infty) = \lim_{n \to \infty} \sum_{k=1}^{n} Q_{k,n}[x, \infty) \quad \text{if } 0 < x \text{ and } \nu\{x\} = 0;$$

$$(16.28) \qquad \begin{aligned} \xi &= \lim_{\varepsilon \searrow 0} \limsup_{n \to \infty} \sum_{k=1}^{n} \int_{(0, \varepsilon]} x \, Q_{k,n}(dx) \\ &= \lim_{\varepsilon \searrow 0} \liminf_{n \to \infty} \sum_{k=1}^{n} \int_{(0, \varepsilon]} x \, Q_{k,n}(dx). \end{aligned}$$

In case these conditions are satisfied, the sequence $(Q_n: n = 1, 2, \ldots)$ *converges to the infinitely divisible distribution on* $\mathbb{R}^+$ *corresponding to* (ξ, ν) *via the Lévy Khinchin Representation Theorem.*

Problem 51. Prove the preceding theorem.

Problem 52. Redo Problem 40 using Theorem 25.

Problem 53. Redo Problem 42 using Theorem 24.

* **Problem 54.** Let $Q_{k,n}$ be the standard exponential distribution with mean $\frac{1}{k+n}$. Decide if

$$\lim_{n\to\infty} (Q_{1,n} * Q_{2,n} * \cdots * Q_{n,n})$$

exists as a distribution on $\mathbb{R}^+$, and if so, find its moment generating function.

Problem 55. Redo Problem 43 and Problem 44 using Theorem 25.

Problem 56. Define $Q_{k,n}$ by $Q_{k,n}\{0\} = \frac{1}{2}$ and

$$Q_{k,n}\Big\{ \frac{1}{(2+(-1)^k)n} \Big\} = \frac{1}{2}.$$

Decide if $\lim(Q_{1,n} * Q_{2,n} * \cdots * Q_{n,n})$ exists, and if so, evaluate it.

Problem 57. For each $\alpha > 0$, check the uan condition and study the limiting behavior of the sequence of row sums of the row-wise independent triangular array of two-sided exponential random variables with mean 0 and the variance of the kth random variable in the nth row equaling $(k+n)^{-\alpha}$.

Problem 58. For $1 \le k \le n^2$, define $Q_{k,n}$ by

$$Q_{k,n}\{0\} = 1 - \frac{k^{1/2}}{2n^3}$$

$$\frac{dQ_{k,n}}{d\lambda}(x) = \Big[-\frac{x}{nk^{1/2}} + \frac{1}{n^2}\Big] \vee 0 \quad \text{if } x > 0,$$

where λ denotes Lebesgue measure. Study the limiting behavior of the sequence $(Q_{1,n} * \cdots * Q_{n^2,n}: n \ge 1)$ as $n \to \infty$. In particular, if the limit Q exists, calculate $Q\{0\}$.

* **Problem 59.** For $1 \le k \le n^2$, define $Q_{k,n}$ by

$$Q_{k,n}\{0\} = 1 - \frac{k^{1/2}}{2n^3}$$

$$\frac{dQ_{k,n}}{d\lambda}(x) = \Big[-\frac{|x|}{2nk^{1/2}} + \frac{1}{2n^2}\Big] \vee 0 \quad \text{if } x \ne 0,$$

where λ denotes Lebesgue measure. Study the limiting behavior of the sequence $(Q_{1,n} * \cdots * Q_{n^2,n}: n \ge 1)$ as $n \to \infty$. In particular, if the limit Q exists, calculate $Q\{0\}$.

Problem 60. Adapt Theorem 25 to the $\overline{\mathbb{R}}^+$-setting in order to give necessary and sufficient conditions for convergence to a limit different from the delta distribution at ∞, including in your statement an identification of the limit when it exists.

Problem 61. Define Q_n by $Q_n\{\frac{1}{n}\} = \frac{n-1}{n}$ and $Q_n\{n\} = \frac{1}{n}$. For the $\overline{\mathbb{R}}^+$-setting decide if there is a distribution Q different from the delta distribution at ∞ such that

$$Q = \lim_{n\to\infty} Q_n^{*n}.$$

If so find Q explicitly.

16.9. † General triangular arrays

Theorem 22, Theorem 24, and Theorem 25 are all special cases of the main result of this section, Theorem 30, although in the cases of Theorem 22 and Theorem 25, some work is needed to extract them as corollaries of the general theorem. This section also contains another special case of the triangular array problem, which we call the Centered Case Lemma. It is called a lemma because its chief purpose is to help us prove the general theorem. The following two problems concern some elementary inequalities which will be used in the proof of the Centered Case Lemma.

Problem 62. Prove that if $w \in [0, 1]$, there exists a constant c depending on w such that

$$\int_0^w \left| u\chi(x) - \sin ux \right| du \leq c \int_0^w (1 - \cos ux) \, du$$

for all $x \in \mathbb{R}$. *Hint:* For $|x| \leq 1$, the integrals can be explicitly calculated, and then bounds from Appendix E can be used. For the remaining values of x, note that for fixed w, the integral on the left is bounded above by a finite constant, and the integral on the right is bounded below by a positive constant.

Problem 63. Prove that for all $w \in [0, 1]$ and all $x \in \mathbb{R}$,

$$\frac{w^3}{12}(x^2 \wedge 1) \leq \int_0^w (1 - \cos ux) \, du \, .$$

Hint: For $x \in (0, \frac{\sqrt{2}}{w}]$, use the inequality

$$\frac{w^3 x^2}{6} - \frac{w^5 x^4}{120} \leq w - \frac{\sin wx}{x} \, ,$$

which may be obtained from an inequality in Appendix E.

Lemma 26. [centered case] *Let $(R_{k,n} : 1 \leq k \leq n, n = 1, 2, \ldots)$ be a uan triangular array of distributions on $\mathbb{R}$ such that*

$$\int_{\mathbb{R}} \chi \, dR_{k,n} = 0 \quad \text{for all } k \text{ and } n \, .$$

For each n, let

$$R_n = R_{1,n} * R_{2,n} * \cdots * R_{n,n} \, .$$

In order that the sequence $(R_n : n = 1, 2, \ldots)$ converge it is necessary and sufficient that there exist a nonnegative number σ and a Lévy measure ν satisfying the following two conditions:

(16.29)
$$\nu(B) = \lim_{n \to \infty} \sum_{k=1}^n R_{k,n}(B) \quad \text{for closed intervals } B$$

$$\text{such that } 0 \notin B \text{ and } \nu(\partial B) = 0;$$

$$\sigma^2 = \lim_{\varepsilon \searrow 0} \limsup_{n \to \infty} \sum_{k=1}^{n} \int_{[-\varepsilon,\varepsilon]} x^2\, R_{k,n}(dx)$$

(16.30)

$$= \lim_{\varepsilon \searrow 0} \liminf_{n \to \infty} \sum_{k=1}^{n} \int_{[-\varepsilon,\varepsilon]} x^2\, R_{k,n}(dx)\,.$$

In case these conditions are satisfied, the sequence $(R_n\colon n = 1,2,\dots)$ converges to the infinitely divisible distribution corresponding to $(0,\sigma,\nu)$ via the Lévy-Khinchin Representation Theorem.

PROOF. Let $\beta_{k,n}$ denote the characteristic function of $R_{k,n}$. The plan is to begin (Steps 1a and 1b of the proof) by showing that the assumption that (R_n) converges and the assumption that both (16.29) and (16.30) hold each lead to the conclusion that

(16.31) $$\sup_n \left\{ \sum_{k=1}^{n} \int (x^2 \wedge 1)\, R_{k,n}(dx)\colon n = 1,2,\dots \right\} < \infty,$$

then (Step 2) use this fact in conjunction with the uan condition to show the logarithm can be replaced by the first term of its expansion, and finally (Step 3) use Theorem 12 as in the proof of Theorem 22.

Step 1a. Suppose that $(R_n\colon n = 1,2,\dots)$ converges to some limit R. Fix $w \in (0,1)$ in the domain of the characteristic exponent of R. By Proposition 19, the characteristic exponent of R_n converges to the characteristic exponent of Q uniformly on $[0,w]$ as $n \to \infty$. In terms of the functions $\beta_{k,n}$, we have

$$\limsup_{n \to \infty} \sup_{0 \le u \le w} \sum_{k=1}^{n} \mathfrak{R}(-\log \circ \beta_{k,n}(u)) < \infty\,.$$

That is,

$$\limsup_{n \to \infty} \sup_{0 \le u \le w} \left\{ \sum_{k=1}^{n} -\log\left(1 - \left[1 - \mathfrak{R}^2(\beta_{k,n}(u)) - \mathfrak{I}^2(\beta_{k,n}(u)) \right] \right) \right\} < \infty\,.$$

Using $-\log(1 - t) \ge t$, $t \in [0,1)$, we see that

$$\limsup_{n \to \infty} \left\{ \sup_{0 \le u \le w} \sum_{k=1}^{n} [1 - \mathfrak{R}^2(\beta_{k,n}(u)) - \mathfrak{I}^2(\beta_{k,n}(u))] \right\} < \infty\,.$$

By Lemma 20, $\mathfrak{R}(\beta_{k,n}(u)) \to 1$ uniformly for $u \in [0,w]$ and $k = 1,\dots,n$ as $n \to \infty$, so

$$\limsup_{n \to \infty} \left\{ \sup_{0 \le u \le w} \sum_{k=1}^{n} [1 - \mathfrak{R}(\beta_{k,n}(u)) - \mathfrak{I}^2(\beta_{k,n}(u))] \right\} < \infty\,.$$

Integration on u gives

(16.32) $$\limsup_{n \to \infty} \left\{ \sum_{k=1}^{n} \int_0^w [1 - \mathfrak{R}(\beta_{k,n}(u)) - \mathfrak{I}^2(\beta_{k,n}(u))]\, du \right\} < \infty\,.$$

Next we use $\int \chi \, dR_{k,n} = 0$ to obtain

(16.33)
$$\int_0^w \mathfrak{J}^2(\beta_{k,n}(u)) \, du$$
$$\leq \max_{0 \leq u \leq w} |\mathfrak{J}(\beta_{k,n}(u))| \int_0^w \int |\sin ux - u\chi(x)| \, R_{k,n}(dx) \, du$$

By Problem 62 and two applications of the Fubini Theorem, there exists a constant c depending only on w such that for $x \in \mathbb{R}$ and $1 \leq k \leq n = 1, 2, \ldots$,

(16.34)
$$\int_0^w \int |\sin ux - u\chi(x)| \, R_{k,n}(dx) \, du$$
$$\leq c \int \int_0^w (1 - \cos ux) \, du \, R_{k,n}(dx)$$
$$= c \int_0^w \left[1 - \mathfrak{R}(\beta_{k,n}(u)) \right] du \, .$$

By Lemma 20, $\mathfrak{J}(\beta_{k,n}(u)) \to 0$ uniformly for $u \in [0, w]$ and $k = 1, \ldots, n$ as $n \to \infty$, so it follows from (16.33) and (16.34) that removing the term $\mathfrak{J}^2(\beta_{k,n}(u))$ from (16.32) does not affect the finiteness of the quantity given there. That is,

$$\limsup_n \left\{ \sum_{k=1}^n \int_0^w \left[1 - \mathfrak{R}(\beta_{k,n}(u)) \right] du \right\} < \infty \, .$$

The inequality in (16.31) now follows from Problem 63 and the Fubini Theorem.

Step 1b. Suppose now that (16.29) and (16.30) hold. From (16.30) it follows that there exists $\varepsilon > 0$ for which

$$\limsup_{n \to \infty} \sum_{k=1}^n \int_{[-\varepsilon, \varepsilon]} x^2 \, R_{k,n}(dx) < \infty \, .$$

From (16.29) we obtain

$$\limsup_{n \to \infty} \sum_{k=1}^n R_{k,n}(\{x : |x| > \varepsilon\}) < \infty \, .$$

Combining these two facts, we obtain (16.31).

Step 2. Using $\int_{\mathbb{R}} \chi \, dR_{k,n} = 0$ we have (for sufficiently large n depending on u)

$$\sum_{k=1}^n -(\log \circ \beta_{k,n})(u)$$
$$= \sum_{k=1}^n -\log \left(1 - \int \left[(1 - \cos ux) + i(-\sin ux + u\chi(x)) \right] R_{k,n}(dx) \right) \, .$$

We want to study the limiting behavior as $n \to \infty$ by replacing the logarithm by the first term of its series. Doing this will be valid provided that we show

(16.35)
$$\lim_{n\to\infty} \sum_{k=1}^{n} \left[\left(\int (1 - \cos ux)\, R_{k,n}(dx) \right)^2 \right.$$
$$\left. + \left(\int (-\sin ux + u\chi(x))\, R_{k,n}(dx) \right)^2 \right] = 0$$

whenever (16.31) holds. Since the functions $x \rightsquigarrow (1 - \cos ux)$ and $x \rightsquigarrow (-\sin ux + u\chi(x))$ are bounded and are continuous at 0 with the value 0 there, the uan condition implies

$$\lim_{n\to\infty} \max_{1 \le k \le n} \left[\int (1 - \cos ux)\, R_{k,n}(dx) + \left| \int (-\sin ux + u\chi(x))\, R_{k,n}(dx) \right| \right] = 0.$$

Therefore, to prove (16.35) we only need prove

$$\lim_{n\to\infty} \sum_{k=1}^{n} \left[\int (1 - \cos ux)\, R_{k,n}(dx) + \left| \int (-\sin ux + u\chi(x))\, R_{k,n}(dx) \right| \right] < \infty.$$

This is a consequence of Step 1 because $1 - \cos ux$ is bounded by a multiple (depending on u) of $x^2 \wedge 1$ and $|-\sin ux + u\chi(x)|$ is bounded by a multiple of $|x|^3 \wedge 1$ and thus by a multiple of $x^2 \wedge 1$.

Step 3. By Step 2, $\sum_{k=1}^{n} -(\log \circ \beta_{k,n}(u))$ has the same limiting behavior as

$$\int \left[(1 - \cos ux) + i(-\sin ux + u\chi(x)) \right] \left(\sum_{k=1}^{n} R_{k,n} \right)(dx)$$
$$= \int \left[1 - e^{iux} + iu\chi(x) \right] \left(\sum_{k=1}^{n} R_{k,n} \right)(dx).$$

This is the characteristic exponent of the infinitely divisible distribution characterized, via its Lévy-Khinchin representation, by the triple $(0, 0, \sum_k R_{k,n})$. The desired result now follows from Theorem 12. $\square$

For the proof of the forthcoming general theorem we will not use the first term of the series for the logarithm. Rather we will use centering to transform the distributions to ones to which the preceding lemma can be applied. The following lemmas will be of use in connection with this centering.

Lemma 27. *For any distribution Q on $\mathbb{R}$, the equation*

$$\int \chi(x - q)\, Q(dx) = 0$$

has at least one solution.

Problem 64. Prove the preceding lemma.

Lemma 28. *Let* $(Q_{k,n}: 1 \leq k \leq n = 1, 2, \ldots)$ *be a uan triangular array of distributions on* $\mathbb{R}$ *and suppose that numbers* $q_{k,n}$ *satisfy*

$$\int \chi(x - q_{k,n}) \, Q_{k,n}(dx) = 0 \,.$$

Then

$$\lim_{n\to\infty} \max_{1\leq k\leq n} |q_{k,n}| = 0 \,.$$

Problem 65. Prove the preceding lemma

The following lemma states that if we use the constants $q_{k,n}$ to center the distributions $Q_{k,n}$, then we obtain a uan triangular array $(R_{k,n})$ which has the property that it satisfies conditions (16.29) and (16.30) of the Centered Case Lemma if and only if $(Q_{k,n})$ satisfies the analogous conditions (16.36) and (16.37) of the general result. The reader should look ahead to Theorem 30 to see precisely what those analogous conditions state before studying the following lemma.

Lemma 29. *Let* $Q_{k,n}$ *and* $q_{k,n}$ *be as in the preceding lemma and define* $R_{k,n}$ *by* $R_{k,n}(B) = Q_{k,n}(B + q_{k,n})$ *for all Borel sets* B. *Then:*

  (i) the triangular array $(R_{k,n})$ *is uan;*
  (ii) for any Lévy measure ν, *the statement* (16.29) *and its*
       analogue (16.36) *are equivalent;*
  (iii) if (16.36) *is true for some Lévy measure* ν, *then*

$$\lim_{n\to\infty} \sum_{k=1}^{n} \left[q_{k,n} - \int \chi(x) \, Q_{k,n}(dx) \right] = 0 \,;$$

  (iv) if (16.36) *is true for some Lévy measure* ν, *then the*
       statement (16.30) *and its analogue* (16.37) *are equivalent.*

Problem 66. Prove the preceding lemma. *Hint:* For the last statement in the lemma prove that if (16.36) holds, then (16.30) and (16.37) are each equivalent to the following statement

$$\sigma^2 = \lim_{\varepsilon \searrow 0} \limsup_{n\to\infty} \sum_{k=1}^{n} \left[\int_{[-\varepsilon,\varepsilon]} x^2 \, Q_{k,n}(dx) - q_{k,n}^2 \right]$$

$$= \lim_{\varepsilon \searrow 0} \liminf_{n\to\infty} \sum_{k=1}^{n} \left[\int_{[-\varepsilon,\varepsilon]} x^2 \, Q_{k,n}(dx) - q_{k,n}^2 \right] \,.$$

Now we are in position to solve the general triangular array problem.

Theorem 30. [general case] *Let* $(Q_{k,n}: 1 \leq k \leq n, n = 1, 2, \dots)$ *be a uan triangular array of distributions on* $\mathbb{R}$. *For each* n, *let*

$$Q_n = Q_{1,n} * Q_{2,n} * \cdots * Q_{n,n}.$$

In order that the sequence $(Q_n: n = 1, 2, \dots)$ *converge it is necessary and sufficient that there exist a real number* η, *a nonnegative number* σ, *and a Lévy measure* ν *satisfying the following three conditions:*

(16.36)
$$\nu(B) = \lim_{n \to \infty} \sum_{k=1}^{n} Q_{k,n}(B) \quad \text{for closed intervals } B$$
$$\text{such that } 0 \notin B \text{ and } \nu(\partial B) = 0;$$

(16.37)
$$\sigma^2 = \lim_{\varepsilon \searrow 0} \limsup_{n \to \infty} \sum_{k=1}^{n} \left(\int_{[-\varepsilon, \varepsilon]} x^2 Q_{k,n}(dx) - \left[\int_{[-\varepsilon, \varepsilon]} x Q_{k,n}(dx) \right]^2 \right)$$
$$= \lim_{\varepsilon \searrow 0} \liminf_{n \to \infty} \sum_{k=1}^{n} \left(\int_{[-\varepsilon, \varepsilon]} x^2 Q_{k,n}(dx) - \left[\int_{[-\varepsilon, \varepsilon]} x Q_{k,n}(dx) \right]^2 \right);$$

(16.38)
$$\eta = \lim_{n \to \infty} \sum_{k=1}^{n} \int \chi(x) Q_{k,n}(dx).$$

In case these conditions are satisfied, the sequence $(Q_n: n = 1, 2, \dots)$ *converges to the infinitely divisible distribution corresponding to* (η, σ, ν) *via the Lévy-Khinchin Representation Theorem.*

PROOF. We first introduce some notation. By Lemma 27, there exist numbers $q_{k,n}$ such that

$$\int_{\mathbb{R}} \chi(x - q_{k,n}) Q_{k,n}(dx) = 0$$

for each k and n. Define distributions $R_{k,n}$ by $R_{k,n}(B) = Q_{k,n}(B + q_{k,n})$. It is clear that $\int_{\mathbb{R}} \chi \, dR_{k,n} = 0$ for each k and n, and by Lemma 29, $(R_{k,n})$ is a uan triangular array so that Lemma 26 applies to it. For each n, let

$$R_n = R_{1,n} * R_{2,n} * \cdots * R_{n,n}.$$

Note that $R_n(B) = Q_n(B + \eta_n)$ for all Borel sets B, where

$$\eta_n = \sum_{k=1}^{n} q_{k,n}.$$

We now prove the sufficiency of (16.36), (16.37), and (16.38). Suppose that (16.36) and (16.37) hold. By Lemma 29, (16.29) and (16.30) hold. By Lemma 26, the sequence $(R_n: n = 1, 2, \dots)$ converges to the infinitely divisible distribution characterized by the triple $(0, \sigma, \nu)$. Suppose that (16.38) also holds. Then (iii) in Lemma 29 implies that $\eta_n \to \eta$ as $n \to \infty$, so the sequence $(Q_n: n =$

$1, 2, \ldots$) converges to the infinitely divisible distribution characterized by the triple (η, σ, ν), as desired.

To prove the necessity of (16.36), (16.37), and (16.38), we introduce the notation $\widehat{T}$ corresponding to any distribution T: $\widehat{T}$ is the symmetrization of T about 0. Suppose that $\lim Q_n = Q$ for some Q. Then since $\widehat{R}_n = \widehat{Q}_n$ for all n,

$$\lim_{n \to \infty} \widehat{R}_n = \lim_{n \to \infty} \widehat{Q}_n = \widehat{Q}.$$

We will show that the convergence of the sequence $(\widehat{R}_n: n = 1, 2, \ldots)$ and the uan condition imply that the sequence $(R_n: n = 1, 2, \ldots)$ has a convergent subsequence $(R_{n_m}: m = 1, 2, \ldots)$. Assuming for the moment that this claim is true, the proof is completed as follows. By the Centered Case Lemma and Lemma 29, (16.36) and (16.37) hold for the corresponding subsequence $(Q_{n_m}: m = 1, 2, \ldots)$ for some Lévy measure ν and positive number σ. By the Convergence of Types Theorem, there exists a real number η such that $\eta_{n_m} \to \eta$ as $m \to \infty$, so (16.38) also holds along the subsequence. By the part of the theorem already proved, the distribution Q, being the limit of the sequence $(Q_{n_m}: m = 1, 2, \ldots)$, is infinitely divisible and determines the triple (η, σ, ν) via the Lévy-Khinchin Representation Theorem. The same argument applies to any subsequence, so (16.36), (16.37), and (16.38) hold along the full sequence.

Thus to complete the proof, it remains to show that the sequence $(R_n: n = 1, 2, \ldots)$ has a convergent subsequence. For each k and n, let $\beta_{k,n}$ be the characteristic function of $R_{k,n}$. By the continuity of the exponential function and Theorem 13 of Chapter 14, it is enough to prove that the sequence of functions

$$\left(\sum_{k=1}^{n} \log \circ \beta_{k,n} : n = 1, 2, \ldots \right)$$

is equicontinuous at 0. In order to do so, we will prove equicontinuity separately for the real and imaginary parts.

We begin with the equicontinuity of the sequence of real parts. Since $|\beta_{k,n}|^2$ is the characteristic function of $\widehat{R}_{k,n}$ and the sequence $(\widehat{R}_n)$ converges, Proposition 19 implies that the sequence of functions

$$\left(\sum_{k=1}^{n} \log \circ (|\beta_{k,n}|^2) : n = 1, 2, \ldots \right)$$

converges uniformly on some open interval containing the origin. As a consequence, this sequence is equicontinuous at 0. Since $\Re(\log \circ \beta_{k,n}) = \log \circ (|\beta_{k,n}|)$ for all k, n, the proof of equicontinuity at 0 of the family of real parts is complete.

The rest of the proof concerns the equicontinuity of the family of imaginary parts. It is sufficient to show that

(16.39) $$\lim_{u \to 0} \sup_{n} \sum_{k=1}^{n} \left| \Im(\log \circ \beta_{k,n}(u)) \right| = 0.$$

Recall that the imaginary part of the logarithm of a point z in the complex plane equals one of the values of $\arg(z)$, a polar coordinates angle of z that is determined only up to constant multiples of 2π. Lemma 20 implies that for large n, the complex number $\beta_{k,n}(u)$ stays close to 1 for $u \in [-1, 1]$ and $k = 1, \ldots, n$, so $\arg(\beta_{k,n}(u))$ stays close to 0. Furthermore, because $\beta_{k,n}(u)$ stays close to 1 for large n, we may choose l sufficiently large so that

$$| \arg(\beta_{k,n}(u))| \le 2|\Im(\beta_{k,n}(u))|$$

for $u \in [-1, 1]$, $n \ge l$, and $k = 1, \ldots, n$. Thus

$$\left|\Im(\log \circ \beta_{k,n}(u))\right| \le 2\left|\Im(\beta_{k,n}(u))\right| \quad \text{for } u \in [-1, 1], n \ge l, \text{ and } k = 1, \ldots, n.$$

We have justified replacing the imaginary part of the logarithm of $\beta_{k,n}$ by the imaginary part of $\beta_{k,n}$ in (16.39). Thus, since $\int \chi(x) R_{k,n}(dx) = 0$ for all k, n, it is enough to show that

$$\lim_{u \to 0} \sup_n \sum_{k=1}^n \left|\int_{\mathbb{R}} (u\chi(x) - \sin ux) R_{k,n}(dx)\right| = 0.$$

For $|x| \le 1$, the absolute value of each integrand is bounded above by $|ux|^3$, which is bounded above by $|u|^3 x^2$. For $|x| > 1$, an upper bound is $|u| + |ux| \wedge 1$, which is bounded above by $2(|ux| \wedge 1)$ for small u. So we will be done once we prove the following two statements:

$$(16.40) \qquad \lim_{u \to 0} \sup_n \sum_{k=1}^n \int_{|x|>1} \left(|ux| \wedge 1\right) R_{k,n}(dx) = 0;$$

and

$$(16.41) \qquad \sup_n \sum_{k=1}^n \int_{[-1,1]} x^2 R_{k,n}(dx) < \infty.$$

To prove (16.40), we use the relation

$$R_{k,n}(x, \infty) R_{k,n}(-\tfrac{1}{2}, \tfrac{1}{2}) \le \widehat{R}_{k,n}(x - \tfrac{1}{2}, \infty),$$

which holds for all $x \in \mathbb{R}$. The uan condition implies that $R_{k,n}(-\tfrac{1}{2}, \tfrac{1}{2}) \ge \tfrac{1}{2}$ for sufficiently large n and $k = 1, \ldots, n$. For such n, we have

$$(16.42) \qquad \sum_{k=1}^n R_{k,n}(x, \infty) \le 2 \sum_{k=1}^n \widehat{R}_{k,n}(x - \tfrac{1}{2}, \infty), \quad x \in \mathbb{R}.$$

Since the sequence of symmetrized distributions $(\widehat{R}_n)$ converges, (16.29) implies that

$$\limsup_{n \to \infty} \sum_{k=1}^n \widehat{R}_{k,n}(x - \tfrac{1}{2}, \infty) \le \nu[x - \tfrac{1}{2}, \infty) < \infty \quad \text{for } x > \tfrac{1}{2},$$

where ν is some Lévy measure. Similar facts hold for intervals of the form $(-\infty, -x)$. It then follows from (16.42) and the Continuity of Measure Theorem that

$$\text{(16.43)} \qquad \sup_n \sum_{k=1}^n R_{k,n}([-1,1]^c) < \infty$$

and

$$\lim_{x \to \infty} \sup_n \sum_{k=1}^n R_{k,n}([-x,x]^c) = 0.$$

The statement in (16.40) now follows easily.

To prove (16.41), we again make a comparison with the array $(\widehat{R}_{k,n})$. Since $\widehat{R}_{k,n}$ is the distribution of the difference of two independent random variables, each having distribution $R_{k,n}$, the Fubini Theorem implies that

$$\int_{[-1,1]} x^2 \, \widehat{R}_{k,n}(dx) = \int_{[-1,1]} \int_{[-1,1]} (x-y)^2 \, R_{k,n}(dx) \, R_{k,n}(dy)$$

$$= 2 \int_{[-1,1]} x^2 \, R_{k,n}(dx) - 2 \left(\int_{[-1,1]} x \, R_{k,n}(dx) \right)^2.$$

Since the symmetrized sequence $(\widehat{R}_n)$ converges, (16.31) applies to the measures $\widehat{R}_{k,n}$, so

$$\sup_n \sum_{k=1}^n \int_{[-1,1]} x^2 \, \widehat{R}_{k,n}(dx) < \infty.$$

Thus, in order to prove (16.41), it suffices to show that

$$\sup_n \sum_{k=1}^n \int_{[-1,1]} x \, R_{k,n}(dx) < \infty.$$

But this last statement follows from (16.43) and the fact that $\int \chi(x) \, R_{k,n}(dx) = 0$. $\square$

Problem 67. Apply Theorem 30 to the triangular array $(Q_{k,n})$ where $Q_{k,n}$ is the delta distribution at $(-1)^k / n^{1/5}$. Show that for this example, (16.37) cannot be replaced by a simpler condition only involving $\int_{[-\epsilon,\epsilon]} x^2 \, Q_{k,n}(dx)$. Also, explain why this example indicates that there was no chance of directly proving Theorem 30 using the first term (or even the first two terms) of the series for the logarithm.

* **Problem 68.** Apply Theorem 30 to the triangular array $(Q_{k,n}: 1 \leq k \leq n = 1, 2, \dots)$ of distributions, where $Q_{k,n}$ is the uniform distribution on the interval $\left[0, \frac{1}{\sqrt{n+k}}\right]$ if k is odd and the uniform distribution on $\left[-\frac{1}{\sqrt{n+k}}, 0\right]$ if k is even.

Problem 69. Apply Theorem 30 to the triangular array of distributions given by

$$Q_{k,n} = \left(1 - \frac{1}{n}\right)\delta_{(-kn^{-2})} + \frac{1}{n}\lambda,$$

where δ_a denotes the delta distribution at a and λ denotes Lebesgue measure on $(0,1)$.

Problem 70. Let $(Y_k : n = 1, 2, \ldots)$ be a sequence of independent random variables having 0 mean and finite variance. Denote the variance of Y_k by σ_k^2 and its distribution by Q_k. Consider the triangular array defined by

$$X_{k,n} = \frac{Y_k}{s_n}, \quad 1 \le k \le n = 1, 2, \ldots,$$

where $s_n = \left(\sum_{k=1}^{n} \sigma_k^2\right)^{1/2}$. Notice that the triangular array is independent within rows and that each row sum S_n has mean 0 and variance 1. Assume that

$$(16.44) \qquad \lim_{n \to \infty} \frac{\max\{\sigma_k^2 : 1 \le k \le n\}}{s_n^2} = 0.$$

Prove that the triangular array is uan. Then prove the Lindeberg-Feller Theorem that a necessary and sufficient condition for $(S_n : n = 1, 2, \ldots)$ to converge to a normal random variable having mean 0 and variance 1 is that

$$(16.45) \qquad \lim_{n \to \infty} \frac{1}{s_n^2} \sum_{k=1}^{n} \int_{|x| > ts_n} x^2 \, Q_k(dx) = 0$$

for every $t > 0$. (Comment: (16.45) implies (16.44).) Also, apply the Lindeberg-Feller Theorem to Problem 68.

Problem 71. Deduce Theorem 22 and Theorem 25 as corollaries of Theorem 30.

Problem 72. Generalize Problem 48.

CHAPTER 17
Stable Distributions as Limits

In Chapter 15, it was seen that if $X_1, X_2, \ldots$, is an iid sequence of $\mathbb{R}$-valued random variables with finite mean and variance, then as a consequence of the Classical Central Limit Theorem, quantities like

$$(17.1) \qquad P[x \leq X_1 + \cdots + X_n \leq y]$$

can be estimated using the normal distribution function. It turns out that for many iid sequences without finite variances, or even without finite means, such a quantity can still be estimated using stable distribution functions. For instance, useful information can be obtained about the distribution of the nth return to 0 of a simple symmetric random walk on $\mathbb{Z}$ (see Problem 20).

After introducing some technical preliminaries in the first section, we characterize the stable distributions. Since every stable distribution is infinitely divisible, this characterization is given in terms of the characteristic exponent in the Lévy-Khinchin Representation Theorem of Chapter 16.

In the third and fourth sections of the chapter, we state and prove the results that are needed for making estimates of quantities like (17.1). As with the Classical Central Limit Theorem, such estimates are most useful when n is large and x and y are chosen in a manner that depends on n so one recasts the problem of estimating (17.1) as a problem involving a limit of the form

$$\lim_{n \to \infty} P\left[x \leq \frac{S_n - b_n}{a_n} \leq y\right],$$

where $S_n = X_1 + \cdots + X_n$. In the case of the Classical Central Limit Theorem, one chooses $a_n = \sqrt{n \operatorname{Var}(X_1)}$ and $b_n = nE(X_1)$. The results given in the present chapter include formulas for the appropriate choices of a_n and b_n. They also tell us the domain of attraction of each stable distribution. In other words, they tell us, in terms of the distribution of X_1, whether or not a given stable distribution is appropriate for estimating (17.1).

17.1. Regular variation

This section, which itself contains no probability, consists of tools useful for identifying all stable distributions and finding their domains of attraction. There may be some discrepancy between what is done here and what might be done in a full treatment of regular variation. Here we strive for only as much generality as is needed.

Definition 1. A function $g: (c, \infty) \to (0, \infty)$, for some c, is said to *vary regularly* or to be of *regular variation at* ∞ if there is a $(0, \infty)$-valued function f for which

$$(17.2) \qquad \lim_{w \to \infty} \frac{g(wy)}{g(w)} = f(y),$$

for all but countably many $y \in (0, \infty)$.

It is easy to extend this definition to other settings, such as regularly varying at $-\infty$ or regularly varying from the left at 0. We will focus here on regular variation at ∞ for two reasons: (i) it is the concept needed for this chapter and (ii) results for the other types of regular variation can then be obtained by a simple change of variables.

Lemma 2. *If g is monotone and regularly varying at ∞ and f is as in Definition 1, then the domain of f can be extended to $(0, \infty)$ so that* (17.2) *holds for every $y \in (0, \infty)$. With this extended domain, f is monotone and continuous and satisfies*

$$f(xy) = f(x)f(y), \quad x > 0, y > 0.$$

PROOF. The domain of f consists of a subset of $(0, \infty)$ having countable (conceivably empty) complement. The monotonicity of g is obviously inherited by f on its domain. If x, y, and xy are in the domain of f, then

$$(17.3) \qquad \begin{aligned} f(xy) &= \lim_{w \to \infty} \frac{g(wxy)}{g(w)} \\ &= \lim_{w \to \infty} \frac{g(wx)}{g(w)} \cdot \lim_{w \to \infty} \frac{g((wx)y)}{g(wx)} = f(x)f(y). \end{aligned}$$

Fix $x > 0$. Suppose that $f(x-) > f(x+)$. By (17.3), $f((xy)-) > f((xy)+)$ for all y in the domain of f for which $f(y) > 0$. Since a monotone function cannot have uncountably many points of discontinuity, the assumption $f(x-) > f(x+)$ is false. A similar argument shows $f(x-) < f(x+)$ to be false. Thus f is continuous at x if defined there and, if f is not defined at x, its domain can be extended to x so that it is continuous there. Now that the domain of f has been extended to all of $(0, \infty)$ so that f is continuous and monotone, we deduce, from the monotonicity of g and f, that (17.2) and therefore (17.3) hold for all x and y in $(0, \infty)$. $\square$

The conditions in the preceding lemma can be written in terms of the function $h = \log \circ f \circ \exp$. Monotonicity and continuity are inherited by h from f, and h satisfies

(17.4) $$h(s + t) = h(s) + h(t), \quad s \in \mathbb{R}, \ t \in \mathbb{R}.$$

Proposition 3. *Suppose that g is monotone and regularly varying at ∞ and that f is given by* (17.2). *Then for some constant κ, $f(y) = y^\kappa$, $y \in (0, \infty)$.*

Problem 1. Prove the preceding proposition. *Hint:* Prove that a continuous monotone function h that satisfies (17.4) necessarily equals the function $s \rightsquigarrow as$ for some constant a.

The exponent κ in the preceding proposition is called the *index* of regular variation of the regularly varying function g. Regularly varying functions of index 0 are said to be *slowly varying*.

Problem 2. Show that the function $\log^c$ is slowly varying at ∞ for every real exponent c.

* **Problem 3.** For which real c is the function $\exp \circ (\log^c)$ slowly varying at ∞? For which other values of c is it regularly varying at ∞ and in those cases what is the index of regular variation?

The next result says that monotone functions that vary regularly with index κ behave somewhat similarly to power functions with exponent κ.

Lemma 4. *Suppose that g is monotone and regularly varying at ∞ with index κ. Then, for any $\varepsilon > 0$,*

$$g(w) = o(w^{\kappa + \varepsilon}) \quad \text{and} \quad w^{\kappa - \varepsilon} = o(g(w)) \quad \text{as } w \to \infty.$$

PROOF. We need only prove the lemma for g increasing, since the result for decreasing g can be obtained immediately from the result for the increasing function $\frac{1}{g}$.

Let $\varepsilon > 0$, and choose $\delta \in (0, \varepsilon)$. By the definition of κ we may fix $x > 0$ so that

(17.5) $$2^{\kappa - \delta} < \frac{g(2y)}{g(y)} < 2^{\kappa + \delta}$$

for $y \geq x$. For each $w > x$ there is a positive integer m such that $x 2^{m-1} < w \leq x 2^m$. From (17.5) we obtain

$$\frac{g(w)}{w^{\kappa + \varepsilon}} \leq \frac{g(x 2^m)}{(x 2^{m-1})^{\kappa + \varepsilon}} \leq \frac{(2^{\kappa + \delta})^m g(x)}{(x 2^{m-1})^{\kappa + \varepsilon}} = \frac{2^{\kappa + \varepsilon} g(x)}{x^{\kappa + \varepsilon}} 2^{-m(\varepsilon - \delta)},$$

which approaches 0 as y and therefore m approach ∞. We also obtain

$$\frac{g(w)}{w^{\kappa-\varepsilon}} \geq \frac{g(x2^{m-1})}{(x2^m)^{\kappa-\varepsilon}} \geq \frac{(2^{\kappa-\delta})^{m-1}g(x)}{(x2^m)^{\kappa-\varepsilon}} = \frac{g(x)}{(2x)^{\kappa-\varepsilon}} 2^{(m-1)(\varepsilon-\delta)},$$

which approaches ∞ as y approaches ∞. $\quad\square$

17.2. The stable distributions

Stable distributions were defined in Chapter 14 and we have already observed that every stable distribution is infinitely divisible. The issues we address in this section are those of identifying which infinitely divisible distributions are stable, in terms of their Lévy-Khinchin representations, and explicitly calculating their characteristic exponents. In the remaining two sections of the chapter we will focus on identifying domains of attraction and strict attraction, as defined in the last section of Chapter 15, but some preliminary work in that direction is in this section.

Let (η, σ, ν) be the characterizing triple, via the Lévy-Khinchin Representation Theorem, of a nondegenerate stable distribution Q on $\mathbb{R}$. To obtain conditions on (η, σ, ν) that are consequences of stability, we let R be any distribution in the domain of attraction of Q. ($R = Q$ would do for this section, but general R is needed for the next one.) Thus, there exist constants $a_n \in (0, \infty)$ and $b_n \in \mathbb{R}$ such that

$$\lim_{n\to\infty} R^{*n}(a_n B + b_n) \to Q(B)$$

for all Borel B for which $Q(\partial B) = 0$. It will be convenient throughout the remainder of this chapter to set $c_n = b_n/n$ and to introduce distributions Q_n defined by

(17.6) $$Q_n(B) = R(a_n B + c_n), \quad B \text{ Borel}.$$

Note that $Q_n^{*n}(B) = R^{*n}(a_n B + b_n)$ for Borel sets B. Thus, we search here for consequences of the assumption $Q_n^{*n} \to Q$ as $n \to \infty$.

Lemma 5. If Q is nondegenerate and $Q_n^{*n} \to Q$ as $n \to \infty$, then $a_n \to \infty$, $\frac{a_{n+1}}{a_n} \to 1$, and $\frac{c_n}{a_n} \to 0$ as $n \to \infty$.

Problem 4. Prove the preceding lemma. *Hint:* Use the last sentence of Theorem 22 of Chapter 16.

Continuing under the assumptions that Q is characterized by the triple (η, σ, ν) and that $Q_n^{*n} \to Q$, we use Corollary 23 of Chapter 16 to obtain

(17.7) $$\lim_{n\to\infty} n \int_{[-y,y]} t^2 Q_n(dt) = \sigma^2 + \int_{[-y,y]\setminus\{0\}} t^2 \nu(dt)$$

for those $y > 0$ for which $\nu\{-y, y\} = 0$. We make the change of variables $t = (s - c_n)/a_n$ in the left side of (17.7) and then use the third conclusion in Lemma 5 in conjunction with $\nu\{-y, y\} = 0$ to obtain

$$(17.8) \qquad \lim_{n \to \infty} \frac{n}{a_n^2} \int_{[-a_n y, a_n y]} s^2 R(ds) = \sigma^2 + \int_{[-y,y] \setminus \{0\}} t^2 \nu(dt).$$

Fix $z > 0$ such that $\nu\{-z, z\} = 0$ and

$$\sigma^2 + \int_{[-z,z] \setminus \{0\}} t^2 \nu(dt) > 0.$$

For $y > 0$ such that $\nu\{-yz, yz\} = 0$, (17.8) and the second conclusion of Lemma 5 yield

$$\lim_{n \to \infty} \frac{\int_{[-a_n yz, a_n yz]} s^2 R(ds)}{\int_{[-a_{n+1} z, a_{n+1} z]} s^2 R(ds)} = \frac{\sigma^2 + \int_{[-yz, yz] \setminus \{0\}} t^2 \nu(dt)}{\sigma^2 + \int_{[-z,z] \setminus \{0\}} t^2 \nu(dt)}$$

and

$$\lim_{n \to \infty} \frac{\int_{[-a_{n+1} yz, a_{n+1} yz]} s^2 R(ds)}{\int_{[-a_n z, a_n z]} s^2 R(ds)} = \frac{\sigma^2 + \int_{[-yz, yz] \setminus \{0\}} t^2 \nu(dt)}{\sigma^2 + \int_{[-z,z] \setminus \{0\}} t^2 \nu(dt)}.$$

By the first conclusion of Lemma 5, every sufficiently large w lies in $[a_n, a_{n+1}]$ for some n. Hence,

$$(17.9) \qquad \lim_{w \to \infty} \frac{\int_{[-wyz, wyz]} s^2 R(ds)}{\int_{[-wz, wz]} s^2 R(ds)} = \frac{\sigma^2 + \int_{[-yz, yz] \setminus \{0\}} t^2 \nu(dt)}{\sigma^2 + \int_{[-z,z] \setminus \{0\}} t^2 \nu(dt)}.$$

Therefore, the function

$$y \rightsquigarrow \int_{[-yz, yz]} s^2 R(ds)$$

is regularly varying at ∞ and, by Proposition 3,

$$(17.10) \qquad \frac{\sigma^2 + \int_{[-yz, yz] \setminus \{0\}} t^2 \nu(dt)}{\sigma^2 + \int_{[-z,z] \setminus \{0\}} t^2 \nu(dt)} = y^\kappa$$

for some $\kappa \geq 0$. Moreover, (17.9) holds for every $z > 0$.

By letting $z \searrow 0$ in (17.10) we see that $\sigma > 0$ implies that $\kappa = 0$ and hence that ν is the 0 measure. In this case, a look at the characteristic function reveals that Q is a normal distribution.

For the remainder of this argument, assume $\sigma = 0$. Set $z = 1$ to obtain

$$(17.11) \qquad \int_{[-y,y] \setminus \{0\}} t^2 \nu(dt) = \lambda y^\kappa$$

for some constant $\lambda > 0$. If $\nu(0, \infty) > 0$, we may repeat the preceding arguments with the lower endpoints of integration replaced by 0; if $\nu(-\infty, 0) > 0$ they can be repeated with the upper endpoints of integration replaced by 0. The result is that there exist positive constants κ^+ and κ^- and nonnegative constants λ^+ and λ^- such that

$$(17.12) \qquad \int_{(0,y]} t^2 \, \nu(dt) = \lambda^+ y^{\kappa^+}$$

and

$$(17.13) \qquad \int_{[-y,0)} t^2 \, \nu(dt) = \lambda^- y^{\kappa^-} \, .$$

The value of κ^+ is arbitrary if $\lambda^+ = 0$; we choose $\kappa^+ = \kappa$ in this case. Similarly, we set $\kappa^- = \kappa$ if $\lambda^- = 0$. With these conventions it is straightforward to use the fact that (17.11), (17.12), and (17.13) hold for all $y > 0$ to conclude first that $\kappa^+ = \kappa^- = \kappa$ and $\lambda = \lambda^+ + \lambda^-$, and second that $\kappa < 2$ (so that ν is a Lévy measure). We may further conclude that ν is given by

$$(17.14) \qquad \nu(y, \infty) = \frac{\lambda^+ \kappa}{2 - \kappa} y^{-(2-\kappa)}, \quad y > 0$$

and

$$(17.15) \qquad \nu(-\infty, x) = \frac{\lambda^- \kappa}{2 - \kappa} |x|^{-(2-\kappa)}, \quad x < 0 \, .$$

We have proved the following proposition.

Lemma 6. *In order that an infinitely divisible distribution corresponding to the triple (η, σ, ν) be a nondegenerate stable distribution it is necessary that exactly one of the following two conditions be satisfied: (i) $\nu = 0$; (ii) $\sigma = 0$ and ν is given by (17.14) and (17.15) for some constants $\kappa \in (0, 2)$, $\lambda^+ \geq 0$, and $\lambda^- \geq 0$, with $\lambda^+ + \lambda^- > 0$.*

In the course of this section it will develop that 'necessary' in the preceding proposition can be replaced by 'necessary and sufficient'. For the $\mathbb{R}^+$-setting, the full story is immediately available.

Problem 5. Use Problem 13 of Chapter 15, the Lévy-Khinchin Representation Theorem for $\mathbb{R}^+$, and Lemma 6 to conclude that the stable moment generating functions are those of the form

$$u \rightsquigarrow e^{-\xi u - c u^\alpha}$$

for some $\alpha \in (0, 1]$, $c \in [0, \infty)$, and $\xi \in [0, \infty)$, and that of these, the strictly stable ones are those with $\xi = 0$. [Note: the correspondence as described is not one-to-one, so for instance, the strictly stable moment generating function obtained by setting $\xi = 0$, $c = 1$, and $\alpha = 1$ can also be obtained by setting $\xi = 1$, $c = 0$, and choosing α arbitrarily.]

In view of (17.14) and (17.15), the change of variables $\alpha = 2 - \kappa$ is natural. With this definition of α, the stable distributions described in case (ii) of Lemma 6 are said to have *index* α, necessarily belonging to $(0, 2)$. A stable distribution for which $\sigma > 0$ (that is, a nondegenerate normal distribution) is said to have *index* 2. The index of the delta distribution at 0 is not defined. Other delta distributions also have no index in the stable setting, but have *index* 1 in the strictly stable setting.

Lemma 6 identifies all the stable distributions in terms of the corresponding triples (η, σ, ν) in the Lévy-Khinchin Representation Theorem. In the remainder of this section we will explicitly calculate the corresponding characteristic exponents and also identify which of these exponents are strictly stable. Before proceeding, however, we record, for use in the next section, necessary conditions for R to be in the domain of attraction of a stable distribution of a particular index, conditions that are contained explicitly or implicitly in the discussion leading to Lemma 6.

Lemma 7. *If a distribution R is in the domain of attraction of a stable distribution distribution of index α, then the function*

(17.16)
$$y \rightsquigarrow \int_{[-y,y]} s^2 \, R(ds)$$

is regularly varying at ∞ of index $2 - \alpha$ and, in case $\alpha \neq 2$,

$$\lim_{y \to \infty} \frac{\int_{[0,y]} s^2 \, R(ds)}{\int_{[-y,y]} s^2 \, R(ds)} = \frac{\lambda^+}{\lambda}$$

and

$$\lim_{y \to \infty} \frac{\int_{[-y,0]} s^2 \, R(ds)}{\int_{[-y,y]} s^2 \, R(ds)} = \frac{\lambda^-}{\lambda}.$$

Problem 6. Identify those aspects of the discussion leading to Lemma 6 that constitute a proof of the preceding lemma.

Problem 7. Prove the following relationships between the constants $\lambda^{\pm}$, λ, and the Lévy measure ν of a stable distribution:

$$\frac{\nu(c, \infty)}{\nu(-\infty, c) + \nu(c, \infty)} = \frac{\lambda^-}{\lambda} \quad \text{and} \quad \frac{\nu(-\infty, -c)}{\nu(-\infty, -c) + \nu(c, \infty)} = \frac{\lambda^-}{\lambda}$$

for $c > 0$.

Problem 8. Let $\beta > \alpha \in (0, 2)$ and suppose that the function defined at (17.16) is regularly varying at ∞ of index $2 - \alpha$. Show that $\int |s|^{\beta} \, R(ds) = \infty$. *Hint:* Use Lemma 4.

* **Problem 9.** Let $0 \leq \beta < \alpha \in (0, 2]$ and suppose that the function defined at (17.16) is regularly varying at ∞ of index $2 - \alpha$. Show that $\int |s|^{\beta} \, R(ds) < \infty$. *Hint:* Use Lemma 4.

Problem 10. Let $\alpha \in (0, 2)$. Show that the assumption that the function defined at (17.16) is regularly varying at ∞ of index $2 - \alpha$ does not enable one to draw a conclusion about the finiteness of $\int |s|^\alpha R(ds)$.

Problem 11. Let $\beta \geq 2$. Show that the assumption that the function defined at (17.16) is slowly varying at ∞ does not enable one to draw a conclusion about the finiteness of $\int |s|^\beta R(ds)$.

The remainder of this section is devoted to the explicit calculation of the stable characteristic exponents. We already mentioned the case $\alpha = 2$ in the preceding discussion, but for the record, we make a formal statement here. This result is also found in Problem 9 of Chapter 16.

Theorem 8. *The stable distributions of index 2 are the normal distributions. The characteristic exponents are of the form $-i\eta u + (\sigma^2/2)u^2$, $\eta \in \mathbb{R}, \sigma^2 > 0$. Of these, the strictly stable distributions are those with mean $\eta = 0$.*

Now we focus on the remaining values of α. In view of (17.14) and (17.15), the density with respect to Lebesgue measure of the Lévy measure of a stable distribution with index $\alpha \in (0, 2)$ is of the form

$$y \rightsquigarrow |y|^{-(\alpha+1)} \left(k^+ I_{(0,\infty)}(y) + k^- I_{(-\infty,0)}(y) \right),$$

where k^+, k^- are nonnegative constants. Since the characteristic exponent is given by

$$\int_{\mathbb{R}\setminus\{0\}} \left(1 - \cos uy - i \sin uy + iu\chi(y) \right) \nu(dy),$$

the following formula is of interest:

(17.17)
$$\int_0^\infty (1 - \cos uy)y^{-(\alpha+1)} \, dy = |u|^\alpha \int_0^\infty \frac{1 - \cos w}{w^{\alpha+1}} \, dw$$
$$= \frac{\pi \csc \frac{\pi\alpha}{2}}{2\Gamma(\alpha + 1)} |u|^\alpha,$$

where Γ denotes the gamma function.

Problem 12. Check the first equality in (17.17) by using substitution and the second by using tables and whatever calculations are necessary in conjunction with the tables. *Hint:* It may be that the tables force consideration of three cases separately: $\alpha \in (0, 1)$, $\alpha = 1$, and $\alpha \in (1, 2)$.

For $0 < \alpha < 1$, the integral of the $-i \sin uy$ term is finite. It can be calculated by replacing α by $\alpha + 1$ in (17.17) and differentiating with respect to u:

(17.18)
$$\int_0^\infty (-i \sin uy) \, y^{-(\alpha+1)} \, dy = -\frac{i\pi \sec \frac{\pi\alpha}{2}}{2\Gamma(\alpha + 1)} |u|^\alpha \operatorname{sgn}(u),$$

valid for $0 < \alpha < 1$. In this case, the integral of the $iu\chi(y)$ term is also finite and can thus be included with the $-i\eta u$ part of the characteristic exponent to give a term $-i\xi u$ for some $\xi \in \mathbb{R}$.

For $1 < \alpha < 2$, the integral of the $-i \sin uy$ term is not finite; the $iu\chi(y)$ term must be included. To obtain a nice formula, it is convenient to add and subtract iuy, and then to split the integral into two pieces. The formula for the first piece is

$$(17.19) \qquad \int_0^\infty i(-\sin uy + uy)\, y^{-(\alpha+1)}\, dy = -\frac{i\pi \sec \frac{\pi\alpha}{2}}{2\Gamma(\alpha+1)} |u|^\alpha \operatorname{sgn}(u)\,,$$

valid for $1 < \alpha < 2$. This formula is proved by replacing α by $\alpha-1$ in (17.17) and integrating with respect to u. Note that the right sides of (17.18) and (17.19) are identical. The second piece is the integral of $iu(\chi(y) - y)$ with respect to ν. It gives a finite multiple of iu and can thus be included with the $-i\eta u$ term to give a term $-i\xi u$ for some $\xi \in \mathbb{R}$.

If we combine (17.17), (17.18), and (17.19), and do a little rearranging of constants, we obtain the following theorem.

Theorem 9. *The stable characteristic exponents of index $\alpha \in (0,1) \cup (1,2)$ are the functions of the form*

$$u \rightsquigarrow -i\xi u + k|u|^\alpha \left(1 - i\gamma(\tan \tfrac{\pi\alpha}{2}) \operatorname{sgn}(u)\right),$$

where $k > 0$, $\xi \in \mathbb{R}$, and $|\gamma| \le 1$. Of these, the strictly stable characteristic exponents are those for which $\xi = 0$.

Problem 13. Prove the preceding theorem. *Hint:* Formulas (17.17), (17.18), and (17.19) involve integrals over $(0,\infty)$. Similar formulas hold for the region $(-\infty, 0)$. The constant γ comes from the fact that the constants in the Lévy measure may be different on these two regions.

For the index $\alpha = 1$, the following integral is relevant:

$$\int_0^\infty \frac{-\sin uy + u\chi(y)}{y^2}\, dy = u \int_0^\infty \frac{-\sin z + |u|\chi(z/|u|)}{z^2}\, dz$$

$$= u \int_0^\infty \frac{-\sin z + \chi(z)}{z^2}\, dz + u \int_0^\infty \frac{|u|\chi(z/|u|) - \chi(z)}{z^2}\, dz\,,$$

the splitting of the integral into two pieces being valid because each of the pieces is finite. The first of the two pieces, when multiplied by i, can be combined with $-i\eta u$ to give a term $-i\xi u$ for some $\xi \in \mathbb{R}$. The second piece can be evaluated explicitly; it equals $u \log |u|$, which we take to equal 0 at $u = 0$.

Theorem 10. *The stable characteristic exponents of index 1 are the functions of the form*

$$u \rightsquigarrow -i\xi u + k|u|\left(1 + i\gamma(\tfrac{2}{\pi} \log |u|) \operatorname{sgn}(u)\right)$$

where $k > 0$, $\xi \in \mathbb{R}$, *and* $|\gamma| \leq 1$. *Of these, the strictly stable characteristic exponents are those for which* $\gamma = 0$. *Moreover, those for which* $k = 0$ *and* $\xi \neq 0$ *are also strictly stable of index* 1.

Problem 14. Prove the preceding theorem.

In Theorem 9 and Theorem 10, parameters k, γ, and ξ appear. The proofs of these theorems show the following relationships of these parameters to $(\eta, 0, \nu)$.

Proposition 11. *For a stable distribution as described in* Theorem 9 *or* Theorem 10, *let* $(\eta, 0, \nu)$ *be the triple for the corresponding Lévy-Khinchin representation. Then* k, γ, *and* ξ, *of those theorems, are given by*

$$k = \frac{\pi \csc \frac{\pi \alpha}{2}}{2\Gamma(\alpha)} \left(\nu(1, \infty) + \nu(-\infty, -1) \right);$$

$$\gamma = \frac{\nu(1, \infty) - \nu(-\infty, -1)}{\nu(1, \infty) + \nu(-\infty, -1)};$$

$$\xi = \eta - \begin{cases} \int_{\mathbb{R} \setminus \{0\}} \chi(y) \, \nu(dy) & \text{if } \alpha \in (0, 1) \\ \int_{\mathbb{R} \setminus \{0\}} [\chi(y) - \sin y] \, \nu(dy) & \text{if } \alpha = 1 \\ \int_{\mathbb{R} \setminus \{0\}} [\chi(y) - y] \, \nu(dy) & \text{if } \alpha \in (1, 2) \end{cases}$$

$$= \eta - \begin{cases} \frac{\nu(1, \infty) - \nu(-\infty, -1)}{1 - \alpha} & \text{if } \alpha \in (0, 1) \cup (1, 2) \\ \left(\nu(1, \infty) - \nu(-\infty, -1) \right) \int_{-\infty}^{\infty} \frac{\chi(y) - \sin y}{y^2} \, dy & \text{if } \alpha = 1. \end{cases}$$

* **Problem 15.** What is the index of a Cauchy distribution that is symmetric about 0?

Problem 16. Show that each strictly stable distribution different from a delta distribution has a continuous density with respect to Lebesgue measure and calculate the value at 0 of that density.

* **Problem 17.** For each strictly stable distribution Q calculate $Q(0, \infty)$. For each fixed index α calculate the maximum value of $Q(0, \infty)$ as the values of the other parameters vary. *Hint:* Use an inversion formula to find an expression for $Q(0, x)$.

17.3. † Domains of attraction

The following theorem identifies the domains of attraction for nondegenerate stable distributions. Only one distribution from each stable type is mentioned since the domains of attraction of all stable distributions of any one type are identical. At some place or other in the theorem a constant, depending on α, must appear. For that purpose we set

$$(17.20) \qquad k_\alpha = \begin{cases} \frac{\pi(2-\alpha)\csc\frac{\pi\alpha}{2}}{2\Gamma(1+\alpha)} & \text{if } \alpha \in (0,2) \\ \frac{1}{2} & \text{if } \alpha = 2\,, \end{cases}$$

where Γ denotes the gamma function. The theorem also gives formulas for relevant normalizing constants. When reading the theorem it should be noticed that the constant γ is irrelevant when $\alpha = 2$.

Theorem 12. *Let* $\alpha \in (0,2]$, $\gamma \in [-1,1]$, *and* Q *be the stable distribution whose characteristic exponent is the function*

$$u \rightsquigarrow \begin{cases} k_\alpha |u|^\alpha \left(1 - i\gamma(\tan\frac{\pi\alpha}{2})\,\mathrm{sgn}(u)\right) & \text{if } \alpha \neq 1 \\ k_1 |u| \left(1 + i\gamma(\frac{2}{\pi}\log|u|)\,\mathrm{sgn}(u)\right) & \text{if } \alpha = 1\,, \end{cases}$$

where k_α *is defined by* (17.20). *Then a distribution* R *is in the domain of attraction of* Q *if and only if the function*

$$(17.21) \qquad y \rightsquigarrow \int_{[-y,y]} s^2\, R(ds)$$

is regularly varying at ∞ *of index* $2 - \alpha$ *and, in case* $\alpha \neq 2$,

$$(17.22) \qquad \lim_{y\to\infty} \frac{\int_{[0,y]} s^2\, R(ds)}{\int_{[-y,y]} s^2\, R(ds)} = \frac{1+\gamma}{2}\,.$$

If R *is in the domain of attraction of* Q, *then* $\lim_{n\to\infty} Q_n^{*n} = Q$, *where* Q_n *is defined by* $Q_n(B) = R(a_n B + c_n)$ *with*

$$(17.23) \qquad a_n = \inf\left\{y > 0 : \frac{1}{y^2}\int_{[-y,y]} s^2\, R(ds) \leq \frac{1}{n}\right\}$$

and

$$(17.24) \qquad c_n = \begin{cases} 0 & \text{if } \alpha < 1 \\ a_n \int \sin\frac{s}{a_n}\, R(ds) & \text{if } \alpha = 1 \\ \int_{(-\infty,\infty)} s\, R(ds) & \text{if } \alpha > 1\,. \end{cases}$$

PARTIAL PROOF. The necessity of the regular variation condition is the content of Lemma 7, and the necessity of (17.22) follows immediately from Proposition 11 and Lemma 7. So, for the remainder of the proof we assume that the function (17.21) is regularly varying at ∞ of some index $2 - \alpha$ with $\alpha \in (0,2]$ and, in case $\alpha \neq 2$, that (17.22) holds for some γ.

For the purposes of obtaining information about the constants a_n and c_n defined in the statement of the theorem, let us consider the properties of the function

$$y \rightsquigarrow \frac{1}{y^2}\int_{[-y,y]} s^2\, R(ds)\,.$$

It is easy to see that this function is right continuous and has limits from the left at each point y. Furthermore, the limit from the left at y is always less than or equal to the value at y. By Lemma 4 the function converges to 0 as $y \to \infty$. It follows from these facts that $a_n \nearrow \infty$ as $n \nearrow \infty$, and that for all sufficiently large n,

$$(17.25) \qquad \frac{n}{a_n^2} \int_{[-a_n, a_n]} s^2 R(ds) = 1.$$

By Problem 9 the definition of c_n is meaningful. Also, $\frac{c_n}{a_n} \to 0$ as $n \to \infty$, a fact that will be used below. We will finish the proof by verifying (16.22), (16.23), and (16.24) in Corollary 23 of Chapter 16, thereby establishing that $Q_n^{*n} \to Q$ as $n \to \infty$ (where Q_n is defined in terms of R, a_n, and c_n as in the statement of the theorem).

With the goal of proving (16.22) we first fix $y > 0$ and, for all sufficiently large n (depending on y), we use (17.25) to obtain

$$nQ_n(y, \infty) = nR(a_n y + c_n, \infty)$$

$$\leq nR((a_n y / 2), \infty)$$

$$= \sum_{m=0}^{\infty} nR(a_n y 2^{m-1}, a_n y 2^m]$$

$$\leq \sum_{m=0}^{\infty} \frac{4n}{(a_n y 2^m)^2} \int_{[0, a_n y 2^m]} s^2 R(ds)$$

$$= \sum_{m=0}^{\infty} \frac{4}{(y 2^m)^2} \frac{\int_{[0, a_n y 2^m]} s^2 R(ds)}{\int_{[-a_n, a_n]} s^2 R(ds)}$$

$$(17.26) \qquad = \sum_{m=0}^{\infty} \frac{4}{(y 2^m)^2} \frac{\int_{[0, a_n y]} s^2 R(ds)}{\int_{[-a_n, a_n]} s^2 R(ds)} \prod_{i=1}^{m} \frac{\int_{[0, a_n y 2^i]} s^2 R(ds)}{\int_{[0, a_n y 2^{i-1}]} s^2 R(ds)}.$$

By the regular variation hypothesis, the factors in the product $\prod_{i=1}^{m}$ in (17.26) approach $2^{2-\alpha}$ uniformly in i, $1 \leq i \leq \infty$, as $n \to \infty$. Therefore, for fixed $\beta \in (0, \alpha)$ and all sufficiently large n, not depending on m, this product is less than $2^{(2-\beta)m}$. Thus, (17.26) is bounded above by

$$(17.27) \quad \frac{\int_{[0, a_n y]} s^2 R(ds)}{\int_{[-a_n, a_n]} s^2 R(ds)} \sum_{m=0}^{\infty} \frac{4}{y^2 2^{\beta m}} = \frac{\int_{[0, a_n y]} s^2 R(ds)}{\int_{[-a_n, a_n]} s^2 R(ds)} y^{-2} \frac{4}{1 - 2^{-\beta}}$$

for sufficiently large n. As $n \to \infty$ the regular variation condition and (17.22) give

$$
(17.28)
\quad
\begin{aligned}
&\lim_{n\to\infty} \frac{\int_{[0,a_n y]} s^2 \, R(ds)}{\int_{[-a_n, a_n]} s^2 \, R(ds)} \\
&= \lim_{n\to\infty} \frac{\int_{[0,a_n y]} s^2 \, R(ds)}{\int_{[-a_n y, a_n y]} s^2 \, R(ds)} \frac{\int_{[-a_n y, a_n y]} s^2 \, R(ds)}{\int_{[-a_n, a_n]} s^2 \, R(ds)} \\
&= \frac{1+\gamma}{2} y^{2-\alpha}.
\end{aligned}
$$

This calculation shows that the quantity in (17.27) goes to 0 as first $n \to \infty$ and then $y \to \infty$. Since a similar argument works for $nQ_n(-\infty, x)$ when $x \to -\infty$, (16.22) holds.

In order to prove (16.24), we first suppose that $\alpha < 2$. By (17.25) and (17.28)

$$
\begin{aligned}
\lim_{n\to\infty} \frac{n}{a_n^2} \int_{[0,a_n y]} s^2 \, R(ds) &= \lim_{n\to\infty} \frac{\int_{[0,a_n y]} s^2 \, R(ds)}{\int_{[-a_n, a_n]} s^2 \, R(ds)} \\
&= \frac{1+\gamma}{2} y^{2-\alpha}.
\end{aligned}
$$

A similar argument for intervals $[a_n x, 0]$ gives $\frac{1-\gamma}{2} |x|^{2-\alpha}$. Thus,

$$
(17.29)
\quad
\begin{aligned}
&\lim_{n\to\infty} \frac{n}{a_n^2} \int_{[a_n x, a_n y]} s^2 \, R(ds) \\
&= \frac{\mathrm{sgn}(y) + \gamma}{2} |y|^{2-\alpha} - \frac{\mathrm{sgn}(x) + \gamma}{2} |x|^{2-\alpha}.
\end{aligned}
$$

for all $x < y$. A slightly different, somewhat simpler argument shows (17.29) to be valid for $\alpha = 2$ provided that neither x nor y equals 0. Using Proposition 11 and a little straightforward calculation to replace the right side of (17.29) by an expression in terms of σ and ν gives

$$
\lim_{n\to\infty} \frac{n}{a_n^2} \int_{[a_n x, a_n y]} s^2 \, R(ds) = \sigma^2 \delta_0([x,y]) + \int_{[x,y]\setminus\{0\}} t^2 \, \nu(dt)
$$

where δ_0 denotes the delta distribution at 0. Since $\frac{c_n}{a_n} \to 0$, the substitution $s = a_n t + c_n$ gives

$$
(17.30)
\quad
\lim_{n\to\infty} n \int_{[x,y]} t^2 \, Q_n(dt) = \sigma^2 \delta_0([x,y]) + \int_{[x,y]\setminus\{0\}} t^2 \, \nu(dt)
$$

provided that neither x nor y equals 0. Thus (16.24) is satisfied.

Before continuing with the proof, it will be useful to mention one consequence of (17.30). Fix an interval $[x, y]$. Then

$$
B \rightsquigarrow n \int_B t^2 \, Q_n(dt)
$$

defines a finite measure μ_n on the Borel subsets of $[x, y]$. It is an easy exercise to extend the notion of convergence in distribution to the sequence $(\mu_n \colon n = 1, 2, \ldots)$, since $\sup_n \mu_n[x, y]$ is finite. We may conclude from (17.30) that if $f \colon [x, y] \to \mathbb{R}$ is continuous and neither x nor y is 0, then

$$(17.31) \qquad \lim_{n \to \infty} n \int_{[x,y]} f(t) t^2 \, Q_n(dt) = \int_{[x,y] \setminus \{0\}} f(t) t^2 \, \nu(dt) \,.$$

We will use this fact several times in the rest of the proof.

It remains to consider (16.23) of Chapter 16. We first treat $\alpha \in (0, 1)$ in which case $\eta = \int \chi \, d\nu$, by Proposition 11. Thus, (16.23) is equivalent to

$$n \int \chi \, dQ_n \to \int \chi \, d\nu \,.$$

In view of the boundedness of the function χ and the fact that (16.22) has already been proved, it is enough to prove that

$$n \int_{[x,y]} \chi(t) \, Q_n(dt) \to \int_{[x,y]} \chi(t) \, \nu(dt)$$

for all real $x < y$. If $0 \notin [x, y]$, then we may multiply and divide by t^2 and apply (17.31) to see that

$$n \int_{[x,y]} \chi \, dQ_n = n \int_{[x,y]} \tfrac{\chi(t)}{t^2} t^2 \, Q_n(dt) \to \int_{[x,y]} \tfrac{\chi(t)}{t^2} t^2 \, \nu(dt) = \int_{[x,y]} \chi \, d\nu \,.$$

We have used the fact that $t \rightsquigarrow \chi(t)/t^2$ is a continuous function on such an interval. To take care of intervals $[x, y]$ containing 0, it is enough to prove that

$$(17.32) \qquad \lim_{y \searrow 0} \limsup_{n \to \infty} \left(n \int_{[-y,y]} t \, Q_n(dt) \right) = 0 \,.$$

The proof of this fact is carried out in two parts, one part for the interval $[0, y]$ and the other for the interval $[-y, 0]$. The arguments are similar for the two parts, so we concentrate on the interval $[0, y]$. Written in terms of R, we need to prove

$$(17.33) \qquad \lim_{y \searrow 0} \limsup_{n \to \infty} \frac{n}{a_n} \int_{[0, a_n y]} s \, R(ds) = 0 \,.$$

For the proof of (17.33) the following equalities and inequalities are useful:

$$\frac{n}{a_n} \int_{[0, a_n y]} s \, R(ds) = \sum_{m=0}^{\infty} \frac{n}{a_n} \int_{(a_n y 2^{-(m+1)}, a_n y 2^{-m}]} s \, R(ds)$$

$$\leq \sum_{m=0}^{\infty} \frac{n 2^{(m+1)}}{a_n^2 y} \int_{[0, a_n y 2^{-m}]} s^2 \, R(ds)$$

$$= \frac{2^{(m+1)}}{y} \frac{\int_{[0, a_n y 2^{-m}]} s^2 \, R(ds)}{\int_{[0, a_n]} s^2 \, R(ds)} \,.$$

It is left for the reader to now prove (17.33).

Next we consider $\alpha \in (1, 2]$. In this case $\eta = \int [\chi(t) - t] \nu(dt)$ by Proposition 11. Thus, (16.23) is equivalent to

$$n \int \chi(t) Q_n(dt) \to \int_{\mathbb{R} \setminus \{0\}} [\chi(t) - t] \nu(dt).$$

The mean of Q_n is 0, so we obtain the following equivalent statement as the one we need to prove:

$$n \int [\chi(t) - t] Q_n(dt) \to \int_{\mathbb{R} \setminus \{0\}} [\chi(t) - t] \nu(dt).$$

As in the previous case, we multiply and divide by t^2. The function $t \rightsquigarrow \frac{\chi(t) - t}{t^2}$, defined at 0 by continuity, is continuous on every bounded interval. It follows from (17.31) that

$$n \int_{[x,y]} [\chi(t) - t] Q_n(dt) \to \int_{[x,y] \setminus \{0\}} [\chi(t) - t] \nu(dt)$$

for real numbers x, y not equal to 0. To treat the behavior near $\pm \infty$, it suffices to show that

(17.34)
$$\lim_{y \nearrow \infty} \limsup_{n \to \infty} n \int_{(y, \infty)} [\chi(t) - t] Q_n(dt) = 0;$$

and a similar assertion near $-\infty$. It is left for the reader to prove (17.34) by a method similar to the proof above that (16.22) holds, and to observe that a similar argument would also work near $-\infty$.

Finally we treat (16.23) for the case $\alpha = 1$. In view of Proposition 11 we want to prove that

$$n \int \chi(t) Q_n(dt) \to \int_{\mathbb{R} \setminus \{0\}} [\chi(t) - \sin t] \nu(dt),$$

which we rewrite as

(17.35)
$$n \int [\chi(t) - \sin t] Q_n(dt) + n \int \sin t \, Q_n(dt)$$
$$\to \int_{\mathbb{R} \setminus \{0\}} [\chi(t) - \sin t] \nu(dt).$$

Since $t \rightsquigarrow \frac{\chi(t) - \sin t}{t^2}$, defined at 0 by continuity, is a continuous function on every bounded interval, we may once again apply (17.31) to obtain

$$n \int_{[x,y]} [\chi(t) - \sin t] Q_n(dt) \to \int_{[x,y] \setminus \{0\}} [\chi(t) - \sin t] \nu(dt)$$

for every finite x and y. Since $t \rightsquigarrow \chi(t) - \sin t$ is a bounded function, (16.22) implies that $[x, y]$ can be replaced by $[x, \infty)$ or $(-\infty, y]$ in this equation.

To show $n \int \sin t \, Q_n(dt) \to 0$, thus finishing the proof, we change variables; the quantity we want to study is

$$n \int \left[\sin \frac{s - c_n}{a_n} \right] R(ds),$$

which equals

$$(17.36) \quad n \left[\cos \frac{c_n}{a_n} \right] \int \left[\sin \frac{s}{a_n} \right] R(ds) - n \left[\sin \frac{c_n}{a_n} \right] \int \left[\cos \frac{s}{a_n} \right] R(ds).$$

We approximate $\cos \frac{c_n}{a_n}$ by 1, $\sin \frac{s}{a_n}$ by $\frac{s}{a_n}$, $\sin \frac{c_n}{a_n}$ by $\frac{c_n}{a_n} = \int \sin \frac{s}{a_n} R(ds)$, and $\cos \frac{s}{a_n}$ by 1. If these approximations were equalities (17.36) would equal 0. It is left for the reader to use Problem 18 below and the fact that the function (17.21) is regularly varying of index 1 to show that the error arising from such approximations goes to 0 as $n \to \infty$. $\square$

Problem 18. Suppose that R is in the domain of attraction of a stable distribution of index α. Let a_n be defined by (17.23). Prove that

$$a_n = o(n^{1/(\alpha - \varepsilon)}) \quad \text{and} \quad n^{1/(\alpha + \varepsilon)} = o(a_n) \quad \text{as } n \to \infty$$

for every $\varepsilon > 0$.

Problem 19. Some steps in the proof of Theorem 12 have been explicitly left for the reader. Supply these missing steps, possibly using the preceding problem.

Example 1. Let us apply Theorem 12 to a distribution R supported by $[0, \infty)$ for which

$$(17.37) \qquad\qquad \lim_{x \to \infty} x^{1/2} (\log x)^{-1} R(x, \infty) = 1.$$

The integral

$$\int_{[0,y]} s^2 \, R(ds) = -\int_0^y s^2 \, dR(s, \infty)$$

$$(17.38) \qquad\qquad = -y^2 R(y, \infty) + 2 \int_0^y s R(s, \infty) \, ds$$

is relevant for applying the theorem. Temporarily we act as if the limiting relation (17.37) is an equality in which case (17.38) becomes

$$-y^{3/2} \log y + 2 \int_0^y s^{1/2} \log s \, ds = \tfrac{1}{3} y^{3/2} \log y - \tfrac{8}{9} y^{3/2}$$

$$\sim \tfrac{1}{3} y^{3/2} \log y \quad \text{as } y \to \infty.$$

It is easy to check that the error in treating (17.37) as an equality rather than as a limiting relation is $o(y^{3/2} \log y)$. Thus,

$$(17.39) \qquad \int_{[0,y]} s^2\, R(ds) \sim \tfrac{1}{3} y^{3/2} \log y \quad \text{as } y \to \infty.$$

It is now clear that the function (17.21) is regularly varying with index $\frac{3}{2} = 2 - \frac{1}{2}$ and that (17.22) is satisfied with $\gamma = 1$.

By Theorem 12, there exist constants a_n such that $Q_n^{*n} \to Q$, where Q_n is defined by $Q_n(B) = R(a_n B)$ and Q is strictly stable of index $\frac{1}{2}$ with characteristic function

$$(17.40) \qquad \begin{aligned} u &\rightsquigarrow \exp\left(-k_{1/2}|u|^{1/2}\left(1 - i\operatorname{sgn}(u)\right)\right) \\ &= \exp\left(-\tfrac{3\sqrt{2\pi}}{2}|u|^{1/2}\left(1 - i\operatorname{sgn}(u)\right)\right). \end{aligned}$$

Let $\varepsilon \in (0,1)$. From (17.23) (or slightly more quickly from (17.25)) and (17.39) we obtain

$$(17.41) \qquad 1 - \varepsilon \le \frac{n \log a_n}{3\sqrt{a_n}} \le 1 + \varepsilon$$

for all sufficiently large n. Of course, we are willing to replace (a_n) by any sequence $(\hat{a}_n)$ for which $\hat{a}_n/a_n \to 1$ as $n \to \infty$. Indeed, we would like to find such numbers $\hat{a}_n$ that can be written in a simple explicit form. Here is a try: set $\varepsilon = 0$ in the above inequality and set the slowly varying function log equal to a convenient constant—say 3 —and solve to obtain $\hat{a}_n = n^2$. Now drop the fiction that $\log \hat{a}_n$ is a constant and replace it by the better approximation $2 \log n$. Then, solve again for $\hat{a}_n$ to obtain

$$(17.42) \qquad \hat{a}_n = \tfrac{4}{9} n^2 \log^2 n.$$

It remains to show that $\hat{a}_n/a_n \to 1$. From (17.41) and (17.42) we obtain

$$\left[(1 - \varepsilon)\frac{2\log n}{\log a_n}\right]^2 \le \frac{\hat{a}_n}{a_n} \le \left[(1 + \varepsilon)\frac{2\log n}{\log a_n}\right]^2$$

for all sufficiently large n. To finish the proof that $\hat{a}_n/a_n \to 1$ we need only show that $(2 \log n)/(\log a_n) \to 1$. But this asymptotic relation is an immediate consequence of Problem 18.

Let us summarize using random variables. Suppose that $(X_n\colon n = 1, 2, \ldots)$ is an iid sequence of nonnegative random variables with common distribution R satisfying (17.37) and let Y denote a strictly stable random variable of index $\frac{1}{2}$ with characteristic function (17.40). Then

$$\frac{\sum_{m=1}^{n} X_m}{\tfrac{4}{9} n^2 \log^2 n} \xrightarrow{\ \mathcal{D}\ } Y$$

as $n \to \infty$.

For the special situation at hand we can give additional information about the limiting distribution Q. By Problem 17, it is supported by $\mathbb{R}^+$. To calculate the moment generating function of Q we use Proposition 11 to obtain the Lévy measure: $\nu(x, \infty) = 3x^{-1/2}$, $x > 0$. Integration by parts in the Lévy-Khinchin representation gives a formula for the moment generating function in terms of the shift ξ:

$$v \rightsquigarrow e^{-\xi v - 3\sqrt{\pi v}}.$$

Since Q is strictly stable (as can be seen by looking at its characteristic function), $\xi = 0$. Comparison with Problem 16 of Chapter 15 shows that the density of Q with respect to Lebesgue measure on $(0, \infty)$ is

$$x \rightsquigarrow \frac{3}{2x^{3/2}} e^{-9\pi/4x}.$$

Remark 1. Another approach to finding the moment generating function of Q in the preceding example is to use the substitution $v = -iu$, say for $u \geq 0$, in the formula for the characteristic function:

$$v \rightsquigarrow \exp\left(-\frac{3\sqrt{2\pi}}{2} v^{1/2} i^{1/2} (1 - i)\right).$$

The simplest way to justify this change of variables is to treat the moment generating function as having domain equal to the right half of the complex plane and using a bit of complex function theory. Continuity considerations force the first-quadrant version of $i^{1/2}$ rather than the third-quadrant version. Thus $i^{1/2} = \frac{1+i}{\sqrt{2}}$, and the formula given above for the moment generating function results.

Problem 20. Let T_n be the time of the nth return to 0 of a simple symmetric random walk on $\mathbb{Z}$. The distribution R of T_1 is given in Example 5 of Chapter 11, and the distribution of T_n is R^{*n} by Theorem 12 of that same chapter. Prove that R is in the domain of attraction of a stable distribution of index $\alpha = \frac{1}{2}$. Use this fact to estimate the quantity $P[T_{100} \leq 10000]$. *Hint:* A simple change of variables allows one to use tables of the normal distribution function for the last part of the problem.

Problem 21. Apply Theorem 12 to the symmetric distribution R given by

$$R(x, \infty) = (16 + x)^{-1/4}, \quad x \geq 0.$$

Problem 22. Apply Theorem 12 to the distribution R given by

$$R(y, \infty) = 2R(-\infty, -y) = (16 + y)^{-1/4}, \quad y \geq 0.$$

Problem 23. Apply Theorem 12 to the distribution with density

$$x \rightsquigarrow \begin{cases} 0 & \text{if } |x| < 1 \\ |x|^{-3} & \text{if } |x| \geq 1. \end{cases}$$

Problem 24. Obtain the Classical Central Limit Theorem as a corollary of Theorem 12.

Problem 25. Use Theorem 12 to redo Problem 7 of Chapter 15.

Problem 26. Apply Theorem 12 to the distribution with density

$$x \rightsquigarrow \begin{cases} 0 & \text{if } x < 1 \\ x^{-2} & \text{if } x \geq 1. \end{cases}$$

Problem 27. Apply Theorem 12 to distributions Q supported by $\mathbb{R}^+$ and satisfying $x^{3/2}Q(x,\infty) \to 5$ as $x \to \infty$.

Problem 28. Apply Theorem 12 to distributions R supported by $(-\infty, 0]$ and satisfying

$$\lim_{x \to -\infty} \left(|x| \log |x|\right)^{1/2} R(-\infty, x] = 5.$$

* **Problem 29.** Apply Theorem 12 to the distribution with density

$$x \rightsquigarrow \begin{cases} 0 & \text{if } |x| < e^2 \\ |x|^{-1}(\log x)^{-2} & \text{if } |x| \geq e^2. \end{cases}$$

Problem 30. Apply Theorem 12 to the symmetric distribution function F given by

$$F(x) = \begin{cases} \frac{1}{|x|(\log |x|)^2} & \text{if } x < -e \\ \frac{1}{e} & \text{if } -e \leq x < 0. \end{cases}$$

Make an interesting comment involving means.

* **Problem 31.** Apply Theorem 12 to distributions R that are symmetric about 0 and satisfy $x^{4/3} \exp(\sqrt{\log x}) R(x,\infty) \to 1$ as $x \to \infty$.

Problem 32. Let $R = R_1 * R_2$, where R_1 is a standard normal distribution and R_2 is a standard symmetric Cauchy distribution. Apply Theorem 12 to R.

In both the last chapter and this one, our goal has been to obtain conditions in terms of distributions because they are usually regarded as the fundamental data. But if one is given a transform of a distribution it is often easier, as in the following problem, to work with the transform rather than Theorem 12.

Problem 33. As seen from Problem 33 of Chapter 5, the function $s \rightsquigarrow 1 - \sqrt{1-s}$ is the probability generating function of a distribution R. Decide if R is in any domain of attraction and, if so, find both the stable type to which R is attracted and appropriate normalizing constants.

17.4. ‡ Domains of strict attraction

In Theorem 12 only one stable distribution appeared of each stable type. For all $\alpha \neq 1$ the representative stable distribution there happens to be strictly stable. Therefore, that theorem goes some distance towards the following result.

Corollary 13. *Let* $\alpha \in (0,1) \cup (1,2]$, $\gamma \in [-1,1]$, *and* Q *be the strictly stable distribution whose characteristic exponent is the function*

$$u \rightsquigarrow k_\alpha |u|^\alpha \left(1 - i\gamma(\tan \tfrac{\pi\alpha}{2}) \operatorname{sgn}(u)\right),$$

where k_α *is defined by* (17.20). *Then a distribution* R *is in the domain of strict attraction of* Q *if and only if the function*

$$y \rightsquigarrow \int_{[-y,y]} s^2 R(ds)$$

is regularly varying at ∞ *of index* $2 - \alpha$ *and, in case* $\alpha \neq 2$,

$$\lim_{y \to \infty} \frac{\int_{[0,y]} s^2 R(ds)}{\int_{[-y,y]} s^2 R(ds)} = \frac{1+\gamma}{2}$$

and, in case $\alpha > 1$, *the mean of* R *is* 0. *If* R *is in the domain of strict attraction of* Q, *then* $\lim_{n\to\infty} Q_n^{*n} = Q$, *where* Q_n *is defined by* $Q_n(B) = R(a_n B)$ *with*

$$a_n = \inf\left\{y > 0: \frac{1}{y^2} \int_{[-y,y]} s^2 R(ds) \leq \frac{1}{n}\right\}.$$

Problem 34. Prove the preceding corollary.

The next theorem treats the domains of strict attraction of the nonconstant strictly stable distributions of index 1.

Theorem 14. *Let* $\xi \in \mathbb{R}$ *and denote by* Q *the strictly stable distribution whose characteristic exponent is the function*

$$u \rightsquigarrow -i\xi u + \frac{\pi}{2}|u|.$$

Then a distribution R *is in the domain of attraction of* Q *if and only if the function*

$$y \rightsquigarrow \int_{[-y,y]} s^2 R(ds)$$

is regularly varying at ∞ *of index* 1,

$$\lim_{y \to \infty} \frac{\int_{[0,y]} s^2 R(ds)}{\int_{[-y,y]} s^2 R(ds)} = \frac{1}{2},$$

and

$$\lim_{n\to\infty} \frac{n}{a_n} \int_{[-a_n,a_n]} sR(ds) = \xi,$$

where

$$a_n = \inf\left\{y > 0 : \frac{1}{y^2}\int_{[-y,y]} s^2\, R(ds) \leq \frac{1}{n}\right\}.$$

If R is in the domain of strict attraction of Q, then $\lim_{n\to\infty} Q_n^{*n} = Q$, where Q_n is defined by $Q_n(B) = R(a_n B)$.

Problem 35. Prove the preceding theorem.

Problem 36. Let Q be as in Theorem 14. Of course, Q is in its own domain of strict attraction. As a check on the correctness of Theorem 14, use it to prove that Q is in its own domain of strict attraction and also to find appropriate scaling constants a_n.

To complete the story concerning domains of strict attraction, it remains to determine the domains of strict attraction of nonzero constants. Such a result might be called a general 'Law of Large Numbers' although many would use that term only with scaling constants $a_n = n$, as in Theorem 2 of Chapter 15.

Theorem 15. *For an arbitrary distribution R, let*

$$(17.43) \qquad a_n = \inf\left\{y > 0 : \frac{1}{y}\left|\int_{[-y,y]} s\, R(ds)\right| \leq \frac{1}{n}\right\}$$

(the infimum of the empty set equaling ∞). Then R is in the domain of attraction of either δ_1 or δ_{-1}, the delta distributions at 1 and -1, if and only if $a_n \in (0,\infty)$ for all sufficiently large n,

$$(17.44) \qquad \lim_{n\to\infty} nR\big((-\infty,-a_n) \cup (a_n,\infty)\big) = 0,$$

and

$$(17.45) \qquad \lim_{n\to\infty} \frac{n}{a_n^2}\int_{[-a_n,a_n]} s^2\, R(ds) = 0.$$

In case these conditions are satisfied either

$$(17.46) \qquad \int_{[-a_n,a_n]} s\, R(ds) > 0$$

for all sufficiently large n and $\lim_{n\to\infty} Q_n^{*n} = \delta_1$, where Q_n is defined by $Q_n(B) = R(a_n B)$, or

$$(17.47) \qquad \int_{[-a_n,a_n]} s\, R(ds) < 0$$

for all sufficiently large n and $\lim_{n\to\infty} Q_n^{*n} = \delta_{-1}$.

PARTIAL PROOF. Suppose that $a_n \in (0, \infty)$ for all large n and that (17.44) and (17.45) hold. Clearly (16.22) and (16.24) in Corollary 23 of Chapter 16 hold with $\sigma = 0$, $\nu = 0$, and Q_n defined as in the theorem at hand.

In order to verify (16.23) with $\eta = \pm 1$, and thus complete the proof of the 'if' portion of the theorem we first study the sequence (a_n). Clearly this sequence is increasing. By the right continuity of $y \rightsquigarrow y^{-1} \left| \int_{[-y,y]} s\, R(ds) \right|$,

$$\frac{n}{a_n} \left| \int_{[-a_n, a_n]} s\, R(ds) \right| \leq 1.$$

Let $\zeta \in (0, 1)$. Then

(17.48)
$$\frac{n}{(1-\zeta) a_n} \left| \int_{[-(1-\zeta) a_n, (1-\zeta) a_n]} s\, R(ds) \right| > 1.$$

Use

(17.49)
$$\frac{n}{(1-\zeta) a_n} \int_{[-a_n, -(1-\zeta) a_n] \cup ((1-\zeta) a_n, a_n]} |s|\, R(ds)$$
$$\leq \frac{n}{(1-\zeta)^2 a_n^2} \int_{[-a_n, a_n]} s^2\, R(ds)$$

in conjunction with (17.45) to enlarge the domain of integration in (17.48) to $[-a_n, a_n]$ and then multiply both sides by $1 - \zeta$ to obtain

$$\liminf_{n \to \infty} \frac{n}{a_n} \left| \int_{[-a_n, a_n]} s\, R(ds) \right| \geq (1 - \zeta).$$

We let $\zeta \searrow 0$ to conclude

(17.50)
$$\lim_{n \to \infty} \frac{n}{a_n} \left| \int_{[-a_n, a_n]} s\, R(ds) \right| = 1.$$

Let $\varepsilon > 0$. Using a slight variation of (17.49) in conjunction with (17.50), we obtain

$$\frac{n+1}{(1+\beta) a_n} \int_{[-(1+\beta) a_n, (1+\beta) a_n]} s\, R(ds) = \frac{1 + o(1)}{1 + \beta} \frac{n}{a_n} \int_{[-a_n, a_n]} s\, R(ds),$$

uniformly for $\beta \in [0, \varepsilon]$ as $n \to \infty$. We conclude that $a_{n+1} \leq (1 + \varepsilon) a_n$ for sufficiently large n and then that either (17.46) holds for all sufficiently large n or that (17.47) holds for all sufficiently large n.

The following calculation, using (17.44), completes the proof of (16.23):

$$\lim_{n \to \infty} \int \chi(t)(n Q_n)(dt) = \lim_{n \to \infty} \int \chi(s/a_n)(n R)(ds)$$
$$= \lim_{n \to \infty} \frac{n}{a_n} \int_{[-a_n, a_n]} s\, R(ds).$$

For the converse, we suppose that, for some sequence $\hat{a}_n$ of positive constants, $\hat{Q}_n^{*n} \to \delta_1$, where $\hat{Q}$ is defined by $\hat{Q}(B) = R(\hat{a}_n B)$, because we will omit the case of convergence to δ_{-1}. Thus, the conditions in Theorem 22 and Corollary 23 of

Chapter 16 hold with $\eta = 1$, $\sigma = 0$, $\nu = 0$, and $\hat{Q}$ in place of Q. By (16.16) and (16.18)

$$(17.51) \qquad \frac{n}{\hat{a}_n} \int_{[-\hat{a}_n, \hat{a}_n]} s\, R(ds) = n \int_{[-1,1]} t\, \hat{Q}_n(dt) \to 1\,.$$

Replacing $\hat{a}_n$ by $\frac{\hat{a}_n}{1+\varepsilon}$ in the definition of $\hat{Q}_n$ changes the value of $\lim \hat{Q}_n^{*n}$ to $\delta_{1+\varepsilon}$. Hence

$$\frac{n}{\frac{\hat{a}_n}{1+\varepsilon}} \int_{[-\frac{\hat{a}_n}{1+\varepsilon}, \frac{\hat{a}_n}{1+\varepsilon}]} s\, R(ds) \to 1 + \varepsilon\,.$$

By using a diagonalization procedure as $n \to \infty$ and $\varepsilon \searrow 0$ we can find a new sequence $(\tilde{a}_n)$, asymptotic to (a_n), such that

$$(17.52) \qquad \frac{n}{\tilde{a}_n} \int_{[-\tilde{a}_n, \tilde{a}_n]} s\, R(ds) = n \int_{[-1,1]} t\, \tilde{Q}_n(dt) \searrow 1 \quad \text{as } n \to \infty$$

and $\tilde{Q}_n^{*n} \to \delta_1$ where $\tilde{Q}_n$ is defined by $\tilde{Q}_n(B) = R(\tilde{a}_n B)$. By (17.52), $\tilde{a}_n \le a_n$ (where a_n is defined by (17.43)). It is left for the reader to show that $a_n < 2\tilde{a}_n$ for all large n, so that in particular $a_n < \infty$.

By (16.16),

$$nR\big((-\infty, -a_n) \cup (a_n, \infty)\big) \le nR\big((-\infty, -\tilde{a}_n) \cup (\tilde{a}_n, \infty)\big) \to 0\,.$$

By (16.24) and (16.16),

$$\frac{n}{a_n^2} \int_{[-a_n, a_n]} s^2\, R(ds)$$

$$\le \frac{n}{\tilde{a}_n^2} \int_{[-\tilde{a}_n, \tilde{a}_n]} s^2\, R(ds) + nR\big((-\infty, -\tilde{a}_n) \cup (\tilde{a}_n, \infty)\big) \to 0\,,$$

thus completing the proof. $\square$

Problem 37. One part of the preceding proof has been explicitly left for the reader. Complete that part.

* **Problem 38.** Let X be a Cauchy random variable symmetric about 0. Decide if $|X|$ is in any domain of attraction and, if so, identify the attracting stable type and find appropriate normalizing constants for convergence. Also, decide if $|X|$ is in any domain of strict attraction and, if so, identify the attracting strict stable type and find appropriate scaling constants for convergence.

Problem 39. Generalize the preceding problem by continuing the assumption of symmetry, but removing the requirement that 0 be the center of symmetry.

Problem 40. Decide which stable distributions belong to which domains of strict attraction, if any.

Problem 41. Show that the slow variation at ∞ of

$$y \rightsquigarrow \int_{[-y,y]} |s| \, R(ds)$$

in combination with the existence of

$$\lim_{y \to \infty} \frac{\int_{[0,y]} s \, R(ds)}{\int_{[-y,y]} |s| \, R(ds)}$$

as a number larger than $\frac{1}{2}$ is sufficient for R to be in the domain of strict attraction of δ_1.

Problem 42. Can Problem 38 be done using the preceding problem? If so, do so.

If the conditions in Problem 41 were necessary for R to be in the domain of attraction of δ_1, these conditions would make a better theorem than Theorem 15, since they do not involve the sequence (a_n). However, the following problem shows that they are not necessary.

Problem 43. Consider the distribution having the following density with respect to Lebesgue measure:

$$y \rightsquigarrow \begin{cases} \frac{k\sqrt{\log |y|}}{y^2} & \text{if } y \leq -1 \\ 0 & \text{if } -1 < y < 1 \\ \frac{k(1+\sqrt{\log y})}{y^2} & \text{if } 1 \leq y, \end{cases}$$

where k is defined by the property that the integral on $\mathbb{R}$ of the density equals 1. Show that this distribution is in the domain of strict attraction of δ_1, but that this fact cannot be deduced from Problem 41.

As indicated by the last few problems, Theorem 15 is not entirely satisfactory. Unfortunately, although there are other results of this type in the literature, the authors of this book have not been able to find anything better for the general case. However, for the nonnegative case, there is a satisfactory result, which we state here without proof.

Theorem 16. *Let R be a distribution on $\mathbb{R}^+$ with distribution function F. Then R is in the domain of strict attraction of δ_1 if and only if*

$$\lim_{y \to \infty} \frac{y(1 - F(y))}{F(y)} = 0,$$

in which case the scaling constants may be chosen as in Theorem 15.

CHAPTER 18
Convergence in Distribution
on Polish Spaces

We want to extend the concept of convergence in distribution to probability spaces other than $(\mathbb{R}, \mathcal{B})$. Certain metric spaces, known as 'Polish spaces', play a central role. Particularly important examples of Polish spaces are the real line, the extended real line, d-dimensional Euclidean space, infinite products of intervals, and spaces of continuous functions. Thus, this chapter may be viewed as a mechanism for extending the concepts and results discussed in Chapter 14 to a wide variety of settings. (Basic facts about metric spaces are treated briefly in Appendix B. Some of the topology in Appendix C is also relevant.)

Particular attention will be given to distributions on $\mathbb{R}^d$. A central limit theorem will be proved by using the 'Cramér-Wold Device' which reduces certain problems for $\mathbb{R}^d$ to problems for $\mathbb{R}$.

18.1. Polish spaces

Much of the theory developed in Chapter 14 can be adapted to a certain type of metric space which we now define.

Definition 1. A *Polish space* is a complete metric space that has a countable dense subset.

The real line $\mathbb{R}$ (with the usual Euclidean metric) is a Polish space, with the rational numbers constituting a countable dense subset. As described in Appendix B, the extended real line $\overline{\mathbb{R}}$ is a complete metric space with the distance between x and y defined as $|\arctan y - \arctan x|$. It is a Polish space because the rational numbers constitute a countable dense set.

Remark 1. Because the definition of Polish space requires completeness of the metric, the correct choice of metric is important. For example, if we used the metric $\rho(x, y) = |\arctan x - \arctan y|$ in $\mathbb{R}$, we would not have completeness, despite the fact that the open sets arising from this metric are the same as those arising from the usual metric.

Another important Polish space is $\mathbb{R}^d$ with the usual (Euclidean) metric: the distance between x and y equals

$$\sqrt{\sum_{j=1}^{d}(y_j - x_j)^2}.$$

The set of points having rational coordinates is a countable dense set. Three other metrics for $\mathbb{R}^d$ that give the same open sets as this metric and also make $\mathbb{R}^d$ into a Polish space are

$$\sum_{j=1}^{d}|y_j - x_j|, \quad \sum_{j=1}^{d}(|y_j - x_j| \wedge 1),$$

and

(18.1) $$\sum_{j=1}^{d}\frac{|y_j - x_j| \wedge 1}{2^j}.$$

Although (18.1) is perhaps the most complicated of the alternate metrics for $\mathbb{R}^d$, it has the advantage that it generalizes easily to $\mathbb{R}^\infty$, as shown by the following example.

Example 1. We use $\mathbb{R}^\infty$ to denote the space of all sequences $(x_1, x_2, \ldots)$ of real numbers, with the product topology. We want to *metrize* this topological space; that is, we want to make it into a metric space in such a way that the metric gives the same open sets as does the product topology. We define the distance between two sequences x and y to be

(18.2) $$\rho(x, y) = \sum_{j=1}^{\infty}\frac{|y_j - x_j| \wedge 1}{2^j}.$$

It is straightforward to check that this definition gives a metric for $\mathbb{R}^\infty$. We will check that this metric gives the same open sets as does the product topology.

We must show (i) if O is an open set in the product topology, then for each $x \in O$, there is an open set U in the topology given by the metric ρ such that $x \in U$ and $U \subseteq O$, and (ii) if U is an open set in the topology given by ρ, then for each $x \in U$, there is an open set O in the product topology such that $x \in O$ and $O \subseteq U$.

Every open set O in the product topology is the union of sets of the form

(18.3) $$\{(x_1, x_2, \ldots) : (x_1, \ldots, x_d) \in O_d\}$$

for some positive integer d, where O_d is an open set in $\mathbb{R}^d$. It follows that in proving (i) and (ii), we may restrict our attention to sets O of the form (18.3). Similarly, every open set U in the topology given by the metric ρ is a union of sets that are open balls in that metric, so in (i) and (ii) we only need to consider open balls U.

We will prove (ii), and leave the proof of (i) to the reader. Let U be an open ball in $\mathbb{R}^\infty$ with the metric ρ, and pick a point $x = (x_1, x_2, \ldots) \in U$. By a standard argument involving the triangle inequality, there exists an $\varepsilon > 0$ such that the open ball $U' = \{y : \rho(x, y) < \varepsilon\}$ is contained in U. We will find a set of the form (18.3) that contains x and is contained in U'. Choose d such that $2^{d-1} > \frac{1}{\varepsilon}$, and let O_d be the open ball centered at the point $(x_1, \ldots, x_d)$ in $\mathbb{R}^d$ with radius $\frac{\varepsilon}{2}$, using the metric (18.1). Let O be defined in terms of O_d as in (18.3). Clearly $x \in O$. Moreover, it is easy to check from the definition of ρ that if $y \in O$, then $\rho(x, y) < \varepsilon$, so $O \subseteq U'$, as desired.

Those sequences containing only rational terms and only finitely many terms different from 0 constitute a countable dense set in $\mathbb{R}^\infty$. The metric space $\mathbb{R}^\infty$ is also complete, a consequence of the second problem below and the completeness of $\mathbb{R}$. Therefore $\mathbb{R}^\infty$ is a Polish space.

Problem 1. Decide if the following sequence in $\mathbb{R}^\infty$ converges and if so to what:

$$\big((1, 0, 0, 0, 0, \ldots), (0, 2, 0, 0, 0, \ldots), (0, 0, 4, 0, 0, \ldots), (0, 0, 0, 8, 0, \ldots), \ldots\big).$$

Problem 2. Let x_n, $n = 1, 2, \ldots$, and y be members of $\mathbb{R}^\infty$. Prove that $x_n \to y$ as $n \to \infty$ if and only if $x_{j,n} \to y_j$ where $x_{j,n}$ is the jth term of x_n and y_j is the jth term of y. (Comment: One says that the topology that we have given $\mathbb{R}^\infty$ is the topology of *coordinate-wise convergence*.)

The method used in Example 1 to find a suitable metric for $\mathbb{R}^\infty$ generalizes easily to countable products of arbitrary Polish spaces.

Proposition 2. *For $j = 1, 2, \ldots$, let (Ψ_j, ρ_j) be a Polish space, and let $\Psi = \bigotimes_{j=1}^\infty \Psi_j$, with the topology on Ψ being the product topology. For $x = (x_1, x_2, \ldots)$ and $y = (y_1, y_2, \ldots)$ in Ψ, define*

$$\rho(x, y) = \sum_{j=1}^\infty \frac{\rho_j(x_j, y_j) \wedge 1}{2^j}.$$

Then (Ψ, ρ) is a Polish space.

In Example 1 and the preceding proposition, the setting is one in which the space of interest already has a natural topology attached to it. Therefore the problem was to construct a metric consistent with the topology so that the resulting metric space is a Polish space. The following problem presents a situation where there is already a natural choice for the metric.

Problem 3. Let $\mathbf{C}[0, 1]$ denote the metric space of continuous $\mathbb{R}$-valued functions on $[0, 1]$ with the distance between two functions f and g being defined as the quantity

(18.4) $$\max\{|f(t) - g(t)| : t \in [0, 1]\}.$$

Prove that $\mathbf{C}[0,1]$ is a Polish space and that the Borel σ-field equals

$$\sigma(\{f\colon f(t) \in B\}\colon t \in [0,1], \text{Borel } B \subseteq \mathbb{R})\,.$$

The following proposition provides further examples of Polish spaces.

Proposition 3. *A closed subset of a Polish space is a Polish space with the inherited metric.*

PROOF. Let (Ψ, ρ) denote the Polish space and C the closed subset. By Problem 1 of Appendix B, (C, ρ) is a metric space.

Consider a Cauchy sequence in C. It converges to a member of ψ of Ψ. This point ψ must be a member of C; otherwise C would not be closed. Thus C is complete.

Let D be a countable dense subset of Ψ. For each positive integer n let D_n consist of those members of D whose distance from C is less than $\frac{1}{n}$. For each member ψ of D_n choose a member of C, conceivably ψ itself, whose distance from ψ is less than $\frac{1}{n}$ and let E_n denote the set of such chosen points. Every point in C is within a distance of $\frac{2}{n}$ of some member of E_n. Hence, $\bigcup_{n=1}^{\infty} E_n$ is a countable dense subset of C. $\square$

Problem 4. Explain why $\{f \in \mathbf{C}[0,1]\colon f(0) = 0\}$ is a Polish space, with the distance between f and g being specified by (18.4).

* **Problem 5.** Give the set $\mathbf{C}[0,\infty)$ of continuous $\mathbb{R}$-valued functions on $[0,\infty)$ a metric so that it becomes a Polish space with the topology of uniform convergence on bounded sets.

Example 2. [Infinite-dimensional cube] Consider the space $[0,1]^\infty$. This is the set of all sequences $(x_1, x_2, \dots)$ of real numbers belonging to the interval $[0,1]$. It is a subset of the space $\mathbb{R}^\infty$ introduced in Example 1, and it is easy to check that it is closed, since it is a product of closed sets. By Proposition 3, it is itself a Polish space, with distance function ρ given by (18.2).

We may also take the point of view that $[0,1]^\infty$ is a countable product of Polish spaces, and hence is itself a Polish space by Proposition 2. And since it is also the product of compact sets, it is compact by the Tychonoff Theorem (Theorem 3 of Appendix C). This fact gives importance to the next result, which says that an arbitrary Polish space is topologically equivalent to a Borel subset of the infinite-dimensional cube.

Lemma 4. *For any Polish space Ψ, there exists a function φ from Ψ onto a Borel subset of $[0,1]^\infty$ such that φ is continuous and one-to-one on Ψ and φ^{-1} is continuous (and one-to-one) on $\varphi(\Psi)$.*

PROOF. Let Ψ be an arbitrary Polish space with metric σ, and let $(\psi_n : n = 1, 2, \ldots)$ be a countable dense subset of Ψ. Consider the function $\varphi : \Psi \to [0, 1]^\infty$ defined by

$$\psi \rightsquigarrow \left(\sigma(\psi, \psi_1) \wedge 1 \,,\, \sigma(\psi, \psi_2) \wedge 1 \,,\, \ldots \right).$$

We claim that φ is one-to-one and that both φ and φ^{-1} (defined on the image of φ) are continuous.

The continuity of φ is an immediate consequence of the continuity of the functions $\psi \rightsquigarrow \sigma(\psi, \psi_n)$ for all n. To prove that φ is one-to-one, let ψ and η be two distinct members of Ψ, and let $\varepsilon = \sigma(\psi, \eta) \wedge 2$. By the definition of a metric, $\varepsilon > 0$. Using the fact that the set $(\psi_n : n = 1, 2, \ldots)$ is dense, choose k so that $\sigma(\psi, \psi_k) < \varepsilon/2$. It follows from the triangle inequality that $\sigma(\eta, \psi_k) > \varepsilon/2$, so $\varphi(\eta)$ and $\varphi(\psi)$ necessarily differ at the kth coordinate. Therefore, $\varphi(\eta) \neq \varphi(\psi)$, and we may conclude that φ is one-to-one.

To show that φ^{-1} is continuous at an arbitrary point $(x_1, x_2, \ldots)$ in $\varphi(\Psi)$, set $\varphi^{-1}(x_1, x_2, \ldots) = \psi$, fix $\varepsilon \in (0, \frac{1}{3})$, and then consider an arbitrary $(y_1, y_2, \ldots) \in \varphi(\Psi)$ for which

$$\rho((x_1, x_2, \ldots), (y_1, y_2, \ldots)) < \frac{\varepsilon}{3 \cdot 2^k} \,,$$

where ρ is given by (18.2) and k is chosen so that $\sigma(\psi, \psi_k) < \varepsilon/3$. Let $\eta = \varphi^{-1}(y_1, y_2, \ldots)$. Note that, by the definition of φ, $\sigma(\eta, \psi_k) = y_k$ and $\sigma(\psi, \psi_k) = x_k$. It follows from the definition of the metric ρ that

$$|\sigma(\eta, \psi_k) - \sigma(\psi, \psi_k)| < 2^k \rho((x_1, x_2, \ldots), (y_1, y_2, \ldots)) < \frac{\varepsilon}{3} \,,$$

so $\sigma(\eta, \psi_k) < 2\varepsilon/3$. Thus

$$\sigma(\psi, \eta) \leq \sigma(\psi, \psi_k) + \sigma(\eta, \psi_k) < \varepsilon \,.$$

The continuity of φ^{-1} follows.

We will prove that $\varphi(\Psi)$ is a Borel set by writing it in terms of countably many operations involving open and closed subsets of $[0, 1]^\infty$. Let D be a countable dense subset of Ψ. For each $d \in D$ and each positive integer k, let $B(d, 1/k)$ be the open ball of radius $1/k$ centered at d. By the continuity of φ^{-1}, the set $\varphi(B(d, 1/k))$ is an open subset of $\varphi(\Psi)$ in the relative topology, so there exists a set $V(d, k)$ which is open in the topology of $[0, 1]^\infty$ such that

$$\varphi(B(d, 1/k)) = V(d, k) \cap \varphi(\Psi).$$

Let $V(k) = \bigcup_{d \in D} V(d, k)$. We claim that

(18.5) $$\varphi(\Psi) = \overline{\varphi(\Psi)} \cap \left(\bigcap_{k=1}^{\infty} V(k) \right),$$

where $\overline{\varphi(\Psi)}$ is the closure in $[0, 1]^\infty$ of $\varphi(\Psi)$. Clearly, (18.5) implies that $\varphi(\Psi)$ is Borel.

Each member of $\varphi(\Psi)$ belongs to each $V(k)$ and thus to the set on the right side of (18.5). It remains to show that if v belongs to the set on the right side of

(18.5), then v belongs to $\varphi(\Psi)$. For each k, the fact that $v \in V(k)$ implies the existence of $d_k \in D$ such that $v \in V(d_k, k)$. Since v is in the closure of $\varphi(\Psi)$, every open neighborhood of v contains a member of $\varphi(\Psi)$. In particular, for each k, we may choose $v_k \in \varphi(\Psi)$ such that

$$v_k \in V(d_1, 1) \cap \cdots \cap V(d_k, k) \cap \{u \in [0,1]^\infty : \rho(u, v) < \tfrac{1}{k}\}.$$

Note that $v_k \to v$ as $k \to \infty$. Also note that $(\varphi^{-1}(v_k) : k = 1, 2, \dots)$ is a Cauchy sequence in Ψ, since for $j, k \geq m$, v_j and v_k are both members of $V(d_m, m)$ and hence $\sigma(\varphi^{-1}(v_j), \varphi^{-1}(v_k)) < \tfrac{2}{m}$. Since Ψ is a Polish space, $\psi = \lim_k \varphi^{-1}(v_k)$ exists. By the continuity of φ, $\varphi(\psi) = \lim_k v_k = v$. Thus $v \in \varphi(\Psi)$, as desired. □

Problem 6. [Infinite-dimensional space-filling curve] It is a well-known fact, often discussed in topology texts, that there exists a continuous function h from $[0, 1]$ onto $[0, 1]^2$. Such a function is a *Peano curve*, named after its discoverer. Use h to construct a continuous function from $[0, 1]$ onto $[0, 1]^\infty$. *Hint:* Since h is a continuous $[0, 1]^2$-valued function, it can be expressed in terms of two continuous $[0, 1]$-valued functions as $h = (h_1, h_2)$. Consider the function

$$(h_1, h_1 \circ h_2, h_1 \circ h_2 \circ h_2, h_1 \circ h_2 \circ h_2 \circ h_2, \dots).$$

18.2. Definition of and criteria for convergence

When we view Polish spaces as measurable spaces we follow our customary convention that the σ-field is the Borel σ-field unless something to the contrary is explicitly stated. In particular, throughout this section this convention is in force. The following definition is motivated by Proposition 7 of Chapter 14.

Definition 5. Let Q and Q_n, $n = 1, 2, \dots$, be probability measures on a Polish space Ψ. Then $(Q_n : n = 1, 2, \dots)$ *converges* to Q, denoted by $Q_n \to Q$ as $n \to \infty$, if, for every $\mathbb{R}$-valued bounded continuous function g on Ψ, $\int g \, dQ_n \to \int g \, dQ$ as $n \to \infty$.

For random variables having values in a Polish space, we say that a sequence (X_n) *converges* to X *in distribution* and write

$$X_n \xrightarrow{\mathcal{D}} X \quad \text{as } n \to \infty,$$

if $Q_n \to Q$, where Q_n and Q denote the distributions of X_n and X, respectively.

The following theorem generalizes Proposition 7 of Chapter 14 to the setting of Polish spaces. The name of the theorem refers to the fact that it contains so many conditions. Note the new condition (vi) and the change in condition (ii). We leave it to the reader to check that the set which appears in condition (vi) is a Borel set. As in Proposition 7 of Chapter 14 we use ∂A for the boundary of a set A.

Theorem 6. [Portmanteau] *Let Q and $Q_n, n = 1, 2, \ldots$, be probability measures on a Polish space Ψ. Then the following conditions are equivalent:*

(i) $Q_n \to Q$ *as* $n \to \infty$;

(ii) $\lim_{n\to\infty} \int g \, dQ_n = \int g \, dQ$ *for each bounded uniformly continuous function g on Ψ;*

(iii) $\limsup_{n\to\infty} Q_n(C) \le Q(C)$ *for each closed subset C of Ψ;*

(iv) $\liminf_{n\to\infty} Q_n(O) \ge Q(O)$ *for each open subset O of Ψ;*

(v) $\lim_{n\to\infty} Q_n(A) = Q(A)$ *for each Borel subset A of Ψ for which $Q(\partial A) = 0$;*

(vi) $\lim_{n\to\infty} \int g \, dQ_n = \int g \, dQ$ *for each bounded measurable function g for which $Q(\{\psi \in \Psi : g$ is discontinuous at $\psi\}) = 0$.*

PROOF. That (i) $\Longrightarrow$ (ii) and (vi) $\Longrightarrow$ (i) are both obvious.

The proof that (ii) $\Longrightarrow$ (iii) is essentially the same as the proof of the corresponding part of Proposition 7 of Chapter 14, since the functions introduced in that proof are uniformly continuous and are defined in a manner that works equally well in a metric space.

The proofs that (iii) $\Longleftrightarrow$ (iv) and $\{$(iii), (iv)$\} \Longrightarrow$ (v) are also the same as the corresponding parts of the proof of Proposition 7 of Chapter 14.

Finally we prove that (v) $\Longrightarrow$ (vi). Let g be a bounded measurable function from Ψ to $\mathbb{R}$, and let

$$D = \{\psi \in \Psi : g \text{ is discontinuous at } \psi\}.$$

Assume that $Q(D) = 0$. For each n, g may be regarded as a bounded random variable from $(\Psi, \mathcal{A}, Q_n)$ to $(\mathbb{R}, \mathcal{B})$, where $\mathcal{A}$ and $\mathcal{B}$ are the respective Borel σ-fields. Let R_n be the distribution of this random variable. Similarly, g is a bounded random variable from $(\Psi, \mathcal{A}, Q)$ to $(\mathbb{R}, \mathcal{B})$ whose distribution will be denoted by R. We will first show that $R_n \to R$ as $n \to \infty$.

Let B be a Borel subset of $\mathbb{R}$ for which $R(\partial B) = 0$. Then $Q(g^{-1}(\partial B)) = 0$. It is easily checked that if $\psi \in \partial g^{-1}(B)$, then either g is discontinuous at ψ, or $g(\psi) \in \partial B$. Hence $\partial g^{-1}(B) \subseteq D \cup g^{-1}(\partial B)$, so that $Q(\partial g^{-1}(B)) = 0$. Thus, $Q_n(g^{-1}(B)) \to Q(g^{-1}(B))$ as $n \to \infty$, or equivalently, $R_n(B) \to R(B)$ as $n \to \infty$. By (v) of Proposition 7 of Chapter 14, $R_n \to R$ as $n \to \infty$.

By (ii) of Proposition 7 of Chapter 14, $\int h \, dR_n \to \int h \, dR$ as $n \to \infty$ for each bounded continuous $\mathbb{R}$-valued function h on $\mathbb{R}$. We apply this fact with

$$h(x) = \begin{cases} x & \text{if } |x| \le c \\ -c & \text{if } x < -c \\ c & \text{if } x > c \end{cases}$$

where $c = \sup\{|g(\psi)| : \psi \in \Psi\}$. Doing so completes the proof since $\int g \, dQ_n = \int h \, dR_n$ and $\int g \, dQ = \int h \, dR$. $\square$

Corollary 7. *Let Q and R be two distributions on a Polish space. If*

(18.6) $$\int g\,dR = \int g\,dQ$$

for each bounded uniformly continuous function g on Ψ then $R = Q$.

PROOF. Suppose that (18.6) holds. By letting $R = Q_n$ for all n in the Portmanteau Theorem one gets $R(O) \geq Q(O)$ for every open set O. Interchanging the roles of Q and R gives $Q(O) \geq R(O)$. Since $R(O) = Q(O)$ for every open set O, it follows by the Uniqueness Theorem that $R = Q$. □

Corollary 8. *Let Q and R be two distributions on a Polish space. If both R and Q are limits of the same sequence of distributions, then $R = Q$.*

PROOF. If R and Q are both limits of the same sequence, then by the Portmanteau Theorem, (18.6) holds. □

It is not immediately apparent from the definition of convergence for a sequence of distributions on a Polish space that a given sequence can converge to only one distribution, but the preceding corollary shows that to be the case. (Of course, a sequence that does not converge might have various subsequences that converge to different limits.)

In Chapter 14 we found a connection between convergence in distribution on $\mathbb{R}$ and convergence in probability of sequences of $\mathbb{R}$-valued random variables. This connection will carry over to Polish spaces, once we have appropriately generalized the definition of convergence in probability.

Let $(X_n : n = 1, 2, \ldots)$ be a sequence of random variables with values in a Polish space (Ψ, ρ). The sequence (X_n) *converge in probability* to a Ψ-valued random variable X if, for every $\varepsilon > 0$,

$$\lim_{n \to \infty} P\big[\rho(X, X_n) > \varepsilon\big] = 0.$$

It is *Cauchy in probability* if, for every $\varepsilon > 0$, there is an integer l such that

$$P\big[\rho(X_n, X_m) > \varepsilon\big] < \varepsilon$$

whenever $m, n \geq l$.

Since a Polish space is a complete metric space, a sequence of random variables with values in a Polish space converges almost surely if and only if it is almost surely Cauchy. The statements and proofs of Theorem 2 and Lemma 24, both of Chapter 12, carry over to the present setting along with Lemma 3 and Problem 39 of that same chapter. Thus, a sequence is Cauchy in probability if and only if it converges in probability. Moreover, almost sure convergence implies convergence in probability. Also, convergence in probability implies almost sure convergence for an appropriate subsequence.

Proposition 9. *Let X and X_n, $n = 1, 2, \ldots$, be random variables on a common probability space having values in a common Polish space. Suppose that the sequence $(X_n \colon n = 1, 2, \ldots)$ converges to X in probability as $n \to \infty$. Then $X_n \xrightarrow{\mathcal{D}} X$ as $n \to \infty$.*

Problem 7. Prove the preceding proposition. *Hint:* Let g be a uniformly continuous bounded $\mathbb{R}$-valued function defined on the Polish space, and show that the sequence $(g \circ X_n \colon n = 1, 2, \ldots)$ of $\mathbb{R}$-valued random variables converges in probability to the $\mathbb{R}$-valued random variable $g \circ X$.

The next result says that if a sequence of probability measures on a Polish space converges in distribution, then so does any sequence induced from it by a continuous function into another Polish space.

Proposition 10. *Let Q and $Q_n, n = 1, 2, \ldots$, be probability measures on a common Polish space Ψ. Let h be a continuous function from Ψ to a Polish space Υ, and let R and R_n be the measures induced by h from Q and Q_n, respectively. If $Q_n \to Q$ as $n \to \infty$, then $R_n \to R$ as $n \to \infty$.*

* **Problem 8.** Prove the preceding proposition.

Problem 9. Let X and X_n, $n = 1, 2, \ldots$, be $\mathbf{C}[0, 1]$-valued random variables and suppose that $X_n \xrightarrow{\mathcal{D}} X$ as $n \to \infty$. Prove that

$$\max\{X_n(t) \colon 0 \le t \le 1\} \xrightarrow{\mathcal{D}} \max\{X(t) \colon 0 \le t \le 1\}$$

as $n \to \infty$.

18.3. Relative sequential compactness

In this section we prove a basis fact about compactness in Polish spaces, introduce the concept of relative sequential compactness for families of probability distributions on a Polish space, and finally prove that any such family is relatively sequentially compact if the Polish space itself is compact.

As described in Appendix B, a set is *totally bounded* if for every $\varepsilon > 0$, it is contained in the union of a finite collection of balls of radius less than ε.

Proposition 11. *A Polish space is compact if and only if it is totally bounded.*

Problem 10. Prove the preceding proposition. *Hint:* Use Proposition 2 and Proposition 3, both of Appendix B.

Problem 11. Either by using the preceding proposition or by a direct argument, show that $\{x \colon |x(t)| \le 1 \text{ for all } t \in [0, 1]\}$ is not a compact subset of the Polish space $\mathbf{C}[0, 1]$ described in Problem 3.

Problem 12. Without using Proposition 11, give a direct proof of the total boundedness of the infinite-dimensional cube defined in Example 2.

Definition 12. A family Q of probability distributions on a Polish space Ψ is *relatively sequentially compact* if every sequence $(Q_n \colon n = 1, 2, \dots)$ of members of Q has a convergent subsequence.

The following lemma constitutes a major step in the identification of the relatively sequentially compact families of distributions on a Polish space.

Lemma 13. *Every family of probability distributions on the infinite-dimensional cube $[0,1]^\infty$ is relatively sequentially compact.*

PROOF. Let $(Q_n \colon n = 1, 2, \dots)$ be a sequence of probability measures on $[0,1]^\infty$. By Problem 6 there exists a continuous function g from $[0,1]$ onto $[0,1]^\infty$. Define $f \colon [0,1]^\infty \to [0,1]$ by

$$f(x) = \inf\{t \in [0,1] \colon g(t) = x\}, \quad x \in [0,1]^\infty.$$

It follows from the continuity of g that $g \circ f$ is the identity function. The measurability of f follows from the fact that $\{x \colon f(x) \leq a\}$ is compact, being the image under g of the compact set $[0,a]$ (see Problem 7 of Appendix B). For $n = 1, 2, \dots$, let R_n be the sequence of probability measures induced on $[0,1]$ by f from Q_n. By Theorem 13 of Chapter 14, there exists a convergent subsequence $(R_{n_k} \colon k = 1, 2, \dots)$ with limit equal to some probability measure R on $[0,1]$. Let Q be the measure induced on $[0,1]^\infty$ by g. Note that for each n, Q_n is the measure induced by g from R_n. By Proposition 10, the subsequence $(Q_{n_k} \colon k = 1, 2, \dots)$ converges to Q. $\square$

The following result is worth remembering, even though it is a special case of the forthcoming Theorem 17.

Proposition 14. *Every family of probability distributions on a compact Polish space is relatively sequentially compact.*

Problem 13. Use Problem 7 of Appendix B and Lemma 4 and Lemma 13 of this chapter to prove the preceding proposition.

We conclude this section with a simple result that is quite useful for proving convergence in distribution in Polish spaces.

Proposition 15. *Let $(Q_n \colon n = 1, 2, \dots)$ be a relatively sequentially compact sequence of probability measures on a Polish space such that every convergent subsequence has the same limiting probability measure Q. Then $Q_n \to Q$ as $n \to \infty$.*

Problem 14. Prove the preceding proposition. *Hint:* See the solution of Proposition 4 of Appendix B.

18.4. Uniform tightness and the Prohorov Theorem

In this section we identify necessary and sufficient conditions for a family of probability measures on a Polish space to be relatively sequentially compact.

Definition 16. A probability distribution on a Polish space Ψ is *tight* if, for every $\varepsilon > 0$, there exists a compact subset K of Ψ such that $Q(K^c) < \varepsilon$. A family Q of probability distributions on Ψ is *uniformly tight* if, for every $\varepsilon > 0$, there exists a compact subset K of Ψ such that $Q(K^c) < \varepsilon$ for every $Q \in Q$.

Some use the term 'tight' for a family to mean 'uniformly tight', but we will not use the abbreviated term.

Theorem 17. [Prohorov] *A family of probability measures on a Polish space Ψ is relatively sequentially compact if and only if it is uniformly tight.*

PROOF. Let Q be a uniformly tight family of probability measures on Ψ, and let $(Q_n : n = 1, 2, \ldots)$ be a sequence of members of Q. By definition, there exists, for each $\varepsilon > 0$, a compact set $K_\varepsilon \subseteq \Psi$ with the property that $Q_n(K_\varepsilon) > 1 - \varepsilon$ for all n. We use the function φ defined in Lemma 4 to transfer everything to the Polish space $[0, 1]^\infty$. Let $C_\varepsilon = \varphi(K_\varepsilon)$. By Problem 7 of Appendix B, each C_ε is a compact subset of $[0, 1]^\infty$. Let $(R_n : n = 1, 2, \ldots)$ be the sequence of probability measures on $[0, 1]^\infty$ induced by φ from the sequence $(Q_n : n = 1, 2, \ldots)$. Note that $R_n(C_\varepsilon) > 1 - \varepsilon$ for all n and ε.

By Lemma 13 there is a subsequence $(R_{n_k} : k = 1, 2, \ldots)$ that converges to a probability measure R on $[0, 1]^\infty$. By the Portmanteau Theorem, $R(C_\varepsilon) \geq 1 - \varepsilon$ for all ε. Since $C_\varepsilon \subseteq \varphi(\Psi)$ for all ε, it follows that $R(\varphi(\Psi)) = 1$. Let Q be the measure induced by φ^{-1} on Ψ from R. It follows from the continuity of φ^{-1} and Proposition 10 that $Q_{n_k} \to Q$ as $k \to \infty$.

To prove the converse, suppose that Q is a relatively sequentially compact family of probability measures on Ψ. Let $(\psi_n : n = 1, 2, \ldots)$ be a countable dense subset of Ψ, and for each $\delta > 0$, let $B(\psi_n, \delta)$ be the open ball of radius δ about the point ψ_n. Let $\overline{B(\psi_n, \delta)}$ be the closure of $B(\psi_n, \delta)$.

We now show that for each δ, there exists an integer $p(\delta)$ such that for all $Q \in Q$,

$$Q\left(\bigcup_{n=1}^{p(\delta)} B(\psi_n, \delta)\right) > 1 - \delta.$$

Suppose that such an integer $p(\delta)$ does not exist for some particular choice of $\delta > 0$. Then for each positive integer m, there exists a probability measure

$Q_m \in \mathcal{Q}$ such that

$$Q_m\left(\bigcup_{n=1}^{m} B(\psi_n, \delta)\right) \leq 1 - \delta.$$

By relative sequential compactness, the sequence $(Q_m\colon m = 1, 2, \ldots)$ has a convergent subsequence with limit equal to some probability measure Q. By the Portmanteau Theorem,

$$Q\left(\bigcup_{n=1}^{m} B(\psi_n, \delta)\right) \leq 1 - \delta$$

for all m. Since the collection of balls $(B(\psi_n, \delta)\colon n = 1, 2, \ldots)$ covers the space Ψ, it follows, by Continuity of Measure, that $Q(\Psi) \leq 1 - \delta$. Since Q is a probability measure, we have derived a contradiction, so the integer $p(\delta)$ must exist for all $\delta > 0$ as asserted.

For each $\delta > 0$, let

$$C_\delta = \bigcup_{n=1}^{p(\delta)} \overline{B(\psi_n, \delta)}.$$

Each set C_δ is the union of a finite number of closed sets, and hence is closed (but not necessarily compact!). Fix $\varepsilon > 0$ and let

$$K = \bigcap_{n=1}^{\infty} C_{\varepsilon/2^n}.$$

Since K is an intersection of closed sets, it is closed. By construction, K is totally bounded, so K is compact by Proposition 11. Since $Q(C_\delta) > 1 - \delta$ for all $\delta > 0$ and all $Q \in \mathcal{Q}$, an elementary calculation shows that $Q(K) > 1 - \varepsilon$ for all $Q \in \mathcal{Q}$. $\square$

Corollary 18. *Every probability measure on a Polish space is tight.*

* **Problem 15.** Prove that if Q is a probability measure on a Polish space Ψ, then for any open set $A \subseteq \Psi$,

$$Q(A) = \sup\{Q(K)\colon K \text{ is compact and } K \subseteq A\}.$$

18.5. Convergence in product spaces

We begin with some terminology. For $j = 1, 2, \ldots$, let (Ψ_j, ρ_j) be Polish spaces, and let (Ψ, ρ) be the corresponding product Polish space (see Proposition 2). Let $A = \{j_1, j_2, \ldots\}$ be any (finite or countably infinite) set of positive integers. For simplicity, assume that $j_1 < j_2 < \ldots$. Let $h_A\colon \Psi \to \Psi_{j_1} \times \Psi_{j_2} \times \ldots$ be the function

$$x = (x_1, x_2, \ldots) \rightsquigarrow (x_{j_1}, x_{j_2}, \ldots),$$

known as the *projection of* Ψ *onto the coordinates indexed by* A. It is easily checked that any such projection is continuous.

If Q is a distribution on the Polish space Ψ, then the measure induced from Q by h_A is called the *marginal of* Q *corresponding to* A. If A is finite, then the measure induced from Q by h_A is known as a *finite-dimensional marginal*. If A has cardinality n, then the corresponding marginal is sometimes called an *n-dimensional marginal*. If $A = \{j\}$, then the corresponding 1-dimensional marginal is called the jth *coordinate marginal*.

It is easy to use the Uniqueness Theorem to show that a probability measure on a countable product of Polish spaces is determined by its finite-dimensional marginals. The next theorem says that convergence in distribution on a countable product of Polish spaces is equivalent to convergence in distribution of each of the finite-dimensional marginal distributions.

Theorem 19. *Let* Q_n, $n = 1, 2, \ldots$, *and* Q *be distributions on the Polish space* $\Psi = \bigotimes_{j=1}^{\infty} \Psi_j$ *that was defined in* Proposition 2. *If* $Q_n \to Q$ *as* $n \to \infty$, *then for each set* $A \subseteq \{1, 2, \ldots\}$, $Q_n^A \to Q^A$ *as* $n \to \infty$, *where* Q_n^A *and* Q^A *are the measures induced from* Q_n *and* Q *by the projection* h_A. *On the other hand, if for all finite sets* A, *there exists a measure* $\tilde{Q}^A$ *such that* $Q_n^A \to \tilde{Q}^A$ *as* $n \to \infty$, *then there exists a measure* Q *such that* $Q_n \to Q$ *as* $n \to \infty$, *and* $Q^A = \tilde{Q}^A$.

PROOF. The first part of the theorem follows immediately from Proposition 10 and the fact that each of the functions h_A is continuous.

For the proof of the second part, assume that for each finite A, there exists a measure $\tilde{Q}^A$ such that $Q_n^A \to \tilde{Q}^A$ as $n \to \infty$. We use the convergence of the 1-dimensional marginals to prove that the sequence (Q_n) is uniformly tight. For each $j = 1, 2, \ldots$, let Q_n^j be the jth coordinate marginal of Q_n. Since $Q_n^j \to \tilde{Q}^{\{j\}}$ as $n \to \infty$, the sequence (Q_n^j) is uniformly tight for each fixed j. Fix $\varepsilon > 0$, and for each j, choose a compact set $K_j \subseteq \Psi_j$ such that $Q_n^j(K_j) > 1 - \varepsilon/2^j$ for all n. Let $K = K_1 \times K_2 \times \ldots$. By the Tychonoff Theorem (see Appendix C), K is a compact subset of Ψ. For each n,

$$Q_n(K) = Q_n\Big(\bigcap_{j=1}^{\infty} \{(x_1, x_2, \ldots): x_j \in K_j\}\Big) \geq 1 - \sum_{j=1}^{\infty} Q_n^j(K_j^c) > 1 - \varepsilon.$$

Thus the sequence (Q_n) is uniformly tight.

Since (Q_n) is uniformly tight, there exists a measure R and a subsequence (Q_{n_k}) such that $Q_{n_k} \to R$ as $k \to \infty$. By the first part of the theorem, the finite-dimensional marginals of the terms in the sequence (Q_{n_k}) converge to the finite-dimensional marginals of R. Thus, the finite-dimensional marginals of R are the measures $\tilde{Q}^A$. Any other subsequential limit must have these same marginals, and thus be equal to R. An application of Proposition 15 completes the proof. $\square$

Problem 16. Let $\Psi = \Psi_1 \times \Psi_2 \times \ldots$ be a countable product of Polish spaces, as in Theorem 19, and let (Q_n) be a sequence of probability measures on Ψ. Show that (Q_n) is uniformly tight if and only if (Q_n^j) is uniformly tight for each j, where Q_n^j is the jth coordinate marginal of Q_n.

Problem 17. Let Ψ be as in the preceding problem, and for each n, let X_n be a Ψ_n-valued random variable. Show that as $n \to \infty$,

$$(X_1, X_2, \ldots, X_n, X_n, X_n, \ldots) \xrightarrow{\ \mathcal{D}\ } (X_1, X_2, X_3, \ldots).$$

Problem 18. Describe how to use the previous result to quickly obtain Theorem 16 of Chapter 9 as a corollary of Theorem 7 of that same chapter in the Polish space setting.

In general, the hypothesis of the second part of Theorem 19 cannot be weakened in any significant way. For example, it would not be enough for all the n-dimensional marginals to converge for some fixed n. But there is one important special case in which the convergence of the 1-dimensional marginals is sufficient.

Theorem 20. *Let J be a countable set, and for $j \in J$, let Ψ_j be a Polish space. Set*

$$\Psi = \bigotimes_{j \in J} \Psi_j,$$

viewed as a Polish space via Proposition 2. For $n = 1, 2, \ldots,$ let

$$Q_n = \bigotimes_{j \in J} Q_n^j$$

be a product measure on Ψ, where for each j and n, Q_n^j is a probability distribution on the Borel subsets of Ψ_j. Then the sequence (Q_n) converges to a distribution Q as $n \to \infty$ if and only if for each j, the sequence (Q_n^j) converges to a distribution Q^j on Ψ_j, in which case

$$Q = \bigotimes_{j \in J} Q^j.$$

PARTIAL PROOF. For the 'only if' aspect, note that for each j and n, the probability measure Q_n^j is the jth coordinate marginal of Q_n. Thus, if $Q_n \to Q$ as $n \to \infty$, it follows from Theorem 19 that for each j, the sequence (Q_n^j) converges as $n \to \infty$ to the jth coordinate marginal of Q, as desired.

For the proof of the 'if' portion we focus on the case $J = \{1, 2\}$ and leave the rest to the reader. Let g be an $\mathbb{R}$-valued bounded continuous function on $\Psi_1 \times \Psi_2$. By the Fubini Theorem, it is enough to show that

$$(18.7) \qquad \lim_{n \to \infty} \int_{\Psi_2} h_n \, dQ_n^2 = \int_{\Psi_2} h \, dQ^2,$$

where

$$h_n(y) = \int_{\Psi_1} g(x,y)\, Q_n^1(dx) \quad \text{and} \quad h(y) = \int_{\Psi_1} g(x,y)\, Q^1(dy).$$

Let $\varepsilon > 0$ and note that

$$\left| \int_{\Psi_2} h_n\, dQ_n^2 - \int_{\Psi_2} h\, dQ^2 \right| \leq \left| \int_{\Psi_2} h\, dQ_n^2 - \int_{\Psi_2} h\, dQ^2 \right|$$

(18.8)
$$+ \int_{\Psi_2 \setminus C} |h_n - h|\, dQ_n^2$$

$$+ \int_C |h_n - h|\, dQ_n^2$$

for any compact $C \subseteq \Psi_2$. There exists l such that the first term on the right is less than $\varepsilon/3$ for $n > l$, since h inherits continuity and boundedness from g and since $Q_n^2 \to Q^2$ as $n \to \infty$. The second term on the right is less than $\varepsilon/3$ for all n and some C, since all $|h_n|$ and $|h|$ inherit a common bound from g and since the sequence (Q_n^2) is uniformly tight.

For the last term on the right of (18.8), cover C by a finite number of sets $B_1, \ldots, B_k$ having the property that $|g(x_1, u) - g(x_1, v)| < \varepsilon/9$ whenever u and v are in a common B_i. Fix $u_i \in B_i$. There exist l_i such that for $n > l_i$, the following calculation is valid for all $v_i \in B_i$:

$$|h_n(v_i) - h(v_i)| < |h_n(v_i) - h_n(u_i)| + |h_n(u_i) - h(u_i)| + |h(u_i) - h(v_i)| < 3(\tfrac{\varepsilon}{9}) = \tfrac{\varepsilon}{3}.$$

Therefore $|h_n(v) - h(v)| < \varepsilon/3$ for $v \in C$ and $n > \max\{l_i : 1 \leq i \leq k\}$, and thus the left side of (18.8) is less than ε for $n > l \vee \max\{l_i : 1 \leq i \leq k\}$. $\square$

Problem 19. Complete the proof of the preceding theorem by first treating the case of $\#J < \infty$ by mathematical induction and then treating the case of J being countably infinite.

18.6. The Continuity Theorem for $\mathbb{R}^d$

The following result generalizes Theorem 15 of Chapter 14.

Theorem 21. [Continuity (for Characteristic Functions in $\mathbb{R}^d$)] *A sequence of probability distributions on $\mathbb{R}$ converges to a probability distribution Q if and only if the sequence of corresponding characteristic functions converges pointwise to a function γ which is continuous at 0, in which case the convergence to γ is uniform on each compact subset of $\mathbb{R}^d$, and γ is the characteristic function of Q.*

PARTIAL PROOF. We leave it as a problem to prove that if $(Q_n : n = 1, 2, \ldots)$ is a sequence of probability measures on $\mathbb{R}^d$ that converges to a probability measure Q, then $(\beta_n : n = 1, 2, \ldots)$ converges to β uniformly on each compact set, where β_n and β are the characteristic functions of Q_n and Q, respectively.

For the converse, assume that the sequence $(\beta_n \colon n = 1, 2, \ldots)$ converges point-wise to a function γ that is continuous at $0 \in \mathbb{R}^d$. It follows from Problem 58 of Chapter 13 that, for each j, the characteristic functions of the jth coordinate marginals converge to a function that is continuous at $0 \in \mathbb{R}$. By the Continuity Theorem for $\mathbb{R}$, the 1-dimensional marginals converge. By Problem 16, the sequence (Q_n) is uniformly tight. By the first part of this theorem, every convergent subsequence of (Q_n) has a limit with characteristic function γ. By Theorem 16 of Chapter 13, these convergent subsequences all have the same limit. An application of Proposition 15 completes the proof. $\square$

Problem 20. Complete the proof of the preceding theorem by doing the first portion of the proof.

Problem 21. Find a sequence $((X_n, Y_n) \colon n = 1, 2, \ldots)$ which does not converge in distribution (as a sequence of $\mathbb{R}^2$-valued random variables), but for which both of the sequences $(X_n \colon n = 1, 2, \ldots)$ and $(Y_n \colon n = 1, 2, \ldots)$ converge in distribution. For your example, find a $w \in \mathbb{R}^2$ for which $E(\exp(i\langle w, (X_n, Y_n)\rangle))$ does not converge.

Problem 22. Let $(X_n \colon n = 1, 2, \ldots)$ be an iid sequence of $\mathbb{R}$-valued random variables. Show that as $n \to \infty$,

$$(X_n, \, X_{n+1} + \tfrac{1}{n} X_n) \xrightarrow{\mathcal{D}} (X_1, X_2).$$

The preceding problem shows that the lack of independence is not necessarily preserved when passing to a limit. Theorem 20 says that independence is preserved.

In the proof of Theorem 21, we found it useful to analyze a sequence of distributions on $\mathbb{R}^d$ in terms of related distributions on $\mathbb{R}$. The following theorem extends this idea.

Theorem 22. [Cramér-Wold Device] *Let $d < \infty$. A sequence $(X_n \colon n = 1, 2, \ldots)$ of $\mathbb{R}^d$-valued random variables converges in distribution to a random variable X if and only if each sequence $(\langle w, X_n \rangle \colon n = 1, 2, \ldots)$, $w \in \mathbb{R}^d$, converges in distribution in which case*

$$(18.9) \qquad\qquad \langle w, X_n \rangle \xrightarrow{\mathcal{D}} \langle w, X \rangle$$

for each $w \in \mathbb{R}^d$.

PROOF. The convergence in (18.9) follows immediately from Proposition 10.

For the converse, suppose that $(\langle w, X_n \rangle)$ converges in distribution for each w. It is clear that if (X_n) converges in distribution or has a convergent subsequence, the limit must have that unique distribution whose characteristic function is

$$w \rightsquigarrow \lim_{n \to \infty} E(e^{i\langle w, X_n \rangle}).$$

In view of the Prohorov Theorem, we only need show uniform tightness of $\{Q_n : n = 1, 2, \dots\}$, where Q_n is the distribution of X_n, because then an appeal to Proposition 15 completes the proof. By setting $w = (0, \dots, 0, 1, 0, \dots, 0)$, with 1 in the jth position, we see that the sequence $(X_{j,n} : n = 1, 2, \dots)$ of jth coordinates converges in distribution. Thus, for each j, the sequence of jth coordinate marginals of (Q_n) is uniformly tight, so (Q_n) is uniformly tight by Problem 16. $\square$

Problem 23. Show that the hypothesis of convergence of $(\langle w, X_n \rangle)$ for all w in the preceding theorem cannot be replaced by the hypothesis of convergence for all w in some basis of $\mathbb{R}^d$.

Theorem 23. [Multi-dimensional Central Limit] *Let d be a positive integer. Let $(X_1, X_2, \dots)$ be an iid sequence of $\mathbb{R}^d$-valued random variables having finite mean vector $\mu = E(X_1)$ and finite covariance matrix*

$$\Sigma = E\big([X_1 - \mu]^T [X_1 - \mu]\big),$$

where $[X_1 - \mu]$ denotes the row matrix corresponding to the vector $X_1 - \mu$ and $[\cdot]^T$ denotes transpose. Then

$$\frac{\sum_{k=1}^n X_k - n\mu}{\sqrt{n}} \xrightarrow{\mathcal{D}} Z,$$

where Z is a normally distributed $\mathbb{R}^d$-valued random variable with mean vector 0 and covariance matrix Σ.

* **Problem 24.** Use the Classical Central Limit Theorem and the Cramér-Wold Device to prove the preceding theorem.

Problem 25. Apply the Multi-dimensional Central Limit Theorem to the iid sequence $(Z_1, Z_2, \dots)$ where each Z_j has the same distribution as the random vector Z of Problem 60 of Chapter 13.

18.7. † The Prohorov metric

Let $\mathcal{Q}(\Psi)$ denote the family of all probability distributions on a Polish space (Ψ, ρ). Our final goal of this chapter is to turn $\mathcal{Q}(\Psi)$ into a Polish space with a metric $\hat{\rho}$ so that convergence of sequences in the Polish space $(\mathcal{Q}(\Psi), \hat{\rho})$ is equivalent to convergence of sequences of distributions on Ψ. The metric $\hat{\rho}$ we will use is called the *Prohorov metric* and is defined by

(18.10) $\hat{\rho}(Q, R) = \inf\{\varepsilon : R(A) \le Q(A_\varepsilon) + \varepsilon \text{ for every Borel } A \subseteq \Psi\},$

where

$$A_\varepsilon = \{x : \rho(x, y) < \varepsilon \text{ for some } y \in A\}.$$

(Note: It is important that A_ε be Borel; indeed, it is open, since it is the union of open balls.)

* **Problem 26.** Choose $\delta, \varepsilon > 0$. Prove that if $R(A) \leq Q(A_\varepsilon) + \delta$ for all Borel sets $A \subseteq \Psi$, then $Q(A) \leq R(A_\varepsilon) + \delta$ for all Borel sets $A \subseteq \Psi$.

Problem 27. Prove that $\hat{\rho}$ defined by (18.10) is a metric.

From the preceding problem we see that $(Q(\Psi), \hat{\rho})$ is a metric space. Let C be a countable dense subset of Ψ. Then it is easy to show that the set $\mathcal{C}$ of probability distributions whose values are rational and whose supports are finite subsets of C is a countable dense subset of $Q(\Psi)$. We have almost proved the following theorem.

Theorem 24. Let (Ψ, ρ) be a Polish space and let $Q(\Psi)$ denote the family of all probability measures on (Ψ, ρ). Define $\hat{\rho}$ by (18.10). Then $(Q(\Psi), \hat{\rho})$ is a Polish space.

PARTIAL PROOF. In view of the discussion preceding the theorem, we need only show that $(Q(\Psi), \hat{\rho})$ is complete. Let $(Q_n \colon n = 1, 2, \dots)$ be an arbitrary Cauchy sequence. To prove the convergence of a Cauchy sequence, it is enough to find a subsequence that converges. By the Prohorov Theorem, it is enough to find a subsequence that is uniformly tight. Using the definition of Cauchy sequence, a routine argument shows that there exists a subsequence $(R_n \colon n = 1, 2, \dots)$ with the property that the $\hat{\rho}(R_n, R_{n+1}) < 1/2^{n+1}$ for $n = 1, 2, \dots$. We will prove that this subsequence is uniformly tight.

Fix $\varepsilon > 0$. We will define a sequence of compact subsets of Ψ. Choose a positive integer l such that $1/2^l < \varepsilon/2$. By Corollary 18 we can find a compact set K such that $R_n(K) > 1 - \varepsilon/2$ for $n = 1, \dots, l$. Let $K_j = K$ for $j = 1, \dots, l$. We now proceed recursively to define compact sets $K_{l+1}, K_{l+2}, \dots$, such that for all $n > l$,

$$R_n(K_n) > 1 - \left(1 - \frac{1}{2^{n+1-l}}\right)\varepsilon$$

and

(18.11) $$K_{n+1} \subseteq (K_n)_{2^{-(n+1)}},$$

where

$$(K_n)_{1/2^{n+1}} = \left\{x \in \Psi \colon \rho(x, y) < \frac{1}{2^{n+1}} \text{ for some } y \in K_n\right\}.$$

We have already defined K_l with the desired properties. Assume that sets $K_l, \dots, K_n$ have been defined with the desired properties for some $n \geq l$. By the definition of $\hat{\rho}$ and our choice of the integer l,

$$R_{n+1}((K_n)_{1/2^{n+1}}) > R_n(K_n) - \frac{1}{2^{n+1}} > 1 - \left(1 - \frac{1}{2^{n+2-l}}\right)\varepsilon.$$

It follows from Problem 15 that there exists a compact set K_{n+1} satisfying (18.11) such that $R(K_{n+1}) > 1 - (1 - 2^{-(n+2-l)})\varepsilon$. The recursive construction of the sequence $(K_n : n = 1, 2, \ldots)$ of compact sets is complete.

Let

$$K = \text{the closure of } \bigcup_{n=1}^{\infty} K_n.$$

Clearly $R_n(K) > 1 - \varepsilon$ for all n. The proof of uniform tightness is completed by showing that K is compact. By Proposition 11, it is enough to prove that K is totally bounded. This final step is left as a exercise. $\square$

Problem 28. Prove that the set K defined in the preceding proof is totally bounded. *Hint:* Show that for each $\varepsilon > 0$, a sufficiently large value of n can be found so that any covering of $K_1 \cup \cdots \cup K_n$ by ε-balls can be 'inflated' to a covering of K by doubling the diameters of each of the balls.

* **Problem 29.** For the metric space of distributions on $\mathbb{R}$, calculate the distance between the uniform distribution on $[0, 1]$ and the uniform distribution on $[a, a+1]$. Also calculate the distance between the uniform distributions on $[0, 8]$ and $[0, 9]$.

Problem 30. For the metric space of distributions on $\mathbb{R}^2$, calculate the distance between the delta distribution at $(0, 0)$ and the uniform distribution on the square region with vertices at $(\pm a, \pm a)$.

Problem 31. For the metric space of distributions on $\mathbb{R}^\infty$, calculate the distance between the delta distribution at $(0, 0, 0, \ldots)$ and the distribution Q_d induced by the uniform distribution on $[-1, 1]$ in $\mathbb{R}$ and the function $\varphi_d : \mathbb{R} \to \mathbb{R}^\infty$ defined by

$$\varphi_d(x) = (0, \ldots, 0, x, 0, 0, \ldots),$$

where x on the right side is the dth term.

Theorem 25. *Let* $Q(\Psi)$ *be the Polish space described in* Theorem 24. *Then* $\hat{\rho}(Q, Q_n) \to 0$ *as* $n \to \infty$ *if and only if* $Q_n \to Q$ *as* $n \to \infty$.

PROOF. Suppose that $\hat{\rho}(Q, Q_n) \to 0$ and let C be a closed set. For each $\varepsilon > 0$ there exists an integer l such that

$$Q_n(C) \le Q(C_\varepsilon) + \varepsilon$$

for $n \ge l$. Hence,

$$\limsup_{n \to \infty} Q_n(C) \le Q(C_\varepsilon) + \varepsilon.$$

Now let $\varepsilon \searrow 0$ through a sequence and use the equation

$$C = \bigcap_{\varepsilon > 0} C_\varepsilon,$$

which is a consequence of the fact that C is closed, to conclude

$$\limsup_{n \to \infty} Q_n(C) \leq Q(C) \,.$$

An appeal to the Portmanteau Theorem completes this half of the proof.

For the converse assume that $\int g \, dQ_n \to \int g \, dQ$ for all continuous bounded functions g. By the Portmanteau Theorem,

$$(18.12) \qquad\qquad \lim_{n \to \infty} Q_n(A) = Q(A)$$

for every Borel set A for which $Q(\partial A) = 0$.

Let $\varepsilon > 0$ and let $(x_j : j = 1, 2, \dots)$ be a dense sequence in Ψ. For each j, let B_j be a ball centered at x_j with radius strictly between $\frac{\varepsilon}{4}$ and $\frac{\varepsilon}{2}$ and having the additional property that $Q(\partial B_n) = 0$. For each j, set

$$C_j = B_j \setminus \bigcup_{i=1}^{j-1} B_i \,.$$

It is clear that:

- $C_j \cap C_i = \emptyset$ if $i \neq j$;
- $\Psi = \bigcup_{j=1}^{\infty} C_j$;
- $1 = \sum_{j=1}^{\infty} Q(C_j)$;
- $Q(\partial C_j) = 0$ for each j;
- the distance between any two members of any one C_j is less than ε.

Choose k so that $\sum_{j=1}^{k} Q(C_j) > 1 - \frac{\varepsilon}{2}$ and then use (18.12) to deduce the existence of l such that

$$Q(C_j) < Q_n(C_j) + \tfrac{\varepsilon}{2k} \,, \quad \text{for } j \leq k, \, n \geq l \,.$$

Let A be any Borel set and denote by A_ε the set of points that each lie no more than distance ε from some point in A. Let $(C_{j_1}, C_{j_2}, \dots, C_{j_r})$ be the subsequence of (C_j) consisting of those C_j for which $j \leq k$ and $C_j \cap A \neq \emptyset$. Then, for $n \geq l$,

$$Q(A) < \frac{\varepsilon}{2} + \sum_{i=1}^{r} Q(C_{j_i}) < \varepsilon + \sum_{i=1}^{r} Q_n(C_{j_i}) \leq \varepsilon + Q_n(A_\varepsilon) \,.$$

Therefore the distance between Q_n and Q is less than ε for $n \geq l$. $\square$

CHAPTER 19
The Invariance Principle and Brownian Motion

In this chapter, we bring together several of the key ideas of previous chapters to construct one of the most important objects in all of probability theory, namely 'Brownian motion'. In order to apply the theory developed in Chapter 18, we will first take the point of view that Brownian motion is a random variable that takes values in the Polish space $\mathbf{C}[0,1]$ and later switch to the Polish space $\mathbf{C}[0,\infty)$. We will find that when a random walk on $\mathbb{R}$ with finite variance is converted to a $\mathbf{C}[0,1]$-valued random variable in a natural way, then it can be centered and scaled so that its distribution approximates that of Brownian motion. As a consequence, much can be learned about the asymptotic properties of random walks by looking at the properties of Brownian motion. We rely frequently on the material concerning Polish spaces in Chapter 18. The Classical Central Limit Theorem (Chapter 15) and the Arzelà-Ascoli Theorem (Theorem 5 of Appendix B) also play important roles.

We include only a small fraction of the many interesting features of Brownian motion on $[0,\infty)$ that have been discovered over the years, but the ones we do include indicate the variety and subtlety of such results. Some of these results involve stopping times, so this chapter also contains the basic theory needed to extend ideas from Chapter 11 about filtrations and stopping times to the continuous time setting.

Random variables X that are $\mathbf{C}[0,1]$-valued play a central role in this chapter. Thus, $X(\omega)$ is a continuous function from $[0,1]$ to $\mathbb{R}$. Its value at $t \in [0,1]$ will be denoted by $X_t(\omega)$. Hence X_t is an $\mathbb{R}$-valued random variable. An alternative notation for X is

$$\omega \rightsquigarrow [t \rightsquigarrow X_t(\omega)].$$

When speaking of events involving either the $\mathbf{C}[0,1]$-valued random variable X or an $\mathbb{R}$-valued random variable X_t we will sometimes suppress ω. The notation $(X^{(n)} : n = 1, 2, \ldots)$ is a typical description of a sequence of $\mathbf{C}[0,1]$-valued random variables.

19.1. Certain sequences of distributions on C[0,1]

Let $(Y_1, Y_2, \ldots)$ be an iid sequence of $\mathbb{R}$-valued random variables having finite mean μ and positive finite variance σ^2. For $n = 1, 2, \ldots$, let $\mathbf{C}[0,1]$-valued random variables $X^{(n)}$ be defined as follows. First, for $m = 0, \ldots, n$, set

$$(19.1) \qquad X^{(n)}_{m/n}(\omega) = \frac{\sum_{k=1}^m (Y_k(\omega) - \mu)}{\sigma \sqrt{n}},$$

and then extend $X^{(n)}(\omega)$ to be continuous on $[0,1]$ by making it linear on each of the intervals $[\frac{m-1}{n}, \frac{m}{n}], m = 1, \ldots, n$. (See Example 2 and Figures 2.1, 2.2, and 2.3, all of Chapter 2.) The following problems and lemma constitute a first step towards the goal of proving that $(X^{(n)} : n = 1, 2, \ldots)$ converges in distribution.

Problem 1. Suppose that $\mathbb{R}^d$-valued random variables Z_n and U_n, $n = 1, 2, \ldots$, have the property that as $n \to \infty$, $Z_n \xrightarrow{D} Z$ for some Z and $U_n \xrightarrow{D} 0$. Prove that $Z_n + U_n \xrightarrow{D} Z$ as $n \to \infty$.

Problem 2. Suppose that a sequence $(Z_n : n = 1, 2, \ldots)$ of $\mathbb{R}$-valued random variables converges in distribution to an $\mathbb{R}$-valued random variable Z and that a sequence $(c_n : n = 1, 2, \ldots)$ of real numbers converges to 1. Prove that $c_n Z_n \xrightarrow{D} Z$ as $n \to \infty$.

Lemma 1. *Let $0 \le t_0 < t_1 < t_2 < \cdots < t_d \le 1$ and let $X^{(n)}$ be defined by the sentence containing (19.1). Then the sequence*

$$\left(\left((X^{(n)}_{t_1} - X^{(n)}_{t_0}), \ldots, (X^{(n)}_{t_d} - X^{(n)}_{t_{d-1}}) \right) : n = 1, 2, \ldots \right)$$

of random vectors converges in distribution to a random vector having independent normally distributed coordinates with the mean and variance of the jth coordinate equaling 0 and $(t_j - t_{j-1})$, respectively.

PROOF. For $0 < j \le d$ and $n = 1, 2, \ldots$, set

$$r_{j,n} = \frac{\lfloor nt_j \rfloor}{n} \quad \text{and} \quad U_{j,n} = X^{(n)}_{t_j} - X^{(n)}_{r_{j,n}}.$$

For $0 \le j < d$ and $n = 1, 2, \ldots$, set

$$s_{j,n} = \frac{\lceil nt_j \rceil}{n} \quad \text{and} \quad V_{j,n} = X^{(n)}_{s_{j,n}} - X^{(n)}_{t_j}.$$

Then for n sufficiently large,

$$t_0 \le s_{0,n} < r_{1,n} \le t_1 \le s_{1,n} < r_{2,n} \le \cdots \le t_{d-1} \le s_{d-1,n} < r_{d,n} \le t_d,$$

and

$$(19.2) \quad \begin{aligned} &((X_{t_1}^{(n)} - X_{t_0}^{(n)}), \ldots, (X_{t_d}^{(n)} - X_{t_{d-1}}^{(n)})) \\ &= ((X_{r_{1,n}}^{(n)} - X_{s_{0,n}}^{(n)}), \ldots, (X_{r_{d,n}}^{(n)} - X_{s_{d-1,n}}^{(n)})) \\ &\quad + (U_{1,n}, \ldots, U_{d,n}) + (V_{0,n}, \ldots, V_{d-1,n}). \end{aligned}$$

The mean of each coordinate in the last two terms is 0 and the variances approach 0 as $n \to \infty$. By the Chebyshev Inequality it follows that each of the last two terms on the right side of (19.2) converges to $(0, \ldots, 0)$ in distribution. Therefore in view of Problem 1 we only need prove that the sequence

$$(19.3) \quad \left(((X_{r_{1,n}}^{(n)} - X_{s_{0,n}}^{(n)}), \ldots, (X_{r_{d,n}}^{(n)} - X_{s_{d-1,n}}^{(n)})) : n = 1, 2, \ldots \right)$$

converges in distribution to the limit described in the lemma.

The coordinates in each term of the sequence (19.3) are independent, and by Theorem 20 of Chapter 18, independence is preserved on passing to the limit. Thus, we only need show for $1 \le j \le d$, that the sequence

$$\left((X_{r_{j,n}}^{(n)} - X_{s_{j-1,n}}^{(n)}) : n = 1, 2, \ldots \right)$$

converges in distribution to a normally distributed random variable with mean 0 and variance $(t_j - t_{j-1})$. By the Classical Central Limit Theorem the sequence

$$\left(\frac{\sqrt{t_j - t_{j-1}}}{\sqrt{r_{j,n} - s_{j-1,n}}} (X_{r_{j,n}}^{(n)} - X_{s_{j-1,n}}^{(n)}) : n = 1, 2, \ldots \right)$$

does have this property. Clearly,

$$\lim_{n \to \infty} \frac{\sqrt{r_{j,n} - s_{j-1,n}}}{\sqrt{t_j - t_{j-1}}} = 1.$$

An appeal to Problem 2 completes the proof. □

A $\mathbf{C}[0,1]$-valued random variable W has *stationary increments* if for $0 \le s < t \le 1$, the distribution of $(W_t - W_s)$ depends only on $t - s$. It has *independent increments* if for d any positive integer and $0 \le t_0 < t_1 < \cdots < t_d \le 1$, the random variables $(W_{t_i} - W_{t_{i-1}})$, $1 \le i \le d$, are independent. The term 'independent increments' is perhaps not ideal, since it does not reflect the fact that independence is only required for time intervals having nonoverlapping interiors.

Lemma 2. *Let $X^{(n)}$ be defined by the sentence containing (19.1). Then any subsequential distributional limit W of the sequence $(X^{(n)} : n = 1, 2, \ldots)$ has the following three properties:*

(i) *for $0 \le t \le 1$, W_t is normally distributed with mean 0 and variance t;*
(ii) *W has stationary increments;*
(iii) *W has independent increments.*

Moreover, if $W^{(1)}$ has the properties (i)-(iii) in common with W, then W and $W^{(1)}$ have the same distribution.

PROOF. Suppose that $X^{(n_k)} \xrightarrow{\mathcal{D}} W$ as $k \to \infty$. Since the function

$$x \rightsquigarrow \left([x(t_1) - x(t_0)], \, [x(t_2) - x(t_1)], \, \ldots, \, [x(t_d) - x(t_{d-1})] \right)$$

from $\mathbf{C}[0,1]$ to $\mathbb{R}^d$ is continuous, it follows from Proposition 10 of Chapter 18 that

$$\left((X_{t_1}^{(n_k)} - X_{t_0}^{(n_k)}), \, \ldots, \, (X_{t_d}^{(n_k)} - X_{t_{d-1}}^{(n_k)}) \right)$$
$$\xrightarrow{\mathcal{D}} \left((W_{t_1} - W_{t_0}), \, \ldots, \, (W_{t_d} - W_{t_{d-1}}) \right).$$

By Lemma 1, W satisfies (i)-(iii).

The distributions of any $\mathbf{C}[0,1]$-valued random variables W and $W^{(1)}$ satisfying (i)-(iii) are supported by $\{x \in \mathbf{C}[0,1]: x(0) = 0\}$ because of (i), and thus, because of (i)-(iii), agree on $\mathcal{E}$, the class of sets of the form

$$\left\{ x \in \mathbf{C}[0,1]: x(0) \in A, \, \left([x(t_1) - x(t_0)], \, \ldots, \, [x(t_d) - x(t_{d-1})] \right) \in B \right\},$$

where d is a positive integer, $0 = t_0 < t_1 < \cdots < t_d \le 1$, A is a Borel subset of $\mathbb{R}$, and B is a Borel subset of $\mathbb{R}^d$. Clearly $\mathcal{E}$ is closed under finite intersections.

It remains to show that $\sigma(\mathcal{E}) = \mathcal{C}$, the Borel σ-field of subsets of $\mathbf{C}[0,1]$. Let us describe $\mathcal{E}$ in a different manner. Since the function

$$(u_0, u_1, u_2, \ldots, u_d) \rightsquigarrow \left(u_0, (u_1 - u_0, \, u_2 - u_1, \, \ldots, \, u_d - u_{d-1}) \right)$$

is continuous and one-to-one from $\mathbb{R}^{d+1}$ onto $\mathbb{R}^{d+1}$ with a continuous inverse, $\mathcal{E}$ may be described as the class of sets of the form

$$(19.4) \qquad \left\{ x \in \mathbf{C}[0,1]: \left(x(0), x(t_1), \, \ldots, \, x(t_d) \right) \in C \right\},$$

where d is a positive integer, $0 < t_1 < \cdots < t_d \le 1$, and C is a Borel subset of $\mathbb{R}^{d+1}$. To show that $\sigma(\mathcal{E}) = \mathcal{A}$, we will show that any set

$$(19.5) \qquad \left\{ x \in \mathbf{C}[0,1]: \rho(y, x) < \varepsilon \right\}, \quad y \in \mathbf{C}[0,1], \, \varepsilon > 0,$$

can be written in terms of countable unions and intersections of sets of the form (19.4). Since $\mathcal{A}$ is the smallest σ-field containing sets of the form (19.5), it will follow that $\sigma(\mathcal{E}) = \mathcal{A}$. Therefore, the following set-theoretic equality completes the proof of uniqueness:

$$\left\{ x \in \mathbf{C}[0,1]: \rho(y, x) < \varepsilon \right\} = \bigcup_{k=1}^{\infty} \bigcap_{m=1}^{\infty} \bigcap_{i=0}^{2^m} \left\{ x: |x(i2^{-m}) - y(i2^{-m})| < \frac{k\varepsilon}{k+1} \right\}. \quad \square$$

Any $\mathbf{C}[0,1]$-valued random variable satisfying properties (i)-(iii) of the preceding lemma is called a *Brownian motion* on $[0,1]$ or a *Wiener process* on $[0,1]$. The lemma asserts that all Wiener processes have the same distribution and that all subsequential limits of $(X^{(n)}: n = 1, 2, \ldots)$ have the same distribution. The common distribution of all Wiener processes on $[0,1]$, if any exist, is called *Wiener measure* on $[0,1]$. We have yet to prove the existence of Wiener measure or of subsequential distributional limits of the sequence $(X^{(n)}: n = 1, 2, \ldots)$.

19.2. The existence of and convergence to Wiener measure

In view of the Prohorov Theorem and Proposition 15 of Chapter 18, we may resolve the issues of the preceding paragraph by proving that the sequence $\{Q_n : n = 1, 2, \ldots\}$ is uniformly tight, where Q_n is the distribution of $X^{(n)}$, which itself is defined by the sentence containing (19.1). The uniform tightness will be established by a sequence of four lemmas. The first lemma concerns the Q_n-probability that $x(s)$ differs from $x(\frac{j}{m})$ by more than some fixed positive quantity z, where s is a time in the interval $[\frac{j}{m}, \frac{j+1}{m}]$.

Lemma 3. *For $z > 0$ and $j = 0, 1, \ldots$,*

(19.6)
$$\lim_{m \to \infty} \limsup_{n \to \infty} \left(m \sup\{Q_n(\{x : |x(s) - x(\tfrac{j}{m})| > z\}) : \tfrac{j}{m} \le s \le \tfrac{j+1}{m}\} \right) = 0.$$

PROOF. We express the probability in (19.6) in terms of the $C[0,1]$-valued random variable $X^{(n)}$:

$$Q_n\left(\{x : |x(s) - x(\tfrac{j}{m})| > z\}\right) = P\left[|X_s^{(n)} - X_{j/m}^{(n)}| > z\right].$$

Let $k_1(n), k_2(n)$ satisfy $\frac{k_1(n)}{n} \le \frac{j}{m} \le \frac{k_1(n)+1}{n}$ and $\frac{k_2(n)-1}{n} \le \frac{j+1}{m} \le \frac{k_2(n)}{n}$. Because the $C[0,1]$-valued random variable $X^{(n)}$ is linear on each interval of the form $[\frac{k-1}{n}, \frac{k}{n}]$, $k = 1, \ldots, n$, the supremum in (19.6) is bounded above by the supremum over s of the form $s = \frac{k}{n}$, $k_1(n) < k \le k_2(n)$. For such s,

$$P\left[|X_s^{(n)} - X_{j/m}^{(n)}| > z\right]$$
$$\le P\left[|X_s^{(n)} - X_{k_1(n)/n}^{(n)}| > \tfrac{z}{2}\right] + P\left[|X_{j/m}^{(n)} - X_{k_1(n)/n}^{(n)}| > \tfrac{z}{2}\right]$$

$$\le P\left[|X_s^{(n)} - X_{k_1(n)/n}^{(n)}| > \tfrac{z}{2} \,;\, |X_{k_2(n)/n}^{(n)} - X_s^{(n)}| \le \tfrac{z}{4}\right]$$
$$+ P\left[|X_s^{(n)} - X_{k_1(n)/n}^{(n)}| > \tfrac{z}{2} \,;\, |X_{k_2(n)/n}^{(n)} - X_s^{(n)}| > \tfrac{z}{4}\right]$$
$$+ P\left[|X_{j/m}^{(n)} - X_{k_1(n)/n}^{(n)}| > \tfrac{z}{2}\right]$$

$$\le P\left[|X_{k_2(n)/n}^{(n)} - X_{k_1(n)/n}^{(n)}| > \tfrac{z}{4}\right]$$
(19.7)
$$+ P\left[|X_s^{(n)} - X_{k_1(n)/n}^{(n)}| > \tfrac{z}{2} \,;\, |X_{k_2(n)/n}^{(n)} - X_s^{(n)}| > \tfrac{z}{4}\right]$$
$$+ P\left[|X_{j/m}^{(n)} - X_{k_1(n)/n}^{(n)}| > \tfrac{z}{2}\right].$$

We wish to obtain upper bounds for the three terms in (19.7). For the first term, we use the fact that $X^{(n)}$ is constructed by taking sums of iid random variables with finite variance. Let Z and G, respectively, denote a standard normally distributed random variable and its distribution function. The Classical

Central Limit Theorem, in combination with Problem 2, tells us that $(X^{(n)}_{k_2(n)/n} - X^{(n)}_{k_1(n)/n}) \xrightarrow{D} \frac{Z}{\sqrt{m}}$ as $n \to \infty$. Thus,

$$\lim_{n\to\infty} P\big[|X^{(n)}_{k_2(n)/n} - X^{(n)}_{k_1(n)/n}| > \tfrac{z}{4}\big] = 2\big(1 - G(\tfrac{z\sqrt{m}}{4})\big).$$

It is easily shown using the l'Hospital Rule that $(1 - G(x)) < x^{-4}$ for sufficiently large x. It follows that for large m,

$$(19.8) \qquad \lim_{n\to\infty} P\big[|X^{(n)}_{k_2(n)/n} - X^{(n)}_{k_1(n)/n}| > \tfrac{z}{4}\big] \leq \tfrac{512}{z^4 m^2}.$$

We use the independence of $(X^{(n)}_s - X^{(n)}_{k_1(n)/n})$ and $(X^{(n)}_{k_2(n)/n} - X^{(n)}_s)$ to factor the second term of (19.7) into the product of two probabilities, and then apply the Chebyshev Inequality to each factor:

$$P\big[|X^{(n)}_s - X^{(n)}_{k_1(n)/n}| > \tfrac{z}{2}\big]\, P\big[|X^{(n)}_{k_2(n)/n} - X^{(n)}_s| > \tfrac{z}{4}\big]$$

$$(19.9) \qquad \leq \left(\frac{\frac{1}{m} + \frac{1}{n}}{\frac{z^2}{16}}\right)^2.$$

The final term in (19.7) goes to 0 as $n \to \infty$, since

$$P\big[|X^{(n)}_{k_1(n)/n} - X^{(n)}_{j/m}| > \tfrac{z}{2}\big] \leq \tfrac{4}{nz^2}$$

by the Chebyshev Inequality and the definition of $k_1(n)$. Using this fact together with (19.8) and (19.9), we conclude that

$$\limsup_{n\to\infty} P\big[|X^{(n)}_s - X^{(n)}_{j/m}| > z\big] \leq \tfrac{c}{z^4 m^2},$$

where c is a constant that does not depend on m. Multiplying by m and letting $m \to \infty$ gives (19.6). $\square$

The next lemma improves on the previous one by moving the supremum over s into the event under consideration.

Lemma 4. For $z > 0$ and $j = 0, 1, \ldots,$

$$\lim_{m\to\infty} \limsup_{n\to\infty} m\, Q_n\Big(\big\{x\colon \sup\{|x(s) - x(\tfrac{j}{m})|\colon \tfrac{j}{m} \leq s \leq \tfrac{j+1}{m}\} > z\big\}\Big) = 0.$$

Problem 3. Prove the preceding lemma. *Hint:* Arrange to use the Etemadi Lemma to bring the supremum of Lemma 3 inside.

We need one further improvement, which is obtained by taking the union over $j = 0, \ldots, m-1$ and then using the triangle inequality, so that the supremum can be taken over pairs of times t, u such that $|t - u| \leq \tfrac{1}{m}$. The resulting probabilistic statement about uniform continuity provides just what we need to prove uniform tightness in Lemma 6.

Lemma 5. *For $z > 0$,*

$$\lim_{m\to\infty} \limsup_{n\to\infty} Q_n\left(\left\{x\colon \sup\{|x(u) - x(t)|\colon |t - u| \leq \tfrac{1}{m}\} > z\right\}\right) = 0.$$

PROOF. Fix $z > 0$. By Lemma 4

$$\lim_{m\to\infty} \limsup_{n\to\infty} Q_n\left(\bigcup_{j=0}^{m-1}\left\{x\colon \max\{|x(s) - x(\tfrac{j}{m})|\colon \tfrac{j}{m} \leq s \leq \tfrac{j+1}{m}\} > z\right\}\right) = 0.$$

The proof is now completed by replacing z by $\frac{z}{3}$ in this last inequality, and noting that if $|x(u) - x(t)| \geq z$ for some t and u such that $|t - u| \leq \frac{1}{m}$, then by the triangle inequality, there exists an integer $j \in \{0, 1, \ldots, m-1\}$ and a real number $s \in [\tfrac{j}{m}, \tfrac{j+1}{m}]$ such that $|x(s) - x(\tfrac{j}{m})| > \frac{z}{3}$. $\square$

Lemma 6. *The sequence $\{Q_n\colon n = 1, 2, \ldots\}$ is uniformly tight.*

PROOF. Fix $\varepsilon > 0$. We need to show the existence of a Borel subset A of $\mathbf{C}[0, 1]$ which has compact closure and for which $Q_n(A^c) < \varepsilon$ for every n. We will only include in A functions whose value at 0 is 0. Then, by the Arzelà-Ascoli Theorem, in order for A to have compact closure it is necessary and sufficient that A be a uniformly equicontinuous set of functions.

By Lemma 5 there exist, for each $k = 1, 2, \ldots$, integers p_k and r_k such that

$$(19.10) \qquad Q_n\left(\left\{x\colon \max\{|x(u) - x(t)|\colon |t - u| \leq \tfrac{1}{m}\} > \tfrac{1}{k}\right\}\right) < \varepsilon 2^{-k}$$

for $m = p_k$ and $n \geq r_k$. By monotonicity in m, (19.10) holds for $m \geq p_k$ and $n \geq r_k$. By the Continuity of Measure Theorem, there exists an integer q_k such that (19.10) also holds for $m \geq q_k$ and $n < r_k$. Set $m_k = p_k \vee q_k$. Then, for each n and k, (19.10) holds whenever $m = m_k$.

Let

$$A = \bigcap_{k=1}^{\infty}\left\{x\colon x(0) = 0 \text{ and } \max\{|x(u) - x(t)|\colon |t - u| \leq \tfrac{1}{m_k}\} \leq \tfrac{1}{k}\right\}.$$

From (19.10), $Q_n(A^c) < \varepsilon$ for every n. It remains to show that A is an equicontinuous set of functions. Let $\gamma > 0$. Choose k so that $\frac{1}{k} < \gamma$ and set $\delta = \frac{1}{m_k}$. Then, for $x \in A$,

$$\max\{|x(u) - x(t)|\colon |t - u| \leq \delta\} < \gamma,$$

as desired. $\square$

The uniform tightness that has now been established implies that every subsequence of (Q_n) has a further subsequence that converges. In view of Lemma 2 the existence of such a convergent subsequence establishes the existence of Wiener measure Q on $\mathbf{C}[0, 1]$ (and, as already contained in Lemma 2, Wiener measure is unique). Also by Lemma 2, all limits of convergent subsequences are identical, and hence by Proposition 15 of Chapter 18, $Q_n \to Q$ as $n \to \infty$. We have thus proved the next two theorems.

Theorem 7. *Wiener measure on* $\mathbf{C}[0,1]$ *exists and is unique.*

Theorem 8. [Donsker Invariance Principle] *Let* $(Y_1, Y_2, \dots)$ *be an iid sequence of* $\mathbb{R}$-*valued random variables having finite mean* μ *and positive finite variance* σ^2. *For* $n = 1, 2, \dots$, *define* $X^{(n)} \in \mathbf{C}[0,1]$ *as follows: for* $m = 0, \dots, n$, *set*

$$X^{(n)}_{m/n} = \frac{\sum_{k=1}^{m}(Y_k - \mu)}{\sigma\sqrt{n}},$$

and then extend $X^{(n)}$ *to all of* $[0,1]$ *by linearity on each of the intervals* $[\frac{m-1}{n}, \frac{m}{n}]$, $m = 1, \dots, n$. *Let* Q_n *denote the distribution of* $X^{(n)}$. *Then* $Q_n \to Q$ *as* $n \to \infty$ *where* Q *is Wiener measure on* $\mathbf{C}[0,1]$.

The term 'invariance' in the name of the preceding theorem refers to fact that the conclusion depends on the distribution of the Y_k only through its mean and variance.

19.3. Some measurable functionals on $\mathbf{C}[0,1]$

Because $(\mathbf{C}[0,1], \mathcal{C})$ is itself a space of functions, we will use the term 'functional' for any $\mathbb{R}$-valued function having $\mathbf{C}[0,1]$ as its domain. We will treat any one such functional as several random variables by placing different probability measures on the measurable space $(\mathbf{C}[0,1], \mathcal{C})$.

One measurable functional we will treat is M defined by

$$M(x) = \max\{x(t) \colon 0 \le t \le 1\}.$$

The distribution R of M under Wiener measure Q equals the distribution of the random variable $M \circ W$, where W is a Wiener process. Similarly, if Q_n denotes the distribution of $X^{(n)}$, where $X^{(n)}$ is defined in terms of a random walk as in the statement of the Donsker Invariance Principle, then the distribution R_n of M under Q_n is the distribution of $M \circ X^{(n)}$. The functional M is continuous, so by Proposition 10 of Chapter 18, $R_n \to R$. (This observation is also made in Problem 9 of Chapter 18.) Thus, knowledge of R yields knowledge of the limiting behavior of the sequence (R_n), and conversely; we will use both directions.

A second functional J defined by

$$J(x) = \int_0^1 x(t)\,dt$$

will be treated in a problem; it is obviously continuous.

A third functional K will also be studied: $K(x)$ equals the Lebesgue measure of $\{t \in [0,1] \colon x(t) > 0\}$. We will prove that K is measurable, although it is not continuous. Despite this lack of continuity, we will show that the sequence of distributions of K under Q_n converges to that of K under Q.

Returning to the functional M we will use the 'reflection principle' to calculate its distribution under Q_n, where Q_n is induced, as in the Donsker Invariance

Principle, from an iid sequence the common distribution of which assigns probability $1/2$ to each of $\{1\}$ and $\{-1\}$. It is clear that Q_n assigns probability 1 to the set of functions x for which $M(x) = x(\frac{m}{n})$ for some integer m depending on x; we will use this fact without comment below.

Fix a positive integer n. For c a positive integral multiple of $\frac{1}{\sqrt{n}}$ and $m = 1, 2, \ldots, n$, set

$$D_m = \{x \in \mathbf{C}[0,1]\colon x(\tfrac{m}{n}) = c > x(\tfrac{i}{n}) \text{ for } i < m\}\,.$$

Then

$$Q_n(\{x\colon M(x) \geq c\}) = Q_n\left(\bigcup_{m=1}^n D_m\right) = \sum_{m=1}^n Q_n(D_m)\,.$$

We also have

$$
\begin{aligned}
Q_n(D_m) &= Q_n\big(D_m \cap \{x\colon x(1) - x(\tfrac{m}{n}) > 0\}\big) \\
&\quad + Q_n\big(D_m \cap \{x\colon x(1) - x(\tfrac{m}{n}) < 0\}\big) \\
&\quad + Q_n\big(D_m \cap \{x\colon x(1) - x(\tfrac{m}{n}) = 0\}\big) \\
&= Q_n(D_m)\, Q_n(\{x\colon x(1) - x(\tfrac{m}{n}) > 0\}) \\
&\quad + Q_n(D_m)\, Q_n(\{x\colon x(1) - x(\tfrac{m}{n}) < 0\}) \\
&\quad + Q_n\big(D_m \cap \{x\colon x(1) - x(\tfrac{m}{n}) = 0\}\big) \\
&= 2Q_n\big(D_m \cap \{x\colon x(1) - x(\tfrac{m}{n}) > 0\}\big) \\
&\quad + Q_n\big(D_m \cap \{x\colon x(1) - x(\tfrac{m}{n}) = 0\}\big)\,.
\end{aligned}
$$

When we sum over m we obtain

$$(19.11)\quad Q_n(\{x\colon M(x) \geq c\}) = 2Q_n(\{x\colon x(1) > c\}) + Q_n(\{x\colon x(1) = c\})\,.$$

Now fix c to be positive and rational, and thus an integral multiple of $\frac{1}{\sqrt{n}}$ for an infinite sequence of positive integers n. Letting $n \to \infty$ through that sequence, the Donsker Invariance Principle implies that the left side of (19.11) converges to $Q(\{x\colon M(x) \geq c\})$. The set $\{x\colon x(1) = c\}$ is closed in $\mathbf{C}[0,1]$. It also equals the boundary in $\mathbf{C}[0,1]$ of the set $\{x\colon x(1) > c\}$. Since $Q(\{x\colon x(1) = c\}) = 0$, the Portmanteau Theorem of Chapter 18 implies that the right side of (19.11) converges to $2Q(\{x\colon x(1) > c\})$. Thus

$$Q_n(\{x\colon M(x) \geq c\}) \to 2Q(\{x\colon x(1) > c\}) = \sqrt{\frac{2}{\pi}} \int_c^\infty e^{-\frac{u^2}{2}}\, du\,.$$

We have thus proved the following two consequences of the Donsker Invariance Principle.

Theorem 9. *Let Q denote Wiener measure on $\mathbf{C}[0,1]$ and set*

$$M(x) = \max\{x(t)\colon 0 \leq t \leq 1\}\,, \quad x \in \mathbf{C}[0,1]\,.$$

Then, for $c \geq 0$,

$$Q(\{x\colon M(x) \leq c\}) = \sqrt{\frac{2}{\pi}} \int_0^c e^{-\frac{u^2}{2}} \, du \, .$$

Theorem 10. *Let $(Y_k\colon k = 1, 2, \dots)$ be an iid sequence of $\mathbb{R}$-valued random variables having finite mean μ and positive finite variance σ^2. Then, for $c \geq 0$,*

$$\lim_{n \to \infty} P\left[\frac{1}{\sigma\sqrt{n}} \max\left\{\sum_{k=1}^m (Y_k - \mu)\colon m = 1, \dots, n\right\} \leq c\right] = \sqrt{\frac{2}{\pi}} \int_0^c e^{-\frac{u^2}{2}} \, du \, .$$

Notice the general scheme underlying the preceding discussion. Start with some easy-to-analyze sequence of iid random variables having finite nonzero variance. Obtain some result for its partial sums. Go to the limit to obtain a corresponding result for the Wiener process. Then obtain a limit result for the sequence of partial sums of an arbitrary iid sequence, with no restriction on the common distribution of the summands other than that of positive finite variance. For some calculations, such as the one in the following problem, there is no need to start with a particular sequence of iid random variables; instead one does the initial computation directly with Wiener measure or a Brownian motion.

* **Problem 4.** Let Q denote Wiener measure. Find the distribution under Q of the functional J, where

$$J(x) = \int_0^1 x(t) \, dt \, .$$

Then apply the Donsker Invariance Principle in connection with this distribution.

In preparation for treating the functional K introduced earlier, we obtain the following result, which is itself quite interesting. It says that the set of times when a Wiener process equals 0 has Lebesgue measure 0 a.s.

Proposition 11. *Let Q denote Wiener measure and let λ denote Lebesgue measure on $[0, 1]$. Then*

$$Q\big(\{x\colon \lambda(\{t \in [0, 1]\colon x(t) = 0\}) = 0\}\big) = 1 \, .$$

PROOF. The function $(t, x) \rightsquigarrow x(t)$ defined on $[0, 1] \times \mathbf{C}_{[0,1]}$ is continuous as a function of t for each fixed x and also as a function of x for each fixed t. By Proposition 9 of Chapter 9 it is measurable as a function of (t, x), and hence the indicator function of $\{(t, x)\colon x(t) = 0\}$ is measurable. Therefore, we may apply the Fubini Theorem to this indicator function. Since nondegenerate normal distributions assign measure 0 to every one-point set, we obtain 0 by integrating first with respect to Q. Thus, we must also get 0 when we integrate first with respect to λ; that is,

$$\int_{\mathbf{C}_{[0,1]}} \left(\int_{[0,1]} I_{\{(t,x)\colon \, x(t)=0\}} \lambda(dt)\right) dQ = 0 \, .$$

Hence, the inside integral must equal 0 for almost every $x \in \mathbf{C}[0,1]$. This fact finishes the proof since the inside integral equals $\lambda(\{t\colon x(t) = 0\})$. $\square$

With λ continuing to denote Lebesgue measure, the definition of the functional K can be written as

$$(19.12) \qquad K(x) = \lambda(\{t\colon x(t) > 0\}),$$

and we also define

$$\widehat{K}(x) = \lambda(\{t\colon x(t) \geq 0\}).$$

Problem 5. Prove that K and $\widehat{K}$ are measurable functionals.

Problem 6. Prove that the boundary in $\mathbf{C}[0,1]$ of $\{x\colon K(x) \leq c\}$ equals

$$\{x\colon K(x) \leq c \leq \widehat{K}(x)\}$$

for each $c \in \mathbb{R}$.

By Proposition 11 and the preceding problem, the Q-measure of the boundary of $\{x\colon K(x) \leq c\}$ equals 0 for every c that is a continuity point of the distribution function of K. By the Portmanteau Theorem,

$$\lim_{n\to\infty} Q_n(\{x\colon K(x) \leq c\}) = Q(\{x\colon K(x) \leq c\})$$

for such c. Accordingly, we may proceed for K as we did for M—first calculating the distribution of K under the measure Q_n induced by an iid sequence the common distribution of which assigns probability $1/2$ to each of $\{1\}$ and $\{-1\}$, and then taking the limit as $n \to \infty$ to obtain a fact about the Wiener process and a limit theorem.

Lemma 12. *For positive even integers k,*

$$\sum_{\substack{j=2 \\ j\,\mathrm{even}}}^{k} \frac{1}{j-1}\binom{j}{j/2}\binom{k-j}{(k-j)/2} = \binom{k}{k/2}.$$

PROOF. Divide both sides by 2^k to obtain the following equivalent statement

$$(19.13) \qquad \sum_{\substack{j=2 \\ j\,\mathrm{even}}}^{k} \Big[\frac{1}{j-1}\binom{j}{j/2}\frac{1}{2^j}\Big]\Big[\binom{k-j}{(k-j)/2}\frac{1}{2^{(k-j)}}\Big] = \binom{k}{k/2}\frac{1}{2^k}.$$

The right side equals the probability that a simple symmetric random walk equals 0 at time k. From Example 5 of Chapter 11, we see that the jth term in the left side of (19.13) equals the probability that a simple symmetric random walk returns to 0 for the first time at time j and also equals 0 at time k (the possibility that $k = j$ not being excluded). By finite additivity of probability measures, (19.13) follows. $\square$

Theorem 13. *Let n be an even positive integer and Q_n the distribution of $X^{(n)}$ as defined in the sentence containing (19.1) with the distribution of each Y_k assigning the value $1/2$ to $\{1\}$ and to $\{-1\}$. Then,*

$$Q_n(\{x\colon K(x) = \tfrac{m}{n}\}) = \begin{cases} \binom{m}{m/2}\binom{n-m}{(n-m)/2}2^{-n} & \text{for } m = 0, 2, 4, \ldots n \\ 0 & \text{otherwise}, \end{cases}$$

where K is defined by (19.12).

PARTIAL PROOF. We will use induction on even positive n. The verification for $n = 2$ is trivial. Suppose that $n \geq 4$ and that the statement of interest is true for every positive even integer less than n. We only consider m for which $0 < m < n$, leaving the two extreme cases $m = 0$ and $m = n$ for the reader. For $x \in \mathbf{C}[0, 1]$, let

$$T(x) = \inf\{t > 0\colon x(t) = 0\}.$$

By finite additivity, the independence of the steps of a random walk, symmetry, and the distribution of T given in Example 5 of Chapter 11,

$$Q_n(\{x\colon K(x) = \tfrac{m}{n}\})$$

$$= \sum_{\substack{j=2 \\ j\,\text{even}}}^{n} Q_n(\{x\colon T(x) = \tfrac{j}{n}, x(\tfrac{1}{n}) > 0, K(x) = \tfrac{m}{n}\})$$

$$+ \sum_{\substack{j=2 \\ j\,\text{even}}}^{n} Q_n(\{x\colon T(x) = \tfrac{j}{n}, x(\tfrac{1}{n}) < 0, K(x) = \tfrac{m}{n}\})$$

$$= \frac{1}{2} \sum_{\substack{j=2 \\ j\,\text{even}}}^{m} Q_n(\{x\colon T(x) = \tfrac{j}{n}\}) Q_{n-j}(\{x\colon K(x) = \tfrac{m-j}{n-j}\})$$

$$+ \frac{1}{2} \sum_{\substack{j=2 \\ j\,\text{even}}}^{n-m} Q_n(\{x\colon T(x) = \tfrac{j}{n}\}) Q_{n-j}(\{x\colon K(x) = \tfrac{m}{n-j}\})$$

$$= \frac{1}{2} \sum_{\substack{j=2 \\ j\,\text{even}}}^{m} \frac{1}{j-1}\binom{j}{j/2}\frac{1}{2^j}\binom{(m-j)}{(m-j)/2}\binom{n-m}{(n-m)/2}\frac{1}{2^{(n-j)}}$$

$$+ \frac{1}{2} \sum_{\substack{j=2 \\ j\,\text{even}}}^{m} \frac{1}{j-1}\binom{j}{j/2}\frac{1}{2^j}\binom{m}{m/2}\binom{n-m-j}{(n-m-j)/2}\frac{1}{2^{(n-j)}}$$

$$= \frac{1}{2^{(n+1)}}\binom{n-m}{(n-m)/2} \sum_{\substack{j=2 \\ j\,\text{even}}}^{m} \frac{1}{j-1}\binom{j}{j/2}\binom{(m-j)}{(m-j)/2}$$

$$+ \frac{1}{2^{(n+1)}}\binom{m}{m/2} \sum_{\substack{j=2 \\ j\,\text{even}}}^{m} \frac{1}{j-1}\binom{j}{j/2}\binom{n-m-j}{(n-m-j)/2},$$

which in combination with Lemma 12 completes the argument for $0 < m < n$. □

Problem 7. From the above presentation it appears that a formula from Example 5 of Chapter 11 has been used twice—once in the proof of the theorem itself and once in the proof of the lemma that was used in the proof of the theorem. However, we did not actually need the formula from Chapter 11. Explain.

* **Problem 8.** Treat the cases $m = 0$ and $m = n$ which were not treated in the partial proof of Theorem 13.

Theorem 14. [Arcsin Law (for Time Spent Positive)] *Let Q denote Wiener measure on $\mathbf{C}[0, 1]$ and define K by (19.12). Then, for $c \in [0, 1]$,*

$$Q(\{x \colon K(x) \leq c\}) = \frac{1}{\pi} \int_0^c \frac{1}{\sqrt{u(1-u)}}\, du = \frac{2}{\pi} \arcsin \sqrt{c}.$$

Problem 9. Prove the preceding theorem.

Corollary 15. *Let $(S_0 = 0, S_1, S_2, \dots)$ be a random walk in $\mathbb{R}$ whose steps have mean 0 and positive finite variance. Then,*

$$\lim_{n \to \infty} P\left[\frac{\#\{m \leq n \colon S_m > 0\}}{n} \leq c\right] = \frac{2}{\pi} \arcsin \sqrt{c}$$

for $0 \leq c \leq 1$.

Problem 10. Prove the preceding corollary. *Hint:* You may find Proposition 11 useful.

* **Problem 11.** For a Wiener process W calculate the expectation of

$$\lambda(\{t \in [0, 1] \colon W_t > 0\}) \vee \lambda(\{t \in [0, 1] \colon W_t < 0\}),$$

where λ denotes Lebesgue measure.

19.4. Brownian motion on $[0, \infty)$

According to Problem 5 of Chapter 18, $\mathbf{C}[0, \infty)$ can be regarded as a Polish space; convergence in it is equivalent to uniform convergence on bounded intervals.

Here is one way to construct a $\mathbf{C}[0, \infty)$-valued random variable that has stationary independent increments and is a Wiener process when restricted to $[0, 1]$. Let $(W^{(n)} \colon n = 0, 1, \dots)$ be an iid sequence of $\mathbf{C}[0, 1]$-valued Wiener processes, and for $n = 0, 1, \dots$ and $t \in [n, n+1)$ define

$$W_t = W_{t-n}^{(n)} + \sum_{k=0}^{n-1} W_1^{(k)}.$$

The random function W thus constructed is called *Brownian motion on* $[0, \infty)$ or a *Wiener process on* $[0, \infty)$. Its distribution is *Wiener measure on* $\mathbf{C}[0, \infty)$, or simply *Wiener measure* when there is no likelihood of confusion with Wiener measure on $\mathbf{C}[0, 1]$. Without regard to the above construction, any $\mathbf{C}[0, \infty)$-valued random variable whose distribution is Wiener measure is called a *Wiener process* or *Brownian motion*, with the modifying phrase "on $[0, \infty)$" being included if necessary. The following problem gives an alternative construction of Brownian motion on $[0, \infty)$.

Problem 12. Let $W^{(1)}$ denote a Wiener process on $[0, 1]$. For $n = 1, 2, \ldots$, set

$$X_t^{(n)} = \sqrt{n}\, W_{[t \wedge n]/n}^{(1)}, \quad t \in [0, \infty).$$

Prove that $X^{(n)} \xrightarrow{\mathcal{D}} W$ as $n \to \infty$, where W is a Wiener process on $[0, \infty)$.

One advantage of extending Brownian motion to $[0, \infty)$ is that Wiener measure on $\mathbf{C}[0, \infty)$ is invariant under a wider variety of transformations than is Wiener measure on $\mathbf{C}[0, 1]$.

Theorem 16. *Let W be a Brownian motion on* $[0, \infty)$. *Then each of the following is also a Brownian motion on* $[0, \infty)$:

(i) $t \rightsquigarrow -W_t$;
(ii) $t \rightsquigarrow (W_{s+t} - W_s)$, *where* $s \geq 0$ *is fixed;*
(iii) $t \rightsquigarrow \sqrt{a}\, W_{t/a}$, *where* $a > 0$ *is fixed;*
(iv)

$$t \rightsquigarrow \begin{cases} tW_{1/t} & \text{if } t > 0 \\ 0 & \text{if } t = 0. \end{cases}$$

PARTIAL PROOF. It is easy to see that each of the first three transformed processes has stationary independent increments and that the distribution at any particular time t is normal with mean 0 and variance t. By Lemma 2, it follows that these three processes are Brownian motions.

Let V be the fourth process obtained from W as described in the theorem. It is easily checked that for each $t \geq 0$, V_t is normally distributed with mean 0 and variance t. The independence of the increments is also obvious. It follows from these two facts that the variance of $(V_t - V_s)$ equals $|t - s|$ and, therefore, that the increments are stationary. It would seem that we have checked the necessary conditions in order to apply Lemma 2, but this is not so, because we still need to show that V is $\mathbf{C}[0, \infty)$-valued (the corresponding fact being obvious for the first three transformations).

Clearly, $t \rightsquigarrow V_t(\omega)$ is continuous on $(0, \infty)$ for each ω. We need to confirm almost sure continuity at 0. Equivalently, we must show that

$$\lim_{t \to \infty} \frac{W_t}{t} = 0 \quad \text{a.s.}$$

Since W has stationary independent increments with mean 0, it follows immediately from the Strong Law of Large Numbers that

$$\lim_{\substack{n \to \infty \\ n \in \mathbb{Z}^+ \setminus \{0\}}} \frac{W_n}{n} = 0 \text{ a.s.}$$

So to complete the proof, it is sufficient to show that

$$\lim_{n \to \infty} \frac{\max_{n \leq t \leq n+1} |W_t - W_n|}{n} = 0 \text{ a.s.}$$

For $c > 0$ and $n = 1, 2, \ldots$, set

$$A_{c,n} = \{\omega : \max_{n \leq t \leq n+1} |W_t - W_n| > cn\}.$$

The desired almost sure convergence is equivalent to $P(\limsup_n A_{c,n}) = 0$ for all $c > 0$. By the Borel Lemma, it is enough to show that $\sum_n P(A_{c,n}) < \infty$ for $c > 0$. Since W has stationary increments,

$$P(A_{c,n}) = P[\max_{0 \leq t \leq 1} |W_t| > cn].$$

It is left for the reader to use Theorem 9 to verify that $\sum_n P(A_{c,n}) < \infty$. $\square$

Problem 13. Make the calculation to which the last sentence of the preceding proof refers.

The transformations in the preceding theorem are sometimes given the following names: (i) *symmetry* or *spatial symmetry*; (ii) *time shift*; (iii) *scaling* or *change of scale*; (iv) *time inversion*. Note that in the course of showing that Brownian motion is invariant under time inversion, we also proved the following:

Proposition 17. [Strong Law for Brownian motion]

$$\lim_{t \to \infty} \frac{W_t}{t} = 0 \quad a.s.$$

Problem 14. Use Theorem 9 and Theorem 16 to prove that a Wiener process W on $[0, \infty)$ has the following two properties with probability 1: (i) There exists a strictly decreasing random sequence $(T_k : k = 1, 2,)$ converging to 0 a.s. such that $W_{T_k(\omega)}(\omega) > 0$ for every k; (ii) There exists a strictly increasing random sequence $(T_k : k = 1, 2,)$ approaching ∞ a.s. such that $W_{T_k(\omega)}(\omega) > 0$ for every k.

Problem 15. Use symmetry and the preceding problem to draw two further conclusions about Wiener processes.

19.5. Filtrations and stopping times

We adapt to the $\mathbf{C}[0, \infty)$-setting some concepts that were introduced for random sequences in Chapter 11.

Definition 18. Let $(\Omega, \mathcal{F})$ denote a measurable space and let $\mathcal{F}_t$ be a sub-σ-field of $\mathcal{F}$ for each $t \in [0, \infty)$. Then $(\mathcal{F}_t : t \in [0, \infty))$ is a *filtration* in $(\Omega, \mathcal{F})$ if $(s < t) \Rightarrow (\mathcal{F}_s \subseteq \mathcal{F}_t)$. Let Y be a $\mathbf{C}[0, \infty)$-valued random variable defined on $(\Omega, \mathcal{F})$. Then Y is *adapted* to a filtration $(\mathcal{F}_t : t \in [0, \infty))$ if, for each t, Y_t is measurable with respect to $\mathcal{F}_t$. Corresponding to a filtration $(\mathcal{F}_t : t \in [0, \infty))$, we use the notation

$$\mathcal{F}_\infty = \sigma(\mathcal{F}_t : t \in [0, \infty)) \,.$$

A $\mathbf{C}[0, \infty)$-valued random variable Y may be adapted to different filtrations $(\mathcal{F}_t)$. The one defined by $\mathcal{F}_t = \sigma(Y_s : 0 \le s \le t)$ is the *minimal filtration* of Y.

The following result is an immediate consequence of Proposition 9 of Chapter 9.

Proposition 19. *Let Y be a $\mathbf{C}[0, \infty)$-valued measurable function defined on a measurable space $(\Omega, \mathcal{F})$ and adapted to a filtration $(\mathcal{F}_t : t \ge 0)$. Then for each t, the function*

$$(\omega, s) \rightsquigarrow Y_s(\omega) \,, \quad (\omega, s) \in \Omega \times [0, t] \,,$$

is measurable with respect to $\mathcal{F}_t \times \mathcal{B}_t$, where $\mathcal{B}_t$ is the Borel σ-field of $[0, t]$.

The measurability property asserted for Y and the filtration in the preceding proposition is called *progressive measurability*. This concept is also used for random variables having values in function spaces other than $\mathbf{C}[0, \infty)$, but for some function spaces, analogues of the preceding proposition may not hold.

Definition 20. Let $(\mathcal{F}_t : t \in [0, \infty))$ be a filtration in a measurable space $(\Omega, \mathcal{F})$. An $\overline{\mathbb{R}}^+$-valued random variable T defined on $(\Omega, \mathcal{F})$ is a *stopping time* with respect to this filtration if $\{\omega : T(\omega) \le t\} \in \mathcal{F}_t$ for every $t \in \mathbb{R}^+$.

Definition and Proposition 21. Let T be a stopping time with respect to a filtration $(\mathcal{F}_t : t \in [0, \infty))$. The collection of events A such that

$$A \cap \{\omega : T(\omega) \le t\} \in \mathcal{F}_t \quad \text{for all } t \in \overline{\mathbb{R}}^+$$

is a σ-field. It is denoted by $\mathcal{F}_T$.

Problem 16. Prove the preceding proposition.

Problem 17. Prove that any stopping time T is measurable with respect to the σ-field $\mathcal{F}_T$.

Problem 18. Let $S \le T$ be stopping times. Prove that $\mathcal{F}_S \subseteq \mathcal{F}_T$.

Proposition 22. *Let* Y *be a* $\mathbf{C}[0,\infty]$-*valued random variable adapted to a filtration* $(\mathcal{F}_t: t \in [0,\infty))$ *and let* Y_∞ *be any* $\mathbb{R}$-*valued,* $\mathcal{F}_\infty$-*measurable function. Then, for any stopping time* T *with respect to this filtration,* $\omega \rightsquigarrow Y_{T(\omega)}(\omega)$ *is measurable with respect to the* σ-*field* $\mathcal{F}_T$.

PROOF. Let B be a Borel subset of $\mathbb{R}$ and $t \in \overline{\mathbb{R}}^+$. We must show

$$\{\omega: Y_{T(\omega)}(\omega) \in B, \, T(\omega) \le t\} \in \mathcal{F}_t.$$

An equivalent statement is

$$\{\omega: Y_{(T(\omega)\wedge t)}(\omega) \in B, \, T(\omega) \le t\} \in \mathcal{F}_t.$$

Since $\{\omega: T(\omega) \le t\} \in \mathcal{F}_t$, we only need show

$$\{\omega: Y_{(T(\omega)\wedge t)}(\omega) \in B\} \in \mathcal{F}_t.$$

We can do this by showing that

$$\omega \rightsquigarrow Y_{(T(\omega)\wedge t)}(\omega)$$

is a measurable function from $(\Omega, \mathcal{F}_t)$ to $(\mathbb{R}, \mathcal{B})$, where $\mathcal{B}$ denotes the Borel σ-field of $\mathbb{R}$. This function is the composition of two functions:

$$\omega \rightsquigarrow (\omega, \, (T(\omega) \wedge t))$$

from $(\Omega, \mathcal{F}_t)$ to $(\Omega \times [0,t], \mathcal{F}_t \times \mathcal{B}_t)$; and

$$(\omega, s) \rightsquigarrow Y_s(\omega)$$

from $(\Omega \times [0,t], \mathcal{F}_t \times \mathcal{B}_t)$ to $(\mathbb{R}, \mathcal{B})$. The second of these two functions is measurable by the progressive measurability of Y. The first coordinate of the first of the two functions is obviously measurable, since $\mathcal{F}_t$ is used for both domain and target. Finally,

$$\{\omega: T(\omega) \wedge t \le s\} = \left\{ \begin{array}{ll} \{\omega: T(\omega) \le s\} & \text{if } s < t \\ \Omega & \text{if } s \ge t \end{array} \right\} \in \mathcal{F}_{s\wedge t} \subseteq \mathcal{F}_t,$$

as desired. □

Problem 19. Explain why the assertion in Problem 17 is a corollary of the preceding proposition.

Problem 20. Show that several other aspects of the theory concerning filtrations and stopping times carry over from Chapter 11 to the $[0,\infty)$-setting.

The *hitting time* of a set A by a $\mathbf{C}[0,\infty)$-valued measurable function Y is

$$\inf\{t \ge 0: Y_t(\omega) \in A\}.$$

Proposition 23. *Let Y be a $\mathbf{C}[0, \infty)$-valued random variable and let C be a closed subset of $\mathbb{R}$. Then the hitting time of C by Y is a stopping time (with respect to any filtration with respect to which Y is adapted).*

PROOF. Let $(\mathcal{F}_t : t \in [0, \infty))$ denote the filtration to which Y is adapted. Let T denote the hitting time of C and, for $m = 1, 2, \ldots$, let

$$C_m = \{r : |r - c| < \tfrac{1}{m} \text{ for some } c \in C\}.$$

For any $t \in \mathbb{R}^+$, the continuity of $s \rightsquigarrow Y_s(\omega)$ for each ω gives

$$\{\omega : T(\omega) \leq t\} = \lim_{m \to \infty} \bigcup_{\substack{s \leq t \\ s \text{ rational}}} \{\omega : Y_s(\omega) \in C_m\} \in \mathcal{F}_t. \quad \square$$

Suppose that $Y_0 = 0$, as is the case for Brownian motion. Let C be a closed subset of $\mathbb{R}$. If $0 \in C$, the hitting time of C is 0. If $0 \notin C$ and C contains at least one positive number, remove all the positive numbers from C except the smallest. If $0 \notin C$ and C contains at least one negative number, remove all the negative numbers from C except the largest. The hitting time of the new set thus obtained is the same as the hitting time of C. The new set is either a two-point set, a one-point set, or empty. The hitting time in the last case is ∞. In the next section, we will find the moment generating functions of the hitting times by Brownian motion of one- and two-point sets, and the distributions themselves for certain two-point sets and all one-point sets.

Hitting times of sets that are not closed may not even be stopping times with respect to minimal filtrations. There exist, for example, members x and $\hat{x}$ of $\mathbf{C}[0, \infty)$ such that $x(s) = \hat{x}(s)$ for $0 \leq s \leq 2$, the hitting time of $(1, \infty)$ by x equals 2, and the hitting time of $(1, \infty)$ by $\hat{x}$ is greater than 2. Sometimes one uses nonminimal filtrations in order that the hitting time of every set be a stopping time. One commonly used approach is to begin with the minimal filtration $(\mathcal{F}_t : t \in [0, \infty))$ of some $\mathbf{C}[0, \infty)$-valued random variable Y and then set

$$(19.14) \qquad\qquad \mathcal{F}_{t+} = \bigcap_{s > t} \mathcal{F}_s.$$

Clearly $\mathcal{F}_t \subseteq \mathcal{F}_{t+}$, so Y is adapted to the filtration $(\mathcal{F}_{t+} : t \in [0, \infty))$.

Problem 21. For Y and $(\mathcal{F}_{t+} : t \in [0, \infty))$ as described in the preceding paragraph, prove that the hitting time of any set by Y is a stopping time.

Problem 22. Show for any filtration $(\mathcal{F}_t : t \in [0, \infty))$, that the filtration $(\mathcal{F}_{t+} : t \in [0, \infty))$ defined by (19.14) is *right-continuous* in the sense that $\mathcal{F}_{t+} = \bigcap_{s > t} \mathcal{F}_{s+}$. It is the *minimal right-continuous filtration* of Y in case $(\mathcal{F}_t : t \in [0, \infty))$ is minimal for Y.

It is conceivable that a $\mathbf{C}[0, \infty)$-valued random variable have some nice properties with respect to its minimal filtration but loses them if that filtration is replaced by the minimal right-continuous filtration. However, we will see in the next section that Brownian motion does not lose its important properties when such a replacement is made.

19.6. Brownian motion, filtrations, and stopping times

By virtue of having stationary independent increments, a Brownian motion W on $[0, \infty)$ has the property that for any fixed s, the $\mathbf{C}[0, \infty)$-valued random variable $t \rightsquigarrow (W_{s+t} - W_s)$ is a Brownian motion that is independent of $\mathcal{F}_s = \sigma(W(\cdot, u): 0 \le u \le s)$. The following result strengthens this assertion.

Proposition 24. *Let $(\mathcal{F}_{t+}: t \in [0, \infty))$ be the minimal right-continuous filtration of a Brownian motion W. Then for each $s \in [0, \infty)$, $t \rightsquigarrow (W_{s+t} - W_s)$ is a Brownian motion that is independent of $\mathcal{F}_{s+}$.*

PROOF. Since $\mathcal{F}_{s+} \subseteq \mathcal{F}_{s+\varepsilon}$ for $\varepsilon > 0$, the $\mathbf{C}[0, \infty)$-valued random variable $t \rightsquigarrow (W_{s+\varepsilon+t} - W_{s+\varepsilon})$ is a Brownian motion independent of $\mathcal{F}_{s+}$. Now let $\varepsilon \searrow 0$ and use continuity in conjunction with Theorem 20 of Chapter 18 to conclude that the Brownian motion $t \rightsquigarrow (W_{s+t} - W_s)$ is independent of every $\mathbb{R}$-valued $\mathcal{F}_{s+}$-measurable random variable and thus of $\mathcal{F}_{s+}$ itself. □

We wish to adapt Corollary 11 of Chapter 11 to the current setting. However, there is no $\mathbf{C}[0, \infty)$-analogue of Proposition 10 of that chapter, so we will need to use a different approach.

Theorem 25. *Let W be a Brownian motion, $(\mathcal{F}_{t+}: t \in [0, \infty))$ the minimal right-continuous filtration of W, and T a stopping time with respect to that filtration. Suppose that $P[T < \infty] = 1$. Then*

(19.15) $$t \rightsquigarrow [W_{T+t} - W_T]$$

is a Brownian motion with respect to the filtration $(\mathcal{F}_{(T+t)+}: t \in [0, \infty))$ and is independent of $\mathcal{F}_{T+}$.

PROOF. From Problem 18 and Proposition 22 we see that $(\mathcal{F}_{(T+t)+})$ is a filtration and that (19.15) is an almost surely defined $\mathbf{C}[0, \infty)$-valued random variable adapted to this filtration. (This observation includes the fact that the set of ω where (19.15) does not apply is $\{\omega: T(\omega) = \infty\}$, which is a member of every σ-field in the filtration $(\mathcal{F}_{(T+t)+})$.)

For each ω, set $T_n(\omega) = \frac{1}{n}\lceil nT \rceil$. For each $k \in \mathbb{Z}^+$,

$$\{\omega: T_n(\omega) \le \tfrac{k}{n}\} = \{\omega: T(\omega) \le \tfrac{k}{n}\} \in \mathcal{F}_{(k/n)+}.$$

Therefore, T_n is a stopping time with respect to the (discrete time) filtration $(\mathcal{F}_{(k/n)+}: k = 0, 1, 2, \dots)$. Also, $(W_{k/n}: k = 0, 1, 2, \dots)$ is a random walk, and so Proposition 24 and an easy adaptation of Corollary 11 of Chapter 11 imply that

$$(19.16) \qquad k \rightsquigarrow [W_{T_n + \frac{k}{n}} - W_{T_n}]$$

is a random walk that is independent of $\mathcal{F}_{T_n+}$. By Problem 18, $\mathcal{F}_{T+} \subseteq \mathcal{F}_{T_n+}$; thus this random walk is independent of $\mathcal{F}_{T+}$.

Let X be the object defined in (19.15). Since X is $\mathbf{C}[0, \infty)$-valued, its values at various times t can be expressed as limits of the values of the random walks defined in (19.16), each of which has independent increments and is independent of $\mathcal{F}_{T+}$. Since independence is preserved in the limit (Theorem 20 of Chapter 18), it follows that X has independent increments and is independent of $\mathcal{F}_{T+}$. It is also easy to check by taking limits of the increments of the random walks that the distribution of $X_t - X_s$ is normal with mean 0 and variance $t - s$ for each $s, t \geq 0$. Thus, X is a Wiener process that is independent of $\mathcal{F}_{T+}$, as desired. $\square$

In view of Theorem 25, Problem 21, and Problem 14, the hitting time of any set by Brownian motion almost surely equals the hitting time of its closure. Thus, the study of distributions of hitting times of arbitrary sets reduces to the study of distributions of hitting times of one- and two-point sets. We begin by treating the one-point sets.

For $b \geq 0$, let $T_b(x)$ equal the hitting time of the singleton $\{b\}$ by $x \in \mathbf{C}[0, \infty]$. Let W denote a Wiener process. By Proposition 23, T_b is a stopping time with respect to any filtration to which W is adapted. We set two goals: to find the distribution of each $\overline{\mathbb{R}}^+$-valued random variable $T_b \circ W$ and to study the random function $b \rightsquigarrow T_b \circ W$. Equivalently, with Q denoting the distribution of W, our goals are to find the distribution of each T_b viewed as a random variable on $(\mathbf{C}[0, \infty), \mathcal{C}, Q)$ and study the random function $x \rightsquigarrow [b \rightsquigarrow T_b(x)]$. As an aid we use M_t defined by

$$M_t(x) = \max\{x(s) \colon 0 \leq s \leq t\}.$$

Clearly,

$$(19.17) \qquad \{x \in \mathbf{C}[0, \infty] \colon T_b(x) \leq x\} = \{x \in \mathbf{C}[0, \infty] \colon M_t(x) \geq b\}.$$

Theorem 9 and Theorem 16 yield the distribution of $M_t \circ W$ for each t. Thus, we can determine the distribution of T_b from (19.17).

Theorem 26. *Let Q denote Wiener measure on $\mathbf{C}[0, \infty)$. For all $b > 0$, T_b, defined on $(\mathbf{C}[0, \infty), \mathcal{C}, Q)$ as the hitting time of $\{b\}$, has a stable distribution of index $\frac{1}{2}$ which has the continuous density*

$$t \rightsquigarrow \begin{cases} \frac{b}{\sqrt{2\pi t^3}} e^{-\frac{b^2}{2t}} & \text{if } t > 0 \\ 0 & \text{if } t \leq 0 \end{cases}$$

and the moment generating function $u \rightsquigarrow \exp(-b\sqrt{2u})$.

PROOF. From Problem 16 of Chapter 15 we see that the density and the moment generating function in the theorem correspond to each other, and that they are stable of index 1/2. Thus, we only need verify the formula for the density. By (19.17) and the invariance of Brownian motion under scaling (Theorem 16),

$$Q[T_b \leq t] = Q[M_t \geq b] = Q[M_1 \geq \tfrac{b}{\sqrt{t}}]$$

for $t > 0$. By Theorem 9, this last quantity equals

$$1 - \sqrt{\frac{2}{\pi}} \int_0^{b/\sqrt{t}} e^{-\frac{u^2}{2}} \, du \, .$$

Differentiation with respect to t gives the formula in the theorem, as desired. □

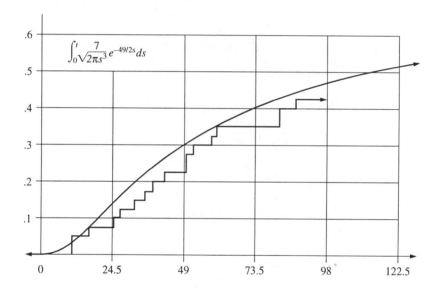

FIGURE 19.1. Stable distribution function of index 1/2 and a related empirical distribution function

Problem 23. The distribution function described in Theorem 26 is illustrated in Figure 19.1 for the case $b = 7$. Also, shown (with jumps filled in as a visual aid) is a portion of the empirical distribution obtained when 40 people each did the following fair-coin-flip experiment. Each flipped until the number of heads exceeded the number of tails by 7 or the total number of flips equaled 98, whichever came first. In the former case the person recorded the total number of flips, whereas in the latter case the person recorded that more than 98 flips would be needed. Discuss reasons for placing the two graphs on the same coordinate system.

Our knowledge of stable distributions shows that $T_b < \infty$ a.s. (This conclusion can also be drawn from the Law of the Iterated Logarithm at ∞ given in the last section of this chapter.) The assumption in the preceding discussion that $b > 0$ was for convenience and, since the distribution of T_{-b} is the same as the distribution of T_b, it entailed no loss of generality. The distribution of T_0 is the delta distribution at 0.

We have been using the terms 'stationary increments', 'independent increments' and 'adapted to a filtration' for $\mathbf{C}[0, 1]$- and $\mathbf{C}[0, \infty)$-valued random variables. However, these terms may equally well be used for random variables taking values in other function spaces, as in the following theorem.

Theorem 27. *Let Q denote Wiener measure on $\mathbf{C}[0, \infty)$. For $b \geq 0$, let $T_b(x)$ denote the hitting time of $\{b\}$ by $x \in \mathbf{C}[0, \infty)$. Then the random function $b \rightsquigarrow T_b$ defined on $(\mathbf{C}[0, \infty), \mathcal{C}, Q)$ has stationary independent increments and is surely left-continuous.*

Problem 24. Prove the preceding theorem. *Hint:* Use Proposition 5 of Chapter 9 and Theorem 25 of this chapter.

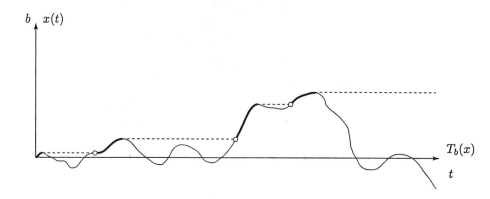

FIGURE 19.2. Hitting time process $b \rightsquigarrow T_b(x)$ for some $x \in \mathbf{C}[0, \infty)$

The function $b \rightsquigarrow T_b(x)$ for a particular x is shown in Figure 19.2 with b on the vertical axis. It is clear that $b \rightsquigarrow T_b(x)$ is not right-continuous. This fact is not inconsistent with the assertion in the following problem that for any fixed b, this function is continuous at b a.s.

Problem 25. Let Q denote Wiener measure on $\mathbf{C}[0, \infty)$ and let $b \geq 0$. Prove that

$$Q(\{x \colon T_b(x) = \lim_{c \searrow b} T_c(x)\}) = 1 .$$

The random function in Theorem 27 is, except for a technicality, an example from the class of random functions treated in Chapter 30. For it to fit exactly into the setting of Chapter 30 it would have to be redefined at its jumps so as to be right-continuous— $\widehat{T}_b(x) = \lim_{c \searrow b} T_c(x)$. The term *first passage time* of b is used for $\widehat{T}_b$. The preceding problem says that, for each fixed b, the first passage time of b and the hitting time of b by a Brownian motion are almost surely equal.

We switch from treating hitting times of one-point sets to studying hitting times of two-point sets.

Theorem 28. *For $-a < 0 < b$ let $T_{a,b}(x)$ denote the hitting time of the two-point set $\{-a, b\}$ by $x \in \mathbf{C}[0, \infty)$. Then the moment generating function of $T_{a,b}$, when governed by Wiener measure on $\mathbf{C}[0, \infty)$, is the function*

$$u \rightsquigarrow \frac{\sinh(a\sqrt{2u}) + \sinh(b\sqrt{2u})}{\sinh((a + b)\sqrt{2u})},$$

where $\sinh$ denotes the hyperbolic sine function.

Problem 26. Prove the preceding theorem. *Hint:* Show that the Wiener measure of the boundary of $\{x : T_{a,b}(x) \le t\}$ equals 0 if $Q(\{x : T_{a,b}(x) = t\}) = 0$. Also, use Example 4 of Chapter 11.

* **Problem 27.** Show that $E_Q(T_{a,b}) = ab$, where $T_{a,b}$ is defined in the preceding theorem and the subscript Q, denoting Wiener measure, indicates that T is regarded as a random variable on $(\mathbf{C}[0, \infty), \mathcal{C}, Q)$.

Problem 28. Continuing with the notation of Theorem 28 show that the density of $T_{b,b}$ with respect to Lebesgue measure on $(0, \infty)$ is

$$t \rightsquigarrow b\sqrt{\frac{2}{\pi t^3}} \sum_{k=0}^{\infty} (-1)^k (2k + 1) e^{-\frac{(2k+1)^2 b^2}{2t}}.$$

Hint: Problem 16 of Chapter 15 may be useful.

19.7. ‡ Characterization of Brownian motion

We will show that condition (i) in Lemma 2 plays a minor role in that lemma. The proof of this fact relies on the convergence theorem for row-wise iid triangular arrays (Theorem 22 of Chapter 16).

Theorem 29. *Let V be a $\mathbf{C}[0, \infty)$-valued random variable such that $V_0 = 0$ a.s. If V has stationary independent increments, then either there exists a constant $b \in \mathbb{R}$ such that $V_t = bt$ a.s., or there exist constants $a > 0$ and $b \in \mathbb{R}$ such that $t \rightsquigarrow (V_t - bt)/a$ is a Brownian motion on $[0, \infty)$.*

PROOF. We begin by proving that for all $t \in [0, \infty)$, V_t is normally distributed (with possibly 0 variance). Fix t and set

$$X_{m,n} = V_{mt/n} - V_{(m-1)t/n}$$

for $m = 1, \ldots, n$ and $n = 1, 2, \ldots$. Since V has stationary independent increments, each row of the resulting triangular array $(X_{m,n})$ is iid. For each fixed n, the row sum is

$$\sum_{m=1}^{n} X_{m,n} = V_t,$$

so the distribution of V_t is obviously infinitely divisible. Let ν be the Lévy measure of the distribution of V. We wish to show that ν is the zero measure.

For each n, let $M_n = \max\{|X_{m,n}| : m = 1, \ldots, n\}$. By Problem 48 of Chapter 16, $M_n \xrightarrow{D} M$ as $n \to \infty$, where M is a nonnegative random variable with distribution function

$$x \rightsquigarrow e^{-\nu(x,\infty) - \nu(-\infty, -x)}, \quad x > 0.$$

On the other hand, since V is $\mathbf{C}[0, 1]$-valued, $s \rightsquigarrow V_s(\omega)$ is a uniformly continuous function on $[0, t]$ for each ω. The definition of uniform continuity implies that for each ω, $M_n(\omega) \to 0$ as $n \to \infty$. Thus $M = 0$ a.s., and so ν is the zero measure. It follows from the Lévy-Khinchin Representation Theorem that each V_t has a (possibly degenerate) normal distribution.

Using the fact that V has stationary independent increments, it is easy to see that

$$\sqrt{\frac{\mathrm{Var}(V_t)}{t}} \quad \text{and} \quad \frac{E(V_t)}{t}$$

do not depend on t, first for rational t and then, by taking limits, for all $t \in (0, 1]$. Denoting these two quantities by a and b, respectively, we see that V_t is normally distributed with mean bt and variance $a^2 t$.

The proof is complete in case $a = 0$. If $a > 0$, let $W_t = (V_t - bt)/a$. It is easily checked that W has stationary independent increments, and that for each $t \in (0, 1]$, W_t is normally distributed with mean 0 and variance t. Thus W is a Brownian motion. $\square$

The terms 'Brownian' and 'Wiener' are often used in conjunction with any $\mathbf{C}[0, 1]$-valued random variable V having stationary independent increments. The process in case $a = 1, b = 0$ is a *standard* Brownian motion, and the cases with $a = 0$ are called *degenerate*. If $a, b \neq 0$, then V is a Brownian motion *with drift*. The preceding sections of this chapter treat standard Wiener measure and standard Brownian motion, although the adjective 'standard' is not used there.

19.8. † Law of the Iterated Logarithm

This section is devoted to the following result which describes more accurately than does Proposition 17 the behavior for large times of the large values of Brownian motion on $[0, \infty)$. A corollary describing the behavior near time 0 is also included.

Theorem 30. [Law of the Iterated Logarithm at ∞] *Let W be a Brownian motion on $[0, \infty)$. Then*

$$\limsup_{t \to \infty} \frac{W_t}{\sqrt{2t \log(\log t)}} = 1 \text{ a.s.}$$

PARTIAL PROOF. Let $\varepsilon > 0$. We can finish the proof by showing that there exist almost surely defined random variables $\widehat{T}$ and $T_1 < T_2 < T_3 < \cdots \to \infty$ such that

(19.18) $\qquad W_t(\omega) < (1 + \varepsilon)\sqrt{2t \log(\log t)} \quad \text{for } t \geq \widehat{T}(\omega),$

and

(19.19) $\qquad W_{T_n(\omega)}(\omega) > (1 - \varepsilon)\sqrt{2T_n(\omega)\log(\log T_n(\omega))}$

for $n = 1, 2, \ldots$.

We can prove (19.18) by showing the existence of $\widehat{T}(\omega)$ such that

(19.20) $\qquad M_t \circ W(\omega) < (1 + \varepsilon)\sqrt{2t \log(\log t)} \quad \text{for } t \geq \widehat{T}(\omega),$

where $M_t(x) = \max\{x(s): 0 \leq s \leq t\}$. Let $c > 1$. Because of the monotonicity of $t \rightsquigarrow M_t(x)$ for each x, (19.20) will follow from $P(\limsup A_n) = 0$, where

$$A_n = \left[M_{c^{n+1}} \circ W \geq (1 + \varepsilon)\sqrt{2c^n \log(\log c^n)} \right].$$

By first using Theorem 9 and part (iii) of Theorem 16 and then using the asymptotic relation (9.12) of Chapter 9 we obtain

$$P(A_n) = \sqrt{\frac{2}{\pi}} \int_{(1+\varepsilon)c^{-1/2}\sqrt{2\log(\log c^n)}}^{\infty} e^{-\frac{u^2}{2}} \, du$$

$$\sim \sqrt{\frac{c}{\pi(1 + \varepsilon)^2}} \cdot \frac{1}{(n \log c)^{\frac{(1+\varepsilon)^2}{c}}\sqrt{\log(n \log c)}} \cdot$$

Since $c > 1$ is, at this point in the argument, arbitrary, we may choose it to be less than $(1 + \varepsilon)^2$. Then $\sum P(A_n) < \infty$, so the Borel Lemma implies that $P(\limsup A_n) = 0$, as desired.

It remains to define $T_1(\omega) < T_2(\omega) < \cdots \to \infty$ that satisfy (19.19). As in the preceding paragraph we introduce $c > 1$, to be further specified later in the argument, with the intention that (T_n) be a random subsequence of $(c^k: k = 1, 2, \ldots)$. Let

$$B_k = \left[W_{c^k} > W_{c^{k-1}} + (1 - \tfrac{\varepsilon}{2})\sqrt{2c^k \log(\log c^k)} \right].$$

Because W has stationary increments,

$$P(B_k) = P\left[W_{(c^k - c^{k-1})} > (1 - \tfrac{\varepsilon}{2})\sqrt{2c^k \log(\log c^k)} \right].$$

It is left to the reader to show that $\sum P(B_k) = \infty$ for sufficiently large c. The independent increment property implies that the events B_k are independent and, therefore, by Borel-Cantelli, that $P(\limsup B_k) = 1$. By the first part of the proof applied to $-W$,

$$(19.21) \qquad W_{c^{k-1}}(\omega) > -(1+\varepsilon)c^{-1/2}\sqrt{2c^k \log(\log c^k)}$$

for all sufficiently large k, depending on ω. For such a k that also satisfies $\omega \in B_k$ we have

$$W_{c^k}(\omega) > \left[(1 - \tfrac{\varepsilon}{2}) - \tfrac{1+\varepsilon}{c^{1/2}}\right]\sqrt{2c^k \log(\log c^k)}.$$

Choose c larger than $\frac{4(1+\varepsilon)^2}{\varepsilon^2}$. Then (19.19) is obtained by defining each $T_n(\omega)$ to equal some c^k, where k is chosen so that $\omega \in B_k$ and (19.21) holds. $\square$

Problem 29. Complete the preceding proof by showing that $\sum P(B_k) = \infty$ for sufficiently large c.

Corollary 31. [Law of the Iterated Logarithm at 0] *Let W denote a Brownian motion on $[0, \infty)$. Then*

$$\limsup_{t \searrow 0} \frac{W_t}{\sqrt{2t \log(|\log t|)}} = 1 \ \ a.s.$$

Problem 30. Prove the preceding corollary. *Hint:* Use Theorem 30 and Theorem 16.

Problem 31. State a Law of the Iterated Logarithm for the simple symmetric random walk in $\mathbb{Z}$. Prove your assertion by using stopping times to 'embed' the random walk in a Brownian motion.

PART 4
Conditioning

If X and Y are two random variables defined on the same probability space, it is often the case that information about the value of X also provides information about the value of Y. The exception to this was studied in Part 2, namely the case in which X and Y are stochastically independent. In general, we will want to develop techniques for studying the way in which knowledge about X affects our assessment of the value of Y. This subject matter comes under the general heading of 'conditioning'. After the introduction of some preliminary concepts from the field of real analysis in Chapter 20, the basic tools of conditioning, namely conditional probabilities and conditional distributions, are introduced in Chapter 21. These tools will enable us to construct several important classes of random sequences, as shown in Chapter 22. In Chapter 23, we study conditional expectations, which may be regarded as the means of conditional probability distributions.

Note: Readers who have skipped Part 3 should read the comment at the end of the introduction to that part concerning notational shortcuts.

CHAPTER 20

Spaces of Random Variables

This chapter is chiefly concerned with two metric spaces consisting of collections of random variables on a probability space $(\Omega, \mathcal{F}, P)$: $\mathbf{L}_1(\Omega, \mathcal{F}, P)$, consisting of all random variables $X\colon \Omega \to \mathbb{R}$ such that $E(|X|) < \infty$, and $\mathbf{L}_2(\Omega, \mathcal{F}, P)$, consisting of those X for which $E(X^2) < \infty$. The space $\mathbf{L}_2(\Omega, \mathcal{F}, P)$ has additional structure which makes it a 'Hilbert space'. General Hilbert spaces are introduced in the first section, and $\mathbf{L}_2(\Omega, \mathcal{F}, P)$ is treated in second section. Basic results from these two sections will play an important role in the definition of conditional probability distributions in Chapter 21. The metric space $\mathbf{L}_1(\Omega, \mathcal{F}, P)$ is discussed briefly in the third section, and the final section of the chapter treats an application of Hilbert space methods to an estimation problem.

20.1. Hilbert spaces

Let $\mathcal{V}$ be a *vector space*. That is, $\mathcal{V}$ is a set of objects for which two operations are defined, *addition* and *scalar multiplication*. The operation of addition assigns to each pair $u, v \in \mathcal{V}$ an element $u + v \in \mathcal{V}$. Addition in $\mathcal{V}$ is associative and commutative; there is an additive identity, denoted by $\mathbf{0}$, which has the property that $u + \mathbf{0} = u$ for all $u \in \mathcal{V}$; and for each $u \in \mathcal{V}$ there is an additive inverse, denoted by $-u$, such that $u + (-u) = 0$. The operation of scalar multiplication assigns to each real number a and each member u of $\mathcal{V}$ an element $au \in \mathcal{V}$. This operation satisfies

$$a(u + v) = au + av \quad \text{for } a \in \mathbb{R} \text{ and } u, v \in \mathcal{V};$$
$$(a + b)v = av + bv \quad \text{for } a, b \in \mathbb{R} \text{ and } v \in \mathcal{V};$$
$$(ab)v = a(bv) \quad \text{for } a, b \in \mathbb{R} \text{ and } v \in \mathcal{V};$$
$$1v = v \quad \text{for } v \in \mathcal{V}.$$

We will be interested in vector spaces $\mathcal{V}$ that have additional structure. Suppose there exists a function that assigns to each pair $(u, v) \in \mathcal{V} \times \mathcal{V}$ a real number

$\langle u, v \rangle$ so that the following properties hold:

$$\langle \mathbf{0}, v \rangle = 0 \quad \text{for } v \in \mathcal{V};$$
$$\langle v, v \rangle > 0 \quad \text{for } v \in \mathcal{V}, v \neq \mathbf{0};$$
$$\langle u, v \rangle = \langle v, u \rangle \quad \text{for } u, v \in \mathcal{V};$$
$$\langle u + v, w \rangle = \langle u, w \rangle + \langle v, w \rangle \quad \text{for } u, v, w \in \mathcal{V};$$
$$a\langle u, v \rangle = \langle au, v \rangle \quad \text{for } a \in \mathbb{R} \text{ and } u, v \in \mathcal{V}.$$

Then $\mathcal{V}$ is called an *inner product space*.

For v a member of an inner product space $\mathcal{V}$, we define the *norm* of v by

$$\|v\| \overset{\text{def}}{=} \sqrt{\langle v, v \rangle}.$$

The *distance* between two members u and v of $\mathcal{V}$ is $\|v - u\|$. It can be checked that this distance is in fact a metric, since the defining properties of a metric hold: $\|v - u\| = \|u - v\| > 0$ if $v \neq u$, $\|v - v\| = 0$, and

$$\|w - u\| \leq \|v - u\| + \|w - v\|.$$

Thus, an inner product space is a metric space with additional structure.

Definition 1. An inner product space is called a *Hilbert space* if it is complete as a metric space.

It is easy to show that every finite-dimensional inner product space is complete, and thus a Hilbert space.

Suppose that $\mathcal{V}$ is a Hilbert space and $\mathcal{U} \subseteq \mathcal{V}$. Then $\mathcal{U}$ is called a *Hilbert subspace* of $\mathcal{V}$ if $\mathcal{U}$ is itself a Hilbert space with respect to the vector operations and inner product inherited from $\mathcal{V}$.

Proposition 2. *Let $\mathcal{V}$ be a Hilbert space and $\mathcal{U}$ a Hilbert subspace of $\mathcal{V}$. Then for each $x \in \mathcal{V}$, there exists a unique $z \in \mathcal{U}$ such that*

$$\|x - z\| = \inf\{\|x - y\| : y \in \mathcal{U}\}.$$

Furthermore, z is the unique vector in $\mathcal{U}$ satisfying

$$\langle y, x - z \rangle = 0$$

for all $y \in \mathcal{U}$.

PROOF. Fix $x \in \mathcal{V}$. Find a sequence (z_n) contained in $\mathcal{U}$ such that

$$\|x - z_n\| \to \inf\{\|x - z\| : z \in \mathcal{U}\}$$

as $n \to \infty$. We wish to prove that this sequence is Cauchy. Suppose that it were not. Then we may assume (by taking a subsequence if necessary) that there exists an $\varepsilon > 0$ such that for all n, $\|z_{n+1} - z_n\| > \varepsilon$. Let $y_n = (z_n + z_{n+1})/2$. Then straightforward algebra with the inner product shows that

$$(20.1) \quad 2\|z_n - x\|^2 + 2\|z_{n+1} - x\|^2 - 4\|y_n - x\|^2 = \|z_{n+1} - z_n\|^2 > \varepsilon^2.$$

On the other hand, in view of the defining property of the sequence (z_n) and the fact that $y_n \in \mathcal{U}$, the limit supremum of the left side of (20.1) is nonpositive. We have arrived at a contradiction, thereby showing that (z_n) is Cauchy. Since $\mathcal{U}$ is complete, the sequence (z_n) converges to a limit $z \in \mathcal{U}$. By the triangle inequality,

$$\|x - z\| \leq \|x - z_n\| + \|z_n - z\|$$

for all n. It follows that no member of $\mathcal{U}$ is closer to x than z is, since $\|z_n - z\| \to 0$ as $n \to \infty$. Thus, we have proved the existence of a member of $\mathcal{U}$ closest to x.

Were there a second member z' of $\mathcal{U}$ as close to x as is z, then we could use (20.1) with $z_n = z$ and $z_{n+1} = z'$ to show that the average of z and z' would be closer to x than z is, and we would have a contradiction. Therefore, z is the unique member of $\mathcal{U}$ that is, among members of $\mathcal{U}$, closest to x.

For $y = \mathbf{0}$, the equality $\langle y, z - x \rangle = 0$ is obvious. To obtain it for other $y \in \mathcal{U}$, consider the quantity

$$(20.2) \qquad \|ay + z - x\|^2 = a^2 \|y\|^2 + 2a\langle y, z - x \rangle + \|z - x\|^2, \quad a \in \mathbb{R}.$$

Since $ay + z \in \mathcal{U}$, we know that the left side, and therefore the right side, of (20.2) has a unique minimum at $a = 0$. On the other hand, from elementary calculus (or standard facts about quadratic polynomials), we know that there is a unique minimum at $a = -\langle y, z - x \rangle / \|y\|^2$. It follows that $\langle y, z - x \rangle = 0$.

Finally, we note that if u is any vector in $\mathcal{U}$ that satisfies $\langle y, u - x \rangle = 0$ for all $y \in \mathcal{U}$, then

$$\langle z - u, z - u \rangle = \langle z - u, z - x + x - u \rangle = \langle z - u, z - x \rangle + \langle z - u, x - u \rangle = 0,$$

since $z \in \mathcal{U}$. It follows that $z = u$. $\square$

The vector z, whose existence and uniqueness is asserted in the preceding proposition, is called the *orthogonal projection* of x onto $\mathcal{U}$, and $|x - z|$ is the *distance* between x and $\mathcal{U}$.

Problem 1. Let $\mathcal{W}$ be a Hilbert space, $\mathcal{V}$ a Hilbert subspace of $\mathcal{W}$, and $\mathcal{U}$ a Hilbert subspace of $\mathcal{V}$. Let $x, y, z \in \mathcal{W}$ be such that y is the projection of x onto $\mathcal{V}$ and z is the projection of y onto $\mathcal{U}$. Show that z is the projection of x onto $\mathcal{U}$.

20.2. The Hilbert space $\mathbf{L}_2(\Omega, \mathcal{F}, P)$

We set the goal of constructing a Hilbert space consisting of the $\mathbb{R}$-valued random variables that have finite second moments on some probability space $(\Omega, \mathcal{F}, P)$.

Problem 2. Prove that the sum of two $\mathbb{R}$-valued random variables has finite second moment if each of the summands has finite second moment. Also, show that the product of a real number and a $\mathbb{R}$-valued random variable X has finite second moment if X itself has finite second moment.

The preceding problem indicates that we can, in fact, form a vector space the members of which are random variables having finite second moments. The operations of multiplication by a scalar and addition are the usual operations with functions, and the appropriate commutative, associative, and distributive properties hold as consequences of the corresponding properties of real numbers. The same can be said for the other properties of vector spaces.

We wish to turn the vector space of the preceding paragraph into an inner product space. However, before doing so, it is necessary to modify the situation slightly by considering two random variables to be equivalent if they are equal almost surely. Thus, rather than considering the space of random variables with finite second moments on some probability space, we instead consider the space of equivalence classes of such random variables. We leave it to the reader to check that the properties required of vector spaces remain valid after this modification.

The preceding paragraph notwithstanding, we often speak of a random variable when we actually mean the equivalence class containing that random variable. For instance, in Definition 3 below, we define the inner product of two random variables X and Y with finite second moments to be the quantity $E(XY)$. The following exercise asks the reader to prove several properties of the expectation that will show that this formula does indeed define an inner product.

Problem 3. Prove or extract from previous work the following facts about real numbers a and b and random variables X, Y, and Z having finite second moments:
  (i) $-\infty < E(XY) < \infty$;
  (ii) $E(XY) = E(YX)$;
  (iii) $E(X(aY + bZ)) = aE(XY) + bE(XZ)$;
  (iv) $E(XX) \geq 0$;
  (v) $E(XX) = 0 \Leftrightarrow X = 0$ a.s.;
  (vi) $E(XY) = E(XZ)$ if $Y = Z$ a.s.

Thus, we may make the following definition.

Definition 3. Let $(\Omega, \mathcal{F}, P)$ denote a probability space. By

$$\mathbf{L}_2(\Omega, \mathcal{F}, P)$$

we denote the inner product space consisting of all equivalence classes of $\mathbb{R}$-valued random variables on $(\Omega, \mathcal{F}, P)$ that have finite second moments, with addition of vectors and multiplication by scalars defined in the usual way for functions, and the inner product of two members X and Y of $\mathbf{L}_2(\Omega, \mathcal{F}, P)$ defined by

$$\langle X, Y \rangle = E(XY).$$

We sometimes write $\mathbf{L}_2$ for $\mathbf{L}_2(\Omega, \mathcal{F}, P)$ when there is no need to explicitly mention the underlying probability space $(\Omega, \mathcal{F}, P)$.

We call the norm associated with $\mathbf{L}_2$ the $\mathbf{L}_2$-*norm* and denote it by

$$\|X\|_2 = \langle X, X \rangle^{1/2}.$$

Thus, the distance between two members X and Y of $\mathbf{L}_2$ is given by

$$d(X, Y) = \|X - Y\|_2.$$

Proposition 4. $\mathbf{L}_2(\Omega, \mathcal{F}, P)$ *is a Hilbert space.*

PROOF. All but completeness has been checked above. Let $(X_1, X_2, \ldots)$ be a Cauchy sequence in $\mathbf{L}_2$. Choose a subsequence $(X_{n_k} : k = 1, 2, \ldots)$ such that

$$\|X_{n_k} - X_{n_{k+1}}\|_2 < 2^{-k}$$

for all k. By the Markov Inequality (Proposition 3 in Chapter 5),

$$P[|X_{n_k} - X_{n_{k+1}}| \geq \varepsilon] \leq \varepsilon^{-2} 2^{-2k}$$

for all $\varepsilon > 0$. By the Borel Lemma,

$$P(\limsup_{k \to \infty} \{\omega : |X_{n_k}(\omega) - X_{n_{k+1}}(\omega)| \geq (2/3)^k\}) = 0.$$

It follows that, for almost all ω,

$$\sum_{k=1}^{\infty} |X_{n_k}(\omega) - X_{n_{k+1}}(\omega)| < \infty.$$

Thus, for almost all ω, the sequence $(X_{n_k}(\omega), k = 1, 2, \ldots)$ is a Cauchy sequence of real numbers. For such ω, let $X(\omega)$ be the limit of this sequence. Hence, X is an almost surely defined $\mathbb{R}$-valued random variable. By the Fatou Lemma and the triangle inequality,

$$(E(X^2))^{1/2} \leq (\liminf_{k \to \infty} \|X_{n_k}\|_2) \leq \left(\|X_{n_1}\|_2 + \sum_{k=1}^{\infty} \|X_{n_k} - X_{n_{k+1}}\|_2 \right) < \infty,$$

so $X \in \mathbf{L}_2$. Similarly,

$$\|X - X_{n_k}\|_2 \leq \sum_{m=k}^{\infty} \|X_{n_m} - X_{n_{m+1}}\|_2 < 2^{-k+1}.$$

It follows that X is the limit, in the metric space $\mathbf{L}_2$, of the sequence (X_{n_k}). Since any subsequential limit of a Cauchy sequence is the limit of the entire sequence, X is also the limit of the sequence (X_n); and so $\mathbf{L}_2$ is complete. $\square$

Problem 4. Discuss the possibility of proving the preceding proposition by showing that the original sequence is Cauchy in probability rather than by showing that some subsequence is Cauchy a.s.

* **Problem 5.** Let $X \in \mathbf{L}_2(\Omega, \mathcal{F}, P)$ and let $\mathcal{U}$ denote the subspace of $\mathbf{L}_2(\Omega, \mathcal{F}, P)$ consisting of all constant random variables. Find the orthogonal projection of X onto $\mathcal{U}$.

We have already studied almost sure convergence and convergence in probability for sequences of random variables (and also convergence in distribution, treated in Chapter 14). We now have another mode of convergence, namely convergence in $\mathbf{L}_2$. With respect to the relationships between these various types of convergence, we already know that almost sure convergence implies convergence in probability. Here is more of the story.

Proposition 5. *Suppose that the sequence* $(X_n \colon n = 1, 2, \ldots)$ *of* $\mathbb{R}$-*valued random variables converges to* X *in* $\mathbf{L}_2$. *Then the following are true:*

(i) $\|X_n\|_2 \to \|X\|_2$ *as* $n \to \infty$;
(ii) (X_n) *converges to* X *in probability;*
(iii) $E(|X_n|) \to E(|X|)$ *as* $n \to \infty$;
(iv) $E(|X - X_n|) \to 0$ *as* $n \to \infty$;
(v) $E(X_n) \to E(X)$ *as* $n \to \infty$;
(vi) *there exists a subsequence of* (X_n) *which converges to* X *almost surely.*

* **Problem 6.** Prove the preceding proposition.

Problem 7. Find a convergent sequence in $\mathbf{L}_2$ of some probability space that does not converge almost surely.

Problem 8. Find a sequence of members of $\mathbf{L}_2$ of some probability space that converges almost surely to a member of $\mathbf{L}_2$, but which does not converge in $\mathbf{L}_2$.

Proposition 6. *Let* X *and* $X_n, n = 1, 2, \ldots$, *be random variables on a probability space* $(\Omega, \mathcal{F}, P)$. *Suppose that* $X_n \to X$ *a.s. as* $n \to \infty$ *and that, for each* n, $X_n \in \mathbf{L}_2$. *Then the following conditions are equivalent:*

(i) *the family* $\{X_n^2 \colon n = 1, 2, \ldots\}$ *is uniformly integrable;*
(ii) $E(X^2) < \infty$ *and* $\lim_{n\to\infty} E((X_n - X)^2) = 0$;
(iii) $\lim_{n\to\infty} E(X_n^2) = E(X^2) < \infty$.

Problem 9. Prove the preceding proposition.

It turns out that most of the results proved so far in this chapter for probability spaces extend to the σ-finite setting. A few comments on how this is done are in order. First note that the Cauchy-Schwarz Inequality can be adapted to the general σ-finite setting, and so the assertions in Problem 2 and Problem 3 apply in the σ-finite setting. As for the proof of Proposition 4, the relevant part of the Borel-Cantelli Lemma is valid more generally, so Proposition 4 extends as well. The upshot is that we can speak of $\mathbf{L}_2$ of any σ-finite measure space, and assert that it is a Hilbert space.

Problem 10. Show that (i), (ii), and (vi) of Proposition 5 still hold if $(\Omega, \mathcal{F}, P)$ is replaced by an arbitrary σ-finite measure space and convergence in probability is replaced by convergence in measure. Give an example that shows that (iii), (iv), and (v) may fail, where, of course, expected values are to be replaced by integrals.

Proposition 5, parts (iii), (iv), and (v), and Proposition 6 do not generalize to σ-finite measure spaces.

20.3. The metric space $\mathbf{L}_1(\Omega, \mathcal{F}, P)$

Before treating an application of $\mathbf{L}_2$, we introduce another metric space of equivalence classes of random variables.

We denote by $\mathbf{L}_1(\Omega, \mathcal{F}, P)$, the space of all equivalence classes of $\mathbb{R}$-valued random variables on $(\Omega, \mathcal{F}, P)$ having finite expectation. It is not a Hilbert space, but a metric space defined via the norm given by $\|X\|_1 = E(|X|)$, called the $\mathbf{L}_1$-*norm* of the member X of $\mathbf{L}_1(\Omega, \mathcal{F}, P)$. The distance between two members X and Y of $\mathbf{L}_1(\Omega, \mathcal{F}, P)$ is $\|Y - X\|_1$. When there is no danger of confusion we write $\mathbf{L}_1$ in lieu of $\mathbf{L}_1(\Omega, \mathcal{F}, P)$.

Problem 11. Prove the assertion just made that $\mathbf{L}_1$ is a metric space.

Proposition 7. *Let $(\Omega, \mathcal{F}, P)$ be a probability space. Then $\mathbf{L}_1(\Omega, \mathcal{F}, P)$ is a complete metric space.*

Problem 12. Prove the preceding proposition.

Problem 13. State and prove a proposition for $\mathbf{L}_1$ analogous to Proposition 5.

Notice that the Uniform Integrability Criterion, Theorem 12 in Chapter 8, contains an analog of Proposition 6 for $\mathbf{L}_1$.

Proposition 8. *For any probability space, $\mathbf{L}_2 \subseteq \mathbf{L}_1$. Moreover, a sequence that converges in $\mathbf{L}_2$ also converges in $\mathbf{L}_1$.*

Problem 14. Prove the preceding proposition.

It is true that the $\mathbf{L}_1$-concept carries over to σ-finite measure spaces, and that even in this more general setting, $\mathbf{L}_1$ is a complete metric space. However, warnings for $\mathbf{L}_1$ in the σ-finite setting analogous to those for $\mathbf{L}_2$ are appropriate. Moreover, Proposition 8 is not valid in infinite measure spaces.

20.4. † Best linear estimator

Let $X, Y_1, Y_2, \ldots, Y_n$ be $\mathbb{R}$-valued random variables with finite second moments. The space

$$\mathcal{V} = \{a_0 + a_1 Y_1 + \cdots + a_n Y_n : a_0, a_1, \ldots, a_n \in \mathbb{R}\}$$

is a finite-dimensional Hilbert subspace of $\mathbf{L}_2(\Omega, \mathcal{F}, P)$, called the *Hilbert space span* of $\{1, Y_1, \ldots, Y_n\}$. Since the Hilbert space $\mathcal{V}$ is finite-dimensional we could also use the term *linear span*. Denote the orthogonal projection of X onto $\mathcal{V}$ by Z. By Proposition 2 we see that Z is the *best linear estimator* of X with respect to the sequence $Y_1, \ldots, Y_n$ in the sense that

$$(20.3) \qquad\qquad E((X - Z)^2) < E((X - Y)^2)$$

for every linear combination Y of $1, Y_1, \ldots, Y_n$ that is not almost surely equal to Z. Because of the inequality (20.3), Z is also called the *least squares estimate*.

We use the tools of linear algebra to do two things: show how to obtain an explicit formula for Z and prove that $E(Z) = E(X)$, thereby indicating, in combination with (20.3), why Z is sometimes called the *minimum variance unbiased linear estimator* of X. An orthonormal basis $\{1, V_1, \ldots, V_m\}$ for $\mathcal{V}$ can be obtained by using the Gram-Schmidt orthonormalization procedure, a procedure which only requires knowledge of the means, variances, and covariances of the random variables $Y_1, \ldots, Y_n$. Then

$$Z = \langle X, 1 \rangle 1 + \langle X, V_1 \rangle V_1 + \cdots + \langle X, V_m \rangle V_m .$$

For $i = 1, \ldots, n$, $E(V_i) = \langle 1, V_i \rangle = 0$. Therefore $E(Z) = \langle X, 1 \rangle = E(X)$.

* **Problem 15.** Let $X, Y_1, \ldots, Y_n$ and Z be as in the preceding example. Show that the random variable $X - Z$ is uncorrelated with each of the random variables $Y_1, \ldots, Y_n$.

Problem 16. Let $X, X_1, X_2, \ldots, X_n$ be independent random variables in $\mathbf{L}_2$ and let $Y_i = X + X_i$ for $i = 1, \ldots, n$. Find the best linear estimator of X with respect to $Y_1, \ldots Y_n$. *Hint:* Use the cases $n = 1, 2$ to guess at the general answer; then check your conjecture.

One can interpret the random variables X_i in the preceding problem as noise. Each time we try to observe X, we instead observe the sum of X and noise. From what we observe, we then try to find a linear combination of the observations that best estimates the true signal X.

CHAPTER 21
Conditional Probabilities

In this chapter we introduce two closely related concepts: conditional probabilities and conditional probability distributions. These two mathematical objects are vital in the construction and analysis of nearly all of the random sequences and stochastic processes that form the chief subject matter of the latter part of this book (Chapters 24 through 33).

We first show how to use Hilbert space ideas to construct conditional probabilities. The concept 'conditional probability' is then extended to 'conditional distribution' and the related 'conditional density'. Conditional distributions do not always exist; a useful result concerning their existence is found in the fourth section. Next, conditional independence is treated. Using the notion of conditional independence, we also define an important class of random sequences known as 'Markov sequences'. The chapter concludes with an illustrative example involving conditional distributions for normally distributed random variables.

21.1. The construction of conditional probabilities

Consider a simple symmetric random walk S in $\mathbb{Z}$, with increments $X_n = S_n - S_{n-1}$. Let A be the event $\{\omega \colon S_5(\omega) > 0\}$. From symmetry considerations we know that $P(A) = 1/2$. Our assessment of the probability changes, however, as we observe the values of S_1, S_2, S_3, and S_4. In particular, if we consider S_4 we find that there are three different situations. If $S_4(\omega) > 0$, then it is necessarily the case that $S_5(\omega) > 0$. Similarly, if $S_4(\omega) < 0$, then $S_5(\omega) < 0$. Finally, if $S_4(\omega) = 0$, then the sign of $S_5(\omega)$ is the same as that of the increment $X_5(\omega)$. We say that the 'conditional probability of A, given S_4' is 1 if $S_4 > 0$, 1/2 if $S_4 = 0$, and 0 if $S_4 < 0$.

We have just described a special case of the following more general situation. We start with a probability space $(\Omega, \mathcal{F}, P)$. The probability measure P represents our initial assessments of the probabilities of the various events in $\mathcal{F}$. We are then given some information about which of the events in a certain collection $\mathcal{G}$ have occurred and which have not. The reader can use the Sierpiński Class

Theorem to see that we might as well assume that $\mathcal{G}$ is a σ-field.

We think of $\mathcal{G}$ as representing information; if $\mathcal{G} = \mathcal{F}$, then $\mathcal{G}$ represents the information possessed by an observer who has seen the outcome of the complete experiment. If $\mathcal{G}$ equals the σ-field $\{\Omega, \emptyset\}$, then it represents the information possessed by one who knows only the probability space for the experiment but has no information about the outcome. Other σ-fields $\mathcal{G}$ such that $\mathcal{G} \subseteq \mathcal{F}$ represent partial information about the outcome of the experiment. Given such a σ-field $\mathcal{G}$ and an event $A \in \mathcal{F}$, we wish to define the 'conditional probability of A given $\mathcal{G}$', which represents our updated assessment of the probability of A, based on the information contained in $\mathcal{G}$. The notation for this conditional probability is '$P(A \mid \mathcal{G})$'. It is not a constant, but a random variable. For the example in the preceding paragraph $\mathcal{G} = \sigma(S_1, S_2, S_3, S_4)$ and

$$
P(A \mid \mathcal{G})(\omega) = \begin{cases} 1 & \text{if } S_4(\omega) > 0 \\ 1/2 & \text{if } S_4(\omega) = 0 \\ 0 & \text{if } S_4(\omega) < 0 \,. \end{cases}
$$

The same formula holds if $\mathcal{G} = \sigma(S_4)$. An interpretation of this fact is that S_4, by itself, gives as much information about the sign of S_5 as does the vector (S_1, S_2, S_3, S_4).

We use the concept of orthogonal projection in a Hilbert space to define conditional probabilities. Let $(\Omega, \mathcal{F}, P)$ be a probability space and $\mathcal{G}$ be a sub-σ-field of $\mathcal{F}$. We wish to think of the Hilbert space $\mathbf{L}_2(\Omega, \mathcal{G}, P)$ as a Hilbert subspace of the Hilbert space $\mathbf{L}_2(\Omega, \mathcal{F}, P)$. Strictly speaking, $\mathbf{L}_2(\Omega, \mathcal{G}, P)$ is not necessarily a subset of $\mathbf{L}_2(\Omega, \mathcal{F}, P)$, since the equivalence class of a random variable X in $\mathbf{L}_2(\Omega, \mathcal{G}, P)$ will typically be smaller than the corresponding equivalence class of X in $\mathbf{L}_2(\Omega, \mathcal{F}, P)$: a random variable in $\mathbf{L}_2(\Omega, \mathcal{G}, P)$ can be almost surely equal to many random variables in $\mathbf{L}_2(\Omega, \mathcal{F}, P)$ that are not $\mathcal{G}$-measurable. Nevertheless, if we identify the equivalence class of X in $\mathbf{L}_2(\Omega, \mathcal{G}, P)$ with the equivalence class of X in $\mathbf{L}_2(\Omega, \mathcal{F}, P)$, then we may identify $\mathbf{L}_2(\Omega, \mathcal{G}, P)$ with a Hilbert subspace of $\mathbf{L}_2(\Omega, \mathcal{F}, P)$. Hereafter we will make this identification without comment.

Definition 1. Let A be an event in a probability space $(\Omega, \mathcal{F}, P)$ and let $\mathcal{G}$ denote a sub-σ-field of $\mathcal{F}$. The *conditional probability of A given $\mathcal{G}$*, denoted by $P(A \mid \mathcal{G})$, is the projection of $I_A \in \mathbf{L}_2(\Omega, \mathcal{F}, P)$ onto $\mathbf{L}_2(\Omega, \mathcal{G}, P)$. That is, $P(A \mid \mathcal{G})$ is the equivalence class of $\mathcal{G}$-measurable random variables X that satisfy the equation

$$
E(XY) = E(I_A Y)
$$

for all $Y \in \mathbf{L}_2(\Omega, \mathcal{G}, P)$.

In practice, it is cumbersome to work with equivalence classes, so we usually speak as if $P(A \mid \mathcal{G})$ denotes a single, arbitrarily chosen $\mathcal{G}$-measurable random variable in the equivalence class obtained by projecting I_A onto $\mathbf{L}_2(\Omega, \mathcal{G}, P)$. Thus, statements of equality of conditional probabilities should technically have the phrase 'almost surely' attached to them, but this phrase is often dropped and one speaks as if conditional probabilities are unique random variables rather than unique equivalence classes of random variables.

The next proposition says that in testing whether a particular random variable equals $P(A \mid \mathcal{G})$, one may focus on its relation to indicator functions of members of $\mathcal{G}$.

Proposition 2. *Let A be an event in a probability space $(\Omega, \mathcal{F}, P)$, X an $\mathbb{R}$-valued random variable, and $\mathcal{G}$ a sub-σ-field of $\mathcal{F}$. Then $X = P(A \mid \mathcal{G})$ a.s. if and only if X satisfies the following two conditions:*

(i) X is $\mathcal{G}$-measurable;
(ii) for all $B \in \mathcal{G}$, $E(XI_B) = P(A \cap B)$.

PROOF. The 'only if' direction of the result is immediate from the definition. For the 'if' direction, let X be any random variable satisfying conditions (i) and (ii). Then $Z = X - P(A \mid \mathcal{G})$ is a $\mathcal{G}$-measurable random variable for which $E(XI_B) = 0$ for every $B \in \mathcal{G}$. It follows that $P(B) = 0$ if B is the set where Z is positive and also if B is the set where Z is negative. Thus, $Z = 0$ a.s., as desired. □

The special case $E(X) = P(A)$ of condition (ii) in Proposition 2 is a fact that is used often without reference to the proposition.

Problem 1. Let P and $\mathcal{G}$ be as in Definition 1 and $A \in \mathcal{G}$. Give two proofs that $P(A \mid \mathcal{G}) = I_A$, one using Definition 1 and the other Proposition 2.

Problem 2. Let P and $\mathcal{G}$ be as in Definition 1, and $\mathcal{H}$ be a sub-σ-field of $\mathcal{G}$. Suppose that $P(A \mid \mathcal{G})$ is $\mathcal{H}$-measurable for some event A. Give two proofs that $P(A \mid \mathcal{G}) = P(A \mid \mathcal{H})$.

Problem 1 can be generalized as follows.

Proposition 3. *Let A and B be events in a probability space $(\Omega, \mathcal{F}, P)$ and suppose that A is a member of a sub-σ-field $\mathcal{G}$ of $\mathcal{F}$. Then*

$$P(A \cap B \mid \mathcal{G}) = P(B \mid \mathcal{G})I_A \ \text{a.s.}$$

* **Problem 3.** Give two proofs of the preceding proposition, one using Definition 1 and the other Proposition 2.

Problem 4. Let A be an event in a probability space $(\Omega, \mathcal{F}, P)$ and suppose that A is independent of a sub-σ-field $\mathcal{G}$ of $\mathcal{F}$. By using Definition 1 and then again by using Proposition 2, prove that $P(A \mid \mathcal{G}) = P(A)$ and, in particular, that the probability of A given $\{\Omega, \emptyset\}$ is the constant random variable $P(A)$.

In the following proposition, which is a generalization of the preceding problem, conditioning with respect to a σ-field $\sigma(\mathcal{G}, \mathcal{H})$ is considered. But, for simplicity, $\mathcal{G}, \mathcal{H}$, rather than $\sigma(\mathcal{G}, \mathcal{H})$ is written, there being no danger of confusion because conditioning is always with respect to a σ-field.

Proposition 4. *Let $\mathcal{G}$ and $\mathcal{H}$ be sub-σ-fields of the σ-field $\mathcal{F}$ of events in a probability space $(\Omega, \mathcal{F}, P)$ and let A be an event. Suppose that $\sigma(A, \mathcal{H})$ and $\mathcal{G}$ are independent. Then $P(A \mid \mathcal{G}, \mathcal{H}) = P(A \mid \mathcal{H})$ a.s.*

PROOF. The plan is to show that $P(A \mid \mathcal{H})$ satisfies the two conditions given in Proposition 2 for a random variable to equal $P(A \mid \mathcal{G}, \mathcal{H})$. It obviously satisfies the first of the two conditions since $P(A \mid \mathcal{H})$, being $\mathcal{H}$-measurable, is $\sigma(\mathcal{G}, \mathcal{H})$-measurable. For the second condition we need to show that

$$E\big(P(A \mid \mathcal{H})I_B\big) = P(A \cap B)$$

for $B \in \sigma(\mathcal{G}, \mathcal{H})$. The collection of those B for which this relation holds is clearly a Sierpiński Class, so we may restrict our attention to B of the form $C \cap D$, where $C \in \mathcal{G}$ and $D \in \mathcal{H}$. We calculate

$$E\big(P(A \mid \mathcal{H})I_{C \cap D}\big) = E\big(P(A \mid \mathcal{H})I_C I_D\big) = E\big(P(A \cap D \mid \mathcal{H})I_C\big),$$

the last equality being a consequence of Proposition 3. Since $\mathcal{G}$ and $\sigma(A, \mathcal{H})$ are independent, $P(A \cap D \mid \mathcal{H})$ is $\mathcal{H}$-measurable, and $C \in \mathcal{G}$, we may continue the calculation as follows:

$$E\big(P(A \cap D \mid \mathcal{H})I_C\big) = E\big(P(A \cap D \mid \mathcal{H})\big) E\big(I_C\big)$$
$$= P(A \cap D)P(C) = P((A \cap D) \cap C) = P(A \cap (C \cap D)),$$

giving us the desired equality. □

The intuition behind Proposition 4 is that if every event formed from A and $\mathcal{H}$ is independent of $\mathcal{G}$, then, once A has been conditioned by $\mathcal{H}$, further conditioning by $\mathcal{G}$ is irrelevant. It is important to note, as indicated by Problem 42, that the conditioning on $\mathcal{G}$ is not necessarily irrelevant if A and $\mathcal{H}$ are only separately independent of $\mathcal{G}$.

The next problem is very important for further developing one's intuition.

* **Problem 5.** Let A and C be events and suppose that $0 < P(C) < 1$. Prove that

(21.1) $$P(A \mid \sigma(C)) = \frac{P(A \cap C)}{P(C)} I_C + \frac{P(A \cap C^c)}{P(C^c)} I_{C^c}.$$

With reference to the preceding problem: if $\omega \in C$, then

$$P(A \mid \sigma(C))(\omega) = \frac{P(A \cap C)}{P(C)},$$

this being the number that one would use in trying to decide whether a proposed bet on A is a good bet in light of the knowledge that C has occurred. We will use the notation $P(A \mid C)$ for the constant $P(A \cap C)/P(C)$ (provided that $P(C) > 0$). Thus, if $0 < P(C) < 1$,

$$P(A \mid \sigma(C))(\omega) = \begin{cases} P(A \mid C) & \text{if } \omega \in C \\ P(A \mid C^c) & \text{if } \omega \notin C. \end{cases}$$

The number $P(A \mid C^c)$ is the number one would want to use for deciding about a bet on the event A given the information that C has not occurred.

One might take the point of view that, in an actual running of a single experiment, one is interested in only one of the two numbers obtained in the preceding paragraph, not in both. In this connection consider a situation in which a unseen machine is flipping a fair coin five times. You are considering a bet that at least three flips have resulted in heads. As you are contemplating, someone who has seen the result tells you that the first two flips were heads. This information would seem to increase the odds of there being at least three heads, but that is not necessarily the case. For instance, he might have intended to tell you the outcomes of the first two flips only in case the last three flips were all tails. Unless either there is a prior algorithm concerning information he will give you or one has actually incorporated his possibly random behavior into the probability space structure, one cannot calculate meaningful conditional probabilities. The mathematician's resolution of this issue is to require all conditioning to be with respect to a σ-field. The relevant σ-field for Problem 5 is $\{C, C^c, \Omega, \emptyset\}$ which corresponds to the information: C occurred or not.

Problem 5 can be generalized. We call a σ-field *purely atomic* if it is generated by a countable (possibly finite) partition $(C_1, C_2, \ldots)$ of the probability space.

Problem 6. Let $X = (X_1, X_2, \ldots)$ be a random sequence of $(\Psi, \mathcal{H})$-valued random variables. Assume that Ψ is a countable set. Prove that each of the σ-fields $\sigma(X_1, X_2, \ldots, X_n)$ is purely atomic. Is the σ-field $\sigma(X)$ necessarily purely atomic?

Proposition 5. *Let $\mathcal{G}$ be a purely atomic σ-field generated by a partition $(C_j : j = 1, 2, \ldots)$ of a probability space $(\Omega, \mathcal{F}, P)$, and let $A \in \mathcal{F}$. Then*

$$(21.2) \qquad P(A \mid \mathcal{G}) = \sum_{j:\, P(C_j)>0} P(A \mid C_j) I_{C_j} \quad a.s.$$

Problem 7. Prove the preceding proposition.

Example 1. Let $(\Omega, \mathcal{F}, P)$ denote a fair-coin-flip space in which -1 is used to denote tails and $+1$ heads; thus $\omega \in \Omega$ can be written as $(\omega_1, \omega_2, \ldots)$ where each ω_n equals -1 or 1. For $n = 0, 1, 2, \ldots$, let

$$S_n(\omega) = \omega_1 + \cdots + \omega_n,$$

where the empty sum $S_0(\omega)$ equals 0. Then $S = (S_0, S_1, \dots)$ is a simple symmetric random walk on $\mathbb{Z}$. We wish to calculate conditional probabilities of arbitrary events given $\mathcal{G} = \sigma(S_4)$.

For $k = -4, -2, 0, 2, 4$, let

$$C_k = \{\omega = (\omega_1, \omega_2, \dots) \in \Omega \colon \omega_1 + \omega_2 + \omega_3 + \omega_4 = k\}.$$

Clearly $\{C_{-4}, C_{-2}, C_0, C_2, C_4\}$ is a partition of Ω, and $\mathcal{G}$ is generated by this partition. Easy calculations give

$$P(C_0) = \frac{6}{16},$$

$$P(C_{-2}) = P(C_2) = \frac{4}{16},$$

$$P(C_{-4}) = P(C_4) = \frac{1}{16}.$$

Thus,

$$P(A \mid \mathcal{G}) = 16P(A \cap C_{-4})I_{C_{-4}} + 4P(A \cap C_{-2})I_{C_{-2}}$$
$$+ \frac{8}{3}P(A \cap C_0)I_{C_0} + 4P(A \cap C_2)I_{C_2} + 16P(A \cap C_4)I_{C_4}.$$

Let us use this formula for $A = \{\omega \colon S_5(\omega) > 0\}$. Then $A \cap C_{-4} = A \cap C_{-2} = \emptyset$ and $C_2 \cup C_4 \subseteq A$. The event $A \cap C_0$ consists of those members of C_0 whose fifth term is $+1$, and so its probability equals $\frac{1}{2}P(C_0)$. The upshot is

$$P(A \mid \mathcal{G}) = \frac{1}{2}I_{C_0} + I_{C_2} + I_{C_4}.$$

In particular, if $\omega = (1, -1, -1, 1, -1, -1, \dots)$ and $\theta = (1, -1, -1, 1, 1, -1, \dots)$, then

$$P(A \mid \mathcal{G})(\omega) = P(A \mid \mathcal{G})(\theta) = \frac{1}{2}$$

even though $\omega \notin A$ and $\theta \in A$. This last sentence does not indicate any error. Rather, the point is that when we are conditioning on $\mathcal{G}$ and wish to evaluate a conditional probability at some sample point, we do not distinguish between two different sample points if they agree in the first four coordinates.

It is often the case when writing expressions for conditional probabilities that two different symbols are needed for generic points in the sample space. For instance, we might write

$$P(\{\omega' \colon S_5(\omega') = 1\} \mid \mathcal{G})(\omega).$$

Of course, the roles of ω and ω' could be reversed, or another letter could be introduced.

* **Problem 8.** For S and $\mathcal{G}$ as in Example 1 compute:
   (i) $P(\{\omega' \colon S_7(\omega') > 0\} \mid \mathcal{G})$,
   (ii) $P(\{\omega' \colon S_2(\omega') = 2\} \mid \mathcal{G})$,

(iii) $P(\{\psi\colon S_6(\psi) = 2\} \mid \mathcal{G})$,

(iv) $P(\{\psi\colon S_2(\psi) = 0,\, S_6(\psi) = 2\} \mid \mathcal{G})$,

(v) $P(\{\omega'\colon S_4(\omega') = 0,\, S_6(\omega') = 2\} \mid \mathcal{G})$.

In particular, evaluate each of these random variables at ω, where $\omega_1 = \omega_3 = 1$ and $\omega_n = -1$ for all $n \neq 1, 3$.

*   Problem 9. Let $(\Omega, \mathcal{F}, P)$ be as in Example 1. Obtain a formula for the conditional probability of an arbitrary event given $\mathcal{H} = \sigma(S_1, S_2, S_3, S_4)$. Repeat Problem 8 with $\mathcal{G}$ replaced by $\mathcal{H}$.

The following result indicates that conditional probabilities have much in common with (unconditional) probabilities.

Proposition 6. *Let $\mathcal{G}$ a sub-σ-field of the σ-field $\mathcal{F}$ of events in a probability space $(\Omega, \mathcal{F}, P)$. Then (i) $P(\Omega \mid \mathcal{G}) = 1$ a.s., (ii) $0 \leq P(A \mid \mathcal{G}) \leq 1$ a.s. for each $A \in \mathcal{F}$, and (iii)*

$$P\left(\bigcup_{n=1}^{\infty} A_n \mid \mathcal{G}\right) = \sum_{n=1}^{\infty} P(A_n \mid \mathcal{G}) \text{ a.s.}$$

for disjoint members $A_1, A_2, \ldots$ of $\mathcal{F}$.

PROOF. To prove (i) we note that the constant function 1 is $\mathcal{G}$-measurable. For any $B \in \mathcal{G}$, $E(1 I_B) = P(B) = P(\Omega \cap B)$. By Proposition 2, 1 is a conditional probability of Ω given $\mathcal{G}$.

To prove (ii) we first let $B = \{\omega\colon P(A \mid \mathcal{G})(\omega) > 1\}$ and, for a proof by contradiction, we suppose that $P(B) > 0$. Then

$$E(P(A \mid \mathcal{G})I_B) > E(I_B) = P(B) \geq P(A \cap B) = E(P(A \mid \mathcal{G})I_B),$$

the last step following from the second condition of Proposition 2. The contradiction we have reached shows $P(A \mid \mathcal{G}) \leq 1$ a.s.. A similar argument shows that $P(A \mid \mathcal{G}) \geq 0$ a.s..

To prove (iii) we show that $\sum P(A_n \mid \mathcal{G})$ satisfies the defining properties of $P(\bigcup A_n \mid \mathcal{G})$. It is certainly measurable with respect to $\mathcal{G}$. To check the second condition of Proposition 2 we let $B \in \mathcal{G}$ and use the Monotone Convergence Theorem in conjunction with the second condition of Proposition 2 for each $E(A_n \mid \mathcal{G})$ to calculate

$$E\left(\left(\sum P(A_n \mid \mathcal{G})\right)I_B\right) = \sum E(P(A_n \mid \mathcal{G})I_B)$$

$$= \sum P(A_n \cap B) = P\left(\left[\bigcup A_n\right] \cap B\right). \quad \square$$

*   Problem 10. [Conditional Borel Lemma] Prove that

$$P(\limsup_{n \to \infty} A_n \mid \mathcal{G})(\omega) = 0$$

for almost every $\omega \in \Omega$ for which $\sum_{n=1}^{\infty} P(A_n \mid \mathcal{G})(\omega) < \infty$. Then conclude that

$$P(\limsup_{n \to \infty} A_n) \leq P\left(\left\{\omega: \sum_{n=1}^{\infty} P(A_n \mid \mathcal{G})(\omega) = \infty\right\}\right).$$

In the following example a conditional probability must be computed without the help of Proposition 5. The technique applied is that of first making an educated guess and then using Proposition 2 to verify that guess.

Example 2. Let X be a random variable uniformly distributed on the interval $[-1, 2]$. Let $\mathcal{G} = \sigma(|X|)$. We set the goal of calculating $P(A \mid \mathcal{G})$, where

$$A = \{\omega: X(\omega) > 0\}.$$

We reason that if $|X| > 1$, we then know that $X > 0$. On the other hand, if $|X| \leq 1$, we feel that the sign of X is determined by a coin flip. Therefore, we conjecture

$$P(A \mid \mathcal{G}) = Y \quad \text{a.s.},$$

where

$$Y = I_{\{\omega: |X(\omega)| > 1\}} + \frac{1}{2} I_{\{\omega: |X(\omega)| \leq 1\}}.$$

Condition (i) of Proposition 2 is clearly satisfied. For condition (ii), let $B \in \mathcal{G}$ and write $B = B_1 \cup B_2$, where

$$B_1 = B \cap \{\omega: |X(\omega)| > 1\} \quad \text{and} \quad B_2 = B \cap \{\omega: |X(\omega)| \leq 1\}.$$

Then

$$E(YI_B) = E(YI_{B_1}) + E(YI_{B_2}) = P(B_1) + \frac{1}{2}P(B_2)$$

$$= P(A \cap B_1) + \frac{1}{2}P[X \in C_2],$$

where C_2 is a certain Borel subset of $[-1, 1]$ having the property that $x \in C_2$ if and only if $-x \in C_2$. Since

$$\frac{1}{2}P[X \in C_2] = P[X \in C_2, X > 0] = P(A \cap B_2),$$

we obtain

$$E(YI_B) = P(A \cap B_1) + P(A \cap B_2) = P(A \cap B),$$

as desired.

Problem 11. Replace the event A in the preceding example by

$$\{\omega: X(\omega) < 1/2\}.$$

* **Problem 12.** Let Ω denote the unit square, $\mathcal{F}$ the Borel σ-field of subsets of Ω, P two-dimensional Lebesgue measure, and $\mathcal{G}$ the σ-field of sets of the form

$$\{(\omega_1, \omega_2) \in \Omega : \omega_1 \in C\},$$

C a Borel subset of the unit interval. Calculate

$$P\big(\{(\omega_1, \omega_2) : \omega_1 > \omega_2\} \mid \mathcal{G}\big).$$

It is often the case that we wish to condition on a σ-field $\mathcal{G}$ of the form $\mathcal{G} = \sigma(X)$ for some random variable X. In this case conditional probabilities have a particularly nice form, as a consequence of the following lemma.

Lemma 7. *Let V be a Ψ-valued random variable and Y a $\sigma(V)$-measurable $\overline{\mathbb{R}}$-valued random variable. Then there exists a measurable function $f \colon \Psi \to \overline{\mathbb{R}}$ such that $Y = f \circ V$.*

PROOF. Let $\mathcal{C}$ be the collection of functions $Y \colon \Omega \to \overline{\mathbb{R}}$ that can be written as $f \circ V$ for some measurable function $f \colon \Psi \to \overline{\mathbb{R}}$. We first show that $\mathcal{C}$ contains all $\overline{\mathbb{R}}$-valued simple functions that are $\sigma(V)$-measurable. Since such functions can be written as a finite sum $\sum_k a_k I_{A_k}$ for constants $a_k \in \overline{\mathbb{R}}$ and pairwise disjoint sets $A_k \in \sigma(V)$, it is enough to show that $\mathcal{C}$ contains all indicator functions of sets in $\sigma(V)$. If A is such a set, then $A = V^{-1}(B)$ for some $B \subseteq \Psi$, so $I_A = I_B \circ V$.

We next show that $\mathcal{C}$ is closed under increasing pointwise limits. Let $(Y_1 \leq Y_2 \leq \ldots)$ be an increasing sequence of members of $\mathcal{C}$, and write $Y_n = f_n \circ V$, for $n = 1, 2, \ldots$, where each f_n is a measurable function from Ψ to $\overline{\mathbb{R}}$. Define $g_n = \max\{f_1, \ldots, f_n\}$ for $n = 1, 2, \ldots$. Each g_n is measurable and $Y_n = g_n \circ V$ since the sequence $(f_n, n = 1, 2, \ldots)$ must be increasing on the image of V. It follows that $\lim_n Y_n = g \circ V$, where $g = \lim_n g_n$. Thus $\mathcal{C}$ is closed under increasing pointwise limits. By the preceding paragraph of this proof and Lemma 13 of Chapter 2, $\mathcal{C}$ contains all $\overline{\mathbb{R}}^+$-valued $\sigma(V)$-measurable functions.

Suppose now that Y is $\overline{\mathbb{R}}$-valued and $\sigma(V)$-measurable. Then $Y = Y^+ - Y^-$, where each of Y^+ and Y^- is $\overline{\mathbb{R}}^+$-valued and $\sigma(V)$-measurable. Thus there exist measurable functions g and h from Ψ to $\overline{\mathbb{R}}^+$ such that

(21.3) $$Y = g \circ V - h \circ V.$$

Since this last expression is meaningful, the set $D = \{\psi : g(\psi) = h(\psi) = \infty\}$ is disjoint from the image of V. Therefore, we may redefine $h(\psi) = 0$ for $\psi \in D$ and maintain the truth of (21.3). Then $Y = (g - h) \circ V$. $\square$

* **Problem 13.** Why near the end of the preceding proof did we not just redefine $h(\psi)$ to equal 0 for ψ not in the image of V.

Proposition 8. *Let V be a $(\Psi, \mathcal{H})$-valued random variable defined on a probability space $(\Omega, \mathcal{F}, P)$ and let $A \in \mathcal{F}$. Then there exists a $[0,1]$-valued measurable function p on $(\Psi, \mathcal{H})$ for which $P(A \mid \sigma(V)) = p \circ V$ a.s.*

Problem 14. Deduce this proposition from Lemma 7, being careful to say why p can be chosen to be $[0,1]$-valued.

The following notation will sometimes be used for the function p described in Proposition 8:

$$P(A \mid V = v) = p(v),$$

a notation that gives rise to the intuitive phrase, "the probability of A given that V equals v".

Problem 15. For the setting of Example 2, find the function

$$x \rightsquigarrow P(A \mid |X| = x).$$

Problem 16. For the setting of Problem 11, find the function

$$x \rightsquigarrow P(A \mid |X| = x).$$

* **Problem 17.** For the setting of Problem 12, find the function

$$v \rightsquigarrow P\big(\{(\omega_1, \omega_2): \omega_1 > \omega_2\} \mid V = v\big),$$

where $V(\omega_1, \omega_2) = \omega_1$.

21.2. Conditional distributions

Conditional probabilities are $[0,1]$-valued random variables. This property corresponds to the fact that (unconditional) probabilities are $[0,1]$-valued. Analogously, we will define 'conditional distributions' so that they are distribution-valued random variables. The following discussion of notation is preparation for this point of view.

Let $(\Omega, \mathcal{F})$ and $(\Psi, \mathcal{H})$ be measurable spaces, and let $\mathcal{Q}$ denote the space of all probability measures on $(\Psi, \mathcal{H})$. To say that Z is a $\mathcal{Q}$-valued function with domain Ω is to say that for each ω, $Z(\omega)$ is a probability measure. We could denote the probability given by $Z(\omega)$ to a measurable set B by $[Z(\omega)](B)$ but will not use this notation. Instead, in order to avoid the clutter of extra parentheses, we express this quantity as

$$Z(\omega, B).$$

When we want to refer to the $[0,1]$-valued function $\omega \rightsquigarrow Z(\omega, B)$, where B is a fixed measurable set, we will often use the notation

$$Z(\cdot, B).$$

When ω is fixed, the probability measure $B \rightsquigarrow Z(\omega, B)$ will sometimes be denoted by $Z(\omega, \cdot)$. It is natural in many cases to suppress one of the arguments, writing

$$Z(B) \text{ for } Z(\cdot, B) \quad \text{and} \quad Z(\omega) \text{ for } Z(\omega, \cdot).$$

The practice of eliminating the ω will be particularly prevalent when we introduce a special notation later for conditional distributions (see the discussion following Theorem 19).

We now define the conditional distribution of a random variable X given a σ-field $\mathcal{G}$ to be a certain distribution-valued function.

Definition 9. Let X be a $(\Psi, \mathcal{H})$-valued random variable defined on a probability space $(\Omega, \mathcal{F}, P)$, $\mathcal{G}$ a sub-σ-field of $\mathcal{F}$, and $\mathcal{Q}$ the space of probability measures on $(\Psi, \mathcal{H})$. A function $Z \colon \Omega \to \mathcal{Q}$ is a *conditional distribution* of X given $\mathcal{G}$ if for each fixed $B \in \mathcal{H}$, the $[0, 1]$-valued function $Z(\cdot, B)$ is a conditional probability of $X^{-1}(B)$ given $\mathcal{G}$.

It is natural for us to think of a conditional distribution Z as a 'random distribution', but technically speaking, in order to do so we need to make the target space $\mathcal{Q}$ into a measurable space. The following definition has been constructed so that if Z is a conditional distribution given some σ-field $\mathcal{G}$, then Z is automatically a $\mathcal{G}$-measurable $\mathcal{Q}$-valued random variable.

Definition 10. Let $(\Psi, \mathcal{H})$ be a measurable space, and denote by $\mathcal{Q}$ the set of probability measures on $(\Psi, \mathcal{H})$. The *measurable space of probability measures* on $(\Psi, \mathcal{H})$ is $(\mathcal{Q}, \mathfrak{H})$, where $\mathfrak{H}$ is the smallest σ-field such that for each $B \in \mathcal{H}$ the function from $\mathcal{Q}$ to $[0, 1]$ defined by $Q \rightsquigarrow Q(B)$ is $\mathfrak{H}$-measurable. A *random distribution on* $(\Psi, \mathcal{H})$ is a measurable function from some probability space $(\Omega, \mathcal{F}, P)$ to $(\mathcal{Q}, \mathfrak{H})$.

Most texts do not involve the measurable space $(\mathcal{Q}, \mathfrak{H})$ in their treatment of conditional distributions. Nevertheless, we feel that coming to grips with Definition 10 will provide the student with a deeper understanding of Definition 9, and so we have provided two problems to aid in that effort. Furthermore, it turns out that $(\mathcal{Q}, \mathfrak{H})$ has some nice properties that will be useful later in the book, particularly in our discussion of Markov sequences in Chapter 26 and Markov processes in Chapter 31. These properties are given in in Lemma 21 and Theorem 22 of the present chapter.

Problem 18. Let $(\Omega, \mathcal{F}, P), (\Psi, \mathcal{H})$, and $(\mathcal{Q}, \mathfrak{H})$ be as in the definition of random distribution. Prove that a function $Z \colon (\Omega, \mathcal{F}, P) \to (\mathcal{Q}, \mathfrak{H})$ is a random distribution if and only if $Z(\cdot, B)$ is a $[0, 1]$-valued random variable for each fixed $B \in \mathcal{H}$.

Problem 19. Let Z be a random distribution on $\mathbb{R}$. Show that the set

$$[Z = \delta_x \text{ for some } x \in \mathbb{R}]$$

is measurable.

Problem 20. Suppose that X is a random variable measurable with respect to a σ-field $\mathcal{G}$. Show that $\omega \rightsquigarrow \delta_{X(\omega)}$ is a conditional distribution of X given $\mathcal{G}$.

Problem 21. Suppose that X is a random variable with distribution Q. Show that X is independent of a σ-field $\mathcal{G}$ if and only if the constant random distribution $Z = Q$ is a conditional distribution of X given $\mathcal{G}$.

Here are some results for using a conditional distribution of one random variable to find conditional distributions of other related random variables.

Proposition 11. Let Z denote a conditional distribution of some $(\Psi, \mathcal{H})$-valued random variable X given a σ-field $\mathcal{G}$, and denote by h a measurable function from $(\Psi, \mathcal{H})$ to a measurable space $(\Xi, \mathcal{K})$. Then

$$(21.4) \qquad C \rightsquigarrow Z(\cdot, h^{-1}(C)), \quad C \in \mathcal{K},$$

is a conditional distribution of the random variable $h \circ X$ given $\mathcal{G}$.

PROOF. That $C \rightsquigarrow Z(\omega, h^{-1}(C))$, $C \in \mathcal{K}$, is a probability measure for each ω follows from the fact that h is measurable and Z takes values in the space of probability measures on $\mathcal{H}$. The measurability of the function (21.4) is a consequence of the measurability of the functions h and $B \rightsquigarrow Z(\cdot, B)$, $B \in \mathcal{H}$.

To check that $Z(\cdot, h^{-1}(C))$ is a conditional probability given $\mathcal{G}$ of $(h \circ X)^{-1}(C)$ for $C \in \mathcal{K}$, we calculate $E\big(Z(\cdot, h^{-1}(C)) I_A\big)$ for $A \in \mathcal{G}$. Since Z itself is a conditional distribution of X given $\mathcal{G}$, the definition implies that $Z(\cdot, h^{-1}(C))$ is a conditional probability of $h^{-1}(C)$ given $\mathcal{G}$, so

$$E\big(Z(\cdot, h^{-1}(C)) I_A\big) = P([X \in h^{-1}(C)] \cap A),$$

which equals $P([h \circ X \in C] \cap A)$, as desired. $\square$

Proposition 12. Let $\mathcal{G}$ be a σ-field and, for $i = 1, 2$, let X_i be a $(\Psi_i, \mathcal{H}_i)$-valued random variable. Suppose that X_1 is $\mathcal{G}$-measurable and that a conditional distribution Z of X_2 given $\mathcal{G}$ exists. Then

$$\delta_{X_1} \times Z$$

is a conditional distribution of the ordered pair (X_1, X_2) given $\mathcal{G}$.

The following is a consequence of the preceding two propositions.

Corollary 13. In the setting of Proposition 12, let $h \colon (\Psi_1, \mathcal{H}_1) \times (\Psi_2, \mathcal{H}_2) \to (\Xi, \mathcal{K})$ be a measurable function, and for $x_1 \in \Psi_1$, let Z_{x_1} be the random distribution induced from Z by the function $x_2 \rightsquigarrow h(x_1, x_2)$. That is,

$$Z_{x_1}(C) = Z(\cdot, \{x_2 \colon h(x_1, x_2) \in C\}), \quad C \in \mathcal{K}.$$

Then Z_{X_1} is a conditional distribution of $h \circ (X_1, X_2)$ given $\mathcal{G}$.

Problem 22. Prove Proposition 12 and Corollary 13.

Problem 23. Let X and Y be $\overline{\mathbb{R}}^+$-valued random variables and suppose that a conditional distribution of Y given $\sigma(X)$ exists. In terms of this conditional distribution, find a conditional distribution of $X + Y$ given $\sigma(X)$.

* **Problem 24.** Let X and Y be independent $\mathbb{R}$-valued random variables. In terms of the distribution of Y, find a conditional distribution of $X \wedge Y$ given $\sigma(X)$.

Example 3. Let $(S_0 = 0, S_1, S_2, \ldots)$ be a random walk in $\mathbb{R}^d$ with distribution Q, and fix a nonnegative integer n. It seems intuitively clear that given $\sigma(S_n)$, a conditional distribution of $(S_n, S_{n+1}, \ldots)$ is Q_{S_n}, where for each $s \in \mathbb{R}^d$, Q_s is the distribution of the random sequence $(s + S_0, s + S_1, s + S_2, \ldots)$. Let us examine how this follows from previous results.

Note that each Q_s is the distribution induced from Q by the function

$$(s_0, s_1, s_2, \ldots) \rightsquigarrow h(s, (s_0, s_1, s_2, \ldots)) \overset{\text{def}}{=} (s + s_0, s + s_1, s + s_2, \ldots).$$

Denoting the steps of the random walk by $X_1, X_2, \ldots$,

$$(S_n, S_{n+1}, S_{n+2}, \ldots) = h(S_n, (0, X_{n+1}, X_{n+1} + X_{n+2}, \ldots)).$$

Since the sequence $(0, X_{n+1}, X_{n+1} + X_{n+2}, \ldots)$ has (unconditional) distribution Q and is independent of $\sigma(S_n)$, its conditional distribution given $\sigma(S_n)$ is also Q, by Problem 21. The desired conclusion follows from Corollary 13.

A good technique for some of the following problems is to guess a correct answer and then verify its correctness.

* **Problem 25.** Let X be an $\mathbb{R}$-valued random variable. Assume that the distribution of X has a density f with respect to a σ-finite measure μ on $\mathbb{R}$, where μ satisfies $\mu(B) = \mu(-B)$ for all Borel sets B. Find a formula for a conditional distribution of X given $\sigma(|X|)$.

Problem 26. Apply the preceding problem to X uniformly distributed on $[-1, 2]$.

Problem 27. Let $X_1, \ldots, X_n$ be iid $\mathbb{R}$-valued random variables and denote the corresponding order statistics by $Y_1, Y_2, \ldots, Y_n$. Find a conditional distribution of $X = (X_1, X_2, \ldots, X_n)$ given $\sigma(Y_1, Y_2, \ldots, Y_n)$.

Problem 28. Let X be any $(\Psi, \mathcal{H})$-valued random variable, and find, in terms of the distribution of X, a conditional distribution of X given $\sigma(\{\omega \colon X(\omega) \in B\})$, where B is a fixed member of $\mathcal{H}$.

Problem 29. Let $X = (X_1, \ldots, X_d)$ be a random vector, the coordinates of which are independent standard gamma random variables with parameters $\gamma_1, \ldots, \gamma_d$, respectively. Find a conditional distribution of X given $\sigma(X_1 + X_2 + \cdots + X_d)$. Discuss relations with Problem 34 of this chapter and Example 2, Problem 28, and Problem 30 of Chapter 10.

21.3. Conditional densities

Just as (unconditional) distributions are often most easily studied via densities, so conditional distributions are often expressed in terms of their densities, when they exist.

Definition 14. Let X be a $(\Psi, \mathcal{H})$-valued random variable defined on a probability space $(\Omega, \mathcal{F}, P)$, μ a σ-finite measure on $(\Psi, \mathcal{H})$, and $\mathcal{G}$ a sub-σ-field of $\mathcal{F}$. A nonnegative measurable function q on $(\Omega \times \Psi, \mathcal{F} \times \mathcal{H})$ is called a *conditional density* of X with respect to μ given $\mathcal{G}$ if

$$B \leadsto \int_B q(\cdot, x)\mu(dx)$$

is a conditional distribution of X given $\mathcal{G}$.

Sometimes the σ-finite measure μ in the preceding definition is not mentioned, especially if it is Lebesgue measure or counting measure.

* **Problem 30.** Let X be exponentially distributed with mean λ. Let $\mathcal{G}$ be the σ-field generated by the event $\{\omega : X(\omega) \geq t\}$, where t is a fixed nonnegative real number. Find a conditional density of X given $\mathcal{G}$.

If two random variables have a joint density with respect to a product measure, then there is a nice formula for a conditional density of one of the two random variables, given the σ-field generated by the other one.

Proposition 15. *For $i = 1, 2$, let X_i be $(\Psi_i, \mathcal{H}_i)$-valued random variables defined on a common probability space $(\Omega, \mathcal{F}, P)$. Assume that the distribution of the random vector (X_1, X_2) has a density f with respect to a σ-finite product measure $\mu = \mu_1 \times \mu_2$ on $\mathcal{H} = \mathcal{H}_1 \times \mathcal{H}_2$. Let*

$$g(x_1) = \int_{\Psi_2} f(x_1, x_2)\,\mu_2(dx_2)\,.$$

Then the function

$$(\omega, x_2) \leadsto \begin{cases} \dfrac{f(X_1(\omega), x_2)}{g(X_1(\omega))} & \text{if } g(X_1(\omega)) > 0 \\ \int_{\Psi_1} f(x_1, x_2)\,\mu_1(dx_1) & \text{if } g(X_1(\omega)) = 0 \end{cases}$$

is a conditional density of X_2 with respect to μ_2 given $\sigma(X_1)$.

Problem 31. Prove the preceding proposition.

Problem 32. Let (X, Y) be an $\mathbb{R}^2$-valued normally distributed random vector with mean $(0, 0)$ and a covariance matrix Γ having positive eigenvalues. Find a conditional density of Y given $\sigma(X)$. Identify the corresponding conditional distribution by name and give a formula for the appropriate 'conditional parameters' (mean and variance).

Problem 33. Discuss the cases in which the covariance matrix in the preceding problem has one or two eigenvalues equal to 0.

* **Problem 34.** Let $X_1, X_2, \ldots, X_d$ be independent positive random variables having gamma densities

$$x \rightsquigarrow \frac{1}{\Gamma(\gamma_i)} x^{\gamma_i - 1} e^{-x}, \quad x > 0,$$

for $i = 1, 2, \ldots, d$. Find an expression for the density of the random vector

$$\left(X_1, X_2, \ldots, X_{d-1}, \sum_{k=1}^{d} X_k \right).$$

Also, find a conditional density of $(X_1, X_2, \ldots, X_{d-1})$ given $\sigma(X_1 + X_2 + \cdots + X_d)$.

Problem 35. Let $X_1, \ldots, X_n$ be iid $\mathbb{R}$-valued random variables, with common density f. Show that for each k, the conditional density of X_k given $\sigma(S)$, where $S = X_1 + \cdots + X_n$, is the function

$$\omega \rightsquigarrow \frac{f(x) f^{*(n-1)}(S(\omega) - x)}{f^{*n}(S(\omega))}.$$

21.4. Existence and uniqueness of conditional distributions

The next proposition says that conditional distributions of $\overline{\mathbb{R}}$-valued random variables always exist regardless of the probability space on which they are defined and regardless of the conditioning σ-field.

Proposition 16. *Let X be an $\overline{\mathbb{R}}$-valued random variable defined on a probability space $(\Omega, \mathcal{F}, P)$ and let $\mathcal{G}$ be a sub-σ-field of $\mathcal{F}$. Then a conditional distribution Z of X given $\mathcal{G}$ exists. Moreover, Z is unique in the sense that if Z' is any other conditional distribution of X given $\mathcal{G}$, then $Z = Z'$ a.s.*

PARTIAL PROOF. For each rational $x \in \mathbb{R}$, set

$$F(\omega, x) = P(\{\omega' : X(\omega') \le x\} \mid \mathcal{G})(\omega).$$

We leave it as an exercise to show that there exists a single null event N such that for $\omega \in N^c$, $F(\omega, \cdot)$ is increasing and right continuous as a function on the rationals, and has limits in $[0, 1]$ at $-\infty$ and at ∞. Thus, for each $\omega \in N^c$, the domain of $F(\omega, \cdot)$ can be extended uniquely to $\mathbb{R}$ so that it is a distribution function for $\overline{\mathbb{R}}$. Let $Z(\omega)$ be the corresponding probability measure on $\overline{\mathbb{R}}$. We take care of $\omega \in N$ by letting $Z(\omega)$ equal the (unconditional) distribution of X for such ω. We will show that Z is a conditional distribution of X given $\mathcal{G}$. By construction, Z is a probability measure for each ω. To finish the proof of existence, we only need show that, for each Borel set B, $Z(\cdot, B)$ is a conditional probability of $X^{-1}(B)$ given $\mathcal{G}$.

It is straightforward to check that the class of those B for which $Z(\cdot, B)$ is a conditional probability of $X^{-1}(B)$ given $\mathcal{G}$ is a Sierpiński class and that it contains the interval $[-\infty, \infty]$ and all intervals of the form $[-\infty, x]$ for x rational. An appeal to the Sierpiński Class Theorem completes the existence proof.

The proof of uniqueness is left to the reader. □

Problem 36. Complete the proof of the preceding proposition by: (i) showing the existence of an event N with the claimed properties, (ii) supplying the details connected with the application of the Sierpiński Class Theorem, and (iii) proving the uniqueness assertion. *Hint:* Use Proposition 6 and, for uniqueness, the observation that Definition 1 entails uniqueness for each fixed x.

In order to generalize Proposition 16, we introduce two definitions.

Definition 17. Two measurable spaces are called *isomorphic* if there exists a bijective function φ between them such that both φ and φ^{-1} are measurable.

Definition 18. A measurable space is called a *Borel space* if it is isomorphic to some $(A, \mathcal{B}(A))$, where A is a Borel set in $[0, 1]$ and $\mathcal{B}(A)$ is the σ-field of Borel subsets of A.

Theorem 19. [Existence and Uniqueness of Conditional Distributions] *Denote by $(\Psi, \mathcal{H})$ a Borel space, by $(\Omega, \mathcal{F}, P)$ a probability space, and by $\mathcal{G}$ a sub-σ-field of $\mathcal{F}$. Then every $(\Psi, \mathcal{H})$-valued random variable defined on $(\Omega, \mathcal{F}, P)$ has a conditional distribution given $\mathcal{G}$. Moreover, such conditional distributions are unique in the same sense as in Proposition 16.*

Problem 37. Use Proposition 16 to prove the preceding theorem.

In view of the uniqueness portion of the preceding theorem we often speak, in the Borel-space-setting, of 'the' conditional distribution For a random variable X having (unconditional) distribution Q on some Borel space, we use the special notation

$$Q(\cdot \mid \mathcal{G})$$

to denote the conditional distribution of X given $\mathcal{G}$. Of course, like other random distributions, this function takes two arguments: a measurable set B and a sample point ω, so that notation like $Q(\cdot \mid \mathcal{G})(\omega)$, $Q(B \mid \mathcal{G})(\cdot)$, and $Q(B \mid \mathcal{G})$, as well as the more explicit

$$\omega \rightsquigarrow \big[B \rightsquigarrow Q(B \mid \mathcal{G})(\omega)\big],$$

will sometimes be used to reflect different points of view. In those cases where $\mathcal{G} = \sigma(Y)$ for some random variable Y, we will often write

$$Q(\cdot \mid Y) \quad \text{for} \quad Q(\cdot \mid \sigma(Y)).$$

When the target space of a random variable X is different from its domain Ω, the notation introduced in the preceding paragraph is distinct from that introduced earlier for conditional probabilities, in spite of the similarities. Even when the target of X is Ω, there is no danger of confusion as long as Q is different from the underlying probability measure P. However, from time to time, it will be convenient to let X be the identity function on $(\Omega, \mathcal{F}, P)$. The distribution of this special random variable is P, so when Ω is itself a Borel space, $P(\cdot \mid \mathcal{G})$ will denote the conditional distribution of X given $\mathcal{G}$. One might worry that the expression $P(A \mid \mathcal{G})$ now denotes two different things: (i) the conditional probability given $\mathcal{G}$ of the event A, and (ii) the value given to A by the conditional distribution of X given $\mathcal{G}$. But the definitions ensure that these two quantities agree almost surely for any fixed event A, so there is no real conflict.

It is fortunate that many of the measurable spaces encountered in probability theory are Borel spaces. The next proposition gives a good idea of the prevalence of Borel spaces. The term 'Polish space' used here is defined in Chapter 18; for those who have not read that chapter, it suffices to mention that $\mathbb{R}$, $\overline{\mathbb{R}}$, $\mathbb{R}^+$, $\overline{\mathbb{R}}^+$ (with their usual topologies), and all countable sets can be regarded as Polish spaces.

Proposition 20. *Every Polish space is a Borel space. A product of a finite or countable number of Borel spaces is itself a Borel space. Every measurable subset A of a Borel space B is itself a Borel space, the measurable subsets in A being those subsets of A that are measurable in B.*

PROOF. The statement that Borel subsets of Borel spaces are themselves Borel spaces is obvious. It is easy to see from the definition of Borel space that any finite or countable product of Borel spaces is isomorphic to a Borel subset of the infinite-dimensional cube $[0, 1]^\infty$, and it follows from Lemma 4 of Chapter 18 that any Polish space is also isomorphic to a Borel subset of the infinite-dimensional cube. Thus, to complete the proof, it is sufficient to prove that $[0, 1]^\infty$ is a Borel space.

By using binary representations of the coordinates of points $x = (x_1, x_2, \ldots) \in [0, 1]^\infty$ (being careful as usual about the ambiguous cases), we can associate to each x a unique doubly indexed sequence of 0's and 1's. A simple relabeling of indices allows us to associate to each x a unique (singly indexed) sequence of 0's and 1's. We leave it to the reader to check that all of this can be done in such a way as to create an isomorphism between the infinite-dimensional unit cube $[0, 1]^\infty$ and a Borel subset of the space $\{0, 1\}^\infty$ of all sequences of 0's and 1's. Thus, it is sufficient to show that $\{0, 1\}^\infty$ is a Borel space. This task is easily accomplished by appropriate use of the binary representation. $\square$

One consequence of Proposition 20 is that conditional distributions of infinite sequences of $\mathbb{R}$- or $\overline{\mathbb{R}}$-valued random variables always exist, not just the conditional distributions of the individual random terms. This fact is useful for the study of random sequences.

Recall from Proposition 8 that a conditional probability given $\sigma(V)$, where V is a random variable, can be written as the composition of V and a measurable $[0,1]$-valued function on the target of V. It is natural to ask if there is a corresponding theorem for conditional distributions. Several exercises, including Problem 23, Problem 25, and Problem 27, indicate that the answer may be 'yes'. If we knew that the distributions of interest formed a Borel space, then we could apply Lemma 7 just as we did to obtain Proposition 8.

Lemma 21. *The measurable space of probability measures on a Borel space is a Borel space.*

Before proving this lemma we record its important consequence, referred to in the discussion preceding the lemma.

Theorem 22. *Let* $X \colon (\Omega, \mathcal{F}, P) \to (\Psi, \mathcal{G})$ *and* $Y \colon (\Omega, \mathcal{F}, P) \to (\Theta, \mathcal{H})$ *be random variables defined on a common probability space, and assume that* $(\Psi, \mathcal{G})$ *is a Borel space. Then the conditional distribution of* X *given* $\sigma(Y)$ *can be written in the form* $R \circ Y$, *where* R *is a measurable function from* $(\Theta, \mathcal{H})$ *to the measurable space of probability measures on* $(\Psi, \mathcal{G})$.

The function R of the preceding theorem is distribution-valued, so our notational conventions dictate that we write $R(y, B)$ for the measure given to the set $B \in \mathcal{G}$ by R when it is evaluated at $y \in \Theta$. There is a special notation for this quantity:

$$P[X \in B \mid Y = y],$$

with corresponding expressions like $P[X \in \cdot \mid Y = y]$ and $P[X \in B \mid Y = \cdot]$ also being used. Do not confuse this notation with expressions like $P[X \in \cdot \mid Y]$, introduced earlier in conjunction with the conditional distribution of X given $\sigma(Y)$. The relationship between $P[X \in \cdot \mid Y]$ and $P[X \in \cdot \mid Y = y]$ is described by Theorem 22.

It remains for us to prove Lemma 21. Those readers who have skipped Chapter 18 may want to omit this proof.

PROOF OF Lemma 21. Let ρ denote the usual metric on $[0,1]$, the distance between two points being the absolute value of their difference. The space $([0,1], \rho)$ is a Polish space. We also call this space $([0,1], \mathcal{B})$, with $\mathcal{B}$ denoting the σ-field of Borel subsets of $[0,1]$, when we want to identify it as a measurable space rather than as a metric space. By Theorem 24 and Theorem 25 of Chapter 18 the set of probability measures on $([0,1], \rho)$ can be turned into a metric space with a metric $\hat{\rho}$ in such a way that convergence in $\hat{\rho}$ corresponds to the standard definition of convergence of sequences of probability measures on a Polish space. Here is a list of the measurable spaces to be used in this proof:

- a Borel space $(\Psi, \mathcal{G})$;
- the measurable space $(\mathcal{Q}, \mathfrak{G})$ of probability measures on $(\Psi, \mathcal{G})$;
- the Polish space $([0,1], \rho)$;

- the measurable space $(\mathcal{R}, \mathfrak{B})$ of probability measures on $([0, 1], \rho)$;
- the Polish space $(\mathcal{R}, \hat{\rho})$ of probability measures on $([0, 1], \rho)$;
- the measurable space $(A, \mathcal{A})$ for A a certain Borel subset of $([0, 1], \mathcal{B})$;
- the measurable space $(\mathcal{R}_A, \mathfrak{B}_A)$ of probability measures on $(A, \mathcal{A})$.

Our goal is to prove that $(\mathcal{Q}, \mathfrak{G})$ is a Borel space.

The first step is to prove that $(\mathcal{R}, \mathfrak{B})$ is a Borel space. We will accomplish this by showing that, when regarded as measurable spaces, $(\mathcal{R}, \hat{\rho})$ and $(\mathcal{R}, \mathfrak{B})$ are the same. Since $(\mathcal{R}, \hat{\rho})$ is a Polish space, the desired conclusion in this first step then follows from Proposition 20.

We know that $\mathfrak{B}$ is the σ-field generated by functions of the form $R \rightsquigarrow R(B)$ for Borel subsets $B \subseteq [0, 1]$. Let us prove that such a function is measurable when its domain is taken to be the Polish space $(\mathcal{R}, \hat{\rho})$. If B is compact, its indicator function I_B is the limit of a decreasing sequence (f_n) of continuous bounded functions. Since each function $R \rightsquigarrow \int f_n \, dR$ is continuous and therefore measurable, the function

$$(21.5) \qquad R \rightsquigarrow \int I_B \, dR = R(B)$$

is measurable. The collection of compact sets is closed under finite intersection and the collection of sets for which (21.5) holds is easily seen to be a Sierpiński class. Therefore, the function (21.5) is measurable for every Borel set B. For the other direction we need to prove that $\mathfrak{B}$ contains all open sets, and therefore all Borel sets, in the Polish space $(\mathcal{R}, \hat{\rho})$. The open sets are inverse images of open sets via functions of the form $R \rightsquigarrow \int f \, dR$, where f is continuous and bounded. To finish this portion of the proof we only need show that such a function is measurable when its domain is regarded as $(\mathcal{R}, \mathfrak{B})$. But this is easy, since f can be approximated by simple functions and $R \rightsquigarrow \int g \, dR$ for g simple is a linear combination of functions of the form (21.5).

Our next task is to prove that $(\mathcal{R}_A, \mathfrak{B}_A)$ is a Borel space for every Borel set $A \subseteq [0, 1]$ and thus, in particular, for the yet-to-be-made choice of A. We will do this by proving that $\mathfrak{B}_A$ consists of exactly those members of $\mathfrak{B}$ that are subsets of $\mathcal{R}_A$ (and thus, in particular, that $\mathcal{R}_A \in \mathfrak{B}$). We know that $\mathfrak{B}_A$ is generated by sets of the form

$$\{R \in \mathcal{R}_A : R(B) \in C\} = \{R \in \mathcal{R} : R(B) \in C\} \cap \{R \in \mathcal{R} : R(A) = 1\} \in \mathfrak{B},$$

where C is Borel set in $[0, 1]$. Hence, every member of $\mathfrak{B}_A$ is a member of $\mathfrak{B}$. For the opposite direction we will prove that every member of $\mathfrak{B}$ is the union of three sets: a member of $\mathfrak{B}_A$, a member of $\mathfrak{B}_{A^c}$, and a set of the form

$$\left\{R \in \mathcal{R} : 0 < R(A) < 1, \frac{1}{R(A)} R|_A \in S_1, \frac{1}{R(A^c)} R|_{A^c} \in S_2\right\},$$

where $S_1 \in \mathfrak{B}_A$, $S_2 \in \mathfrak{B}_{A^c}$, and $R|_A$ and $R|_{A^c}$ denote R restricted to A and to A^c, respectively. This is the case because a collection of sets generating $\mathfrak{B}$ is included in this list and the sets in this list constitute a σ-field. But the sets in this list that are subsets of $\mathcal{R}_A$ are clearly members of $\mathfrak{B}_A$. We have shown that $(\mathcal{R}_A, \mathcal{B}_A)$ is a Borel space.

Since $(\Psi, \mathcal{G})$ is a Borel space there is a one-to-one function φ from Ψ onto a Borel subset A of $[0,1]$ that induces a one-to-one correspondence between $\mathcal{G}$ and $\mathcal{A}$. We will show that φ induces a one-to-one function $\hat{\varphi}$ from $\mathcal{Q}$ to $\mathcal{R}_A$ which itself induces a one-to-one correspondence between $\mathfrak{G}$ and $\mathfrak{B}_A$; then the lemma follows because $(\mathcal{R}_A, \mathfrak{B}_A)$ is known to be a Borel space, by the preceding paragraph. For $Q \in \mathcal{Q}$, define $\hat{\varphi}(Q)$ by

$$(\hat{\varphi}(Q))(B) = Q(\varphi^{-1}(B))$$

for B a Borel subset of A. Clearly $\hat{\varphi}$ is one-to-one because φ is, and also the image of $\hat{\varphi}$ is $\mathcal{R}_A$. Fix a measurable subset D of Ψ and the corresponding Borel subset $B = \varphi(D)$ of A. Then consider the functions g and h defined by $g(Q) = Q(D)$ on $\mathcal{Q}$ and $h(R) = R(B)$ on $\mathcal{R}_A$. Since $g = h \circ \hat{\varphi}$ and both $\hat{\varphi}$ and $\hat{\varphi}^{-1}$ are measurable, $\hat{\varphi}$ induces a one-to-one correspondence between $\mathfrak{G}$ and $\mathfrak{B}_A$, as desired. $\square$

21.5. Conditional independence

We begin with the following natural definition.

Definition 23. Let $(\Omega, \mathcal{F}, P)$ be a probability space and $\mathcal{F}_1$, $\mathcal{F}_2$, and $\mathcal{G}$ sub-σ-fields of $\mathcal{F}$. The σ-fields $\mathcal{F}_1$ and $\mathcal{F}_2$ are *conditionally independent* given $\mathcal{G}$ if

$$P(A_1 \cap A_2 \mid \mathcal{G}) = P(A_1 \mid \mathcal{G}) P(A_2 \mid \mathcal{G}) \text{ a.s.}$$

for all $A_i \in \mathcal{F}_i$, $i = 1, 2$.

We leave it to the reader to extend the notion of conditional independence to to more than two σ-fields and to obtain conditional analogues of Proposition 3 and Proposition 4, both of Chapter 9 .

There are many results that show how conditional independence can be used in ways that parallel uses of (unconditional) independence. The first and third problems below describe two such results. It is left to the interested reader to work out others.

Problem 38. [Conditional Borel-Cantelli Lemma] Let $(\Omega, \mathcal{F}, P)$ be a probability space, $\mathcal{G}$ a sub-σ-field of $\mathcal{F}$, and $(A_n : n = 1, 2, \dots)$ a sequence of events conditionally independent given $\mathcal{G}$. Prove that for almost all $\omega \in \Omega$,

$$P(\limsup_{n \to \infty} A_n \mid \mathcal{G})(\omega) \begin{cases} = 1 \\ = 0 \end{cases}$$

according as

$$\sum_{n=1}^{\infty} P(A_n \mid \mathcal{G})(\omega) \begin{cases} = \infty \\ < \infty . \end{cases}$$

Then conclude that

$$P(\limsup_{n\to\infty} A_n) = P\left(\left\{\omega: \sum_{n=1}^{\infty} P(A_n \mid \mathcal{G})(\omega) = \infty\right\}\right).$$

Problem 39. Make a direct confirmation of the result in Problem 38 for the special case where

$$\mathcal{G} = \sigma(A_n : n = 1, 2, \dots).$$

Problem 40. Let X and Y be two random variables that are conditionally independent given a σ-field $\mathcal{G}$ and that have targets that are Borel spaces. Prove that the conditional distribution of the ordered pair (X, Y) is the product measure of the conditional distribution of X and that of Y.

Notice that the preceding problem contains information about random product measures, whereas the definition of conditional independence only involves products of numbers.

The following proposition shows that conditioning can turn dependence into independence.

Proposition 24. Let $(\Omega, \mathcal{F}, P)$ be a probability space. Let $\mathcal{G}_1$ and $\mathcal{G}_2$ and $\mathcal{H}$ be sub-σ-fields of $\mathcal{F}$ and suppose that $\mathcal{G}_2 \subseteq \mathcal{H}$. Then $\mathcal{G}_1$ and $\mathcal{G}_2$ are conditionally independent given $\mathcal{H}$.

Problem 41. Prove this proposition. *Hint:* Use Proposition 3 and Problem 1.

The next problem shows that conditioning can turn independence into dependence.

Problem 42. Consider the experiment of flipping two fair coins. Let A be the event that the first coin comes up heads, B the event that the second coin comes up heads, and C the event that the two coins agree. Show that any two of the events A, B, and C are independent, but that they are not conditionally independent given the σ-field generated by the remaining event. Explain the connection between this example and the discussion following the proof of Proposition 4.

The preceding problem shows that the interplay between conditioning and independence involving more than two σ-fields can be quite interesting. The same can be said of the following propositions, problems, and corollary.

Problem 43. For each proposition, problem, and corollary between here and Theorem 29, intuitively describe the conclusion.

Proposition 25. *Let $\mathcal{G}$, $\mathcal{H}$, and $\mathcal{K}$ be σ-fields of events in a probability space. If $\mathcal{G}$ and $\mathcal{H}$ are conditionally independent given $\mathcal{K}$, then $\mathcal{G}$ and $\sigma(\mathcal{H}, \mathcal{K})$ are conditionally independent given $\mathcal{K}$.*

PROOF. Denote the underlying probability space by $(\Omega, \mathcal{F}, P)$. We need to show

$$P(A \cap B \mid \mathcal{K}) = P(A \mid \mathcal{K}) P(B \mid \mathcal{K})$$

for arbitrary $A \in \mathcal{G}$ and $B \in \sigma(\mathcal{H}, \mathcal{K})$. For fixed A, Proposition 6 implies that the collection of B for which this equality holds is a Sierpiński class. Thus, we may restrict our attention to B of the form $C \cap D$, where $C \in \mathcal{H}$ and $D \in \mathcal{K}$. For such B, the following calculation that first uses Proposition 3, then the conditional independence of $\mathcal{G}$ and $\mathcal{H}$ given $\mathcal{K}$, and finally Proposition 3 a second time completes the proof:

$$P(A \mid \mathcal{K})P(C \cap D \mid \mathcal{K}) = P(A \mid \mathcal{K})P(C \mid \mathcal{K})I_D$$
$$= P(A \cap C \mid \mathcal{K})I_D = P(A \cap (C \cap D) \mid \mathcal{K}). \quad \square$$

Proposition 26. *Let $\mathcal{G}$ and $\mathcal{H}$ be two σ-fields of events in a probability space and let $\mathcal{G}_1$ and $\mathcal{H}_1$ be sub-σ-fields of $\mathcal{G}$ and $\mathcal{H}$, respectively. Suppose that $\mathcal{G}$ and $\mathcal{H}$ are independent. Then $\mathcal{G}$ and $\mathcal{H}$ are conditionally independent given $\sigma(\mathcal{G}_1, \mathcal{H}_1)$.*

PROOF. Let $A \in \mathcal{G}$ and $B \in \mathcal{H}$. Clearly,

$$P(A \mid \mathcal{G}_1, \mathcal{H}_1)P(B \mid \mathcal{G}_1, \mathcal{H}_1)$$

is $\sigma(\mathcal{G}_1, \mathcal{H}_1)$-measurable. To show that it satisfies the second condition in Proposition 2 characterizing $P(A \cap B \mid \mathcal{G}_1, \mathcal{H}_1)$, we first simplify it using Proposition 4:

$$P(A \mid \mathcal{G}_1)P(B \mid \mathcal{H}_1).$$

Consider a member $C \cap D$ of $\sigma(\mathcal{G}_1, \mathcal{H}_1)$, where $C \in \mathcal{G}_1$ and $D \in \mathcal{H}_1$. Then

$$E(P(A \mid \mathcal{G}_1)P(B \mid \mathcal{H}_1)I_C I_D)$$

is the expectation of the product of a bounded $\mathcal{G}$-measurable random variable and a bounded $\mathcal{H}$-measurable random variable. Therefore it equals

$$E(P(A \mid \mathcal{G}_1)I_C)E(P(B \mid \mathcal{H}_1)I_D) = P(A \cap C)P(B \cap D) = P((A \cap B) \cap (C \cap D)).$$

We complete the proof by using the Sierpiński Class Theorem to assert that the equality so far obtained, namely

$$E(P(A \mid \mathcal{G}_1)P(B \mid \mathcal{H}_1)I_{C \cap D}) = P((A \cap B) \cap (C \cap D)),$$

holds with $C \cap D$ replaced by an arbitrary member of $\sigma(\mathcal{G}_1, \mathcal{H}_1)$. $\quad \square$

---

* **Problem 44.** Show that $\sigma(\mathcal{G}_1, \mathcal{H}_1)$ in the preceding proposition cannot be replaced by an arbitrary sub-σ-field of $\sigma(\mathcal{G}, \mathcal{H})$.

Proposition 27. *Let $\mathcal{G} \subseteq \mathcal{H}$ and $\mathcal{K}$ be three σ-fields of events in a probability space. Then $\mathcal{H}$ and $\mathcal{K}$ are conditionally independent given $\mathcal{G}$ if and only if, for every $C \in \mathcal{K}$,*

$$P(C \mid \mathcal{G}) = P(C \mid \mathcal{H}) \quad a.s.$$

PROOF. Denote the probability space by $(\Omega, \mathcal{F}, P)$. Suppose that $\mathcal{H}$ and $\mathcal{K}$ are conditionally independent given $\mathcal{G}$ and let $C \in \mathcal{K}$. We check that $P(C \mid \mathcal{G})$ satisfies the two conditions in Proposition 2 characterizing $P(C \mid \mathcal{H})$. Since $\mathcal{G} \subseteq \mathcal{H}$, $P(C \mid \mathcal{G})$ is $\mathcal{H}$-measurable. To verify condition (ii) of Proposition 2, we let $B \in \mathcal{H}$ and use Definition 1 and the fact that $P(C \mid \mathcal{G}) \in \mathbf{L}_2(\Omega, \mathcal{G}, P)$ to obtain

$$E(P(C \mid \mathcal{G}) I_B) = E(P(C \mid \mathcal{G}) P(B \mid \mathcal{G})).$$

From the conditional independence of $\mathcal{H}$ and $\mathcal{K}$ we deduce

$$E(P(C \mid \mathcal{G}) P(B \mid \mathcal{G})) = E(P(C \cap B \mid \mathcal{G})) = P(C \cap B).$$

From the preceding two calculations we see that $P(C \mid \mathcal{G})$ satisfies the second of the two characterizing conditions of $P(C \mid \mathcal{H})$.

For the converse we assume $P(C \mid \mathcal{G}) = P(C \mid \mathcal{H})$ for all $C \in \mathcal{K}$. Then we fix $C \in \mathcal{K}$ and $B \in \mathcal{H}$ and set

$$Y(\omega) = P(B \mid \mathcal{G})(\omega) \, P(C \mid \mathcal{G})(\omega).$$

We complete the proof by using Definition 1 to show that $Y = P(B \cap C \mid \mathcal{G})$ a.s. It is clear that $Y \in \mathbf{L}_2(\Omega, \mathcal{G}, P)$. Let $Z \in \mathbf{L}_2(\Omega, \mathcal{G}, P)$. Then

$$E\big((Y - I_B I_C)Z\big) = E\big((P(B \mid \mathcal{G}) - I_B)P(C \mid \mathcal{G})Z\big) + E\big((P(C \mid \mathcal{G}) - I_C)I_B Z\big)$$
$$= 0 + E\big((P(C \mid \mathcal{H}) - I_C)I_B Z\big) = 0,$$

since $P(C \mid \mathcal{G})Z$ is $\mathcal{G}$-measurable and $I_B Z$ is $\mathcal{H}$-measurable. $\square$

Corollary 28. *Let $\mathcal{G} \subseteq \mathcal{H}$ and $\mathcal{K}$ be three σ-fields of events in a probability space. Suppose that $\mathcal{H}$ and $\mathcal{K}$ are independent. Then $\mathcal{H}$ and $\sigma(\mathcal{G}, \mathcal{K})$ are conditionally independent given $\mathcal{G}$.*

PROOF. We apply Proposition 27 with $\mathcal{G}$, $\mathcal{H}$, and $\sigma(\mathcal{G}, \mathcal{K})$ of the corollary playing the roles of $\mathcal{G}$, $\mathcal{H}$, and $\mathcal{K}$, respectively, in that proposition. According to the proposition we only need show $P(C \mid \mathcal{G}) = P(C \mid \mathcal{H})$ for every $C \in \sigma(\mathcal{G}, \mathcal{K})$. By Proposition 6, the collection of those C for which this equality holds is a Sierpiński class, so it suffices to prove that $P(D \cap E \mid \mathcal{G}) = P(D \cap E \mid \mathcal{H})$ whenever $D \in \mathcal{G}$ and $E \in \mathcal{H}$. Since $D \in \mathcal{G}$ and, therefore, also $D \in \mathcal{H}$,

(21.6) $$P(D \cap E \mid \mathcal{G}) = I_D P(E \mid \mathcal{G});$$
(21.7) $$P(D \cap E \mid \mathcal{H}) = I_D P(E \mid \mathcal{H}).$$

To complete the proof that the right sides and therefore the left sides of (21.6) and (21.7) equal each other, we need only observe that since $\mathcal{H}$ and $\mathcal{K}$ are independent, it follows that $\mathcal{G}$ and $\mathcal{K}$ are independent and, therefore,

$$P(E \mid \mathcal{G}) = P(E) = P(E \mid \mathcal{H}) \text{ a.s.} \quad \square$$

Problem 45. In view of Proposition 27 one is tempted to make a conjecture that treats $\mathcal{H}$ and $\mathcal{K}$ symmetrically. Suppose that, besides $\mathcal{G}$, $\mathcal{H}$, and $\mathcal{K}$ as given in that proposition, a fourth σ-field $\mathcal{L} \subseteq \mathcal{K}$ is involved. Then one might conjecture that if

$$P(C \mid \mathcal{G}) = P(C \mid \mathcal{H}) \text{ a.s.}$$

for every $C \in \mathcal{K}$ and

$$P(B \mid \mathcal{L}) = P(B \mid \mathcal{K}) \text{ a.s.}$$

for every $B \in \mathcal{H}$, then $\mathcal{H}$ and $\mathcal{K}$ are conditionally independent given $\sigma(\mathcal{G}, \mathcal{L})$. Use the probability space for some fair coins to show that this conjecture is not true.

We now have a tool to prove that the past and future of a random walk are conditionally independent given the present.

Theorem 29. Let $(S_0 = 0, S_1, \dots)$ be a random walk. Then, for every n, the vector $(S_0, S_1, \dots, S_n)$ and the sequence $(S_n, S_{n+1}, \dots)$ are conditionally independent given $\sigma(S_n)$.

PROOF. Denote the steps of the random walk by $X_1, X_2, \dots$, and set

$$\mathcal{G} = \sigma(S_n), \quad \mathcal{H} = \sigma(X_1, X_2, \dots, X_n), \quad \mathcal{K} = \sigma(X_{n+1}, X_{n+2}, \dots).$$

It is clear that $\mathcal{G}$, $\mathcal{H}$, and $\mathcal{K}$ satisfy the hypotheses of Corollary 28. From that result we conclude that $\mathcal{H}$ and $\sigma(\mathcal{G}, \mathcal{K})$ are conditionally independent given $\mathcal{G}$. This conclusion and the observations that $(S_0, \dots, S_n)$ is $\mathcal{H}$-measurable and $(S_n, S_{n+1}, \dots)$ is $\sigma(\mathcal{G}, \mathcal{K})$-measurable completes the proof. $\square$

The preceding theorem gives another illustration (besides the rather trivial Proposition 24) that conditional independence does not imply independence. The property obtained for random walks in the theorem is quite important. Many interesting random sequences, other than random walks, have this property.

Definition 30. A random sequence $(Y_m \colon m \geq 0)$ adapted to a filtration $(\mathcal{F}_m \colon m \geq 0)$ is a *Markov sequence with respect to* that filtration if, for each nonnegative integer n, $\mathcal{F}_n$ and $\sigma(Y_m \colon m \geq n)$ are conditionally independent given $\sigma(Y_n)$.

For the preceding definition, the minimal filtration is implicit if no filtration is mentioned explicitly. According to Theorem 29, a random walk is a Markov sequence. Regarding n in the definition as indicating the present, a Markov sequence is one in which the past (represented by $\mathcal{F}_n$) and the future (represented by $\sigma(Y_m : m \geq n)$) are conditionally independent given the present. The next proposition describes an equivalent view: a Markov sequence is one in which the conditional probability of a future event given the past equals its conditional probability given the present.

Proposition 31. *Let $Y = (Y_m : m \geq 0)$ be a random sequence adapted to a filtration $(\mathcal{F}_m : m \geq 0)$. Then the following two conditions are equivalent:*

(i) *Y is Markov with respect to $(\mathcal{F}_n : n \geq 0)$;*
(ii) *for each n and each $C \in \sigma(Y_n, Y_{n+1}, \dots)$,*

$$P(C \mid \sigma(Y_n)) = P(C \mid \mathcal{F}_n) \ a.s.$$

Problem 46. Prove the preceding proposition. *Hint:* Use Proposition 27.

Problem 47. Show that Proposition 31 remains true if, in condition (ii), we only require that $C \in \sigma(Y_{n+1}, Y_{n+2}, \dots)$. *Hint:* Use Proposition 25.

We will have more to say about Markov sequences in Chapters 22 and 26.

21.6. ‡ Conditional distributions of normal random vectors

We conclude this chapter with an illustrative example.

Example 4. Let $X = (X_1, \dots, X_d)$ be an $\mathbb{R}^d$-valued normal random variable, with mean vector $\mathbf{0}$ and covariance matrix Σ, and let $Y_1 = (X_1, \dots, X_k)$ and $Y_2 = (X_{k+1}, \dots, X_d)$, where k is an integer strictly between 0 and d. We wish to find the conditional distribution of Y_1 given Y_2. For simplicity, we will assume that the covariance matrix of Y_2 is strictly positive definite, and thus invertible.

Let us write the covariance matrix of X as

$$\Sigma = \begin{pmatrix} \Sigma_{11} & \Sigma_{12} \\ \Sigma_{21} & \Sigma_{22} \end{pmatrix}.$$

Regarding Y_1 and Y_2 as random row matrices, we have the following formula for the matrices Σ_{ij}:

$$\Sigma_{ij} = E(Y_i^T Y_j), \quad i, j = 1, 2,$$

where the expectation is taken term by term. In particular, Σ_{ii} is the covariance matrix of Y_i, $i = 1, 2$.

Let $Z_1 = Y_1 - \Sigma_{12}\Sigma_{22}^{-1}Y_2$. Since X is normal, the $\mathbb{R}^d$-valued random variable (Z_1, Y_2) is also normal. Note that

$$E(Z_1^T Y_2) = \Sigma_{12} - \Sigma_{12}\Sigma_{22}^{-1}\Sigma_{22} = \mathbf{0},$$

where $\mathbf{0}$ stands for the $k \times (d-k)$ zero matrix. Thus, the characteristic function of (Z_1, Y_2) factors into the product of the characteristic functions of Z_1 and Y_2 and hence Z_1 and Y_2 are independent.

By Proposition 12, the conditional distribution of (Z_1, Y_2) given Y_2 is the product of δ_{Y_2} and the unconditional distribution of Z_1. Since $Y_1 = Z_1 + \Sigma_{12}\Sigma_{22}^{-1}Y_2$, it follows from Proposition 11 that the conditional distribution of Y_1 given Y_2 is normal with mean vector $\Sigma_{12}\Sigma_{22}^{-1}Y_2$ and covariance matrix equal to the covariance matrix of Z_1:

$$E(Z_1^T Z_1)$$
$$= E(Y_1^T Y_1) - E(\Sigma_{12}\Sigma_{22}^{-1}Y_2^T Y_1) - E(Y_1^T Y_2 \Sigma_{22}^{-1}\Sigma_{21})$$
$$\qquad + E(\Sigma_{12}\Sigma_{22}^{-1}Y_2^T Y_2 \Sigma_{22}^{-1}\Sigma_{21})$$
$$= \Sigma_{11} - \Sigma_{12}\Sigma_{22}^{-1}\Sigma_{21} .$$

In particular, the conditional distribution of Y_1 given Y_2 is normal.

Problem 48. Work out the formulas for the mean and variance of the conditional distribution in the preceding example for the special case in which $d = 2, k = 1$, and
$$\Sigma = \begin{pmatrix} 1 & \rho \\ \rho & 1 \end{pmatrix} ,$$
with $|\rho| < 1$. Also discuss the situation where $\rho = \pm 1$.

Problem 49. For the special case $k = 1$ of Example 4, show that the variance of the conditional distribution is less than or equal to the variance of Y_1.

Problem 50. (intended for readers with sufficient knowledge of linear algebra) The formulas for the mean and variance of the conditional distribution in the example are still valid when Σ_{22} is not invertible, provided we replace Σ_{22}^{-1} by a matrix $\widetilde{\Sigma}_{22}$, known as the *pseudoinverse* of Σ_{22}. To define this matrix, write $\Sigma_{22} = O^T D O$, where O is an orthogonal matrix and D is a diagonal matrix. Let $\widetilde{D}$ be the matrix obtained by replacing each nonzero element of D by its reciprocal, and let $\widetilde{\Sigma}_{22} = O^T \widetilde{D} O$. Show that in the invertible case, the pseudoinverse equals the inverse. Then show that the modified formula is valid for the noninvertible case. *Hint:* A look at the proof shows that it suffices to show $\Sigma_{12} = \Sigma_{12}\widetilde{\Sigma}_{22}\Sigma_{22}$.

CHAPTER 22

Construction of Random Sequences

There are situations in which it is natural to construct probability spaces by first specifying certain conditional distributions and then showing that there is a unique underlying (unconditional) distribution consistent with those specifications. The tool for doing this is Theorem 3. Several specific examples are included along with three general classes of examples: exchangeable sequences, Markov sequences, and Polya urns.

22.1. The basic result

We begin with an elementary example.

Example 1. Consider two urns labelled H and T. In urn H there are 3 identical blue balls and 5 identical green balls. In urn T there are 7 identical blue balls and 2 identical green balls. A fair coin is flipped. If it shows heads, a ball is drawn at random from urn H. If it shows tails, a ball is drawn at random from urn T.

Let X_0 be the outcome of the coin toss (H or T), and let X_1 be the color of the ball drawn (B or G). We consider X_0 and X_1 to be two random variables defined on an appropriate probability space. The distribution of X_0 is clear from the description: $P[X_0 = H] = P[X_0 = T] = \frac{1}{2}$. But the distribution of X_1 is not directly available from the description. Instead, it is the conditional distribution of X_1 given $\sigma(X_0)$ that is easily obtained:

$$P[X_1 = B \mid \sigma(X_0)](\omega) = \begin{cases} \frac{3}{8} & \text{if } X_0(\omega) = H \\ \frac{7}{9} & \text{if } X_0(\omega) = T \end{cases}$$

and

$$P[X_1 = G \mid \sigma(X_0)](\omega) = \begin{cases} \frac{5}{8} & \text{if } X_0(\omega) = H \\ \frac{2}{9} & \text{if } X_0(\omega) = T. \end{cases}$$

The (unconditional) distribution of the pair (X_0, X_1) can now be obtained from condition (ii) in Proposition 2 of Chapter 21. We must have

$$P[X_0 = H, X_1 = B] = \frac{1}{2} \cdot \frac{3}{8} = \frac{3}{16},$$
$$P[X_0 = H, X_1 = G] = \frac{1}{2} \cdot \frac{5}{8} = \frac{5}{16},$$
$$P[X_0 = T, X_1 = B] = \frac{1}{2} \cdot \frac{7}{9} = \frac{7}{18},$$
$$P[X_0 = T, X_1 = G] = \frac{1}{2} \cdot \frac{2}{9} = \frac{2}{18}.$$

Thus, the unconditional distribution of X_0 together with the conditional distribution of X_1 given $\sigma(X_0)$, both of which arise naturally from the description of the experiment, uniquely determine the unconditional distribution of (X_0, X_1).

Here is a result that generalizes the construction described in the preceding example.

Lemma 1. *Let $(\Psi_0, \mathcal{G}_0)$ and $(\Psi_1, \mathcal{G}_1)$ be two measurable spaces, let R_0 denote a probability measure on $(\Psi_0, \mathcal{G}_0)$, and let $x_0 \rightsquigarrow R_1(x_0, \cdot)$ be a random distribution on $(\Psi_1, \mathcal{G}_1)$ whose domain is the probability space $(\Psi_0, \mathcal{G}_0, R_0)$. Then there is a unique distribution Q on $(\Psi_0 \times \Psi_1, \mathcal{G}_0 \times \mathcal{G}_1)$ such that if $X = (X_0, X_1)$ is any $(\Psi_0 \times \Psi_1)$-valued random variable having distribution Q, then R_0 is the distribution of X_0 and R_1 is a conditional distribution of X_1 given $\sigma(X_0)$. Moreover, Q is given by*

$$(22.1) \qquad Q(A) = \int_{\Psi_0} \int_{\Psi_1} I_A(x_0, x_1) \, R_1(x_0, dx_1) \, R_0(dx_0)$$

for $A \in \mathcal{G}_0 \times \mathcal{G}_1$.

PROOF. We begin by showing that a probability measure Q can be defined by (22.1). The interior integral in (22.1) is well-defined and nonnegative since, for each x_0, the function $I_A(x_0, \cdot)$ is nonnegative and measurable.

To prove that the interior integral is a measurable function of x_0 for all $A \in \mathcal{G}_0 \times \mathcal{G}_1$, consider the collection of all A's for which this measurability property holds. Clearly, this collection contains sets of the form $A_0 \times A_1$, where $A_j \in \mathcal{G}_j$. A straightforward application of the Sierpiński Class Theorem now gives the desired measurability for general A.

Notice that Q is nonnegative and $Q(\Psi_0 \times \Psi_1) = 1$. That Q is countably additive follows from the linearity of integration and the Monotone Convergence Theorem. Hence Q is a probability measure on $(\Psi_0 \times \Psi_1, \mathcal{G}_0 \times \mathcal{G}_1)$. Therefore Q is also the distribution of the random variable X defined by $X(\psi_0, \psi_1) = (\psi_0, \psi_1)$ on the probability space $(\Psi_0 \times \Psi_1, \mathcal{G}_0 \times \mathcal{G}_1, Q)$.

That the distribution of X_0 is R_0 follows from setting $A = A_0 \times \Psi_1$ in (22.1) for an arbitrary $A_0 \in \mathcal{G}_0$.

We now check that $(\psi_0, \psi_1) \rightsquigarrow R_1(X_0(\psi_0, \psi_1), \cdot)$ is a conditional distribution of X_1 given $\sigma(X_0)$, or equivalently, that for each $B \in \mathcal{G}_1$, $(\psi_0, \psi_1) \rightsquigarrow R_1((X_0(\psi_0, \psi_1), B)$ is a conditional probability of $X_1^{-1}(B)$ given $\sigma(X_0)$. Clearly, such a function is $\sigma(X_0)$-measurable. To check the second condition in Proposition 2 of Chapter 21, choose $C \in \sigma(X_0)$. Then

$$C = \{(\psi_0, \psi_1) \colon X_0(\psi_0, \psi_1) \in D\}$$

for some $D \in \mathcal{G}_0$. Thus,

$$\int_{(\Psi_0, \Psi_1)} R_1((X_0(\psi_0, \psi_1), B) I_C(\psi_0, \psi_1) \, P(d(\psi_0, \psi_1))$$

$$= \int_{\Psi_0} R_1(x_0, B) I_D(x_0) \, R_0(dx_0)$$

$$= \int_{\Psi_0} \int_{\Psi_1} I_{D \times B}(x_0, x_1) R_1(x_0, dx_1) \, R_0(dx_0)$$

$$= P(D \times B) = P(C \cap \{(\psi_0, \psi_1) \colon X_1(\psi_0, \psi_1) \in B\}),$$

as desired.

Now let $X = (X_0, X_1)$ be any random vector having the property that R_0 is the distribution of X_0 and R_1 is a conditional distribution of X_1 given $\sigma(X_0)$. By condition (ii) of Proposition 2 of Chapter 21, its distribution $\widehat{Q}$ agrees with Q, defined by (22.1), for A of the form $A_0 \times A_1$, where $A_j \in \mathcal{G}_j$. Now the Uniqueness of Measure Theorem (Theorem 3 of Chapter 7) implies that $\widehat{Q}$ and Q are the same probability measure on $\mathcal{G}_0 \times \mathcal{G}_1$. $\square$

It is not hard to see how to extend the preceding lemma inductively to give a method for constructing finite random sequences $(X_0, \ldots, X_n)$ of arbitrary length in terms of conditional distributions. For the inductive step, one is given the unconditional distribution of $(X_0, \ldots, X_k)$ and a conditional distribution of X_{k+1} given $\sigma(X_0, \ldots, X_k)$. Then one applies the lemma with the random variable $(X_0, \ldots, X_k)$ playing the role of X_0 and X_{k+1} playing the role of X_1 in that lemma, thus obtaining the unconditional distribution of $(X_0, \ldots, X_{k+1})$.

In order to construct infinite sequences using conditional distributions, we will mimic the construction of infinite product measure in Chapter 9. We need the following generalization of the Fubini Theorem, which follows from Lemma 1.

Theorem 2. [Conditional Fubini] *Let* $(\Psi_0, \mathcal{G}_0)$ *and* $(\Psi_1, \mathcal{G}_1)$ *be two measurable spaces and let*

$$(\Omega, \mathcal{F}) = (\Psi_0, \mathcal{G}_0) \times (\Psi_1, \mathcal{G}_1).$$

Let R_0, R_1, *and* Q *be as in* Lemma 1. *If* f *is an* $\overline{\mathbb{R}}$-*valued measurable function defined on* $(\Omega, \mathcal{F}, Q)$ *whose integral with respect to* Q *exists, then the function*

$$x_0 \rightsquigarrow \int_{\Psi_1} f(x_0, x_1) R_1(x_0, dx_1)$$

is an R_0-*almost surely defined* $\mathcal{G}_0$-*measurable function, and*

(22.2) $$\int_{\Omega} f \, dQ = \int_{\Psi_0} \int_{\Psi_1} f(x_0, x_1) \, R_1(x_0, dx_1) \, R_0(dx_0).$$

Problem 1. Prove the preceding lemma.

We might write (22.2) as

$$(22.3) \qquad E(f \circ (X_0, X_1)) = E\left(\int f(X_0, x_1)\, R_1(X_0, dx_1)\right).$$

The right side can be viewed as the iteration of two expectations, the 'interior expectation' being the 'conditional expectation' of $f \circ (X_0, X_1)$ given X_0. It is easy to extend Theorem 2 inductively to treat integration on products of finitely many measurable spaces.

Here is the construction result for infinite sequences.

Theorem 3. *Let $((\Psi_n, \mathcal{G}_n), n \geq 0)$ be a sequence of measurable spaces. Let R_0 be a probability measure on $\mathcal{G}_0$, and for each $n \geq 0$, let R_{n+1} be a measurable function from $(\Psi_0, \mathcal{G}_0) \times \cdots \times (\Psi_n, \mathcal{G}_n)$ to the measurable space of probability measures on $(\Psi_{n+1}, \mathcal{G}_{n+1})$. Then there exists a probability space $(\Omega, \mathcal{F}, P)$ and a random sequence $X = (X_0, X_1, \dots)$ defined on that space such that the distribution of X_0 is R_0 and, for $n \geq 0$, a conditional distribution of X_{n+1} given $\sigma(X_0, \dots, X_n)$ is given by*

$$\omega \rightsquigarrow R_{n+1}\big((X_0(\omega), \dots, X_n(\omega)), \cdot\big).$$

The distribution of X is uniquely determined by the relations

$$
\begin{aligned}
&P[(X_0, \dots, X_n) \in A_n] \\
(22.4) \quad &= \int_{\Psi_0} \cdots \int_{\Psi_n} I_{A_n}(x_0, \dots, x_n)\, R_n\big((x_0, \dots, x_{n-1}),\, dx_n\big) \dots R_0(dx_0),
\end{aligned}
$$

$n \in \mathbb{Z}^+$ and $A_n \in \mathcal{G}_0 \times \cdots \times \mathcal{G}_n$.

Problem 2. Prove the preceding theorem. *Hint:* Look at the proof of Theorem 16 of Chapter 9.

Problem 3. In the setting of Theorem 3, show that for $n = 0, 1, \dots$, and $k = 0, \dots, n-1$,

$$
\begin{aligned}
&P[(X_{k+1}, \dots, X_n) \in A \mid \sigma(X_0, \dots, X_k)] \\
&= \int_{\Psi_{k+1}} \cdots \int_{\Psi_n} I_A(x_{k+1}, \dots, x_n)\, R_n\big((X_0, \dots, X_k, x_{k+1}, \dots, x_{n-1}),\, dx_n\big) \\
&\qquad\qquad\qquad\qquad \dots R_{k+1}\big((X_0, \dots, X_k), dx_{k+1}\big),
\end{aligned}
$$

$A \in \mathcal{G}_{k+1} \times \cdots \times \mathcal{G}_n$. *Hint:* In checking the second of the two conditions that characterize conditional probabilities, use the Conditional Fubini Theorem repeatedly to calculate the relevant expected value.

We often take the point of view that a random sequence X represents the sequence of random values taken by some system that evolves in time, with X_n being the value at time n. The preceding result then says that the distribution of X can be specified by giving the 'initial distribution' (the distribution of X_0) and, for each time n, giving the conditional distribution of the 'present' value X_n in terms of the 'past' values $X_0, \ldots, X_{n-1}$.

Problem 4. [Destructive Random Walk] Show that there exists a sequence of $\mathbb{Z}^d$-valued random variables which fits the following informal description. Let X_n be the position at time $n \geq 0$ of a particle that starts at the origin at time 0. The particle moves somewhat like a simple symmetric random walk, except that it is not allowed to return to a previously visited point. One may think of the points of $\mathbb{Z}^d$ as locations which are destroyed when visited by the particle. If the particle is at location x, it moves with equal probability to any undestroyed location that is distance 1 from x. If there are no such locations, the particle stops moving forever. (Note that the sequence described is actually not a random walk.)

Problem 5. [Random Walk with Reinforcement] Let G be the complete graph on three vertices. In other words, G is the graph that looks like a triangle; it contains three vertices and three edges, with one edge connecting each pair of vertices. Show that there exists a sequence $(X_0, X_1, \ldots)$ whose terms take values in the set of vertices of G and which fits the following informal description. Each edge of G is initially labeled by some positive integer and X_0 equals each of the three vertices with probability 1/3. At each stage, the particle moves from its current position to one of the other two vertices along the corresponding connecting edge. The probability that it chooses to travel along a given edge is proportional to the integer that labels that edge. Each time the particle travels along an edge, the integer that labels that edge is increased by 1. (Note that the sequence described is actually not a random walk.)

22.2. Construction of exchangeable sequences

In this section, we construct a certain rather special but important class of random sequences. It turns out that in this case, we only need Lemma 1 rather than the more powerful Theorem 3.

Example 2. [Random iid sequences] Let Ψ denote a Borel space with σ-field $\mathcal{G}$, and let $(\mathcal{Q}, \mathfrak{H})$ denote the measurable space of probability measures on $(\Psi, \mathcal{G})$. Let R_0 be a probability measure on $(\mathcal{Q}, \mathfrak{H})$. By Lemma 1, there exists a pair (Q, X) of random variables defined on some probability space $(\Omega, \mathcal{F}, P)$ such that the distribution of the random variable Q is R and the conditional distribution of X given $\sigma(Q)$ is

$$\bigotimes_{n=1}^{\infty} Q \, ,$$

the infinite product of Q with itself. Note that Q is a random distribution on $(\Psi, \mathcal{G})$ and X is a $(\Psi, \mathcal{G})^\infty$-valued random variable, or in other words, $X = (X_1, X_2, \dots)$ is a random sequence of $(\Psi, \mathcal{G})$-valued random variables.

It is easy to check that X is a conditionally iid sequence given $\sigma(Q)$. One thinks of a two-stage experiment in which one first chooses a distribution Q on $(\Psi, \mathcal{G})$ at random, and then forms an iid sequence X with common distribution Q. For this reason, X is sometimes called a 'random iid sequence', even though this term may be somewhat misleading. The sequence X is not independent, but is instead only conditionally independent (given $\sigma(Q)$).

Problem 6. Describe the distribution R_0 of Example 2 for the experiment of flipping a coin an infinite sequence of times, in a setting where the probability of heads is not known, but rather uniformly distributed on $[0, 1]$.

Problem 7. Describe the distribution R_0 of Example 2 for the experiment of observing an infinite sequence of normally distributed random variables, the mean and standard deviation of which are not known, but rather distributed like a pair $(Y, |Z|)$, where (Y, Z) is an independent pair of standard normal random variables.

Definition 4. Let $\mathcal{I}$ be a countable set. A sequence $(X_i : i \in \mathcal{I})$ (finite or infinite) of random variables on a probability space $(\Omega, \mathcal{F}, P)$ is *exchangeable* if, for every permutation ρ of $\mathcal{I}$, the distributions of $(X_{\rho(i)} : i \in \mathcal{I})$ and $(X_i : i \in \mathcal{I})$ are identical.

The reader can check that to verify exchangeability one only need verify the condition in Definition 4 for permutations ρ having the property that $\rho(i) = i$ for all but finitely many i. Notice that a finite or infinite iid sequence is exchangeable. The next problem generalizes this fact to conditionally iid sequences.

Problem 8. Show that any sequence that is conditionally iid given some σ-field is exchangeable.

Problem 9. An urn contains r red balls, w white balls, and y yellow balls. The experiment consists of randomly drawing balls one at a time from the urn without replacement until all the balls have been drawn, yielding a random sequence $X = (X_1, X_2, \dots, X_n)$, where $n = r + w + y$. Prove that the sequence X is exchangeable.

For the Borel space setting it will be shown in Chapter 27 that every exchangeable infinite sequence is conditionally iid with respect to some σ-field, and that in fact, every such sequence can be constructed in the manner illustrated by Example 2.

Here is a simple example of a finite exchangeable sequence that is not conditionally iid with respect to any σ-field. The example consists of two random variables X_1 and $X_2 = -X_1$, where X_1 equals ± 1 according to a fair coin flip.

* Problem 10. Show that the sequence (X_1, X_2) defined in the preceding paragraph is not conditionally iid and that it also is not the beginning of a three-term exchangeable sequence.

Definition 5. Let $\mathcal{I}$ be a countable set. A sequence $(X_i: i \in \mathcal{I})$ (finite or infinite) of random variables having values in a Borel space is *conditionally exchangeable* given a σ-field $\mathcal{G}$ if, for every permutation ρ of $\mathcal{I}$, the conditional distributions of $(X_i: i \in \mathcal{I})$ given $\mathcal{G}$ and $(X_{\rho(i)}: i \in \mathcal{I})$ given $\mathcal{G}$ are the same.

The next proposition generalizes Problem 8. It says that conditional exchangeability implies exchangeability.

Proposition 6. *A finite or infinite sequence of random variables that is conditionally exchangeable given some σ-field is (unconditionally) exchangeable.*

* Problem 11. Prove the preceding proposition.

Problem 12. In connection with Definition 5, comment on the modification of Problem 9 in which the numbers of balls of each color are random and not necessarily independent.

Problem 13. Let $(X_1, X_2, \ldots, X_n)$ be an exchangeable sequence of $\mathbb{R}^d$-valued random variables. Let $S = \sum_{j=1}^{n} X_j$. Prove that the sequence $(X_1, X_2, \ldots, X_n)$ is conditionally exchangeable given $\sigma(S)$.

* Problem 14. Consider a simple random walk $S = (S_0 = 0, S_1, S_2, \ldots)$ in $\mathbb{Z}$ whose step distribution assigns probability p to $\{1\}$ and probability $(1-p)$ to $\{-1\}$, where $0 < p < 1$. Use Problem 13 to find the conditional distribution given $\sigma(S_n)$ of the vector consisting of the first n steps of S.

Problem 15. The formulas for the conditional probabilities in the solution of the preceding problem do not depend on p. Discuss this fact. Also, discuss the modifications, if any, needed in that problem to accommodate the cases $p = 1$ and $p = 0$.

* Problem 16. Let U be distributed according to the standard beta distribution with parameters α and β. Find the (unconditional) distribution of the first term of a sequence that is conditionally iid given $\sigma(U)$, where the common conditional distribution is Bernoulli with parameter U. Also find the joint distribution of the first two terms of the infinite sequence.

Problem 17. Repeat the preceding problem for a sequence of conditionally iid normal random variables with variance the constant σ^2 and random mean Θ, where Θ has the standard normal distribution.

22.3. Construction of Markov sequences

Theorem 3 indicates a general method of constructing Markov sequences. Suppose that in that theorem the functions $R_{n+1}, n \geq 1$, have the property that, for each fixed x_n and $B \in \mathcal{G}_{n+1}$,

$$(x_0, \ldots, x_{n-1}) \rightsquigarrow R_{n+1}\big((x_0, \ldots, x_{n-1}, x_n), B\big)$$

is a constant function. Then it follows from Problem 3 that

$$P[(X_{n+1}, \ldots, X_{n+k}) \in B_{n+1} \times \cdots \times B_{n+k} \mid \sigma(X_0, \ldots, X_n)]$$

is a measurable function of X_n for $B_{n+j} \in \mathcal{G}_{n+j}$, $j = 1, \ldots, k$. A straightforward Sierpiński class argument now gives that for any $B \in \mathcal{G}_{n+1} \times \mathcal{G}_{n+2} \times \ldots$, $P[(X_{n+1}, X_{n+2}, \ldots) \in B \mid \sigma(X_0, \ldots, X_n)]$ is also a measurable function of X_n, and thus equal to $P[(X_{n+1}, X_{n+2}, \ldots) \in B \mid \sigma(X_n)]$. By Problem 47 of Chapter 21, X is a Markov sequence.

Under the circumstances just described, it is natural to write

$$R_{n+1}(x_n, \cdot) \quad \text{in lieu of} \quad R_{n+1}\big((x_0, x_1, \ldots, x_n), \cdot\big).$$

The distribution R_0 is called the *initial distribution* of the Markov sequence and, for $n \geq 1$, R_n is called the nth *transition function*. For fixed x, $R_n(x, \cdot)$ is a *transition distribution*.

The preceding paragraph characterizes all Markov sequences in Borel spaces (see Problem 18). Throughout this book all Markov sequences will have values in Borel spaces. When all of the transition functions R_n are identical, the corresponding Markov sequence is said to be *time-homogeneous*, and the function $R_1 = R_2 = \ldots$ is *the* transition function.

Problem 18. Show that if X is a random sequence of random variables that take values in a Borel space $(\Psi, \mathcal{G})$, then X is Markov if and only if for each integer $n \geq 0$ there exists a measurable function $R_{n+1} \colon (\Psi, \mathcal{G}) \to (\mathcal{Q}, \mathfrak{H})$, where $(\mathcal{Q}, \mathfrak{H})$ is the measurable space of probability measures on $(\Psi, \mathcal{G})$, such that $\omega \rightsquigarrow R_{n+1}(X_n(\omega), \cdot)$ is the conditional distribution of X_{n+1} given $\sigma(X_0, \ldots, X_n)$. Further show that this condition is equivalent to the following:

$$(22.5) \qquad P[X_{n+1} \in B \mid \sigma(X_0, \ldots, X_n)] = P[X_{n+1} \in B \mid \sigma(X_n)] \text{ a.s.}$$

for $n \geq 0$ and $B \in \mathcal{G}$. *Hint:* Theorem 22 of Chapter 21 will be needed for the first part.

Problem 19. Add one green ball to the contents of the urn of Problem 9, and introduce the condition that whenever the green ball is drawn it is returned to the urn. Let Z_n denote the contents of the urn after n balls have been drawn (counting draws of the green ball), with Z_0 denoting the initial contents of the urn. Find the transition function of the time-homogeneous Markov sequence thus defined.

A random walk $(S_0, S_1, \dots)$ in either $\mathbb{R}$ or $\overline{\mathbb{R}}^+$ is a time-homogeneous Markov sequence. Its transition function R_1 is related to the step distribution Q of the random walk in a simple manner (see Problem 23 of Chapter 21):

$$(22.6) \qquad R_1(x, B) = Q(\{y : x + y \in B\}).$$

If we are taking the point of view that $S_0 = 0$, then its initial distribution R_0 equals the delta distribution at $\{0\}$. If R_0 is arbitrary, the term 'random walk' may still be used.

22.4. Polya urns

We now introduce a model that will serve many purposes. In particular, it will illustrate some aspects of preceding sections.

Example 3. [Polya urns] An urn contains a finite number of balls, each of which is either blue or orange; there is at least one ball of each of the two colors. The contents of the urn are changed according to the following procedure. A ball is drawn at random from the urn, its color is observed, and then that ball is placed back into the urn along with c more balls of the same color, where c is a fixed positive integer. This procedure is repeated infinitely often. (It is assumed that there is an unlimited supply of balls of both colors outside the urn, and that the urn is large enough to contain an arbitrary finite number of balls.)

We represent the states of the urn by members of $(\mathbb{Z}^+ \setminus \{0\})^2$, the first coordinate indicating the number of blue balls and the second the number of orange balls. The preceding section makes it clear that there is a time-homogeneous Markov sequence $((X_n, Y_n): n = 0, 1, 2, \dots)$ for which the 'Markov-type' description just given is appropriate. (Alternatively, one could construct a probability space especially suited to the example at hand; its members could be all infinite sequences of B's and O's.) The initial (possibly random) state of the urn is represented by the distribution R_0 of (X_0, Y_0) and the transition function R_1 is given by

$$R_1((x, y), \cdot) = \frac{x}{x + y} \delta_{(x+c, y)} + \frac{y}{x + y} \delta_{(x, y+c)},$$

where δ_a denotes the delta distribution at a.

For $n = 1, 2, \dots$, let

$$I_n(\omega) = \begin{cases} 1 & \text{if } X_n(\omega) > X_{n-1}(\omega) \\ 0 & \text{otherwise.} \end{cases}$$

Thus, $I_n(\omega)$ equals 1 or 0 according to whether a blue ball or orange ball is drawn at time n. For each finite sequence $(\varepsilon_1, \varepsilon_2, \dots, \varepsilon_k)$ of 0's and 1's, we will calculate

$$(22.7) \qquad P\big[(I_1, \dots, I_k) = (\varepsilon_1, \dots, \varepsilon_k)\big]$$

and obtain a formula that depends only on the numbers of 0's and 1's in the sequence $(\varepsilon_1, \ldots, \varepsilon_k)$ and not on their order.

We begin with the following equalities among events:

$$\bigcap_{j=1}^{k} [I_j = \varepsilon_j]$$

$$= \bigcap_{j=1}^{k} [(X_j, Y_j) = (X_{j-1}, Y_{j-1}) + c(\varepsilon_j, 1 - \varepsilon_j)]$$

(22.8)
$$= \bigcap_{j=1}^{k} \left[(X_j, Y_j) = (X_0, Y_0) + c \sum_{i=1}^{j} (\varepsilon_i, 1 - \varepsilon_i) \right].$$

By iterating the Conditional Fubini Theorem, we obtain the probability of the event in (22.8):

$$\int \prod_{j=1}^{k} R_1 \left((x_0, y_0) + c \sum_{i=1}^{j-1} (\varepsilon_i, 1 - \varepsilon_i) ; \{ (x_0, y_0) + c \sum_{i=1}^{j} (\varepsilon_i, 1 - \varepsilon_i) \} \right) R_0(d(x_0, y_0))$$

$$= \int \prod_{j=1}^{k} \frac{\varepsilon_j (x_0 + c \sum_{i=1}^{j-1} \varepsilon_i) + (1 - \varepsilon_j)(y_0 + c \sum_{i=1}^{j-1} (1 - \varepsilon_i))}{x_0 + y_0 + c(j-1)} R_0(d(x_0, y_0))$$

$$= \sum_{x_0=1}^{\infty} \sum_{y_0=1}^{\infty} \frac{\prod_{l=1}^{r} [x_0 + (l-1)c] \prod_{m=1}^{k-r} [y_0 + (m-1)c]}{\prod_{j=1}^{k} [x_0 + y_0 + (j-1)c]} R_0\{(x_0, y_0)\},$$

where r denotes the number of 1's in the sequence $(\varepsilon_1, \ldots, \varepsilon_k)$ (and, of course, $k - r$ equals the number of 0's). It follows that $(I_n : n = 1, 2, \ldots)$ is an exchangeable sequence.

Problem 20. For the preceding example, calculate

$$P(\{\omega' : I_2(\omega') = 1\} \mid \sigma(I_9)).$$

Hint: Use exchangeability to avoid extensive calculation.

* **Problem 21.** In Example 3 suppose that there are x_0 blue balls and y_0 orange balls in the urn at time 0. (That is, suppose that R_0 is the delta distribution at (x_0, y_0).) Calculate the correlation of I_m and I_n as a function of x_0, y_0, c, m and n. Find the limits (if they exist) of the correlation as $x_0 \to \infty$ (as a function of y_0, c, m, and n) and as $(x_0, y_0) \to (\infty, \infty)$ (as a function of c, m and n) and as $c \to \infty$ (as a function of x_0, y_0, m, and n). Give intuitive explanations of the results of your limit calculations.

* **Problem 22.** For Example 3 show that the events

$$\limsup_{n \to \infty} [I_n = 1] \quad \text{and} \quad \limsup_{n \to \infty} [I_n = 0]$$

both have probability 1.

Problem 23. Continuing with the notation of Example 3, let

$$J = \inf\{j: X_j > X_0, Y_j > Y_0\}$$

For $c = 1$ and R_0 the delta distribution at $(1,1)$, decide whether $E(J) < \infty$ or $E(J) = \infty$. Does your answer remain the same for other c and R_0?

22.5. † Coupon collecting

Suppose a person is trying to collect a certain class of objects (such as stamps, baseball cards, movie posters). We call these objects 'coupons', and denote by m the number of different coupons contained in a 'complete' collection. The collector receives coupons of random type one by one, continuing until a complete collection is achieved.

We model the situation just described as follows. Let $(Y_1, Y_2, \dots)$ be an iid sequence of random variables, each of which is uniformly distributed on the finite set $\{1, \dots, m\}$. These represent the coupon types received by the collector during the course of trying to build a complete collection. When n coupons (not necessarily all different from each other) have been received, the state of the collection is

$$X_n = \{Y_1\} \cup \{Y_2\} \cup \cdots \cup \{Y_n\},$$

$n = 0, 1, 2, \dots$, where it is understood that $X_0 = \emptyset$. It is easily checked that the random sequence $X = (X_n: n = 0, 1, 2, \dots)$ is a Markov sequence that takes values in the space of all subsets of $\{1, \dots, m\}$, whose initial distribution is $\delta_\emptyset$ and whose transition function $R_1 = R_2 = \dots$ is given by

$$R_1(x_n, \cdot) = \frac{\#x_n}{m}\delta_{x_n} + \sum_{a \notin x_n}\frac{1}{m}\delta_{x_n \cup \{a\}}.$$

(In fact, the sequence X is a random walk on the semigroup of subsets of $\{1, \dots, m\}$ with the semigroup operation being set union. See Chapter 11.)

For $0 \le j \le m$, set

$$N_j(\omega) = \inf\{n: \#(X_n(\omega)) = j\}.$$

Our goal is to study N_m, the time at which the full collection is completed. Of course $N_0 = 0$ and hence

$$(22.9) \qquad N_m = \sum_{j=1}^{m}[N_j - N_{j-1}].$$

Our first step will be to show that the summands in (22.9) are independent and that $[N_j - N_{j-1}]$ is geometrically distributed with support $\mathbb{Z} \setminus \{0\}$, mean

$$E(N_j - N_{j-1}) = \frac{m}{m-j+1},$$

variance

$$\mathrm{Var}(N_j - N_{j-1}) = \frac{m(j+1)}{(m-j+1)^2},$$

and probability generating function

$$s \rightsquigarrow \frac{(m-j+1)s}{m-(j-1)s}.$$

For x a subset of $\{1, \ldots, m\}$, we let $x^{\cup}$ denote the collection of all subsets of $\{1, \ldots, m\}$ that contain every member of x and exactly one more element. If $\#x = j - 1$, then for integers $p \geq (j-1)$ and $n > 0$,

$$P\big[(N_j - N_{j-1}) = n \mid (N_1 - N_0), \ldots, (N_{j-1} - N_{j-2}), X_{j-1} = x, N_{j-1} = p\big]$$
$$= R_1(x, \{x\})^{n-1} R_1(x, x^{\cup}) = \left(\frac{j-1}{m}\right)^{n-1} \left(\frac{m-j+1}{m}\right).$$

Now multiply by $P[X_{j-1} = x, N_{j-1} = p]$ and sum over p and x to obtain

$$P\big[(N_j - N_{j-1}) = n \mid (N_1 - N_0), \ldots, (N_{j-1} - N_{j-2})\big]$$
(22.10)
$$= \left(\frac{j-1}{m}\right)^{n-1} \left(\frac{m-j+1}{m}\right).$$

Since this conditional probability is a constant, it follows from Problem 21 of Chapter 21 that $(N_j - N_{j-1})$ is independent of $\sigma\big((N_1 - N_0), \ldots, (N_{j-1} - N_{j-2})\big)$. The independence of the m random variables $(N_j - N_{j-1})$, $1 \leq j \leq m$, follows from Proposition 5 of Chapter 9, and the advertised geometric distribution becomes apparent from (22.10).

From (22.9) we obtain

(22.11)
$$E(N_m) = m \sum_{j=1}^{m} \frac{1}{m-j+1} = m \sum_{i=1}^{m} \frac{1}{i}.$$

and

$$\mathrm{Var}(N_m) = \sum_{j=1}^{m} \frac{m(j-1)}{(m-j+1)^2}$$
$$= m \sum_{i=1}^{m} \frac{m-i}{i^2}$$
$$= m^2 \left(\sum_{i=1}^{m} \frac{1}{j^2} - \frac{1}{m} \sum_{i=1}^{m} \frac{1}{i}\right).$$

The random variable N_m/m represents the average amount of time that one waits for a coupon different from those that one already has. Its mean can be obtained by dividing both sides of (22.11) by m:

(22.12)
$$E\left(\frac{N_m}{m}\right) = \sum_{j=1}^{m} \frac{1}{j},$$

which approaches ∞ as $m \to \infty$. By dividing $\mathrm{Var}(N_m)$ by m^2 we obtain

$$(22.13) \qquad \mathrm{Var}\left(\frac{N_m}{m}\right) = \sum_{i=1}^{m} \frac{1}{i^2} - \frac{1}{m} \sum_{i=1}^{m} \frac{1}{i},$$

which, unlike the mean, approaches the finite limit $\sum_{i=1}^{\infty} i^{-2} = \pi^2/6$ as $m \to \infty$.

We turn to the issue of explicitly calculating the distribution of N_m. The case $m = 1$ is trivial; we take $m > 1$ in what follows. The probability generating function ρ of N_m is the product of the probability generating functions of the appropriate random variables of geometric type:

$$\rho(s) = \prod_{k=0}^{m-1} \frac{(m-k)s}{m-ks}.$$

Using partial fractions we obtain

$$(22.14) \qquad \rho(s) = -\frac{m! \, s^m}{m^{m-1}} \sum_{k=0}^{m-1} \left(k^{m-1} \prod_{\substack{l=0 \\ l \neq k}}^{m} \frac{1}{k-l} \right) \frac{m-k}{m-ks}.$$

We recognize that $\frac{m-k}{m-ks}$ is the probability generating function of a geometric distribution whose density with respect to counting measure on $\mathbb{Z}^+$ is

$$n \leadsto \left(1 - \frac{k}{m}\right)\left(\frac{k}{m}\right)^n, \quad n = 0, 1, 2, \dots.$$

Thus, $\frac{s^m(m-k)}{m-ks}$ is the probability generating function of the distribution whose density with respect to counting measure is

$$n \leadsto \begin{cases} (1 - \frac{k}{m})(\frac{k}{m})^{n-m} & \text{if } n \geq m \\ 0 & \text{if } n < m. \end{cases}$$

Taking account of the coefficients and summation in (22.14) we obtain

$$P[N_m = n] = -\frac{m!}{m^{m-1}} \sum_{k=0}^{m-1} \left(k^{m-1} \prod_{\substack{l=0 \\ l \neq k}}^{m} \frac{1}{k-l} \right) \left(1 - \frac{k}{m}\right) \left(\frac{k}{m}\right)^{n-m}$$

$$= \frac{m!}{m^n} \sum_{k=0}^{m-1} \left(k^{n-1} \prod_{\substack{l=0 \\ l \neq k}}^{m-1} \frac{1}{k-l} \right)$$

$$(22.15) \qquad = \frac{1}{m^{n-1}} \sum_{k=0}^{m-1} \binom{m-1}{k} (-1)^{m-1-k} k^{n-1}.$$

* **Problem 24.** Simplify the formula for $P[N_m = n]$ obtained above by writing k^{n-1} in terms of falling factorials and Stirling numbers.

Problem 25. Let $S(n, k)$ denote a Stirling number of the second kind. Use the answer to the preceding problem to prove that $k!S(n, k)$ equals the number of ways of placing n distinguishable objects into k distinguishable urns so that no urn is empty. Also, show that $S(n, k)$ equals the number of ways of placing n distinguishable objects into k indistinguishable urns so that no urn is empty.

Problem 26. As an alternative approach to obtaining the formula (22.15), use inclusion-exclusion to calculate $P[N_m \geq n]$.

Problem 27. We have kept m fixed for the coupon collecting problem, except when describing the limiting behavior of the mean and variance. If one wants to treat m as a variable throughout, then double subscripts are appropriate. We let $N_{j,m}$ denote the first time j types are obtained when the total number of types is m, and ask ourselves about convergence in distribution of the sequence

$$\left(\frac{N_{m,m} - E(N_{m,m})}{\sqrt{\mathrm{Var}(N_{m,m})}} : m = 1, 2, \ldots \right).$$

Is Chapter 16 useful for solving this problem? If so, use it. If not, approach the problem in some other fashion.

CHAPTER 23
Conditional Expectations

Integration with respect to conditional distributions gives conditional expectations. A precise definition is given in the first section of this chapter, after which several equivalent formulations are given. An interesting sidelight is the proof, at the end of the first section, of the Radon-Nikodym Theorem. The remaining sections are devoted to various formulas and properties, some of which are analogous to properties obtained in Chapters 4, 5, and 8 for (unconditional) expectations. Conditional variances are also treated and a useful formula relating conditional and unconditional variances is proved.

23.1. Definition of conditional expectation

The relationship between (unconditional) expectations and (unconditional) distributions is our guide for the following definition.

Definition 1. Let X be an $\overline{\mathbb{R}}$-valued random variable defined on a probability space $(\Omega, \mathcal{F}, P)$, Q the distribution of X, and $\mathcal{G}$ a sub-σ-field of $\mathcal{F}$. The *conditional expectation* of X given $\mathcal{G}$, denoted by

$$E(X \mid \mathcal{G}),$$

is the function

$$\omega \rightsquigarrow \int_{\mathbb{R}} x \, Q(dx \mid \mathcal{G})(\omega),$$

where it is to be understood that, evaluated at any particular value of ω, the conditional expectation can be any member of $\overline{\mathbb{R}}$ or be undefined.

By Proposition 16 of Chapter 21, conditional distributions of $\overline{\mathbb{R}}$-valued random variables are determined up to a null event. Therefore conditional expectations are also determined up to a null event. We have used this fact implicitly in the preceding definition when using the word 'the' before the term 'conditional expectation'.

Unless otherwise stated, if we make an assertion concerning an equality that involves one or more conditional expectations, we are in particular asserting that for almost all ω, either the quantities on both sides of the equality are defined and equal, or neither side is defined. Usually, the existence of a null event where the equality fails is to be understood, even when the modifier 'a.s.' has been omitted.

In case $\mathcal{G} = \sigma(Y)$ for some random variable Y, we often write

$$E(X \mid Y) \quad \text{for} \quad E(X \mid \sigma(Y)).$$

The reader can observe that the following is a consequence of Lemma 7 of Chapter 21.

Proposition 2. *Let X be a $\overline{\mathbb{R}}$-valued random variable and Y a $(\Psi, \mathcal{H})$-valued random variable, with X and Y both defined on a common probability space. Then there exists a measurable function $q \colon \Psi \to [\overline{\mathbb{R}} \cup \{\text{undefined}\}]$ such that $E(X \mid Y) = q \circ Y$ a.s.*

Definition 1 has the advantage of being quite general and intuitive. It makes sense for any $\mathbb{R}$-valued random variable X and any conditioning σ-field $\mathcal{G}$, and is particularly useful when one has a formula for the conditional distribution. But otherwise it can be difficult to use, and it is quite helpful to have some other equivalent versions of the definition. The remainder of this section is mainly concerned with three results, each of which provides a different characterization of conditional expectation and could therefore be taken as an alternative to Definition 1, although in each case, an additional assumption is required for equivalence. The following proposition and its consequences will be useful in obtaining the first characterization.

Proposition 3. *Let X be a $(\Psi, \mathcal{H})$-valued random variable on a probability space $(\Omega, \mathcal{F}, P)$ and suppose that a conditional distribution Z of X given $\mathcal{G}$ exists, where $\mathcal{G}$ is a sub-σ-field of $\mathcal{F}$. Let φ denote a measurable $\overline{\mathbb{R}}$-valued function on $(\Psi, \mathcal{H})$. Then*

$$E(\varphi \circ X \mid \mathcal{G}) = \int_{\Psi} \varphi(x) \, Z(dx) \ \text{a.s.}$$

PROOF. Apply Proposition 11 of Chapter 21. □

Problem 1. Show that if X is $\overline{\mathbb{R}}$-valued, then

$$E(X \mid \mathcal{G}) = E(X^{+} \mid \mathcal{G}) - E(X^{-} \mid \mathcal{G}),$$

with the usual understanding that the equality holds almost surely on the set where either side is defined. *Hint:* Use the preceding proposition with $\varphi(x) = x^{+}$ and $\varphi(x) = x^{-}$.

Problem 2. [Positivity of conditional expectation] Show that conditional expectation has the same kind of positivity as unconditional expectation, in the sense that if X is $\overline{\mathbb{R}}^{+}$-valued, then $E(X \mid \mathcal{G})$ is almost surely $\overline{\mathbb{R}}^{+}$-valued.

Problem 3. [Linearity of conditional expectation] Show that conditional expectation is linear, in the sense that (i) $E(aX \mid \mathcal{G}) = aE(X \mid \mathcal{G})$ for real a and (ii) $E(X + Y \mid \mathcal{G}) = E(X \mid \mathcal{G}) + E(Y \mid \mathcal{G})$. In (i), it is to be understood that $0 \cdot (\text{undefined}) = 0$ and $a \cdot (\text{undefined}) = \text{undefined}$ if $a \neq 0$. In (ii), the equality is required to hold almost surely on the set where each of the two conditional expectations on the right side and their sum are defined, and is not required to hold elsewhere, even if the left side is defined. *Hint:* Let $Z = (X, Y)$ and apply Proposition 3 with $\varphi((x, y)) = x + y$, with $\varphi((x, y)) = x$, and with $\varphi((x, y)) = y$.

Problem 4. Show that $E(I_A \mid \mathcal{G}) = P(A \mid \mathcal{G})$. Use this fact in conjunction with Proposition 3 to show that if $X \geq 0$, then there exist nonnegative simple random variables $X_n, n \geq 1$, such that $X_n \nearrow X$ a.s. and $E(X_n \mid \mathcal{G}) \nearrow E(X \mid \mathcal{G})$ a.s. as $n \to \infty$. Conclude from this result and the previous problem that if X is $\overline{\mathbb{R}}$-valued, then $E(X \mid \mathcal{G})$ is $\mathcal{G}$-measurable, in the sense that each of the following sets is in $\mathcal{G}$:

$$\{\omega\colon -\infty < E(X \mid \mathcal{G})(\omega) \leq x\}, \; x \in \mathbb{R},$$
$$\{\omega\colon E(X \mid \mathcal{G})(\omega) = -\infty\},$$
$$\{\omega\colon E(X \mid \mathcal{G})(\omega) = +\infty\},$$
$$\{\omega\colon E(X \mid \mathcal{G})(\omega) \text{ is undefined}\}.$$

Also, give a direct proof of this last fact using the Conditional Fubini Theorem.

Problem 5. Let $(X_1, \ldots, X_n)$ be a finite exchangeable sequence of $\mathbb{R}$-valued random variables, and let $S = X_1 + \cdots + X_n$. Prove that for $k = 1, 2, \ldots, n$,

$$E(X_k \mid S) = \frac{S}{n}$$

almost surely on the set where $E(X_k \mid S)$ exists. *Hint:* Use Problem 13 of Chapter 22.

Problem 6. Let $X_1, \ldots, X_n$ be iid $\mathbb{R}$-valued random variables with common density f with respect to Lebesgue measure. Assume that $f(x)|x|$ is bounded for $x \in \mathbb{R}$. Show that $E(X_k \mid S) = S/n$ a.s. *Hint:* Use the formula in Problem 35 of Chapter 21.

Problem 7. Let Y be an $\mathbb{R}^+$-valued random variable with infinite mean. Let (X_1, X_2) be a random pair that equals $(Y, -Y)$ or $(-Y, Y)$ each with probability $1/2$. Discuss the relevance of this example to Problem 5.

Our first characterization of conditional expectation is analogous to the characterization of conditional probabilities in Proposition 2 of Chapter 21. In order to properly understand its statement, we need to make two conventions concerning the value 'undefined'. The first concerns the expectation of a random variable Y that takes values in the set $\overline{\mathbb{R}} \cup \{\text{undefined}\}$. If $P[Y = \text{undefined}] > 0$, then $E(Y)$ is undefined. Otherwise, $E(Y) = E(\tilde{Y})$, where $\tilde{Y}(\omega) = Y(\omega)$ if $Y(\omega) \in \overline{\mathbb{R}}$ and $\tilde{Y}(\omega) = 0$ if $Y(\omega) = \text{undefined}$. The second convention was introduced in Problem 3. Both conventions play a role when we consider something like $E(Y; B) = E(Y I_B)$.

Theorem 4. *Let X be an $\overline{\mathbb{R}}$-valued random variable and let Y be a random variable which takes values in the space $\overline{\mathbb{R}} \cup \{undefined\}$. If $Y = E(X \mid \mathcal{G})$ a.s. then*

(i) Y is measurable with respect to $\mathcal{G}$;
(ii) $E(Y\,;\,B) = E(X\,;\,B)$ for all $B \in \mathcal{G}$ such that $E(X\,;\,B)$ exists.

On the other hand, if $E(X)$ exists and Y satisfies (i) and (ii), then Y is defined and equal to $E(X \mid \mathcal{G})$ almost surely.

PROOF. For the proof that $E(X \mid \mathcal{G})$ is $\mathcal{G}$-measurable, refer to Problem 4. To complete the proof of the first half of the theorem, it remains to show that (ii) holds with $Y = E(X \mid \mathcal{G})$. First consider the case in which X is $\overline{\mathbb{R}}^+$-valued. By Problem 2, $E(X \mid \mathcal{G})$ is also $\overline{\mathbb{R}}^+$-valued, so both expectations in (ii) exist for all $B \in \mathcal{G}$. To calculate $E(E(X \mid \mathcal{G})I_B)$, we use the Conditional Fubini Theorem, with $X_0 = I_B, X_1 = X$, and $f(x_0, x_1) = x_0 x_1$ (see (22.3) of Chapter 22 for the appropriate version of the formula in that theorem). Let Q denote the distribution of X. We obtain

$$E(E(X \mid \mathcal{G})I_B) = E\left(I_B \int x\, Q(dx \mid \mathcal{G})\right)$$
$$= E\left(\int I_B x\, Q(dx \mid \mathcal{G})\right) = E(I_B X)$$

as desired. In this calculation, the first equality follows from the definition of conditional expectation, the second from the fact that I_B does not depend on x, and the third is the one that uses the Conditional Fubini Theorem.

Now consider $\overline{\mathbb{R}}$-valued X. If $E(X\,;\,B)$ exists, then at least one of the two quantities $E(X^+\,;\,B)$ and $E(X^-\,;\,B)$ is finite. From the preceding paragraph we obtain $E(X^+\,;\,B) = E(E(X^+ \mid \mathcal{G})\,;\,B)$ and similarly for X^-. The desired result now follows from Problem 1 and the linearity of expectation.

For the proof of the second half, we assume that $E(X)$ exists, so that $E(X\,;\,B)$ exists for all $B \in \mathcal{G}$. Thus, if Y satisfies (i) and (ii), $E(Y\,;\,B)$ exists and is determined for all $B \in \mathcal{G}$, so Y is almost surely defined. By the first part of the theorem, to complete the proof it is enough to show that if Y and Z are almost surely defined $\mathcal{G}$-measurable random variables such that $E(Y\,;\,B)$ and $E(Z\,;\,B)$ exist and are equal for all $B \in \mathcal{G}$, then $Y = Z$ a.s. Let $W = Y - Z$. Then W is $\mathcal{G}$-measurable, and $E(W\,;\,B) = 0$ for all $B \in \mathcal{G}$. By letting $B = [W > 0]$ and $B = [W < 0]$ in this equation, we see that $W = 0$ a.s., as desired. $\square$

Problem 8. Show that if $E(X)$ exists, then $E(X \mid \mathcal{G})$ is almost surely defined, and $E(E(X \mid \mathcal{G})) = E(X)$. Conclude that if $E(X)$ is finite, then $E(X \mid \mathcal{G})$ is almost surely finite.

Problem 9. Let X have a standard Cauchy distribution and set $\mathcal{G} = \sigma(|X|)$. Show that $E(X \mid \mathcal{G}) = 0$ a.s. and that there exists a $B \in \mathcal{G}$ for which $E(X\,;\,B)$ does not exist, but $E(E(X \mid \mathcal{G}); B)$ does exist. (This example shows that the assumption that $E(X)$ exists in the second half of Theorem 4 cannot be eliminated entirely.)

The next set of problems contains our second and third characterizations of conditional expectation. The second characterization requires that X have a finite second moment. The third requires that X be finite almost surely.

Problem 10. Let $X \in \mathbf{L}_2(\Omega, \mathcal{F}, P)$ and let $\mathcal{G}$ be a sub-σ-field of $\mathcal{F}$. Show that $E(X \mid \mathcal{G})$ is the orthogonal projection of X onto $\mathbf{L}_2(\Omega, \mathcal{G}, P)$. *Hint:* Use Theorem 4.

* **Problem 11.** Suppose that X and V are $\mathbb{R}$-valued with finite second moments and that $E(X \mid V) = q \circ V$ for some decreasing q. Prove that $\mathrm{Cov}(X, V) \leq 0$.

Problem 12. Use the Radon-Nikodym Theorem to construct $E(X \mid \mathcal{G})$ if X is almost surely finite. *Hint:* First assume that X is nonnegative and bounded. Consider the measurable space $(\Omega, \mathcal{G})$ and let μ equal the restriction of P to $\mathcal{G}$. Define $\nu(A) = E(XI_A)$ for all $A \in \mathcal{G}$. Show that $\nu \ll \mu$ and then check that $\frac{d\nu}{d\mu}$ satisfies the conditions of Theorem 4.

The preceding problem shows that the Radon-Nikodym Theorem, which was introduced without proof in Chapter 8, can be used to construct conditional expectations. We will now reverse the viewpoint of this problem and use conditional expectations to prove the Radon-Nikodym Theorem, under the additional assumption that both measures are finite. The reader can do the extension to the σ-finite case.

PROOF OF RADON-NIKODYM THEOREM IN FINITE CASE. Let μ and ν be finite measures on a measurable space $(\Omega, \mathcal{F})$ having the property that $\nu \ll \mu$. The existence of $\frac{d\nu}{d\mu}$ is obvious if $\nu(\Omega) = 0$, so hereafter we assume $\nu(\Omega) > 0$ and therefore $\mu(\Omega) > 0$.

Let $\Psi = \Omega \times \{1, 2\}$ and denote by $\mathcal{H}$ the collection of subsets of Ψ having the form $(A \times \{1\}) \cup (B \times \{2\})$, where $A, B \in \mathcal{F}$. We define P by

$$P\big((A \times \{1\}) \cup (B \times \{2\})\big) = \tfrac{1}{2\mu(\Omega)}\, \mu(A) + \tfrac{1}{2\nu(\Omega)}\, \nu(B).$$

Clearly, $(\Psi, \mathcal{H}, P)$ is a probability space. On this probability space we introduce the random variable X equal to the indicator function of $\Omega \times \{2\}$, and we condition with respect to the σ-field $\mathcal{G}$ consisting of sets of the form $A \times \{1, 2\}$, where $A \in \mathcal{F}$. For such a set,

(23.1) $\qquad E\big(X\,;\, [A \times \{1, 2\}]\big) = E\big(E(X \mid \mathcal{G})\,;\, [A \times \{1, 2\}]\big),$

by Theorem 4. The left side equals $\tfrac{1}{2\nu(\Omega)}\, \nu(A)$. We note that there exists a measurable function g defined on Ω such that $E(X \mid \mathcal{G})(\omega, k) = g(\omega)$ for every ω and both k. Hence the right side of (23.1) equals

$$\int_A g\, d\big(\tfrac{1}{2\mu(\Omega)}\, \mu + \tfrac{1}{2\nu(\Omega)}\, \nu\big).$$

From (23.1) and the discussion following it we obtain

$$(23.2) \qquad \eta(A) \overset{\text{def}}{=} \int_A (1 - g)\, dv = \int_A \frac{\nu(\Omega)}{\mu(\Omega)}\, g\, d\mu\,.$$

Letting $A = \{\omega\colon g(\omega) = 1\}$ in (23.2) we see that $\mu(\{\omega\colon g(\omega) = 1\}) = 0$. Since $\nu \ll \mu$ we also obtain $\nu(\{\omega\colon g(\omega) = 1\}) = 0$. The Reciprocal Rule (Problem 33 of Chapter 8) thus applies to give $\nu \ll \eta$ and $\frac{dv}{d\eta} = \frac{1}{1-g}$ (when $g \neq 1$ and defined to equal any constant when $g = 1$). We also have $\frac{d\eta}{d\mu} = \frac{\nu(\Omega)}{\mu(\Omega)} g$, by (23.2). By the Chain Rule (Proposition 20 of Chapter 8), $\frac{dv}{d\mu}$ exists (equaling $\frac{\nu(\Omega)}{\mu(\Omega)} \frac{g}{1-g}$ almost everywhere with respect to μ). $\square$

Problem 13. Use the validity of the Radon-Nikodym Theorem for the case where both measures are finite to prove it in case both measures are σ-finite.

23.2. Conditional versions of unconditional theorems

As has already been done with positivity and linearity, certain other properties and results concerning expectation will be now be adapted to the conditional setting. The choices made here are not necessarily based on importance, but rather on various phenomena they highlight. The first example illustrates how a 'big' space like $\mathbb{R}^\infty$ can be of use even in a rather simple situation.

Example 1. [Conditional Monotone Convergence Theorem] Let $0 \leq X_1 \leq X_2 \leq \ldots$ be an increasing sequence of $\overline{\mathbb{R}}$-valued random variables defined on a probability space $(\Omega, \mathcal{F}, P)$. Then $X = (X_1, X_2, \ldots)$ is a random sequence that takes values in the space A of increasing sequences of members of $\overline{\mathbb{R}}$. It is easily shown that A is a Borel subset of $\overline{\mathbb{R}}^\infty$, so A is a Borel space by Proposition 20 of Chapter 21. Let Q be the distribution of X. It follows from the existence theorem for conditional distributions that, for any sub-σ-field $\mathcal{G}$ of $\mathcal{F}$, the random sequence X has a conditional distribution $Q(\cdot \mid \mathcal{G})$ given $\mathcal{G}$. For $x = (x_1, x_2, \ldots) \in A$, let $\varphi(x) = x_n, n = 1, 2, \ldots$, and $\varphi(x) = \lim_{n\to\infty} x_n$. Since the functions φ_n and φ are measurable, we have by Proposition 3 that

$$E(X_n \mid \mathcal{G}) = \int \varphi_n(x)\, Q(dx \mid \mathcal{G}) \text{ a.s.}, n = 1, 2, \ldots,$$

and

$$E(\lim_{n\to\infty} X_n \mid \mathcal{G}) = \int \varphi(x)\, Q(dx \mid \mathcal{G}) \text{ a.s.}$$

Since $(\varphi_n\colon n = 1, 2, \ldots)$ is an increasing sequence of $\overline{\mathbb{R}}^+$-valued functions on A, it follows from the (unconditional) Monotone Convergence Theorem that

$$(23.3) \qquad \lim_{n\to\infty} E(X_n \mid \mathcal{G}) = E(\lim_{n\to\infty} X_n \mid \mathcal{G}),$$

giving us a Conditional Monotone Convergence Theorem.

Problem 14. Provide an alternative proof of the Conditional Monotone Convergence Theorem based on Theorem 4.

Problem 15. [Conditional Dominated Convergence Theorem] State and prove a Conditional Dominated Convergence Theorem. Then notice that, as a consequence of Problem 8, the same conclusion holds if the dominating random variable Y has finite (unconditional) expectation.

Problem 16. Find an example for which the Conditional Dominated Convergence Theorem can be applied for ω in a set of positive probability, but not for almost every ω.

* **Problem 17.** Find an example for which the Conditional Dominated Convergence Theorem can be applied for almost every ω, but for which the (unconditional) Dominated Convergence Theorem is not applicable.

Problem 18. [Conditional Uniform Integrability Criterion] State and prove a Conditional Uniform Integrability Criterion.

The following example shows that unconditional uniform integrability does not imply conditional uniform integrability, so the conclusion of the Conditional Uniform Integrability Criterion does not necessarily hold for uniformly integrable sequences.

(Counter)example 2. Let P denote Lebesgue measure on the unit square $\Omega = (0,1)^2$ and let $\mathcal{F}$ denote the Borel field of Ω. Let $\mathcal{G}$ denote the σ-field generated by events of the form

$$\{(\omega_1, \omega_2) \in \Omega : \omega_1 \in B\}$$

for Borel subsets B of $(0,1)$. Let $(B_1, B_2, \dots)$ denote a sequence of Borel subsets of $(0,1)$ such that the one-dimensional Lebesgue measure of B_n equals $1/n$ and $\limsup B_n = (0,1)$. (We leave it as an exercise to show that such a sequence exists.) Let

$$X_n(\omega_1, \omega_2) = I_{B_n}(\omega_1) Z_n(\omega_2),$$

where Z_n is a nonnegative random variable yet to be specified. We want to choose Z_n so that the sequence $(X_1, X_2, \dots)$ converges almost surely and is uniformly integrable, but with probability 1 fails to be conditionally uniformly integrable. We leave it to the reader to arrange for:

(i) $Z_n(\omega_2) \to 0$ for almost every ω_2 ;

(ii) $\int_{(0,1)} Z_n(\omega_2)\, d\omega_2 \to \infty$;

(iii) $\dfrac{\int_{(0,1)} Z_n(\omega_2)\, d\omega_2}{n} \to 0$.

From condition (i) it follows that $X_n \to 0$ a.s. and so we are in a situation where the Uniform Integrability Criterion is relevant. The quotient in condition (iii) is just the expectation of the nonnegative random variable X_n and so, by condition (iii), all of the equivalent conditions in the Uniform Integrability Criterion hold. It is easy to calculate the conditional distribution of each X_n given $\mathcal{G}$ and thus the corresponding conditional expectation. The result is

$$(23.4) \qquad E(X_n \mid \mathcal{G})(\omega_1, \omega_2) = I_{B_n}(\omega_1) \int_{(0,1)} Z_n(\alpha) \, d\alpha \,,$$

which, for each (ω_1, ω_2), approaches ∞ on the sequence of those n for which $\omega_1 \in B_n$. Thus, according to the Conditional Uniform Integrability Criterion, the sequence $(X_1, X_2, \dots)$ fails with probability 1 to be conditionally uniformly integrable.

Problem 19. Show that the Borel sets $(B_1, B_2, \dots)$ can be chosen as described in the preceding example. Also show that conditions (i), (ii), and (iii) can be satisfied. Finally, make the calculation required to establish (23.4).

Example 3. [Conditional Jensen Inequality] Let X be an $\mathbb{R}$-valued random variable defined on a probability space $(\Omega, \mathcal{F}, P)$, Q the distribution of X, and J an interval that supports Q. Let φ be an $\mathbb{R}$-valued convex function defined on J and let $\mathcal{G}$ be a sub-σ-field of $\mathcal{F}$. Since J is a Borel space, the conditional distribution of X given $\mathcal{G}$ exists. Therefore, by the (unconditional) Jensen Inequality,

$$(23.5) \qquad \int_I \varphi(x) \, Q(dx \mid \mathcal{G})(\omega) \geq \varphi\left(\int_I x \, Q(dx \mid \mathcal{G})(\omega) \right)$$
$$= \left(\varphi \circ E(X \mid \mathcal{G}) \right)(\omega)$$

for almost all ω such that $E(X \mid \mathcal{G})(\omega)$ exists and is finite. Applying Proposition 3 to the left side of this last expression, we obtain, for such ω,

$$(23.6) \qquad E(\varphi \circ X \mid \mathcal{G})(\omega) \geq \left(\varphi \circ E(X \mid \mathcal{G}) \right)(\omega) \,,$$

which is the desired conditional form of the Jensen Inequality. In particular, if X has finite expectation, then it follows from Problem 8 that $E(X \mid \mathcal{G})$ is a.s. finite, and so (23.6) holds for almost every ω.

For a further interesting inequality we assume that $E(X)$ is finite. By taking expectations of both sides of (23.6) we obtain the first inequality below, and by applying the (unconditional) Jensen Inequality to the random variable $\omega \rightsquigarrow \varphi \circ E(X \mid \mathcal{G})(\omega)$ and then using Problem 8 we obtain the second inequality:

$$(23.7) \qquad E(\varphi \circ X) \geq E(\varphi \circ E(X \mid \mathcal{G})) \geq \varphi(E(X)) \,.$$

Problem 20. State (23.6) and (23.7) for the special case $\varphi(x) = |x|$. For each of the three inequalities obtained, decide whether it is an inequality previously known to you and, if so, on what basis.

Problem 21. State conditional versions of some of the main results in Part 2 concerning independent random variables. Possibilities include conditional versions of the Strong Law of Large Numbers, the Kolmogorov 0-1 Law, and the Etemadi Lemma. Review the proofs of the unconditional versions of these theorems to verify that they can be adapted to the conditional setting.

23.3. Formulas for conditional expectations

There are various special cases in which one can make explicit calculations involving conditional expectations. The following are designed to introduce the reader to the more useful and instructive of these.

Problem 22. From which problem in Chapter 21 does it follow that $E(X \mid \mathcal{G}) = X$ if X is $\mathcal{G}$-measurable?

* **Problem 23.** From which problem in Chapter 21 does it follow that $E(X \mid \mathcal{G}) = E(X)$ if X and $\mathcal{G}$ are independent?

Problem 24. Let $\mathcal{G}$ be a purely atomic σ-field generated by a partition $(C_j : j = 1, 2, \dots)$ of a probability space $(\Omega, \mathcal{F}, P)$, and let $X : \Omega \to \overline{\mathbb{R}}$ be a random variable. Then

$$E(X \mid \mathcal{G}) = \sum_{j \,:\, P(C_j) > 0} E(X \mid C_j) I_{C_j} \text{ a.s.} ,$$

where $E(X \mid C) = E(X \,;\, C)/P(C)$ for events C such that $P(C) > 0$.

Problem 25. Let X be a Cauchy random variable. Find a purely atomic sub-σ-field $\mathcal{G}$ of $\sigma(X)$ such that each of the following four events has positive probability:
(i) $\{\omega : E(X \mid \mathcal{G})(\omega) \in \mathbb{R}\}$;
(ii) $\{\omega : E(X \mid \mathcal{G})(\omega) = \infty\}$;
(iii) $\{\omega : E(X \mid \mathcal{G})(\omega) = -\infty\}$;
(iv) $\{\omega : E(X \mid \mathcal{G})(\omega) \text{ is undefined}\}$.

The following result is a nontrivial improvement of Problem 22. See also Problem 27, which is an important special case.

Proposition 5. For $i = 1, 2$, let X_i be a $(\Psi_i, \mathcal{H}_i)$-valued random variable defined on a probability space $(\Omega, \mathcal{F}, P)$ and let $\mathcal{G}$ be a sub-σ-field of $\mathcal{F}$ such that X_2 is measurable with respect to $\mathcal{G}$. Suppose that each $(\Psi_i, \mathcal{H}_i)$ is a Borel space. Let φ be a measurable $\overline{\mathbb{R}}$-valued function defined on $(\Psi_1, \mathcal{H}_2) \times (\Psi_2, \mathcal{H}_2)$. If Q_1 is the distribution of X_1, then

$$(23.8) \qquad E(\varphi \circ (X_1, X_2) \mid \mathcal{G})(\omega) = \int_{\Psi_1} \varphi(x, X_2(\omega)) \, Q_1(dx \mid \mathcal{G})(\omega)$$

for almost every ω, in the sense that the set of ω for which one side exists but the other does not is a null event.

Problem 26. Prove the preceding proposition. *Hint:* Use Proposition 12 or Corollary 13 of Chapter 21.

Problem 27. Let X and Y be $\mathbb{R}$-valued random variables on a probability space $(\Omega, \mathcal{F}, P)$ and suppose that Y is measurable with respect to a sub-σ-field $\mathcal{G}$ of $\mathcal{F}$. Prove that

$$E(XY \mid \mathcal{G}) = Y E(X \mid \mathcal{G}) \text{ a.s.},$$

making sure that your proof shows that the set on which one side exists and the other does not is a null event.

Problem 28. Suppose that X and Y are two $\overline{\mathbb{R}}$-valued random variables that are conditionally independent given some σ-field $\mathcal{G}$. Show that

$$E(XY \mid \mathcal{G})(\omega) = E(X \mid \mathcal{G})(\omega) E(Y \mid \mathcal{G})(\omega)$$

for almost every ω for which both factors on the right side are finite.

Problem 29. For conditional expectations, state and prove an analogue of Theorem 9 of Chapter 4, the result that describes basic properties of expectations such as linearity. Note that linearity and positivity have already been covered in Problem 2 and Problem 3.

* **Problem 30.** Let the random vector (X, Y) be uniformly distributed on the triangle $\{(x, y): 0 < x < y < 1\}$. Fix a number b between 0 and 1. Let $A_b = \{\omega: X(\omega) < b < Y(\omega)\}$. Calculate

$$E(Y - X \mid A_b)$$

in the following two ways. One way is to find the conditional distribution of (X, Y) given A_b and integrate $y - x$ with respect to it. A second way is to interpret (X, Y) as order statistics and use linearity of conditional expectation either in the form

$$E(Y - X \mid A_b) = E(b - X \mid A_b) + E(Y - b \mid A_b)$$

or in the form

$$E(Y - X \mid A_b) = E(Y \mid A_b) - E(X \mid A_b).$$

Compare your answers for various values of b and also with the (unconditional) expectation $E(Y - X)$. Discuss intuitively.

We conclude this section with an important formula involving iterated conditional expectations.

Proposition 6. *Let X be an $\overline{\mathbb{R}}$-valued random variable on $(\Omega, \mathcal{F}, P)$ and denote by $\mathcal{G}$ and $\mathcal{H}$ two sub-σ-fields of $\mathcal{F}$, with $\mathcal{H} \subseteq \mathcal{G}$. If $E(X \mid \mathcal{H})$ exists a.s., then $E(X \mid \mathcal{G})$ exists a.s., and*

$$E(E(X \mid \mathcal{G}) \mid \mathcal{H}) = E(X \mid \mathcal{H}) = E(E(X \mid \mathcal{H}) \mid \mathcal{G}) \text{ a.s.}$$

Problem 31. Prove the preceding proposition. *Hint:* If $E(X^2) < \infty$, the result is a consequence of Problem 1 of Chapter 20 and Problem 10. Use the Conditional Monotone Convergence Theorem to extend to nonnegative X, then (carefully) use linearity.

Problem 32. Let $(\Omega, \mathcal{F}, P)$ be a probability space, and let $\mathcal{G}, \mathcal{H}$ be σ-fields such that $\mathcal{H} \subseteq \mathcal{G} \subseteq \mathcal{F}$. Show that for any event $A \in \mathcal{F}$, $E(P(A \mid \mathcal{G}) \mid \mathcal{H}) = P(A \mid \mathcal{H})$.

* **Problem 33.** Let $\mathcal{H} \subseteq \mathcal{G}$ and $B \in \mathcal{G}$. Use Proposition 6 to prove

(23.9) $$E(E(X \mid \mathcal{G})I_B \mid \mathcal{H})(\omega) = E(XI_B \mid \mathcal{H})(\omega)$$

for almost all ω for which the right side exists. In particular, $E(X \mid \mathcal{G})$ exists almost surely on the set where $E(X \mid \mathcal{H})$ exists.

Problem 34. Let $X = (X_n : n = 0, 1, 2, \dots)$ be a Markov sequence taking values in a Borel space Ψ. Show that for any measurable function $f \colon \Psi \to \overline{\mathbb{R}}^+$ and nonnegative integers k and $m \leq n$,

$$E(E(f \circ Y_{n+k} \mid Y_n) \mid Y_m) = E(f \circ Y_{n+k} \mid Y_m) \text{ a.s.}$$

23.4. Conditional variance

In this section we will see that, on average, conditioning lowers variances.

Definition 7. Let X be an $\overline{\mathbb{R}}$-valued random variable defined on a probability space $(\Omega, \mathcal{F}, P)$, Q the distribution of X, and $\mathcal{G}$ a sub-σ-field of $\mathcal{F}$. The *conditional variance* of X given $\mathcal{G}$ is defined by

$$\operatorname{Var}(X \mid \mathcal{G})(\omega) = \int_{\mathbb{R}} (x - E(X \mid \mathcal{G})(\omega))^2 Q(dx \mid \mathcal{G})(\omega),$$

for every ω for which $E(X \mid \mathcal{G})(\omega)$ is finite.

Proposition 8. *For X and $\mathcal{G}$ as in the preceding definition,*

(23.10) $$\operatorname{Var}(X \mid \mathcal{G}) = E(X^2 \mid \mathcal{G}) - [E(X \mid \mathcal{G})]^2 \text{ a.s.}$$

Problem 35. Prove the preceding proposition, making sure to prove that the set where one side exists and the other does not is a null event.

By Proposition 5, another way to write (23.10) is

$$\operatorname{Var}(X \mid \mathcal{G}) = E((X - E(X \mid \mathcal{G}))^2 \mid \mathcal{G}),$$

provided that $E(X \mid \mathcal{G})$ is almost surely defined.

By Problem 8, the conditional expectation of a random variable is finite almost surely if the (unconditional) expectation is finite. The following theorem implies that the same is true for variances.

Theorem 9. *If X is any $\mathbb{R}$-valued random variable with finite second moment, then*
$$\mathrm{Var}(X) = E(\mathrm{Var}(X \mid \mathcal{G})) + \mathrm{Var}(E(X \mid \mathcal{G}))$$
for any conditioning σ-field $\mathcal{G}$.

PROOF. Taking note of the fact that the hypothesis implies that the random variable $E(X \mid \mathcal{G})$ is finite almost surely and has finite expectation and, as a consequence, has a variance (not known initially to be finite), we calculate

$$
\begin{aligned}
E(\mathrm{Var}(X \mid \mathcal{G})) &= E\big(E(X^2 \mid \mathcal{G}) - [E(X \mid \mathcal{G})]^2\big) \\
&= E\big(E(X^2 \mid \mathcal{G})\big) - E\big([E(X \mid \mathcal{G})]^2\big) \\
&= E(X^2) - \big[E(E(X \mid \mathcal{G}))\big]^2 - \mathrm{Var}(E(X \mid \mathcal{G})) \\
&= E(X^2) - [E(X)]^2 - \mathrm{Var}(E(X \mid \mathcal{G})) \\
&= \mathrm{Var}(X) - \mathrm{Var}(E(X \mid \mathcal{G})),
\end{aligned}
$$

the linearity used for the second equality being valid because the first term on its right is finite by hypothesis. □

Problem 36. To what familiar fact does Theorem 9 reduce in case X is the sum of two independent random variables having finite second moment and $\mathcal{G}$ is the σ-field generated by one of the two summands.

Problem 37. Theorem 9 asserts that the expectation of a conditional variance is typically smaller than the (unconditional) variance. Give an intuitive explanation.

Problem 38. Let $Z = X + Y$, where X and Y are iid with finite second moment, and set $\mathcal{G} = \sigma(Z)$. For X and $\mathcal{G}$ as defined here calculate each of the three terms in Theorem 9 in terms of quantities associated with the distribution of Z.

Problem 39. Suppose that X and Y in the preceding problem are normally distributed. Then the random variable $\mathrm{Var}(X \mid \mathcal{G})$ is a constant a.s.—Why? Give an example that shows that this conclusion can fail if the normality assumption is dropped.

Problem 40. Let $(Y_1, \ldots, Y_d)$ have a Dirichlet distribution. Use Theorem 9 to calculate $\mathrm{Var}(E(Y_1 \mid Y_2, \ldots, Y_d))$.

Problem 41. Let $(X_k : k = 1, 2, \ldots)$ be a sequence of independent $\mathbb{R}$-valued random variables each having zero mean and finite variance. Let

$$S_n = \sum_{k=1}^{n} X_k$$

and

$$\mathcal{F}_n = \sigma(X_1, \ldots, X_n)$$

for $n \geq 0$. Show that $E(S_{n+1} \mid \mathcal{F}_n) = S_n$ and $\mathrm{Var}(S_{n+1} \mid \mathcal{F}_n) = \mathrm{Var}(X_{n+1})$ for each n.

A random sequence (S_n) having the property mentioned in the preceding problem that $E(S_{n+1} \mid \mathcal{F}_n) = S_n$ is called a *martingale*. Thus, $\mathbb{R}$-valued random walks with expected step equal to 0 are martingales. Martingales will be treated in the next chapter.

We have already seen that many results for ordinary expectations generalize to conditional expectations. A similar statement holds true for variances. In most cases, such generalizations are rather straightforward, both in statement and in proof. Here is an exception, the proof being slightly subtle.

Proposition 10. [Conditional Chebyshev Inequality] *Let X be an $\mathbb{R}$-valued random variable defined on a probability space $(\Omega, \mathcal{F}, P)$, and let $\mathcal{G}$ be a sub-σ-field of $\mathcal{F}$. Then, for $z > 0$ and almost every ω for which $E(X \mid \mathcal{G})(\omega)$ is finite,*

$$P\big[|X - E(X \mid \mathcal{G})| \geq z \,\big|\, \mathcal{G}\big](\omega) \leq \frac{\operatorname{Var}(X \mid \mathcal{G})(\omega)}{z^2}.$$

PROOF. Fix ω such that $E(X \mid \mathcal{G})(\omega)$ is finite. Letting Q be the distribution of X, apply the (unconditional) Chebyshev Inequality to the probability measure $Q(\cdot \mid \mathcal{G})$ to obtain

$$\operatorname{Var}(X \mid \mathcal{G})(\omega) \geq z^2 Q\big(\{x \colon |x - E(X \mid \mathcal{G})(\omega)| \geq z\} \mid \mathcal{G}\big).$$

Rewrite the right side as

$$z^2 \int I_B\big(x, E(X \mid \mathcal{G})(\omega)\big) \, Q(dx \mid \mathcal{G})(\omega),$$

where $B = \{(x, y) \colon |x - y| \geq z\}$. By Proposition 5, this last expression equals

$$z^2 E\big(I_B \circ (X, E(X \mid \mathcal{G})) \,\big|\, \mathcal{G}\big)(\omega)$$
$$= z^2 P\big[|X - E(X \mid \mathcal{G})| \geq z \,\big|\, \mathcal{G}\big](\omega). \quad \square$$

* **Problem 42.** Let Y be a random variable taking values between 0 and 1. Let $(X_1, X_2, \dots)$ be a random sequence whose conditional distribution given Y is that of a sequence of Bernoulli trials with parameter Y. Show that for $\varepsilon > 0$,

$$P\left[\left|\frac{X_1 + \cdots + X_n}{n} - Y\right| > \varepsilon\right] \leq \frac{E(Y(1 - Y))}{n\varepsilon^2}.$$

Problem 43. Show that the statement obtained by replacing $E(X \mid \mathcal{G})$ by $E(X)$ in the Conditional Chebyshev Inequality can be false.

Problem 44. Show that the statement obtained by replacing $E(X \mid \mathcal{G})$ by $E(X)$ and $\operatorname{Var}(X \mid \mathcal{G})$ by $\operatorname{Var}(X)$ in the Conditional Chebyshev Inequality can be false.

Problem 45. [Conditional Markov Inequality] Let X, Y be $\mathbb{R}$-valued random variables defined on a probability space $(\Omega, \mathcal{F}, P)$, and let $f \colon \mathbb{R} \times \mathbb{R} \to (0, \infty)$ be a measurable function which is increasing in the second variable, in the sense that $f(x, y_1) \leq f(x, y_2)$ if $y_1 \leq y_2$. Show that if $\mathcal{G}$ is a sub-σ-field of $\mathcal{F}$ such that X is $\mathcal{G}$-measurable, then for all $z \in \mathbb{R}$ and almost every ω,

$$P\big[Y \geq z \mid \mathcal{G}\big](\omega) \leq \frac{E(f \circ (X, Y) \mid \mathcal{G})(\omega)}{f(X(\omega), z)}.$$

Problem 46. [Conditional Cauchy-Schwarz Inequality] State and prove the Conditional Cauchy-Schwarz Inequality.

PART 5
Random Sequences

In the previous parts of the book, we have laid the theoretical foundations of modern probability theory. In the remaining parts, we build on these foundations to study what are known as 'stochastic processes'. This part is concerned with discrete-time stochastic processes, also known as random sequences, and Part 6 is concerned with stochastic processes in continuous time.

We have chosen to give a thorough introduction to five types of random sequences: martingales (Chapter 24), renewal sequences (Chapter 25), Markov sequences (Chapter 26), exchangeable sequences (Chapter 27), and stationary sequences (Chapter 28). We have already seen examples of each of these types, often without making explicit mention of the fact. For instance, every iid sequence is Markov, exchangeable, and stationary, and iid sequences of 0's and 1's are renewal sequences. Random walks are Markov, and when the steps have mean 0, they are martingales.

Thus, much of what we have to say will be concerned with generalizing certain of the properties of iid sequences and random walks. But the theory goes far beyond mere generalization. By taking the various points of view represented by the five chapters in this part of the book, we will be led to ask and answer questions that did not arise naturally in the context of iid sequences or random walks. And we will encounter random sequences that have interesting behavior that is not possible for iid sequences or random walks.

CHAPTER 24
Martingales

We will treat what many would regard as the most important type of random sequence, for it is both intrinsically natural and also a tool for treating other topics in probability. Martingales are particularly important in the study of Markov sequences and Markov processes, as will be seen later in this book.

Martingales can be used to model 'fair games'. In this context, an important theorem to be proved in this chapter says that the fairness of a game is not affected by a broad range of strategies that might be employed by a player. Other theorems in this chapter, treating the issue of convergence as time gets large, have the flavor of laws of large numbers.

To illustrate the power of the theory, we give applications to a variety of models, including random walks, Polya urns, and some random sequences relevant to gambling theory. In an extended example at the end of the chapter, we determine the optimal strategy in a game known as 'Red and Black'.

24.1. Basic definitions

It is reasonable to regard the following definition as a description of a sequence of fair games. This interpretation will become clearer as we present various examples throughout the chapter.

Definition 1. A random sequence $X = (X_0, X_1, X_2, \dots)$ of $\mathbb{R}$-valued random variables with finite mean is a *martingale* with respect to a filtration $(\mathcal{F}_n, n \geq 0)$ to which it is adapted if

$$(24.1) \qquad E(X_{n+1} \mid \mathcal{F}_n) = X_n \text{ a.s.}$$

for all $n \geq 0$.

By relaxing the equality (24.1) to one-sided inequalities, we obtain the following two related definitions.

Definition 2. A random sequence $X = (X_0, X_1, X_2, \dots)$ of $\mathbb{R}$-valued random variables with finite mean is a *supermartingale* with respect to a filtration $(\mathcal{F}_n, n \geq 0)$ to which it is adapted if

$$(24.2) \qquad\qquad E(X_{n+1} \mid \mathcal{F}_n) \leq X_n \text{ a.s.}$$

for all $n \geq 0$.

Definition 3. A random sequence $X = (X_0, X_1, X_2, \dots)$ of $\mathbb{R}$-valued random variables with finite mean is a *submartingale* with respect to a filtration $(\mathcal{F}_n, n \geq 0)$ to which it is adapted if

$$(24.3) \qquad\qquad E(X_{n+1} \mid \mathcal{F}_n) \geq X_n \text{ a.s.}$$

for all $n \geq 0$.

If the filtration, in any of the above definitions, is the minimal filtration for X, we often drop the phrase "with respect to the filtration ... ". According to the preceding definition, an $\mathbb{R}$-valued random sequence X adapted to a filtration $(\mathcal{F}_n, n \geq 0)$ is a submartingale with respect to that filtration if and only if for all n,

$$E\big((X_{n+1} - X_n) \mid \mathcal{F}_n\big) \geq 0 \text{ a.s.}$$

and $E(|X_n|) < \infty$. By Theorem 4 of Chapter 23 this is equivalent to

$$(24.4) \qquad\qquad E\big((X_{n+1} - X_n); A\big) \geq 0$$

for all $A \in \mathcal{F}_n$.

A random sequence X is a supermartingale if and only if the random sequence $-X$ is a submartingale, and X is a martingale if and only if it is both a supermartingale and a submartingale. Because of the simple relationship that exists between sub- and supermartingales, we will usually state results for one of these two types of sequences (whichever seems more natural) and leave it to the reader to formulate the analogous result for the other type. Of course, every result concerning sub- or supermartingales applies to martingales.

Problem 1. Show that if X is a submartingale with respect to some filtration, then it is a submartingale with respect to the minimal filtration. Give an example of a random sequence that is a martingale with respect to the minimal filtration but is not a martingale with respect to some other filtration to which it is adapted.

* **Problem 2.** Let X be a random sequence of $\mathbb{R}$-valued random variables having finite mean. Prove that X is a submartingale with respect to a filtration $(\mathcal{F}_n, n \geq 0)$ to which it is adapted if and only if for all $m, n \geq 0$,

$$E(X_{n+m} \mid \mathcal{F}_n) \geq X_n \text{ a.s.}$$

Problem 3. Prove that $E(X_n) \leq E(X_{m+n})$ for all submartingales X and $m, n \geq 0$.

The inequality in the preceding exercise will often be used without comment, especially in the form of an equality in the case of martingales.

24.2. Examples

Martingales, submartingales, and supermartingales appear in a variety of contexts, as illustrated by the following examples and exercises. More examples will appear in later sections.

Example 1. [Polya urns, continued] The random sequences X and Y in Example 3 of Chapter 22 can be used in a natural way to define a martingale. Let

$$V_n = \frac{X_n}{X_n + Y_n},$$

the proportion of blue balls in the urn after n steps. It is easy to obtain the conditional distribution of V_{n+1} given $\mathcal{F}_n$: it assigns probability

$$\frac{X_n(\omega)}{X_n(\omega) + Y_n(\omega)} \quad \text{to the value} \quad \frac{X_n(\omega) + c}{X_n(\omega) + Y_n(\omega) + c}$$

and probability

$$\frac{Y_n(\omega)}{X_n(\omega) + Y_n(\omega)} \quad \text{to the value} \quad \frac{X_n(\omega)}{X_n(\omega) + Y_n(\omega) + c}.$$

By using this conditional distribution, we obtain the following expression for the conditional expectation of V_{n+1} given $\mathcal{F}_n$:

$$\frac{X_n[X_n + c] + Y_n X_n}{[X_n + Y_n][X_n + Y_n + c]} = V_n.$$

Thus the random sequence V is a martingale.

Problem 4. Let $S = (S_0 = 0, S_1, \dots)$ be an $\mathbb{R}$-valued random walk. Assume that $m = E(S_1)$ exists and is finite. Show that S is a supermartingale, martingale or submartingale with respect to the minimal filtration, depending on whether m is $\leq, =,$ or ≥ 0.

Historically, the study of martingales was motivated by questions concerning gambling. In modern times, gambling theory has developed into an important area within probability theory. The following example introduces some of the questions that interest gambling theorists.

Example 2. [Red and Black] A roulette wheel in Las Vegas typically has 38 positions. Two of these are colored green, eighteen of them are colored black, and the remaining eighteen are colored red. On any given play, the wheel is given a spin, then allowed to turn freely until it comes to a rest, at which point there is a mechanism that marks one of the 38 positions. The game is repeated indefinitely, and we will assume that the marked positions form an iid sequence, each member of which is uniformly distributed among the 38 possibilities.

One of the options available to a gambler is to bet an amount of money, known as a 'stake', on the black numbers. We denote the stake by s, and the amount of money available to the gambler for betting, known as her 'fortune', by f. The rules of casinos require that $0 \leq s \leq f$. If a black position comes up, the gambler receives back the stake plus as much again. Otherwise, she loses the stake. Thus if she wins, her fortune becomes $f + s$, and if she loses it becomes $f - s$. For the roulette wheel described above, these two possibilities have probability $18/38$ and $20/38$, respectively. Since there is nothing special about these quantities from a mathematical point of view, we will allow the probability of winning to be any number p in $[0, 1]$. Typically, $p < 1/2$.

In this simple model, we assume that the gambler has decided not to make any other types of bets or to play any other games. Thus, if we denote by $Q(f, s, \cdot)$ the distribution of the gambler's fortune after betting a stake s from a fortune f, then

$$Q(f, s, \cdot) = p\delta_{f+s} + (1 - p)\delta_{f-s},$$

where δ_x denotes the delta distribution at x.

We wish to define a random sequence which represents the sequence of fortunes of a gambler whose is repeatedly betting on the black numbers, as described above. The only choice being made by the gambler is how much to bet each time. In our model, we want this choice to depend only on the history of outcomes of previous bets. To make this last requirement more precise, we consider a sequence of functions that constitute an allowable 'strategy' for this game.

For each $n \geq 0$, let γ_{n+1} be a Borel measurable function from $[0, \infty)^{n+1}$ to $[0, \infty)$ with the property that

$$\gamma_{n+1}(f_0, f_1, \ldots, f_n) \leq f_n$$

for all $f_0, f_1, \ldots, f_n \in [0, \infty)$. The sequence $\gamma = (\gamma_1, \gamma_2, \ldots)$ is called a *strategy*. The quantity $\gamma_{n+1}(f_0, \ldots, f_n)$ represents the stake that will be bet on the $(n+1)^{\text{st}}$ play of the game, given that the 'initial fortune' was f_0 and, for $k = 1, \ldots, n$, the fortune after the kth play was f_k. Let

$$(\Omega, \mathcal{F}) = ([0, \infty), \mathcal{B})^\infty,$$

where $\mathcal{B}$ is the Borel σ-field of subsets of $[0, \infty)$. For $\omega = (f_0, f_1, \ldots) \in \Omega$ and $n \geq 0$, let $X_n(\omega) = f_n$. By Theorem 3 of Chapter 22 there exists, for each initial fortune f_0 and strategy γ, a probability measure P on $(\Omega, \mathcal{F})$ such that

$$P[X_0 = f_0] = 1$$

and, for each $n \geq 0$, $Q(X_n, \gamma(X_0, \ldots, X_n), \cdot)$ is a conditional distribution of X_{n+1} given $\sigma(X_0, \ldots, X_n)$.

We now have a precise formulation of the situation we wanted to model. Note that we have actually defined a family of random sequences, one for each initial fortune and strategy. This family is referred to collectively as *Red and Black*.

Suppose $p \leq 1/2$. Then for any initial fortune f_0 and strategy γ, the mean of the conditional distribution given above is less than or equal to X_n, so that the sequence $X = (X_0, X_1, \dots)$ is a supermartingale. When $p < 1/2$, this mean is strictly less than X_n, and one could say that the game is 'unfair', in the sense that it favors the casino over the gambler. Supermartingales are often used to model games which favor the casino rather than the gambler. When $p = 1/2$, the sequence X is a martingale. Martingales are often called 'fair games'.

One of the chief concerns of gambling theory is that of finding an 'optimal strategy'. The meaning of the word 'optimal' depends on the goals of the gambler. Suppose that she has chosen a certain amount $g > 0$, and that her only goal is to achieve a fortune which is not less than g. Let

$$N(\omega) = \inf\{n \geq 0 : X_n(\omega) = 0\}$$

and

$$G(\omega) = \inf\{n \geq 0 : X_n(\omega) \geq g\}.$$

These random variables are stopping times for the minimal filtration of X. No nonzero stakes may be bet after time N, so $G = \infty$ if $G > N$.

The 'gambler's ruin problem' is to determine for each fixed initial fortune and strategy, the probability of 'ruin', which is the probability that $G > N$. A related optimization problem is to find a strategy which maximizes the probability that $G < N$. After we have developed some tools, we will solve the gambler's ruin problem for the case of constant bets, and in the last section of the chapter, we will solve the optimization problem just mentioned.

Problem 5. In real gambling casinos, gamblers have more choices than our gambler in Red and Black. For example, in roulette, a gambler may bet a nonnegative amount on each position. For betting correctly on a given position, the gambler receives back the stake plus 35 times that amount. The bet made in Red and Black is equivalent to placing stakes of size $s/18$ on each of the black positions. One could also imagine other types of roulette wheels. Generalize the construction of Red and Black to take some of these possibilities into account.

We continue with more examples of martingales.

Problem 6. Let $f : [0, 1) \to \mathbb{R}$ be a measurable function whose integral with respect to Lebesgue measure λ on $[0, 1)$ is finite. For $n \geq 0$, let $\mathcal{F}_n$ be the σ-field generated by the collection of intervals of the form $[(k-1)/2^n, k/2^n)$, $k = 1, 2, \dots, 2^n$. Let $(\Omega, \mathcal{F}, P) = ([0, 1), \mathcal{B}, \lambda)$, and define

$$X_n = E(f \mid \mathcal{F}_n).$$

Show that the random sequence X is a martingale with respect to the filtration $(\mathcal{F}_n, n \geq 0)$. Find an explicit formula for each random variable X_n in terms of f.

The next exercise generalizes the example in the preceding problem, and in so doing, also makes it simpler in a sense.

Problem 7. Let Z be an $\mathbb{R}$-valued random variable with finite mean, defined on a probability space $(\Omega, \mathcal{F}, P)$, and let $(\mathcal{F}_n, n \geq 0)$ be a filtration of sub-σ-fields of $\mathcal{F}$. Prove that the random sequence

$$(E(Z \mid \mathcal{F}_0), E(Z \mid \mathcal{F}_1), \dots)$$

is a martingale with respect to the filtration $(\mathcal{F}_n, n \geq 0)$.

We will find later that every uniformly integrable martingale takes the form given in Problem 7 (see Theorem 20).

* **Problem 8.** Let $(S_0 = 0, S_1, S_2, \dots)$ be an $\mathbb{R}$-valued random walk and let φ be the characteristic function of the step distribution. ('Characteristic function' is defined in Chapter 13.) Fix $u \in \mathbb{R}$ and define

$$Y_n = \exp(iuS_n)/(\varphi(u))^n$$

for $n \geq 0$. Show that the real and imaginary parts of the random sequence $(Y_n : n \geq 0)$ are martingales.

Example 3. Let S be an $\mathbb{R}$-valued random walk starting at 0 with steps having finite expectation. For $n \geq 1$ define

$$\mathcal{G}_n = \sigma\Big(\frac{S_n}{n}, \frac{S_{n+1}}{n+1}, \dots\Big).$$

Note that, for all n,

$$\mathcal{G}_n = \sigma(S_n, X_{n+1}, X_{n+2}, \dots),$$

where $(X_1, X_2, \dots)$ is the sequence of steps of S. It follows from Proposition 4 of Chapter 21 and Problem 5 of Chapter 23 that

$$E\Big(\frac{S_n}{n} \mid \mathcal{G}_{n+1}\Big) = \frac{1}{n}\sum_{k=1}^{n} E(X_j \mid \mathcal{G}_{n+1}) = \frac{1}{n}\sum_{k=1}^{n} \frac{S_{n+1}}{n+1} = \frac{S_{n+1}}{n+1} \quad \text{a.s.}$$

If we let $Z_n = (S_n/n)$ in the preceding example, then the last equation becomes $E(Z_n \mid \mathcal{G}_{n+1}) = Z_{n+1}$, which is like a reversed version of the defining property of martingales. We are motivated to make some definitions. A decreasing sequence of σ-fields is a *reverse filtration*. A random sequence $Z = (Z_1, Z_2, \dots)$ is called a *reverse supermartingale* with respect to a reverse filtration $(\mathcal{G}_n, n \geq 1)$ if, for all n, Z_n has finite mean, is measurable with respect to $\mathcal{G}_n$, and satisfies

$$E(Z_n \mid \mathcal{G}_{n+1}) \leq Z_{n+1} \quad \text{a.s.}$$

A random sequence Z is called a *reverse submartingale* if $-Z$ is a reverse super-martingale. A *reverse martingale* is a random sequence which is both a reverse supermartingale and a reverse submartingale. We have shown that the sequence $(S_1, S_2/2, S_3/3, \dots)$ is a reverse martingale with respect to the sequence of σ-fields $(\mathcal{G}_n, n \geq 1)$. This fact will be used later in this chapter to give a nice proof of the Strong Law of Large Numbers.

24.3. Doob decomposition

The next problem shows that every random sequence of random variables having finite mean can be changed into a martingale by conditional centering.

Problem 9. Let $X = (X_0, X_1, \dots)$ be a sequence of $\mathbb{R}$-valued random variables with finite mean, adapted to a filtration $(\mathcal{F}_n, n \geq 0)$. For $n > 0$, define

$$U_n = E(X_n - X_{n-1} \mid \mathcal{F}_{n-1})$$

and

$$Y_n = X_n - \sum_{k=1}^{n} U_k \,.$$

Let $Y_0 = X_0$. Show that the random sequence $Y = (Y_0, Y_1, \dots)$ is a martingale with respect to the filtration $(\mathcal{F}_n, n \geq 0)$.

Theorem 4. [Doob Decomposition] *Let X be a submartingale with respect to a filtration $(\mathcal{F}_n, n \geq 0)$. There exist unique random sequences Y and V such that*

  (i) for all $n \geq 0$, $X_n = Y_n + V_n$;
  (ii) Y is a martingale with respect to the filtration $(\mathcal{F}_n, n \geq 0)$;
  (iii) $0 = V_0 \leq V_1 \leq V_2 \leq \dots$;
  (iv) for all $n > 0$, V_n is measurable with respect to $\mathcal{F}_{n-1}$.

Because of property (iv), we say that the random sequence V in the preceding proposition is *previsible* with respect to the filtration $(\mathcal{F}_n, n \geq 0)$. Thus, the Doob Decomposition Theorem states that any submartingale can be written as the sum of a martingale and an increasing previsible random sequence.

* **Problem 10.** Prove Theorem 4. Also state and prove an analogous fact concerning supermartingales. *Hint:* Use Problem 9.

* **Problem 11.** Let S be a random walk on $\mathbb{R}$ starting at 0. Assume that $E(S_1) = 0$ and $\mathrm{Var}(S_1) < \infty$. Show that the random sequence

$$S^2 = (S_0^2, S_1^2, S_2^2, \dots)$$

is a submartingale. Find an increasing previsible sequence V such that $S^2 - V$ is a martingale.

24.4. Transformations of submartingales

We turn now to two results concerning different ways in which submartingales may be transformed into other submartingales. As usual, there are analogous results for supermartingales. The first result, Proposition 5, states that a convex, increasing function of a submartingale is itself a submartingale. Proposition 6 says that both the sum and the maximum of two submartingales are submartingales.

Proposition 5. *Let $X = (X_0, X_1, X_2, \ldots)$ be an $\mathbb{R}$-valued random sequence adapted to a filtration $(\mathcal{F}_n : n \geq 0)$. Let φ be an $\mathbb{R}$-valued function which is convex on an interval that contains the supports of the distributions of the random variables $X_n, n \geq 0$. Assume that $E(|\varphi \circ X_n|) < \infty$ for all n. Let Y be the random sequence*

$$(\varphi \circ X_0, \, \varphi \circ X_1, \, \varphi \circ X_2, \, \ldots).$$

If X is a martingale with respect to $(\mathcal{F}_n : n \geq 0)$, or if X is a submartingale with respect to $(\mathcal{F}_n : n \geq 0)$ and φ is also increasing, then Y is a submartingale with respect to $(\mathcal{F}_n : n \geq 0)$.

Problem 12. Use the Conditional Jensen Inequality (Example 3 in Chapter 23) to prove the preceding proposition.

Problem 13. Repeat the first part of Problem 11, using Proposition 5.

Proposition 6. *Let X and Y be submartingales with respect to a filtration $(\mathcal{F}_n, n \geq 0)$. Then the sequences*

$$(X_0 + Y_0, X_1 + Y_1, X_2 + Y_2, \ldots)$$

and

$$(X_0 \vee Y_0, X_1 \vee Y_1, X_2 \vee Y_2, \ldots)$$

are submartingales with respect to the filtration $(\mathcal{F}_n, n \geq 0)$.

Problem 14. Prove the preceding proposition.

Problem 15. Let $X = (X_1, X_2, \ldots)$ be a random sequence of independent $\mathbb{R}$-valued random variables with common distribution function F. For $n \geq 1$ let F_n be the empirical distribution function based on n observations $X_1, \ldots, X_n$, and let

$$Y_n = \sup\{|F_n(x) - F(x)| : x \in \mathbb{R}\}.$$

Show that the random sequence $Y = (Y_n : n \geq 1)$ is a reverse submartingale with respect to an appropriate filtration. *Hint:* First, let $\varphi(y) = |y|$ and apply a reverse martingale version of Proposition 5 to the sequence $(F_n(x) - F(x), n \geq 1)$ for fixed x. Then extend and modify Proposition 6 to apply to the supremum of countably many reverse submartingales. Then complete the proof.

24.5. Another transformation: optional sampling

A central feature of martingale theory concerns random subsequences obtained from martingales by 'sampling' them at an increasing sequence of stopping times. If X is a random sequence adapted to a filtration $(\mathcal{F}_n, n \geq 0)$, and if T is an almost surely finite stopping time for the same filtration, then according to Proposition 9 of Chapter 11, X_T is a random variable which is measurable with respect to $\mathcal{F}_T$. We may regard X_T as the value obtained by sampling the sequence X at the random time T. The σ-field $\mathcal{F}_n$ is often interpreted as the information available at time n, so the assumption that T be a stopping time means that the decision ('option') to sample at time n must be based on information available at time n. If we sample X at successive random times $T_0 \leq T_1 \leq T_2 \leq \ldots$, we obtain a new random sequence. The Optional Sampling Theorem, to be stated and proved below, implies that if the original sequence is a submartingale, then the new sequence obtained by sampling is also a submartingale, provided each of the sampling times satisfies certain conditions.

Definition 7. Let $X = (X_0, X_1, \ldots)$ be a random sequence of $\mathbb{R}$-valued random variables with finite mean. A $\overline{\mathbb{Z}}^+$-valued random variable T that is a.s. finite satisfies the *sampling integrability conditions* for X if the following two conditions hold:

(i) $E(|X_T|) < \infty$;
(ii) $\liminf_{m \to \infty} E(|X_m|\,; [T > m]) = 0$.

Theorem 8. [Optional Sampling] *Let $T_0 \leq T_1 \leq T_2 \leq \ldots$ be an increasing sequence of stopping times for a filtration $(\mathcal{F}_m, m \geq 0)$, and let $X = (X_m : m \geq 0)$ be a submartingale with respect to the same filtration. Assume that each T_n is finite almost surely and satisfies the sampling integrability conditions for X. Define*

$$Y_n(\omega) = X_{T_n(\omega)}(\omega),$$

and

$$\mathcal{G}_n = \mathcal{F}_{T_n}.$$

Then the random sequence $Y = (Y_0, Y_1, \ldots)$ is a submartingale with respect to the filtration $(\mathcal{G}_n, n \geq 0)$.

PROOF. By Proposition 9 of Chapter 11, Y is an $\mathbb{R}$-valued random sequence which is adapted to the filtration $(\mathcal{G}_n, n \geq 0)$. Each random variable in the sequence Y has finite mean by the sampling integrability condition (i). Thus, it remains to show that

$$E(Y_{n+1} \mid \mathcal{G}_n) \geq Y_n,$$

or equivalently that

$$E(Y_{n+1}\,; A) \geq E(Y_n\,; A)$$

for all $A \in \mathcal{G}_n$. Fix $n \geq 0$ and $A \in \mathcal{G}_n$. For $m = 0, 1, 2, \ldots$, let

$$B_m = A \cap [T_n = m].$$

By hypothesis, T_n is finite almost surely, so A is the disjoint union of the sets B_m and a null set. On the set B_m, $Y_n = X_m$. Thus, by the Dominated Convergence Theorem, it is enough to prove that

$$E(Y_{n+1} ; B_m) \geq E(X_m ; B_m)$$

for all m. For an arbitrary integer $p > m$,

$$E(Y_{n+1} ; B_m)$$

(24.5)
$$= E(X_m ; B_m) + \sum_{l=m}^{p-1} E\big((X_{l+1} - X_l) ; B_m \cap [T_{n+1} > l]\big)$$
$$+ E\big((Y_{n+1} - X_p) ; B_m \cap [T_{n+1} > p]\big) .$$

For each l,
$$[T_{n+1} > l] = [T_{n+1} \leq l]^c \in \mathcal{F}_l .$$

Also $B_m \in \mathcal{F}_m \subseteq \mathcal{F}_l$ for $l \geq m$. For such l it thus follows from (24.4) that

$$E\big((X_{l+1} - X_l) ; B_m \cap [T_{n+1} > l]\big) \geq 0 .$$

From (24.5) we then obtain

(24.6) $$E(Y_{n+1} ; B_m) \geq E(X_m ; B_m) + E\big((Y_{n+1} - X_p) ; [T_{n+1} > p]\big) .$$

By the sampling integrability condition (ii), we can let $p \to \infty$ along an appropriate subsequence so that

$$E(|X_p| ; B_m \cap [T_{n+1} > p]) \leq E(|X_p| ; [T_{n+1} > p]) \to 0 .$$

Since
$$I_{[T_{n+1} > p]} \to 0 \text{ a.s.}$$

as $p \to \infty$, it follows from sampling integrability condition (i) and the Dominated Convergence Theorem that

$$\lim_{p \to \infty} E(Y_{n+1} ; B_m \cap [T_{n+1} > p]) = 0 .$$

Thus the second expectation on the right of (24.6) converges to 0 as $p \to \infty$ along an appropriate subsequence, and the proof is complete. $\square$

In view of Problem 3 we see that if $S \leq T$ are stopping times that satisfy the sampling integrability conditions for a submartingale X, then $E(X_S) \leq E(X_T)$. This fact will be used without comment, often with $S = 0$.

Problem 16. Show that the Optional Sampling Theorem can be improved somewhat, in that $|X_m|$ can be replaced by X_m^+ in the sampling integrability condition (ii). State and prove a version of the Optional Sampling Theorem for supermartingales, and incorporate a similar improvement into your hypotheses.

Sampling integrability condition (i) is clearly necessary in the Optional Sampling Theorem since submartingales are, by definition, sequences of random variables with finite means. The following example shows that condition (ii) cannot be eliminated entirely.

Example 4. [Double or Nothing] Let X be a random sequence for which the conditional distribution of X_{n+1} given $\sigma(X_0, \ldots, X_n)$ is the uniform distribution on the two-point set $\{0, 2X_n\}$. Let $X_0 = 1$. Such a sequence is a special case of Red and Black. It represents the fortunes of a gambler who starts with 1 dollar, then stakes her entire fortune at each step on a fair bet. After each bet, the fortune either doubles or vanishes, each possibility occurring with probability $\frac{1}{2}$. It is easily checked that X is a martingale. Let

$$T(\omega) = \inf\{m > 0\colon X_m(\omega) = 0\}.$$

We leave it to the reader to check that T is of geometric type and, in particular, that T is finite almost surely. Since $E(X_T) = E(0) = 0 < 1 = E(X_0)$, T does not satisfy the sampling integrability conditions for X. It must be condition (ii) that fails, because condition (i) is clearly satisfied.

Problem 17. Show directly that sampling integrability condition (ii) fails in the preceding example.

The sampling integrability conditions are somewhat technical in nature. They are chosen to make the proof of the Optional Sampling Theorem work. Here are some results and exercises concerning sets of conditions which imply the sampling integrability conditions. They are often easier to check than the sampling integrability conditions themselves, and they apply in a number of important situations.

Proposition 9. If T is an almost surely bounded stopping time, then T satisfies the sampling integrability conditions for any sequence X of $\mathbb{R}$-valued random variables with finite mean.

PROOF. Suppose that $T \leq a$ a.s. for some real number a. Then condition (i) follows from the following inequality:

$$|X_T| \leq \sum_{i=0}^{a} |X_i| \text{ a.s.}$$

Condition (ii) follows from the fact that, for $m \geq a$, $P[T > m] = 0$. $\square$

Problem 18. Let X be a submartingale and T a stopping time, both with respect to a filtration $(\mathcal{F}_n, n \geq 0)$. Define a random sequence Y by

$$Y_n(\omega) = X_{T(\omega) \wedge n}(\omega).$$

By the Proposition 9 and the Optional Sampling Theorem, Y is a submartingale with respect to the filtration $(\mathcal{F}_{T \wedge n}, n \geq 0)$. Prove that for all integers $p \geq 0$ and all sets $A \in \mathcal{F}_{T \wedge p}$,

$$E(Y_p \,;\, A) \leq E(X_p \,;\, A).$$

If, in addition, X is a martingale with respect to the filtration $(\mathcal{F}_n, n \geq 0)$, show that

$$E(|Y_p| \,;\, A) \leq E(|X_p| \,;\, A).$$

Problem 19. Let $X = (X_0, X_1, \dots)$ be an $\mathbb{R}^\infty$-valued random variable, and T an almost surely finite $\mathbb{Z}^+$-valued random variable. Show that if there exists a real number c such that for almost every ω,

$$\max\{|X_0(\omega)|, |X_1(\omega)|, \dots, |X_{T(\omega)}(\omega)|\} \leq c,$$

then T satisfies the sampling integrability criterion for X.

* **Problem 20.** Show that if $0 \leq X_0 \leq X_1 \leq X_2 \dots$, then (i) implies (ii) in the sampling integrability conditions.

Proposition 10. *Let X be a random sequence of $\mathbb{R}$-valued random variables with finite mean, adapted to a filtration $(\mathcal{F}_n, n \geq 0)$, and let T be an almost surely finite stopping time for the same filtration. Suppose that for each $n > 0$, there exists a finite constant m_n such that*

$$E\big(|X_n - X_{n-1}| \,\big|\, \mathcal{F}_{n-1}\big)(\omega) \leq m_n$$

for almost all ω in the set $[T \geq n]$. Let

$$f(n) = \sum_{i=1}^n m_i.$$

If $E(f \circ T) < \infty$, then T satisfies the sampling integrability conditions for X.

PROOF. For $n \geq 0$, let

$$Y_n = |X_0| + \sum_{i=1}^n |X_i - X_{i-1}|.$$

Then $Y_n \geq |X_n|$ for all n, so it is enough to show that T satisfies the sampling integrability conditions for the random sequence $Y = (Y_0, Y_1, \dots)$. By the Monotone Convergence Theorem,

(24.7) $$E(Y_T) = E(|X_0|) + \sum_{n=1}^\infty E(|X_n - X_{n-1}| \,;\, [T \geq n]).$$

Since $[T \geq n] = [T \leq n - 1]^c \in \mathcal{F}_{n-1}$, it follows from Theorem 4 of Chapter 23 that

$$E(|X_n - X_{n-1}| \, ; \, [T \geq n]) \leq m_n P[T \geq n] \, .$$

Sampling integrability condition (i) follows from $E(|X_0|) < \infty$, (24.7), and the following computation:

$$\sum_{n=1}^{\infty} m_n P[T \geq n] = \sum_{n=1}^{\infty} \sum_{k=n}^{\infty} m_n P[T = k] = \sum_{k=1}^{\infty} \sum_{n=1}^{k} m_n P[T = k]$$

$$= \sum_{k=1}^{\infty} f(k) P[T = k] = E(f \circ T) < \infty \, .$$

Since Y is a nonnegative increasing random sequence, condition (ii) now follows from Problem 20. $\square$

Problem 21. Show that the hypotheses of the preceding proposition are satisfied if $E(T) < \infty$, $E(|X_0|) < \infty$, and $|X_n - X_{n-1}| \leq b$ a.s. for some finite b and all $n > 0$.

Problem 22. Let S be a random walk on $\mathbb{R}$ whose steps have finite mean. Prove that if T is a stopping time for which $E(T) < \infty$, then T satisfies the sampling integrability conditions for S.

Proposition 9 and Proposition 10 both impose conditions on T that ensure that the sequence X does not get sampled too late. It turns out that if the sequence X is a uniformly integrable submartingale and if T is an almost surely finite stopping time, then T cannot be too late. In order to prove this result, we need a lemma concerning uniformly integrable collections of random variables.

Lemma 11. *Let* $\{X_t, t \in \mathcal{T}\}$ *be a uniformly integrable collection of* $\mathbb{R}$-*valued random variables defined on a common probability space, and let* $(A_1, A_2, \dots)$ *be a sequence of events which converges to the empty set. Then*

$$(24.8) \qquad \lim_{n \to \infty} \sup_{t \in \mathcal{T}} E(|X_t| \, ; \, A_n) = 0 \, .$$

PROOF. Fix $\varepsilon > 0$. By the definition of uniform integrability, there exists a finite constant m such that, for all $t \in \mathcal{T}$,

$$E(|X_t| \, ; \, B_t) \leq \varepsilon \, ,$$

where $B_t = [|X_t| \geq m]$. Thus

$$E(|X_t| \, ; \, A_n) \leq \varepsilon + m P(A_n \cap B_t^c) \leq \varepsilon + m P(A_n)$$

for all $t \in \mathcal{T}$. Since $A_n \to \emptyset$, it now follows from the Continuity of Measure Theorem that $E(|X_t| \, ; \, A_n) < 2\varepsilon$ for all sufficiently large n and all $t \in \mathcal{T}$. Since ε is an arbitrary positive quantity, the proof is complete. $\square$

Theorem 12. *Let X be a uniformly integrable submartingale and T an almost surely finite stopping time, both with respect to a filtration $(\mathcal{F}_n, n \geq 0)$. Then T satisfies the sampling integrability conditions for X.*

PROOF. Sampling integrability condition (ii) follows from Lemma 11 with $A_n = [T > n]$.

In order to verify condition (i), we define a random sequence Y by letting $Y_n = X_{T \wedge n}$ for $n \geq 0$. By Proposition 9 and the Optional Sampling Theorem, Y is a submartingale with respect to the filtration $(\mathcal{F}_{T \wedge n}, n \geq 0)$. It is enough to show that Y is uniformly integrable, since condition (i) then follows immediately from the Uniform Integrability Criterion and the fact that $Y_n \to X_T$ a.s. as $n \to \infty$. Use the Doob Decomposition Theorem to write $X = Z + V$, where Z is a martingale and V is an increasing previsible random sequence, both with respect to the filtration $(\mathcal{F}_n, n \geq 0)$. Let $V_\infty = \lim_n V_n$. By the Monotone Convergence Theorem,

$$
\begin{aligned}
E(V_\infty) = \lim_{n \to \infty} E(V_n) &= \lim_{n \to \infty} E(X_n - Z_n) \\
&= \lim_{n \to \infty} [E(X_n) - E(Z_0)] \leq E(|Z_0|) + \sup_n E(|X_n|) \,.
\end{aligned}
$$

By the uniform integrability of the sequence X, this last quantity is finite. It follows from the Uniform Integrability Criterion that the sequence V is uniformly integrable. Therefore $Z = X - V$ is also uniformly integrable. Since

$$
V_T = \lim_{n \to \infty} V_{T \wedge n} \text{ a.s.}\,,
$$

and since $E(V_T) \leq E(V_\infty)$, it also follows from the Monotone Convergence Theorem and the Uniform Integrability Criterion that the sequence $(V_{T \wedge n}, n \geq 0)$ is uniformly integrable. In view of the fact that $Y_n = Z_{T \wedge n} + V_{T \wedge n}$, we only need prove that the sequence $(Z_{T \wedge n} : n \geq 0)$ is uniformly integrable to finish the proof.

Let $z > 0$. By Problem 18,

$$
\begin{aligned}
E(|Z_{T \wedge n}| \,;\, [|Z_{T \wedge n}| \geq z]) &\leq E(|Z_n| \,;\, [|Z_{T \wedge n}| \geq z]) \\
&\leq E(|Z_n| \,;\, [|Z_n| \geq z]) + E(|Z_n| \,;\, [|Z_T| \geq z]) \,.
\end{aligned}
$$

By the definition of uniform integrability and Lemma 11, this last expression converges to 0 uniformly in n as $z \to \infty$, as desired. $\square$

While proving Theorem 12, we have shown the following fact which will be useful later.

Proposition 13. *Let X be a uniformly integrable submartingale. Suppose that $X = Z + V$, where Z is a martingale and V is a monotone previsible random sequence. Then Z and V are uniformly integrable.*

* **Problem 23.** Let X be a submartingale and T an almost surely finite stopping time, both with respect to a filtration $(\mathcal{F}_n, n \geq 0)$. Show that if $E(|X_n|^p)$ is uniformly bounded in n for some $p > 1$, then T satisfies the sampling integrability conditions for X.

24.6. Applications of optional sampling

We have obtained several results which aid in verifying the conditions of the Optional Sampling Theorem. It is time to look at some of the ways in which that theorem may be applied.

Problem 24. Let $S = (0, S_1, S_2, \dots)$ denote a random walk on $\mathbb{R}$ whose steps have mean 0. Set

$$T = \inf\{n \colon S_n > 0\}.$$

Show that $E(T) = \infty$. Hint: Assume that $E(T) < \infty$, then apply Problem 22 and the Optional Sampling Theorem to obtain a contradiction.

Problem 25. Let S be a random walk on $\mathbb{Z}$ starting at 0. Assume that $0 < E(|S_1|) < \infty$. Let T be the time of the first return to 0. Prove that $E(T) = \infty$. Hint: Let $\mu = E(S_1)$ and consider the martingale

$$\big((S_n - \mu n) \colon n \geq 0\big).$$

The preceding problem may also be relevant.

* **Problem 26.** Let X be a supermartingale with respect to some filtration. Assume that $X_0 = f_0$ and that $0 \leq X_n \leq g$ for all n. Show that for any almost surely finite stopping time T,

$$P[X_T = g] \leq \frac{f_0}{g}.$$

The result of the preceding exercise is given an interpretation in the following example.

Example 5. [Gambler's ruin] Let X be a sub- or supermartingale. Assume that $X_0 = f_0$. We imagine that X_n represents the fortune of a gambler after n bets. We will be mostly interested in the case in which X is a supermartingale (the game is unfavorable). As in Example 2, the gambler has the goal of achieving a fortune of at least g before going broke. We will assume that the gambler never bets more money than he has, so that $X_n \geq 0$ for all n. We will also assume that he does not bet more than is needed to reach the goal, so that $X_n \leq g$ for all n. Thus, X is bounded and the Optional Sampling Theorem can be used with any almost surely finite stopping time. Let

$$T = \inf\{n \geq 0 \colon X_n = 0 \text{ or } g\}.$$

Because of the assumptions made on the types of bets allowed, the gambler will stop betting at time T, which is a stopping time. He reaches the goal if and only if T is finite and $X_T = g$. Depending on the strategy chosen, T may or may not be almost surely finite. For example, the gambler may choose a strategy in which all but finitely many of the stakes are of size 0. Nevertheless, in the supermartingale case, we can, for each n, apply the preceding problem to the bounded stopping time $T \wedge n$ to obtain

$$(24.9) \qquad p(f_0, g) = \lim_{n \to \infty} P[X_{T \wedge n} = g] \le \frac{f_0}{g},$$

where $p(f_0, g)$ is the probability that the gambler reaches the goal g when the initial fortune is f_0. Note that this bound depends neither on the strategy employed by the gambler nor on the particular game being played, provided that the sequence X is a supermartingale. Thus, it places an *a priori* limit on the best that the gambler can do in an unfavorable game.

We will now show how to compute $p(f_0, g)$ for the case when the game is Red and Black, g and f_0 are integers (of course, satisfying $0 \le f_0 \le g$), and the strategy is to bet 1 any time the fortune is strictly between 0 and g. Thus, up to the stopping time T, the sub- or supermartingale X describing the sequence of fortunes is a simple random walk, where the probability that any step equals 1 is the probability p of winning a bet. We leave it to the reader to check that, for the strategy just described, T is almost surely finite and, therefore,

$$(24.10) \qquad p(f_0, g) = P[X_T = g] = \frac{E(X_T)}{g}.$$

If $p = \frac{1}{2}$, X is a martingale and the Optional Sampling Theorem applies (for the same reason as in the solution of Problem 26) to give $E(X_T) = f_0$. It then follows from (24.10) that $p(f_0, g) = f_0/g$. Therefore, in this case the *a priori* upper bound in (24.9) is achieved and the strategy of always betting 1 is seen to be optimal.

When $p \ne \frac{1}{2}$, we must use a different method to calculate $p(f_0, g)$, one which appears, in the present context, to be merely a clever trick. We will see in Section 3 of Chapter 26 that this trick is part of a general theory for computing certain probabilities for Markov sequences. We omit the trivial cases $p = 0$ and $p = 1$ and consider the sequence $Y = (Y_n : n \ge 0)$ defined by

$$Y_n = (q/p)^{X_n},$$

where $q = 1 - p$. Let $(\mathcal{F}_n, n \ge 0)$ be the minimal filtration for X. The sequence Y is adapted to this filtration, and, for each n, the increment $X_{n+1} - X_n$ is independent of $\mathcal{F}_n$. It follows from Problem 27 and Proposition 5, both in Chapter 23, that

$$E(Y_{n+1} \mid \mathcal{F}_n) = E\left((q/p)^{[X_{n+1} - X_n]}\right) Y_n = Y_n.$$

Thus Y is a bounded martingale. By the Optional Sampling Theorem,

$$E(Y_T) = E(Y_0) = (q/p)^{f_0}.$$

We may also compute $E(Y_T)$ in terms of $p(f_0, g)$:

$$E(Y_T) = p(f_0, g)(q/p)^g + (1 - p(f_0, g)).$$

Together, these two equalities imply that

$$p(f_0, g) = \frac{(q/p)^{f_0} - 1}{(q/p)^g - 1}.$$

Note that even though this expression is undefined for $p = \frac{1}{2}$, it converges as $p \to \frac{1}{2}$ to the correct value for $p = \frac{1}{2}$ —namely, f_0/g.

Problem 27. What general conclusion can be drawn from the preceding example about the optimization problem for Red and Black?

Problem 28. For the preceding example, in the case that X is a simple (but not necessarily symmetric) random walk, calculate

$$\lim_{g \to \infty} p(f_0, g)$$

and interpret the results.

The next result is useful in computing expected values connected with stopping times for random walks.

Theorem 14. [First Wald Identity] *Let* $S = (S_0 = 0, S_1, S_2, \dots)$ *be a random walk on* $\mathbb{R}$ *whose steps have finite mean. If* T *is a stopping time that satisfies the sampling integrability conditions for* S, *then*

$$(24.11) \qquad\qquad E(S_T) = E(S_1)E(T),$$

where we interpret $0 \cdot \infty$ *as* 0.

PROOF. Let $\mu = E(S_1)$ and, for all n,

$$Y_n = S_n - n\mu.$$

The random sequence $Y = (Y_n \colon n \geq 0)$ is a random walk whose step distribution has mean 0, so it is a martingale. It follows from Proposition 9 and the Optional Sampling Theorem that, for all n,

$$0 = E(Y_{T \wedge n}) = E(S_{T \wedge n}) - \mu E(T \wedge n).$$

By the Monotone Convergence Theorem,

$$(24.12) \qquad\qquad \mu E(T) = \mu \lim_{n \to \infty} E(T \wedge n) = \lim_{n \to \infty} E(S_{T \wedge n}).$$

By the first sampling integrability condition, $E(|S_T|) < \infty$. Thus, second sampling integrability condition (ii) and the Dominated Convergence Theorem imply that

$$\liminf_{n\to\infty} E(|S_{T\wedge n} - S_T|) \leq \liminf_{n\to\infty} E(|S_n| + |S_T|\,;\,[T > n])$$
$$\leq \liminf_{n\to\infty} E(|S_n|\,;\,[T > n]) + \limsup_{n\to\infty} E(|S_T|\,;\,[T > n]) = 0\,.$$

Equation (24.11) follows from this equality and (24.12). □

Remark 1. One might be tempted to prove the preceding theorem by applying the Optional Sampling Theorem to the martingale Y and stopping time T. However, it is not easy to directly deduce that T satisfies the sampling integrability conditions for Y as a consequence of the fact that it satisfies them for S, although, as the reader is asked to show below, such an implication can be proved now that the above proof is complete.

Problem 29. Use the First Wald Identity to show that if T satisfies the sampling integrability conditions for S having steps with a nonzero finite mean, then $E(T) < \infty$. Then prove the implication mentioned in the last sentence of the preceding remark.

The First Wald Identity is not useful for computing $E(T)$ when $E(S_1) = 0$. In this case, the following result may help.

Theorem 15. [Second Wald Identity] *Let* $S = (S_0 = 0, S_1, S_2, \dots)$ *be a random walk on* $\mathbb{R}$ *whose steps have mean 0 and finite variance. If T is a stopping time that satisfies the sampling integrability conditions for the random sequence* $(S_m^2 : m = 0, 1, 2, \dots)$*, then*

$$(24.13) \qquad\qquad \mathrm{Var}(S_T) = \mathrm{Var}(S_1)E(T)\,.$$

PARTIAL PROOF. Note first that the hypotheses imply that T satisfies the sampling integrability conditions for the sequence S. By the First Wald Identity, $E(S_T) = 0$ and hence $\mathrm{Var}(S_T) = E(S_T^2)$. Let

$$Z_n = S_n^2 - n\sigma^2\,, \quad n = 0, 1, 2, \dots\,.$$

By Problem 41 of Chapter 23,

$$E\big(Z_{n+1} \mid \sigma(S_0, \dots, S_n)\big) = Z_n\,,$$

and so $(Z_0, Z_1, Z_2, \dots)$ is a martingale with respect to the minimal filtration of S. The rest of the proof is similar to the proof of the First Wald Identity. □

Problem 30. Complete the proof of the preceding theorem.

Problem 31. Compute $E(T)$ for Example 5 (the Gambler's ruin example) in the case that X is a simple random walk on $\mathbb{Z}$, with $p \in [0,1]$. Compute the limiting value of your answer as $g \to \infty$ and interpret the result.

Problem 32. Let S be a random walk on $\mathbb{R}$ starting at 0, and let T be a stopping time. Assume that $E(T)$ and $E((S_1)^2)$ are both finite. Let $\mu = E(S_1)$. Prove that

$$E([S_T - \mu T]^2) = \text{Var}(S_1)E(T).$$

* **Problem 33.** Let $(S_n : n = 1, 2, \dots)$ be a random walk in $\mathbb{Z}^+$ with step distribution Q given by $Q\{0\} = Q\{1\} = \frac{1}{2}$. Set $T = \inf\{n : S_n = 1\}$, and $T_n = T \wedge n$. Calculate $E([S_{T_n} - \frac{1}{2}T_n]^2)$ and $\text{Var}(S_{T_n})$ and describe the behavior of these quantities as a function of n. Discuss the relevance to Theorem 15. Also, calculate

$$E\big(\text{Var}(S_{T_n} \mid T_n)\big).$$

Repeat the calculations with T_n replaced by T.

Problem 34. Let $S = (S_0 = 0, S_1, S_2, \dots)$ be a random walk on $\mathbb{R}$, and let T be a stopping time. Prove that if $E(|S_1|) < \infty$ and $E(|S_T|) = \infty$, then $E(T) = \infty$. Also, show that if $E(S_1) = 0$, $E(S_1^2) < \infty$, and $E(S_T^2) = \infty$, then $E(T) = \infty$.

Problem 35. Show that the hypothesis on the stopping time T in the Second Wald Identity can be replaced by the hypothesis that $E(T) < \infty$, and that the two hypotheses are equivalent if $\text{Var}(S_1) > 0$.

24.7. Inequalities and convergence results

We now turn our attention to some results that are similar to many of the properties discussed in Part 2 for sums of independent random variables. The Optional Sampling Theorem will play an important role in many of the proofs. We start with a result which can be used in much the same way as the Etemadi Inequality.

Proposition 16. *Let X be a submartingale. Then, for $z > 0$,*

$$P\Big[\max_{0 \le k \le n} X_k > z\Big] \le \frac{1}{z}E(|X_n|).$$

PROOF. Let

$$A = \Big[\max_{0 \le k \le n} X_k > z\Big].$$

Note that

$$A = [X_{T \wedge n} > z],$$

where

$$T(\omega) = \inf\{k \ge 0 : X_k(\omega) > z\}.$$

Then

(24.14) $$zP(A) \le E(X_{T \wedge n} ; A) \le E((X_{T \wedge n})^+).$$

By Proposition 6, $(X_k^+ : k \geq 0)$ is a submartingale, so by the Optional Sampling Theorem,

$$E((X_{T \wedge n})^+) \leq E(X_n^+) \leq E(|X_n|),$$

which, in combination with (24.14), finishes the proof. $\square$

The previous proposition has many variations, of which the following is particularly useful:

Corollary 17. [Kolmogorov Inequality] *Let X be a martingale. Then for all $z > 0$ and $p \geq 1$,*

$$P[\max_{0 \leq k \leq n} |X_k| > z] \leq \frac{1}{z^p} E(|X_n|^p).$$

Problem 36. Prove the preceding corollary.

Problem 37. Let X be a martingale with respect to a filtration $(\mathcal{F}_n, n \geq 0)$. Assume that

$$\sup_{n \geq 0} E(X_n^2) < \infty.$$

Prove that the sequence X converges almost surely and in $\mathbf{L}_2$ to a random variable Y. Also show that $E(Y \mid \mathcal{F}_n) = X_n$ a.s. for all $n \geq 0$. *Hint:* First prove that X is Cauchy in $\mathbf{L}_2$ by showing that

$$E(|X_n - X_m|^2) = E(X_n^2) - E(X_m^2).$$

Then imitate the proof of Proposition 25 of Chapter 12 to get almost sure convergence. Problem 12 of Chapter 8 may be useful.

One of the conclusions in the preceding exercise motivates us to extend the definition of martingale. We speak of a *martingale* $(X_n : n = 0, 1, 2, \ldots, \infty)$ adapted to a filtration $(\mathcal{F}_n : n = 0, 1, 2, \ldots, \infty)$ where, in addition to the usual conditions, we require that X_∞ be $\mathcal{F}_\infty$-measurable with finite mean and that, for each n, both $\mathcal{F}_n \subseteq \mathcal{F}_\infty$ and

(24.15) $E(X_\infty \mid \mathcal{F}_n) = X_n$ a.s.

The extended definitions of *submartingale* and *supermartingale* are similar, with the equality in (24.15) being replaced by $\geq$ and $\leq$, respectively.

In Problem 37, we obtain such a martingale by setting $X_\infty = Y$. The assertion in that problem that $X_n \to Y$ as $n \to \infty$ is a 'martingale convergence theorem'. We will strengthen this result to a version in which there is no $\mathbf{L}_2$-hypothesis.

In order to obtain such an improvement, we will need the notion of 'upcrossing'. Let X be a random sequence of $\mathbb{R}$-valued random variables. For $a < b$ define

$$T_0(\omega) = \inf\{k \geq 0 : X_k(\omega) \leq a\},$$

and having defined T_{2n} for $n \geq 0$, let

$$T_{2n+1}(\omega) = \inf\{k > T_{2n}(\omega) : X_k(\omega) \geq b\}$$

and

$$T_{2n+2}(\omega) = \inf\{k > T_{2n+1}(\omega)\colon X_k(\omega) \le a\}\,.$$

Informally speaking, the times T_n are the successive times at which the sequence X crosses the interval $[a, b]$. When n is odd, T_n is the time at which an *upcrossing*, is completed. For $p > 0$, let

$$U_p = \sharp\{n\colon n \text{ is odd and } T_n \le p\}\,.$$

Thus, U_p, illustrated in Figure 24.1 for $p = 15$, is the number of upcrossings of the interval $[a, b]$ made by the finite sequence $(X_0, \ldots, X_p)$.

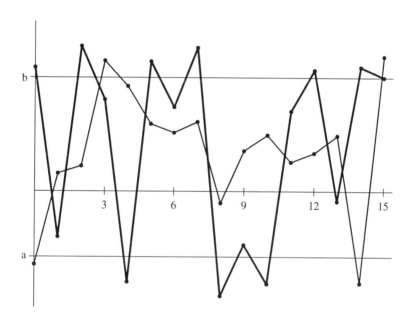

FIGURE 24.1. Instances of exactly two upcrossings by time 15 ($U_{15} = 2$)

Lemma 18. [Doob Upcrossing] *Let X be a submartingale. Fix real numbers $b > a$. Then for all integers $p > 0$,*

$$E(U_p) \le \frac{E((X_p - a)^+)}{b - a}\,.$$

PROOF. It is easily checked that the crossing times T_n defined above are stopping times with respect to the minimal filtration of X. Fix $p > 0$ and let $Y_n = X_{T_n \wedge p}$ for all $n \ge 0$. If $k < U_p$, then $Y_{2k-1} \ge b$ and $Y_{2k} \le a$, so $(b - a) \le Y_{2k-1} - Y_{2k}$. If $k = U_p$, then $Y_{2k-1} \ge b$ and either $Y_{2k} \le a$ or

$Y_{2k} = X_p > a$. In either case, $(b - a) \leq Y_{2k-1} - Y_{2k} + (X_p - a)^+$. If $k > U_p$, then $Y_{2k-1} = Y_{2k} = X_p$, so $Y_{2k-1} - Y_{2k} = 0$. Thus,

$$(b - a)U_p \leq (X_p - a)^+ + \sum_{k=1}^{\infty}(Y_{2k-1} - Y_{2k}) = (X_p - a)^+ + \sum_{k=1}^{p}(Y_{2k-1} - Y_{2k}).$$

By the Optional Sampling Theorem, the expected value of each term in the sum is nonpositive. Thus, the lemma follows by taking the expected value of both sides. $\square$

Theorem 19. [Submartingale Convergence] *Let X be a submartingale. If*

$$\liminf_{n \to \infty} E(|X_n|) < \infty,$$

then there exists a random variable Y such that $E(|Y|) < \infty$ and

$$\lim_{n \to \infty} X_n = Y \text{ a.s.}$$

Problem 38. Prove the previous theorem. *Hint:* Let $Y = \liminf X_n$ and $Z = \limsup X_n$. Use the Doob Upcrossing Lemma to show that for all real numbers $a < b$,

$$P[Y \leq a \text{ and } b \leq Z] = 0.$$

Conclude that $Y = Z$ a.s. Then use the Fatou Lemma to prove that $E(|Y|) < \infty$.

When Theorem 19 is applied to martingales it is usually called the Martingale Convergence Theorem.

Problem 39. Prove that if X is a nonnegative supermartingale, then there exists a random variable Y with finite mean such that $Y = \lim X_n$ a.s.

Problem 40. Find an example to show that it need not be the case that $E(Y) = \lim E(X_n)$ in Theorem 19. *Hint:* Look back at the Double or Nothing example.

The hypotheses in Theorem 19 say nothing about second moments, so, unlike Problem 37, there is no $\mathbf{L}_2$ conclusion. One might have expected an $\mathbf{L}_1$ conclusion, but the preceding exercise shows that an additional hypothesis is needed to get such a conclusion.

Theorem 20. *Let X be a submartingale with respect to a filtration $(\mathcal{F}_n, n = 0, 1, 2, \ldots)$. Then X is uniformly integrable if and only if there exists a random variable Y such that*

(24.16) $$\lim_{n \to \infty} E(|X_n - Y|) = 0.$$

Furthermore, when (24.16) holds,

$$Y = \lim_{n \to \infty} X_n \text{ a.s.}$$

and, with X_∞ *defined to equal* Y, $(X_n: n \in \overline{\mathbb{Z}}^+)$ *is a submartingale with respect to* $(\mathcal{F}_n: n = 0, 1, 2, \ldots, \infty)$, *where* $\mathcal{F}_\infty = \sigma(\mathcal{F}_n, n \geq 0)$.

PROOF. First assume that (24.16) holds for some random variable Y. For sufficiently large n,

$$E(|Y|) \leq E(|X_n - Y|) + E(|X_n|) < \infty,$$

so $E(|Y|) < \infty$. Since

$$E(|X_n|) \leq E(|X_n - Y|) + E(|Y|),$$

it follows that X satisfies the hypotheses of the Submartingale Convergence Theorem. Thus, there exists a random variable Z such that $Z = \lim X_n$ a.s. Since X_n converges to Y in $\mathbf{L}_1$, we must have $Y = Z$ a.s. It now follows from the Uniform Integrability Criterion that X is uniformly integrable.

Now assume that X is a uniformly integrable submartingale. Then $E(|X_n|)$ must be bounded in n, so the conditions of the Submartingale Convergence Theorem hold. Thus, there exists a random variable Y with finite mean such that $Y = \lim X_n$ a.s. By the Uniform Integrability Criterion, $E(|X_n - Y|) \to 0$.

It remains to prove that $E(Y \mid \mathcal{F}_n) \geq X_n$ when the equivalent conditions of uniform integrability and (24.16) hold. Fix $A \in \mathcal{F}_n$ for some $n \geq 0$. Then $E(|Y - X_m|; A) \to 0$ as $m \to \infty$, and so

$$E(E(Y \mid \mathcal{F}_n); A) = E(Y; A) = \lim_{m \to \infty} E(X_m; A)$$
$$= \lim_{m \to \infty} E(E(X_m \mid \mathcal{F}_n); A) \geq E(X_n; A).$$

It follows from Problem 30 of Chapter 8 that $E(Y \mid \mathcal{F}_n) \geq X_n$ a.s., and the proof is complete. $\square$

Corollary 21. *A sequence* X *of* $\mathbb{R}$-*valued random variables is a uniformly integrable martingale with respect to a filtration* $(\mathcal{F}_n, n \geq 0)$ *if and only if there exists a random variable* Y *having finite expectation such that*

$$E(Y \mid \mathcal{F}_n) = X_n \ a.s.$$

for all $n \in \mathbb{Z}^+$. *Furthermore, when this condition holds, then*

$$E(Y \mid \mathcal{F}_\infty) = \lim_{n \to \infty} X_n \ a.s.,$$

where

$$\mathcal{F}_\infty = \sigma(\mathcal{F}_n, n \geq 0).$$

* **Problem 41.** Prove the preceding corollary.

* **Problem 42.** Apply Theorem 20 to the martingale in Example 1.

Example 6. Let X be a sub- or supermartingale. Let us imagine that X represents the fortune of a gambler. In the real world, fortunes are measured in some indivisible monetary unit (such as pennies). Actual fortunes are also bounded by some number b representing the total amount of wealth in the world. Thus, even if we allow the gambler to borrow money, we may assume that X is a sequence of random variables which take integer values in the interval $[-b, b]$ for some $b > 0$. Since X is bounded, it is uniformly integrable. By Theorem 20, $\lim X_n = Y$ a.s. for some random variable Y. Since X is $\mathbb{Z}$-valued, it follows that for almost all ω, there exists a positive integer $N(\omega)$ such that

$$Y(\omega) = X_n(\omega)$$

for all $n \geq N(\omega)$. In other words, if the game can be represented as a martingale, submartingale, or supermartingale (that is, the game is either fair, favorable, or unfavorable), then no matter what strategy is employed, there must be a last play, provided that bets of size 0 are prohibited. The gambler cannot play forever. In practice, this last play occurs when the gambler cannot borrow any more money or the casino goes broke or the gambler 'chooses' to stop gambling (because of death or lack of time, for instance). If X is a supermartingale, then Theorem 20 implies that the expected value of the final fortune is less than or equal to the expected value of the initial fortune, just as we might expect.

Problem 43. In what way must the preceding discussion be modified if we drop the assumption that money comes in indivisible units, but retain the assumption that fortunes are bounded by some finite quantity?

Problem 44. For $n \geq 0$ and $x \in [0, 1)$, let $I(n, x)$ be the unique interval of the form

$$[(k-1)/2^n, k/2^n)$$

that contains x. Let λ denote Lebesgue measure on $[0, 1)$. Prove that for any measurable $\mathbb{R}$-valued function f defined on $[0, 1)$,

$$\lim_{n \to \infty} \frac{1}{\lambda(I(n, x))} \int_{I(n,x)} f \, d\lambda = f(x)$$

for λ-almost all $x \in [0, 1)$ (see Problem 6).

Theorem 22. [Reverse Submartingale Convergence] *Let* $(Z_1, Z_2, \dots)$ *be a reverse submartingale. Then there exists an* $\overline{\mathbb{R}}$-*valued random variable* Y *such that*

$$\lim_{n \to \infty} Z_n = Y \text{ a.s.}$$

PROOF. For $n \geq 0$, let U_n be the number of upcrossings of the interval $[a, b]$ by the sequence $(Z_0, \ldots, Z_n)$, as in the statement of the Doob Upcrossing Lemma. This differs by at most one from the number of upcrossings of the interval $[a, b]$ by the sequence $(X_{-n}, X_{-n+1}, \ldots, X_0)$, where $X_{-k} = Z_k$ for all $k \geq 0$. Let $\mathcal{F}_{-k} = \mathcal{G}_k$ for $k \geq 0$. By the definition of reverse submartingale,

$$E(X_{-n+1} \mid \mathcal{F}_{-n}) \geq X_{-n} \text{ a.s.}$$

for all $n > 0$. Thus, except for the indexing set, the sequence X behaves like a submartingale, and we can apply the Doob Upcrossing Lemma to obtain

$$E(U_n) \leq \frac{1}{b - a} E((X_0 - a)^+).$$

The right side of this expression is independent of n. Now follow the rest of the proof of convergence in the proof of the Submartingale Convergence Theorem (see Problem 38). $\square$

* **Problem 45.** Construct an example of a reverse submartingale whose limit is finite a.s. but does not have finite expectation.

The next lemma shows that a reverse martingale cannot constitute a solution of the preceding exercise.

Lemma 23. *A reverse martingale is uniformly integrable.*

PROOF. Let $(Z_n : n \in \mathbb{Z}^+)$ be a reverse martingale with respect to a reverse filtration $(\mathcal{G}_n : n \in \mathbb{Z}^+)$. For each n let $Y_n = |Z_n|$. We will complete the proof by showing that this nonnegative reverse submartingale is uniformly integrable.

For each $n \in \mathbb{Z}^+$ and each $r \in \mathbb{R}^+$, let $A_{n,r} = [Y_n \geq r]$. Since $Y_n \leq E(Y_0 \mid \mathcal{G}_n)$ and $A_{n,r} \in \mathcal{G}_n$, we have, for any $m > 0$,

$$
\begin{aligned}
E(Y_n ; A_{n,r}) &\leq E\big(E(Y_0 \mid \mathcal{G}_n) ; A_{n,r}\big) = E(Y_0 ; A_{n,r}) \\
&\leq m P(A_{n,r}) + E(Y_0 ; [Y_0 > m]).
\end{aligned}
$$

(24.17)

By the Dominated Convergence Theorem, the second term on the right goes to 0 as $m \to \infty$. Thus, to show that the right side (and therefore the left side) of (24.17) goes to 0 uniformly in n as $r \to \infty$, we only need show that $P(A_{n,r}) \to 0$ as $r \to \infty$ uniformly in n. By the Markov Inequality,

$$P(A_{n,r}) \leq r^{-1} E(Y_n) \leq r^{-1} E(Y_0),$$

the right hand inequality being a consequence of the reverse submartingale property. The rightmost term does not depend on n and goes to 0 as $r \to \infty$, as desired. $\square$

Theorem 24. [Reverse Martingale Convergence] *Let $(Z_n : n \geq 0)$ be a reverse martingale with respect to a reverse filtration $(\mathcal{G}_n : n \geq 0)$. Then there exists a random variable Z_∞ such that the sequence $(Z_n : n \geq 0)$ converges to Z_∞ both in $\mathbf{L}_1$ and a.s. Moreover, $(Z_n : n = 0, 1, 2, \ldots, \infty)$ is a reverse martingale with respect to the filtration $(\mathcal{G}_n : n = 0, 1, 2, \ldots, \infty)$, where*

$$\mathcal{G}_\infty = \bigcap_{n=1}^{\infty} \mathcal{G}_n .$$

Problem 46. Prove the preceding theorem. *Hint:* Use Lemma 23.

Problem 47. Use the preceding theorem to prove the Strong Law of Large Numbers in the finite mean case. *Hint:* Refer back to Example 3.

24.8. † Optimal strategy in Red and Black

We complete this chapter by solving the optimization problem for the game Red and Black, Example 2. Denote by p the probability that the gambler wins a given bet. In the original example, $p = \frac{18}{38}$, but any value of p will do for the first portion of this section and any value between 0 and $\frac{1}{2}$ will do for the latter portion. We assume that the gambler's only goal is to achieve a fortune of at least g, where g is some positive number. The probability that she reaches this goal is a function

$$(x, \gamma) \rightsquigarrow \Pi(x, \gamma)$$

of the initial fortune x and the strategy γ that she employs. We wish to find a strategy γ which is optimal, in the sense that it maximizes $\Pi(x, \gamma)$ for all x. More precisely, an optimal strategy γ is one which has the property that for any nonnegative x and any strategy τ,

$$\Pi(x, \tau) \leq \Pi(x, \gamma) .$$

An important class of strategies is that consisting of all *stationary strategies*, namely those strategies γ such that

$$\gamma_{n+1}(f_0, f_1, \ldots, f_n) = \gamma_1(f_n)$$

for all $n \geq 0$. The stationary strategies are those for which the $(n+1)^{\text{st}}$ bet depends only on the fortune after n bets. It should not be too surprising that in many settings optimal strategies are stationary.

Problem 48. Let X be a random sequence of fortunes corresponding to some initial fortune f_0 and some stationary strategy γ in the game of Red and Black. Prove that the sequence

$$(\Pi(X_0, \gamma), \Pi(X_1, \gamma), \Pi(X_2, \gamma), \ldots)$$

is a martingale with respect to the minimal filtration of X.

We will find that there is a stationary optimal strategy in the game of Red and Black. The following result, which is a special case of what is sometimes known as the Fundamental Theorem of Gambling, provides us with a criterion for recognizing such a strategy.

Theorem 25. *Let γ be a stationary strategy with the property that $\gamma_1(x) = 0$ for all $x \geq g$. Then γ is optimal for the goal of achieving a fortune which is at least g in the game of Red and Black if and only if for all nonnegative x and all $s \in [0, x]$,*

$$(24.18) \qquad \Pi(x, \gamma) \geq p\Pi(x + s, \gamma) + (1 - p)\Pi(x - s, \gamma).$$

PROOF. First suppose that (24.18) fails for some x and s. Let τ be the strategy (which is not stationary) defined by

$$\tau_1(x) = s \quad \text{and} \quad \tau_1(y) = \gamma_1(y), \, y \neq x,$$

and $\tau_n = \gamma_n$ for $n > 1$. Let X be the random sequence of fortunes corresponding to the initial fortune x and the strategy τ in the game of Red and Black. Since the strategies τ and γ are identical after the first bet,

$$\Pi(x, \tau) = p\Pi(x + s, \gamma) + (1 - p)\Pi(x - s, \gamma) > \Pi(x, \gamma).$$

Thus, γ is not optimal.

Now suppose that (24.18) holds for all nonnegative x and all $s \in [0, x]$. Let X be a random sequence of fortunes corresponding to an initial fortune of f_0 and an arbitrary (not necessarily stationary) strategy τ, and let $(\mathcal{F}_n, n \geq 0)$ be the minimal filtration of X. By (24.18),

$$E(\Pi(X_{n+1}, \gamma) \mid \mathcal{F}_n)) = p\Pi(X_n + S_n, \gamma) + (1 - p)\Pi(X_n - S_n, \gamma)$$
$$\leq \Pi(X_n, \gamma),$$

where $S_n = \tau_{n+1}(X_0, \ldots, X_n)$. Thus, the sequence $(\Pi(X_n, \gamma), n \geq 0)$ is a supermartingale. By the Optional Sampling Theorem,

$$(24.19) \qquad \Pi(f_0, \gamma) \geq E(\Pi(X_{T \wedge n}, \gamma))$$

for all $n \geq 0$, where

$$T(\omega) = \inf\{m \geq 0 \colon X_m(\omega) \geq g\}.$$

Clearly, $E(\Pi(X_{T \wedge n}, \gamma)) \geq P[X_{T \wedge n} \geq g]$, so, by (24.19),

$$\Pi(f_0, \gamma) \geq P[X_{T \wedge n} \geq g].$$

Let $n \to \infty$ to obtain $\Pi(f_0, \gamma) \geq \Pi(f_0, \tau)$, as desired. $\quad \square$

Problem 49. Generalize the preceding theorem to roulette games in which there is a greater variety of bets available to the gambler (see Problem 5).

If we are given a strategy γ, Theorem 25 provides us with a way to check whether or not γ is optimal. Unfortunately, it does not tell us how to construct an optimal strategy. For that task, we must make an educated guess. If our goal is to achieve a fortune of at least g in the game of Red and Black with $p < \frac{1}{2}$, our intuition tells us that a strategy calling for many small stakes is not likely to be optimal. We might guess that an optimal strategy involves only large stakes. Of course, we are not allowed to bet stakes which are larger than our fortune, and it seems reasonable that we would not want to bet more than is needed to reach our goal. These considerations lead to the stationary strategy γ known as *bold play*:

$$\gamma_{n+1}(f_0, \ldots, f_n) = \begin{cases} f_n \wedge (g - f_n) & \text{if } 0 \le f_n < g \\ 0 & \text{if } f_n \ge g. \end{cases}$$

Theorem 26. *In the supermartingale versions of Red and Black, bold play is an optimal strategy for achieving a fortune of at least g .*

PROOF. By a change of scale, if necessary, we may assume that $g = 1$. Since the case $p = 0$ is trivial, we may also assume that $p > 0$. Let $h(x) = \Pi(x, \gamma)$, where γ denotes bold play. In other words, $h(x)$ is the probability that the gambler who starts with an initial fortune x reaches the goal by following bold play. By Theorem 25, to show the optimality of bold play it suffices to show that

$$(24.20) \qquad h(x) \ge ph(x + s) + (1 - p)h(x - s), \quad 0 \le s \le x \le 1.$$

Let us first determine some properties of the function h. It is obvious that h takes values in $[0, 1]$ and that $h(0) = 0$ and $h(1) = 1$. If the gambler has a fortune between 0 and $\frac{1}{2}$ and plays boldly, her fortune doubles with probability p and vanishes with probability $1 - p$. From this and a similar consideration involving fortunes between $\frac{1}{2}$ and 1, we deduce that

$$(24.21) \qquad h(x) = \begin{cases} ph(2x) & \text{if } 0 \le x \le \frac{1}{2} \\ p + (1 - p)h(2x - 1) & \text{if } \frac{1}{2} \le x \le 1. \end{cases}$$

We will now prove by induction on $n \ge 0$ that (24.20) holds for $x \in [0, 1]$ of the form

$$x = \frac{k}{2^n}.$$

The case $n = 0$ is easily checked. Now consider the case $n = 1$. We must show that for all $s \in [0, \frac{1}{2}]$,

$$h(\tfrac{1}{2}) \ge ph(\tfrac{1}{2} + s) + (1 - p)h(\tfrac{1}{2} - s).$$

By (24.21)

$$h(\tfrac{1}{2}) - ph(\tfrac{1}{2} + s) - (1 - p)h(\tfrac{1}{2} - s)$$
$$= p - p[p + (1 - p)h(2s)] - (1 - p)ph(1 - 2s)$$
$$= p(1 - p)\left[1 - \left(h(2s) + h(1 - 2s)\right)\right].$$

That this last expression is nonnegative follows from the fact in Problem 26 that $h(x) \leq x$ for $x \geq 0$.

We are now ready for the inductive step. Fix $n > 1$ and let $x = k/2^n$ for some positive integer $k < 2^n$. Also, choose $s \in [0, x]$. We will consider several cases. First, assume that $x + s \leq \tfrac{1}{2}$. Then (24.21) and the inductive hypothesis imply that

$$ph(x + s) + (1 - p)h(x - s) = p(ph(2x + 2s)) + (1 - p)(ph(2x - 2s))$$
$$\leq ph(2x) = h(x).$$

The case in which $x - s \geq \tfrac{1}{2}$ is treated in a similar fashion. Now assume that $x \leq \tfrac{1}{2}$ and $x + s \geq \tfrac{1}{2}$. Then by (24.21)

$$h(x) - ph(x + s) - (1 - p)h(x - s)$$
$$= ph(2x) - p[p + (1 - p)h(2x + 2s - 1)] - (1 - p)ph(2x - 2s)$$
$$= p[h(2x) - p - (1 - p)h(2x + 2s - 1) - (1 - p)h(2x - 2s)].$$

In this case, it is necessarily true that $x \geq \tfrac{1}{4}$, since otherwise $s \leq \tfrac{1}{4}$ and then $x + s \leq \tfrac{1}{2}$. Therefore, $2x \geq \tfrac{1}{2}$, so further applications of (24.21) imply that the right side of the last expression equals

$$p[p + (1 - p)h(4x - 1) - p - (1 - p)h(2x + 2s - 1) - (1 - p)h(2x - 2s)]$$
$$= p(1 - p)[h(4x - 1) - h(2x + 2s - 1) - h(2x - 2s)]$$
$$= (1 - p)\left[h\left(2x - \tfrac{1}{2}\right) - ph\left((2x - \tfrac{1}{2}) + (2s - \tfrac{1}{2})\right) - ph\left((2x - \tfrac{1}{2}) - (2s - \tfrac{1}{2})\right)\right].$$

Since $p \leq \tfrac{1}{2}$, the right side of this expression is bounded below by

$$(1 - p)\left[h\left(2x - \tfrac{1}{2}\right) - ph\left((2x - \tfrac{1}{2}) + (2s - \tfrac{1}{2})\right) - (1 - p)h\left((2x - \tfrac{1}{2}) - (2s - \tfrac{1}{2})\right)\right],$$

which is nonnegative, by the inductive hypothesis. The verification of (24.20) is similar for the case in which $x \geq \tfrac{1}{2}$ and $x - s \leq \tfrac{1}{2}$.

Thus we have checked (24.20) for x in a set which is dense in $[0, 1]$, so in order to show that (24.20) holds for all $x \in [0, 1]$, it is enough to prove that h is continuous. To show this, we will use the fact (Problem 50 following this proof) that h is increasing. In particular, at each x, the right and left limits $h(x+)$ and $h(x-)$ exist. Since we know that (24.20) holds for x in a dense subset of $[0, 1]$, we can take limits from the left to obtain

$$h(x-) \geq ph((x + s)-) + (1 - p)h((x - s)-)$$

for all $0 \le s \le x \le 1$. Now let $s \searrow 0$ to see that

$$h(x-) \ge ph(x+) + (1-p)h(x-).$$

Since $h(x+) \ge h(x-)$ and $p > 0$, it follows immediately that $h(x+) = h(x-)$, so h is continuous. $\square$

Problem 50. Prove that the function h of the preceding proof is increasing. *Hint:* Compare the fortunes of two gamblers who have different initial fortunes and who simultaneously follow the bold play strategy at the same roulette wheel.

CHAPTER 25
Renewal Sequences

The main subject of this chapter is a class of random sequences that are defined in terms of random walks $T = (T_m \colon m = 0, 1, 2, \ldots)$ in $\overline{\mathbb{Z}}^+$ satisfying $T_{m+1}(\omega) \geq 1 + T_m(\omega)$ (with the understanding that $\infty \geq 1 + \infty$). The random sequence $X = (X_n \colon n \in \mathbb{Z}^+)$, defined by

$$X_n(\omega) = \begin{cases} 1 & \text{if } n = T_m(\omega) \text{ for some } m \geq 0 \\ 0 & \text{otherwise} \end{cases}$$

is known as the *renewal sequence* corresponding to T. The correspondence between T and X is one-to-one; the random walk T corresponding to a renewal sequence X is given by $T_0(\omega) = 0$ and, for $m > 0$,

(25.1) $$T_m = \inf\{n > T_{m-1} \colon X_n = 1\},$$

where, as usual, the infimum of the empty set is understood to be $+\infty$.

In this chapter we will see that renewal sequences are important tools for studying random walks. In Chapter 26, they will be used in the analysis of Markov sequences. In physical applications, renewal sequences are used to model situations in which a task is repeated indefinitely, with the quantities T_n representing the (integer) times at which the repetitions of the task begin. Our assumption that T be a random walk means that the lengths of time taken to complete the repetitions are iid. The integers $n > 0$ for which $X_n = 1$ are called *renewal times,* consistent with practice in renewal theory of using the noun 'time' for the target of the random walk and thus for the domain of the renewal sequence. (Notice that we do not call 0 a renewal time even though $X_0 = 1$ a.s.) If $T_m < \infty$, then it is the mth renewal time and the time difference $T_m - T_{m-1}$ is the mth *waiting time.* The distribution of the first renewal time T_1 is known as the *waiting time distribution.* This distribution is, of course, the same as the step distribution of the random walk T.

25.1. Basic criterion

Suppose that X is a random sequence of 0's and 1's. One might ask if it is a renewal sequence, or equivalently, whether the sequence T defined by $T_0 = 0$ and (25.1) is a random walk. The following proposition gives a useful criterion for a random sequence to be a renewal sequence, a criterion that does not involve the corresponding random walk. Roughly speaking, it says that renewal sequences are sequences of 0's and 1's that start over independently after each occurrence of a 1; that is, they 'regenerate' themselves.

Proposition 1. *A random sequence $X = \{X_0, X_1, X_2, \ldots\}$ of 0's and 1's is a renewal sequence if and only if $X_0 = 1$ a.s. and*

$$
\begin{aligned}
(25.2) \quad & P[X_n = x_n \text{ for } 0 < n \le r + s] \\
& = P[X_n = x_n \text{ for } 0 < n \le r]\, P[X_{n-r} = x_n \text{ for } r < n \le r + s]
\end{aligned}
$$

for all positive integers r and s and sequences $(x_1, \ldots, x_{r+s})$ of 0's and 1's such that $x_r = 1$.

* **Problem 1.** Prove the preceding proposition.

Note that if $x_r = 0$, (25.2) does not necessarily hold for a renewal sequence $(X_0, X_1, X_2, \ldots)$.

If X is a renewal sequence, the random set

$$\Sigma_X = \{n \colon X_n = 1\},$$

is called a *regenerative set in* $\mathbb{Z}^+$. The study of renewal sequences is equivalent to the study of regenerative sets in $\mathbb{Z}^+$.

The following result describes many regenerative sets connected with an arbitrary random walk in $\mathbb{R}$. Figure 25.1 illustrates two of these sets, one located along the horizontal axis and the other located along the vertical axis in the graph of a random walk.

Proposition 2. *Let $S = (S_0 = 0, S_1, S_2, \ldots)$ be a random walk in $\mathbb{R}$. Then each of the following conditions defines a regenerative set in $\mathbb{Z}^+$:*

$$
\begin{aligned}
(25.3) \quad & \{n \colon S_n = 0\}\,; \\
(25.4) \quad & \{n \colon S_n > S_k \text{ for } 0 \le k < n\}\,; \\
(25.5) \quad & \{n \colon S_n < S_k \text{ for } 0 \le k < n\}\,; \\
(25.6) \quad & \{n \colon S_n \ge S_k \text{ for } 0 \le k < n\}\,; \\
(25.7) \quad & \{n \colon S_n \le S_k \text{ for } 0 \le k < n\}.
\end{aligned}
$$

If S is a random walk in $\mathbb{Z}$, then each of the following conditions also defines a regenerative set in $\mathbb{Z}^+$:

(25.8) $$\{S_n : S_n > S_k \text{ for } 0 \leq k < n\};$$
(25.9) $$\{-S_n : S_n < S_k \text{ for } 0 \leq k < n\}.$$

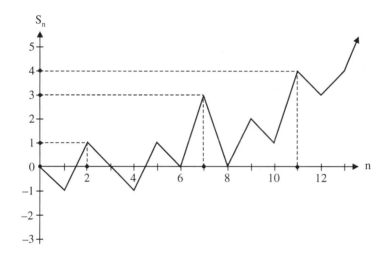

FIGURE 25.1. The regenerative sets $\{n : S_n > S_k, 0 \leq k < n\}$ and $\{S_n : S_n > S_k, 0 \leq k < n\}$.

Problem 2. Check that Theorem 12 of Chapter 11 applies directly to show that (25.3) is a regenerative set. Then prove that one of (25.4), (25.5), (25.6), and (25.7) is a regenerative set. Finally, show that one of (25.8) and (25.9) is a regenerative set. (The rest of the proof of the proposition will be omitted.)

Later in this chapter, we will analyze some of the regenerative sets in Proposition 2.

25.2. Renewal measures and potential measures

Let $(X_0, X_1, X_2, \dots)$ denote a renewal sequence. The random σ-finite measure N given by

$$N(B) = \sum_{n \in B} X_n$$

is called the *renewal measure* of the renewal sequence $X = (X_0, X_1, X_2, \dots)$; it describes the *number of renewals* occurring during a time set B. By the

Monotone Convergence Theorem, the function

$$B \rightsquigarrow E(N(B)) = E\left(\sum_{n \in B} X_n\right)$$

is a (nonrandom) σ-finite measure; it is the *potential measure* of the renewal sequence X. The density of the potential measure U of X with respect to counting measure is the *potential sequence* of X. Thus, the potential sequence $(u_0, u_1, u_2, \ldots)$ satisfies $u_0 = 1$ and

$$u_n = U\{n\} = P[X_n = 1].$$

Let R be a waiting time distribution for a renewal sequence X corresponding to a random walk T in $\overline{\mathbb{Z}}^+$. For any set $B \subseteq \mathbb{Z}^+$ and any nonnegative integer m, it follows from the definitions that

$$R^{*m}(B) = P[T_m \in B],$$

where R^{*0} is the delta distribution at 0. This observation will be used in the proof of the following result that relates the potential measure of a renewal sequence to the waiting time distribution.

Proposition 3. *The potential measure U and the waiting time distribution R of a renewal sequence are related via*

$$U(B) = \sum_{m=0}^{\infty} R^{*m}(B), \quad B \subseteq \mathbb{Z}^+,$$

where R^{*0} is the delta distribution at 0.

PROOF. By the Fubini Theorem,

$$U(B) = \sum_{n \in B} E(X_n) = \sum_{n \in B} E\left(\sum_{m=0}^{\infty} I_{[T_m = n]}\right)$$

$$= \sum_{m=0}^{\infty} \sum_{n \in B} E(I_{[T_m = n]})$$

$$= \sum_{m=0}^{\infty} \sum_{n \in B} R^{*m}\{n\} = \sum_{m=0}^{\infty} R^{*m}(B)$$

for all $B \subseteq \mathbb{Z}^+$. □

We will explore further the relationship between a potential measure U and the corresponding waiting time distribution R by using generating functions. Consider a distribution Q on $\mathbb{Z}^+$. Recall that the probability generating function ρ of Q is given by

(25.10) $$\rho(s) = \sum_{n=0}^{\infty} Q\{n\}s^n, \quad 0 \le s \le 1.$$

In order to extend the definition (25.10) to certain measures other than probability measures, we remove 1 from the domain. The *measure generating function* of a measure U on $\mathbb{Z}^+$ that has a bounded density $(u_0, u_1, \ldots)$ with respect to counting measure is the function

$$(25.11) \qquad\qquad s \rightsquigarrow \sum_{n=0}^{\infty} u_n s^n, \quad 0 \le s < 1.$$

The following two problems extend essential parts of the theory of probability generating functions to measure generating functions.

Problem 3. Suppose that U and V are measures on $\mathbb{Z}^+$ with bounded densities. Show that if the measure generating functions of U and V are equal, then $U = V$.

Problem 4. Let U and V be measures on $\mathbb{Z}^+$. Suppose that U has a bounded density $(u_0, u_1, \ldots)$ and that V is a finite measure. Define a measure $U * V$ by

$$(U * V)\{n\} = \sum_{k=0}^{n} u_k v_{n-k},$$

where $(v_0, v_1, \ldots)$ denotes the density of V. Prove that $U * V$ has a bounded density whose measure generating function is the product of the measure generating functions of U and V.

The measure $U * V$, described in the preceding problem is called the *convolution* of U and V; it can also be written with the finite measure first: $V * U$.

When we speak of the measure generating function of a measure that happens to be a probability measure it is to be understood that the domain is $[0, 1)$ rather than $[0, 1]$ and that the probability attached to the one-point set $\{\infty\}$ does not enter the formula for the generating function. The following theorem would fail to be true in general were we to allow $s = 1$.

Theorem 4. *Let R and U be the waiting time distribution and potential measure, respectively, of a renewal sequence and let φ denote the measure generating function of R. Then the measure generating function of U is $1/(1 - \varphi)$.*

* **Problem 5.** Prove the preceding theorem.

Theorem 5. *Let R and U be the waiting time distribution restricted to $\mathbb{Z}^+$ and potential measure, respectively, of a renewal sequence. Then*

$$U = (U * R) + \delta_0,$$

where δ_0 denotes the standard delta distribution.

PROOF. By Problem 3, it suffices to prove equality of the corresponding measure generating functions. By Theorem 4 the measure generating function of U is $1/(1 - \varphi)$, where φ denotes the measure generating function of R. So, the measure generating function of $U * R$ is $\varphi/(1 - \varphi)$. Since the measure generating function of δ_0 is the constant function 1, our task is to show that

$$\frac{1}{1 - \varphi} = \frac{\varphi}{1 - \varphi} + 1.$$

But this follows immediately by simple algebra. □

Problem 6. Rewrite Theorem 5 using the densities of U and R rather than U and R themselves.

Theorem 6. *For $i = 1, 2$, let R_i and U_i denote the waiting time distribution and potential measure of a renewal sequence, and let φ_i denote the measure generating function of R_i. Then the following three conditions are equivalent:*

   (i) $R_1 = R_2$;
   (ii) $U_1 = U_2$;
   (iii) $\varphi_1 = \varphi_2$.

PROOF. By Proposition 3, the first equality implies the second. By Theorem 4, the second equality implies

$$(1 - \varphi_1)^{-1} = (1 - \varphi_2)^{-1},$$

and hence $\varphi_1 = \varphi_2$. That the third equality implies the first is the content of Problem 3. □

25.3. Examples

We begin with a series of exercises intended to help the reader become familiar with the concepts introduced so far.

Problem 7. For $p \in [0, 1]$ let X be a renewal sequence whose waiting time distribution R is given by

$$R(\{2\}) = p = 1 - R(\{1\}).$$

Calculate the potential sequence of X.

* **Problem 8.** Show that the sequence $(1, 0, \frac{1}{4}, \frac{1}{4}, \frac{1}{4}, \dots)$ is a potential sequence and find the corresponding waiting time distribution.

Problem 9. For which values of q is the sequence $(1, 0, q, q, q, \dots)$ a potential sequence? For each such q find the density of the corresponding waiting time distribution.

Problem 10. Describe all potential sequences beginning with two 1's: $(1, 1, \dots)$.

Problem 11. For $q \in [0, 1]$ let R be the waiting time distribution defined by $R(\{1\}) = q = 1 - R(\{\infty\})$. Find the corresponding potential sequence. Describe the corresponding regenerative set, minimizing the use of formulas in the description.

Problem 12. For $p \in (0, 1)$ let X be a renewal sequence with waiting time density $n \rightsquigarrow (1 - p)p^{n-1}$, $n > 0$. Calculate the corresponding potential sequence.

Problem 13. Use a sequence of independent Bernoulli random variables to create the renewal sequence described in the preceding exercise.

* **Problem 14.** Let $(X_0 = 1, X_1, X_2, \dots)$ be the renewal sequence described in Problem 12. Show that $(1, 1 - X_1, 1 - X_2, \dots)$ is a renewal sequence and calculate its potential sequence.

* **Problem 15.** For $p \in (0, 1)$ let $(Y_n : n = 1, 2, \dots)$ be an iid sequence of random variables with

$$P(\{\omega : Y_n(\omega) = 1\}) = 1 - P(\{\omega : Y_n(\omega) = -1\}) = p.$$

Let $(S_n : n = 0, 1, \dots)$ denote the corresponding sequence of partial sums. In each of (i)-(vii) below, we define a random set Σ in $\mathbb{Z}^+$. Determine in each case whether Σ is a regenerative set. For those cases in which the answer is affirmative, calculate the corresponding potential sequence and waiting time distribution.

(i) $\{0\} \cup \{n : Y_n = 1\}$;
(ii) $\{0\} \cup \{n : n = 1 \text{ and } Y_1 = 1, \text{ or } n > 1 \text{ and } Y_n = Y_{n-1} = 1\}$;
(iii) $\{0\} \cup \{n : n = 1 \text{ and } Y_1 = 1 \text{ or } n > 1 \text{ and } Y_n \neq Y_{n-1}\}$;
(iv) $\{0\} \cup \{n > 1 : -Y_{n-1} = Y_n = 1\}$;
(v) $\{0\} \cup \{n : n = 2 \text{ and } Y_2 = Y_1 = 1 \text{ or } n > 2 \text{ and } Y_n = Y_{n-1} \neq Y_{n-2}\}$;
(vi) $\{0\} \cup \{n > 2 : -1 = Y_n = Y_{n-1} \neq Y_{n-2}\}$;
(vii) $\{0\} \cup \{n > 0 : S_{n-1} < 0 \text{ and } S_n = 0\}$.

Hint: Partial fractions and the Binomial Theorem are useful tools when working with generating functions.

The following theorem identifies another way in which renewal sequences arise—namely as term-by-term products of independent renewal sequences.

Theorem 7. Let $X = (X_0, X_1, \dots)$ and $Y = (Y_0, Y_1, \dots)$ be independent renewal sequences. Then the term-by-term product $XY = (X_0 Y_0, X_1 Y_1, \dots)$ is a renewal sequence the potential sequence of which is the term-by-term product $(u_0 v_0, u_1 v_1, \dots)$ of the potential sequences $(u_0, u_1, \dots)$ and $(v_0, v_1, \dots)$ of X and Y.

PROOF. To show that XY is a renewal sequence we will apply Proposition 1. Thus, for positive integers r and s and a sequence $(z_1, \dots, z_{r+s})$ of 0's and 1's such that $z_r = 1$, we calculate

(25.12) $$P[X_n Y_n = z_n \text{ for } 0 < n \leq r + s] =$$
$$\sum P[X_r = Y_r = 1 \text{ and } X_n = x_n, Y_n = y_n \text{ for } 0 < n \leq r + s, n \neq r],$$

where the sum is over all choices of sequences (x_n) and (y_n) whose terms equal 0 or 1 and which have the property that $z_n = x_n y_n$. Since X and Y are independent, each term in the sum can be rewritten as a product:

$$P[X_r = 1 \text{ and } X_n = x_n \text{ for } 0 < n \leq r + s, n \neq r]$$
$$\cdot P[Y_r = 1 \text{ and } Y_n = y_n \text{ for } 0 < n \leq r + s, n \neq r].$$

Since each of X and Y is a renewal sequence, this product of two factors can, according to Proposition 1, be rewritten as a product of four factors:

$$P[X_r = 1 \text{ and } X_n = x_n \text{ for } 0 < n < r] \, P[X_{n-r} = x_n \text{ for } r < n \leq r + s]$$
$$\cdot P[Y_r = 1 \text{ and } Y_n = y_n \text{ for } 0 < n < r] \, P[Y_{n-r} = y_n \text{ for } r < n \leq r + s].$$

Using the independence of X and Y again, we combine the first and third factors and the second and fourth factors to obtain

$$P[X_r = Y_r = 1 \text{ and } X_n = x_n, Y_n = y_n \text{ for } 0 < n < r]$$
$$\cdot P[X_{n-r} = x_n, Y_{n-r} = y_n \text{ for } r < n \leq r + s].$$

Inserting this product into the summation (25.12) and then summing gives

$$P[X_r Y_r = 1 \text{ and } X_n Y_n = z_n \text{ for } 0 < n < r]$$
$$\cdot P[X_{n-r} Y_{n-r} = z_n \text{ for } r < n \leq r + s],$$

thus completing the argument showing that XY is a renewal sequence.

The value at n of the potential sequence of the renewal sequence XY equals

$$P[X_n = Y_n = 1] = P[X_n = 1] \, P[Y_n = 1] = u_n v_n,$$

as desired. □

Problem 16. Restate the preceding theorem using the language of regenerative sets rather than that of renewal sequences.

Problem 17. Apply Theorem 7 to a variety of pairs of independent renewal sequences.

Problem 18. Theorem 7 asserts that the term-by-term minimum of two independent renewal sequences is a renewal sequence. Show that the term-by-term maximum of two independent renewal sequences is not necessarily a renewal sequence.

25.4. Renewal theory: a first step

Our goal in this section is to obtain information about the random variable $N(B)$ for various sets B. In all cases, B will be a set of consecutive nonnegative integers, so $N(B)$ will equal the number of renewals that occur during some time interval.

Our first result concerns the case in which B is of the form $\{k+1, \ldots, k+l\}$ for some integers $k, l \geq 0$, and gives a formula for $P[N(B) > 0]$, which is the probability that a renewal occurs between the times $k+1$ and $k+l$.

Theorem 8. *Let R, $(u_n : n = 0, 1, 2, \ldots)$, and N denote the waiting time distribution, the potential sequence, and the renewal measure, respectively, of some renewal sequence. For k and l nonnegative integers,*

$$P\big[N[k+1, k+l] > 0\big] = \sum_{n=0}^{k} R[k+1-n, k+l-n] u_n$$

$$= 1 - \sum_{n=0}^{k} R[k+l+1-n, \infty] u_n$$

$$= \sum_{n=k+1}^{k+l} R[k+l+1-n, \infty] u_n \, .$$

PARTIAL PROOF. We will only prove the first of the three equalities. Denoting the random walk corresponding to X by T, we have

$$P\big[N[k+1, k+l] > 0\big] = \sum_{j=0}^{\infty} P[T_j \leq k, \, k+1 \leq T_{j+1} \leq k+l]$$

$$= \sum_{j=0}^{\infty} \sum_{n=0}^{k} R[k+1-n, \, k+l-n] \, R^{*j}\{n\}$$

$$= \sum_{n=0}^{k} R[k+1-n, \, k+l-n] \, u_n \, ,$$

the last step being based on the Fubini Theorem. $\square$

Problem 19. Complete the proof of the preceding theorem by proving the other two equalities.

* **Problem 20.** Apply Theorem 8 to renewal sequences with geometrically distributed waiting times. Then arrive at the same conclusion by using Problem 13.

Problem 21. Apply Theorem 8 to the renewal sequence of Problem 11.

Problem 22. Prove the following corollary.

Corollary 9. *For each* $k \in \mathbb{Z}^+$,

$$\sum_{n=0}^{k} R[k+1-n, \infty] u_n = 1.$$

The preceding result will be used in the proofs of Theorem 15 and Theorem 16 in this chapter.

We now turn our attention to the quantity $N(\mathbb{Z}^+)$, which equals the total number of renewals made by the renewal sequence. The following result, which generalizes Corollary 13 of Chapter 11, shows how to determine the distribution of $N(\mathbb{Z}^+)$ when either $U(\mathbb{Z}^+)$ or $R\{\infty\}$ is known. The proof is requested in Problem 23.

Theorem 10. *Let X be a renewal sequence with potential measure U and waiting time distribution R. If $U(\mathbb{Z}^+) < \infty$, then the distribution of $N(\mathbb{Z}^+)$ is of geometric type, supported by $\{1, 2, \dots\}$, with mean $U(\mathbb{Z}^+)$. If $U(\mathbb{Z}^+) = \infty$, then $N(\mathbb{Z}^+) = \infty$ a.s. In any case,*

$$U(\mathbb{Z}^+) = \frac{1}{R\{\infty\}},$$

where it is to be understood that $1/0 = \infty$.

According to the preceding result, the random variable $N(\mathbb{Z}^+)$ either is finite a.s. or else it is infinite a.s. The following definition introduces terminology for these two cases.

Definition 11. A renewal sequence and the corresponding regenerative set are *transient* if the corresponding renewal measure is finite almost surely and they are *recurrent* otherwise.

Problem 23. Prove Theorem 10.

Problem 24. Find a formula for the distribution function of the maximum member of a transient regenerative set in terms of its potential measure and waiting time distribution. In particular, check that your formula correctly gives the probability that the maximum equals 0.

Our next theorem concerns the asymptotic properties of the random variable $N(\{0, 1, \dots, n\})$ as $n \to \infty$.

Theorem 12. [Strong Law for Renewal Sequences] *Let N, U, and $\mu \in [1, \infty]$ denote the renewal measure, the potential measure, and the mean waiting time, respectively, of some renewal sequence. Then*

$$\lim_{n \to \infty} \frac{U(\{0, 1, \dots, n\})}{n} = \lim_{n \to \infty} \frac{N(\{0, 1, \dots, n\})}{n} = \frac{1}{\mu} \quad a.s.$$

PROOF. Since $N(\{0,1,\ldots,n\}) \le n$, the first equality is a consequence of the second equality and the Bounded Convergence Theorem.

To prove the second equality let T be the random walk corresponding to U. Then

$$n \in [T_{m-1}(\omega), T_m(\omega)) \implies \frac{N(\{0,1,\ldots,n\})(\omega)}{n} = \frac{m}{n}.$$

By the Strong Law of Large Numbers, $T_m/m \to \mu$ a.s. as $m \to \infty$. The second equality now follows easily. $\square$

In terms of renewal sequences X, the preceding theorem says that

$$\lim_{n \to \infty} \frac{\sum_{k=0}^{n} X_k}{n} = \frac{1}{\mu} \text{ a.s.},$$

a fact justifying the term 'Strong Law'.

If X is transient, we already know from Theorem 10 that $N(\mathbb{Z}^+) < \infty$ a.s. and $\mu = \infty$ (since $R\{\infty\} > 0$), so Theorem 12 does not tell us anything new in that case.

If X is recurrent, Theorem 12 tells us that except for sample points ω in some null event, the sequence $X(\omega)$ has a well-defined 'frequency' of 1's. This frequency is 0 if the mean waiting time is infinite; otherwise, it is the positive real number $1/\mu$. It is useful to have terminology to distinguish these last two cases.

Definition 13. A recurrent renewal sequence is *null recurrent* if the mean waiting time is infinite; otherwise, it is *positive recurrent*.

After we prove the Renewal Theorem later in this chapter, we will obtain a criterion in terms of the potential measure U for distinguishing null recurrence and positive recurrence.

25.5. Delayed renewal sequences

Let $S = (S_0 = 0, S_1, S_2, \ldots)$ be a random walk in $\mathbb{Z}$ and for a fixed integer z consider the random sequence $X = (X_0, X_1, X_2, \ldots)$ defined by

$$X_n = I_{\{z\}} \circ S_n.$$

Thus, the sequence X marks the visits made by S to z. If $z \ne 0$, X is not a renewal sequence. However, once the point z is reached by S, then thereafter X behaves like a renewal sequence.

To make this idea precise and to introduce an appropriate name we consider an arbitrary random sequence X of 0's and 1's, and let $T = \inf\{n \ge 0 \colon X_n = 1\}$. Then X is called a *delayed renewal sequence* if either $P[T = \infty] = 1$, or $P[T = \infty] < 1$ and conditioned on the event $[T < \infty]$, the sequence

$$Y = (X_{T+n} \colon n = 0, 1, 2, \ldots)$$

is a renewal sequence. The random variable T is called the *delay time*, and the distribution of T is the *delay distribution* of X. The waiting time distribution of

Y is also the *waiting time distribution* of X. Clearly a delayed renewal sequence X is a renewal sequence if and only if $P[X_0 = 1] = 1$.

Problem 25. For a simple random walk S in $\mathbb{Z}$, let X be the delayed renewal sequence defined by $X_n = I_{\{z\}} \circ S_n$ for some $z \in \mathbb{Z}$. Find the probability generating function of the delay distribution of X and, from it, the probability that the delay equals ∞.

Problem 26. State and prove a version of Theorem 12 for delayed renewal sequences.

Problem 27. Prove the following analogue of Proposition 1: Let $W = (W_n : n \geq 0)$ be a renewal sequence. A random sequence $X = \{X_0, X_1, X_2, \ldots\}$ of 0's and 1's is a delayed renewal sequence with the same waiting time distribution as W if and only if

$$P[X_n = x_n \text{ for } 0 \leq n \leq r + s]$$
$$= P[X_n = x_n \text{ for } 0 \leq n \leq r] \, P[W_{n-r} = x_n \text{ for } r < n \leq r + s]$$

for all positive integers r and s and sequences $(x_0, \ldots, x_{r+s})$ of 0's and 1's such that $x_r = 1$.

There is an important technique, known as *coupling* that can be used to compare two delayed renewal sequences, provided they have the same waiting time distribution. Previously in this book, versions of this technique occurred in Theorem 12 of Chapter 11 and Proposition 5 of Chapter 14, although they were not identified as such.

Let X and Y be independent delayed renewal sequences, both with waiting time distribution R. Let $M = \inf\{n \geq 0 : X_n = Y_n = 1\}$, and define

$$Z_n = \begin{cases} X_n & \text{if } n \leq M \\ Y_n & \text{if } n > M. \end{cases}$$

We will prove that the random sequence $Z = (Z_n : n \geq 0)$ has the same distribution as X. This result is particularly useful when $M < \infty$ a.s., because in that case, the two sequences Y and Z clearly have the same asymptotic behavior as $n \to \infty$, allowing us to conclude that X and Y have the same asymptotic behavior.

Proposition 14. *Let X, Y, and Z be as above. Then X and Z have the same distribution.*

PROOF. Fix an arbitrary finite sequence $x_0, x_1, \ldots, x_n$ of 0's and 1's. We must show that

$$P[X_0 = x_0, \ldots, X_n = x_n] = P[Z_0 = x_0, \ldots, Z_n = x_n].$$

Let $M' = n \wedge \inf\{k \colon Y_k = x_k = 1\}$. Since X and Y are independent, X and (Y, M') are independent. It is easily checked that for $\omega \in [Z_0 = x_0, \ldots, Z_n = x_n]$, $M'(\omega) = n \wedge M(\omega)$. These two facts imply that

$$P[Z_0 = x_0, \ldots, Z_n = x_n]$$

$$= \sum_{k=0}^{n} P[M' = k, Z_0 = x_0, \ldots, Z_n = x_n]$$

$$= \sum_{k=0}^{n} P[M' = k, X_0 = x_0, \ldots, X_k = x_k, Y_{k+1} = x_{k+1}, \ldots, Y_n = x_n]$$

$$= P[X_0 = x_0, \ldots, X_k = x_k]\left(\sum_{k=0}^{n} P[M' = k, Y_{k+1} = x_{k+1}, \ldots, Y_n = x_n]\right).$$

Note that the event $[M' = k]$ is empty for $k < n$ such that $x_k = 0$. For such k we have by Problem 27 that

(25.13)
$$\begin{aligned} P[M' = k, Y_{k+1} &= x_{k+1}, \ldots, Y_n = x_n] \\ &= P[M' = k]P[W_1 = x_{k+1}, \ldots, W_{n-k} = x_n], \end{aligned}$$

where $W = (W_n \colon n \geq 0)$ is a (nondelayed) renewal sequence with waiting time distribution R. This equality also holds trivially when $k = n$, since then it reduces to $P[M' = n] = P[M' = n]$. Substituting (25.13) into the expression preceding it gives

$$P[Z_0 = x_0, \ldots, Z_n = x_n]$$

$$= \sum_{k=0}^{n}\left(P[M' = k]P[X_0 = x_0, \ldots, X_k = x_k]P[W_0 = x_k, \ldots, W_{n-k} = x_n]\right).$$

By Problem 27, this last expression equals

$$\sum_{k=0}^{n} P[M' = k]P[X_0 = x_0, \ldots, X_n = x_n] = P[X_0 = x_0, \ldots, X_n = x_n],$$

as desired. □

The previous result will be used in the next section in the proof of the Renewal Theorem. In the proof of the positive recurrent case of that result, the special delayed renewal sequences identified in the following theorem are particularly important.

Theorem 15. *Let R be a waiting distribution with finite mean μ and let X be a delayed renewal sequence corresponding to R and the delay distribution D defined by*

$$D\{n\} = \frac{1}{\mu} R[n+1, \infty), \quad n \in \mathbb{Z}^+.$$

Then $P[X_n = 1] = 1/\mu$ for all nonnegative integers n.

PROOF. By Corollary 20 of Chapter 4, D is a distribution, as asserted in the theorem.

Let $(u_0, u_1, u_2, \dots)$ be the potential sequence corresponding to R (and thus to the corresponding undelayed renewal sequence) and let T_0 have distribution D. By definition, $P[X_n = 1 \mid T_0 = k] = u_{n-k}$ for $k \le n$ and $= 0$ for $k > n$. Hence,

$$P[X_n = 1] = E\big(P(X_n = 1 \mid T_0)\big) = \sum_{k=0}^{n} u_{n-k} P[T_0 = k] = \frac{1}{\mu} \sum_{k=0}^{n} u_{n-k} R[k+1, \infty),$$

which by Corollary 9 and the fact that $R\{\infty\} = 0$, equals $1/\mu$, as desired. $\quad\square$

Problem 28. Let $X = (X_0, X_1, X_2 \dots)$ be a delayed renewal sequence of the form described in Theorem 15. Prove that the various sequences $(X_k, X_{k+1}, X_{k+2}, \dots)$ all have the same distribution. [Thus X is an example of a 'stationary sequence', to be defined in Chapter 28. It is a 'stationary renewal sequence'.]

* **Problem 29.** For a delayed renewal sequence of the form described in Theorem 15, find a formula for the mean of the delay in terms of the mean (assumed to be finite) and variance (not necessarily finite) of the waiting time distribution.

Problem 30. [Bus-stop paradox] Let $X = (X_0, X_1, X_2, \dots)$ be a delayed renewal sequence as described in Theorem 15, and set $V = \min\{n > 0 : X_n = 1\}$. Suppose that X represents the times at which buses come to a particular bus stop. Then V can be interpreted as the time one must wait for the next bus provided that one arrives at the bus stop just after time 0. Find necessary and sufficient conditions on the mean and variance of the waiting time distribution in order that

$$E(V \mid X_0 = 0) > E(V \mid X_0 = 1).$$

Thus when the requested conditions hold, the conditional mean of the time to wait for the next bus is less if one arrives just as a bus is pulling away from the bus stop than it is if one arrives with no bus in sight. Give an intuitive explanation that agrees with the mathematics and makes this seemingly paradoxical conclusion appear natural. If one goes to the bus stop just after some fixed positive (integral) time rather than just after time 0, are the conclusions the same?

25.6. The Renewal Theorem

Recall, from Theorem 12, that the potential measure U of a renewal sequence satisfies

$$\lim_{n \to \infty} \frac{U(\{0, 1, \dots, n\})}{n} = \frac{1}{\mu},$$

where $\mu \in [1, \infty]$ denotes the mean of the waiting time distribution. Thus, relative to n, $(n-1)/\mu$ and n/μ are close to $U(\{0, 1, \dots, n-1\})$ and $U(\{0, 1, \dots, n\})$, respectively, when n is large. By subtracting we arrive at the conjecture that $1/\mu$ is a good approximation of $u_n = U(\{n\})$ when n is large. Although this conjecture is false as it stands, the next theorem gives a precise statement in this direction. For the statement we need a concept. The *period* of a renewal sequence is the greatest common divisor of the support of the corresponding potential measure, with the greatest common divisor being defined to equal ∞ in case the support equals $\{0\}$. It is easy to see that the period is also the greatest common divisor of the support of the waiting time distribution. A renewal sequence is *aperiodic* if its period is 1.

Theorem 16. [Renewal] *Let $\mu \in [1, \infty]$ and $(u_n : n \geq 0)$ be the mean waiting time and potential sequence, respectively, of a renewal sequence with finite period γ. Then*

$$\lim_{k \to \infty} u_{k\gamma} = \frac{\gamma}{\mu}.$$

BEGINNING OF PROOF. In the transient case $\mu = \infty$; also $\sum u_n < \infty$ by Theorem 10 and, hence, $u_n \to 0$. This completes the proof for the transient case and we hereafter restrict consideration to the recurrent case.

Treating multiples of γ is equivalent to treating all members of $\mathbb{Z}^+$ for the renewal distribution $\hat{R}$ defined by $\hat{R}(B) = R(\gamma B)$ and having mean μ/γ. Therefore, we may, without loss of generality, assume aperiodicity throughout the remainder of the proof.

Clearly $u_n > 0$ if and only if n is a linear combination of members of the support of R with $\mathbb{Z}^+$-valued coefficients. It follows from the forthcoming Lemma 18 that $u_n > 0$ for all sufficiently large n. $\square$

For the remainder of the proof, we consider the positive recurrent and null recurrent cases separately.

END OF PROOF IN POSITIVE RECURRENT CASE. By Theorem 15 there exists a delayed renewal sequence $Y = (Y_n : n = 0, 1, 2, \dots)$ independent of X having waiting time distribution R and satisfying $P[Y_n = 1] = 1/\mu$ for all n. Set

$$M(\omega) = \inf\{n : X_n(\omega) = Y_n(\omega) = 1\},$$

and define $Z = (Z_n : n = 0, 1, 2, \dots)$ by

$$Z_n(\omega) = \begin{cases} X_n(\omega) & \text{if } n \leq M(\omega) \\ Y_n(\omega) & \text{if } n > M(\omega). \end{cases}$$

By Proposition 14, X and Z have the same distribution.

Our goal is to prove that $M < \infty$ a.s., for then it will follow that $Y_n - Z_n \to 0$ and hence that $E(Y_n) - E(Z_n) = \frac{1}{\mu} - E(X_n) \to 0$, thus finishing the proof in the positive recurrent case. Let $K = \min\{k : Y_k = 1 \text{ and } u_k > 0\}$, a random variable that is finite almost surely because X is recurrent and $u_k > 0$ for all sufficiently large k. By considering the random walk and delay time corresponding to the sequence Y, it is easy to see that X, K, and the random sequence $(Y_{n+K} : n \geq 0)$ are independent, and that $(Y_{n+K} : n \geq 0)$ has the same distribution as X. Thus, by Theorem 7, the random sequence $(X_n Y_{n+K} : n \geq 0)$ is a renewal sequence that is independent of K, and the corresponding potential sequence is $\{u_n^2 : n \geq 0\}$. The positive recurrence of X and Theorem 12 imply that $u_n \not\to 0$ and hence that $\sum u_n^2 = \infty$, so this renewal sequence is recurrent by Theorem 10.

In other words, with probability 1 there exist infinitely many times n such that $X_n = Y_{n+K} = 1$. Let $S_0 = 0 < S_1 < S_2, \ldots$ be that random sequence of times. Since this sequence is strictly increasing, $S_{2kj} + k < S_{2k(j+1)}$ for all $k > 0$ and $j \geq 0$. Therefore, repeated applications of Proposition 1 imply that for each positive integer k, the events

$$[X_{S_{2kj}+k} = 1], \quad j = 1, 2, \ldots$$

are independent and have probability u_k. Since K is independent of X, the events

$$[X_{S_{Kj}+K} = 1], \quad j = 1, 2, \ldots$$

are conditionally independent given K, with conditional probability u_K. By the definition of K, $u_K > 0$, so the conditional version of the Borel-Cantelli Lemma implies that with probability 1, infinitely many of these events occur. Each such occurrence corresponds to a time at which the sequences X and Y are simultaneously equal to 1, so $M < \infty$ a.s. as desired. $\square$

END OF PROOF IN NULL RECURRENT CASE. For a proof by contradiction, suppose that for some $\varepsilon > 0$ there are infinitely many n for which $u_n > \varepsilon$.

Since $\mu = \infty$, Corollary 20 of Chapter 4 implies the existence of an integer q such that

$$(25.14) \qquad\qquad \sum_{n=1}^{q+1} R[n, \infty) > \frac{2}{\varepsilon}.$$

Consider $q + 1$ independent delayed renewal sequences $Y^{(r)}, 0 \leq r \leq q$, with $Y^{(0)} = X$ and, for $1 \leq r \leq q$, $Y^{(r)}$ having a nonrandom delay equal to r and having the same waiting time distribution as X. Loosely mimicking the proof for the case of positive recurrence, we set

$$M(\omega) = \inf\{n : Y_n^{(r)}(\omega) = 1 \text{ for } 0 \leq r \leq q\}$$

and define

$$Z_n^{(r)}(\omega) = \begin{cases} Y_n^{(r)}(\omega) & \text{if } n \leq M(\omega) \\ X_n(\omega) & \text{if } n \geq M(\omega). \end{cases}$$

As for the case of positive recurrence, here each $Z^{(r)}$ has the same distribution as the corresponding $Y^{(r)}$. Since $u_n > \varepsilon$ for infinitely many n, $\sum u_n^{(q+1)} = \infty$. Hence, $P[M < \infty] = 1$, as can be seen by using an argument similar to that used for the case of positive recurrence. Therefore, there exists a positive integer $k > q$ such that

$$u_k = P[X_k = 1] > \varepsilon \quad \text{and} \quad P[Z_k^{(r)} = 1] > \frac{\varepsilon}{2} \quad \text{for } 1 \leq r \leq q.$$

It follows that

$$P[Y_k^{(r)} = 1] > \frac{\varepsilon}{2} \quad \text{for } 0 \le r \le q;$$

that is,

(25.15) $$u_{k-r} > \frac{\varepsilon}{2} \quad \text{for } 0 \le r \le q.$$

From (25.14) and (25.15) we obtain

$$\sum_{n=1}^{q+1} u_{k+1-n} R[n, \infty) > 1.$$

We have reached the desired contradiction since this last inequality is inconsistent with Corollary 9, as can be seen by a change of variables. □

The following is an immediate consequence of Theorem 10 and the Renewal Theorem.

Corollary 17. *Let X be a renewal sequence with potential sequence $(u_n : n \ge 0)$. Then X is recurrent if and only if $\sum_n u_n = \infty$, in which case X is null recurrent if and only if $\lim_{n \to \infty} u_n = 0$.*

We conclude this section with a proof of the number-theoretical fact used near the beginning of the proof of the Renewal Theorem.

Lemma 18. *Let B be a set of positive integers having greatest common divisor equal to 1. Then there exists a positive integer k such that every integer $n > k$ can be written as a linear combination of members of B with coefficients belonging to $\mathbb{Z}^+$.*

PROOF. Let H denote the set of linear combinations of members of B having coefficients in $\mathbb{Z}^+$, and set $G = \{n - k : k, n \in H\}$. It is easy to check that G also equals the set of all linear combinations of members of B having coefficients in $\mathbb{Z}$. The Euclidean algorithm gives a method of representing the greatest common divisor of two positive integers as a linear combination of those integers with coefficients in $\mathbb{Z}$. Repeated use (needed only finitely many times even if $\sharp(B) = \infty$) of this fact shows that 1 is a linear combination of members of B having coefficients in $\mathbb{Z}$. That is, $1 \in G$ and so $1 = q - p$ for some members p and q of H.

We will finish the proof by using mathematical induction to show that every $n \ge pq$ can be represented as a linear combination of p and q having coefficients in $\mathbb{Z}^+$. It is clearly true if $n = pq$. Suppose, as the induction hypothesis, that $k = ap + bq \ge pq$, and $a, b \in \mathbb{Z}^+$. Then

$$k + 1 = (a - 1)p + (b + 1)q,$$

thus finishing the induction argument in case $a \neq 0$. The following calculation showing that the case $a = 0$ can be reduced to the case $a \neq 0$ completes the proof:

$$k = bq = qp + (b - p)q \,,$$

which is a linear combination of p and q with positive coefficient on p and nonnegative coefficient on q. $\square$

25.7. † Applications to random walks

We have already seen that there are several regenerative sets that are useful for understanding random walks in $\mathbb{R}$. In this section, we will investigate properties of some of these sets.

We begin by treating the zero set of an arbitrary random walk $S = (S_n : n \in \mathbb{Z}^+)$ in $\mathbb{Z}$, that is, by treating the regenerative set defined in (25.3). The corresponding renewal sequence X is defined by $X_n = I_{\{0\}} \circ S_n$. The random walk S is called *transient, positive recurrent,* or *null recurrent* according to which phrase applies to X.

The potential sequence $(u_n : n \geq 0)$ of X is easily seen to be related to the step distribution Q of S by $u_n = Q^{*n}\{0\}$. By Theorem 13 of Chapter 13 we obtain

$$u_n = \frac{1}{2\pi} \int_{-\pi}^{\pi} \beta^n(v)\, dv \,,$$

where β is the characteristic function of Q. Multiplying by s^n and summing gives

$$\frac{1}{1 - \varphi(s)} = \sum_{n=0}^{\infty} s^n u_n = \frac{1}{2\pi} \int_{-\pi}^{\pi} \frac{1}{1 - s\beta(v)}\, dv \,,$$

where φ is the measure generating function of the waiting time distribution of X. We state this conclusion formally. (The essence of this result is also found in Section 10 of Chapter 13. However, the corollary after the result adds something new that is not found in Chapter 13.)

Theorem 19. *Let* $S = (S_0 = 0, S_1, S_2, \dots)$ *be a random walk in* $\mathbb{Z}$*, and let* $T_1 = \inf\{n > 0 : S_n = 0\}$*. Then the distribution of* T_1 *has measure generating function* φ *given by*

$$\varphi(s) = 1 - \left[\frac{1}{2\pi} \int_{-\pi}^{\pi} \frac{1}{1 - s\beta(v)}\, dv \right]^{-1} \,,$$

where β *is the characteristic function of the step distribution of* S.

Corollary 20. *Except for the identically zero random walk, every random walk on* $\mathbb{Z}$ *is either transient or null recurrent.*

Problem 31. Prove the preceding corollary.

Problem 32. Apply Theorem 19 to each simple random walk in $\mathbb{Z}$ to determine transience and null recurrence, and to calculate the distribution of the first return time to 0 of the random walk. (Of course, the results of this problem have also been obtained earlier by other means. See, for example, Problem 33 of Chapter 5.)

We now consider the regenerative sets (25.4) and (25.7) identified in Proposition 2 as arising from a random walk $(S_0 = 0, S_1, S_2, \dots)$ in $\mathbb{R}$:

$$\{n \colon S_n > S_k \text{ for } 0 \le k < n\},$$

the set of *strict ascending ladder times;* and

$$\{n \colon S_n \le S_k \text{ for } 0 \le k < n\},$$

the set of *weak descending ladder times.* Our first goal is to obtain a relation between the probability generating functions φ^{++} and φ^- of the waiting time distributions for the strict ascending ladder times and the weak descending ladder times. Here is the key tool.

Lemma 21. *Let* $n \ge 0$*. For a random walk in* $\mathbb{R}$ *the probability that* n *is a strict ascending ladder time equals the probability that there is no positive weak descending ladder time less than or equal to* n*.*

PROOF. The result is obvious for $n = 0$. Fix $n > 0$. Denote the partial sums and the steps of the random walk by S_k and Y_k, respectively. The event that n is a strict ascending ladder time equals

$$\left[\sum_{k=m}^{n} Y_k > 0 \quad \text{for } m = 1, 2, \dots, n \right].$$

By using the fact that the random variables Y_k are independent and identically distributed we see that we can replace Y_k by Y_{n+1-k} for $k = 1, 2, \dots, n$ without changing the probability. Thus the probability that n is a strict ascending ladder time equals

$$P\left[\sum_{k=1}^{m} Y_k > 0 \quad \text{for } m = 1, 2, \dots, n \right],$$

which is the probability that none of the integers $1, 2, \dots, n$ is a weak descending ladder time. $\square$

Theorem 22. *The probability generating functions* φ^{++} *and* φ^- *of the waiting time distributions for the regenerative sets of strict ascending ladder times and weak descending ladder times of a random walk in* $\mathbb{R}$ *satisfy the relation*

$$[1 - \varphi^{++}(s)][1 - \varphi^-(s)] = 1 - s, \quad 0 \le s \le 1.$$

PROOF. The equality is clear for $s = 1$, so we take $s \in [0,1)$. For $n = 0, 1, 2, \ldots$, let u_n^{++} denote the probability that n is a strict ascending ladder time and r_n^- the probability that n is the smallest positive weak ascending ladder time. Since $[1 - \varphi^{++}]^{-1}$ is the measure generating function of the sequence $(u_n^{++}, n = 0, 1, \ldots)$ and, according to the preceding lemma, $u_n^{++} = 1 - \sum_{k=1}^{n} r_k^-$, we conclude that

$$
\frac{1}{1 - \varphi^{++}(s)} = \sum_{n=0}^{\infty} \left[1 - \sum_{k=1}^{n} r_k^- \right] s^n
$$

$$
= \frac{1}{1-s} - \sum_{k=1}^{\infty} \sum_{n=k}^{\infty} r_k^- s^n
$$

$$
= \frac{1}{1-s} - \sum_{k=1}^{\infty} r_k^- \frac{s^k}{1-s} = \frac{1 - \varphi^-(s)}{1-s}. \quad \square
$$

Corollary 23. *Either the set of strict ascending ladder times and the set of weak descending ladder times are both null recurrent or one of these sets is transient and the other is positive recurrent.*

Problem 33. Prove the preceding corollary.

Problem 34. Formulate results analogous to the preceding theorem and corollary for the regenerative sets (25.5) and (25.6) identified in Proposition 2: the sets of strict descending ladder times and of weak ascending ladder times.

Problem 35. For random walks in $\mathbb{R}$ other than those are almost surely identically zero prove that the set of strict ascending ladder times and the set of weak ascending ladder times are both transient or both null recurrent or both positive recurrent.

In case the step distribution of a random walk in $\mathbb{R}$ assigns zero probability to each one-point set, each weak ascending ladder time is, with probability one, a strict ascending ladder time; and similarly for descending ladder times. In this case we have

$$
[1 - \varphi^{++}(s)] [1 - \varphi^{--}(s)] = 1 - s,
$$

where φ^{--} is the probability generating function of the waiting time distribution for the regenerative set of strict descending ladder times. Since $\varphi^{++} = \varphi^{--}$ for a symmetric random walk we obtain the following fact, remarkable in that the conclusion does not depend on further properties of the step distribution of the random walk.

Corollary 24. *The probability generating function of the waiting time distribution for the regenerative set of strict (or weak) ascending ladder times of a symmetric random walk whose step distribution assigns zero probability to each one-point set is the function $s \rightsquigarrow 1 - \sqrt{1-s}$.*

* **Problem 36.** Calculate the potential measure and waiting time distribution for the regenerative set described in Corollary 24.

Example 1. Suppose that the support of the step distribution of the random walk $S = (S_n : n \geq 0)$ consists of integers smaller than 2. Thus,

$$P[S_1 = j] = q_j \quad \text{for } j = \ldots, -2, -1, 0, 1,$$

and

$$\sum_{j=-\infty}^{1} q_j = 1.$$

Let

$$h(s) = \sum_{k=0}^{\infty} q_{1-k} s^k, \quad 0 \leq s \leq 1,$$

and let φ^{++} be as in Theorem 22. For $0 \leq s < 1$,

$$\varphi^{++}(s) = \sum_{n=0}^{\infty} r_n^{++} s^n,$$

where, for $n = 1, 2, \ldots$, the number r_n^{++} denotes the probability that the smallest positive strict ascending ladder time equals n.

We want to relate φ^{++} to h. For $m = 1, 2, \ldots$, let $r_{n,m}^{++}$ denote the probability that n is the mth smallest positive strict ascending ladder time. Then, for $0 \leq s < 1$,

$$[\varphi^{++}(s)]^m = \sum_{n=0}^{\infty} r_{n,m}^{++} s^n,$$

valid even for $m = 0$ with the conventions $r_{0,0}^{++} = 1$ and $r_{n,0}^{++} = 0$ for $n > 0$.

If $S_1(\omega) = 1 - m$ for some $m > 0$, then in order that $n > 1$ be the first positive strict ascending ladder time, it is necessary and sufficient that $n - 1$ be the mth positive strict ascending ladder time for the random walk with steps $(S_2 - S_1), (S_3 - S_2), \ldots$, a random walk that is independent of S_1. Therefore,

$$r_n^{++} = \sum_{m=0}^{\infty} q_{1-m} r_{n-1,m}^{++}, \quad n > 0.$$

Multiply by s^n and sum to obtain

$$\varphi^{++}(s) = \sum_{m=0}^{\infty} q^{1-m} [\varphi^{++}(s)]^m = sh(\varphi^{++}(s)).$$

Thus, we confront the issue of whether, for each $s \in [0, 1]$, the equation $x = sh(x)$ has a unique solution in $[0, 1]$, which would then necessarily be the value of $\varphi^{++}(s)$. We note that $h(1) = h(1-) = 1$. Thus, for $s < 1$, the function $x \leadsto sh(x)$ is concave up, continuous, nonnegative at 0, and less than 1 at 1. It follows that the equation $x = sh(x)$ has a unique solution for each $s \in [0, 1)$, and that solution equals $\varphi^{++}(s)$.

The equation $x = h(x)$ is of interest. Of course, the solution $x = 1$ is the value of $\varphi^{++}(1)$, but in case of more than one solution, it is the smallest of the solutions that necessarily equals $\varphi^{++}(1-)$. So, $h(x) = x$ has a unique solution, namely $x = 1$, if and only if the set of strict ascending ladder times is recurrent. It is easy to see that there is a unique solution if and only if $h(0) > 0$ and $h'(1) \leq 1$. Since $h'(1) = 1 - E(S_1)$, we conclude that the set of strict ascending ladder times is recurrent if $E(S_1) \geq 0$ and $P[S_1 = 0] < 1$, as was expected. In the case of transience, the probability that there is at least one positive ascending ladder time equals the smallest solution of $x = h(x)$.

It is worth noting that only when $h(x) = x$ for all $x \in [0, 1]$, a trivial case, does the equation $h(x) = x$ have more than two solutions.

Problem 37. For a random walk S as described in the preceding example with $E(S_1) < 0$, calculate the distribution of $\sup\{S_n : n \in \mathbb{Z}^+\}$.

Problem 38. For a random walk S as described in Example 1 with $E(S_1) \geq 0$, find the probability that there is at least one positive weak descending ladder time.

* **Problem 39.** For each simple random walk in $\mathbb{Z}$, find the waiting time distributions and potential measures for the following four regenerative sets: the random sets of strict ascending ladder times, of strict descending ladder times, of weak ascending ladder times, and of weak descending ladder times.

Problem 40. Let $q \in (0, 1)$ and $c \in [0, 1 - q]$. Follow the instructions of the preceding problem for the random walk whose step distribution is given by

$$P[S_1 = -k] = cq^k, \quad k = 0, 1, 2, \ldots$$
$$P[S_1 = 1] = \frac{1 - q - c}{1 - q}.$$

For each q find those c for which the regenerative sets of all four types of ladder times are recurrent.

CHAPTER 26
Time-homogeneous Markov Sequences

Markov sequences are so important that several chapters could be devoted to their study. We will mainly limit our coverage to two kinds of topics: (i) those that involve instructive applications of material presented earlier in this book and (ii) those that lay the groundwork for material on continuous-time Markov processes presented later in the book.

26.1. Transition operators and discrete generators

Most of this section is concerned with new terminology and notation. If $X = (X_n : n = 0, 1, 2, \ldots)$ is a Markov sequence of Ψ-valued random variables, then the space Ψ is called the *state space* of X. Throughout this chapter, all state spaces will be Borel spaces. We identify the subscript n on X_n with *time*, so that an event of the form $[X_n \in B]$ might be read as "the state of the Markov sequence X at time n is a member of B" or "X_n is in B at time n".

Recall from Chapter 22 that the distribution of a time-homogeneous Markov sequence $(X_0, X_1, \ldots)$ in a Borel space Ψ is characterized by an initial distribution R_0 and a transition function R satisfying

$$P[X_{n+1} \in B \mid (X_0, \ldots, X_n)] = R(X_n, B).$$

In this chapter, we often write Q_n for the distribution of X_n, $n \geq 0$, so that the initial distribution will usually be denoted by Q_0 rather than R_0. Also, all Markov sequences treated in this chapter are time-homogeneous, so we will usually drop the adjective 'time-homogeneous'.

We now introduce some new objects that will take over the role of the transition function R.

Definition 1. Let Ψ be a Borel space on which a collection of distributions $\mathcal{M} = (\mu_x : x \in \Psi)$ is defined. Assume that for each measurable set $B \subseteq \Psi$, the function $x \rightsquigarrow \mu_x(B)$ is measurable. The *left transition operator* T corresponding to the collection $\mathcal{M}$ of *transition distributions* operates on the space of bounded measurable $\mathbb{R}$-valued functions on Ψ and is given by

$$(T(f))(x) = \int_\Psi f(y)\,\mu_x(dy)\,.$$

The corresponding *right transition operator,* also denoted by T, operates on the space of probability measures on Ψ and is given by

$$((Q_0)T)(B) = \int_\Psi \mu_x(B)\,Q_0(dx)\,.$$

To avoid a clutter of symbols, parentheses will usually be omitted from the notation introduced in the preceding definition. Thus, one might write $(Tf)(x)$ or $Tf(x)$, and similarly, $(Q_0T)(B)$ or even $Q_0T(B)$.

We usually refer to a *transition operator* without the modifiers 'left' or 'right' since an explicit or implicit identification of domain serves to distinguish left from right.

The connection between transition functions and transition operators is as follows. Given a transition function R, the transition distributions are given by $\mu_x = R(x,\cdot), x \in \Psi$. The collection of transition distributions uniquely determines the corresponding left and right transition operators T, so a transition function R uniquely determines a corresponding transition operator T.

On the other hand, since $\mu_x(B) = TI_B(x)$ for $x \in \Psi$ and measurable sets B, it is clear that a left transition operator T uniquely determines a corresponding collection of transition distributions, which in turn determines a transition function R. Similarly a right transition operator T determines the collection of transition distributions and the transition function since $\mu_x = \delta_x T$. Thus, there is a one-to-one correspondence between transition operators (either left or right) and transition functions R. We are led to reformulate the definition of time-homogeneous Markov sequence in the following equivalent way.

Definition 2. Let $(\Psi,\mathcal{G})$ be a Borel space and T a transition operator for Ψ, defined in terms of a collection $(\mu_x : x \in \Psi)$ of probability measures on Ψ. A random sequence X of Ψ-valued random variables adapted to a filtration $(\mathcal{F}_n : n \geq 0)$ is a *time-homogeneous Markov sequence* with respect to that filtration, having *state space* Ψ and *transition operator* T, if for all $n \geq 0$, μ_{X_n} is the conditional distribution of X_{n+1} given $\mathcal{F}_n$.

Given a transition operator T and initial distribution Q_0, the existence of a corresponding Markov sequence (with respect to the minimal filtration) is guaranteed by an easy application of Theorem 3 of Chapter 22, and the distribution of X is uniquely determined by T and Q_0. However, there is no uniqueness in the converse direction (see Problem 7).

Problem 1. Prove that the transition operator T takes bounded measurable functions to bounded measurable functions and probability measures to probability measures.

Problem 2. If ν is a probability measure on a Borel space $(\Psi, \mathcal{G})$ and $f \colon \Psi \to \mathbb{R}$ is a bounded measurable function, let νf denote $\int f \, d\nu$. With this notation, show that if T is a transition operator, then $(Q_0 T)f = Q_0(Tf)$.

Problem 3. Let T be an operator that takes the space of bounded measurable functions on a Borel space Ψ to itself. Show that T is a left transition operator if and only if: (i) T is linear, (ii) $Tf \geq 0$ whenever $f \geq 0$, (iii) $T1 = 1$, where 1 denotes the function that is identically equal to 1; (iv) $\lim_{n \to \infty} T f_n(x) = 0$ for every x whenever $(f_n \colon n = 1, 2,)$ is a decreasing sequence satisfying $\lim_{n \to \infty} f_n(x) = 0$ for every x.

Problem 4. Let T be a measurable operator that takes the Borel space of probability measures on a Borel space Ψ to itself. Show that T is a right transition operator if and only if $(b\nu_1 + (1-b)\nu_2)T = b(\nu_1 T) + (1-b)(\nu_2 T)$ whenever $0 \leq b \leq 1$.

* **Problem 5.** Let $(X_n \colon n \geq 0)$ be a Markov sequence with respect to a filtration $(\mathcal{F}_n \colon n \geq 0)$, with transition operator T. Let Q_n denote the distribution of X_n, and f a bounded measurable function on the state space. Show that, for all $n \geq 0$,

$$Q_n T = Q_{n+1} \quad \text{and} \quad E(f \circ X_{n+1} \mid \mathcal{F}_n) = (Tf) \circ X_n \text{ a.s.}$$

Problem 6. Show that if T is a transition operator, then T^k is a transition operator for all $k \geq 1$, where T^k denotes the k-fold composition of T with itself. Then generalize Problem 5 to show that, for all $n \geq 0$,

$$Q_n T^k = Q_{n+k} \quad \text{and} \quad E(f \circ X_{n+k} \mid \mathcal{F}_n) = (T^k f) \circ X_n \text{ a.s.}$$

Problem 7. For the state space $\Psi = \{0, 1\}$, find two transition operators T and $\widehat{T}$ such that the (deterministic) sequence $X = (0, 0, 0, \dots)$ is a Markov sequence with transition operator T and also a Markov sequence with transition operator $\widehat{T}$.

Technically speaking, a time-homogeneous Markov sequence consists of a random sequence, a state space, a filtration, a transition operator, and the distribution of the first term of the sequence. The state space is often implicit, and an unmentioned filtration is usually the minimal filtration. The transition operator is usually mentioned explicitly, and if one speaks of a particular random sequence in connection with two different transition operators, as in the preceding exercise, then one is speaking of two different Markov sequences.

While both the transition operator T and the initial distribution Q_0 are needed to uniquely determine the distribution of a Markov sequence, we will see that many of the most important properties of a given sequence have more to do with properties of T than with properties of Q_0. Thus it is common to simultaneously consider all of the Markov sequences corresponding to a given choice of T. In doing so, it is convenient to use a single underlying measurable space. Given a Borel space $(\Psi, \mathcal{G})$, the corresponding *canonical sample space* is

$$\Omega = \bigotimes_{n=0}^{\infty} \Psi \,,$$

to which we assign the usual product σ-field, denoted by $\mathcal{F}$. Thus each sample point $\omega \in \Omega$ is a sequence $(\omega_0, \omega_1, \dots)$ of members of Ψ. In this context, we will always use the symbol X to denote the random sequence $(X_0, X_1, \dots)$, where for each $n \geq 0$,

$$(26.1) \qquad\qquad\qquad X_n(\omega) = \omega_n \,.$$

Fix a transition operator T. It and an initial measure Q_0, specified to be the distribution of X_0, determine a unique measure P^{Q_0} on $(\Omega, \mathcal{F})$. The slightly abbreviated notation P^x is used for P^{δ_x}. When X is governed by P^x, we say that x is the *initial state* of X. The expectation operators associated with P^x and P^{Q_0} are denoted by E^x and E^{Q_0} respectively.

The transition operator T determines and is determined by the family of distributions $(P^x \colon x \in \Psi)$, called the *Markov family*, with transition operator T, and this Markov family determines the probability measures P^{Q_0} via the formula in Problem 8.

The following exercises are intended to give the reader some practice with the notation just introduced.

Problem 8. Show that for a given transition operator T, the probability measures $P^x, x \in \Psi$, determine the probability measures P^{Q_0} for arbitrary Q_0 by way of

$$P^{Q_0}(A) = \int_{\Omega} P^x(A)\, Q_0(dx)$$

for all $A \in \mathcal{F}$.

Problem 9. Show that

$$E^x(f \circ X_k) = T^k f(x) \,, \quad k \geq 0 \,,$$

for all bounded measurable $f \colon \Psi \to \mathbb{R}$.

Problem 10. Show that for a Markov sequence with transition operator T, the conditional distribution of the random sequence $(X_k, X_{k+1}, X_{k+2}, \dots)$ given X_k is P^{X_k}. Generalize to allow k to be a stopping time.

We conclude this section with one further definition.

Definition 3. The *discrete generator* of a Markov sequence and of its transition operator T is the operator $G = T - I$, where I is the identity operator.

The connection between T and G is so simple that it hardly seems worthwhile to define G. However, one of our purposes here is to prepare the reader for Markov processes in continuous time, for which generators are nontrivial and very useful.

26.2. Examples

When Markov sequences are studied in elementary courses, their state spaces are usually finite or countable. In such cases, transition operators can be represented by matrices.

Example 1. [Finite state space] Let Ψ be the finite set $\{1, 2, \ldots, m\}$. In this case, the $\mathbb{R}$-valued functions defined on Ψ can be identified with vectors in $\mathbb{R}^m$. By convention, we represent such a vector f by the column matrix with entries $f(x)$, $x = 1, \ldots, m$. Then a transition operator T can be regarded as an $m \times m$ matrix $(T(x, y): 1 \le x, y \le m)$ in which each entry is nonnegative and each row sum satisfies

$$\sum_{y=1}^{m} T(x, y) = 1 \quad \text{for } x = 1, 2, \ldots, m.$$

The condition that the row sums equal 1 arises from the relation $\mu_x\{y\} = T(x, y)$, where μ_x, $1 \le x \le m$, are the transition distributions. The matrix T operates on column matrices f by multiplication on the left, so that the notation Tf used earlier is consistent with matrix multiplication.

Similarly, matrix multiplication is useful when regarding T as a right transition operator. We identify each probability measure Q_0 on Ψ with the row matrix whose entries are the quantities $Q_0(\{x\})$, $x = 1, 2, \ldots, m$. It is easily checked that when this identification is made, the probability measure $Q_0 T$ is identified with the matrix $Q_0 T$.

Here is a specific example. Let $n = 5$ and

$$T = \begin{pmatrix} 1 & 0 & 0 & 0 & 0 \\ \frac{2}{3} & 0 & \frac{1}{3} & 0 & 0 \\ 0 & \frac{2}{3} & 0 & \frac{1}{3} & 0 \\ 0 & 0 & \frac{2}{3} & 0 & \frac{1}{3} \\ 0 & 0 & 0 & 0 & 1 \end{pmatrix}.$$

We let our Markov sequence start at 3. Then the sequence corresponds to the gambler's ruin problem, in which the gambler starts with 3 dollars, wins a dollar each time with probability $\frac{1}{3}$, loses a dollar each time with probability $\frac{2}{3}$, and quits as soon as the fortune reaches either 1 dollar or 5 dollars. The gambler's fortune at time 2 is a random variable whose distribution is the row matrix $(0, 0, 1, 0, 0)T^2 = (\frac{4}{9}, 0, \frac{4}{9}, 0, \frac{1}{9})$.

We may generalize the matrix representation in the preceding example to arbitrary countable state spaces. The following result uses the notation of Problem 2.

Proposition 4. *If T is a transition operator for a countable state space Ψ, then T may be represented by a matrix, also denoted by T, with entries $T(x,y)$ given by the formula*

$$T(x,y) = \delta_x T f_y = P^x(X_1 = y), \quad x, y \in \Psi,$$

where δ_x is the point mass at x and f_y is the indicator function of the singleton $\{y\}$. This representation has the property that for any distribution Q_0 and bounded function f on Ψ, the distribution $Q_0 T$ and the function Tf are represented by the respective matrix products $Q_0 T$ and Tf, and for any nonnegative integer k, the operator T^k is represented by the matrix product T^k.

For countable state spaces Ψ, any matrix $T = (T(x,y) \colon x, y \in \Psi)$ with nonnegative entries and with row sums equal to 1 is called a *transition matrix*. On any countable state space, there is a one-to-one correspondence between the set of transition matrices and the set of transition operators.

Problem 11. Prove Proposition 4.

Problem 12. Make and interpret some calculations for the transition matrix

$$T = \begin{pmatrix} 1 & 0 & 0 & 0 & 0 & \cdots \\ \frac{2}{3} & 0 & \frac{1}{3} & 0 & 0 & \cdots \\ 0 & \frac{2}{3} & 0 & \frac{1}{3} & 0 & \cdots \\ 0 & 0 & \frac{2}{3} & 0 & \frac{1}{3} & \cdots \\ 0 & 0 & 0 & \frac{2}{3} & 0 & \cdots \\ 0 & 0 & 0 & 0 & \frac{2}{3} & \cdots \\ 0 & 0 & 0 & 0 & 0 & \cdots \\ \vdots & \vdots & \vdots & \vdots & \vdots & \ddots \end{pmatrix},$$

similar to those made in Example 1.

Problem 13. Let X be a random walk on $\mathbb{Z}$. Show that X is a Markov sequence, and determine the corresponding transition matrix and discrete generator in terms of the step distribution of the random walk.

Example 2. [Birth-death sequences] A *birth-death sequence* is a Markov sequence $(X_n \colon n \geq 0)$ with state space $\mathbb{Z}^+$ whose transition probability measures μ_x satisfy

$$\mu_x(\{x - 1, x, x + 1\}) = 1 \quad \text{for } x > 0$$

and $\mu_0(\{0, 1\}) = 1$. In terms of the corresponding transition matrix T, the condition is that $T(x,y) = 0$ if $|x - y| > 1$. It follows from this condition that $|X_n - X_{n-1}| \leq 1$ a.s. for all $n > 0$. If $X_n = X_{n-1} + 1$, we say that a *birth* occurs at time n. If $X_n = X_{n-1} - 1$, we say a *death* occurs at time n. Note that unless all of the diagonal entries of T are 0, it is not necessarily true that all transitions are births or deaths.

Birth-death sequences can be regarded as simple models of changing population sizes. In such a model, the population size can only increase or decrease by at most 1 during each time step. When thinking about such a model, it is sometimes best not to think of the index n in X_n as marking the passage of time intervals of uniform length, but rather as indicating the times at which changes in the population might occur.

Example 3. [Branching processes] Let μ be a probability distribution on $\mathbb{Z}^+$. A *branching process* with *branching distribution* μ is a Markov sequence on $\mathbb{Z}^+$ having transition probability measures

$$\mu_x = \mu^{*x}.$$

The branching distribution is also called the *offspring distribution*.

Like birth-death sequences, branching processes can be interpreted as population models. When the branching process is in state x, we think of the population as having x members, each of which independently goes through a reproductive cycle. During each cycle the member dies and produces a random number of offspring. The distribution μ gives the probabilities of the various outcomes: $\mu\{k\}$ is the probability that there are k offspring from a particular individual. Thus, the convolution μ^{*x} is the distribution of the upcoming population, given that the population is currently x.

Problem 14. Suppose that $\mu\{0\} = \mu\{2\} = \frac{1}{2}$ in the preceding example. Calculate μ_2 and μ_3.

Example 4. Let μ be a distribution on $\overline{\mathbb{Z}}^+ \setminus \{0\}$. For $x \in \overline{\mathbb{Z}}^+ \setminus \{0\}$, set

$$\mu_x = \begin{cases} \mu & \text{if } x = 1 \\ \delta_{x-1} & \text{if } 1 < x < \infty \\ \delta_\infty & \text{if } x = \infty. \end{cases}$$

Let $(X_n \colon n = 0, 1, \dots)$ be a Markov sequence with initial state 1 and transition distributions μ_x, $x \in \overline{\mathbb{Z}}^+ \setminus \{0\}$. Set $Y_n = I_{[X_n = 1]}$. It is rather obvious that $(Y_n \colon n \geq 0)$ is a renewal sequence with waiting time distribution μ. The correspondence thus obtained between renewal sequences and certain Markov sequences is illustrated in Figure 26.1. If the initial distribution δ_1 for the Markov sequence is replaced by an arbitrary distribution Q_0, the sequence (Y_n) is a delayed renewal sequence. A special Q_0 is given in the next problem.

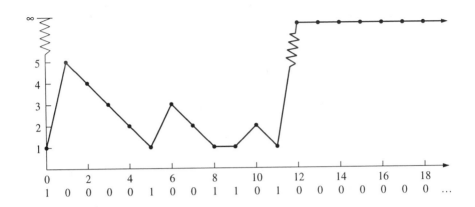

FIGURE 26.1. Correspondence between a renewal sequence and a certain Markov sequence

Problem 15. Suppose that the waiting time distribution μ in the preceding example has finite mean. Let $(X_n : n \geq 0)$ be the corresponding Markov sequence with initial distribution Q_0 defined by

$$(26.2) \qquad Q_0\{n\} = \frac{\mu[n, \infty)}{\sum_{j=1}^{\infty} \mu[j, \infty)} .$$

Show that distribution of each X_n is Q_0. Also show that the corresponding delayed renewal sequence (Y_n) is 'stationary', in the sense that it has the same distribution as the sequence (Z_n) defined by $Z_n = Y_{n+1}, n = 0, 1, 2, \ldots$. Compare with Theorem 15 of Chapter 25, resolving any apparent differences between formulas.

Problem 16. Let μ be as in Example 4, and define transition distributions μ_x for a Markov sequence $(X_n : n \geq 0)$ on $\mathbb{Z}^+ \setminus \{0\}$ by

$$\mu_x\{1\} = \frac{\mu\{x\}}{\mu[x, \infty]} = 1 - \mu_x\{x + 1\} .$$

Show that if the initial state is 1, then the random sequence (Y_n) defined by $Y_n = I_{[X_n = 1]}$ is a renewal sequence with waiting time distribution μ. See Figure 26.2 for an illustration of the correspondence identified in this problem.

Example 5. [Network walks] Let G be a graph, meaning that G consists of a set of points called 'vertices' and a collection of 'edges' connecting certain pairs of vertices. Two vertices are neighbors if they are connected by an edge. We assume that every vertex in G has at least one neighbor, and also that G is locally finite, meaning that each vertex has only finite many neighbors. Let Ψ be

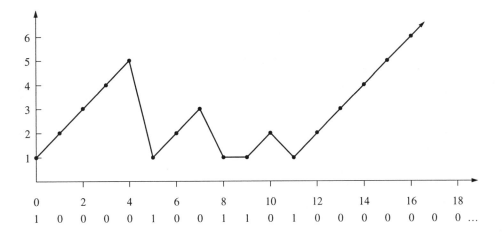

FIGURE 26.2. Correspondence between a renewal sequence and a second Markov sequence

the vertex set of G, and for each $x \in \Psi$, let n_x denote the number of neighbors of x. Define a transition matrix T by

$$T(x, y) = \begin{cases} \frac{1}{n_x} & \text{if } x \text{ and } y \text{ are neighbors} \\ 0 & \text{otherwise}. \end{cases}$$

Here is an intuitive description of a Markov sequence X with transition operator T. Given that X is in state x at time n, then at time $n + 1$, it will be at one of the neighbors of x, each with equal probability. Sometimes X is called a 'random walk on G', although according to the definition of random walk given in Chapter 11, X is not a random walk unless G is very special.

Example 6. [Probabilistic cellular automata] Let $\Psi = \{0, 1\}^{\mathbb{Z}}$. For this more complicated state space, it will be convenient for us to change our usual notation. We will write ψ for a typical state, and use the letter z for a typical member of $\mathbb{Z}$. Each state ψ is a sequence of 0's and 1's indexed by the integers. We write $\psi(z)$ for the element in the sequence ψ that is indexed by the integer z. A state ψ is called a *configuration* in this context. We think of ψ as the 'global' state of a one-dimensional array of 2-state components or *cells*. We call z the *site* of the cell located at z, and $\psi(z)$ is the 'local state' at the site z.

For each integer z, let $f_z : \Psi \to \mathbb{R}$ be defined by

$$f_z(\psi) = \psi(z).$$

We will start to define an operator T by saying how T acts on the functions f_z. For each $\eta \in \{0, 1\} \times \{0, 1\}$, let S_η be a transition operator on the state space

$\{0,1\}$, and let g be the identity function on $\{0,1\}$. Define

$$Tf_z(\psi) = S_{(\psi(z-1),\psi(z+1))}g(\psi(z)).$$

We extend T to products of functions f_z by defining

(26.3)
$$T\left(\prod_{z \in A} f_z\right) = \prod_{z \in A} Tf_z$$

for all finite sets $A \subseteq \mathbb{Z}$. Now extend T to finite linear combinations of such products by linearity. We leave it as an exercise for the reader to extend T to bounded measurable functions by taking limits and then showing that T is, in fact, a transition operator.

We can describe Markov sequences X with transition operator T as follows. Suppose the sequence is in state ψ at time n. Now consider a particular site z. The local state at the cell at z will make a transition according to the transition matrix S_η, where $\eta = (\psi(z-1), \psi(z+1))$. Thus, the local transition probability at z depends on the states of the cells at the sites $z-1$ and $z+1$. We say that the cell at z *updates* according to S_η, and, as a consequence of (26.3), conditioned on the state ψ, all of the cells update independently. We have described the *cellular automaton* with local update rules $\{S_\eta, \eta \in \{0,1\} \times \{0,1\}\}$. Continuous-time versions of cellular automata form the subject matter of Chapter 32.

Problem 17. Fill in the details for the construction of the transition operator T in the preceding example.

26.3. Martingales and the strong Markov property

There is an important connection between Markov sequences and martingales.

Definition 5. Let G be an operator with domain and target equal to the set of bounded measurable functions on some Borel space Ψ, and let $(\mathcal{F}_n : n \geq 0)$ be a filtration in a probability space $(\Omega, \mathcal{F}, P)$. A sequence X of Ψ-valued random variables on $(\Omega, \mathcal{F}, P)$ is a *solution to the martingale problem* for G and $(\mathcal{F}_n : n \geq 0)$ if for all bounded measurable functions $f : \Psi \to \mathbb{R}$, the sequence Y defined by

$$Y_n = f \circ X_n - \sum_{k=0}^{n-1} (Gf) \circ X_k$$

is a martingale with respect to the filtration $(\mathcal{F}_n : n \geq 0)$.

Theorem 6. *Let G, Ψ, x, $(\mathcal{F}_n : n \geq 0)$, and $(\Omega, \mathcal{F}, P)$ be as in the preceding definition. Then a sequence X of Ψ-valued random variables is a Markov sequence with respect to $(\mathcal{F}_n : n \geq 0)$ and having discrete generator G if and only if X is a solution to the martingale problem for G and $(\mathcal{F}_n : n \geq 0)$.*

Problem 18. Prove the preceding theorem.

* **Problem 19.** Suppose that $X = (X_n : n \geq 0)$ is a $[0, 1]$-valued random sequence that is both a time-homogeneous Markov sequence and a submartingale with respect to some filtration $(\mathcal{F}_n : n \geq 0)$. Obtain the Doob decomposition of X as a corollary of Theorem 6.

Theorem 6 is useful in two ways. First, it gives us a large collection of martingales that can be used to analyze a given Markov sequence. Second, it provides us with a way to recognize that a random sequence is Markov. This second use is particularly valuable in the continuous-time version of this theory, but is usually ignored in the case of Markov sequences. Nevertheless, we will show how it can be applied by using it to prove the *strong Markov property*.

Theorem 7. [Strong Markov property] *Let X be a Markov sequence with respect to a filtration $(\mathcal{F}_n : n \geq 0)$, with transition operator T. Let τ be an almost surely finite stopping time relative to the filtration $\{\mathcal{F}_n\}$. Then the random sequence Y defined by*

$$Y_n = X_{\tau+n}, \quad n \geq 0,$$

is a Markov sequence with transition operator T that is adapted to the filtration $\{\mathcal{F}_{\tau+n} : n \geq 0\}$.

PROOF. This result follows immediately from Theorem 6 and the Optional Sampling Theorem. □

Problem 20. Prove the strong Markov property without using martingales.

Let G be a discrete generator on a state space Ψ and let A be a Borel subset of Ψ. A bounded measurable function $f \colon \Psi \to \mathbb{R}$ is *G-subharmonic on A* if $Gf(x) \geq 0$ for all $x \in A$. It is *G-superharmonic on A* if $-f$ is G-subharmonic on A. It is *G-harmonic on A* if it is both G-subharmonic and G-superharmonic on A. If $A = \Psi$, then the phrase 'on A' is often dropped. The modifier 'G-' is also often omitted when only one discrete generator is being considered.

Problem 21. Let G be the discrete generator of a Markov sequence X. Let A be a Borel set and $\tau = \inf\{n \colon X_n \in A^c\}$. If f is a bounded measurable function on the state space that is G-subharmonic on A, show that the random sequence Y defined by

$$Y_n = f \circ X_{n \wedge \tau}$$

is a submartingale with respect to the same filtration to which X is adapted.

Problem 22. Let X be a simple random walk with step distribution $p\delta_{\{1\}} + (1 - p)\delta_{\{-1\}}$. Find the discrete generator G of X and show that the function

$$f(x) = \left(\frac{1-p}{p}\right)^x$$

is G-harmonic. (Refer back to Chapter 24 and our treatment of the gambler's ruin problem for an application of this fact.)

26.4. Hitting times and return times

Let T be a transition operator with state space Ψ and discrete generator G. For each Borel set $B \subseteq \Psi$, set

$$\pi_{x,B} = P^x[X_n \in B \text{ for some } n \geq 0],$$

called the *hitting probability of B starting at x*, and

$$\pi_{x,B}^+ = P^x[X_n \in B \text{ for some } n > 0].$$

Note that $\pi_{x,B} = 1$ for $x \in B$ and $\pi_{x,B} = \pi_{x,B}^+$ for $x \in B^c$. For $x \in B$, $\pi_{x,B}^+$ is called the *return probability to B starting at x*. We often write π_{xy} and π_{xy}^+ for $\pi_{x,B}$ and $\pi_{x,B}^+$ when $B = \{y\}$.

Problem 23. Prove that

$$\pi^+(\cdot, B) = T\pi(\cdot, B)$$

for any transition operator and every Borel subset B of the state space.

Problem 24. Let G denote the discrete generator of some Markov sequence X, and B a measurable subset of its state space Ψ. Let $f(x) = \pi_{x,B}$ and $g(x) = \pi_{x,B}^+$ for $x \in \Psi$. Prove that f and g are G-superharmonic on Ψ and that f is G-harmonic on B^c. Also show that $\lim_{n\to\infty} f \circ X_n$ and $\lim_{n\to\infty} g \circ X_n$ exist almost surely.

The preceding problem constitutes part of the proof of the following theorem, which characterizes the function $x \rightsquigarrow \pi_{x,B}$.

Theorem 8. *Let G be the discrete generator of a Markov sequence, and B a measurable subset of its state space. Then $x \rightsquigarrow \pi_{x,B}$ is the minimal bounded function h having the following properties:*

    (i) $h(x) = 1$ for $x \in B$,
    (ii) h is G-harmonic on B^c,
    (iii) $h(x) \geq 0$ for all x,

where the word 'minimal' is used in the sense that if h is any other function satisfying properties (i)-(iii), then $\pi_{x,B} \leq h(x)$ for all x.

PROOF. That the function $x \rightsquigarrow \pi_{x,B}$ has properties (i) and (iii) is obvious. Property (ii) is contained in Problem 24.

To complete the proof, suppose that h satisfies (i), (ii), and (iii). Set $\tau = \inf\{n: X_n \in B\}$ and define $Y = (Y_1, Y_2, \ldots)$ by $Y_n = h \circ X_{n \wedge \tau}$. By Problem 21, Y is a bounded martingale. Let Y_∞ denote its limit. Clearly, $Y_\infty(\omega) \geq 0$ for all ω and equals 1 if $\tau(\omega) < \infty$. Therefore,

$$\pi_{x,B} \leq E^x(Y_\infty) = E^x(Y_0) = E^x(h \circ X_0) = h(x). \quad \square$$

Example 7. For the state space $\mathbb{R}^+ \setminus \{0\}$ consider a Markov sequence with transition distributions $\mu_x = \frac{1}{2}\delta_{x/2} + \frac{1}{2}\delta_{4x}$, where δ_y denotes the delta distribution at y. We set $B = (0, 1]$ with the goal of calculating the function $x \rightsquigarrow \pi_{x,B}$.

In order to apply Theorem 8 we need the following formula:

$$Gh(x) = \frac{1}{2}h(x/2) - h(x) + \frac{1}{2}h(4x).$$

A change of variables is helpful: set $g(y) = h(2^y)$, $y \in \mathbb{R}$. The requirement (ii) of Theorem 8 becomes

(26.4) $$\frac{1}{2}g(y-1) - g(y) + \frac{1}{2}g(y+2) = 0, \quad y > 0.$$

Let us begin by restricting attention to $y \in \mathbb{Z}^+ \setminus \{0\}$. The equation (26.4) is what is known as a third-order homogeneous linear difference equation. We make the educated guess that there is a solution of the form k^y, and obtain the following condition on k:

$$\frac{1}{2} - k + \frac{1}{2}k^3 = 0.$$

Solving for the three values of k and putting the solutions together just as one does for differential equations, we conclude that every solution of (26.4) is of the form

$$a\left(\frac{-1+\sqrt{5}}{2}\right)^y + b + c\left(\frac{-1-\sqrt{5}}{2}\right)^y, \quad y \in \mathbb{Z}^+.$$

By (iii) of Theorem 8, applied for large odd y if $c \geq 0$ and large even y if $c \leq 0$, we conclude that $c = 0$. By (i) of that theorem, we obtain $a + b = 1$. Finally, to get a minimal solution without violating condition (iii) we take $b = 0$. Thus,

$$g(y) = \left(\frac{-1+\sqrt{5}}{2}\right)^y, \quad y \in \mathbb{Z}^+.$$

A little thought shows that $g(y) = g(\lceil y \rceil)$ for all $y > 0$ and $g(y) = 1$ for $y \leq 0$. Therefore,

$$\pi_{x,B} = \begin{cases} 1 & \text{if } 0 < x \leq 1 \\ \left(\frac{-1+\sqrt{5}}{2}\right)^m & \text{if } 2^{m-1} < x \leq 2^m, \ m = 1, 2, \ldots. \end{cases}$$

Problem 25. Let $\Psi = \{0, 1, 2, \ldots, g\}$ for some positive integer g. Fix $p \in (0, 1)$ and for $x \in \Psi$, let μ_x be the uniform distribution on Ψ if $x = 0$, the delta distribution δ_g if $x = g$, and $p\delta_{x+1} + (1-p)\delta_{x-1}$ for other x. Calculate the functions $x \rightsquigarrow \pi_{x0}$ and $x \rightsquigarrow \pi_{x0}^+$.

Problem 26. For all simple random walks in $\mathbb{Z}$, calculate the functions $x \rightsquigarrow \pi_{x0}$ and $x \rightsquigarrow \pi_{x0}^{+}$.

Problem 27. For a Markov sequence on $\mathbb{Z}^{+} \setminus \{0\}$ with transition operator T given by

$$Tf(x) = \frac{x}{x+1}f(x+1) + \frac{1}{x+1}f(1)$$

calculate the function $x \rightsquigarrow \pi_{x1}$. Discuss the relation between this problem and Problem 16.

* **Problem 28.** For a Markov sequence on $\mathbb{Z}^{+} \setminus \{0\}$ with transition operator T given by

$$Tf(x) = \frac{(x-1)(x+2)}{x(x+1)}f(x+1) + \frac{2}{x(x+1)}f(1)$$

calculate the function $x \rightsquigarrow \pi_{x1}$. Discuss the relation between this problem and Problem 16.

* **Problem 29.** For a Markov sequence on $\mathbb{Z}$ with transition operator T given by

$$Tf(x) = \tfrac{1}{2}f(x-1) + \tfrac{1}{2}f(x+2)$$

calculate the function $x \rightsquigarrow \pi_{x0}$.

Problem 30. For a Markov sequence on $\mathbb{Z}^{+}$ with transition operator T given by

$$Tf(x) = \begin{cases} \tfrac{1}{2}f(x-1) + \tfrac{1}{2}f(x+2) & \text{if } x > 0 \\ f(x) & \text{if } x = 0 \end{cases}$$

calculate the function $x \rightsquigarrow \pi_{x,\{0,3\}}$.

* **Problem 31.** For the birth-death sequence with transition distributions μ_x given by $\mu_x\{x+1\} = \frac{x}{x+1}$ and for $x > 0$, $\mu_x\{x-1\} = \frac{1}{x+1}$ calculate the function $x \rightsquigarrow \pi_{x0}$.

Theorem 9. *Let ρ denote the generating function of the offspring distribution of a branching process. Then for this process, $\pi_{x0} = c^x$, where c is the smallest root in $[0,1]$ of the equation $\rho(s) = s$.*

PROOF. A branching process with initial state $x > 1$ is constructed from x independent processes with initial state 1. Once any of these independent process hits 0 it stays at 0. Hence, the branching process with initial state x hits 0 if and only if all x of the independent processes hit 0. We conclude that $p_{x0} = (p_{10})^x$.

Using the fact that $x \rightsquigarrow (p_{10})^x$ is harmonic we obtain

$$\sum_{y=0}^{\infty} (p_{10})^y \, \mu^{*x}\{y\} - (p_{10})^x = 0.$$

This condition simplifies to

$$\left[\rho(p_{10})\right]^x = (p_{10})^x,$$

from which it follows that p_{10} is a solution of the equation $\rho(s) = s$. Moreover, for any solution c of this equation, the function $x \rightsquigarrow c^x$ is harmonic. The desired conclusion follows from the minimality assertion in Theorem 8. $\square$

Problem 32. Let $p \in (0,1)$. For a branching process with branching distribution $x \rightsquigarrow p\delta_2 + (1-p)\delta_0$, calculate the function $x \rightsquigarrow \pi_{x0}$.

Problem 33. Let $q \in (0,1)$ and $\mu\{x\} = (1-q)q^x$, $x \in \mathbb{Z}^+$. Calculate $x \rightsquigarrow \pi_{x0}$ for a branching process with branching distribution μ.

26.5. Renewal theory and Markov sequences

We have already seen some connections between Markov sequences and renewal sequences in Example 4 and Problem 16. In this section we exploit another such connection. Although some of the results apply generally, we will restrict our attention to Markov sequences with countable state spaces. In this context, a transition operator T is represented by a transition matrix $T = (T(x,y): x,y \in \Psi)$, as described in Proposition 4. We will be interested in the asymptotic behavior of T^k as $k \to \infty$. Denote the entries of the matrix T^k by $T^k(x,y)$ for nonnegative integers k. It follows from Proposition 4 and Problem 6 that $T^k(x,y) = P^x[X_k = y]$, so the results below will give information about the asymptotic behavior of X_k as $k \to \infty$.

Let $X = (X_n : n \geq 0)$ be a time-homogeneous Markov sequence with state space Ψ, let y be a fixed state in Ψ, and define $Y = (Y_n : n \geq 0)$ by the formula

$$Y_n = I_{\{y\}} \circ X_n.$$

Thus, Y is a random sequence of 1's and 0's, with the 1's indicating the times when the Markov sequence X visits the state y. It is a straightforward consequence of the strong Markov property that Y is a delayed renewal sequence.

Given a transition operator T for a countable state space Ψ, we have defined a delayed renewal sequence Y on the probability space $(\Omega, \mathcal{F}, P^x)$ for each $x \in \Psi$. When $x = y$, this delayed renewal sequence is a renewal sequence, so there is a renewal sequence corresponding to each state. We use the classification of renewal sequences in Chapter 25 to classify the states in Ψ. A state y is *transient, null recurrent,* or *positive recurrent* according to the classification of its corresponding renewal sequence. The period γ_y of this renewal sequence is called the *period* of the state y, and y is *aperiodic* if $\gamma_y = 1$. We denote the mean waiting time of the renewal sequence corresponding to y by m_y. In the context of Markov sequences, m_y is called the *mean return time* of y.

We will use the Renewal Theorem to analyze the behavior of the sequence $(T^k(x,y): k \geq 0)$ for fixed states x,y. In terms of the random sequence Y, $T^k(x,y) = P^x[Y_k = 1]$, so $(T^k(y,y): k \geq 0)$ is the potential sequence for the renewal sequence corresponding to y. It follows immediately from the Renewal Theorem that

$$\lim_{k\to\infty} T^{k\gamma_y}(y,y) = \frac{\gamma_y}{m_y},$$

provided that $\gamma_y < \infty$. By the definition of 'period', $T^k(y,y) = 0$ if k is not an integer multiple of γ_y, so

(26.5)
$$\lim_{k\to\infty} \left(\frac{1}{\gamma_y} \sum_{i=0}^{\gamma_y - 1} T^{k+i}(y,y) \right) = \frac{1}{m_y},$$

whether or not $\gamma_y < \infty$. We have almost proved the following result.

Theorem 10. [Renewal for Markov sequences] *Let T be a transition operator for a countable state space Ψ. Then for all states $x, y \in \Psi$,*

(26.6)
$$\lim_{k\to\infty} \left(\frac{1}{\gamma_y} \sum_{i=0}^{\gamma_y - 1} T^{k+i}(x,y) \right) = \frac{\pi_{xy}}{m_y}.$$

PROOF. Since $\pi_{yy} = 1$, the case $x = y$ follows from (26.5). The case $\gamma_y = \infty$ is trivial, since at most 1 visit to the state is possible and $m_y = \infty$ for that case.

It remains to consider general x and y with $\gamma_y < \infty$. Let $\tau = \inf\{n \geq 0 : X_n = y\}$. By the strong Markov property,

$$T^{k+i}(x,y) = \sum_{n=0}^{k+i} P^x[X_{k+i} = y \text{ and } \tau = n]$$

$$= \sum_{n=0}^{k+i} P^y[X_{k+i-n} = y]\, P^x[\tau = n]$$

$$= \sum_{n=0}^{\infty} I_{\{m\,:\, m\leq k+i\}}(n)\, T^{k+i-n}(y,y)\, P^x[\tau = n].$$

We substitute this last expression into (26.6) and then use the Dominated Convergence Theorem for sums to justify bringing the limit inside the infinite sum. Since

$$\pi_{xy} = \sum_{n=0}^{\infty} P^x[\tau = n],$$

the desired result now follows from the case for which $x = y$. □

Problem 34. Prove that every time-homogeneous Markov sequence with finite state space has at least one recurrent state.

Problem 35. Use the Strong Law for Renewal Sequences to determine the asymptotic behavior of the number of visits to a state y made by a time-homogeneous Markov sequence with initial state x.

26.6. Irreducible Markov sequences

We introduce some terminology that will be useful in further analyzing the asymptotic behavior of Markov sequences. Although we do not assume in this section that the state spaces are countable, the terminology is generally useful only in the countable case.

If $\pi_{xy} > 0$, then we say that the state y is *accessible from* the state x. We call a transition operator and corresponding Markov sequence *irreducible* if all states are accessible from one another.

It is an immediate consequence of the forthcoming Problem 39 that when the transition operator is irreducible, all states have the same period and are of the same type as far as recurrence is concerned. Thus one can speak of the *period* of an irreducible transition operator, identify it as *aperiodic* or not, and classify it as *positive recurrent, null recurrent,* or *transient.* Of course we can apply the same terminology to the irreducible Markov sequence itself.

Problem 36. Prove that a necessary and sufficient condition for a state x to be transient is that there exists a state $y \neq x$ that is accessible from x and for which $\pi_{yx} < 1$.

Problem 37. Using the necessary and sufficient conditions for transience given in the preceding problem, write a statement that gives necessary and sufficient conditions for a state to be recurrent.

Problem 38. Prove that if x and y are recurrent states, then either $\pi_{xy} = \pi_{yx} = 1$ or $\pi_{xy} = \pi_{yx} = 0$. Also, prove that $\pi_{xz} = 0$ if x is recurrent and z is transient.

* **Problem 39.** Show that if two states are each accessible from one another, then they both have the same period and are of the same type—positive recurrent, null recurrent, or transient.

Problem 40. Call recurrent states x and y *equivalent* if $\pi_{xy} = 1$. Prove that this relationship is, in fact, an equivalence relation among the recurrent states.

The equivalence classes induced by the equivalence relation of the preceding problem are called *irreducible recurrence classes*. A Markov sequence is sometimes analyzed according to the following scheme: (i) The irreducible recurrence classes are identified; (ii) for each transient state the probability is calculated—using Theorem 8 for instance—of hitting each irreducible recurrence class and also of staying forever in the transient states; (iii) a study is made of each of the irreducible Markov sequences obtained by considering the irreducible recurrence classes to be separate state spaces, a legitimate procedure in view of Problem 38.

Problem 41. Show that if y and z are in the same irreducible recurrence class, then $\pi_{xy} = \pi_{xz}$ for all states x.

Problem 42. For each of the Markov sequences of Example 4 there is a unique subset Υ of the state space such that the Markov sequence restricted to Υ is irreducible. Describe Υ in terms of the distribution μ of that example. Also, for each $\gamma \in \mathbb{Z}^+ \setminus \{0\}$ find three distributions μ for which $\#\Upsilon = \infty$ and the corresponding transition operator has period γ: one for which the transition operator is positive recurrent, one for which it is null recurrent, and one for which it is transient.

* **Problem 43.** Show that if T is a transition operator with finite state space, then T is irreducible and aperiodic if and only if for some positive integer k, all entries of the matrix T^k are positive.

Problem 44. Show that if T is a transition operator with countable state space, then T is irreducible and aperiodic if and only if for every pair of states x, y, there exists a positive integer k such that $T^n(x, y) > 0$ for all $n \geq k$.

Problem 45. Let T be an irreducible transition operator with countable state space and suppose that $T(x, x) > 0$ for some x. Prove that T is aperiodic. Also show that it is possible for T to be aperiodic even if $T(x, x) = 0$ for all x.

Problem 46. Let T be an irreducible recurrent transition operator with countable state space and period k. Prove that T^k is an aperiodic transition operator with no transient states and k irreducible recurrence classes.

26.7. Equilibrium distributions

Let T be a transition operator with state space Ψ. If Q_0 is a probability distribution on Ψ, then we call Q_0 an *equilibrium distribution* for T if $Q_0 T = Q_0$. The following fact is easily proved:

Definition and Proposition 11. If Q_0 is an equilibrium for T and if X is a Markov sequence with transition operator T and initial distribution Q_0, then Q_0 is the distribution of X_k for all $k \geq 0$, and X is called a *stationary Markov sequence*.

Problem 47. Let Q_0 be an equilibrium distribution for a transition operator T. Show that if x is a transient or null recurrent state, then $Q_0\{x\} = 0$.

Problem 48. Let T be a transition operator with state space Ψ. Assume that Ψ is a compact Polish space. (Polish spaces are introduced and treated in Chapter 18.) Show that there exists at least one equilibrium distribution for T. *Hint:* Let X be any Markov sequence with transition operator T, and let Q_n be the distribution of X_n, $n \geq 0$. For $n \geq 0$, define probability distributions R_n by

$$R_n(A) = \frac{1}{n+1} \sum_{k=0}^{n} Q_n(A).$$

Show that every limit point of the sequence $(R_n : n \geq 0)$ is an equilibrium distribution for T.

We now focus on equilibrium distributions in the setting of countable state spaces. There is a strong connection between these distributions and the behavior of corresponding Markov sequences $(X_0, X_1, \dots)$ for large time. The distribution of X_n is represented by the matrix $Q_0 T^n$, where Q_0 is the row matrix that represents the distribution of X_0 and T is the transition matrix. Thus we will be interested in the behavior of the powers of T. The Renewal Theorem for Markov sequences will be our main tool. An immediate consequence of that theorem is

$$(26.7) \qquad \lim_{n \to \infty} \frac{1}{n} \sum_{k=1}^{n} T^k(x, y) = \frac{\pi_{xy}}{m_y} \overset{\text{def}}{=} A(x, y)$$

for all $x, y \in \Psi$.

Proposition 12. *For the matrix A given in (26.7), all rows corresponding to states in the same irreducible recurrence class are identical, and all entries in columns corresponding to null recurrent and transient states equal 0. Moreover,*

- *if x is positive recurrent, then $A(x, y) = 1/m_y$ for y in the irreducible recurrence class of x and $A(x, y) = 0$ for other y;*
- *if x is null recurrent, then $A(x, y) = 0$ for all y;*
- *if x is transient, then row x is a linear combination of rows corresponding to various irreducible recurrence classes, where the weight corresponding to a given recurrence class is π_{xy} for any choice of y from that class.*

Problem 49. Prove the preceding proposition. *Hint:* Use Problem 38, Problem 39, and Problem 41.

Proposition 13. *For T a transition matrix, let the matrix A be defined by (26.7). Then $TA = AT = A^2 = A$. Moreover, a probability distribution Q_0 is an equilibrium distribution for T if and only if $Q_0 A = Q_0$.*

PROOF. By treating each row of T as a probability distribution, we can apply the Bounded Convergence Theorem to obtain

$$TA = \lim_{n \to \infty} \frac{1}{n} \sum_{k=1}^{n} T^{k+1} = \lim_{n \to \infty} \frac{1}{n+1} \sum_{j=2}^{n+1} T^j = \lim_{n \to \infty} \frac{1}{n+1} \sum_{j=1}^{n+1} T^j = A.$$

By the Fatou Lemma, every entry of the matrix AT is less than or equal to the corresponding entry of the matrix A and the sum of the entries in any row of A is finite (in fact, less than or equal to 1). We finish the argument that $AT = A$ by the following calculation showing that the sum of the entries in any row of AT equals the sum of the entries in the corresponding row of A:

$$\sum_z \sum_y A(x, y) T(y, z) = \sum_y \sum_z A(x, y) T(y, z) = \sum_y A(x, y).$$

From the preceding paragraph, we obtain $A = AT = AT^2 = \ldots$ and thus

$$A = A\Big(\frac{1}{n}\sum_{k=1}^{n}T^k\Big).$$

By treating each row of A as a finite measure, we can apply the Bounded Convergence Theorem to obtain $A = A^2$.

Suppose that Q_0 is an equilibrium distribution. Then, viewing T as a right transition operator,

$$Q_0 = Q_0 T = Q_0 T^2 = \ldots ,$$

so

$$Q_0 = Q_0\Big(\frac{1}{n}\sum_{k=1}^{n}T^k\Big).$$

By the Bounded Convergence Theorem, $Q_0 A = Q_0$.

Conversely, suppose that $Q_0 = Q_0 A$. Operate on the right by T to obtain $Q_0 T = (Q_0 A)T = Q_0(AT) = Q_0 A = Q_0$. $\square$

Proposition 12 shows that, in some sense, the interesting rows of A are those that correspond to positive recurrent states since all other rows can be written as (possibly trivial) linear combinations of those rows. The following theorem connects this fact to the equilibrium distributions for T.

Theorem 14. *For a transition matrix T, let the matrix A be defined by (26.7). The equilibrium distributions of T are the countable convex combinations of the rows of A that correspond to positive recurrent states. The row corresponding to any particular positive recurrent state is the unique equilibrium distribution supported by the irreducible recurrence class containing that state.*

PARTIAL PROOF. Let x be a positive recurrent state. Since $A = A^2$, the entry in position (x, x) of the matrix A equals the corresponding entry of A^2. Because of Proposition 12 this equality can be written as

$$(26.8) \qquad \frac{1}{m_x} = \sum_y A(x, y)\, A(y, x).$$

By Proposition 12, the first factor in any of the summands is nonzero only if y is in the same irreducible recurrence class as x in which case the second factor equals $1/m_x$. Hence (26.8) becomes

$$\frac{1}{m_x} = \frac{1}{m_x}\sum_y A(x, y).$$

Multiply both sides by the finite number m_x to obtain $\sum_y A(x, y) = 1$.

We leave the rest of proof to the reader. $\square$

Problem 50. Use Proposition 12 and Proposition 13 to finish the proof of Theorem 14.

The following is a corollary of the preceding theorem and the Renewal Theorem for Markov sequences.

Corollary 15. *Let T be an irreducible transition matrix on a countable state space Ψ. Then there exists an equilibrium distribution for T if and only if T is positive recurrent in which case there is a unique equilibrium distribution Q_0 given by $Q_0\{x\} = 1/m_x$. If in addition T is aperiodic, then for any Markov sequence X with transition matrix T,*

$$\lim_{n \to \infty} P[X_n = x] = \frac{1}{m_x}.$$

Example 8. Let T be a transition matrix with countable state space Ψ. A state $x \in \Psi$ is called *absorbing* if $T(x, x) = 1$. Suppose that all of the states in Ψ are either absorbing or transient, and denote the number of absorbing states by a and the number of transient states by t. (By Problem 34, $a > 0$ if Ψ is finite.) Label the absorbing states (if any) by the integers $1, \ldots, a$ and the transient states by $a + 1, \ldots, a + t$. Then

$$T = \begin{pmatrix} I_a & \mathbf{0} \\ B & C \end{pmatrix}$$

where for any positive integer l, I_l is the $l \times l$ identity matrix, $\mathbf{0}$ is a zero matrix, and B and C are, respectively, $t \times a$ and $t \times t$ matrices. Since each transient state can only be visited finitely many times with probability 1, it is easy to see that the entries of C^k converge to 0 as $k \to \infty$. Elementary matrix calculations then give that

$$\lim_{k \to \infty} T^k = \begin{pmatrix} I_a & \mathbf{0} \\ M & \mathbf{0} \end{pmatrix},$$

where M is the matrix $(I_t - C)^{-1} B$. In making this calculation, we have used the formula $(I_t - C)^{-1} = I_t + C + C^2 + \ldots$. The entries of the $t \times a$ matrix M are the *absorption probabilities*. Thus, $M(i, j)$ is the probability that the sequence with transition matrix T and transient initial state i eventually hits (and gets stuck at) the absorbing state j. Also,

$$1 - \sum_{j=1}^{a} M(i, j)$$

equals the probability that the sequence starting in the transient state i stays among the transient states forever. Of course, this probability equals 0 if $t < \infty$.

Problem 51. Calculate the absorption probabilities for the transition matrix

$$\begin{pmatrix} 1 & 0 & 0 & 0 & 0 \\ 0 & 1 & 0 & 0 & 0 \\ \frac{1}{3} & 0 & \frac{1}{3} & \frac{1}{3} & 0 \\ \frac{1}{2} & \frac{1}{2} & 0 & 0 & 0 \\ 0 & 0 & \frac{1}{4} & \frac{1}{2} & \frac{1}{4} \end{pmatrix}.$$

* **Problem 52.** Calculate the absorption probabilities for the transition matrix

$$\begin{pmatrix} 1 & 0 & 0 & 0 & 0 & 0 & 0 & \cdots \\ 0 & 1 & 0 & 0 & 0 & 0 & 0 & \cdots \\ \frac{1}{3\cdot2^0-1} & 0 & 0 & \frac{3\cdot2^0-2}{3\cdot2^0-1} & 0 & 0 & 0 & \cdots \\ 0 & \frac{1}{3\cdot2^1-1} & 0 & 0 & \frac{3\cdot2^1-2}{3\cdot2^1-1} & 0 & 0 & \cdots \\ \frac{1}{3\cdot2^2-1} & 0 & 0 & 0 & 0 & \frac{3\cdot2^2-2}{3\cdot2^2-1} & 0 & \cdots \\ 0 & \frac{1}{3\cdot2^3-1} & 0 & 0 & 0 & 0 & \frac{3\cdot2^3-2}{3\cdot2^3-1} & \cdots \\ \vdots & \vdots & \vdots & \vdots & \vdots & \vdots & \vdots & \ddots \end{pmatrix}$$

starting at the state represented by the third row and column of the matrix. Also, find the probability of staying among the transient states forever starting at that state.

Problem 53. Explain how the ideas in Example 8 can be used to help calculate the rows of the matrix A in Theorem 14 that correspond to transient states. *Hint:* Think of each irreducible recurrence class as a kind of absorbing state.

Problem 54. Consider a birth-death sequence for which the transition distributions μ_x satisfy $\mu_x\{x-1\} > 0$ for $x \neq 0$. Prove that there is an equilibrium distribution if and only if

$$r \stackrel{\text{def}}{=} \sum_{x=0}^{\infty} \left(\prod_{z=1}^{x} \frac{\mu_{z-1}\{z\}}{\mu_z\{z-1\}} \right) < \infty,$$

in which case there is a unique equilibrium distribution Q_0 given by

$$Q_0\{x\} = \frac{1}{r} \left(\prod_{z=1}^{x} \frac{\mu_{z-1}\{z\}}{\mu_z\{z-1\}} \right).$$

* **Problem 55.** [Ehrenfest urn sequence] In this model, there are two urns containing a total of b balls. The state of the system is the number of balls in the first urn, so the state space is $\{0, \ldots, b\}$. At each time unit, a ball is chosen at random from among all of the balls, and then the chosen ball is moved from the urn that contains it to the other urn. Thus, the entries in the transition matrix are given by

$$T(x, x-1) = \frac{x}{b}, \qquad 0 < x \leq b,$$

$$T(x, x+1) = \frac{b-x}{b}, \qquad 0 \leq x < b,$$

$$T(x, y) = 0, \qquad y \neq (x-1), (x+1).$$

Find the unique equilibrium distribution for T. *Hint:* Use the preceding problem.

CHAPTER 27
Exchangeable Sequences

Recall from Definition 4 of Chapter 22 that a finite or infinite sequence is exchangeable if its distribution is invariant under permutations of its terms. Our main goal in this chapter is to develop some simple ways to describe all exchangeable sequences. We will see that an infinite sequence is exchangeable if and only if it is conditionally iid, and that finite exchangeable sequences can be described in terms of a certain urn model. The characterization results, known as De Finetti Theorems, were foreshadowed in Example 2 and Problem 9 of Chapter 22. Problem 10 of Chapter 22 provides an example of a finite exchangeable sequence that is not conditionally iid, thereby showing that the finite case cannot be derived from the infinite case. We focus on exchangeable sequences whose terms take values in a finite set, and describe how some of the formulas obtained for such sequences can be used to obtain information about unknown parameters. Later we generalize to sequences whose terms take values in a Borel space. In the last section, we introduce some important distributions that are naturally connected with infinite exchangeable sequences and an urn model.

27.1. Finite exchangeable sequences

The following result describes how to form a finite exchangeable sequence from an arbitrary finite sequence of random variables. The construction is rather natural in settings in which there is no intrinsic order among the indices of the random variables.

Proposition 1. *Let n be a positive integer and let $(Z_1, \ldots, Z_n)$ be a sequence of random variables defined on a common probability space and taking values in a common Borel space. Let Π be a random permutation of the set $\{1, 2, \ldots, n\}$ with each permutation having probability $1/n!$. Furthermore, suppose that Π is independent of the sequence $(Z_1, Z_2, \ldots, Z_n)$. Then*

$$(27.1) \qquad (Z_{\Pi(1)}, Z_{\Pi(2)}, \ldots, Z_{\Pi(n)})$$

is a finite exchangeable sequence.

PROOF. Let ρ be a fixed permutation of $\{1, 2, \ldots, n\}$. Then the distribution of $\rho \circ \Pi$ is uniform over the set of permutations; that is, the distribution of $\rho \circ \Pi$ is the same as the distribution of Π. Also, each of Π and $\rho \circ \Pi$ is independent of $(Z_1, \ldots, Z_n)$. Therefore the distribution of the vector $(Z_{\rho(\Pi(1))}, \ldots, Z_{\rho(\Pi(n))})$ is the same as the distribution of the vector (27.1). $\square$

For the next two sections and the remainder of this section we consider exchangeable sequences having values in a finite set, and after that we return to general Borel spaces. The names of the members of the finite set are fundamentally irrelevant, but it will usually be notationally useful to call them $1, 2, \ldots, d$ for some $d \in \mathbb{Z}^+ \setminus \{0\}$.

Definition 2. Let $X = (X_1, \ldots, X_n)$ be a finite exchangeable sequence taking values in $\{1, 2, \ldots, d\}$. The *empirical density* of X is the random vector

$$Y = \frac{1}{n} \sum_{k=1}^{n} \mathbf{e}_{X_k},$$

where $\mathbf{e}_i, 1 \leq i \leq d$, are the standard basis vectors in $\mathbb{R}^d$. The *De Finetti measure* of X is the distribution of Y.

Let Y^i denote the ith coordinate of Y. The word 'density' is justified in the preceding definition by the observation that the function $i \rightsquigarrow Y^i$ is the density with respect to counting measure on $\{1, \ldots, d\}$ of the random distribution corresponding to the empirical distribution function defined in the paragraph preceding Theorem 15 of Chapter 12.

The values of an empirical density belong to

$$\Delta^{(d-1)} \stackrel{\text{def}}{=} \{(y_1, \ldots, y_d): y_1 + \cdots + y_d = 1, \, y_i \geq 0 \text{ for each } i\},$$

called the *standard* $(d-1)$-*simplex,* and the De Finetti measure is a distribution on $\Delta^{(d-1)}$.

Problem 1. Let X_k denote $1+$ the indicator random variable of 'heads on flip k' of a sequence of n independent flips of a fair coin. Find the De Finetti measure of the sequence $(X_1, X_2, \ldots, X_n)$.

* **Problem 2.** Let X_1 be the result of the roll of a fair die and $X_2 = 7 - X_1$. Show that (X_1, X_2) is exchangeable and find its De Finetti measure.

Problem 3. Fix an integer $n \in [2, 6]$. Consider an infinite sequence of independent rolls of a fair die. Let X_1 be the result of the first roll, and for $2 \leq m \leq n$, let X_m be the result of the first roll that is different from X_k, $1 \leq k < m$. Find the De Finetti measure of $(X_1, \ldots, X_n)$.

* **Problem 4.** Here is a way of generating an exchangeable sequence of 96 $\mathbb{Z}$-valued random variables. For $1 \leq x \leq 12$, place 8 balls labelled x in an urn and draw the resulting 96 balls out of the urn one at a time without replacement, noting the label on each ball as it is drawn. Find the De Finetti measure.

Problem 5. Suppose that 48 balls are drawn at random from a pile of the 96 balls described in the preceding exercise and that these 48 balls are then placed in an urn from which 48 drawings are made one at a time without replacement. Find the De Finetti measure for the resulting exchangeable sequence of length 48.

* **Problem 6.** Four balls numbered 1, 2, 3, and 4 are placed in an urn and then 4 balls are drawn with or without replacement according as a single independent flip of a coin is heads or tails. Is the random sequence of length 4 thus obtained an exchangeable sequence? If so, find its De Finetti measure.

Problem 7. Modify the preceding problem by replacing the single flip by three independent flips, one after each of the first three draws to determine whether the ball drawn on that flip is to be replaced.

Problem 8. Let $(Z_1, \ldots, Z_n)$ be a sequence of $\{1, \ldots, d\}$-valued random variables. In terms of the distribution of $(Z_1, \ldots, Z_n)$, describe the De Finetti measure of the exchangeable sequence constructed from $(Z_1, \ldots, Z_n)$ as in Proposition 1.

A probability measure on Δ^1 is a distribution on the set of pairs of the form $(p, (1-p))$, $0 \le p \le 1$. Since the second coordinate of such an ordered pair is determined by the first coordinate, a distribution on Δ^1 may also be regarded as a distribution on $[0, 1]$. Whenever we wish to adopt this point of view, we indicate our desire by a change of notation, in that we replace an exchangeable sequence of $\{1, 2\}$-valued random variables with the equivalent exchangeable sequence of Bernoulli (that is, $\{0,1\}$-valued) random variables obtained by subtracting 1 from each member of the sequence. We also replace the De Finetti measure of the original sequence by the distribution on $[0, 1]$ obtained by taking the first-coordinate marginal of the original De Finetti measure. We will distinguish the viewpoint of this paragraph from that arising from Definition 2 by saying whether the De Finetti measure is a distribution on $[0, 1]$ or on Δ^1, or alternatively whether the empirical density is $[0, 1]$-valued or Δ^1-valued. Note that the De Finetti measure on $[0, 1]$ equals the distribution of $\frac{1}{n} \sum_{k=1}^{n} X_k$, where $(X_1, \ldots, X_n)$ is the exchangeable sequence of Bernoulli random variables.

Problem 9. Treat Problem 1 along the lines suggested by the preceding paragraph, and express the De Finetti measure as a distribution on $[0, 1]$.

* **Problem 10.** In case $(Z_1, Z_2, \ldots, Z_n)$ is a sequence of standard Bernoulli random variables, describe the answer to Problem 8 as a distribution on $[0, 1]$.

Problem 8 and Problem 10 might make one ask: What role is exchangeability playing in the definition of 'empirical density' and 'De Finetti measure'? One answer, as we will soon see, is that under the assumption of exchangeability, the distribution of a sequence is determined by its De Finetti measure, which would not be the case without this assumption.

From Definition 2, it is clear that the De Finetti measure of an exchangeable length-n sequence of $\{1, \ldots, d\}$-valued random variables is supported by the set of points in $\Delta^{(d-1)}$ having coordinates that are integral multiples of $1/n$. This assertion and its converse constitute part of the following result.

Theorem 3. [De Finetti (finite case, special version)] *The distribution of a length-n exchangeable sequence of $\{1, \ldots, d\}$-valued random variables is uniquely determined by its De Finetti measure, and the family of such De Finetti measures consists of the probability measures on the standard $(d-1)$-simplex that are supported by the set of points whose coordinates are integral multiples of $1/n$.*

PROOF. As mentioned in the paragraph preceding the theorem, all De Finetti measures of the exchangeable sequences of the theorem have the property described in the theorem.

Let μ denote a probability measure that is supported by the set of points in $\Delta^{(d-1)}$ whose coordinates are integral multiples of $1/n$, and let Y be a random vector with distribution μ. Let Π be a uniformly distributed random permutation of $\{1, 2, \ldots, n\}$ that is independent of Y. There exist $\{1, \ldots, d\}$-valued random variables $Z_1 \leq Z_2 \leq \cdots \leq Z_n$ such that

$$Y = \frac{1}{n} \sum_{k=1}^{n} e_{Z_k}.$$

Set $X_m = Z_{\Pi(m)}$ for $m = 1, 2, \ldots, n$. By Proposition 1, $(X_1, X_2, \ldots, X_n)$ is exchangeable. It is clear that Y is the empirical density of $(X_1, \ldots, X_n)$.

For the uniqueness, let Y denote the empirical density of an exchangeable sequence $(X_1, \ldots, X_n)$. Since Y is unchanged under permutations of $(X_1, \ldots, X_n)$, in order that

$$P[Y = y, X_m = x_m \text{ for } 1 \leq m \leq n]$$

be unchanged under such permutations it is necessary that

$$P[X_m = x_m \text{ for } 1 \leq m \leq n \mid Y = y]$$

be unchanged under these permutations whenever $P[Y = y] > 0$. Thus this conditional probability must be uniform on the set of $(x_m : 1 \leq m \leq n)$ for which

$$y = \frac{1}{n} \sum_{m=1}^{n} e_{x_m}.$$

Therefore the distribution of an exchangeable sequence $(X_1, \ldots, X_n)$ is determined by the distribution of its empirical density (that is, by its De Finetti measure). □

The following problem shows that the construction in the preceding proof is equivalent to a certain urn model.

Problem 11. Discuss the following urn model in connection with Theorem 3. For $1 \leq i \leq d$, nY^i balls of color i are placed in an urn, and then the urn is emptied one ball at a time. (Here, Y^i denotes the ith coordinate of the random vector Y.)

For the forthcoming collection of problems involving explicit formulas, the following notation is useful:

$$\begin{pmatrix} n \\ r_1 \ r_2 \ \cdots \ r_d \end{pmatrix} \overset{\text{def}}{=} \frac{n!}{r_1! r_2! \cdots r_d!} \quad \text{with } n = \sum_{i=1}^d r_i .$$

This, the coefficient of $\alpha_1^{r_1} \alpha_2^{r_2} \cdots \alpha_d^{r_d}$ in the expansion of $(\alpha_1 + \alpha_2 + \cdots + \alpha_d)^n$, is called a *multinomial coefficient*.

Problem 12. Let Y, n, and $(X_1, \ldots, X_n)$ be as in Theorem 3 and its proof, and denote the ith coordinate of Y by Y^i. Show that

$$P[X_m = x_m \text{ for } 1 \leq m \leq n \mid Y]$$

(27.2)
$$= \begin{cases} \left(\begin{matrix} n \\ _{nY^1} \ \cdots \ _{nY^d} \end{matrix} \right)^{-1} & \text{if } \#\{m : x_m = i\} = nY^i, \ 1 \leq i \leq d \\ 0 & \text{otherwise} \end{cases}$$

for $x_1, \ldots, x_n \in \{1, \ldots, d\}$.

Problem 13. Fix $m \geq 1$. Adjoin $w_i = \#\{k \leq m : x_k = i\}$ for $1 \leq i \leq d$ to the notation in the preceding problem. Show that

$$P[X_k = x_k \text{ for } 1 \leq k \leq m \mid Y]$$

$$= \left(\begin{matrix} n \\ _{nY^1} \ \cdots \ _{nY^d} \end{matrix} \right)^{-1} \left(\begin{matrix} (n - m) \\ (nY^1 - w_1) \ \cdots \ (nY^d - w_d) \end{matrix} \right)$$

if

$$w_i \leq nY^i, \quad 1 \leq i \leq d,$$

and

$$P[X_k = x_k \text{ for } 1 \leq k \leq m \mid Y] = 0$$

otherwise.

Problem 14. Let μ be the De Finetti measure of a length-n exchangeable sequence $(X_1, \ldots, X_n)$ of $\{1, 2, \ldots, d\}$-valued random variables, denote the corresponding empirical density by Y, and let $W_i = \#\{k \leq m : X_k = i\}$. For $1 \leq m \leq n$ and $w_i \in \mathbb{Z}^+$ for which $\sum_{i=1}^d w_i = n$, prove that

$$P[nY^i = w_i \text{ for } 1 \leq i \leq d \mid X_1, \ldots, X_m]$$

$$= \frac{\frac{w_1! \cdots w_d!}{(w_1 - W_1)! \cdots (w_d - W_d)!} \mu\left\{ \left(\frac{w_1}{n}, \ldots, \frac{w_d}{n} \right) \right\}}{\displaystyle\sum_{\substack{z_i \geq W_i \\ z_1 + \cdots + z_d = n}} \frac{z_1! \cdots z_d!}{(z_1 - W_1)! \cdots (z_d - W_d)!} \mu\left\{ \left(\frac{z_1}{n}, \ldots, \frac{z_d}{n} \right) \right\}}$$

if $w_i \geq W_i$ for $1 \leq i \leq d$, and that this conditional probability equals 0 otherwise. The case of a 0 denominator requires a comment.

* **Problem 15.** Let $(X_1, \ldots, X_n)$ be an exchangeable sequence of Bernoulli random variables whose De Finetti measure is the uniform distribution on $\{\frac{0}{n}, \frac{1}{n}, \ldots, \frac{n}{n}\}$. Calculate the distribution of X_1, and in case $n > 1$, the distribution of the pair (X_1, X_2) and the conditional distribution of $\frac{1}{n} \sum_{m=1}^{n} X_m$ given $\sigma(X_1, X_2)$.

Problem 16. Generalize Problem 15 to $\mathbb{R}^d$ for a De Finetti measure that is uniform on the set of all members of $\Delta^{(d-1)}$ whose coordinates are integral multiples of $1/n$.

Problem 17. Let $X = (X_1, X_2, \ldots, X_n)$ be an exchangeable sequence of Bernoulli random variables, and suppose that $n \geq 3$. Calculate the distributions of X_1, (X_1, X_2), and (X_1, X_2, X_3) in terms of the De Finetti measure of X. Present your answer in two forms depending on whether the De Finetti measure is viewed as a distribution on Δ^1 or on the interval $[0, 1]$.

Since the first m terms, $m \leq n$, of an exchangeable sequence of length n constitute an exchangeable sequence of length m, it is natural to ask how the De Finetti measure of the shorter sequence is related to that of the longer sequence. The answer is presented in the following result in which the natural extension of 'martingale' to finite sequences of $\mathbb{R}^d$-valued random variables plays a role. Later such an extension will be used for infinite sequences.

Proposition 4. *Let μ denote the De Finetti measure for a length-n exchangeable sequence $(X_1, \ldots, X_n)$ of $\{1, \ldots, d\}$-valued random variables. For $m \leq n$, set*

$$(27.3) \qquad Y_m = \frac{1}{m} \sum_{k=1}^{m} \mathbf{e}_{X_k},$$

and denote the distribution of Y_m by ν_m. Then ν_m is the De Finetti measure of the exchangeable sequence $(X_1, X_2, \ldots, X_m)$ and is related to μ via the formula

$$
(27.4) \quad
\begin{aligned}
&\nu_m\{\tfrac{1}{m}(w_1, \ldots, w_d)\} \\
&= \sum_{\substack{z_i \geq w_i \\ z_1 + \cdots + z_d = n}} \frac{\binom{m}{w_1 \ \cdots \ w_d} \binom{(n-m)}{(z_1 - w_1) \ \cdots \ (z_d - w_d)}}{\binom{n}{z_1 \ \cdots \ z_d}} \mu\{\tfrac{1}{n}(z_1, \ldots, z_d)\}
\end{aligned}
$$

for $w_1, \ldots, w_d \in \mathbb{Z}^+$ for which $\sum_{i=1}^{d} w_i = m$. Moreover, $(Y_1, Y_2, \ldots, Y_n)$ is a reverse martingale with respect to the reverse filtration $(\mathcal{G}_m : m = 1, 2, \ldots)$, where

$$\mathcal{G}_m = \sigma(Y_m, Y_{m+1}, \ldots, Y_n).$$

PROOF. By definition, ν_m is the De Finetti measure of $(X_1, X_2, \ldots, X_m)$. Clearly, $mY_m \leq nY_n$ coordinate by coordinate. Thus

$$
\begin{aligned}
&\nu_m\{\tfrac{1}{m}(w_1, \ldots, w_d)\} \\
&= \sum_{\substack{z_i \geq w_i \\ z_1 + \cdots + z_d = n}} P[mY_m = (w_1, \ldots, w_d) \mid nY_n = (z_1, \ldots, z_d)] \, \mu\{\tfrac{1}{n}(z_1, \ldots, z_d)\}.
\end{aligned}
$$

The conditional probability in this formula is equal to a product one factor of which is the number of ways of choosing $(x_1, \ldots, x_m)$ so that $(w_1, \ldots, w_d)$, defined in Problem 13, equals mY_m and the other factor of which is the conditional distribution obtained in that problem for any such $(x_1, \ldots, x_m)$. The formula in the proposition follows.

By linearity

$$(27.5) \qquad \sum_{k=1}^{m} E(e_{X_k} \mid \mathcal{G}_m) = E(mY_m \mid \mathcal{G}_m) = mY_m \,.$$

Because $Y_m, \ldots, Y_n$ all remain unchanged by permutations of $(X_1, \ldots, X_m)$, conditional exchangeability of the sequence $(X_1, \ldots, X_m)$ given $\mathcal{G}_m$ follows from the (unconditional) exchangeability of $(X_1, \ldots, X_m)$. Hence the terms in the sum on the left side of (27.5) are all equal, and each one of them equals Y_m. Therefore for $2 \le m \le n$,

$$E(Y_{m-1} \mid \mathcal{G}_m) = \frac{1}{m-1} \sum_{k=1}^{m-1} E(e_{X_k} \mid \mathcal{G}_m) = Y_m \,,$$

thus showing that $(Y_1, Y_2, \ldots, Y_n)$ is a reverse martingale with respect to the reverse filtration $(\mathcal{G}_1, \mathcal{G}_2, \ldots, \mathcal{G}_n)$. $\square$

Problem 18. Let $(X_1, \ldots, X_n)$ be an exchangeable sequence of $\{1, \ldots, d\}$-valued random variables. Show that

$$(e_{X_1}, e_{X_2}, \ldots, e_{X_n})$$

is an exchangeable sequence of multivariate Bernoulli random vectors. Restate some of the results of this section using multivariate Bernoulli terminology.

27.2. Infinite exchangeable sequences

We need a variation on the last assertion in Proposition 4.

Lemma 5. *Let X be an infinite exchangeable sequence of $\{1, \ldots, d\}$-valued random variables, and for each m, define Y_m by (27.3) and $\mathcal{G}_m$ by*

$$(27.6) \qquad \mathcal{G}_m = \sigma(Y_m, Y_{m+1}, Y_{m+2}, \ldots) \,.$$

Then the infinite sequence $(Y_1, Y_2, \ldots)$ is a reverse martingale with respect to the reverse filtration $(\mathcal{G}_1, \mathcal{G}_2, \ldots)$.

Problem 19. Prove the preceding lemma.

Definition and Proposition 6. Let $X = (X_1, X_2, \ldots)$ be an infinite exchangeable sequence of $\{1, \ldots, d\}$-valued random variables. Then

$$(27.7) \qquad\qquad Y = \lim_{n \to \infty} \frac{1}{n} \sum_{m=1}^{n} e_{X_m}$$

exists almost surely. The $\Delta^{(d-1)}$-valued random variable Y is the *limiting empirical density* of X, and the distribution of Y is the *De Finetti measure* of X.

PROOF. The result is an immediate consequence of Lemma 5 and the Reverse Martingale Convergence Theorem. □

Problem 20. Suppose a fair coin is flipped once. Let $X_m = 2$ for all m if the result of the flip is heads, and $X_m = 1$ for all m if the result is tails. Show that the sequence $(X_m : m = 1, 2, \ldots)$ is exchangeable and calculate its De Finetti measure on Δ^1.

Problem 21. Suppose that a fair die is rolled, and that based on the outcome U, first a coin is chosen whose probability of heads is $U/7$ and then it is flipped infinitely many times. Find the De Finetti measure of the infinite sequence of coin flips. Express your answer as a distribution on Δ^1 and also as a distribution on $[0, 1]$.

As an aid for its proof, the following theorem contains a more complicated notation for a limiting empirical density than might seem necessary.

Theorem 7. [De Finetti (infinite case, special version)] *Every probability distribution μ on the standard $(d-1)$-simplex is the De Finetti distribution of an infinite exchangeable sequence of $\{1, \ldots, d\}$-valued random variables whose distribution is determined uniquely by μ via the two facts that (i) it is conditionally iid given its limiting empirical density Y_∞ and (ii) the conditional distribution of each term in the sequence is Y.*

PROOF. In view of Example 2 of Chapter 22, we only need prove that given its limiting empirical density Y_∞, every exchangeable sequence $(X_1, X_2, X_3 \ldots)$ is conditionally iid with common conditional distribution Y_∞, the ith coordinate of which we denote by Y_∞^i.

Let Y_n denote the empirical density of the finite sequence $(X_1, \ldots, X_n)$ and Y_n^i its ith coordinate. Fix m and a sequence $(x_1, \ldots, x_m)$ of members of $\{1, \ldots, d\}$. For $1 \leq i \leq d$, let $w_i = \sharp\{k \leq m : x_m = i\}$. For $n \geq m$, we obtain from

Problem 13 that

$$P[X_k = x_k \text{ for } 1 \le k \le m \mid Y_n]$$

(27.8)
$$= \left(\begin{matrix} n \\ nY_n^1 \ \dots \ nY_n^d \end{matrix} \right)^{-1}$$
$$\cdot \left(\begin{matrix} (n-m) \\ (nY_n^1 - w_1) \ \dots \ (nY_n^d - w_d) \end{matrix} \right)$$

if $nY_n^i \ge w_i$ for each i, and that otherwise this conditional probability equals 0. Moreover, a modification of the solution of Problem 13 shows that the left side of (27.8) can be replaced by

$$P[X_k = x_k \text{ for } 1 \le k \le m \mid Y_n, Y_{n+1}, \dots, Y_\infty]$$

and therefore by

$$P[X_k = x_k \text{ for } 1 \le k \le m \mid Y_n, Y_\infty].$$

Since $Y_n \to Y_\infty$ a.s. as $n \to \infty$, an easy limiting argument applied to the right side of (27.8) gives

$$\lim_{n \to \infty} P[X_k = x_k \text{ for } 1 \le k \le m \mid Y_n, Y_\infty] = [Y_\infty^1]^{w_1} \dots [Y_\infty^d]^{w_d}.$$

Take the conditional expectation given Y_∞ of both sides and use the Conditional Bounded Convergence Theorem to obtain

$$P[X_k \in B_k \text{ for } 1 \le k \le m \mid Y_\infty] = [Y_\infty^1]^{w_1} \dots [Y_\infty^d]^{w_d}.$$

This formula characterizes the property of being conditionally iid given Y_∞ with common conditional density $i \rightsquigarrow Y_\infty^i$. $\square$

Problem 22. Find a single probability measure μ that can serve as a De Finetti measure for exchangeable sequences of lengths 2, 4, and ∞, such that different distributions are obtained for X_1 in the three different settings.

The following problem describes an explicit construction of infinite exchangeable sequences of Bernoulli random variables.

Problem 23. Let μ and λ, respectively, denote an arbitrary probability measure and Lebesgue measure on $[0, 1]$. Set

$$(\Omega, \mathcal{F}, P) = ([0, 1], \mathcal{A}, \mu) \times \bigotimes_{m=1}^{\infty} ([0, 1], \mathcal{A}, \lambda),$$

where $\mathcal{A}$ is the Borel σ-field on $[0, 1]$. Write a typical member ω of Ω as $\omega = (\omega_0, \omega_1, \omega_2, \dots)$, where each $\omega_m \in [0, 1]$. For $m = 1, 2, \dots$, let

$$X_m(\omega) = \begin{cases} 1 & \text{if } \omega_m \le \omega_0 \\ 0 & \text{if } \omega_m > \omega_0. \end{cases}$$

Show that $(X_m \colon m = 1, 2, \dots)$ is an infinite exchangeable sequence of Bernoulli random variables whose De Finetti measure is μ.

Problem 24. Let Y denote the limiting empirical density of an infinite exchangeable sequence $(X_1, X_2, \dots)$ of $\{1, 2, \dots, d\}$-valued random variables. Fix m and set $W_i = \sharp\{k \leq m \colon X_k = i\}$. Prove that the conditional distribution η_m of Y given $(X_1, \dots, X_m)$ has a Radon-Nikodym derivative with respect to the De Finetti measure μ of X given by

$$\frac{d\eta_m}{d\mu}(p) = \frac{p_1^{W_1} \cdots p_d^{W_d}}{\int_{\Delta^{(d-1)}} q_1^{W_1} \cdots q_d^{W_d} \, \mu(dq)}$$

where, of course, p_i and q_i are the ith coordinates of p and q, respectively, and $p \in \Delta^{(d-1)}$. The case of a 0 denominator requires a comment.

Problem 25. Find the (random) vector p that maximizes the Radon-Nikodym derivative in the preceding problem.

Problem 26. Let μ denote the De Finetti measure of an infinite exchangeable sequence $(X_1, X_2, \dots)$ of $\{1, \dots, d\}$-valued random variables. For $m < \infty$, let ν_m be the De Finetti measure of the finite exchangeable sequence $(X_1, X_2, \dots, X_m)$. Show that

$$\nu_m\{\tfrac{1}{m}(w_1, \dots, w_d)\} = \binom{m}{w_1 \ \cdots \ w_d} \int_{\Delta^{(d-1)}} y_1^{w_1} \cdots y_d^{w_d} \, \mu(dy)$$

for $w_1, \dots, w_d \in \mathbb{Z}^+$ for which $\sum_{i=1}^d w_i = m$.

Problem 27. If you have studied Chapter 14, continue the preceding problem by showing that $\nu_m \to \mu$ as $m \to \infty$.

Problem 28. At least at the intuitive level, relate Proposition 4 and Problem 26 by letting $n \to \infty$ and each $z_i \to \infty$ in (27.4) so that $\frac{z_i}{n} \to y_i$.

Problem 29. Let $X = (X_1, X_2, \dots, X_n)$ be a finite exchangeable Bernoulli sequence with De Finetti measure μ on $[0, 1]$. Prove that a necessary condition for the distribution of X to be the distribution of the first n terms of an infinite exchangeable sequence is that $\mu\{\frac{w+1}{n}\} + \mu\{\frac{w-1}{n}\} - 2\mu\{\frac{w}{n}\} \geq 0$ for $1 \leq w \leq n - 1$.

27.3. Posterior distributions

Problem 24 and to a lesser extent Problem 14 contain important results for a certain type of application. With respect to Problem 24, an experimenter views the limiting empirical density Y as the quantity of interest. From the perspective of the experimenter, Y is random with a distribution μ based on relevant factors such as the scientific judgement of people with experience. The distribution μ is called the *prior distribution*. In an actual real-world situation, Y is of course not random; it has a certain value. It may for instance, equal the probability that a certain asymmetrical coin comes up 'heads'. If one were to flip this coin infinitely many times, one would, by the Strong Law of Large Numbers, learn the value of Y. In practice one does only finitely many such experiments, and on the basis of the results of these, obtains the distribution η_m of Y described in Problem 24. Since η_m is calculated on the basis of the experimental results, it is called a *posterior distribution* of Y. The adjective 'Bayesian' is attached to statisticians who favor the methodology: get good prior densities or distributions and then calculate posterior densities or distributions based on results of experiments.

The formula obtained in Problem 14 may be regarded as a formula for the posterior distribution of the contents of an urn in terms of the prior distribution and some observations based on sampling without replacement.

Problem 30. An urn contains 6 balls, each of which is colored red, yellow, or blue. Show that there are 28 possible color arrangements, including those that do not use all 3 colors. Suppose that the prior distribution assigns probability $1/28$ to each of the 28 arrangements. Then three balls are drawn without replacement and one ball of each color is obtained. Use Problem 14 to calculate the posterior probabilities of the various color arrangements of the 6 balls originally in the urn.

* **Problem 31.** Suppose that the prior distribution of a conditionally iid Bernoulli sequence is standard beta with parameters α and β, when viewed as a measure on $[0, 1]$. Show that any posterior distribution is also beta, and find a formula for the parameters of the posterior distribution in terms of α, β, and the observations.

* **Problem 32.** For the case of exchangeable Bernoulli sequences, extend the result of Problem 24 by describing the conditional distribution of (Y, X_{m+1}, X_{m+2}) given $(X_1, \ldots, X_m)$. Use your answer to calculate the conditional expectation of X_{m+1} given $(X_1, \ldots, X_m)$ for the case where the De Finetti measure is Lebesgue measure on $[0, 1]$.

Problem 33. Let $X = (X_1, X_2, \ldots)$ be an infinite exchangeable sequence with De Finetti measure μ. In terms of μ and $(X_1, \ldots, X_m)$, describe the conditional distribution of X given $(X_1, \ldots, X_m)$.

The following four problems treat infinite exchangeable Bernoulli sequences having De Finetti measures that are beta distributions. In this special setting martingales appear, in addition to the reverse martingales that arise generally.

Problem 34. Let $X = (X_1, X_2, \ldots)$ be an infinite exchangeable Bernoulli sequence the De Finetti measure of which is a standard beta distribution with parameters α and β. For $n = 0, 1, 2, \ldots$, let $S_n = \sum_{m=1}^{n} X_m$ and $\mathcal{F}_n = \sigma(X_1, X_2, \ldots, X_n)$, where, according to standard conventions, $S_0 = 0$ and $\mathcal{F}_0$ is the trivial σ-field. Show that the sequence

$$\left(\frac{\alpha + S_n}{\alpha + \beta + n} : n = 0, 1, 2, \ldots \right)$$

is a martingale with respect to the filtration $(\mathcal{F}_n : n = 0, 1, \ldots)$.

Problem 35. Let Y denote the limiting empirical distribution of the exchangeable sequence X of the preceding problem. Find a formula for $E(Y \mid \mathcal{F}_n)$, where $\mathcal{F}_n = \sigma(X_1, \ldots, X_n)$. *Hint:* Copious calculations are not required.

Problem 36. Suppose that α and β in Problem 34 and Problem 35 are positive integers and relate these exercises to Polya urns. In particular, calculate the distribution of the limiting proportion of blue balls in the urn in case the urn initially has one blue ball and one orange ball and one additional ball of the drawn color is inserted after each draw.

Problem 37. Modify the description of Polya urns so that the preceding problem becomes interesting even when the positive numbers α and β are not necessarily integers. *Hint:* Imagine that the urn contains α ounces of blue sand and β ounces of green sand.

27.4. † Generalization to Borel spaces

We set the goal of generalizing earlier concepts and results to the setting of exchangeable sequences of random variables taking values in a Borel space Ψ. Proposition 1 asserting that finite exchangeable sequences can be obtained by randomly permuting the order of an arbitrary sequence of random variables is stated at the desired level of generalization.

All probability measures on the finite set $\{1, \ldots, d\}$ have densities with respect to counting measure. For this reason we were able to focus on densities rather than distributions in Definition 2. Now we must focus on distributions. This change forces a change in the definition of 'De Finetti measure'—a technical change, not just a straight generalization.

Definition 8. Let $X = (X_1, \ldots, X_n)$ be a finite exchangeable sequence of random variables taking values in a Borel space $(\Psi, \mathcal{G})$. The *empirical distribution* of X is the random distribution Y defined by

$$Y(B) = \frac{1}{n} \sum_{k=1}^{n} I_B \circ X_k, \quad B \in \mathcal{G},$$

where I_B denotes the indicator function of B. The *De Finetti measure* of X is the distribution of Y.

Here the De Finetti measure is a distribution on the space $\mathcal{Q}$ of distributions on Ψ, whereas in Definition 2 it is a distribution on the space of densities on $\Psi = \{1, \ldots, d\}$, viewed as members of $\Delta^{(d-1)}$.

Problem 38. Find the De Finetti measure of an iid sequence of 2 random variables uniformly distributed on $[0, 1]$.

* **Problem 39.** For $n = 2$ apply Proposition 1 in case Z_1 and Z_2 are independent standard exponentially distributed random variables with means 1 and 2, respectively. For the corresponding exchangeable sequence (X_1, X_2), calculate the densities of X_1, X_2, and (X_1, X_2). Also calculate the De Finetti measure.

Problem 40. On intuitive grounds guess whether the correlation of X_1 and X_2 in the preceding problem is positive, negative, or 0. Then calculate the correlation.

Here is our desired generalization of Theorem 3.

Theorem 9. [De Finetti (finite case)] *Let $\mathcal{Q}$ denote the space of probability measures on a Borel space Ψ. The distribution of a length-n exchangeable sequence of Ψ-valued random variables is uniquely determined by its De Finetti measure, and the family of such De Finetti measures consists of the probability measures on $\mathcal{Q}$ that assign probability 1 to the set of distributions of the form*

$$(27.9) \qquad\qquad \frac{1}{n} \sum_{k=1}^{n} \delta_{x_k} \, ,$$

where δ_x denotes the delta distribution at x.

Much of the proof of Theorem 3 carries over to form a portion of the proof of Theorem 9. However, the uniqueness aspect of the proof of Theorem 3 does not in itself establish the uniqueness asserted in Theorem 9. What it does do is give enough information so that the following lemma applies to give the desired uniqueness.

Lemma 10. *Let U and V be two finite or infinite equally long sequences of random variables having values in a Borel space Ψ. Let n be no larger than the length of the sequences. Suppose that for every finite measurable partition $(C_1, \ldots, C_d)$ of Ψ,*

$$(27.10) \qquad P[U_m \in B_m \text{ for } 1 \le m \le n] = P[V_m \in B_m \text{ for } 1 \le m \le n]$$

for all choices of $B_m \in \{C_1, \ldots, C_d\}$. Then the distributions of $(U_1, \ldots, U_n)$ and $(V_1, \ldots, V_n)$ are identical. If U and V are infinite sequences and (27.10) holds for every n, then U and V have the same distribution.

Problem 41. Prove the preceding lemma.

Problem 42. Check the assertion made in the paragraph preceding Lemma 10 that the proof of Theorem 9 is now essentially complete.

Problem 43. Create an urn model relevant for Theorem 9.

For treating infinite sequences we need a variation of Lemma 5.

Lemma 11. *Let X be an infinite exchangeable sequence of random variables taking values in a Borel space Ψ, and for each m, define Y_m by*

$$Y_m(B) = \frac{1}{m} \sum_{k=1}^{m} I_B \circ X_k$$

and $\mathcal{G}_m$ by

$$\mathcal{G}_m = \sigma(Y_m, Y_{m+1}, Y_{m+2}, \dots).$$

Then for each measurable B, the sequence $(Y_1(B), Y_2(B), \dots)$ is a reverse martingale with respect to the reverse filtration $(\mathcal{G}_1, \mathcal{G}_2, \dots)$.

Problem 44. Prove the preceding lemma, taking care that your proof uses the specified filtration, not just the minimal filtration of $(Y_1(B), Y_2(B), \dots)$.

As for the special case of $\{1, \dots, d\}$-valued random variables, the general definition of 'De Finetti measure' for infinite exchangeable sequences is based on a fact that must be proved.

Definition and Proposition 12. Let $(\Psi, \mathcal{G})$ be a Borel space and X an infinite exchangeable sequence of Ψ-valued random variables. Then there exists a random distribution Y on $(\Psi, \mathcal{G})$ such that for each $B \in \mathcal{G}$,

$$(27.11) \qquad\qquad Y(B) = \lim_{n \to \infty} \frac{1}{n} \sum_{m=1}^{n} I_B \circ X_m$$

almost surely. The distribution of Y is the *De Finetti measure* of X, and Y itself is the *limiting empirical distribution* of X

PARTIAL PROOF. We assume that $\Psi = \mathbb{R}$, without loss of generality. (Why can we do this?) Let Y_m and $\mathcal{G}_m$ be as in Lemma 11, and set $\mathcal{G}_\infty = \bigcap_{m=1}^{\infty} \mathcal{G}_m$. Fix a measurable set B. By Lemma 11 and the Reverse Martingale Convergence Theorem, the sequence $(Y_m(B) \colon m \geq 1)$ converges almost surely to a random variable $Y_\infty(B)$, and

$$(27.12) \qquad\qquad Y_\infty(B) = E(Y_n(B) \mid \mathcal{G}_\infty) \text{ a.s.},$$

for all m. By working with $Y_\infty((-\infty, y])$ for y rational and then using limits from the right we obtain a random distribution function. Let Y denote the corresponding random distribution.

We leave it for the reader to prove that the set $\{(\omega, y) \colon Y(\omega)\{y\} > 0\}$ is a measurable subset of $\Omega \times \mathbb{R}$, where Ω denotes the underlying probability space. By the Fubini Theorem the integral of its indicator function with respect to $P \times \lambda$, where λ denotes Lebesgue measure, can be calculated via iterated integrals in either order. Integration first with respect to λ gives 0 for every $\omega \in \Omega$. Then integration with respect to P gives 0 as the value of the iterated integral. Hence, when the iteration is done in the other order, it must be that for almost every

$y \in \mathbb{R}$, the integral with respect to P must equal 0. Therefore, there is a (nonrandom) dense subset D of $\mathbb{R}$ such that $Y\{y\} = 0$ a.s. for $y \in D$. It is clear that $Y_\infty(-\infty, y] = Y(-\infty, y]$ a.s. for $y \in D$. Also, $Y_\infty(\mathbb{R}) = Y(\mathbb{R})$. To show that

$$(27.13) \qquad Y_\infty(B) = Y(B) \text{ a.s.}$$

for each Borel $B \subseteq \mathbb{R}$ and thereby finish the proof, we only need show that the collection of B for which (27.13) holds is a Sierpiński class, for then the Sierpiński Class Theorem applies.

The collection of B for which (27.13) holds is clearly closed under proper differences. That it is closed under increasing limits follows from (27.12) and the Conditional Monotone Convergence Theorem. $\square$

Problem 45. Complete the proof of Proposition 6 by showing the measurability of $\{(\omega, y): Y(\omega)\{y\} > 0\}$.

Problem 46. Show that the conclusion of Proposition 6 cannot be strengthened to the assertion that almost surely, (27.11) holds for every Borel set B.

Theorem 13. [De Finetti (infinite case)] *Let Q denote the space of probability measures on a Borel space Ψ. The distribution of an infinite exchangeable sequence of Ψ-valued random variables is uniquely determined by its De Finetti measure via the two facts that it is (i) conditionally iid given its limiting empirical distribution Y and (ii) the conditional distribution of each term in the sequence is Y. The family of such De Finetti measures consists of all probability measures on Q.*

PROOF. In view of Example 2 of Chapter 22, we only need prove that each exchangeable sequence $(X_1, X_2, X_3 \ldots)$ is conditionally iid given its limiting empirical distribution Y and that the common conditional distribution is Y. This conclusion follows from Theorem 7 and Lemma 10. $\square$

Example 1. Denote by μ the distribution of the random distribution Y defined by

$$Y(\omega)\{i\} = (1 - V(\omega))\, V^{i-1}(\omega), \quad i = 1, 2, \ldots,$$

where V is uniformly distributed on the interval $(0, 1)$. Let us study an infinite exchangeable sequence $(X_1, X_2, \ldots)$ having De Finetti measure μ.

For $v \in [0, 1]$ and positive integers $x_1, x_2, \ldots, x_m$ we calculate

$$P[V \le v, X_1 = x_1, \ldots, X_m = x_m]$$
$$(27.14) \qquad = \int_0^v \left[\prod_{k=1}^m (1 - u) u^{x_k - 1} \right] du = \int_0^v (1 - u)^m u^{s-m}\, du,$$

where $s = \sum_{k=1}^{m} x_k$. By using our knowledge of the beta distribution, we can calculate the integral explicitly in case $v = 1$ to obtain

$$P[X_1 = x_1, \ldots, X_m = x_m] = \frac{m! \, (s-m)!}{(s+1)!}.$$

Given $X_1, \ldots, X_m$, we are interested in the conditional distribution of Y, which we might view as an unknown distribution, in which case $X_1, \ldots, X_m$ are m 'independent' observations based on the unknown distribution Y. Since Y is determined by the $\mathbb{R}$-valued random variable V, we focus on the conditional distribution of V given $X_1, \ldots, X_m$:

$$P[V \leq v \mid X_1 = x_1, \ldots, X_m = x_m] = \frac{P[V \leq v, X_1 = x_1, \ldots, X_m = x_m]}{P[X_1 = x_1, \ldots, X_m = x_m]}.$$

From (27.14) we see that the conditional distribution of V given $(X_1, \ldots, X_m)$ is standard beta with parameters $1 + m$ and $1 - m + \sum_{k=1}^{m} X_k$.

* **Problem 47.** Mimic the preceding example in case the values of Y are exponential distributions having support $[0, \infty)$, with the reciprocal of the random mean being itself exponentially distributed with mean 1.

Problem 48. For the preceding problem and for Example 1, calculate the (unconditional) expectation of X_1. Comment on your answers.

Problem 49. Modify Example 1 by replacing the uniform distribution on V by an arbitrary, but fixed, standard beta distribution. For which values of the beta parameters does X_1 have finite (unconditional) expectation? What about finite (unconditional) variance?

Problem 50. Modify Problem 47 by replacing the exponential distribution for the reciprocal of the (random) mean by an arbitrary standard gamma distribution having support $[0, \infty)$. For which values of the gamma parameter does X_1 have finite expectation? What about finite variance?

Problem 51. Mimic Example 1 in case the values of Y are beta distributions having support $[0, 1]$ and density there with respect to Lebesgue measure λ given by:

$$\frac{dY(\omega)}{d\lambda}(x) = \frac{\Gamma(U(\omega) + V(\omega))}{\Gamma(U(\omega)) \, \Gamma(V(\omega))} x^{U(\omega) - 1} (1 - x)^{V(\omega) - 1},$$

where $U + V$ is exponentially distributed on $[0, \infty)$ with mean 1, the conditional distribution of U given $U + V$ is uniform on the interval $(0, U + V)$, and Γ denotes the gamma function.

Example 2. [Broken stick as probability distribution] Let $V = (V_1, V_2, \ldots)$ be an independent sequence of random variables each of which is uniformly distributed on the interval $(0,1)$. Let Y be the random measure defined by

$$Y\{i\} = V_i \prod_{j=1}^{i-1}(1 - V_j), \quad x = 1, 2, \ldots,$$

where, as is usual, an empty product equals 1. Mathematical induction shows that

$$Y(\{1, 2, \ldots, i\}) = 1 - \prod_{j=1}^{i}(1 - V_j),$$

which approaches 1 a.s. as $i \to \infty$. Thus, with probability 1, Y is a probability measure on the set of positive integers, a measure closely related to the stick-breaking random walk of Problem 17 in Chapter 12.

Let $(X_m : m \geq 1)$ be an infinite exchangeable sequence with De Finetti measure μ equal to the distribution of Y. Even in this situation where the support of μ is large, we can make explicit calculations. For positive integers $x_1, x_2, \ldots, x_m$, we set $w_i = \#\{k \leq m : x_k = i\}$ for $i = 1, 2, \ldots$ and calculate

$P[X_k = x_k \text{ for } 1 \leq k \leq m]$

$$= E\left(\prod_{k=1}^{m}\left[V_{x_k} \prod_{j=1}^{x_k-1}(1 - V_j)\right]\right) = E\left(\prod_{i=1}^{\infty} V_i^{w_i}(1 - V_i)^{(w_{i+1}+w_{i+2}+\cdots)}\right)$$

$$= \prod_{i=1}^{\infty}\left(\int_0^1 u^{w_i}(1 - u)^{(w_{i+1}+w_{i+2}+\cdots)}\, du\right) = \prod_{i=1}^{\infty} \frac{w_i! \, (\sum_{j=i+1}^{\infty} w_j)!}{[1 + \sum_{j=i}^{\infty} w_j]!}$$

$$= \frac{1}{(m+1)!} \prod_{i=1}^{\infty} \frac{w_i!}{1 + \sum_{j=i+1}^{\infty} w_j},$$

where all but finitely many of the factors in the infinite product equal 1.

Since Y is defined in terms of V we are interested in the conditional distribution function of V given $(X_1, X_2, \ldots, X_m)$. The preceding calculation gives the denominator of the conditional distribution function, and the numerator is obtained from the preceding calculation by replacing the interval $[0,1]$ of integration on u by intervals of the form $[0, v]$. Thus, the random sequence V is conditionally independent given $(X_1, \ldots, X_m)$ and the conditional distribution of the ith member of V is a standard beta distribution with parameters $1 + W_i$ and $1 + \sum_{j=i+1}^{\infty} W_j$, where $W_i(\omega) = \#\{k \leq m : X_k(\omega) = i\}$.

---

* **Problem 52.** Give an intuitive explanation of the conditional independence of the sequence V described in the preceding example.

Problem 53. [GEM distributions] Modify Example 2 by replacing the (unconditional) uniform distribution of each V_r by a beta distribution with density

$$v \rightsquigarrow \theta(1-v)^{\theta-1}, \quad 0 < v < 1,$$

where θ is a fixed positive constant. The distribution of the random measure Y is called the *GEM distribution* with parameter θ.

27.5. ‡ Ferguson distributions and Blackwell-MacQueen urns

Suppose that the De Finetti measure for an infinite exchangeable sequence of Bernoulli random variables is a beta distribution. We saw in Problem 31 that the posterior distributions of the Bernoulli parameter are also beta distributions. For this reason beta distributions are important De Finetti measures. As the following example shows, Dirichlet distributions have a similar property for exchangeable sequences of $\{1, 2, \ldots, d\}$-valued random variables.

Proposition 14. *Let* $(X_1, X_2, X_3, \ldots)$ *be an infinite exchangeable sequence of* $\{1, \ldots, d\}$-*valued random variables whose De Finetti measure is Dirichlet with parameters* γ_i, $1 \leq i \leq d$, *when regarded as a distribution on the standard* $(d-1)$-*simplex. For members* $x_1, \ldots, x_m$ *of* $\{1, \ldots, d\}$, *set* $w_i = \sharp\{k \leq m : x_k = i\}$. *Then*

$$P[X_k = x_k \text{ for } 1 \leq k \leq m] = \frac{\prod_{i=1}^d (\gamma_i)^{\uparrow}_{w_i}}{\left(\sum_{i=1}^d \gamma_i\right)^{\uparrow}_m},$$

where $(a)^{\uparrow}_b$ *denotes the rising factorial* $\prod_{c=0}^{b-1}(a+c)$. *Also, the posterior distribution of* Y *based on* m *observations is Dirichlet with parameters* $(\gamma_i + W_i)$, $1 \leq i \leq d$, *where* $W_i = \sharp\{k \leq m : X_k = i\}$.

Problem 54. Prove the preceding proposition. *Hint:* Use Problem 24 for part of the proof.

We have seen in Problem 37 that an infinite exchangeable sequence of Bernoulli random variables having a specified beta distribution for its De Finetti measure can be constructed via a 'generalized Polya urn model' in which it is not required that the initial contents be described by integers. The following example extends this construction to accommodate Dirichlet distributions.

Example 3. [Blackwell-MacQueen urns, special case] Consider an urn containing sands of different colors: γ_i units of color i, $1 \leq i \leq d$, where each $\gamma_i > 0$. We imagine that the 'grains' of sand are infinitesimal in size so that there is no divisibility requirement on the numbers γ_i. At each time $m = 1, 2, \ldots$, a grain of sand is drawn, the probability of each color being proportional to amount of sand of that color, and then it is returned to the urn (actually not relevant since it is infinitesimal in size) and one additional unit of sand of its color is poured into the urn. Thus, after the mth stage, the amount of sand in the urn equals $m + \sum_{i=1}^{d} \gamma_i$. For $m = 1, 2, \ldots$, let X_m denote the color— $1, 2, \ldots$, or d —drawn at time m. An induction argument shows that

$$P[(X_1, \ldots, X_m) = (x_1, \ldots, x_m)] = \frac{\prod_{i=1}^{d} (\gamma_i)_{w_i}^{\uparrow}}{\left(\sum_{i=1}^{d} \gamma_i\right)_{m}^{\uparrow}},$$

where $w_i = \sharp\{m \le k : x_m = i\}$.

By comparing this formula with Proposition 14, we see that every Dirichlet distribution can be viewed as the De Finetti measure of an exchangeable sequence arising from an appropriate Blackwell-MacQueen urn.

* **Problem 55.** Carry out the induction mentioned in the preceding example.

Problem 56. Consider a Blackwell-MacQueen urn with initial contents being three units of red sand, one unit of brown sand, and one unit of gray sand. Calculate the probability that in the long run the contents of the urn is more than 50% red.

Problem 57. As $\gamma \searrow 0$ the gamma distribution with parameter γ and scaling parameter a (not depending on γ), approaches the delta distribution at 0. Thus the delta distribution at 0 may be called a gamma distribution with parameter 0. Use this fact to generalize the definition of 'Dirichlet distribution' to accommodate parameters $(\gamma_1, \ldots, \gamma_d)$ in which some but not all of the coordinates might equal 0. Then comment on corresponding generalizations of Proposition 14 and Example 3.

We will continue the process of generalization by moving from exchangeable sequences of random variables having values in a finite set to those having values in an arbitrary Borel space. As we generalize we want to focus on De Finetti measures having the nice properties that Dirichlet distributions have—namely those described in Proposition 14 of this chapter and Problem 31 of Chapter 10. We use the indirect route of first generalizing a model rather than a definition.

To generalize Example 3 we replace the set $\{1, \ldots, d\}$ of colors by a Borel space Ψ of colors, and the vector $(\gamma_1, \ldots, \gamma_d)$ describing the color of the initial contents of the urn by a nonzero (meaning not identically zero) finite measure ξ on Ψ. The probability that the color of the first grain drawn from the urn belongs to a Borel set B is $\frac{\xi(B)}{\xi(\Psi)}$. As in the above example a unit amount of sand of the same color as the drawn grain is added to the urn at each time $m = 1, 2, \ldots$. The preceding sentences describe the *Blackwell-MacQueen urn* with initial measure ξ.

Definition and Proposition 15. Let ξ be a nonzero finite measure on a Borel space Ψ, and denote by $\mathcal{Q}(\Psi)$ the Borel space of distributions on Ψ. The sequence of colors drawn from a Blackwell-MacQueen urn with initial measure ξ is an exchangeable sequence, the De Finetti measure μ of which is the *Ferguson distribution* with parameter ξ. The measure μ is the unique probability measure on $Q(\Psi)$ that, for every finite measurable partition $(B_1, \ldots, B_d)$ of Ψ, induces the Dirichlet distribution on $\mathbb{R}^d$ with parameter $(\xi(B_1), \ldots, \xi(B_d))$, via the mapping $y \rightsquigarrow (y(B_1), \ldots, y(B_d))$ from $Q(\Psi)$ to $\mathbb{R}^d$.

PROOF. Consider the Blackwell-MacQueen urn with initial measure ξ. By renaming all the colors in B_r as r, $1 \leq r \leq d$, we obtain a Blackwell-MacQueen urn as described in Example 3 with initial parameter $(\xi(B_1), \ldots, \xi(B_d))$. From that example we see that the sequence of drawn colors from the Blackwell-MacQueen urn with initial measure ξ is exchangeable and that the corresponding De Finetti measure induces the desired Dirichlet distribution.

For the proof of uniqueness, let λ be a probability measure on $Q(\Psi)$ that induces the Dirichlet distribution with parameter $(\xi(B_1), \ldots, \xi(B_d))$ on $\mathbb{R}^d$ via the mapping $y \rightsquigarrow (y(B_1), \ldots, y(B_d))$ from $Q(\Psi)$ to $\mathbb{R}^d$. We will prove that $\lambda = \mu$ by proving that corresponding exchangeable sequences $(X_{\lambda,1}, X_{\lambda,2}, \ldots)$ (not necessarily arising from a Blackwell-MacQueen urn) and $(X_{\mu,1}, X_{\mu,2}, \ldots)$ have the same distribution. Fix a measurable partition $(B_1, \ldots, B_d)$; for any finite sequence $(A_1, \ldots, A_k)$ where each A_m equals some B_r, we have

$$P[X_{\lambda,m} \in A_m \, , \, 1 \leq m \leq k] = P[X_{\mu,m} \in A_m \, , \, 1 \leq m \leq k] \, .$$

From this it follows that the distributions of the two exchangeable sequences are identical and thus, by the definition of De Finetti measure, that $\lambda = \mu$. $\square$

Problem 58. Prove that with probability 1, infinitely many different colors are drawn from any Blackwell-MacQueen urn whose initial measure ξ has the property that there exists an infinite partition $\{B_j : j = 1, 2, \ldots\}$ such that $\xi(B_j) > 0$ for every j.

Problem 59. For a Blackwell-MacQueen urn with initial measure ξ on a Borel space Ψ, let Ξ_m, for $m = 0, 1, 2, \ldots$, denote the measure describing the color distribution of the sand in the urn immediately after time m. [Notice that $\Xi_0 = \xi$ and $\Xi_m(\Psi) = m + \xi(\Psi)$.] The description of the urn makes it clear that the random sequence $(\Xi_m : m \geq 0)$ is a time-homogeneous Markov sequence in the space of finite measures on Ψ. Find the corresponding transition distributions.

Problem 60. For a sequence $(\Xi_m : m \geq 0)$ of random measures constructed on Ψ as in the preceding problem and any measurable $B \subseteq \Psi$, show that

$$\left(\frac{\Xi_m(B)}{\Xi_m(\Psi)} : m \geq 0 \right)$$

is a martingale, the limit of which has a standard beta distribution with parameters $\xi(B)$ and $\xi(\Psi) - \xi(B)$. Relate this problem to Problem 56.

CHAPTER 28

Stationary Sequences

Many of the random sequences studied so far in this book are related in some significant way to 'stationary sequences'. Informally speaking, a random sequence is stationary if it models the successive states of some system that is in equilibrium. Thus, an important example of a stationary sequence is a Markov sequence whose initial distribution is an equilibrium distribution. Exchangeable sequences are also stationary. And there are many other important examples, some of which will be introduced in this chapter.

After providing basic definitions and examples in the first three sections, we turn our attention in the fourth section to the Birkhoff Ergodic Theorem, which is a remarkable generalization of the Strong Law of Large Numbers to sequences with stationary (but not necessarily independent) increments. Section 5 introduces the concept of 'ergodicity': a stationary sequence is ergodic if its distribution cannot be decomposed into a nontrivial mixture of distributions of stationary sequences. Important criteria for ergodicity are given. In Section 6, we improve the Strong Law of Large Numbers even further with a very useful result known as the Kingman-Liggett Subadditive Ergodic Theorem. In an entirely different direction, in the final section of the chapter, we take an introductory look at the 'spectral analysis' of stationary sequences.

28.1. Definitions

A random sequence $X = (X_0, X_1, X_2, \ldots)$ is *stationary* if it has the same distribution as the random sequence $(X_1, X_2, X_3, \ldots)$. We may also have doubly infinite stationary sequences: a random sequence $(\ldots, X_{-1}, X_0, X_1, \ldots)$ is *stationary* if the distribution of the sequence $(X_k, X_{k+1}, X_{k+2}, \ldots)$ does not depend on $k \in \mathbb{Z}$. Sometimes, in order to distinguish these two types of stationary sequences, we use the term *one-sided* for a sequence $X = (X_0, X_1, X_2, \ldots)$ and the term *two-sided* for a sequence $(\ldots, X_{-1}, X_0, X_1, \ldots)$. For simplicity, we mostly concern ourselves with one-sided stationary sequences X whose terms are $(\Psi, \mathcal{G})$-

valued for some Borel space $(\Psi, \mathcal{G})$. Thus X itself is $(\Theta, \mathcal{H})$-valued, where

$$(28.1) \qquad\qquad (\Theta, \mathcal{H}) = \prod_{n=0}^{\infty} (\Psi, \mathcal{G}) = (\Psi, \mathcal{G})^{\infty}.$$

Define the *shift transformation* $\tau \colon \Theta \to \Theta$ by

$$\tau((\psi_0, \psi_1, \psi_2, \dots)) = (\psi_1, \psi_2, \psi_3, \dots).$$

We use the notation τ^k to denote the k-fold composition of τ with itself, with τ^0 denoting the identity function. Thus, for example,

$$\tau^3((\psi_0, \psi_1, \psi_2, \dots)) = (\psi_3, \psi_4, \psi_5, \dots).$$

In order that a random sequence X of Ψ-valued random variables be stationary, it is obviously necessary and sufficient that X and $\tau \circ X$ have the same distribution. If Q is the distribution of such a stationary sequence X, then τ is a *measure-preserving transformation* on the probability space $(\Theta, \mathcal{H}, Q)$ in the sense that

$$Q(\tau^{-1}(A)) = Q(A) \quad \text{for all } A \in \mathcal{H}.$$

Thus stationary sequences can be treated by studying 'ergodic theory', the theory of measure-preserving transformations. The following problem shows also that questions in ergodic theory can be viewed as questions about stationary sequences.

Problem 1. Let $(\Omega, \mathcal{F}, P)$ be an arbitrary probability space, and let $\tau \colon \Omega \to \Omega$ be a measurable function. Let X_0 be the identity function on Ω, and for $n = 1, 2, \dots$, define random variables X_n inductively by setting $X_n = \tau \circ X_{n-1}$. Show that the sequence $X = (X_0, X_1, X_2, \dots)$ is stationary if and only if τ is measure-preserving.

Problem 2. For the particular case of the preceding problem in which $(\Omega, \mathcal{F})$ equals $(\Theta, \mathcal{H})$ as defined in (28.1) and τ equals the shift transformation, write an explicit formula for each X_n in terms of the coordinates of members of Θ.

Problem 3. Show that a one-sided sequence X is stationary if and only if for all nonnegative integers k and n, $(X_0, X_1, \dots, X_k)$ has the same distribution as $(X_n, X_{n+1}, \dots, X_{n+k})$. Use this fact to prove that every exchangeable sequence is stationary.

* **Problem 4.** Let $(X_0, X_1, \dots)$ be a stationary sequence, and fix integers $m \geq 0$ and $k > 0$. Prove that the sequence $(X_m, X_{m+k}, X_{m+2k}, \dots)$ is a stationary sequence.

Problem 5. Let X be a one-sided stationary sequence of Ψ-valued random variables, τ the shift transformation on $\Theta = \Psi^{\infty}$, and $g \colon \Theta \to \Psi$ a measurable function. Show that the sequence

$$(g \circ X, \, g \circ \tau \circ X, \, g \circ \tau^2 \circ X, \, \dots)$$

is stationary.

*   **Problem 6.** Let $(\Psi, \mathcal{G})$ be a Borel space. Show that if $X = (X_0, X_1, X_2, \dots)$ is a one-sided stationary sequence of Ψ-valued random variables, then there exists a two-sided stationary sequence $Y = (\dots, Y_{-1}, Y_0, Y_1, \dots)$ such that X has the same distribution as the one-sided sequence $(Y_0, Y_1, Y_2, \dots)$.

28.2. Notation

Throughout this chapter, it will be useful to agree on some standard notation and assumptions. The space Ψ in which the terms of random sequences typically take their values is a Borel space, and the corresponding σ-field, which will often not be mentioned, is denoted by $\mathcal{G}$. The Borel space Θ and σ-field $\mathcal{H}$ are defined by (28.1), with $\mathcal{H}$ often omitted. The symbol τ is reserved for the shift transformation on Θ. With respect to a probability measure Q on Θ, τ may or may not be measure-preserving. Correspondingly, we say that Q is or is not *shift-invariant*. We denote by $\mathcal{M}$ the collection of all shift-invariant probability measures on Θ, and a member of $\mathcal{M}$ will often be called a distribution since it may be the distribution of a stationary sequence or, as in Problem 2, the distribution of the first term of a stationary sequence.

The σ-field

$$S = \{A \in \mathcal{H} : A = \tau^{-1}(A)\},$$

is called the *shift-invariant σ-field* of $\mathcal{H}$, with the phrase "of $\mathcal{H}$" often being dropped.

There is another sub-σ-field of $\mathcal{H}$ that is useful in the context of stationary sequences. Even though τ might not be one-to-one, τ^{-1} may be viewed as a function from $\mathcal{H}$ to $\mathcal{H}$. As such it can be composed with itself. For positive integers n, let τ^{-n} denote n-fold composition of τ^{-1} with itself. Set

$$\mathcal{H}_n = \{\tau^{-n}(A) : A \in \mathcal{H}\},$$

and

$$\mathcal{T} = \bigcap_{n=1}^{\infty} \mathcal{H}_n.$$

We call $\mathcal{T}$ the *tail σ-field* of $\mathcal{H}$. The reader may recall that our definition of tail σ-field in Chapter 12 was more general than the one given here. In this chapter, the tail σ-field will always be defined in terms of the σ-fields $\mathcal{H}_n$ as above, and not in terms of other natural sequences of σ-fields, such as $\sigma(X_n, X_{n+1}, \dots)$. Also note that we have not bothered to define the tail σ-field for the two-sided case. In that context, our definition of $\mathcal{H}_n$ is inappropriate, since it gives $\mathcal{H}_n = \mathcal{H}$. Nevertheless, the reader who is interested in two-sided sequences should have no difficulty in making appropriate modifications.

Example 3 of the next section shows that the invariant and tail σ-fields are not necessarily the same. But there is a relation between them as indicated in the following problem.

Problem 7. Prove that $\mathcal{S} \subseteq \mathcal{T}$.

Problem 8. Let X be a stationary sequence taking values in some Borel space. Denote the conditional distribution of X given the σ-field $X^{-1}(\mathcal{S})$ by $\omega \rightsquigarrow R^\omega$. Prove that for P-almost every ω, R^ω is the distribution of a stationary sequence.

Problem 9. Let X be an $\mathbb{R}$-valued stationary sequence and A a member of $X^{-1}(\mathcal{S})$. Show that the sequence $(X_n I_A : n = 0, 1, 2, \ldots)$ is stationary.

Problem 10. Show that the set $\mathcal{M}$ of all shift-invariant measures is a convex set of measures. That is, show that if $R, S \in \mathcal{M}$, then for any $t \in [0, 1]$, $tR + (1-t)S \in \mathcal{M}$.

28.3. Examples

We have already seen that exchangeable sequences (including, of course, iid sequences) are stationary. The following example uses such sequences to illustrate some of the concepts of the preceding section.

Example 1. Let X be an exchangeable sequence in a Borel space $(\Psi, \mathcal{G})$. Let Y denote its limiting empirical distribution (see Chapter 27), and set

$$R(\omega, \cdot) = \bigotimes_{k=0}^{\infty} Y(\omega, \cdot).$$

Our goal is to show that R is a version of the conditional distribution introduced in Problem 8. By Proposition 8 of Chapter 27, R is $X^{-1}(\mathcal{S})$-measurable.

To complete the argument we need to show

$$(28.2) \qquad E(R(B)I_C) = P([X \in B] \cap C),$$

for $B \in \mathcal{H}$ and $C \in X^{-1}(\mathcal{S})$. Write $C = X^{-1}(A)$ for $A \in \mathcal{S}$. The desired relation (28.2) can be written as

$$(28.3) \qquad E(R(B)I_A(X)) = P[X \in B \cap A].$$

The left side of (28.3) equals

$$E\Big(E(R(B)I_A(X) \mid Y)\Big) = E\Big(R(B)E(I_A(X) \mid Y)\Big)$$

$$(28.4) \qquad\qquad\qquad = E\Big(R(B)\,R(A)\Big).$$

By the Kolmogorov 0-1 Law, $R(\omega, \cdot)$, being a product measure on Θ for each ω, assigns probability 0 or 1 to each member of $\mathcal{T}$. By Problem 7, $A \in \mathcal{T}$, and therefore for every ω, $R(\omega, A) = 0$ or $= 1$. Thus

$$R(\omega, A)\,R(\omega, B) = R(\omega, A \cap B)$$

for all ω. This equality enables us to conclude that (28.4) equals $P(X \in B \cap A)$, as desired.

Our next example indicates which Markov sequences are stationary.

Example 2. Let $X = (X_0, X_1, X_2, \dots)$ be a Markov sequence with transition operator T and initial distribution Q_0. If Q_0 is an equilibrium distribution for T, then the sequence $Y = (X_1, X_2, X_3, \dots)$ has the same initial distribution and transition operator as X, so it has the same distribution as X. Thus, X is a stationary sequence. We will sometimes use the term *stationary Markov sequence* to describe a sequence like X.

Example 3. [Rotations of the circle] Let X_0 be a random variable that is uniformly distributed on $[0, 1)$, and let a be an arbitrary real number. Set $X_n = an + X_0 \pmod 1, n \in \mathbb{Z}^+$. Then the sequence $X = (X_0, X_1, X_2, \dots)$ is stationary. It is not hard to prove this fact directly, but we will prove it using some of the theory that has been developed so far.

Clearly X is a Markov sequence with state space $[0, 1)$ (the transition operator takes a function $x \rightsquigarrow f(x)$ to the function $x \rightsquigarrow f(x + a \pmod 1)$). It is also easy to see that the uniform distribution on $[0, 1)$ is an equilibrium distribution, using the translation invariance of Lebesgue measure. Since X_0 has this distribution, X is a stationary sequence by Example 2.

Any sequence distributed like X is called a *stationary rotation of the circle*. The quantity $2\pi a$ is known as the *rotation angle*.

The fact that $x \rightsquigarrow x + a \pmod 1$ is a one-to-one function having a measurable inverse implies that the tail σ-field $\mathcal{T}$ equals the σ-field $\mathcal{H}$ of measurable sets of infinite sequences of members of $[0, 1)$. The character of the shift-invariant σ-field $\mathcal{S}$ depends on the rotation angle $2\pi a$. If a is rational, let q denote its denominator when written in lowest terms with positive denominator. The set $[0, 1)$ viewed as a circle is the union of equivalence classes each of which contains q equally spaced members. In order for a measurable subset of $[0, 1)$ to belong to $X^{-1}(\mathcal{S})$ it is necessary and sufficient that it contain all members of an equivalence class whenever it contains any one of them. A similar statement is true in case a is irrational, except that each equivalence class is a countable set dense in $[0, 1)$.

Example 4. Let X_0 be as in the preceding example, and let $X_n = 2X_{n-1} = 2^n X_0 \pmod 1$ for $n = 1, 2, \dots$. In this example as in the preceding one, the entire sequence X is determined by its first term X_0. However, this sequence is unlike the preceding example, in that the knowledge of the sequence for large n does not determine X_0. In fact, we will argue that in this example, the tail σ-field is very small; that each of its members has probability 0 or 1.

By writing each member of $[0, 1)$ in binary, we can view X as an infinite sequence, each of whose members is itself an infinite sequence of 0's and 1's. The distribution of the infinite sequence X_0 is the fair-coin-flip measure, and X_n is obtained from X_{n-1} by discarding the first term of X_{n-1} and then subtracting

1 from the indices on the remaining terms. Clearly X_n does not determine X_0 if $n \geq 1$.

Example 5. [Stationary Gaussian sequences] Let $U = (U_1, U_2, \dots)$ be an iid sequence of random variables uniformly distributed on $[0, 2\pi)$, and let $Z = (Z_1, Z_2, \dots)$ be an iid sequence of random variables with the standard normal distribution. Assume that the pair (U, Z) is independent. Also, let $(a_1, a_2, \dots)$ and $(b_1, b_2, \dots)$ be two sequences of real constants. Set

$$X_n = \sum_{k=1}^{\infty} b_k Z_k \cos(na_k + U_k), \quad n \in \mathbb{Z}.$$

By the Kolmogorov Three-Series Theorem, this sum converges a.s. if and only if $\sum_n b_n^2 < \infty$, in which case the sequence $X = (X_0, X_1, X_2, \dots)$ is a.s. well-defined. The reader is asked in Problem 14 to show that the sequence X is stationary and 'Gaussian'. (A random sequence of $\mathbb{R}$-valued random variables is *Gaussian* if all of its finite-dimensional marginal distributions are normal distributions.) More will be said about Gaussian sequences at the end of this chapter.

Problem 11. [Moving averages] Let $X = (X_0, X_1, X_2, \dots)$ be a stationary sequence of $\mathbb{R}$-valued random variables. Fix a positive integer k, and let

$$Y_n = \frac{X_n + \cdots + X_{n+k-1}}{k}, \quad n \in \mathbb{Z}^+.$$

Show that $Y = (Y_0, Y_1, Y_2, \dots)$ is stationary.

Problem 12. Let X and Y be stationary sequences of $\mathbb{R}$-valued random variables. Show that if X and Y are independent, then the sequence $X+Y$ is stationary. Also, show by example that the independence assumption cannot be entirely eliminated.

Problem 13. [Stationary renewal sequences] Show that the delayed renewal sequence defined in Theorem 15 of Chapter 25 is a stationary sequence.

Problem 14. Show that the sequence X defined in Example 5 is stationary and Gaussian. *Hint:* Example 2, Example 3, and Problem 12 are all helpful in this argument.

28.4. The Birkhoff Ergodic Theorem

This section is devoted to the following result.

Theorem 1. [Birkhoff Ergodic] *Let $X = (X_n : n = 0, 1, 2, \dots)$ be an $\mathbb{R}$-valued stationary sequence, and let $S_n = X_0 + X_1 + \cdots + X_{n-1}$. If $E(X_0 \mid X^{-1}(\mathcal{S}))$ is a.s.-defined, then*

$$(28.5) \qquad \lim_{n \to \infty} \frac{S_n}{n} = E(X_0 \mid X^{-1}(\mathcal{S})) \text{ a.s.},$$

and if $E(|X_0|) < \infty$, then

(28.6) $$\lim_{n\to\infty} E\left(\left|\frac{S_n}{n} - E(X_0 \mid X^{-1}(\mathcal{S}))\right|\right) = 0.$$

PROOF. We first treat the special case in which $E(X_0 \mid X^{-1}(\mathcal{S})) = 0$ a.s. and $E(X_0)$ exists, in which case $E(X_0)$ necessarily equals 0. For this case our goal is to prove that as $n \to \infty$, $S_n/n \to 0$ a.s. and in $\mathbf{L}_1$.

Fix $\varepsilon > 0$ and set

$$A = \left\{\omega : \limsup_{n\to\infty} \frac{S_n(\omega)}{n} > \varepsilon\right\}.$$

We will prove that $P(A) = 0$. Once we prove this fact, the proof of the desired almost sure convergence is completed by applying the same argument to the sequence $-X$.

Note that $A \in X^{-1}(\mathcal{S})$, so $E(X_0 I_A) = E\big(E(X_0 \mid X^{-1}(\mathcal{S}))I_A\big) = 0$. Thus,

$$E((X_0 - \varepsilon)I_A) = -\varepsilon P(A).$$

Therefore, in order to show that $P(A) = 0$, it suffices to prove that

(28.7) $$E((X_0 - \varepsilon)I_A) \geq 0.$$

For $n = 1, 2, 3, \ldots$, set $S'_n = S_{n+1} - X_0$. Let

$$M_n = \max\{(S_1 - \varepsilon), (S_2 - 2\varepsilon), \ldots, (S_n - n\varepsilon)\},$$
$$M'_n = \max\{(S'_1 - \varepsilon), (S'_2 - 2\varepsilon), \ldots, (S'_n - n\varepsilon)\}.$$

By the definition of S'_n,

$$(X_0 - \varepsilon) + (0 \vee M'_n) \geq M_n, \quad n = 0, 1, 2, \ldots.$$

Since $X_0 = S_1$, it follows that

$$X_0 - \varepsilon \geq M_n - (0 \vee M'_n),$$

so

$$E\big((X_0 - \varepsilon)I_A\big) \geq E\big([M_n - (0 \vee M'_n)]I_A\big)$$
$$= E\big([(0 \vee M_n) - (0 \vee M'_n)]I_A\big) + E\big([M_n - (0 \vee M_n)]I_A\big),$$

for $n = 0, 1, 2, \ldots$. Since X is stationary and $A \in X^{-1}(\mathcal{S})$, the random variables $(0 \vee M_n)I_A$ and $(0 \vee M'_n)I_A$ have the same distribution. Thus, the first expectation on the right side of the preceding inequality is 0, so that inequality becomes

(28.8) $$E((X_0 - \varepsilon)I_A) \geq E((M_n - (0 \vee M_n))I_A), \quad n = 0, 1, 2, \ldots.$$

By the definitions of M_n and A, $\lim_n M_n(\omega) \geq 0$ for all $\omega \in A$, so

(28.9) $$\lim_n [M_n - (0 \vee M_n)I_A] = 0.$$

By the definition of M_n, $(X_0 - \varepsilon) \le M_n$, so $|X_0 - \varepsilon| \ge |M_n - (0 \vee M_n)|$. Since we have assumed that $E(|X_0|) < \infty$, the Dominated Convergence Theorem, (28.8), and (28.9) imply that

$$E\big((X_0 - \varepsilon)I_A\big) \ge \lim_{n \to \infty} E\big((M_n - (0 \vee M_n))I_A\big) = 0,$$

proving (28.7).

For the special case thus far treated, the proof of (28.6) is essentially the same as the proof of the corresponding fact in the Strong Law of Large Numbers (see the proof of Lemma 13 of Chapter 12).

Next we drop the assumption that $E(X_0 \mid X^{-1}(\mathcal{S})) = 0$ a.s., but continue to assume $E(|X_0|) < \infty$. Thus $E(X_0 \mid X^{-1}(\mathcal{S}))$ is finite a.s. The sequence $(X_n - E(X_0 \mid X^{-1}(\mathcal{S})): n = 0, 1, 2, \dots)$, which is the same as the sequence $(X_n - E(X_n \mid X^{-1}(\mathcal{S})): n = 0, 1, 2, \dots)$, is stationary. Apply what has already been proved to this stationary sequence to obtain (28.5) and (28.6) whenever $E(|X_0|) < \infty$.

Finally we drop the assumption that $E(|X_0|) < \infty$, assuming instead only that $E(X_0 \mid X^{-1}(\mathcal{S}))$ is a.s. defined. To finish the proof we only need prove (28.5) in this situation. This is easily done by treating the stationary sequences $(X_n^+ \wedge c_1: n = 0, 1, \dots)$ and $(X_n^- \wedge c_2: n = 0, 1, \dots)$ and letting $c_1 \to \infty$ and $c_2 \to \infty$. $\square$

Problem 15. [Maximal Ergodic Lemma] Let X be an $\mathbb{R}$-valued stationary sequence such that $E(|X_0|) < \infty$. Let $S_n = X_0 + \cdots + X_{n-1}$ and $M_n = \max\{S_1, S_2, \dots, S_n\}$. Prove that

$$E(X_0 \,;\, [M_n > 0]) \ge 0.$$

Hint: Use ideas from the proof of the Birkhoff Ergodic Theorem.

Problem 16. Let X be a stationary sequence of $(\Psi, \mathcal{G})$-valued random variables. Fix a measurable set $A \in \mathcal{G}$, and define

$$T_n = \begin{cases} 0 & \text{for } n = 0 \\ \inf\{k > T_{n-1}: X_k \in A\} & \text{for } n > 0. \end{cases}$$

Show that

$$\lim_{n \to \infty} \frac{T_n}{n} = \frac{1}{P[X_0 \in A \mid X^{-1}(\mathcal{S})]} \quad \text{a.s.},$$

where it is understood that $1/0 = \infty$.

28.5. Ergodicity

In Chapter 12 we saw examples of σ-fields containing events other than $\emptyset$ and Ω but which nevertheless have the property that each member is an event of probability 0 or 1. We will call such σ-fields *0-1 trivial*. Of course, the trivial σ-field is also 0-1 trivial.

It is apparent from the statement and proof of the Birkhoff Ergodic Theorem that the shift-invariant σ-field S plays an important role for stationary sequences, somewhat reminiscent of the role of the tail σ-field in the study of iid sequences. However, for general stationary sequences, there is no 0-1 law for S, that is, S is not necessarily a 0-1 trivial σ-field. And when S is not 0-1 trivial, the limit in the Birkhoff Ergodic Theorem need not be a constant. The following definition gives us terminology for describing whether or not a 0-1 law is in force for a given stationary sequence.

Definition 2. A stationary sequence X with distribution Q is *ergodic* if the shift-invariant σ-field S is 0-1 trivial under Q. In this case, we also say that the distribution Q is *ergodic*.

If X is an ergodic sequence defined on some probability space $(\Omega, \mathcal{F}, P)$, then the sub-σ-field $X^{-1}(S)$ of $\mathcal{F}$ is 0-1 trivial under P. Thus if $E(X_0 \mid X^{-1}(S))$ is defined with positive probability, then it is a.s.-defined and equals a constant a.s. In this case therefore, the limit in the Birkhoff Ergodic Theorem is a constant, just as in the Strong Law of Large Numbers.

By the Kolmogorov 0-1 Law and Problem 7, any iid sequence is ergodic. We will also see in the exercises that a stationary Markov sequence is ergodic if its initial distribution is not a mixture of two or more equilibrium distributions, and an infinite exchangeable sequence is ergodic if and only if it is iid. Many stationary sequences are not ergodic. In general, the most that can be said for a stationary sequence X taking values in a Borel space is that the conditional distribution of X given $X^{-1}(S)$ is ergodic almost surely.

Problem 17. Let X_0 be uniformly distributed on $[-1, 1]$ and for $n > 0$, set $X_n = (-1)^n X_0$. Clearly the sequence $(X_0, X_1, \dots)$ is stationary. Show that it is not ergodic but that the limit in the Birkhoff Ergodic Theorem is a constant a.s.

Problem 18. Let Q be an ergodic distribution on $(\Theta, \mathcal{H})$. Prove that the sequence $(g, g \circ \tau, g \circ \tau^2, \dots)$ is ergodic under Q. Restate the conclusion of this problem using the notation of Problem 5.

Problem 19. Let Q be the distribution of a stationary sequence X whose terms take values in a Borel space. Show that X is ergodic if and only if the conditional distribution of X given $X^{-1}(S)$ is almost surely equal to Q.

Problem 20. Let X be an infinite exchangeable sequence of random variables taking values in some Borel space. Show that X is ergodic if and only if X is iid.

We wish to develop some useful criteria for ergodicity. For our first criterion, we need a definition.

Definition 3. Let $\mathcal{P}$ be a convex set of distributions on some measurable space. A measure $Q \in \mathcal{P}$ is *extremal* in $\mathcal{P}$ if Q cannot be written in the form $Q = tR + (1-t)S$ for some $t \in (0,1)$ and distinct measures $R, S \in \mathcal{P}$.

Another way of stating the previous definition is as follows: Q is extremal in $\mathcal{P}$ if the equation $Q = tR + (1-t)S$ implies $Q = R = S$ whenever $t \in (0,1)$ and $R, S \in \mathcal{P}$.

Theorem 4. *Let $\mathcal{M}$ be the set of shift-invariant distributions on $(\Theta, \mathcal{H})$. A measure $Q \in \mathcal{M}$ is ergodic if and only if it is extremal in $\mathcal{M}$.*

PROOF. Suppose that Q is an ergodic member of $\mathcal{M}$ and $Q = tR + (1-t)S$ for some $t \in (0,1)$ and $R, S \in \mathcal{M}$. It follows easily from the definition of ergodicity that R and S are ergodic. Let B be any member of $\mathcal{H}$. By Problem 5 and Problem 18, the sequence

$$Y = (I_B, \, I_B \circ \tau, \, I_B \circ \tau^2, \, \ldots)$$

is an $\mathbb{R}$-valued ergodic stationary sequence. By the Birkhoff Ergodic Theorem,

$$\lim_{n \to \infty} \frac{Y_0 + Y_1 + \ldots Y_{n-1}}{n} = Q(B) \quad Q\text{-a.s.}$$

Since $Q = tR + (1-t)S$ and R and S are ergodic, the Birkhoff Ergodic Theorem also implies that the limit equals $R(B)$ with Q-probability t and $S(B)$ with Q-probability $(1-t)$. Thus $Q(B) = R(B) = S(B)$. Since B is an arbitrary member of $\mathcal{H}$, it follows that $Q = R = S$, so Q is extremal.

Now suppose Q is not ergodic. Let A be a member of $\mathcal{S}$ such that $0 < Q(A) < 1$. Let $R(B) = Q(B \cap A)/Q(A)$ and $S(B) = Q(B \cap A^c)/Q(A^c)$ for all $B \in \mathcal{H}$. Since $A \in \mathcal{S}$, it is easily checked that R and S are shift-invariant distributions. Since $Q = tR + (1-t)S$ with $t = Q(A)$, Q is not extremal. $\square$

* **Problem 21.** Show that if R and S are distinct ergodic measures in $\mathcal{M}$, then R and S are mutually singular. *Hint:* Look at the proof of Theorem 4.

Problem 22. Let T be a Markov transition operator, and let $\mathcal{N}$ be the collection of equilibrium distributions for T. Show that a stationary Markov sequence with transition operator T is ergodic if and only if its initial distribution is extremal in $\mathcal{N}$. Use this fact to show that a stationary Markov sequence with countable state space is ergodic if its transition operator is irreducible.

* **Problem 23.** Let X be a stationary rotation of the circle through angle $2\pi a$. (As indicated in Example 3, the initial distribution of a stationary rotation is uniform, by definition.) Show that X is ergodic if and only if a is irrational. As a consequence, deduce the Weyl Equidistribution Theorem: If a is irrational,

$$\lim_{n\to\infty} \frac{I_A \circ X_0 + \cdots + I_A \circ X_{n-1}}{n} = \lambda(A) \text{ a.s.},$$

where A is any Borel subset of $[0, 2\pi)$ and λ is Lebesgue measure. *Hint:* When a is irrational, show that there is only one shift-invariant distribution for the relevant shift transformation on the appropriate product space.

Problem 24. Let $\mathcal{E}$ be the set of ergodic shift-invariant distributions on $(\Theta, \mathcal{H})$. Show that for any $Q \in \mathcal{M}$, there exists a random $\mathcal{E}$-valued distribution R such that $Q(B) = E(R(B))$ for all $B \in \mathcal{H}$. Explain how this formula may be viewed as giving a decomposition of Q into a 'convex combination' of ergodic measures.

* **Problem 25.** Let X be a Markov sequence on $\mathbb{Z}^+$ with transition matrix $T = (T(i, j): i, j \geq 0)$. Suppose T is irreducible and positive recurrent, with equilibrium distribution Q_0. For each state i, let f_i be the indicator function of the set $\{i\}$. For each pair of states i, j, calculate in terms of Q_0 and T the almost sure limit as $n \to \infty$ of

$$\frac{1}{n} \sum_{k=0}^{n-1} f_i(X_k) f_j(X_{k+1}).$$

Hint: First consider the case in which the distribution of X_0 is Q_0.

We turn our attention now to a second criterion for ergodicity, one which establishes a relationship between the ergodicity of a stationary sequence X and the amount of dependence that exists between the individual random variables in the sequence. Before stating the criterion, we introduce some terminology.

Definition 5. A $(\Theta, \mathcal{H})$-valued stationary sequence X and its distribution Q are *weakly mixing* if

$$\lim_{n\to\infty} \frac{1}{n} \sum_{k=0}^{n-1} Q(A \cap \tau^{-k}(B)) = Q(A) Q(B) \quad \text{for all } A, B \in \mathcal{H},$$

mixing if

$$\lim_{n\to\infty} Q(A \cap \tau^{-n}(B)) = Q(A) Q(B) \quad \text{for all } A, B \in \mathcal{H},$$

and *strongly mixing* if

$$\lim_{n\to\infty} \sup_{B \in \mathcal{H}} |Q(A \cap \tau^{-n}(B)) - Q(A) Q(B)| = 0 \quad \text{for all } A \in \mathcal{H}.$$

Roughly speaking, mixing is a kind of asymptotic independence. For example, with $(\Theta, \mathcal{H}) = (\Psi, \mathcal{G})^\infty$, a mixing stationary sequence $(X_n: n = 0, 1, \dots)$ satisfies

$$\lim_{n\to\infty} P([X_0 \in A] \cap [X_n \in B]) = P[X_0 \in A] P[X_0 \in B]$$

for $A, B \in \mathcal{G}$. From this same point of view, strongly mixing is uniform asymptotic independence, and weakly mixing could be understood as asymptotic independence 'on the average'. Clearly, strongly mixing $\Longrightarrow$ mixing $\Longrightarrow$ weakly mixing.

Theorem 6. *A stationary sequence X is ergodic if and only if it is weakly mixing.*

PROOF. Suppose X is weakly mixing. Choose $A \in \mathcal{S}$. For all $k \in \mathbb{Z}^+$, $Q(A \cap \tau^{-k}(A)) = Q(A)$, since $\tau^{-k}(A) = A$. So the definition of weakly mixing implies that $Q(A) = Q(A)^2$. Thus $Q(A) = 0$ or 1, and it follows that X is ergodic.

Now suppose X is ergodic. Choose $A, B \in \mathcal{H}$ and, for $n = 0, 1, 2, \ldots$, let $Y_n = I_B \circ \tau^n \circ X$. The sequence $Y = (Y_n : n = 0, 1, 2, \ldots)$ is stationary by Problem 5 and ergodic by Problem 18. We now calculate:

$$\lim_{n \to \infty} \frac{1}{n} \sum_{k=0}^{n-1} Q(A \cap \tau^{-k}(B)) = \lim_{n \to \infty} E\left((I_A \circ X)\frac{Y_0 + \cdots + Y_{n-1}}{n}\right) = Q(A)\, Q(B),$$

the last equality following from the Birkhoff Ergodic Theorem and the Bounded Convergence Theorem. $\square$

Problem 26. Show that the stationary rotation of the circle through angle $2\pi a$ is not mixing for any a. What is the story with respect to weak mixing? *Hint:* Problem 23 may be useful.

Problem 27. Let X and Y be independent stationary sequences of $\mathbb{R}$-valued random variables. Show that $X + Y$ is ergodic if and only if both X and Y are ergodic.

* **Problem 28.** Show that X is strongly mixing if and only if the tail field $\mathcal{T}$ is 0-1 trivial under the distribution of X.

28.6. † The Kingman-Liggett Subadditive Ergodic Theorem

The main result of this section is a useful generalization of the Birkhoff Ergodic Theorem. The hypotheses may seem strange at first glance, so we give an example to show how these hypotheses can arise in a natural way.

Example 6. Let $X = (X_n : n = 0, 1, 2, \ldots)$ be a stationary sequence of $\mathbb{R}^d$-valued random variables. For $n > 0$, let $S_n = X_0 + \cdots + X_{n-1}$, and for $0 \le m < n$, set

$$R_{m,n} = \sharp\{S_k : m < k \le n\}$$

and

$$R_n = R_{0,n}.$$

In the special case that X is an iid sequence, the sequence (S_n) is a random walk, and the sequence $R = (R_n : n = 0, 1, 2, \ldots)$ is the subject of Theorem 28 in Chapter 12.

It will be seen later in this section that the doubly indexed sequence $\widetilde{R} = (R_{m,n} : m, n \geq 0)$ can be used to analyze the asymptotic behavior of the terms in the sequence R. We give here the properties of $\widetilde{R}$ that make such an analysis possible.

First, we note the following obvious inequality:

$$R_{0,n} \leq R_{0,m} + R_{m,n} \quad \text{for all } n > m > 0.$$

This relationship is known as 'subadditivity'. Next we note that by Problem 4 and Problem 5, $(R_{0,k}, R_{k,2k}, R_{2k,3k}, \ldots)$ is a stationary sequence for any positive integer k. Also, the stationarity of X implies that for any $k \geq 0$, the sequence $(R_{k,k}, R_{k,k+1}, R_{k,k+2}, \ldots)$ has the same distribution as R. These three conditions are the first three conditions in the Kingman-Liggett Ergodic Theorem. The fourth and final condition of that theorem involves moment assumptions which are also satisfied by $\widetilde{R}$. The conclusion is that R_n/n converges a.s. as $n \to \infty$. If X is ergodic, then the a.s.-limit is a constant.

Theorem 7. [Kingman-Liggett Ergodic] *Let $(Z_{m,n} : 0 \leq m < n < \infty)$ be a doubly indexed sequence of $\mathbb{R}$-valued random variables satisfying the following four conditions:*

(i) $Z_{0,n} \leq Z_{0,m} + Z_{m,n}$ for $0 < m < n$;

(ii) for each $k = 1, 2, 3, \ldots$, the sequence $(Z_{nk,(n+1)k} : n = 0, 1, 2, \ldots)$ is stationary;

(iii) the sequence $(Z_{k,k+n} : n = 1, 2, 3, \ldots)$ has the same distribution for all $k = 0, 1, 2, \ldots$;

(iv) there exists a constant $c > 0$ such that $E(|Z_{0,n}|) \leq cn$ for $n = 1, 2, 3, \ldots$.

Then there exists an $\mathbb{R}$-valued random variable L such that

(28.10)
$$L = \lim_{n \to \infty} \frac{Z_{0,n}}{n} \quad \text{a.s. and in } \mathbf{L}_1$$

and

(28.11)
$$E(L) = \inf_n \frac{E(Z_{0,n})}{n}.$$

Furthermore, if the stationary sequences in condition (ii) are all ergodic, then

(28.12)
$$L = E(L) \text{ a.s.}$$

Remark 1. If the random variables $Z_{m,n}$ are nonnegative, then (iv) in the Kingman-Liggett Subadditive Ergodic Theorem can be replaced by $E(Z_{0,1}) < \infty$, since this inequality and (i) together imply (iv).

PROOF. Let

$$\gamma = \inf_{n \geq 1} \frac{E(Z_{0,n})}{n}$$

$$\overline{L} = \limsup_{n \to \infty} \frac{Z_{0,n}}{n}$$

$$\underline{L} = \liminf_{n \to \infty} \frac{Z_{0,n}}{n}.$$

We break the proof into five steps. In the first step, $|\gamma| < \infty$ is proved. Then in the second, we obtain

(28.13) $$E(\overline{L}) \leq \gamma.$$

In the third step, we show that

(28.14) $$\limsup_{n \to \infty} \frac{E(Z_{0,n})}{n} \leq E(\underline{L}).$$

These three parts of the proof immediately give the existence of an $\mathbb{R}$-valued random variable L satisfying (28.11) and the 'almost sure' aspect of (28.10) [by Property (v) in Theorem 9 of Chapter 4]. The fourth step shows how (28.12) is obtained as a by-product of the second and third steps. In the fifth and final step of the proof, we use uniform integrability to obtain $\mathbf{L}_1$ convergence at (28.10).

Step 1. Clearly $|E(Z_{0,n})| \leq E(|Z_{0,n}|) \leq cn$, from which it follows that $|\gamma| \leq c$.

Step 2. In order to prove (28.13), we set

$$S_n^{(k)} = \sum_{m=0}^{n-1} Z_{mk,(m+1)k}$$

for positive integers k and n. By condition (ii) and the Birkhoff Ergodic Theorem,

$$L^{(k)} \stackrel{\text{def}}{=} \lim_{n \to \infty} \frac{S_n^{(k)}}{nk}$$

exists almost surely, and

(28.15) $$E(L^{(k)}) = \frac{E(Z_{0,k})}{k}.$$

From $n + 1$ applications of condition (i) we have

$$Z_{0,nk+l} \leq S_n^{(k)} + Z_{nk,nk+l}$$

for integers $k, n \geq 1$ and $0 \leq l < k$, so

(28.16) $$\limsup_{n \to \infty} \frac{Z_{0,nk+l}}{nk+l} \leq L^{(k)} + \limsup_{n \to \infty} \frac{Z_{nk,nk+l}}{nk+l} \quad \text{a.s.}$$

for $k \geq 1$. By conditions (iii) and (iv), the random variables $Z_{nk,nk+l}$, $n = 1, 2, \ldots$, are identically distributed and have finite mean, so $Z_{nk,nk+l}/(nk+l) \to$

0 a.s. as $n \to \infty$, by Problem 9 of Chapter 12. It follows from this fact and (28.16) that

$$(28.17) \qquad \overline{L} \leq L^{(k)} \text{ a.s.}$$

for $k = 1, 2, \ldots$, from which it follows by the Fatou Lemma and (ii) that $E(\overline{L}) \leq \gamma$.

Step 3. We next wish to prove that

$$(28.18) \qquad E(\underline{L}) \geq \limsup_{n \to \infty} \frac{E(Z_{0,n})}{n}.$$

(This part of the proof is considered to be the hardest part.) For each $k = 1, 2, \ldots$, let U_k be a random variable that is uniformly distributed on $\{1, 2, \ldots, k\}$ and independent of the collection $(Z_{m,n})$. For $n = 0, 1, 2, \ldots$, let

$$Y_n^{(k)} = Z_{0,n+U_k} - Z_{0,n+U_k-1},$$

where $Z_{0,0}$ is defined to be identically 0, and let

$$Y^{(k)} = (Y_n^{(k)} : n = 0, 1, 2, \ldots)$$

be the corresponding sequence. Choose a subsequence (k_i) of the positive integers so that

$$(28.19) \qquad \lim_{i \to \infty} E(Y_0^{(k_i)}) = \limsup_{k \to \infty} E(Y_0^{(k)})$$

(possible by the definition of the limit supremum), and so that

$$Y^{(k_i)} \xrightarrow{D} Y \quad \text{as } i \to \infty$$

for some random sequence

$$Y = (Y_n : n = 0, 1, 2, \ldots)$$

of $\overline{\mathbb{R}}$-valued random variables (possible since the space $\overline{\mathbb{R}}^\infty$ is compact).

We wish to show that Y is stationary. By the definition of convergence in distribution,

$$(28.20) \qquad \lim_{i \to \infty} E(f \circ Y^{(k_i)}) = E(f \circ Y)$$

for any bounded continuous function $f \colon \overline{\mathbb{R}}^\infty \to \mathbb{R}$. By the definition of $Y^{(k)}$,

$$(28.21) \quad E(f \circ Y^{(k)}) = E(f \circ \tau^{U_k-1} \circ Y^{(1)}) = \frac{1}{k} \sum_{l=0}^{k-1} E(f \circ \tau^l \circ Y^{(1)}).$$

Both (28.20) and (28.21) also hold with f replaced by $f \circ \tau$ since τ is continuous, so we have

$$E(f \circ Y - f \circ \tau \circ Y) = \lim_{i \to \infty} \frac{1}{k_i} \left[E(f \circ \tau^{k_i} \circ Y^{(1)} - f \circ Y^{(1)}) \right].$$

Since f is bounded, the limit on the right of this last expression is 0, implying that $E(f \circ Y) = E(f \circ \tau \circ Y)$. Since f is an arbitrary bounded continuous function, it follows that Y and $\tau \circ Y$ have the same distribution, so Y is stationary.

Our next immediate goal is to prove the inequality (28.25) below. By the definition of $Y_0^{(k)}$ and condition (i),

$$(28.22) \qquad (Y_0^{(k)})^+ = (Z_{0,U_k} - Z_{0,U_k-1})^+ \le (Z_{U_k-1,U_k})^+ .$$

By condition (iii), the right side of this equation has the same distribution as $Z_{0,1}^+$, and by condition (iv), $E(Z_{0,1}^+) < \infty$. Thus the family $\{(Y_0^{(k)})^+ : k = 1, 2, 3, \dots\}$ is uniformly integrable. By the Uniform Integrability Criterion for convergence in distribution (Problem 27 of Chapter 14),

$$E(Y_0^+) = \lim_{i \to \infty} E[(Y_0^{(k_i)})^+] \le E[(Z_{0,1})^+] < \infty .$$

By the Fatou Lemma for convergence in distribution,

$$E[(Y_0)^-] \le \liminf_{i \to \infty} E[(Y_0^{(k_i)})^-] .$$

Combining these last two inequalities with (28.19) gives

$$(28.23) \qquad \limsup_{k \to \infty} E(Y_0^{(k)}) \le E(Y_0) < \infty .$$

By the definition of $Y^{(k)}$,

$$(28.24) \qquad E(Y_0^{(k)}) = E(Z_{0,U_k} - Z_{0,U_k-1})$$

$$= \frac{1}{k} \sum_{l=1}^{k} E(Z_{0,l} - Z_{0,l-1}) = \frac{E(Z_{0,k})}{k} ,$$

so by (28.23) and the definition of γ,

$$(28.25) \qquad E(Y_0) \ge \limsup_{k \to \infty} \frac{E(Z_{0,k})}{k} .$$

Thus, to complete this part of the proof, it is enough to prove that

$$(28.26) \qquad E(\underline{L}) \ge E(Y_0) .$$

We have shown that Y is a stationary sequence and that $\gamma \le E(Y_0) < \infty$, from which it follows that the random variables in the sequence Y are almost surely $\mathbb{R}$-valued. By the Birkhoff Ergodic Theorem,

$$L' = \lim_{n \to \infty} \frac{Y_0 + \cdots + Y_{n-1}}{n}$$

exists a.s. and $E(L') = E(Y_0)$, so (28.26) is equivalent to $E(\underline{L}) \ge E(L')$. We will prove this last inequality by comparing the two sequences that produce $\underline{L}$ and L'.

We need some terminology. A function $f : \mathbb{R}^\infty \to \mathbb{R}$ is *increasing* if

$$f(x_0, x_1, \dots) \le f(y_0, y_1, \dots) \quad \text{whenever } x_n \le y_n \text{ for all } n = 0, 1, 2, \dots .$$

If $R^{(1)}$ and $R^{(2)}$ are two $\mathbb{R}^\infty$-valued random variables (not necessarily defined on the same probability space), $R^{(1)}$ *stochastically dominates* $R^{(2)}$ if $E(f \circ R^{(1)}) \geq E(f \circ R^{(2)})$ for every increasing bounded continuous function $f \colon \mathbb{R}^\infty \to \mathbb{R}$.

For each k, denote by $T^{(k)}$ the sequence of partial sums of the sequence $Y^{(k)}$. Then by the definition of $Y^{(k)}$ and conditions (i) and (iii),

$$
\begin{aligned}
& E(f \circ T^{(k)}) \\
&= E\big[f\big((Z_{0,U_k} - Z_{0,U_k-1}), (Z_{0,U_k+1} - Z_{0,U_k-1}), (Z_{0,U_k+2} - Z_{0,U_k-1}), \ldots\big)\big] \\
&\leq E\big(f(Z_{U_k-1,U_k}, Z_{U_k-1,U_k+1}, Z_{U_k-1,U_k+2}, \ldots)\big) \\
&= E\big(f(Z_{0,1}, Z_{0,2}, Z_{0,3}, \ldots)\big) \\
&= E(f \circ T^{(1)})
\end{aligned}
$$

for increasing bounded continuous f, so $T^{(1)}$ stochastically dominates $T^{(k)}$ for all $k = 1, 2, 3, \ldots$.

Now let T denote the sequence of partial sums of the sequence Y. Convergence in distribution implies that

$$
E(f \circ T) = \lim_{i \to \infty} E(f \circ T^{(k_i)})
$$

for all continuous bounded functions $f \colon \mathbb{R}^\infty \to \mathbb{R}$. Thus $T^{(1)}$ stochastically dominates T. Since $T^{(1)} = (Z_{0,n} \colon n = 0, 1, 2, \ldots)$ and $E(L') = E(Y_0)$, (28.26) is now a consequence of Problem 29, which follows this proof.

Step 4. If each of the stationary sequences in condition (ii) is ergodic, then for each k, $L^{(k)} = E(Z_{0,k}/k)$ a.s.. By (28.17) and the definition of γ, $\overline{L} \leq \gamma$ a.s. In Step 3 we showed that $E(\underline{L}) \geq \gamma$. The desired conclusion follows from Property (v) in Theorem 9 of Chapter 4.

Step 5. The second and third steps permit us to write

$$
L = \underline{L} = \overline{L}.
$$

Now let

$$
R_n = \sum_{l=0}^{n-1} (Z_{l,l+1})^+.
$$

By the Birkhoff Ergodic Theorem, R_n/n converges a.s. as $n \to \infty$ to a random variable R with finite mean, and $E(|R - (R_n/n)|) \to 0$. By the Uniform Integrability Criterion, the family $\{R_n/n \colon n = 1, 2, \ldots\}$ is uniformly integrable. Repeated applications of (i) show that

$$
(Z_{0,n})^+ \leq R_n,
$$

so the family $\{(Z_{0,n})^+/n \colon n = 1, 2, \ldots\}$ is also uniformly integrable. Thus,

$$
\lim_{n \to \infty} E\Big(\frac{(Z_{0,n})^+}{n}\Big) = E(L^+).
$$

It follows from this equation, (28.13), and (28.14) that

$$\lim_{n\to\infty} E\left(\left|\frac{Z_{0,n}}{n}\right|\right) = E(|L|) .$$

The Uniform Integrability Criterion now implies $\mathbf{L}_1$ convergence at (28.10). $\square$

Problem 29. Let $R = (R_n : n = 1, 2, \dots)$ and $S = (S_n : n = 1, 2, \dots)$ be sequences of $\mathbb{R}$-valued random variables, and suppose that R stochastically dominates S. Show that for any bounded increasing measurable function $f \colon \mathbb{R}^\infty \to \mathbb{R}$,

$$E(f \circ R) \geq E(f \circ S),$$

provided either the expectation on the left is $> -\infty$ or the expectation on the right is $< \infty$. Note: In the proof of the Kingman-Liggett Subadditive Ergodic Theorem, this fact is used with $f(x_1, x_2, \dots) = \liminf_{n\to\infty}(x_n/n)$.

Example 7. [First-passage percolation] For an arbitrary pair of points $x, y \in \mathbb{Z}^2$, a *path* from x to y is any finite sequence $(x_0, \dots, x_n)$ of points in $\mathbb{Z}^2$ such that $x_0 = x$, $x_n = y$, and $|x_j - x_{j-1}| = 1$ for $j = 1, \dots, n$. Let $\{T_{x,y} : |x - y| = 1\}$ be an iid collection of nonnegative random variables, indexed by the 'nearest neighbor' pairs in $\mathbb{Z}^2$, and for any path $\pi = (x_0, \dots, x_n)$, let

$$U(\pi) = \sum_{j=1}^{n} T_{x_{j-1}, x_j} .$$

For an arbitrary pair of points $x, y \in \mathbb{Z}^2$ (not necessarily nearest neighbors), define

$$M_{x,y} = \inf U(\pi) ,$$

where the infimum is taken over all paths from x to y. The random variable $M_{x,y}$ is called the *first-passage time* from x to y. We assume that $E(T_{x,y}) < \infty$ for each x and y that are 'nearest neighbors' from which it follows that $E(M_{x,y}) < \infty$ for all x and y.

Fix a vector $v \in \mathbb{Z}^2$, and let

$$Z_{m,n} = M_{mv,nv} , \quad 0 \leq m < n < \infty .$$

In view of Remark 1 it is easily checked that the collection $(Z_{m,n})$ satisfies conditions (i)-(iv) of Theorem 7. The reader is asked in Problem 30 to show that the ergodicity condition is also satisfied. Thus

$$\lim_{n\to\infty} \frac{M_{0,nv}}{n} = C(v)$$

exists a.s. and in $\mathbf{L}_1$, where $C(v)$ is not random and is known as the *time constant*, although this constant depends on v.

* **Problem 30.** Prove that the collection $(Z_{m,n})$ in Example 7 satisfies the ergodicity condition in the Kingman-Liggett Subadditive Ergodic Theorem.

Problem 31. Let $X = (X_n : n \geq 0)$ be a stationary sequence of random variables taking values in a Hilbert space with inner product $\langle \cdot, \cdot \rangle$. Show that

$$\lim_{n \to \infty} \frac{\langle X_0 + \cdots + X_n, X_0 + \cdots + X_n \rangle}{(n+1)^2}$$

exists almost surely.

Problem 32. [Wiener sausage] Let U, V, W be independent standard Wiener processes on $[0, \infty)$ (as defined in Chapter 19), and for $t \geq 0$, let

$$S_t = \bigcup_{0 \leq s \leq t} ((U_s, V_s, W_s) + A),$$

where A is a Borel subset of $\mathbb{R}^3$. (Thus, S_t is the random tube or 'sausage' swept out by the set A as it follows along with the randomly moving point (U_s, V_s, W_s), $0 \leq s \leq t$.) Let λ denote Lebesgue measure in $\mathbb{R}^3$. Show that

$$\lim_{t \to \infty} \frac{\lambda(S_t)}{t} = C(A)$$

exists a.s., where $C(A)$ is a constant depending on A, known as the 'Newtonian capacity' of A. Prove that $C(A) > 0$ if A has positive Lebesgue measure. (Note that $C(A)$ can be positive even if A has Lebesgue measure 0. For instance, let A be the surface of a sphere.)

Problem 33. [Products of random matrices] Let $(A_n : n \geq 0)$ be an ergodic stationary sequence of $d \times d$ matrices with positive entries, and for $n \geq 0$, let B_n be the product $A_0 A_1 \cdots A_n$. Show that for $1 \leq i, j \leq d$,

$$\lim_{n \to \infty} \frac{\log B_n(i,j)}{n} = C(i,j) \text{ a.s.},$$

where $C(i,j)$ is a constant.

28.7. ‡ Spectral analysis of stationary sequences

Throughout this section, we will restrict our attention to sequences of $\mathbb{R}$-valued random variables that have finite second moments. Such sequences are called *second-order sequences*.

For a second-order sequence $X = (X_0, X_1, \dots)$, we define, as in Chapter 5, the *mean vector* m_X and *covariance matrix* Σ_X. Here m_X has infinitely many coordinates and Σ_X has infinitely many rows and columns:

$$m_X(k) = E(X_k), \quad k = 0, 1, 2, \dots$$

and

$$\Sigma_X(j,k) = \mathrm{Cov}(X_j, X_k), \quad j, k = 0, 1, 2, \dots.$$

A second-order sequence X is *second-order stationary* if it has the same mean vector and covariance matrix as the shifted sequence $\tau \circ X$, that is, if $m_X(k)$ does not depend on k and $\Sigma_X(j, k) = \Sigma_X(0, |j - k|)$ for all j and k.

The most important type of second-order sequences are Gaussian sequences, defined in Example 5 to be those random sequences whose finite-dimensional marginal distributions are normal. Since any normal distribution is uniquely determined by its mean vector and covariance matrix (see Theorem 20 of Chapter 13), it is clear that the distribution of a Gaussian sequence X is uniquely determined by m_X and Σ_X. It follows from this fact that a Gaussian sequence is stationary if and only if it is second-order stationary.

> **Problem 34.** Prove that the mean vector and covariance matrix of any second-order stationary sequence are also the mean vector and covariance matrix of some stationary Gaussian sequence.

In general, a second-order stationary sequence need not be stationary. For instance, the forthcoming Example 8 shows how to construct a certain class of second-order stationary sequences, and Problem 36 indicates precisely which sequences of that class are not stationary. It is remarkable that the main result of this section, which is a representation theorem, can be stated in a way that applies to general second-order stationary sequences. However, for simplicity we will only state and prove this result for the Gaussian case.

Example 8. Let Z_1, Z_2 be uncorrelated $\mathbb{R}$-valued random variables, each having mean 0 and finite variance σ^2, and let λ be a real number. Then the random sequence X defined by

$$X_k = Z_1 \cos \lambda k + Z_2 \sin \lambda k, \quad k = 0, 1, 2, \ldots$$

is easily seen by direct computation to be a second-order stationary sequence with mean vector $m_X = (0, 0, \ldots)$ and covariance matrix Σ_X satisfying

$$\Sigma_X(j, k) = \sigma^2 \cos \lambda(k - j), \quad j, k \geq 0.$$

This sequence is a second-order stationary *pure-tone* sequence with *frequency* $\lambda/2\pi$ and *random coefficients* Z_1, Z_2.

In applications in the 'real world', one usually considers the extension of such sequences to continuous time. Thus, a pure-tone sequence X becomes a pure-tone 'random signal':

$$X_t = Z_1 \cos \lambda t + Z_2 \sin \lambda t, \quad t \geq 0.$$

This continuous-time stochastic process is a sine wave. Its frequency is deterministic, but its amplitude and phase angle are random. The extension to continuous time is not unique, since replacing λ by $\lambda + 2\pi p$ for some integer p changes the continuous-time extension without changing the original random sequence.

A main result of this section is that an arbitrary second-order stationary Gaussian sequence can be decomposed into a mixture of second-order stationary pure-tone sequences with different frequencies, so it follows that a second-order stationary sequence can be naturally extended to a continuous-time stochastic process by replacing each of its component pure-tone sequences by the corresponding pure-tone random signal.

Problem 35. Let X be a second-order stationary pure-tone sequence. Show that if the pair of random coefficients (Z_1, Z_2) is normally distributed, then X can be written in the form

$$X_k = \sigma Z \cos(\lambda k + U),$$

where Z is standard normal, U is uniformly distributed on $[0, 2\pi)$, and (Z, U) is an independent pair. (Compare with Example 5.)

Problem 36. Show that a second-order stationary pure-tone sequence X with frequency $\lambda/2\pi$ is stationary if and only if the distribution of the random vector (Z_1, Z_2) of random coefficients is invariant under a rotation of λ radians about the origin in $\mathbb{R}^2$.

Let $X^{(1)}, \ldots, X^{(m)}$ be second-order stationary pure-tone sequences defined on a common probability space. If the random coefficients of $X^{(j)}$ are uncorrelated with the random coefficients of $X^{(k)}$ for $1 \le j, k \le m$, then it is easily checked that $a_1 X^{(1)} + \cdots + a_m X^{(m)}$ is a second-order stationary sequence; it is a mixture of finitely many pure-tone sequences.

In order to be able to represent all second-order stationary sequences as mixtures of pure-tone sequences, we need to be able to make sense out of mixing uncountably many pure-tone sequences. While it is possible to do so for the general case, there are some technicalities involved that we wish to avoid. Matters are much simpler in the Gaussian case, as illustrated by the following example.

Example 9. Let $F \colon [-\pi, \pi] \to \mathbb{R}^+$ be an increasing right-continuous function that satisfies $F(-\pi) \ge 0$. Extend F to $(-\infty, -\pi)$ by setting $F(\lambda) = 0$ for $\lambda < -\pi$. Also let V and W be independent standard Wiener processes on $[0, \infty)$. For $n \ge 0$, define

$$(28.27) \quad X_n = \int_{(-\pi)-}^{\pi} \cos \lambda n \, d(V \circ F)(\lambda) + \int_{(-\pi)-}^{\pi} \sin \lambda n \, d(W \circ F)(\lambda),$$

where the integrals are understood to be Riemann-Stieltjes integrals, the existence of which follows from Problem 14 of Appendix D. Consider Riemann-Stieltjes sums for the two integrals in (28.27), using the same point partition for both. It is easily checked that when these two sums are added together, one obtains a stationary Gaussian sequence that is a mixture of finitely many pure-tone sequences. It follows from a straightforward limiting argument that X is a stationary Gaussian sequence.

By using integration by parts (Proposition 2 of Appendix D), the integrals in (28.27) can be written as:

(28.28)
$$\int_{(-\pi)-}^{\pi} \cos \lambda n \, d(V \circ F)(\lambda)$$
$$= (V \circ F)(\pi)(\cos \pi n) + n \int_{-\pi}^{\pi} (V \circ F)(\lambda) \sin \lambda n \, d\lambda \,,$$

(28.29)
$$\int_{(-\pi)-}^{\pi} \sin \lambda n \, d(W \circ F)(\lambda) = -n \int_{-\pi}^{\pi} (W \circ F)(\lambda) \cos \lambda n \, d\lambda \,.$$

By setting $n = 0$ in (28.28) and (28.29) and taking expected values we get $E(X_0) = 0$ and hence conclude that the mean vector is $(0, 0, \dots)$. We can use these same two formulas in conjunction with the Fubini Theorem to obtain

$$\mathrm{Cov}(X_0, X_n) = E\big[(V \circ F)^2(\pi)\big](\cos \pi n) + n \int_{-\pi}^{\pi} E\big[(V \circ F)(\pi)(V \circ F)(\lambda)\big] \sin \lambda n \, d\lambda \,,$$

where we have used the fact that V and W are independent in order to eliminate one term. [The Fubini Theorem cannot be used directly with (28.27) since the Riemann-Stieltjes integrals there are not measure-theoretic integrals or even differences of such integrals unless F is very special.] The first expectation on the right side is the variance of the value taken by a Wiener process at time $F(\pi)$ so it equals $F(\pi)$. The expectation inside the integral is the covariance of a Wiener process at time $F(\pi)$ with the same Wiener process at time $F(\lambda)$. Therefore

$$\mathrm{Cov}(X_0, X_n) = F(\pi) \cos \pi n + n \int_{-\pi}^{\pi} [F(\pi) \wedge F(\lambda)] \sin \lambda n \, d\lambda$$
$$= \int_{(-\pi)-}^{\pi} \cos \lambda n \, dF(\lambda) \,,$$

the last equality resulting from integration by parts. Thus, the covariance matrix Σ_X is given by

(28.30)
$$\Sigma_X(m, n) = \int_{(-\pi)-}^{\pi} \cos \lambda (m - n) \, dF(\lambda) \,.$$

We call the function F in the preceding example the *spectral distribution function* of the random sequence X. It can be regarded as the distribution function of a finite measure μ on $[-\pi, \pi]$ called the *spectral measure;* that is,

$$F(\lambda) = \mu[-\pi, \lambda] \,, \quad \lambda \in [-\pi, \pi] \,.$$

We may view X as a mixture of uncountably many pure-tone sequences, with μ determining the contribution of each individual frequency. The support of μ is the *spectrum* of X. The points in the spectrum of X at which F is continuous constitute the *continuous spectrum* of X, and if F is continuous, X has a *pure continuous spectrum*. The discontinuity points of F constitute the *point spectrum* of X. If the spectral measure of X assigns measure 0 to the complement of some countable set, then X has a *pure point spectrum*.

We have seen how to use a finite measure μ to construct a stationary Gaussian sequence X. The covariance matrix and hence the distribution of X are determined by μ in this construction. It is easy to see, however, that more than one measure μ can lead to the same covariance matrix. Consider, for example, a Gaussian pure-tone sequence with covariance matrix given by $\Sigma(m,n) = \cos(m-n)$. Any probability measures that is a convex combination of δ_1 and δ_{-1} can be used for a spectral measure. Thus, we have been somewhat imprecise in talking about 'the' spectral measure and spectral distribution function.

There are several common conventions for eliminating this lack of uniqueness. Two of these will be useful to us. The first is to consider only measures μ that are symmetric about 0. This option is particularly useful in proofs in which complex exponentials are used to simplify calculations involving cos and sin, and it will also be useful in Example 10. The second is to consider only measures μ that are supported by $[0, \pi]$. This will typically be our choice when trying to make the formulas look as simple as possible, such as in the next lemma and theorem.

Lemma 8. [Herglotz] *A matrix Σ with entries $\Sigma(m,n)$, $m,n \geq 0$, is the covariance matrix of a second-order stationary sequence X if and only if there exists a finite measure μ on $[0, \pi]$ such that*

$$(28.31) \qquad \Sigma(m,n) = \int_{[0,\pi]} \cos \lambda(m-n) \, \mu(d\lambda)$$

for $m,n \geq 0$. Furthermore, the correspondence between Σ and μ is one-to-one.

We omit the proof of the preceding lemma because it is quite similar to the proof of Theorem 13 of Chapter 13.

The following theorem is the main result of this section. It says that every stationary Gaussian sequence can be uniquely represented as a mixture with respect to its spectral measure of pure tone sequences. The result is an immediate consequence of the argument given in Example 9, the Herglotz Lemma, and the fact that the distribution of a Gaussian vector is determined by its mean vector and covariance matrix.

Theorem 9. *Let V, W be independent Wiener processes on $[0, \infty)$ and let $F: [0, \pi] \to \mathbb{R}^+$ be an increasing right continuous function that satisfies $F(0) \geq 0$, and extend F to $(-\infty, 0)$ by setting $F(\lambda) = 0$ for $\lambda < 0$. For $n \geq 0$, define*

$$(28.32) \qquad X_n = \int_{0-}^{\pi} \cos \lambda n \, d(V \circ F)(\lambda) + \int_{0-}^{\pi} \sin \lambda n \, d(W \circ F)(\lambda),$$

where the two integrals are Riemann-Stieltjes integrals. Then $X = (X_n: n \geq 0)$ is a stationary Gaussian sequence with mean vector $(0, 0, \dots)$ and covariance matrix Σ_X given by

$$\Sigma_X(m, n) = \int_{0-}^{\pi} \cos \lambda (m - n) \, dF(\lambda), \quad m, n \geq 0.$$

Moreover, if $Y = (Y_n: n \geq 0)$ is any stationary Gaussian sequence with spectral distribution function F, then the sequence $(Y_n - EY_0: n \geq 0)$ has the same distribution as X.

Problem 37. Let X be a stationary Gaussian sequence with spectral measure μ. Suppose that μ can be written as the sum of pairwise mutually singular measures $\mu_1, \dots, \mu_k$. Let $X^{(1)}, \dots, X^{(k)}$, be independent stationary Gaussian sequences such that for $j = 1, \dots, k$, $X^{(j)}$ has spectral measure μ_j. Prove that X has the same distribution as $X^{(1)} + \cdots + X^{(k)}$.

Problem 38. Show that a stationary Gaussian sequence has the same distribution as the sum of two appropriate independent Gaussian sequences $X^{(1)}$ and $X^{(2)}$, with $X^{(1)}$ having a pure continuous spectrum and $X^{(2)}$ a pure point spectrum.

Problem 39. Show that if a stationary Gaussian sequence X is ergodic then it has a pure continuous spectrum. *Hint:* Use the preceding two problems and Problem 27.

Problem 40. Let X be a stationary Gaussian sequence with spectral measure μ. Show that if μ is absolutely continuous with respect to Lebesgue measure, then X is mixing and therefore ergodic. *Hint:* First consider the case in which the density of μ is continuous, and use the continuity to prove that $\text{Cov}(X_0, X_n) \to 0$ as $n \to \infty$.

Problem 41. [Second-order ergodic theorem] Let $X = (X_n: n \geq 0)$ be a (not necessarily Gaussian) second-order stationary sequence, and for $n > 0$, let $S_n = X_0 + \cdots + X_{n-1}$. Show that there exists a random variable L with finite second moment such that S_n/n converges to L in $\mathbf{L}_2$ as $n \to \infty$. In other words, show that

$$\lim_{n \to \infty} E\left[\left(\frac{S_n}{n} - L \right)^2 \right] = 0.$$

For the Gaussian case, show that the convergence is also almost sure. *Hint:* First use the Birkhoff Ergodic Theorem for the Gaussian case. To generalize, show that the property of a sequence being Cauchy in $\mathbf{L}_2$ depends only on the mean vector and covariance matrix.

It turns out that one can explicitly evaluate the limit in the preceding problem in the Gaussian case. Here is a brief sketch of the method. We use the formulas

$$\sum_{k=0}^{n-1} \cos \lambda k = \Re\left(\frac{1 - e^{i\lambda n}}{1 - e^{i\lambda}} \right) \quad \text{and} \quad \sum_{k=0}^{n-1} \sin \lambda k = \Im\left(\frac{1 - e^{i\lambda n}}{1 - e^{i\lambda}} \right),$$

which are proved by summing the appropriate finite geometric series of complex exponentials. By Theorem 9,

$$
\frac{S_n}{n} = E(X_0)
$$

(28.33)
$$
+ \int_{0-}^{\pi} \Re\left(\frac{1 - e^{i\lambda n}}{n(1 - e^{i\lambda})}\right) d(V \circ F)(\lambda)
$$
$$
+ \int_{0-}^{\pi} \Im\left(\frac{1 - e^{i\lambda n}}{n(1 - e^{i\lambda})}\right) d(W \circ F)(\lambda).
$$

The integrals in (28.33) are Riemann-Stieltjes integrals, not measure-theoretic integrals. Accordingly, tools such as the Dominated Convergence Theorem are not likely to be useful. Nevertheless, it is reasonable to make a guess of a limiting random variable by taking limits inside of the integrals. The integrand in the first integral converges to the indicator function of the singleton $\{0\}$ as $n \to \infty$, and the integrand in the second integral converges to 0. Thus, the guess is that the limit as $n \to \infty$ of the entire expression is $E(X_0) + (V \circ F)(0)$. In particular, the limit is constant if and only if $F(0) = \mu\{0\} = 0$, in which case the limit is $E(X_0)$. Once this guess has been made, it can be verified by calculating some means and second moments, as in the following problem.

Problem 42. Let J_n and K_n be the two integrals in (28.33). Show that $E(K_n^2)$ and $E[(J_n - (V \circ F)(0))^2]$ converge to 0 as $n \to \infty$. Then explain why these facts imply that $S_n/n \to E(X_0) + (V \circ F)(0)$ a.s. as $n \to \infty$. *Hint:* Use Riemann-Stieltjes sums to calculate the means and second moments of J_n and K_n. Also, note that $V \circ F(0)$ is independent of 'most' of the integral J_n, except for the part that involves integration near 0.

Problem 43. Let $X = (X_n : n \geq 0)$ be an ergodic stationary Gaussian sequence, with $E(X_0) = 0$. Fix an integer $j \geq 0$. Show that

$$
\lim_{n \to \infty} \frac{1}{n} \sum_{k=0}^{n-1} X_k X_{k+j}
$$

exists a.s., and find a simple expression for this limit.

The following random sequences (some of which have continuous-time analogues mentioned in Chapter 33) are stationary, Markov, and Gaussian.

Example 10. [Ornstein-Uhlenbeck sequence] Let $c \in (0, \infty)$ and consider the symmetric spectral measure on $[-\pi, \pi]$ with density

$$
\lambda \rightsquigarrow \sum_{k=-\infty}^{\infty} \frac{c}{\pi[c^2 + (\lambda + 2\pi k)^2]}.
$$

Using (28.30), the Dominated Convergence Theorem, a change of variables, and the periodicity of cos, we obtain a formula for the covariance function:

$$\Sigma(m,n) = \sum_{k=-\infty}^{\infty} \int_{-\pi}^{\pi} \cos \lambda(m-n) \frac{c}{\pi[c^2 + (\lambda + 2\pi k)^2]}$$

$$= \sum_{k=-\infty}^{\infty} \int_{\pi(-1+2k)}^{\pi(1+2k)} \cos y(m-n) \frac{c}{\pi[c^2 + y^2]} dy = \int_{-\infty}^{\infty} \cos y(m-n) \frac{c}{\pi[c^2 + y^2]},$$

which we recognize as the characteristic function of a Cauchy distribution. Thus it equals $e^{-c|m-n|}$. Letting $\rho = e^{-c}$, we see that for every $\rho \in (0,1)$ there is a stationary Gaussian sequence with correlation function given by $(m,n) \rightsquigarrow \rho^{|m-n|}$. It is easily checked that if X is such a sequence, then the sequence $(X_0, -X_1, X_2, \dots)$ is stationary Gaussian with correlation function $(m,n) \rightsquigarrow (-\rho)^{|m-n|}$.

To accommodate $\rho = \pm 1$ and $\rho = 0$, we note that a spectral measure supported by $\{0\}$ gives the correlation function $(m,n) \rightsquigarrow 1^{|m-n|}$, and a spectral measure equal to any multiple of Lebesgue measure on $[-\pi, \pi]$ gives the correlation function $(m,n) \rightsquigarrow 0^{|m-n|}$.

The stationary sequences with correlation functions of the form $(m,n) \rightsquigarrow \rho^{|m-n|}$, $\rho \in [-1,1]$, are *Ornstein-Uhlenbeck sequences*. It develops that these are exactly the stationary Gaussian sequences that are also Markov. The argument in one direction is given below. Problem 45 addresses the other direction.

Let $X = (X_0, X_1, \dots)$ be a stationary Markov Gaussian sequence, and set $\rho = \mathrm{Corr}(X_0, X_1)$. With no loss of generality, we assume that $\mathrm{Var}(X_n) = 1$ and $E(X_n) = 0$ for each n. In this situation we see from Problem 48 of Chapter 21 that for any m and n,

$$E(X_n \mid X_m) = \mathrm{Corr}(X_m, X_n) X_m.$$

We will prove that $E(X_n \mid X_0) = \rho^n X_0$, from which it will then immediately follow by stationarity that the covariance function of X is $(m,n) \rightsquigarrow \rho^{|m-n|}$. That $E(X_1 \mid X_0) = \rho X_0$ is true from the definition of ρ and the relation between correlations and conditional expectations. For an induction proof we suppose that $E(X_{n-1} \mid X_0) = \rho^{n-1} X_0$. By stationarity, $E(X_n \mid X_1) = \rho^{n-1} X_1$. Since X is Markov, we then obtain

$$E(X_n \mid (X_1, X_0)) = \rho^{n-1} X_1.$$

Take conditional means of both sides to get $E(X_n \mid X_0) = \rho^n$, as desired.

Problem 44. For each $\rho \in [-1, 0)$, find a spectral measure corresponding to the correlation function $(m,n) \rightsquigarrow \rho^{|m-n|}$.

* **Problem 45.** Prove that a stationary Gaussian sequence with correlation function $(m,n) \rightsquigarrow \rho^{|m-n|}$ is Markov. *Hint:* One approach is to use Example 4 of Chapter 21.

PART 6
Stochastic Processes

Although the term 'stochastic process' has been given a variety of meanings in the literature, we use it in this book to refer to any collection of random variables X_t defined on a common probability space Ω, taking values in a common 'state space' Ψ, and indexed by the continuous-time parameter $t \in [0, \infty)$. Thus, stochastic processes are continuous-time analogues of random sequences. Often the shorter term *process* is used in context. It should come as no surprise that we will now encounter continuous-time versions of many of the types of random sequences introduced in previous parts. Our emphasis will be on concepts and types of behavior that are either more natural in or unique to the continuous-time setting.

We focus our attention on four important classes of stochastic processes: Lévy processes (Chapter 30), pure-jump Markov processes with bounded rates (Chapter 31), interacting particle systems with bounded, finite-range rates (Chapter 32), and 1-dimensional diffusions with 'nice' coefficients (Chapter 33). In each case, we provide a completely rigorous construction procedure.

The main ingredient in the construction of the first three classes of processes is a type of random set known as a 'Poisson point process' (Chapter 29). Poisson point processes are not really stochastic processes in our sense of the term, since they do not involve the time parameter. However, they are important in their own right, and our use of them shows that they are closely related to stochastic processes.

The construction of diffusions requires the 'stochastic calculus'. In Chapter 33, this important concept is introduced, and then diffusions are seen to be solutions of 'stochastic differential equations' that involve Brownian motion in a manner that is analogous to the way ordinary differential equations involve the time variable.

All of the stochastic processes mentioned above are 'Markov processes'. They also are closely related to many continuous-time martingales. A brief introduction to the general theory of Markov processes and their relationship to martingales is provided in Chapter 31.

CHAPTER 29
Point Processes

Loosely speaking, a point process is a random 'discrete' set of points in some Polish space. Thus, one could use a point process to model experiments like throwing grains of sand onto the floor and noting their locations, or pointing an astronomical telescope in a random direction and noting the positions of the stars seen in the field of view. A mathematical example would be the random set of values taken by a finite sequence of random variables. This latter example makes it clear that we may want to generalize the notion of sets to allow a given point to appear more than once. It turns out that there is a nice mathematical way to accommodate the generalization using a certain class of $\overline{\mathbb{Z}}^+$-valued measures. The relevant definitions and basic facts are given in the first section. The most important point processes are 'Poisson point processes', which are characterized by the property that their intersections with disjoint subsets of the underlying Polish space are independent. These are treated in Sections 3 and 4. An important tool for studying the distributions of point processes is introduced in the fourth section. This tool is needed in the final two sections of the chapter, where various operations on point processes are studied. In particular, the convergence in distribution of point processes is considered in the final section. One nice result from that section is that the Poisson point processes arise as limits of certain naturally defined sequences.

29.1. Point processes as random Radon measures

We begin by describing the appropriate setting for point processes. A Polish space Ψ is *locally compact* if for every $x \in \Psi$, there exists $\varepsilon > 0$ such that the closed ball $\overline{B}(x, \varepsilon)$ is compact. Equivalently, local compactness means that every point $x \in \Psi$ has an open neighborhood with compact closure.

A measure μ on a locally compact Polish space is a *Radon measure* if $\mu(C)$ is finite for every compact set C. Since we wish to define a point process to be a certain type of random Radon measure, we need to make the collection of Radon measures on a given locally compact Polish space into a measurable space. The following lemma makes this task easy to do.

Lemma 1. *Let* Ψ *be a locally compact Polish space. Then there exist open balls* $B(y_1, \varepsilon_1), B(y_2, \varepsilon_2), \ldots$ *such that*

$$\Psi = \bigcup_{n=1}^{\infty} B(y_n, \varepsilon_n)$$

and for each n, *the closed ball* $\overline{B}(y_n, \varepsilon)$ *is compact.*

PROOF. Let $(y_1, y_2, \ldots)$ be a dense sequence of points in Ψ. For $n = 1, 2, \ldots$, set

$$(29.1) \qquad \varepsilon_n = 1 \wedge \tfrac{1}{2} \sup\{\varepsilon \colon \overline{B}(y_n, \varepsilon) \text{ is compact}\},$$

which is positive since Ψ is locally compact. Moreover, $\overline{B}(y_n, \varepsilon_n)$ is compact for every n.

To show that $\Psi \subseteq \cup_{n=1}^{\infty} B(y_n, \varepsilon_n)$, we let x be an arbitrary member of Ψ and choose $\varepsilon \in (0, 3]$ so that $\overline{B}(x, \varepsilon)$ is compact. Fix n so that the distance between x and y_n is less than $\frac{\varepsilon}{3}$. Then $\overline{B}(y_n, \frac{2\varepsilon}{3})$ is a closed subset of the compact set $\overline{B}(x, \varepsilon)$, and is thus compact. By (29.1), $\varepsilon_n \geq \frac{\varepsilon}{3}$ and thus $x \in B(y_n, \varepsilon_n)$, as desired. $\square$

Let Ψ be a locally compact Polish space and $\mathcal{M}$ the collection of all Radon measures on Ψ. We will place the same type of measurability structure on $\mathcal{M}$ that we did on the space of probability measures in Chapter 21. Denote by $\mathfrak{H}$ the smallest σ-field of subsets of $\mathcal{M}$ such that for every Borel set $B \subseteq \Psi$, the function from $\mathcal{M}$ to $\overline{\mathbb{R}}^+$ defined by $\mu \rightsquigarrow \mu(B)$ is $\mathfrak{H}$-measurable.

Proposition 2. *The space* $(\mathcal{M}, \mathfrak{H})$ *defined above is a Borel space.*

PROOF. According to Lemma 1, the locally compact Polish space Ψ is the union of countably many compact sets $C_1, C_2, \ldots$. For each $n \geq 1$, let

$$A_n = C_n \setminus \left(\bigcup_{k=1}^{n-1} C_k \right),$$

and let $\mathcal{M}_n$ be the collection of all finite measures on the Borel subsets of A_n. Note that $(A_1, A_2, \ldots)$ is a measurable partition of Ψ. Since each set A_n is a subset of a compact set, the restriction to A_n of any Radon measure on Ψ is a finite measure on the Borel subsets of A_n. We leave it to the reader to use this fact to show that there is a Borel isomorphism between $(\mathcal{M}, \mathfrak{H})$ and the infinite product space

$$\bigotimes_{n=1}^{\infty} (\mathcal{M}_n, \mathfrak{H}_n).$$

where for each n, $\mathfrak{H}_n$ is defined in the same manner as $\mathfrak{H}$.

Thus, it is enough to show that each space $(\mathcal{M}_n, \mathfrak{H}_n)$ is a Borel space. Identify each nonzero finite measure $\mu \in \mathcal{M}_n$ with the pair $(\mu(A_n), \mu/\mu(A_n))$. This identification provides an obvious Borel isomorphism between the nonzero finite measures on A_n and the product of two Borel spaces, namely $(0, \infty)$ and the space of probability measures on A_n. The desired result follows immediately. □

We will be particularly interested in Radon measures that take values in $\overline{\mathbb{Z}}^+$, so the following result is relevant.

Corollary 3. *Let Ψ be a locally compact Polish space. The collection of all $\overline{\mathbb{Z}}^+$-valued Radon measures on Ψ is a measurable subset of the Borel space of all Radon measures on Ψ, and thus is itself a Borel space.*

Problem 1. Prove the preceding corollary.

Problem 2. Show that every Radon measure on a locally compact Polish space is σ-finite.

We are now prepared to define point processes.

Definition 4. Let Ψ be a locally compact Polish space. A *point process* on Ψ is a random variable taking values in the space of $\overline{\mathbb{Z}}^+$-valued Radon measures on Ψ.

We often use the letter X to denote a point process, and $X(B)$ to denote the measure assigned to a Borel set B by the random measure X. It is clear from the definition of the space $(\mathcal{M}, \mathfrak{H})$ that if A is a Borel subset of Ψ and X is a point process on Ψ, then the set $[X(A^c) = 0]$ is an event. If this event has probability 1, then we sometimes say that X is a *point process on A*.

It is not obvious from Definition 4 why the objects defined are named as they are. The 'points' in question have to do with the support of the random measure. We will see that the support of a $\overline{\mathbb{Z}}^+$-valued Radon measure μ consists precisely of those 'locations' x such that $\mu\{x\}$ is a positive integer. We take the point of view that μ represents a collection of points that occupy the locations in its support. Any location x such that $\mu\{x\} \geq 2$ is 'multiply occupied'.

To be more precise, we introduce some more terminology. A closed subset A of a Polish space Ψ is *discrete* if for every $x \in A$ there is an open set B such that $A \cap B = \{x\}$. Equivalently, a closed set A is discrete if no point $x \in A$ is the limit of a convergent sequence of points in $A \setminus \{x\}$. The following result implies that a point process is a random $\overline{\mathbb{Z}}^+$-valued Radon measure with discrete support.

Proposition 5. *Let C be the support of a $\overline{\mathbb{Z}}^+$-valued Radon measure μ on a locally compact Polish space Ψ. Then*

$$(29.2) \qquad\qquad C = \{x \in \Psi : \mu\{x\} \geq 1\}$$

and C is discrete.

PROOF. Because of Lemma 1, there is no loss of generality in assuming that Ψ is compact and μ is finite. Let x be a point in Ψ. By the Continuity of Measure Theorem,

$$\mu\{x\} = \lim_{\varepsilon \searrow 0} \mu(B(x,\varepsilon)).$$

Since μ is $\mathbb{Z}^+$-valued, there exists an $\varepsilon > 0$ such that $\mu\{x\} = \mu(B(x,\varepsilon))$. If $\mu(B(x,\varepsilon)) = 0$, then x is not in the support, since the support of a measure is a closed set. Otherwise, $\mu\{x\} \geq 1$ and x is clearly in the support of μ. We have proved (29.2).

To show that C is discrete, let x be a member of C. As in the preceding paragraph, there exists an $\varepsilon > 0$ such that $\mu\{x\} = \mu(B(x,\varepsilon))$. It follows from (29.2) that $C \cap B(x,\varepsilon) = \{x\}$, as desired. $\square$

Let μ be a $\overline{\mathbb{Z}}^+$-valued Radon measure on a locally compact Polish space Ψ. Since each singleton $\{x\}$ is a compact set, μ assigns a finite integer value to each singleton. The *multiplicity* (with respect to μ) of a point x is the integer $\mu\{x\}$. The preceding proposition implies that μ is uniquely determined by the multiplicities of the points in its support. In fact, we have the following formula:

$$\mu = \sum_{x \in C} \mu\{x\}\delta_x.$$

where C is the support of μ and δ_x denotes the delta distribution at x. This formula is the reason that many of the examples of random and nonrandom $\overline{\mathbb{Z}}^+$-valued Radon measures in this chapter are expressed as linear combinations of delta distributions.

A set whose members have been assigned positive integer multiplicities is called a *multiset*. Thus, there is a natural one-to-one correspondence between $\overline{\mathbb{Z}}^+$-valued Radon measures on a locally compact Polish space Ψ and discrete multisets of points in Ψ. The notation used for multisets is similar to that used for sets, except that elements with multiplicities greater than 1 are repeated according to their multiplicities. Thus $\{a, a, b\}$ is the multiset in which the element a has multiplicity 2 and the element b has multiplicity 1. As with ordinary set notation, the ordering is not important, so $\{a, b, a\}$ is another way of writing the same multiset. Expressed as a Radon measure, this multiset is $2\delta_a + \delta_b$.

Thus, a point process can be viewed as a 'random discrete multiset', and sometimes it is convenient to express point processes in this manner. For example, in this introduction, we mentioned that one way to construct a point process is to use the values of a finite sequence $(Y_1, \ldots, Y_n)$ of random variables. Expressed as a random Radon measure, the resulting point process is

$$\delta_{Y_1} + \cdots + \delta_{Y_n}.$$

As a random multiset, the point process is $\{Y_1, \ldots, Y_n\}$. However, the set notation for a multiset should only be used when accompanied by words that are indicative of a multiset. Thus for instance, in the last section of Chapter 12 where the size of the image of a random walk is treated and no mention of multisets is made, repeated occurrences of a value are not to be counted.

From the multiset point of view, $X(B)$ is the number of members of X that lie in B, counting multiplicities, and for ψ a member of Ψ, $X\{\psi\}$ is the multiplicity of ψ, possibly 0.

Problem 3. Prove that a discrete subset of a compact Polish space is finite, and that a discrete subset of a locally compact Polish space is countable. Thus, point processes have countable support a.s.

Problem 4. Let Y_j, $1 \le j \le n$, be iid random variables taking values in a locally compact Polish space Ψ. Let X be the point process defined by $X = \sum_{k=1}^{n} \delta_{Y_k}$. For arbitrary Borel subsets A and B of Ψ calculate the distribution of the random number $X(A)$ and the distribution of the random pair $(X(A), X(B))$ in terms of the common distribution of the random variables Y_j.

* **Problem 5.** Consider an urn with balls of $m \ge 3$ colors and denote by r_j the number of balls of color j, $1 \le j \le m$. Fix an integer $n \le \sum_{j=1}^{m} r_j$ and consider the point process X that assigns to each color the number of balls of that color drawn when a total of n balls are drawn without replacement. Calculate the distributions of the random number $X\{1\}$, the random pair $(X\{1\}, X\{2\})$, and the random triple $(X\{1\}, X\{2\}, X\{3\})$.

Problem 6. Let $\Psi = \{1, 2, \ldots, n\}$ and fix an integer $r \in [0, n]$. Consider the point process X for which the random variables $X\{k\}$ sum to r and constitute an exchangeable Bernoulli sequence of length n. Describe the distributions of the random n-tuple $(X\{1\}, X\{2\}, \ldots, X\{n\})$ and the random numbers $X(B)$, $B \subseteq \Psi$.

Problem 7. Modify the preceding problem by the deleting the parameter r and changing "exchangeable" to "iid with success probability $\frac{1}{2}$".

* **Problem 8.** Let $\Psi = \{1, 2, \ldots, n\}$ and $r \in \mathbb{Z}^+$. Consider the point process X defined by

$$P[X\{k\} = r_k, 1 \le k \le n] = \begin{cases} \dfrac{r!}{n^r \prod_{k=1}^{n} r_k!} & \text{if } r_1 + \cdots + r_n = r \\ 0 & \text{otherwise}. \end{cases}$$

Calculate the distribution of $X(B)$ for each $B \subseteq \Psi$.

Problem 9. At times we extend the use of the term 'point process' to a random variable which with probability 1 equals a $\overline{\mathbb{Z}}^+$-valued Radon measure, but which may have some other value or values on a null set. This problem illustrates such an extended use. Let $(S_n : n = 0, 1, 2, \ldots)$ be a $\mathbb{Z}^+$-valued random walk with a step distribution whose support is the two-point set $\{0, 1\}$. Describe the relevance of the opening sentence of this problem to the random Radon measure defined by $R\{y\} = \sharp\{n : S_n = y\}$.

We conclude this section with a criterion for checking that two point processes have the same distribution. Since the σ-field $\mathfrak{H}$ on which the distribution Q of a point process is defined is somewhat complicated, we seek a relatively small subcollection $\mathcal{E}$ of $\mathfrak{H}$ such that the values of Q on $\mathcal{E}$ determine the values of Q on $\mathfrak{H}$. The following result is a straightforward consequence of the Uniqueness of Measure Theorem in Chapter 7.

Proposition 6. *Let Q and R be probability measures on the measurable space of $\overline{\mathbb{Z}}^+$-valued Radon measures on a locally compact Polish space Ψ. If*

$$Q\{\mu \colon \mu(B_i) = z_i,\, 1 \le i \le m\} = R\{\mu \colon \mu(B_i) = z_i,\, 1 \le i \le m\}$$

for every finite collection $\{B_1, \ldots, B_m\}$ of disjoint compact subsets of Ψ and nonnegative integers z_i, then $Q = R$.

29.2. Intensity measures

In this section, we introduce a concept for point processes that plays a role similar to that played by expectations for $\mathbb{R}$-valued random variables.

Definition 7. Let X be a point process on a locally compact Polish space Ψ with Borel σ-field $\mathcal{A}$. The set function $A \rightsquigarrow E(X(A))$ from $\mathcal{A}$ to $\overline{\mathbb{R}}^+$ is the *intensity measure* of X.

Note that if X is a point process on Ψ whose intensity measure assigns the value 0 to A^c for some Borel set A, then X is a point process on A. The following proposition shows that intensity measures are indeed measures.

Proposition 8. *The intensity measure of a point process is a measure (not necessarily Radon nor even σ-finite).*

Problem 10. Prove the preceding proposition twice: by using the Monotone Convergence Theorem and again by using the Fubini Theorem.

Problem 11. Let $X_1, X_2, \ldots, X_n$ be iid random variables taking values in a locally compact Polish space Ψ. Show that the intensity measure of the point process $\delta_{X_1} + \delta_{X_2} + \cdots + \delta_{X_n}$ equals nQ, where Q denotes the common distribution of the random variables X_k.

Problem 12. Let $X_1, X_2, \ldots$ be iid random variables taking values in a locally compact Polish space Ψ. Let N be a $\mathbb{Z}^+$-valued random variable independent of the sequence $(X_1, X_2, \ldots)$. Express the intensity measure of the point process $\delta_{X_1} + \delta_{X_2} + \cdots + \delta_{X_N}$ in terms of the distributions of N and X_1.

* **Problem 13.** Explain how a renewal sequence (defined in Chapter 25) may be regarded as a point process. Express its intensity measure in terms of a quantity or quantities associated with the renewal sequence.

29.3. Poisson point processes

In this section we will see how to construct point processes X having a given intensity measure and also having the property that $X(A)$ and $X(B)$ are independent random variables for disjoint A and B. Each random variable $X(A)$ will have a standard Poisson distribution, with the understanding that the standard Poisson distributions with mean 0 and ∞ are, respectively, δ_0 and δ_∞. We begin with a problem showing that only Radon intensity measures should be considered, and then treat finite intensity measures before considering the general situation.

Problem 14. Let X be a point process on a locally compact Polish space Ψ. Suppose for each Borel subset $A \subseteq \Psi$, that $X(A)$ has a standard Poisson distribution. Explain why the intensity measure of X is a Radon measure.

Lemma 9. *For every finite measure ν on a locally compact Polish space Ψ, there exists a point process X with intensity measure ν such that for each finite measurable partition $\{A_1, \ldots, A_d\}$ of Ψ, the random variables $X(A_1), \ldots, X(A_d)$ are independent and Poisson distributed.*

Proof. Let ν be a finite measure on Ψ. We may assume that $\nu(\Psi) > 0$ since the case $\nu(\Psi) = 0$ is trivial. Let $(Y_1, Y_2, \ldots)$ be an iid sequence having common distribution $\frac{1}{\nu(\Psi)}\nu$, and N a Poisson random variable that is independent of $(Y_1, Y_2, \ldots)$ and has mean $\nu(\Psi)$. For $n = 0, 1, 2, \ldots$, introduce point processes

$$X_n = \delta_{Y_1} + \cdots + \delta_{Y_n},$$

with the understanding that X_0 equals the zero measure. We want to prove that by using N to randomize the number of terms in the sum, we obtain a point process having the desired properties.

Let $\{A_1, \ldots, A_d\}$ be a measurable partition of Ψ, and for each n set

$$Z_n = \big(X_n(A_1), \ldots, X_n(A_d)\big).$$

Clearly, the distribution of Z_n is multinomial with parameters $\frac{\nu(A_i)}{\nu(\Psi)}$, $1 \le i \le d$. By Problem 18 of Chapter 10, the coordinates of Z_N are independent and the ith coordinate of Z_N is Poisson with mean $\frac{\nu(A_i)}{\nu(\Psi)}\nu(\Psi) = \nu(A_i)$. Therefore the point process X_N has the desired properties. $\square$

Theorem 10. *For every Radon measure ν on a locally compact Polish space Ψ, there exists a point process X with intensity measure ν such that for each finite measurable partition $\{A_1, \ldots, A_d\}$ of Ψ, the random variables $X(A_1), \ldots, X(A_d)$ are independent and Poisson distributed. Moreover, any two such point processes with the same intensity measure have the same distribution.*

PROOF. If $\nu(\Psi) < \infty$ the existence result is contained in the preceding lemma, so we assume that $\nu(\Psi) = \infty$. As in the proof of Proposition 2, choose a countably infinite measurable partition $(\Psi_1, \Psi_2, \dots)$ of Ψ such that $0 < \nu(\Psi_k) < \infty$ for each k. Let ν_k be the restriction of ν to Ψ_k.

Construct independent point processes X_k corresponding to ν_k as in the preceding lemma, and set $X = \sum_{k=1}^{\infty} X_k$. To prove that X has the desired property we let $\{A_1, \dots, A_d\}$ be a measurable partition of Ψ. For $1 \le i \le d$,

$$(29.3) \qquad\qquad X(A_i) = \sum_{k=1}^{\infty} X_k(A_i).$$

This sum of independent Poisson random variables is Poisson and its mean, being the sum of the means, equals $\sum_{k=1}^{\infty} \nu_k(A_i) = \sum_{k=1}^{\infty} \nu(A_i \cap \Psi_k) = \nu(A_i)$. From the preceding lemma and the fact that the point processes X_k, $k \ge 1$, are independent, it follows that the random variables $X_k(A_i)$, $1 \le i \le d$, $k \ge 1$, are independent. Hence the random variables $X(A_i)$, $1 \le i \le d$, are independent. The final assertion of the theorem is an immediate consequence of Proposition 6. $\square$

A point process having the properties given in Theorem 10 is called a *Poisson point process* with intensity measure ν. The proofs of that theorem and Lemma 9 give a method for constructing Poisson point processes with arbitrary Radon intensity measures.

The most important Poisson point processes are those for which the intensity measure gives each one-point set measure 0. Proposition 12 below states that if X is such a Poisson point process, then X is a random set (rather than a random multiset) with probability 1. A key step in the proof of this result is the following fact about locally compact Polish spaces.

Lemma 11. *Let Ψ be a locally compact Polish space, and let μ be a Radon measure on Ψ. Then for all $\delta > 0$, there exists a countable measurable partition of Ψ such that each set in the partition has measure less than δ.*

PROOF. It follows from Proposition 2 that we may restrict our attention to the case in which Ψ is compact. Since each point x in Ψ has measure 0, the proof of Proposition 5 shows that there exists an $\varepsilon_x > 0$ such that $\mu(B(x, \varepsilon_x)) < \delta$. The collection of balls $B(x, \varepsilon_x)$ forms an open covering of the compact set Ψ, so there is a finite subcovering. Each member of the finite subcovering has measure less than δ. It is now an easy matter to construct the desired measurable partition of Ψ. $\square$

Proposition 12. *Let X be a Poisson point process whose intensity measure assigns the value 0 to each one-point set. Then with probability 1, every member of the support of X has multiplicity 1.*

When X satisfies the conditions of the preceding proposition, it is natural to call X a random discrete set rather than a random discrete multiset.

Problem 15. Prove Proposition 12. *Hint:* (i) Show that one only need consider the case where the underlying Polish space Ψ is compact and the intensity measure is finite. (ii) Use Lemma 11 to show that for each $\delta > 0$, Ψ has a measurable partition consisting of sets to which the intensity measure assigns value less than δ. (iii) Find an upper bound for the probability that the point process assigns value greater than 1 to at least one of these sets.

Problem 16. Give an example of point process (not Poisson) for which the intensity measure assigns value 0 to each one-point set, but which with positive probability, is a random multiset with at least one member having multiplicity greater than 1.

Problem 17. [Uniqueness of the Poisson distribution] Let X be a point process whose intensity measure is a Radon measure that assigns the value 0 to each one-point set. Use the Central Limit Theorem for triangular arrays (given in Chapter 16) to prove that if the random variables $X(A_1), \ldots, X(A_n)$ are independent for each finite sequence $A_1, \ldots, A_n$ of pairwise disjoint measurable sets, then $X(A)$ has a compound Poisson distribution for each measurable set A. Also show that if X is a.s. a random set (rather than a random multiset), then $X(A)$ has a Poisson distribution for each measurable A.

29.4. Examples of Poisson point processes

A variety of interesting calculations arise in connection with specific Poisson point processes.

Example 1. Let X be the Poisson point process the intensity measure of which is counting measure on $\mathbb{Z}$, set $V = \inf\{n \geq 0 : X\{n\} \geq 1\}$, and set $U = \sup\{n \leq 0 : X\{n\} \geq 1\}$. We will calculate the distribution of $Z = V - U$, the length of the longest open interval in $\mathbb{R}$ that contains the origin but no member of the support of X.

Of course, the distribution of $-U$ is the same as the distribution of V. Thus, if U and V were independent we would obtain the answer by finding the distribution of V and then convolving it with itself. But U and V are not independent since one of them equals 0 if and only if the other one equals 0.

It is easy to see that $-U$ and V are conditionally independent given $\sigma(D_0)$, where $D_0 = \{\omega : X(\omega, \{0\}) \geq 1\}$, and that they have identical conditional distributions. The conditional distribution of V given $\sigma(D_0)$ is given by

$$P(V = v \mid \sigma(D_0))(\omega) = \begin{cases} 1 & \omega \in D_0, \ v = 0 \\ 0 & \omega \in D_0, \ v = 1, 2, \ldots \\ 0 & \omega \notin D_0, \ v = 0 \\ (e-1)e^{-v} & \omega \notin D_0, \ v = 1, 2, \ldots . \end{cases}$$

Taking the convolution of this conditional distribution with itself gives

$$P(Z = z \mid \sigma(D_0))(\omega) = \begin{cases} 1 & \omega \in D_0, \ z = 0 \\ 0 & \omega \in D_0, \ z = 1, 2, \ldots \\ 0 & \omega \notin D_0, \ z = 0 \\ (e-1)^2(z-1)e^{-z} & \omega \notin D_0, \ z = 1, 2, \ldots \end{cases}$$

For each fixed z we take the expected value of this random variable in order to obtain the (unconditional) probability that $Z = z$:

$$P(Z = z) = \begin{cases} 1 - e^{-1} & z = 0 \\ (e-1)^2(z-1)e^{-z-1} & z = 1, 2, \ldots \end{cases}$$

* **Problem 18.** Show that the distribution of the random variable V in the preceding example is geometric and given by $P[V = v] = (1 - e^{-1})e^{-v}$ for $v = 0, 1, 2, \ldots$.

Problem 19. Find the distribution of $\inf\{n \geq 0: X\{n\} \geq 2\}$ for X defined as in Example 1.

Problem 20. Let $c \in (0, \infty)$. For the Poisson point process X on $\mathbb{R}^d$ with intensity measure equal to c^d times Lebesgue measure, calculate the distribution function of $\inf\{|y|: y \in X\}$, with X viewed as a random set.

Problem 21. For the Poisson point process X on $\mathbb{R}$ with intensity measure equal to c times Lebesgue measure, calculate the distribution, expectation, and variance of $V - U$, where $V = \inf\{y \geq 0: X\{y\} \geq 1\}$ and $U = \sup\{y \leq 0: X\{y\} \geq 1\}$.

Problem 22. Let X denote the Poisson point process the intensity measure of which is Lebesgue measure on $\mathbb{R}$. For each $t \in \mathbb{R}$, let V_t be the smallest member of the support of X that is no smaller than t and let U_t be the largest member of the support of X that is no larger than t. Calculate the distributions of $(V_t - t, t - U_t)$, $V_t - U_t$, and $(V_t - U_t, t - U_t)$. Also, calculate the correlations $\mathrm{Corr}(V_t - t, V_s - s)$ and $\mathrm{Corr}(V_t - U_t, V_s - U_s)$ and calculate the limits of these functions as $s \to t$.

* **Problem 23.** Show that if Y is a Poisson point process on $\mathbb{R}^+ \setminus \{0\}$ with intensity measure $c\lambda$, where λ is Lebesgue measure, then the distances between successive members of $Y \cup \{0\}$ are iid and exponentially distributed with mean c^{-1}. Equivalently, let W be the image of a random walk with exponentially distributed steps having mean c^{-1}, and show that $W \setminus \{0\}$ is a Poisson point process on $\mathbb{R}^+ \setminus \{0\}$, with intensity measure $c\lambda$.

* **Problem 24.** Let Z be a Poisson point process on $(0, \infty) \times \Psi$ with intensity measure $\lambda \times \mu$, where Ψ is a locally compact Polish space, λ is Lebesgue measure on $(0, \infty)$, and μ is a probability measure on Ψ. Show that with probability 1, Z can be written as $Z = \{(U_n, V_n): n = 1, 2, \ldots\}$, in such a way that the sequence $U = (U_0 = 0 < U_1 < U_2, \ldots)$ is a random walk on $\mathbb{R}^+$ with exponentially distributed steps having mean 1, the sequence $V = (V_1, V_2, \ldots)$ is independent of U, and V is iid with common distribution μ.

Example 2. For the Poisson point process X on $\mathbb{R}^2$ with intensity measure equal to c^2 times Lebesgue measure, let us calculate the expectation of the area U of the random region consisting of those points in $\mathbb{R}^2$ which are closer to the origin than to any member of support of X. Denote the underlying probability space by $(\Omega, \mathcal{F}, P)$.

We first note that the set of points closer to the origin than to some particular other point is the set of points on one side of the perpendicular bisector of the line connecting the origin to the other point, a set known as a 'half-plane'. Thus the random region of points closer to the origin than to any member of the support of X is the intersection of half-planes, a set known to be convex. Since, with probability one, it contains the origin, its boundary can be represented in polar coordinates by a random radial function $\omega \rightsquigarrow [\theta \rightsquigarrow R(\theta, \omega)]$ of the polar angle θ.

A standard polar-coordinate formula gives

$$E(U) = \frac{1}{2} \int_\Omega \left(\int_0^{2\pi} R^2(\theta, \omega) \, d\theta \right) P(d\omega) ,$$

an iterated integral. It will develop, for fixed θ, that $R^2(\theta, \cdot)$ is measurable — and it is clear, for fixed ω, that $R^2(\cdot, \omega)$ is continuous. From Proposition 9 of Chapter 9 it will then follow that $(\theta, \omega) \rightsquigarrow R(\theta, \omega)$ is measurable and thus that the Fubini Theorem can be applied to give

$$E(U) = \frac{1}{2} \int_0^{2\pi} \left(\int_\Omega R^2(\theta, \omega) \, P(d\omega) \right) d\theta = \frac{1}{2} \int_0^{2\pi} E(R^2(\theta, \cdot)) \, d\theta .$$

In order to establish the preceding formula we need to show that $R^2(\theta, \cdot)$ is measurable, which we will do simultaneously with calculating the distribution of $R(\theta, \cdot)$, and then we will use this distribution to calculate $E(R^2(\theta, \cdot))$. Before proceeding we note that a calculation valid for one value of θ is, with the obvious rotational changes, valid for all values of θ, so we need only consider $\theta = 0$. Then the above expression for $E(U)$ reduces to

$$(29.4) \qquad E(U) = \pi E(R^2(0, \cdot)) .$$

For any $r > 0$, the probability that $R(0, \cdot) \geq r$ is the probability that there is no point of the support of X that has the property that the perpendicular bisector of the segment between it and the origin intersects the 0 angle polar ray at a distance less than r from the origin. Let κ_r denote the Lebesgue measure of the set of points B_r in $\mathbb{R}^2$ having this property. Then

$$(29.5) \qquad P[R(0) \geq r] = e^{-c^2 \kappa_r} .$$

Consider a point with polar coordinate description $[s, \varphi]$, $s > 0$ and $|\varphi| \leq \pi$. The midpoint of the line segment between it and the origin is $[\frac{s}{2}, \varphi]$. In order that $[s, \varphi] \in B_r$ the point $[\frac{s}{2}, \varphi]$ must be the vertex of the right angle in the right triangle having angle $|\varphi|$ at the origin and hypotenuse of length less than r lying along a portion of the positive horizontal axis. This condition is equivalent to the inequality $s < 2r \cos \varphi$, from which it follows that

$$
\kappa_r = \int_{-\pi/2}^{\pi/2} \int_0^{2r \cos \varphi} s \, ds \, d\varphi = 4r^2 \int_0^{\pi/2} \cos^2 \varphi \, d\varphi
$$

$$
= 2r^2 \int_0^{\pi/2} (1 + \cos 2\varphi) \, d\varphi = \pi r^2 .
$$

From Proposition 19 of Chapter 4 we see that

$$
E(R^2(0)) = 2 \int_0^\infty r P[R(0) \geq r] \, dr .
$$

So, from (29.4) and (29.5) we conclude that

$$
E(U) = 2\pi \int_0^\infty r e^{-c^2 \pi r^2} \, dr = c^{-2} .
$$

Problem 25. Calculate the expectation of the perimeter of the random region treated in the preceding example.

* **Problem 26.** For the Poisson point process X on $\mathbb{R}^3$ with intensity measure equal to c^3 times Lebesgue measure, calculate the expectation of the volume of the random region in $\mathbb{R}^3$ consisting of those points that are closer to the origin than to any member of the support of X.

Problem 27. Generalize the preceding problem by replacing "3" by d.

Problem 28. Let X be the Poisson point process on $\mathbb{R}$ with intensity measure equal to Lebesgue measure, and let S consist of those real numbers which lie midway between two consecutive members of the support of X. Let I be an interval of length l. Show that the probability that I contains no member of S equals

(29.6) $2l^2 e^{-2l} + \frac{1}{2}(e^{-2l} + e^{-4l}) .$

Show that the intensity measure of S is Lebesgue measure and then use (29.6) to show that it is not a Poisson point process.

In Example 6 of Chapter 7 we introduced the measure space $(\mathcal{L}, \mathcal{A}, \nu)$, where $\mathcal{L}$ is the space of all lines in $\mathbb{R}^2$. To obtain the σ-field $\mathcal{A}$ and the measure ν we made use of a one-to-one correspondence between $\mathcal{L}$ and $\{(s, \varphi) : s \in \mathbb{R}, 0 \leq \varphi < \pi\}$. The members of $\mathcal{A}$ are the sets which correspond to Borel subsets of $\mathbb{R} \times [0, \pi)$ and the measure ν is induced by Lebesgue measure on $\mathbb{R} \times [0, \pi)$. We may call a particular (s, φ) the coordinates of the corresponding member of $\mathcal{L}$, and in fact, we will speak of the line (s, φ). Recall that $|s|$ is the distance in $\mathbb{R}^2$ from the origin to the line (s, φ) and φ or $\varphi + \pi$ denotes the polar angle of the point on the line closest to the origin, with a suitable interpretation in case $s = 0$. The metric

$$\rho((s_1, \varphi_1), (s_2, \varphi_2))$$
$$= \sqrt{[s_2 - s_1]^2 + [(\varphi_2 - \varphi_1) \wedge (\varphi_2 - \varphi_1 + \pi) \wedge (\varphi_2 - \varphi_1 - \pi)]^2}$$

gives a locally compact Polish structure on $\mathcal{L}$. We will omit the straightforward arguments that $\mathcal{A}$ is the Borel σ-field and ν is a Radon measure. Notice that $\nu\{l\} = 0$ for every $l \in \mathcal{L}$. By Proposition 12, a point process corresponding to ν assigns, with probability 1, measure ≤ 1 to every one-point set. Thus, it is natural to call such a point process a 'random set of lines'. Figure 29.1 is relevant for some of the following problems.

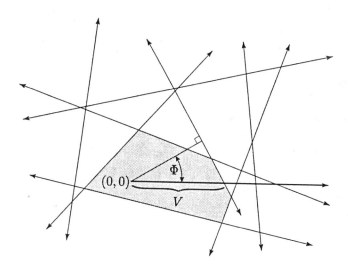

FIGURE 29.1. Poisson point process of lines

* **Problem 29.** Let X be a random set of lines in $\mathbb{R}^2$ having a Poisson distribution with intensity measure ν as described above. Let $D(\omega)$ denote the region consisting of those points in $\mathbb{R}^2$ having the following property: segments from the origin to them do not intersect any members of $X(\omega)$. Explain why D is, with probability 1, an open bounded polygonal region. Calculate the means of the perimeter and area of D. *Hint:* Use Problem 25 of Chapter 7.

Problem 30. For the point process X described in the preceding problem let $V(\omega)$ be that point on the positive horizontal axis that is closest to the origin among all such points lying on a member of $X(\omega)$ and let $\Phi(\omega)$ be the polar coordinate angle of that (unique with probability one) member of $X(\omega)$. Prove that V and Φ are independent, that V has a standard exponential distribution with mean $\frac{1}{2}$, and that Φ has density

$$\varphi \rightsquigarrow \frac{|\cos\varphi|}{2}, \quad 0 \le \varphi < \pi.$$

Problem 31. For the region D described in Problem 29, let R denote the function that describes its boundary in polar coordinates, with distance from the origin being given as a function of angle. For each θ calculate the distribution of the random pair $(R(\theta), R'(\theta))$.

In many situations there is attached to a Polish space a group of measurable permutations of that space. For the space $\mathcal{L}$ of Problem 29 the usual group G of permutations consists of the isometries of $\mathbb{R}^2$. Each isometry of $\mathbb{R}^2$ takes a member of $\mathcal{L}$ to a member of $\mathcal{L}$. The measure ν of Problem 29 is a natural measure to be used in conjunction with G since it is G-*invariant* (or just *invariant* if G is understood), that is, $\nu(A) = \nu(g^{-1}(A))$ for every $g \in G$ and every Borel set A. It can be shown that multiples of ν are the only G-invariant Radon measures on $\mathcal{L}$.

The isometries of $\mathbb{R}^d$ constitute a natural group H to be associated with the Polish space $\mathbb{R}^d$. The multiples of d-dimensional Lebesgue measure are the only H-invariant Radon measures on $\mathbb{R}^d$.

In general, a finite point process X on a locally finite Polish space to which is attached a group G of measurable permutations is G-*invariant* if $\hat{g} \circ X$ has the same distribution as X for every $g \in G$.

Proposition 13. *The intensity measure of an invariant point process is invariant. A Poisson point process with an invariant intensity measure is invariant.*

Problem 32. Prove the preceding proposition.

Problem 33. Give an example of a noninvariant point process whose intensity measure is an invariant Radon measure.

In view of Proposition 13, the Poisson point process of lines illustrated by Figure 29.1 is invariant under the isometries of $\mathbb{R}^2$.

29.5. † Probability generating functionals

The *probability generating functional* of a point process X on a locally compact Polish space Ψ is the functional

$$h \rightsquigarrow E\Big(\prod_{\psi \in \Psi} [h(\psi)]^{X\{\psi\}}\Big),$$

defined for continuous $[0, 1]$-valued functions h for which the set $\{x : h(x) < 1\}$ is relatively compact. The infinite product above has only finitely many factors different from 1 and the factors that equal 1 can be ignored. In terms of the distribution Q of X, the probability generating functional is given by

$$h \rightsquigarrow \int \prod_{\psi \in \Psi} [h(\psi)]^{\mu\{\psi\}} \, Q(d\mu) = \int e^{-\int_\Psi \log(1/h) \, d\mu} \, Q(d\mu) \,,$$

where the integration is taken over the space of $\overline{\mathbb{Z}}^+$-valued Radon measures on Ψ, $\log(1/0) = \infty$, $0 \cdot \infty = 0$, and $e^{-\infty} = 0$.

Example 3. The functional

$$h \rightsquigarrow \sum_{j=0}^{\infty} 2^{-(j+1)} \prod_{i=1}^{j} h(i)$$

is the probability generating functional of the random subset of $\overline{\mathbb{Z}}^+ \setminus \{0\}$ that equals $\emptyset$ with probability $\frac{1}{2}$, $\{1\}$ with probability $\frac{1}{4}$, $\{1, 2\}$ with probability $\frac{1}{8}$, and so forth. This random set may be regarded as the set of times of tail flips preceding the first head flip in an infinite sequence of fair coin flips.

* **Problem 34.** Calculate the probability generating functional for each of the cases $r = 1$ and $r = n - 1$ in Problem 6.

Problem 35. Calculate the probability generating functional for the case $r = 3, n = 4$ in Problem 8.

Problem 36. Decide if the functional

$$h \rightsquigarrow \sum_{j=1}^{\infty} \frac{h(j)h(j+1)}{j(j+1)}$$

is the probability generating functional of some point process on $\mathbb{Z}^+ \setminus \{0\}$. If so, describe such a point process.

Problem 37. Follow the instructions of the preceding problem for the functional

$$h \rightsquigarrow \sum_{j=1}^{\infty} \frac{[h(j)]^2}{j(j+1)} \,.$$

Theorem 14. *Point processes that have the same probability generating functional have the same distribution.*

PROOF. Let Q and R be as in Proposition 6, and suppose that their probability generating functionals are the same. Let B_i, $1 \le i \le m$, be as in that proposition. Fix numbers $s_i \in [0, 1]$ and set

$$g = \sum_{i=1}^{m} s_i I_{B_i} \,.$$

Choose a decreasing sequence of functions h_n in the domain of the probability generating functional of Q (and thus of R) such that, for each $\psi \in \Psi$, $h_n(\psi) \to g(\psi)$ as $n \to \infty$. By the Dominated Convergence Theorem we obtain

$$\int \prod_{i=1}^{m} s_i^{\mu(B_i)} \, Q(d\mu) = \int \prod_{i=1}^{m} s_i^{\mu(B_i)} \, R(d\mu) \, .$$

Now treat $s_1, \ldots, s_m$ as variables. For arbitrary $z_i \in \mathbb{Z}^+$, $1 \le i \le m$, we equate the coefficients of $\prod_{i=1}^{m} s_i^{z_i}$ in these equal Taylor series to obtain

$$Q\{\mu \colon \mu(B_i) = z_i \, , \, 1 \le i \le m\} = R\{\mu \colon \mu(B_i) = z_i \, , \, 1 \le i \le m\} \, .$$

An appeal to Proposition 6 completes the proof. $\square$

Proposition 15. *The probability generating functional of a Poisson point process on a locally compact Polish space having intensity measure ν is given by*

$$h \rightsquigarrow e^{-\int (1-h) \, d\nu} \, .$$

PARTIAL PROOF. Let X denote a Poisson point process having intensity measure ν on a locally compact Polish space Ψ. We first focus on $[0,1]$-valued functions h (not necessarily continuous) with finite image $\{h_1, \ldots, h_m\}$, such that $\{\psi \colon h(\psi) < 1\}$ is relatively compact. For each i, let $A_i = \{y \colon h(y) = h_i\}$. Then

$$E\Big(\prod_{\psi \in \Psi} [h(\psi)]^{X\{\psi\}} \Big) = E\Big(\prod_{i=1}^{m} h_i^{X(A_i)} \Big) = \prod_{i=1}^{m} E\big(h_i^{X(A_j)} \big)$$

$$= \prod_{i=1}^{m} e^{-\nu(A_i)(1-h_i)} = e^{-\int (1-h) \, d\nu} \, ,$$

as desired. It is left to reader to treat arbitrary $[0,1]$-valued continuous h for which $\{\psi \colon h(\psi) < 1\}$ is relatively compact. $\square$

Problem 38. Finish the proof of the preceding proposition.

* **Problem 39.** Calculate the probability generating functionals of Poisson point processes whose intensity measures are counting measures on countable sets.

Problem 40. Construct a metric that makes $(0, \infty)$ into a locally compact Polish space. Then show that there is a Poisson point process X on $(0, \infty)$ whose probability generating functional has the form

$$h \rightsquigarrow e^{-\int_0^\infty h'(t) \log t \, dt}$$

for those h in its domain satisfying the additional condition that h' exists and is continuous. Then calculate the distributions of $\inf\{t > 1 \colon X\{t\} \ge 1\}$ and $\sup\{t < 1 \colon X\{t\} \ge 1\}$.

We will not fully address the question of which functionals are probability generating functionals of some point process, nor will we address the general existence issue of a point process when probabilities of events of the form

$$\{\mu \colon \mu(B_i) = z_i\,,\ 1 \le i \le m\}$$

are specified. However, the following proposition provides a necessary condition for a functional to be a probability generating functional.

Proposition 16. *Let $\mathfrak{F}$ be a probability generating functional of some point process. Let h_m, $m = 1, 2, \ldots$, and h be in the domain of $\mathfrak{F}$, and suppose that $h_m(\psi) \to h(\psi)$ as $m \to \infty$ for every ψ, and $\bigcup_{m=1}^{\infty}\{\psi \colon h_m(\psi) < 1\}$ is relative compact. Then $\mathfrak{F}(h_m) \to \mathfrak{F}(h)$ as $m \to \infty$.*

Problem 41. Prove Proposition 16, making sure that your proof shows where the relative compactness hypothesis is used. Show that a false statement is obtained if this hypothesis is removed.

Problem 42. Proposition 16 might be stated concisely as: All probability generating functionals are 'continuous'. If you like topology, describe the neighborhood structure of the weakest topology on the domain of a probability generating functional consistent with this concise version of Proposition 16.

29.6. ‡ Operations on point processes

There is a natural addition for $\overline{\mathbb{Z}}^{+}$-valued Radon measures on a locally compact Polish space: $(\mu_1 + \mu_2)(B) = \mu_1(B) + \mu_2(B)$. (The collection of $\overline{\mathbb{Z}}^{+}$-valued Radon measures forms a commutative semigroup under this operation with the zero measure being the identity.) In terms of multisets this operation corresponds to 'union' with an appropriate interpretation regarding multiplicities.

Theorem 17. *Let X and Y be independent point processes on a locally compact Polish space Ψ. The probability generating functional of $X+Y$ is the product of the probability generating functionals of X and Y.*

* **Problem 43.** Prove the preceding theorem.

Problem 44. Let $(Z_k \colon k = 1, 2, \ldots)$ be an iid sequence of random variables taking values in a locally compact Polish space. Show that the probability generating functional of the point process $\delta_{Z_1} + \cdots + \delta_{Z_n}$ is the functional $h \rightsquigarrow [E(h \circ Z_1)]^n$.

Let Ψ and Φ be locally compact Polish spaces and let $g \colon \Psi \to \Phi$ be a function satisfying: $g^{-1}(C)$ is compact if C is a compact subset of Φ. (Such a function is known to be continuous.) The function g induces a function $\hat{g}$ from the space of $\overline{\mathbb{Z}}^{+}$-valued Radon measures on Ψ to the space of $\overline{\mathbb{Z}}^{+}$-valued Radon measures on Φ: $(\hat{g}(\mu))(A) = \mu(g^{-1}(A))$ for Borel $A \subseteq \Phi$.

Proposition 18. *The function $\hat{g}$ defined above is a measurable function.*

Problem 45. Prove the preceding proposition.

For locally compact Polish spaces Ψ and Φ, Proposition 18 implies that a function $g\colon \Psi \to \Phi$ for which $g^{-1}(C)$ is compact whenever C is compact transforms a point process X in Ψ into a point processes $\hat{g} \circ X$ on Φ.

Problem 46. For X and $\hat{g} \circ X$ as above, find a relation between their intensity measures and also between their probability generating functionals. Also, prove that $\hat{g} \circ X$ is Poisson if X is. Finally, give an example that shows that $\hat{g} \circ X$ might be Poisson even if X is not.

Problem 47. Let X be a point process having intensity measure equal to Lebesgue measure in $\mathbb{R}^d$. For $x \in \mathbb{R}^d$, let $g(x) = |x|$. Calculate the intensity measure of the point process $\hat{g} \circ X$.

29.7. ‡ Convergence in distribution for point processes

A sequence $(\mu_n\colon n = 1, 2, \ldots)$ of Radon measures on a locally compact Polish space Ψ is said to *converge* to a Radon measure μ if $\int_\Psi f\, d\mu_n \to \int_\Psi f\, d\mu$ as $n \to \infty$ for every continuous function $f\colon \Psi \to \mathbb{R}$ for which the set $\{x\colon f(x) \neq 0\}$ is relatively compact. It can be shown that there exists a countable sequence $(f_m\colon m = 1, 2, \ldots)$ of such functions having the property: $\mu_n \to \mu$ if and only if $\int_\Psi f_m\, d\mu_n \to \int_\Psi f_m\, d\mu$ for every m. It can then be shown that the function

$$(29.7) \qquad (\mu, \lambda) \rightsquigarrow \sum_{m=1}^{\infty} \frac{1 \wedge |\int_\Psi f_m\, d\lambda - \int_\Psi f_m\, d\mu|}{2^m}$$

turns the space of Radon measures into a Polish space in which convergence is the same concept as that introduced in the first sentence of this section. This Polish space and corresponding Borel σ-field is the same as $(\mathcal{M}, \mathfrak{H})$, introduced in Section 1. Moreover, the measurable subset of $\overline{\mathbb{Z}}^+$-valued Radon measures is a closed subset of the Polish space of all Radon measures, and thus is itself a Polish space.

The comments above indicate that we may speak of *almost sure convergence*, *convergence in probability*, and *convergence in distribution* for sequences of point processes.

Problem 48. Supply some of the proofs relevant to the preceding discussion.

Theorem 19. *A family $\mathcal{Q}$ of distributions on the space of $\mathbb{Z}^+$-valued Radon measures on some locally compact Polish space Ψ is relatively sequentially compact if and only if for every compact $C \subseteq \Psi$,*

$$(29.8) \qquad\qquad Q\{\mu\colon \mu(C) > z\} \to 0 \quad \text{as } z \to \infty,$$

uniformly in $Q \in \mathcal{Q}$.

PARTIAL PROOF. We leave to the reader the proof that relative sequential compactness implies the uniform convergence in (29.8). For the opposite direction we assume that (29.8) holds uniformly for each compact C.

Let $\varepsilon > 0$. Choose f_m, $m = 1, 2, \ldots$, as in the paragraph leading to (29.7), and denote the compact closure of $\{\psi\colon f_m(\psi) \neq 0\}$ by C_m. For each m choose z_m so that $Q\{\mu\colon \mu(C_m) > z_m\} < \varepsilon 2^{-m}$ for all $Q \in \mathcal{Q}$. Let

$$\Lambda = \{\mu\colon \mu(C_m) \leq z_m \,,\, m = 1, 2, \ldots\}.$$

Clearly, $Q(\Lambda) > 1 - \varepsilon$ for $Q \in \mathcal{Q}$.

For any m and any sequence in Λ, there is a subsequence $(\mu_n\colon n = 1, 2, \ldots)$ and an integer $z \leq z_m$ such that $\mu_n(C_m) = z$ for all n and either $z = 0$ or the sequence $(z^{-1}\mu_n\colon n = 1, 2, \ldots)$ of probability measures on C_m converges. In either case the sequence $(\int f_m \, d\mu_n\colon n = 1, 2, \ldots)$ is Cauchy. In view of (29.7) we can use the Cantor diagonalization procedure to show that any sequence in Λ has a convergent subsequence. Hence Λ is compact. So, $\mathcal{Q}$ is uniformly tight, and therefore it is relatively sequentially compact by the Prohorov Theorem. $\square$

Problem 49. Complete the proof of Theorem 19 by showing that relative sequential compactness implies (29.8) uniformly.

The next theorem is our motivation for having only continuous functions in the domain of probability generating functionals.

Theorem 20. [Continuity (for Probability Generating Functionals)] *A sequence of point processes converges in distribution to a point processes if and only if the corresponding sequence of probability generating functionals converges pointwise (that is, function-wise) to a functional $\mathfrak{F}$ that satisfies: if h_m is in the domain of $\mathfrak{F}$ for each m, $\bigcup_{m=1}^{\infty}\{\psi\colon h_m(\psi) < 1\}$ is relatively compact, and $h_m(\psi) \to 1$ as $m \to \infty$ for each ψ, then $\mathfrak{F}(h_m) \to 1$ as $m \to \infty$. In this case $\mathfrak{F}$ is the probability generating functional of the limiting point process.*

* **Problem 50.** Prove Theorem 20.

Corollary 21. *A sequence of Poisson point processes on a locally compact Polish space converges in distribution if and only if the corresponding sequence of intensity measures converges to a Radon measure ν, in which case the limiting point process is Poisson with intensity measure ν.*

Problem 51. Prove the preceding corollary.

We conclude this chapter with a nice application of Theorem 20. The following theorem can be described informally as follows: given an intensity measure, 'randomly place' n points in a set of 'size' n; for large n, the resulting point process is approximately Poisson with the given intensity measure. It is interesting to note that the independence that is inherent in the limiting Poisson point process X is absent from each of the point processes in the sequence that converges to X.

Theorem 22. *Let ν be an infinite Radon measure on a locally compact Polish space Ψ and suppose there exist open sets $A_1 \subseteq A_2 \subseteq \cdots \nearrow \Psi$ with $\nu(A_n) = n$ for each n. For each $n = 1, 2, \ldots$, let $(Y_{k,n} : 1 \leq k \leq n)$ be an iid sequence with common distribution ν_n on Ψ defined to equal $\frac{1}{n}$ times the restriction of ν to A_n. Define point processes X_n on Ψ by*

$$X_n = \sum_{k=1}^{n} \delta_{Y_{k,n}} .$$

Then the sequence $(X_n : n = 1, 2, \ldots)$ converges in distribution to the Poisson point process X having intensity measure ν.

PROOF. Let h be in the domain of the probability generating functional $\mathfrak{F}$ of a Poisson point process having intensity measure ν. Then $\{\psi : h(\psi) < 1\} \subset A_n$ for all sufficiently large n because $\{\psi : h(\psi) < 1\}$ is relatively compact. For such n, we use Problem 44 to obtain formulas for the probability generating functional of X_n evaluated at h:

$$\left(\int_\Psi h \, d\nu_n \right)^n = \left(1 - \int_\Psi [1 - h] \, d\nu_n \right)^n = \left(1 - \frac{1}{n} \int_{A_n} [1 - h] \, d\nu \right)^n .$$

Let $n \to \infty$ to obtain $\mathfrak{F}(h)$. An appeal to Theorem 20 completes the proof. $\square$

Problem 52. For the case where ν is Lebesgue measure on $\mathbb{R}^d$, describe appropriate choices for A_n in Theorem 22 and find the corresponding ν_n.

CHAPTER 30
Lévy Processes

In this chapter we treat continuous-time analogues of random walks, while restricting ourselves to the state spaces $\overline{\mathbb{R}}^+$ and $\mathbb{R}$. In the $\overline{\mathbb{R}}^+$-setting, these 'Lévy processes' can be constructed using Poisson point processes. Matters are slightly more complicated in the $\mathbb{R}$-setting; in addition to Poisson point processes, Brownian motion and a limiting procedure are needed. The main results of this chapter show that there is a Lévy process for each infinitely divisible distribution. After these results are proved, we extend to the continuous-time setting some of the theory that was developed earlier for random walks. The chapter concludes with a section in which a few 'sample function properties' of Lévy processes are discussed.

30.1. Measurable spaces of right-continuous functions

The basic random variables in this chapter are function-valued. The domain of these functions is $[0, \infty)$. Often we will use 'time' to refer to this domain, as well as to an individual member of the domain.

It turns out that the spaces of all $\overline{\mathbb{R}}^+$- and $\mathbb{R}$-valued functions on $[0, \infty)$ are quite cumbersome. Therefore, we restrict our attention to certain nice subsets of these spaces. In the $\overline{\mathbb{R}}^+$-setting, we consider the space $\mathbf{D}^+[0, \infty)$ of increasing, right-continuous functions from $[0, \infty)$ to $\overline{\mathbb{R}}^+$. For the $\mathbb{R}$-setting, we introduce a new term: a function $f \colon [0, \infty) \to \mathbb{R}$ is *cadlag* if f is right-continuous, and for every $t \in (0, \infty)$, the limit $f(t-) = \lim_{s \nearrow t} f(t)$ exists. (The term 'cadlag' is an acronym for the French phrase 'continues à droite, limites à gauche' which means 'continuous on the right, limits on the left'.) The space of all $\mathbb{R}$-valued cadlag functions is denoted by $\mathbf{D}[0, \infty)$. (Note that the existence of left limits is automatic for the functions in $\mathbf{D}^+[0, \infty)$, since they are increasing.)

We regard both of these spaces as measurable spaces, with corresponding σ-fields generated by sets of the form $\{f \colon f(t) \in B\}$ for $t \in [0, \infty)$ and Borel $B \subseteq \mathbb{R}$. The σ-field will usually not be mentioned explicitly; rather it is implicit in a phrase such as 'the measurable space $\mathbf{D}[0, \infty)$'. The notation f_t can be used in place of $f(t)$; in fact, when the function itself is random we will prefer the subscript position for the time variable. The following two problems treat issues of pointwise convergence for functions in $\mathbf{D}[0, \infty)$ and $\mathbf{D}^+[0, \infty)$.

* **Problem 1.** Give an example to show that the pointwise limit of a sequence of functions in $\mathbf{D}[0, \infty)$ need not be a member of $\mathbf{D}[0, \infty)$.

Problem 2. Prove that if a sequence of functions in $\mathbf{D}[0, \infty)$ converges uniformly on every bounded set, then the limit is a member of $\mathbf{D}[0, \infty)$.

Problem 3. Prove that $\mathbf{C}[0, \infty)$ is a measurable subset of $\mathbf{D}[0, \infty)$, and comment on the significance of this fact in conjunction with Problem 3 of Chapter 18.

30.2. Definition of Lévy process

Recall from Chapter 19 that a Wiener process on $[0, \infty)$ is a $\mathbf{C}[0, \infty)$-valued random variable with stationary independent increments. The following definition, inspired by Proposition 3 of Chapter 11 and the discussion preceding that proposition, says that a Lévy process is a $\mathbf{D}[0, \infty)$-valued or $\mathbf{D}^+[0, \infty)$-valued random variable with stationary independent increments. Wiener processes are examples of Lévy processes.

Definition 1. A *Lévy process* in $\mathbb{R}$ or $\overline{\mathbb{R}}^+$, respectively, is a $\mathbf{D}[0, \infty)$- or $\mathbf{D}^+[0, \infty)$-valued random variable Y for which $Y_0 = 0$ a.s. and

$$(30.1) \qquad P[Y_{t_j} - Y_{t_{j-1}} \in C_j, j = 1, 2, \ldots, n] = \prod_{j=1}^{n} P[Y_{t_j - t_{j-1}} \in C_j]$$

whenever $C_1, C_2, \ldots, C_n$ are Borel subsets of $\mathbb{R}$ and

$$0 = t_0 < t_1 < t_2 \cdots < t_n.$$

In the $\overline{\mathbb{R}}^+$-setting it is understood that the undefined quantity $\infty - \infty$ is not a member of any Borel subset of $\mathbb{R}$. A Lévy process in $\overline{\mathbb{R}}^+$ is also called a *subordinator*.

Just as with random walks in $\mathbb{R}$, we have defined Lévy processes so that they always start at the value 0 at time $t = 0$. Sometimes the term 'Lévy process' (or the equivalent term 'process with stationary independent increments') is used more generally to include processes Y satisfying $\widehat{Y}_t = Y_t + X$, where Y is a Lévy process in the sense of Definition 1, and X is a $\mathbb{R}$-valued random variable that is independent of Y.

Problem 4. Let Y be a Lévy process in $\mathbb{R}$ and let $s \in [0, \infty)$. Prove that the process

$$t \rightsquigarrow Y_{s+t} - Y_s$$

is a Lévy process that has the same distribution as Y itself and is independent of $\sigma(Y_r : r \leq s)$.

Problem 5. Let Y be a Lévy process and fix $a, b > 0$. Show that the random sequence $(Y_{a+bn} - Y_{bn} : n = 0, 1, 2, \ldots)$ is a stationary sequence (defined in Chapter 28). Calculate the mean vector and covariance matrix of this stationary sequence for the case in which Y_1 has finite mean μ and finite variance σ^2.

Problem 6. Let X be a Poisson point process on $(0, \infty)$ with intensity measure $\kappa \lambda$, where κ is a positive real constant and λ is Lebesgue measure. Define a $\mathbf{D}^+[0, \infty)$-valued random variable Y by $Y_t = yX(0, t]$ for $t \geq 0$, where y is a constant in $\overline{\mathbb{R}}^+$ and it is to be understood in case $y = \infty$ that $\infty \cdot 0 = 0$. Show that Y is a Lévy process in $\overline{\mathbb{R}}^+$, and that for each $t \geq 0$, the distribution of Y_t is that of the product of y and a Poisson random variable with mean κt.

Problem 7. Show that the sum of two independent Lévy processes is a Lévy process, both in the $\mathbb{R}$- and in the $\overline{\mathbb{R}}^+$-setting. (Take care that your proof accommodates the process constructed in the preceding problem for the case $y = \infty$.)

The preceding exercise shows that Poisson point processes on $(0, \infty)$ can be used to construct certain simple Lévy processes. The following example shows that by using Poisson point processes on $(0, \infty) \times \mathbb{R}$, a much larger class of Lévy processes can be constructed. This example is a special case of the main representation theorem for $\mathbb{R}$-valued Lévy processes, to be proved in the next section.

Example 1. [Compound Poisson processes] Let Q be an arbitrary distribution on $\mathbb{R}$, and let X be a Poisson point process on $(0, \infty) \times \mathbb{R}$ with intensity measure $\kappa(\lambda \times Q)$, where $\kappa > 0$ and λ denotes Lebesgue measure on $[0, \infty)$. We define $Y = (Y_t : t \in [0, \infty))$ by

$$Y_t = \int_{(0,t] \times \mathbb{R}} x \, X(d(s, x)) = \int x \, X((0, t] \times dx) = \sum_{x \in \mathbb{R}} x X((0, t] \times \{x\}),$$

where $(0, t] = \emptyset$ in case $t = 0$, and the last summation is meaningful because with probability 1, all but finitely many terms in it equal 0. It is now easy to see that with probability one, $[t \rightsquigarrow Y_t]$ is cadlag, has its discontinuities limited to a discrete set of times, and is constant on intervals between discontinuities.

That Y has independent increments follows immediately from the independence properties of Poisson point processes. That it has stationary increments follows from Proposition 13 of Chapter 29 and the invariance of $\kappa(\lambda \times Q)$ under translations in the direction of the first coordinate axis. Therefore (30.1) is satisfied, and hence Y is a Lévy process.

According to Proposition 12 of Chapter 29, X can be viewed as a random discrete set rather than a random Radon measure, because the intensity measure of X assigns the value 0 to every one-point set. From this perspective, the formula for Y may be rewritten as

$$Y_t = \sum_{(s,x) \in X \cap ((0,t] \times \mathbb{R})} x .$$

That is, to find Y_t, sum the second coordinates of the points in X whose first coordinates are $\leq t$. Inversion of this formula gives an expression for the random set X in terms of Y:

$$X = \{(s,x) : Y_s - Y_{s-} = x\} .$$

We conclude this example by determining the distribution of Y_t for each t. In the definition of Y_t, only the restriction of X to $(0,t] \times \mathbb{R}$ is relevant. Denote this restriction by X_t which is a Poisson point process having a finite intensity measure. From the proof of Lemma 9 of Chapter 29, it is clear that this random finite set is distributed like a random set of the form $\{(T_1, V_1), \ldots, (T_N, V_N)\}$, where N is Poisson with mean κt, and $(T_1, V_1), (T_2, V_2), \ldots$, is an iid sequence of $(0,t] \times \mathbb{R}$-valued random variables that is independent of N and has common distribution equal to the intensity measure of X_t divided by κt —that is $\lambda \times Q$. Therefore Y_t has the same distribution as $V_1 + \cdots + V_N$, which we recognize from Example 2 of Chapter 16 as a compound Poisson random variable with Lévy measure $\kappa t Q$.

The preceding example is easily modified to accommodate the $\overline{\mathbb{R}}^+$-setting. In both settings, the resulting processes are called *compound Poisson processes*. For the case where Q is a delta distribution, the adjective 'compound' is omitted. The *standard Poisson processes* are obtained by setting $Q = \delta_1$.

Remark 1. If Q in Example 1 has the property that $Q\{0\} \in (0,1)$, then if one introduces $\hat{\kappa} = \kappa Q(\mathbb{R} \setminus \{0\})$ and $\widehat{Q} : B \rightsquigarrow \frac{1}{Q(\mathbb{R}\setminus\{0\})} \widehat{Q}(B \setminus \{0\})$, one obtains a pair $(\hat{\kappa}, \widehat{Q})$ that is different from (κ, Q) but which gives a compound Poisson process with the same distribution. Therefore, in order that there be a one-to-one correspondence between the set of pairs (κ, Q) and the set of distributions of compound Poisson processes, it is sometimes required for the construction of Example 1 that Q be a measure on $\mathbb{R} \setminus \{0\}$. Similarly, for the $\overline{\mathbb{R}}^+$-setting one might require that Q be a probability measure on $(0, \infty]$, not just on $[0, \infty]$.

Problem 8. [Alternate construction of compound Poisson processes] Let $t \rightsquigarrow N_t$ be a standard Poisson process. Independent of N let $(V_1, V_2, \ldots)$ be an iid sequence of $\mathbb{R}$-valued random variables with common distribution Q. Set

$$Z_t = \sum_{k=1}^{N_t} V_k .$$

Show that $t \rightsquigarrow Z_t$ is a compound Poisson process in $\mathbb{R}$, and that the distribution of Z_t is compound Poisson with Lévy measure $E(N_t)Q = tE(N_1)Q$.

Problem 9. Let Z be a compound Poisson process in $\mathbb{R}$, and let $T_1 < T_2 < \ldots$ be the (random) times at which the random function Z has discontinuities. Show that the random variables $T_1, T_2 - T_1, T_3 - T_2, \ldots$ are iid exponentially distributed random variables, and find their common mean in terms of the parameter κ of Example 1.

* **Problem 10.** In the $\overline{\mathbb{R}}^+$-setting, obtain the moment generating function of Y_t, constructed as in Example 1, by applying the formula for the probability generating functional of the Poisson point process X (Proposition 15 in an optional section of Chapter 29) to appropriate functions in its domain. Use the same method to calculate the characteristic function of Y_t in the $\mathbb{R}$-setting.

As the following proposition shows, the appearance of an infinitely divisible distribution in Example 1 was to have been expected.

Proposition 2. *Let Z be a Lévy process in either $\mathbb{R}$ or $\overline{\mathbb{R}}^+$. Then Z_t is infinitely divisible for every t.*

Problem 11. Prove the preceding proposition. *Hint:* Care is needed for treating the $\overline{\mathbb{R}}^+$-setting.

30.3. Construction of Lévy processes

In view of Proposition 2, the question arises: For every infinitely divisible distribution R does there exist a Lévy process whose distribution at time 1 equals R? We will build on Example 1 to give affirmative answers for both the $\overline{\mathbb{R}}^+$- and $\mathbb{R}$-settings. Since we will be using Lévy measures as intensity measures of Poisson point processes, we need to make $\overline{\mathbb{R}}^+ \setminus \{0\}$ and $\mathbb{R} \setminus \{0\}$ into locally compact Polish spaces in such a way that all Lévy measures are Radon measures and the Borel sets are the usual measurable sets.

For the $\overline{\mathbb{R}}^+$-setting, define the distance between $w, x \in \overline{\mathbb{R}}^+ \setminus \{0\}$ to be $|\frac{1}{x} - \frac{1}{w}|$ (where, of course, $\frac{1}{\infty} = 0$). Notice that any compact set in $\overline{\mathbb{R}}^+ \setminus \{0\}$ is a subset of $[\varepsilon, \infty]$ for some $\varepsilon > 0$ and is thus assigned a finite value by each Lévy measure for $\overline{\mathbb{R}}^+$.

For the $\mathbb{R}$-setting, we define the distance between $w, x \in \mathbb{R} \setminus \{0\}$ to be

$$\left| \frac{1}{x} - \frac{1}{w} \right| + |x - w|.$$

Since any compact subset of $\mathbb{R} \setminus \{0\}$ is a subset of $(-\infty, -\varepsilon] \cup [\varepsilon, \infty)$ for some $\varepsilon > 0$, Lévy measures for $\mathbb{R}$ are Radon measures on $\mathbb{R} \setminus \{0\}$.

Problem 12. Why would it not work to define the distance between x and w in $\mathbb{R} \setminus \{0\}$ to equal $|\frac{1}{x} - \frac{1}{w}|$?

Theorem 3. [Itô Representation ($\overline{\mathbb{R}}^+$-version)] Let $\xi \geq 0$ and X be a Poisson point process in $(0, \infty) \times (0, \infty]$ whose intensity measure is $\lambda \times \nu$, where λ denotes Lebesgue measure and ν is a Lévy measure for $\mathbb{R}^+$. Then Z defined by

$$(30.2) \qquad Z_t = \xi t + \int x\, X((0, t] \times dx)$$

is a Lévy process in $\overline{\mathbb{R}}^+$, and the distribution of Z_t is the infinitely divisible distribution corresponding to $(t\xi, t\nu)$ via the Lévy-Khinchin Representation Theorem for the $\overline{\mathbb{R}}^+$-setting.

PROOF. We first assume that $\nu\{\infty\} = 0 = \xi$. For $\varepsilon > 0$, set

$$\nu_\varepsilon(B) = \nu([\varepsilon, \infty) \cap B), \quad \text{Borel } B \subseteq (0, \infty).$$

An application of Example 1 to the intensity measure

$$\frac{1}{\nu([\varepsilon, \infty))}(\lambda \times \nu_\varepsilon)$$

shows that

$$Z_t^\varepsilon \stackrel{\text{def}}{=} \int_{[\varepsilon, \infty)} x\, X((0, t] \times dx)$$

is compound Poisson with Lévy measure $t\nu_\varepsilon$. As $\varepsilon \searrow 0$, $Z_t^\varepsilon \nearrow Z_t$ a.s., whether finite or infinite. Thus, we also have convergence in distribution, which in terms of moment generating functions can be written as

$$\lim_{\varepsilon \searrow 0} \exp\left(-t \int_{[\varepsilon, \infty)} \left(1 - e^{-uz}\right) \nu(dz)\right) = E\left(e^{-uZ_t}\right).$$

Thus Z_t is finite a.s. and is infinitely divisible with Lévy measure $t\nu$ and shift 0.

Since $Z_s - Z_s^\varepsilon \leq Z_t - Z_t^\varepsilon$ for $s < t$, the convergence $Z^\varepsilon \to Z$ is uniform on bounded intervals. By Problem 2, Z is cadlag a.s. The stationary increment property is clearly preserved under passage to the limit, so the theorem has now been proved under the assumption that $\xi = 0$ and $\nu\{\infty\} = 0$.

In case $\nu\{\infty\} > 0$, Z may be written as the sum of random functions $\widehat{Z}$ and V, where

$$\widehat{Z}_t = \int_{[0, \infty)} x\, X((0, t] \times dx) \quad \text{and} \quad V = \infty \cdot X((0, t] \times \{\infty\}).$$

(As usual, we understand $\infty \cdot 0$ to equal 0.) By the part of the theorem already proved, $\widehat{Z}$ is a Lévy process, and for each $t \geq 0$, $\widehat{Z}_t$ is infinitely divisible with shift equal to 0 and Lévy measure equal to the restriction of $t\nu$ to $[0, \infty)$. By Problem 6, V is also a Lévy process, and for $t \geq 0$, the distribution of V_t is described in that problem. The independence properties of Poisson point processes imply that $\widehat{Z}$ and V are independent. By Problem 7, their sum Z is a Lévy process. It is easily checked from the properties of $\widehat{Z}$ and V that Z_t has the desired distribution for each $t \geq 0$. It is also easy to see that the addition of a deterministic linear function $t \rightsquigarrow \xi t$ accommodates arbitrary ξ. $\square$

The Lévy measure ν and shift ξ of the $\overline{\mathbb{R}}^+$-valued random variable Z_1 in the preceding theorem are also called the *Lévy measure* and *drift*, respectively, of the Lévy process Z.

* **Problem 13.** Let ν be the Lévy measure of a Lévy process Z in $\overline{\mathbb{R}}^+$. For each $t \in (0, \infty)$ and $y \in (0, \infty]$ calculate the probability, in terms of ν, that there exists $s \in (0, t]$ such that $Z_s - Z_{s-} \geq y$. Interpret the limit of your answer as $y \searrow 0$.

Problem 14. [Gamma processes] Let a and c be positive constants. For the Lévy process Z whose drift equals 0 and whose Lévy measure has density

$$y \rightsquigarrow \frac{ce^{-ay}}{y}, \quad y \in (0, \infty),$$

with respect to Lebesgue measure, show that for $t > 0$, the $\mathbb{R}^+$-valued random variable Z_t has the gamma density

$$y \rightsquigarrow \frac{a^{ct} y^{ct-1} e^{-ay}}{\Gamma(ct)}.$$

Problem 15. Let Z be the gamma process described in the preceding problem for $c = 1$, and define V by

$$V_t = \frac{Z_t}{Z_1}, \quad 0 \leq t \leq 1.$$

(Using a natural variation of notation previously introduced, we say that V is a $\mathbf{D}^+[0, 1]$-valued random variable.) For each $t \in [0, 1]$, identify the distribution of V_t. Find a formula for the function $(s, t) \rightsquigarrow \text{Corr}(V_s, V_t)$ for $(s, t) \in (0, 1)^2$, and discuss the limiting behavior of this function at the boundary (a square) of its domain. If you have become familiar with Dirichlet distributions (described in an optional section of Chapter 11), discuss why the term 'Dirichlet process' would be appropriate for the random function V had that term not already been appropriated as a synonym for 'Ferguson distribution', as noted in the section of Appendix G related to Chapter 27.

* **Problem 16.** (This problem requires familiarity with Dirichlet distributions, which are introduced in an optional section of Chapter 11.) Let Z be the gamma process defined in Problem 14 with $c = 1$, and let R denote the distribution of the corresponding random function V described in Problem 15. Show that

$$B \rightsquigarrow R(\{v \in \mathbf{D}^+[0, 1]: yv \in B\})$$

is the conditional distribution given $Z_1 = y$ of $Z\big|_{[0,1]}$, the restriction of Z to $[0, 1]$.

Problem 17. [Negative binomial processes] Let a and c be positive constants. For the Lévy process Z whose drift equals 0 and whose Lévy measure has density

$$y \rightsquigarrow \frac{ce^{-ay}}{y}, \quad y \in \mathbb{Z}^+ \setminus \{0\},$$

with respect to counting measure, calculate the distribution of each $\mathbb{Z}^+$-valued random variable Z_t.

Problem 18. Comment on similarities and differences between the gamma and negative binomial processes defined in Problem 14 and Problem 17. Give some attention to small times and also comment on large times.

For the $\mathbb{R}$-setting we have the following companion of Theorem 3.

Theorem 4. [Itô Representation ($\mathbb{R}$-version)] Let (X, W) be an independent pair, where W is a standard Wiener process and X is a Poisson point process in $(0, \infty) \times (\mathbb{R} \setminus \{0\})$ whose intensity measure is $\lambda \times \nu$, where λ denotes Lebesgue measure and ν is a Lévy measure for $\mathbb{R}$. Let $\eta \in \mathbb{R}$, let $\sigma \in [0, \infty)$, and let χ be the function described by Figure 16.1 of Chapter 16. Then there exists a decreasing sequence $(\varepsilon_k : k = 1, 2, \dots)$ of positive numbers converging to 0 such that

$$t \rightsquigarrow \eta t + \sigma W_t$$

$$+ \lim_{k \to \infty} \left[\int_{(-\infty, -\varepsilon_k] \cup [\varepsilon_k, \infty)} x \, X((0, t] \times dx)) - t \int_{(-\infty, -\varepsilon_k] \cup [\varepsilon_k, \infty)} x \, \nu(dx) \right]$$

defines a Lévy process in $\mathbb{R}$ whose value at t has the infinitely divisible distribution corresponding to $(t\eta, t\sigma, t\nu)$ via the Lévy-Khinchin Representation Theorem.

PARTIAL PROOF. In order to focus on the crux of the proof we will assume that $\eta = \sigma = \nu(-\infty, 0) = 0 < \nu(0, \varepsilon)$ for every $\varepsilon > 0$, leaving it to the reader to remove these restrictions.

Recall, from (16.6) of Chapter 16, that $\int_{(0,1]} x^2 \, \nu(dx) < \infty$. For $k \in \mathbb{Z}^+$, set

$$\varepsilon_k = \sup\{\varepsilon : \int_{(0, \varepsilon)} y^2 \, \nu(dy) \leq 8^{-k}\}.$$

Clearly, $\varepsilon_k \searrow 0$ as $k \to \infty$.

Define $Y^{(k)}$ by

(30.3) $$Y_t^{(k)} = \int_{[\varepsilon_k, \infty)} x \, X((0, t] \times dx)) - t \int_{[\varepsilon_k, \infty)} \chi(x) \, \nu(dx).$$

Applying Example 1 to the first term on the right and using the fact that the second function is a nonrandom constant times t, we conclude that with probability 1, $Y^{(k)}$ is a Lévy process. Moreover, the characteristic function of $Y_t^{(k)}$ equals

$$v \rightsquigarrow \exp\left(-\int_{[\varepsilon_k, \infty)} \left(1 - e^{ivx} - iv\chi(x)\right) \nu(dx)\right).$$

The limit as $k \to \infty$ equals the infinitely divisible characteristic function that corresponds to $(0, 0, \nu)$ via its Lévy-Khinchin representation.

In view of Problem 2 we can finish the proof by showing almost sure uniform convergence on bounded intervals by the sequence (Y^k). We write

$$Y^{(k)} = Y^{(0)} + \sum_{j=0}^{k-1} [Y^{(j+1)} - Y^{(j)}].$$

It is enough to show that for each $t < \infty$,

$$(30.4) \qquad \sum_{j=0}^{\infty} \sup_{s \le t} \left| Y_s^{(j+1)} - Y_s^{(j)} \right| < \infty \text{ a.s.}$$

By right continuity the term indexed by j equals

$$\lim_{m \to \infty} \sup_{i \le 2^m} \left| Y_{i2^{-m}t}^{(j+1)} - Y_{i2^{-m}t}^{(j)} \right|.$$

By the Etemadi Inequality (Lemma 21 of Chapter 12) and the Continuity of Measure Theorem, the probability that this quantity is larger than 2^{-j} is bounded above by

$$(30.5) \qquad \begin{aligned} &4 \lim_{m \to \infty} \sup_{i \le 2^m} P\left[\left| Y_{i2^{-m}t}^{(j+1)} - Y_{i2^{-m}t}^{(j)} \right| > 2^{-j-2} \right] \\ &= 4 \sup_{s \le t} P\left[\left| Y_s^{(j+1)} - Y_s^{(j)} \right| > 2^{-j-2} \right]. \end{aligned}$$

For j large enough so that $\varepsilon_j < 1$, we obtain from Problem 14 of Chapter 16 and our choice of (ε_k) that

$$E\left(Y_s^{(j+1)} - Y_s^{(j)} \right) = 0;$$

$$\mathrm{Var}\left(Y_s^{(j+1)} - Y_s^{(j)} \right) = \int_{[\varepsilon_{j+1}, \varepsilon_j)} x^2 \, \nu(dx) \le 8^{-j}.$$

By the Chebyshev Inequality, (30.5) is bounded by 2^{4-j}, which when summed over j gives a convergent series. By the Borel Lemma, only finitely many of the events

$$\left[\sup_{s \le t} \left| Y_s^{(j+1)} - Y_s^{(j)} \right| \ge 2^{-j} \right]$$

occur, with probability 1. Therefore the series (30.4) is a.s. finite. $\square$

Problem 19. Complete the proof of Theorem 4 by removing the restrictions on η, σ, and ν imposed in the partial proof given above.

From a Lévy processes in $\mathbb{R}$ constructed via its Itô representation, it is easy to recover the corresponding point process X, viewed as a random set:

$$X = \{ (s, x) \colon Y_s - Y_{s-} = x \}.$$

We have shown in this section that if Q is an infinitely divisible distribution on $\mathbb{R}$ or $\overline{\mathbb{R}}^+$, then there is a Lévy process Z such that Z_1 has distribution Q. In most cases, if the distribution Q has a name, then the same name is applied to Z. Thus, for example, the Lévy processes whose distributions at time 1 (and therefore at any positive time) are stable or strictly stable are said to be *stable* or *strictly stable*, respectively. And a *standard Cauchy process* is also known as a *strictly stable process with index 1*. Other examples of this terminology are found in Problem 14 and Problem 17. A major exception to this usage is the Wiener process, which is not called a 'normal process'.

30.4. Filtrations and stopping times

Filtrations indexed by nonnegative integers were introduced in Chapter 11. Indexing by nonnegative real numbers was treated in Chapter 19, but the random functions treated there were assumed to be continuous.

> **Problem 20.** Prove that Proposition 19 of Chapter 19 can be generalized to random variables that are $\mathbf{D}[0, \infty)$-valued, thus showing that Lévy processes are progressively measurable.

In this section we will focus on Lévy processes in $\mathbb{R}$, although some of what is done also applies to the $\overline{\mathbb{R}}^+$.

Suppose that $(\mathcal{F}_t : t \in [0, \infty))$ is the minimal filtration for a Lévy process Y. By Problem 4, the $\mathbf{D}[0, \infty)$-valued random variable

$$(30.6) \qquad\qquad t \rightsquigarrow Y_{s+t} - Y_s$$

is a Lévy process that has the same distribution as Y itself and is independent of $\mathcal{F}_s$. For a filtration that is not minimal, the preceding statement may or may not be true. In case Y is adapted to $(\mathcal{F}_t : t \in [0, \infty))$ and the process (30.6) is independent of $\mathcal{F}_s$ for each s, we say that Y is a *Lévy process with respect to* $(\mathcal{F}_t : t \in [0, \infty))$. This terminology is also used in case the Lévy process has a more specific name. Thus we may speak of a Wiener process or strictly stable process with respect to a filtration.

> **Proposition 5.** *A Lévy process in $\mathbb{R}$ with respect to a filtration $(\mathcal{F}_t : t \in [0, \infty))$ is also a Lévy process with respect to the filtration $(\mathcal{F}_{t+} : t \in [0, \infty))$.*

PROOF. The proof of Proposition 24 of Chapter 19 applies: continuity is assumed there but only right-continuity is used; and minimality of filtration is assumed but, in the terminology introduced above, only the fact that the Wiener process there is a Wiener process with respect to the filtration $(\mathcal{F}_t : t \in [0, \infty))$ is used. □

The following corollary is a special case of a more general result with the same name.

> **Corollary 6.** [Blumenthal 0-1 Law] *Let $(\mathcal{F}_{t+} : t \in [0, \infty))$ denote the minimal right-continuous filtration of a Lévy process in $\mathbb{R}^+$. Then every event in $\mathcal{F}_{0+}$ has probability 1 or 0.*

PROOF. Denote the Lévy process by Y and let $B \in \mathcal{F}_{0+}$. Because of the hypothesis of a minimal filtration,

$$B \in \sigma(Y_t : t \geq 0) = \sigma((Y_t - Y_0) : t \geq 0).$$

By Proposition 5, with $t = 0$, events that belong to this σ-field are independent of events belonging to $\mathcal{F}_{0+}$. Since B belongs to both of these σ-fields, $P(B \cap B) = P(B)P(B)$ from which the desired conclusion follows. $\square$

In the terminology introduced in Chapter 12, the conclusion of Corollary 6 can be restated as: The σ-field $\mathcal{F}_{0+}$ is 0-1 trivial.

Problem 21. Identify some interesting events that have probability 0 or 1 on the basis of the Blumenthal 0-1 Law.

The following result says that Lévy processes start over at stopping times. Thus Lévy processes are 'strong Markov processes'.

Theorem 7. *Let Y be a Lévy process in $\mathbb{R}$ with respect to a filtration $(\mathcal{F}_t : t \geq 0)$. Let T be an almost surely finite stopping time with respect to this filtration. Then the σ-field $\mathcal{F}_T$ is independent of the $\mathbf{D}[0, \infty)$-valued random variable $t \rightsquigarrow Y_{T+t} - Y_T$, $t \geq 0$, which is a Lévy process with respect to the filtration $(\mathcal{F}_{T+t} : t \in [0, \infty))$ having the same distribution as Y itself.*

Problem 22. Prove the preceding theorem by adapting the proof of Theorem 25 of Chapter 19.

30.5. † Subordination

Let Y be a Lévy process in $\mathbb{R}$ and Z a subordinator that is independent of Y and whose Lévy measure assigns the value 0 to the one-point set $\{\infty\}$. Define $\widehat{Y}$ by $\widehat{Y}_\tau = Y_{Z_\tau}$. It is straightforward to use Theorem 7 to show that $\widehat{Y}$ is a Lévy process in $\mathbb{R}$. It is said to be *subordinate to* Y and it arises from Y by *subordination* using the subordinator Z. Let us use Proposition 5 of Chapter 23 to calculate the characteristic function of $\widehat{Y}_\tau$ in terms of functions associated with Y and Z:

$$(30.7) \qquad E(e^{iu\widehat{Y}_\tau}) = E\big(E(e^{iuY_{Z_\tau}} \mid Z_\tau)\big) = E\big(e^{-Z_\tau \psi(u)}\big) = e^{-\tau \varphi(\psi(u))},$$

where ψ denotes the characteristic exponent of Z_1 and

$$\varphi(v) = -\log E\big(e^{-vT_1}\big).$$

Example 2. Let us find the Lévy process that is subordinate to standard Brownian motion using a strictly stable subordinator of index $\frac{1}{2}$. (See Problem 13 in Chapter 15.) For some $a > 0$ (depending on which stable subordinator is used), the formula for the characteristic exponent in (30.7) becomes

$$a\left|\frac{u^2}{2}\right|^{1/2} = \frac{a}{\sqrt{2}}|u|,$$

the characteristic exponent of a symmetric Cauchy process. Even though Brownian motion has no jumps and finite expectation at each fixed time, a Lévy process with jumps and undefined expectations at each time has been created from Brownian motion by 'sampling' it at a random set of times.

Problem 23. Show that the strictly stable Lévy processes that are not subordinate to any Lévy process of a different strict stable type are exactly those of index greater than or equal to 1 whose Lévy measure assigns the value 0 to $(0, \infty)$ or $(-\infty, 0)$. *Hint:* See Problem 13 of Chapter 15, where the moment generating functions of all strictly stable distributions in $\mathbb{R}^+$ are identified.

Problem 24. Suppose that a Lévy process $\widehat{Y}$ is subordinate to a Lévy process Y using a subordinator Z. Let $\widehat{Q}_\tau$, Q_t, and R_τ denote the distributions of $\widehat{Y}_\tau$, Y_t, and Z_τ, respectively. Show that

$$(30.8) \qquad \widehat{Q}_\tau(B) = \int_{[0,\infty)} Q_t(B) \, R_\tau(dt) \, .$$

Apply this formula in the case that Y is a standard Wiener process and Z is a strictly stable subordinator of index $\frac{1}{2}$ as described in Problem 13 and Problem 16, both of Chapter 15. Confirm that your answer is consistent with Example 2.

* **Problem 25.** Apply (30.8) to find Q_τ in case Y is a standard Poisson process and Z is a standard gamma process, giving your answer in terms of the means of Y_1 and Z_1.

30.6. ‡ Local-time processes and regenerative subsets of $[0, \infty)$

We begin with an example in order to illustrate several related concepts.

Example 3. Figure 30.1 illustrates, with short-dashed line segments, a standard *two-sided Poisson process* Y corresponding to the triple $\left(0, 0, \frac{1}{2}[\delta_1 + \delta_{-1}]\right)$ via the Lévy-Khinchin Representation Theorem. The function L_0, shown with the solid line segments, is called the 'local-time process at 0' of Y. Its value $L_0(t)$ at any particular t equals the amount of time less than or equal to t that Y is at 0. Similarly for each $y \in \mathbb{Z}$, the 'local-time process at y' is denoted by $L_y(t)$ and equals the Lebesgue measure of $\{s \leq t \colon Y_s = y\}$. For $C \subseteq \mathbb{Z}$, the random function

$$t \rightsquigarrow \sum_{y \in C} L_y(t)$$

is the 'occupation-time process' of C; its value at t is the Lebesgue measure of $\{s \leq t \colon Y_s \in C\}$. In Figure 30.1, the function shown with dots is L_{-1} of the particular ω being illustrated, and the function shown with long dashes is the occupation-time process of $\{-1, 0\}$. For each fixed t an 'occupation-measure

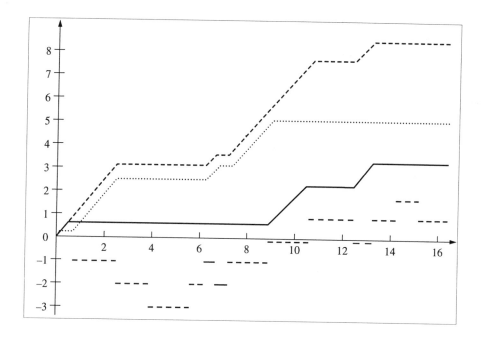

FIGURE 30.1. Local-time and occupation-time processes for the zero set of a two-sided Poisson process

process' is obtained by letting C vary. Its density with respect to counting measure is the random function $y \leadsto L_y(t)$.

Let

$$\Sigma(\omega) = \{t \in [0, \infty) \colon Y_t(\omega) = 0\},$$

and for each ω denote its indicator function by $t \leadsto X_t(\omega)$. We call X a 're-newal process'. and Σ a 'regenerative set'. Clearly $X_t(\omega) = 1$ if and only if $L_0(\omega, t + \gamma) > L_0(\omega, t)$ for all positive γ. It is straightforward to use the independent increment and stationary increment properties of Y to show that X has a property analogous to that in Proposition 1 of Chapter 25:

$$(30.9) \quad \begin{aligned} &P[X_{t_n} = x_n \text{ for } 0 < n \le r + s] \\ &= P[X_{t_n} = x_n \text{ for } 0 < n \le r] \, P[X_{t_n - t_r} = x_n \text{ for } r < n \le r + s], \end{aligned}$$

for all positive integers r and s, sequences $(x_1, \ldots, x_{r+s})$ of 0's and 1's such that $x_r = 1$, and $t_1 < t_2 < \cdots < t_{r+s}$.

Unlike the situation in Chapter 25, there is no increasing random walk associated with L_0, Σ, and X. Instead there is a subordinator. The subordinator Z is the 'right-continuous' inverse function of L_0; its image is Σ. It is shown in Figure 30.1 with the vertical axis as its domain, its graph being obtained by removing the horizontal line segments from the graph of L_0. To prove that Z is a subordinator one needs the fact that Y has the strong Markov property with respect to the minimal right-continuous filtration (Theorem 7 and Proposition 5) as well as stationary independent increments. The reason that the strong Markov property is needed is that a fixed time τ (on the vertical axis in Figure 30.1) in the domain of Z corresponds to the random time

$$\inf\{t\colon L_0(t) > \tau\},$$

which is a stopping time with respect to the minimal right-continuous filtration of Y.

The random σ-finite measure N on $[0, \infty)$ given by

$$(30.10) \qquad N(B) = \int_B X_t\, \lambda(dt) = \lambda(\Sigma \cap B),$$

where λ denotes Lebesgue measure, is the 'renewal measure' of the renewal process $t \rightsquigarrow X_t$. The 'potential measure' of $t \rightsquigarrow X_t$ is the (nonrandom) measure U defined by

$$(30.11) \qquad U(B) = E(N(B))$$
$$= \int_B P[X_t = 1]\,\lambda(dt) \quad \text{(by the Fubini Theorem).}$$

The following relations are clear and give the impression that we have introduced redundant concepts and notation:

$$N[0, t] = L_0(t)\,;$$
$$U[0, t] = E(L_0(t))\,;$$
$$(30.12) \qquad U(B) = \int_B P[Y_t = 0]\,\lambda(dt)\,.$$

The density $t \rightsquigarrow P[X_t = 0]$ of U with respect to λ is the 'potential function' of the renewal process $t \rightsquigarrow X_t$. Problem 5 of Chapter 16 gives an explicit formula for the potential function:

$$(30.13) \qquad t \rightsquigarrow e^{-t} I_0(t)\,,$$

where I_0 is a modified Bessel function of the first kind.

From Figure 30.1 we see that

$$(30.14) \qquad U(B) = E\big(\lambda(\{\tau\colon Z_\tau \in B\})\big) = E\big(\lambda(\Sigma \cap B)\big)\,.$$

By the Fubini Theorem,

$$(30.15) \qquad U(B) = \int_0^\infty R_\tau(B)\,d\tau\,,$$

where R_τ denotes the distribution of Z_τ. Another application of the Fubini Theorem gives us a relation between the 'Laplace-Stieltjes transform' of U and the pair (ξ, ν), where ξ and ν are the drift and Lévy measure of Z:

(30.16)
$$\int_{[0,\infty)} e^{-vt} U(dt)$$
$$= \int_0^\infty \exp\left[-\tau\left(\xi v + \int_{\mathbb{R}^+} (1 - e^{-vt})\, \nu(dt)\right)\right] d\tau$$
$$= \frac{1}{\left(\xi v + \int_{\mathbb{R}^+} (1 - e^{-vt})\, \nu(dt)\right)}, \quad v > 0.$$

We are able to calculate the Laplace-Stieltjes transform of U using (30.12), (30.13), and either the defining series (16.4) of I_0 or a table of Laplace transforms:

$$\int_{[0,\infty)} e^{-vt} U(dt) = \frac{1}{\sqrt{v^2 + 2v}}, \quad v > 0.$$

Hence

(30.17)
$$\xi v + \int_{\mathbb{R}^+} (1 - e^{-vt})\, \nu(dt) = \sqrt{v^2 + 2v}.$$

To calculate ξ and ν from (30.17) we first let $v \searrow 0$ to obtain $\nu\{\infty\} = 0$. Then we write the integral as a Riemann-Stieltjes integral, integrate by parts, and divide throughout by v to obtain

$$\xi + \int_0^\infty e^{-vt} \nu(t, \infty)\, dt = \sqrt{1 + (2/v)}.$$

Let $v \to \infty$ to obtain $\xi = 1$, a conclusion that we could have also obtained from Figure 30.1. We thus obtain a formula for the (ordinary) Laplace transform of the function $t \rightsquigarrow \nu(t, \infty)$:

$$v \rightsquigarrow \sqrt{1 + (2/v)} - 1.$$

By using a power series expansion or a table of inverse Laplace transforms we obtain

$$\nu(t, \infty) = e^{-t}[I_0(t) + I_1(t)].$$

The fact $\xi = \nu(0, \infty]$ implies that whenever Y is at 0, the amount of additional time it spends there before leaving has mean 1 (and is of course exponentially distributed). The function

(30.18)
$$t \rightsquigarrow \frac{\nu(0, t]}{\nu(0, \infty]} = 1 - e^{-t}[I_0(t) + I_1(t)]$$

is the distribution function of each of the independent random variables $(T_1 - S_1), (T_2 - S_2), \ldots$ that we now define using $T_0 = 0$:

$$S_n(\omega) = \inf\{t > T_{n-1}(\omega) \colon Y_t(\omega) \neq 0\}\,;$$
$$T_n(\omega) = \inf\{t > S_n(\omega) \colon Y_t(\omega) = 0\}\,.$$

Thus S_n is the sum of $2n - 1$ independent random variables, n of which are exponential with mean 1 and $n-1$ of which have the distribution function (30.18).

Similarly T_n is the sum of $2n$ independent random variables, n of which are exponential with mean 1 and n of which have the distribution function (30.18).

Problem 26. Summarize the various concepts and interconnections described in the preceding example. In doing so make sure you identify which objects are random objects and which are descriptive quantities connected with random objects. Also, say which conclusions have potential for generalization beyond two-sided Poisson processes and which conclusions are specific to these particular processes.

Problem 27. Mimic Example 3 for Y a compound Poisson with Lévy measure $p\delta_1 + (1-p)\delta_{-1}$. Also, find the distribution of the Lebesgue measure of Σ. *Hint:* You may find your solution of Problem 7 of Chapter 16 useful.

Problem 28. Mimic Example 3 for Y a compound Poisson with Lévy measure $\frac{1}{3}[\delta_1 + \delta_{-1} + \delta_\pi]$. Also, find the distribution of the Lebesgue measure of Σ. Check for consistency with the formulas in Example 3.

The random set Σ of Example 3 satisfies (30.9) for t_n and x_n as described there and has the property that both it and its complement in $\mathbb{R}^+$ are unions of half-open intervals, each of which contains its own left endpoint. Any random set satisfying these conditions is called a *regenerative set in* $\mathbb{R}^+$, and its (random) indicator function is a *renewal process*. [See Appendix G for a comment on these and the forthcoming definitions.] Equations (30.10) and (30.11) of Example 3 are the definitions of the *renewal measure* N and *potential measure* U of a regenerative set Σ and corresponding renewal process X. The random function $t \rightsquigarrow L(t) \overset{\text{def}}{=} N[0,t]$ is the *local-time process* of Σ. As in Example 3, its 'right-continuous inverse' Z is a subordinator with drift 1 and finite Lévy measure. The renewal process X can be recovered from Z: $X_t(\omega) = 1$ if and only if t is in the image of $Z(\omega)$ and $t < \infty$.

The potential measure U is related, as at (30.15), to the distributions R_τ of the $\mathbb{R}^+$-valued random variables Z_τ :

$$U(B) = \int_0^\infty R_\tau(B)\, d\tau\,.$$

The density of U with respect to Lebesgue measure is the function $t \rightsquigarrow P[X_t = 1]$, called the *potential function* of X. The calculation (30.16) is valid in general, so

$$(30.19) \qquad \int_{[0,\infty)} e^{-vt}\, U(dt) = \frac{1}{v + \int_{\mathbb{R}^+}(1 - e^{-vt})\, \nu(dt)}\,, \qquad v > 0\,.$$

Returning to Example 3, we note that L_y for $y \neq 0$ is not the local-time process of a regenerative set. It is the local-time process of a *delayed regenerative set in* $\mathbb{R}^+$, the formal definition of which we omit since it is very similar to that of a delayed regenerative set in $\mathbb{Z}^+$. For any compound Poisson process Y, $(t,y) \rightsquigarrow L_y(t)$ can be defined as it was for the two-sided compound Poisson process in Example 3. For a Borel subset C of $\mathbb{R}$, $t \rightsquigarrow \sum_{y \in C} L_y(t)$ is the *occupation-time process* of C.

Problem 29. Characterize those regenerative sets in $\mathbb{R}^+$ that have the property that their complements are delayed regenerative sets.

Theorem 8. *Let* $(X_t : t \in [0, \infty))$ *be a renewal process, and denote by* ν *the Lévy measure of the corresponding subordinator with drift 1. Then*

$$(30.20) \qquad \lim_{t \to \infty} P[X_t = 1] = \frac{1}{1 + \int_{(0,\infty]} y\, \nu(dy)}.$$

Also, $X_t = 0$ *for all sufficiently large* t *(depending on* ω*) if and only if* $\nu\{\infty\} > 0$.

PARTIAL PROOF. The last assertion of the theorem is obvious. There are two aspects of the first assertion: the existence of a limit and its value.

For the existence we let Z denote the subordinator corresponding to X and use Problem 13 to obtain

$$(30.21) \qquad \begin{aligned} P[X_t = 1] &\geq P[X_s = 1 \text{ for } s \in [0, t]] \\ &= P[Z_t = t] = e^{-t\nu(0,\infty]} > 0. \end{aligned}$$

Now we fix $b > 0$ and consider the sequence $(X_{bn} : n = 0, 1, 2, \dots)$. It is easy to see that it is a renewal sequence. By (30.21) it has period 1. So by the Renewal Theorem (Theorem 16 of Chapter 25), there exists a constant $c \in [0, 1]$ such that

$$\lim_{n \to \infty} P[X_{bn} = 1] = c.$$

Consider t between bn and $b(n + 1)$. By (30.9) and (30.21),

$$\begin{aligned} P[X_t = 1] &\geq P[X_{bn} = 1]\, P[X_{t-bn} = 1] \\ &\geq P[X_{bn} = 1]e^{-(t-bn)\nu(0,\infty]} \\ &\geq P[X_{bn} = 1]e^{-b\nu(0,\infty]}, \end{aligned}$$

and by a similar sequence of inequalities,

$$P[X_{b(n+1)} = 1] \geq P[X_t = 1]e^{-b\nu(0,\infty]}.$$

Hence

$$ce^{-b\nu(0,\infty]} \leq \liminf_{t \to \infty} P[X_t = 1] \leq \limsup_{t \to \infty} P[X_t = 1] \leq ce^{b\nu(0,\infty]}.$$

Now let $b \searrow 0$ to obtain $\lim_{t \to \infty} P[X_t = 1] = c$.

To evaluate c we rely in part on the problem following this proof which gives

$$(30.22) \qquad \lim_{v \searrow 0} v \int_0^\infty e^{-vt} P[X_t = 1]\, dt = c.$$

By (30.19), this is equivalent to

(30.23)
$$\lim_{v \searrow 0} \frac{v}{v + \int_{(0,\infty]} (1 - e^{-vt}) \nu(dt)} = c.$$

It is left for the reader to now show that c equals the right side of (30.20). □

Problem 30. Prove (30.22), and then finish the preceding proof by showing that the left side of (30.23) equals the right side of (30.20).

In view of Theorem 8 and the definitions in Chapter 25, we say that a renewal process is *positive recurrent* if $\int y \, \nu(dy) < \infty$, *null recurrent* if $\int y \, \nu(dy) = \infty$ but $\nu\{\infty\} = 0$, and *transient* if $\nu\{\infty\} = \infty$.

30.7. ‡ Sample function properties of subordinators

It is natural to ask for the probability that a given random function, such as a Lévy process, has certain properties like continuity or differentiability. For Lévy processes, there is a vast literature treating questions of this sort. In view of the prevalence of 0-1 laws, it is not surprising that for many properties, the probability is 0 or 1. For illustrative purposes we give a few such properties for subordinators in this section.

Theorem 9. *Let Z be a subordinator with drift ξ. Then*

$$\frac{dZ_t}{dt}\bigg|_{t=0} = \xi \ a.s.$$

* **Problem 31.** Prove the preceding theorem according to the following steps:
    (i) Show that without loss of generality one may assume $\xi = 0$; then use this assumption in the remaining steps.
    (ii) Show that without loss of generality one may assume that the Lévy measure ν satisfies $\nu(1, \infty] = 0$; then use this assumption in the remaining steps.
    (iii) Show that $t^{-1} Z_t \to 0$ i.p. as $t \searrow 0$.
    (iv) Show that $E(Z_t) < \infty$ for all t.
    (v) Show that $n \rightsquigarrow 2^n Z_{2^{-n}}$ is a martingale.
    (vi) Complete the proof.

* **Problem 32.** For any subordinator Z that never takes the value ∞ and any $s \in [0, \infty)$, show that $\frac{dZ_t}{dt}\big|_{t=s} = \xi$ a.s., where ξ denotes the drift of Z. One might carelessly conclude from this fact that $Z_t = \xi t$ a.s. Identify the carelessness that is likely to be involved.

Theorem 10. *Let ν denote the Lévy measure of a subordinator Z having 0 drift, and let g be a function from $\mathbb{R}^+$ to $\mathbb{R}^+$ satisfying: $g(0) = g(0+) = 0$ and $x \rightsquigarrow g(x)/x$ is increasing on $(0, \infty)$. Then*

$$\limsup_{t \searrow 0} \frac{Z_t}{g(t)} = \begin{cases} 0 \text{ a.s.} & \text{if } \int_0^1 \nu[g(t), \infty] \, dt < \infty \\ \infty \text{ a.s.} & \text{otherwise}. \end{cases}$$

Note that in Theorem 10, only the behavior of g near 0 is relevant.

Problem 33. Prove Theorem 10 according to the following steps:
  (i) Let X denote the Poisson point process in the Itô representation of Z.
  (ii) Assume that $\int_0^1 \nu[g(t), \infty] \, dt = \infty$, let $c \in (0, \infty)$, and show that the intensity measure of X assigns infinite measure to $\{(s, x): x \geq cg(s), s \leq t\}$ for every $t > 0$.
  (iii) Use the preceding step to deduce that

$$\limsup_{t \searrow 0} \frac{Z_t - Z_{t-}}{g(t)} \geq c \text{ a.s.}$$

  (iv) Note that the preceding inequality remains true if Z_{t-} is removed from the numerator. Then let $c \to \infty$ to complete half of the proof.
  (v) Assume that $\int_0^1 \nu[g(t), \infty] \, dt < \infty$, and let $b \in (0, \infty)$.
  (vi) Define $\hat{\nu}$ by $\hat{\nu}[x, \infty] = \nu[bg(x), \infty]$ for $x \in (0, \infty)$, and prove that $\hat{\nu}$ is a Lévy measure for $\overline{\mathbb{R}}^+$.
  (vii) Define a point process $\widehat{X}$ by $\widehat{X}([0, t] \times [x, \infty]) = X([0, t] \times [g(x), \infty])$, and say why it is Poisson with intensity measure $\lambda \times \hat{\nu}$, where λ denotes Lebesgue measure.
  (viii) Define $\widehat{Z}$ via the Itô Representation Theorem using the Poisson point process $\widehat{X}$ and drift 0. Then show that $Z(t) \leq g^{-1}(\widehat{Z}_t)$ for all t.
  (ix) Complete the proof by using Theorem 9 and then letting $b \searrow 0$.

Problem 34. For a gamma process Z find which numbers $\beta \in [1, \infty)$ have the property that $Z_t/t^\beta \to 0$ a.s. as $t \searrow 0$ and which have the property that

$$\limsup_{t \searrow 0} \frac{Z(t)}{t^\beta} = \infty \text{ a.s.}$$

* **Problem 35.** Repeat the preceding problem with the gamma process replaced by a strictly stable subordinator of index $\alpha \in (0, 1]$.

It turns out that the image of any subordinator with 0 drift has Lebesgue measure 0. In the other direction, the image is also uncountable unless the Lévy measure is finite. We want a method to distinguish the various images that have zero Lebesgue measure.

Definition and Proposition 11. Let $h: [0, \infty) \to [0, \infty)$ be a continuous strictly increasing function satisfying $h(0) = 0$. The set function h-meas defined on the Borel σ-field of $\mathbb{R}$ by

$$(30.24) \qquad h\text{-meas}(B) = \lim_{\varepsilon \searrow 0} \inf_{\substack{|J_k| < \varepsilon \\ B \subseteq \bigcup_{k=1}^{\infty} J_k}} \sum_{k=1}^{\infty} h(|J_k|),$$

where each J_k is an interval, possibly empty, and $|J_k|$ denotes the length of J_k, is a measure called *Hausdorff h-measure*.

We omit the proof that h-meas is a measure. The usual Cantor set C can be covered by 2^n intervals each of length 3^{-n}. Thus, if $h(x) = x^{\log 2 / \log 3}$, the sum in (30.24) equals 1; in fact, it can be proved that $h\text{-meas}(C) = 1$.

More generally, let $h_\beta(x) = x^\beta$ for some $\beta > 0$. It is easy to show that for any Borel set B in $\mathbb{R}$, there exists a unique $\alpha \in [0, \infty)$ such that

$$h_\beta\text{-meas}(B) = \begin{cases} \infty & \text{if } 0 < \beta < \alpha \\ 0 & \text{if } \beta > \alpha. \end{cases}$$

Both Definition 16 and this fact can easily be generalized to $\mathbb{R}^d$ with convex sets replacing intervals and diameter replacing length. The unique number α is called the *Hausdorff dimension* of B. By working with the function h_d one easily shows that the Hausdorff dimension of every Borel subset of $\mathbb{R}^d$ is less than or equal to d, and those that have positive d-dimensional Lebesgue measure have Hausdorff dimension d. The preceding paragraph shows that the usual Cantor set has dimension $\log 2 / \log 3$.

There are several possibilities for Borel sets B in $\mathbb{R}$ of Hausdorff dimension $\alpha < 1$:

- $h_\alpha\text{-meas}(B) = 0$,
- $0 < h_\alpha\text{-meas}(B) < \infty$,
- $h_\alpha\text{-meas}(B) = \infty$ and there exist $B_1, B_2, \ldots$ such that $B = \bigcup_{n=1}^{\infty} B_n$ and $h_\alpha\text{-meas}(B_n) < \infty$ for all n,
- $h_\alpha\text{-meas}(B) = \infty$ and there do not exist $B_1, B_2, \ldots$ such that $B = \bigcup_{n=1}^{\infty} B_n$ and $h_\alpha\text{-meas}(B_n) < \infty$ for all n.

If B falls into one of the middle two of these four cases, we regard h_α as 'an appropriate' function for 'measuring' B. In the other two cases it is natural to look for a function other than a power function to 'measure' B. The purpose of the preceding discussion is to set the background for the following theorem which we give without proof.

Theorem 12. *The image of a strictly stable subordinator Z of index α has Hausdorff dimension α. Set $h(0) = 0$ and $h(x) = x^\alpha \log^{1-\alpha}(\log(1/x))$ for $x \in (0, 1)$. Then h is an appropriate function for measuring the image of Z.*

CHAPTER 31
Introduction to Markov Processes

In many ways, continuous time is a more natural setting than discrete time for the study of random time evolutions. Fortunately, there is a close connection between the two settings, particularly in the case of Markovian time evolutions. In this chapter, we build on much of what was done for Markov sequences in Chapter 26 in order to develop the basic theory of Markov processes. Our main goal in doing so is to prepare the way for the final two chapters in which two of the most important classes of Markov processes are studied.

31.1. Cadlag space

In the modern study of Markov processes, it is common to view a Markov process as a random variable X, where for each ω, $X(\omega)$ is a function from the 'time-line' $[0, \infty)$ to a 'state space' Ψ. Thus, we need to define a measurable space consisting of such functions. Since convergence issues will play an important role, we will restrict our attention to state spaces that are Polish. (The following definition generalizes the space $\mathbf{D}[0, \infty)$, described in Chapter 30, to the Polish space setting.)

Definition 1. Let (Ψ, ρ) be a Polish space. A function $\varphi \colon [0, \infty) \to \Psi$ is *cadlag* if it is right continuous and the limit

$$\lim_{s \nearrow t} \varphi(t)$$

exists for all $t \in (0, \infty)$. The space of all cadlag functions $\varphi \colon [0, \infty) \to \Psi$ is denoted by $\mathbf{D}([0, \infty), \Psi)$ and called *cadlag space*.

For $t \in [0, \infty)$ and $\varphi \in \mathbf{D}([0, \infty), \Psi)$, we commonly write φ_t for $\varphi(t)$. This notation is consistent with that used earlier in Chapters 19 and 30. We make $\mathbf{D}([0, \infty), \Psi)$ into a measurable space by introducing the σ-field $\mathcal{H}$ generated by sets of the form

$$\{\varphi \colon \varphi_t \in B\},$$

for Borel $B \subseteq \Psi$ and $t \in [0, \infty)$. Usually, this σ-field will not be mentioned explicitly. A $\mathbf{D}([0, \infty), \Psi)$-valued random variable X is a *random cadlag function*, and X_t is the *state* of X at *time* t.

One reason for working with the space $\mathbf{D}([0, \infty), \Psi)$ is the following theorem.

Theorem 2. *Let (Ψ, ρ) be a Polish space and $(\mathbf{D}([0, \infty), \Psi), \mathcal{H})$ the corresponding measurable cadlag space. Then there exists a metric $\hat{\rho}$ on $\mathbf{D}([0, \infty), \Psi)$ such that $(\mathbf{D}([0, \infty), \Psi), \hat{\rho})$ is a Polish space and $\mathcal{H}$ is the σ-field of Borel sets in $\mathbf{D}([0, \infty), \Psi)$.*

We omit the highly technical proof of this result. Its main importance is that it allows us to study the convergence of sequences of distributions on $\mathbf{D}([0, \infty), \Psi)$. This fact is particularly important when constructing a relatively complicated stochastic process by taking the limit of a sequence of simpler stochastic processes. However, we will not need to use the theorem in this way. Our only application of it is the following.

Corollary 3. *Let X be a random cadlag function defined on a probability space $(\Omega, \mathcal{F}, P)$. If $\mathcal{G}$ is a sub-σ-field of $\mathcal{F}$, then there exists a conditional distribution of X given $\mathcal{G}$, and this conditional distribution is unique in the sense given in* Proposition 16 *of Chapter 21.*

Problem 1. Show that the distribution of a random cadlag function X uniquely determines and is uniquely determined by its *finite-dimensional distributions*, which are the distributions of finite sequences of the form $(X_{t_1}, \ldots, X_{t_n})$, for positive integers n and times $t_1, \ldots, t_n \in [0, \infty)$.

* **Problem 2.** Let X be a random cadlag function defined on a probability space $(\Omega, \mathcal{F}, P)$ and $\mathcal{G}$ a sub-σ-field of $\mathcal{F}$. Let Q be the conditional distribution of X given $\mathcal{G}$, and for each $t \in [0, \infty)$, let Q_t be the marginal of Q corresponding to t; that is, Q_t is the random distribution on Ψ defined by $Q_t(\cdot) = Q[X_t \in \cdot]$. Show that for each $t \in [0, \infty)$, Q_t is the conditional distribution of X_t given $\mathcal{G}$, and $Q_u \to Q_t$ a.s. as $u \searrow t$.

31.2. Markov, strong Markov, and Feller processes

In Chapter 26 we defined a Markov sequence in terms of a transition operator. In continuous time, a single transition operator is insufficient.

Definition 4. Let Ψ be a Polish space. A *transition semigroup* for Ψ is a collection $(T_t \colon t \in [0, \infty))$ of transition operators for Ψ satisfying the following properties:

(i) $T_0 f = f$ for all bounded measurable $f \colon \Psi \to \mathbb{R}$;
(ii) $T_s T_t = T_{s+t}$ for all $s, t \in [0, \infty)$;

(iii) $\lim_{t \searrow 0} T_t f(x) = f(x)$ for all bounded continuous $f \colon \Psi \to \mathbb{R}$ and $x \in \Psi$.

Since transition operators are defined in terms of transition distributions, a transition semigroup is necessarily associated with a collection $(\mu_{x,t} \colon x \in \Psi, t \in [0, \infty))$ of probability measures on Ψ, with

$$(31.1) \qquad T_t f(x) = \int f(y) \, \mu_{x,t}(dy)$$

for all $x \in \Psi, t \in [0, \infty)$. These probability measures are the *transition distributions* associated with the transition semigroup $(T_t \colon t \in [0, \infty))$.

* **Problem 3.** Find necessary and sufficient conditions on $(\mu_{x,t} \colon x \in \Psi, t \in [0, \infty))$ so that $(T_t \colon t \in [0, \infty))$ defined by (31.1) is a transition semigroup.

We are now ready to define time-homogeneous Markov processes, distributions, and families.

Definition 5. Let Ψ be a Polish space and $(T_t \colon t \in [0, \infty))$ a transition semigroup for Ψ, with associated transition distributions $\mu_{x,t}$, $x \in \Psi, t \in [0, \infty)$. A $\mathbf{D}([0, \infty), \Psi)$-valued random variable X adapted to a filtration $(\mathcal{F}_t \colon t \in [0, \infty))$ is a *time-homogeneous Markov process* with respect to that filtration, having *state space* Ψ and *transition semigroup* $(T_t \colon t \in [0, \infty))$, if for all $s, t \in [0, \infty)$ the conditional distribution of X_{s+t} given $\mathcal{F}_s$ is $\mu_{X_s, t}$. The distribution of such a process is a *time-homogeneous Markov distribution* with transition semigroup $(T_t \colon t \in [0, \infty))$.

We will often omit the adjective 'time-homogeneous', since we will not consider other types of Markov processes or distributions in this book. If a stochastic process X is called a Markov process without any reference to a filtration, then the minimal filtration of X is implied.

Definition 6. A *Markov family* of processes is a collection $\{X^x \colon x \in \Psi\}$ of Markov processes with common transition semigroup $(T_t \colon t \in [0, \infty))$, such that $X_0^x = x$ a.s. for every $x \in \Psi$. The corresponding collection $(Q^x \colon x \in \Psi)$ of Markov distributions is a *Markov family* of distributions with transition semigroup $(T_t \colon t \in [0, \infty))$.

Often one is able to construct a collection of distributions $(Q^x \colon x \in \Psi)$ on $\mathbf{D}([0, \infty), \Psi)$ that is a candidate for being a Markov family, but the construction is made in the absence of a transition semigroup. The following simple but useful result tells us when there exists a transition semigroup that makes the collection of probability measures into a Markov family.

Proposition 7. *Let Ψ be a Polish space, and for $t \in [0, \infty)$, set*

$$\mathcal{H}_t = \sigma \left(\varphi_s \colon s \in [0, t], \varphi \in \mathbf{D}([0, \infty), \Psi) \right).$$

Then a family $(Q^x: x \in \Psi)$ of distributions on $(\mathbf{D}([0,\infty), \Psi), \mathcal{H})$ is a Markov family if and only if the following three conditions are satisfied:

(i) For all $x \in \Psi$, $Q^x\{\varphi: \varphi_0 = x\} = 1$;

(ii) For all $t \in [0, \infty)$, $x \rightsquigarrow Q^x\{\varphi: \varphi_t \in \cdot\}$ is a measurable function from Ψ to the measurable space of probability measures on Ψ;

(iii) For all $x \in \Psi$, $s, t \in [0, \infty)$, and Borel sets $A \subseteq \Psi$,

$$Q^x[\varphi_{s+t} \in \cdot \mid \mathcal{H}_s](\theta) = Q^{\theta_s}[\varphi \in \cdot]$$

for Q^x-a.e. $\theta \in \mathbf{D}([0,\infty), \Psi)$.

Under these conditions, the transition semigroup $(T_t: t \in [0,\infty))$ of the Markov family is defined by

$$T_t f(x) = \int f(\varphi_t) \, Q^x(d\varphi)$$

for $x \in \Psi$, $t \in [0, \infty)$, and bounded measurable $f: \Psi \to \mathbb{R}$.

Problem 4. Prove the preceding proposition.

We have described a one-to-one correspondence between a certain collection of transition semigroups and the collection of all Markov families $(Q^x: x \in \Psi)$. We will not deal with the general question of which transition semigroups correspond to a Markov family, although we will give (without proof) a sufficient condition at the end of this section. Problem 6 below shows that an individual Markov distribution can correspond to more than one transition semigroup.

The *initial distribution* of a Markov process X is the distribution of X_0. If the initial distribution of X is a delta distribution δ_x for some x in the state space, then we call x the *initial state* of X. A Markov distribution is uniquely determined by the corresponding initial distribution and transition semigroup (see Problem 5 below).

Let $(Q^x: x \in \Psi)$ be a Markov family with transition semigroup $(T_t: t \in [0,\infty))$ and Q_0 a probability distribution on Ψ. Then

$$(31.2) \qquad C \rightsquigarrow \int_\Psi Q^x(C) Q_0(dx), \quad C \in \mathcal{H},$$

is the Markov distribution having initial distribution Q_0 and transition semigroup $(T_t: t \in [0,\infty))$.

The function $\varphi \rightsquigarrow \varphi_t$ from $\mathbf{D}([0,\infty), \Psi)$ to Ψ is measurable. This function and the Markov distribution (31.2) induce a distribution on Ψ —namely, the distribution of the state of the Markov process at time t when the initial distribution is Q_0. If this induced distribution is also Q_0 for every t, we say that Q_0 is an *equilibrium distribution* for the Markov family, and also for the corresponding transition semigroup.

Problem 5. Show that the distribution of a Markov process is uniquely determined by its initial distribution and transition semigroup.

Problem 6. Show that the delta distribution at the identically zero function in $\mathbf{D}([0,\infty),\mathbb{R})$ is a Markov distribution with (at least) two different transition semigroups $(T_t : t \in [0,\infty))$ and $(\widehat{T}_t : t \in [0,\infty))$ given by

$$T_t f(x) = f(x) \quad \text{and} \quad \widehat{T}_t f(x) = f(x + t \operatorname{sgn} x),$$

where $\operatorname{sgn} x$ is defined to equal 1, 0, or -1 according as $x > 0$, $x = 0$, or $x < 0$.

The proof of the following proposition is requested in Problem 7. To prepare for its statement, we remark that a function f from a Polish space Ψ (in fact, any topological space) to $\mathbb{R}$ is said to *vanish at ∞* if for every $\varepsilon > 0$ there exists a compact subset C of Ψ such that $|f(x)| < \varepsilon$ for $x \in \Psi \setminus C$. Notice that continuous functions that vanish at ∞ are necessarily bounded and uniformly continuous.

Proposition 8. *Let Ψ be a Polish space, $(T_t : t \in [0,\infty))$ a transition semigroup for Ψ, X a $\mathbf{D}([0,\infty), \Psi)$-valued random variable, and $(\mathcal{F}_t : t \in [0,\infty))$ a filtration to which X is adapted. The following are equivalent:*

(i) X is Markov with respect to $(\mathcal{F}_t : t \in [0,\infty))$;

(ii) For all times $s, t \in [0,\infty)$ and bounded measurable functions $f: \Psi \to \mathbb{R}$,

$$E(f \circ X_{s+t} \mid \mathcal{F}_s) = T_t f(X_s);$$

(iii) The equation in (ii) holds for all times $s, t \in [0,\infty)$ and all continuous functions $f: \Psi \to \mathbb{R}$ that vanish at ∞;

(iv) For all $t > 0$, the random sequence $(X_0, X_t, X_{2t}, \ldots)$ is Markov with respect to the filtration $(\mathcal{F}_{nt} : n = 0, 1, 2, \ldots)$, with transition operator T_t.

Problem 7. Prove Proposition 8. *Hint:* For the equivalence of (i) and (iii), use Corollary 7 of Chapter 18. For the equivalence of (i) and (iv), use Problem 2.

Problem 8. Show that the three conditions in Definition 4 are 'necessary', in the sense that if any one of them fails, a corresponding Markov family could not exist. *Hint:* For the second property, take expectations on both sides of the equation in (ii) of Proposition 8.

* **Problem 9.** Show that any Lévy process is a Markov process with state space $\mathbb{R}$ and initial state 0, and find the corresponding transition semigroup (or transition distributions). What is the appropriate Markov family associated with this transition semigroup?

Problem 10. [Chapman-Kolmogorov equations] Let X be a Markov process with state space Ψ, initial distribution Q_0, and transition distributions $\mu_{x,t}$. Show that for all positive integers n, bounded measurable functions $f\colon \Psi^n \to \mathbb{R}$, and times $t_1, \ldots, t_n$,

$$E\Big[f(X_{t_1}, X_{t_1+t_2}, \ldots, X_{t_1+\cdots+t_n})\Big]$$
$$= \int \Big(\int \cdots \Big(\int f(x_1, \ldots, x_n)\, \mu_{x_{n-1},t_n}(dx_n)\Big) \cdots \mu_{x_0,t_1}(dx_1)\Big) Q_0(dx_0).$$

Hint: One easy way to do this is to show that for any infinite sequence $t_1, t_2, \ldots$ of times, the random sequence $(X_0, X_{t_1}, X_{t_1+t_2}, \ldots)$ is a Markov sequence (not necessarily time-homogeneous), with transition functions $R_k(x, \cdot) = \mu_{x,t_k}(\cdot)$, and then use Theorem 3 or Problem 3, both of Chapter 22.

Problem 11. Let X be a Markov process with respect to a filtration $(\mathcal{F}_t\colon t \in [0, \infty))$, and suppose that a Markov family $(Q^x\colon x \in \Psi)$ exists corresponding to the transition semigroup of X. Fix $t \in [0, \infty)$, and define Y by $Y_s = X_{s+t}$. Show that the conditional distribution of Y given $\mathcal{F}_t$ is Q^{X_t}.

If we replace the time s in Definition 5 by an a.s.-finite stopping time S, we obtain the definition of a strong Markov process. The reader may want to compare this definition with the 'strong Markov property' that was proved for Lévy processes in Chapter 30.

Definition 9. Let X be a Markov process with respect to a filtration $(\mathcal{F}_t\colon t \in [0, \infty))$, with transition distributions $\mu_{x,t}$, $x \in \Psi, t \in [0, \infty)$). Then X is *strong Markov* with respect to the filtration $(\mathcal{F}_t\colon t \in [0, \infty))$ if for each a.s.-finite stopping time S with respect to that filtration and all $t \in [0, \infty)$, the conditional distribution of X_{S+t} given $\mathcal{F}_S$ is $\mu_{X_S,t}$. The distribution of a strong Markov process is a *strong Markov* distribution. Markov families whose members are all strong Markov are *strong Markov families*.

In the discrete-time setting, Markov and strong Markov are equivalent. Unfortunately (see Problem 13), the same does not hold true in continuous time.

Problem 12. State and prove strong Markov analogues of Proposition 7, of the equivalence among (i), (ii), and (iii) in Proposition 8, and of the result in Problem 11.

Problem 13. Let Ψ be the following subset of $\mathbb{R}^2$:

$$\{(x,y)\colon y = 0 \text{ or } x^2 + (y-1)^2 = 1\}.$$

Thus, Ψ is the union of the x-axis and the circle with radius 1 and center $(0,1)$. Define $\varphi\colon \mathbb{R} \to \Psi$ by

$$\varphi(x) = \begin{cases} (x,0) & \text{if } x \leq 0 \\ (\sin x, 1 - \cos x) & \text{if } 0 < x < 2\pi \\ (x - 2\pi, 0) & \text{otherwise}. \end{cases}$$

Let W be a Wiener process on $[0, \infty)$, and define X by

$$X_t = \varphi(W_t + \pi).$$

Show that X is Markov, but not strong Markov. *Hint:* To show that X is Markov, use the fact that φ is invertible except at $(0,0)$. To show that X is not strong Markov, consider the hitting time of the point $(0,0)$.

We conclude this section with a discussion of an important special class of transition semigroups. This class has two nice properties: (i) every member of it corresponds to a Markov family and (ii) every such Markov family is strong Markov.

Definition 10. A transition semigroup $(T_t \colon t \in [0, \infty))$ is a *Feller semigroup* if for all times $t \in [0, \infty)$ and continuous functions $f \colon \Psi \to \mathbb{R}$ that vanish at ∞, $T_t f$ is a continuous function that vanishes at ∞. A Markov process is a *Feller process* if its transition semigroup is a Feller semigroup.

Theorem 11. *Let X be a Markov process with respect to a filtration $(\mathcal{F}_t \colon t \in [0, \infty))$. If X is a Feller process, then X is strong Markov with respect to the filtration $(\mathcal{F}_{t+} \colon t \in [0, \infty))$.*

PROOF. By the strong Markov version of the equivalence between (i) and (iii) in Proposition 8 (see Problem 12), it is enough to prove that

$$(31.3) \qquad E(f \circ X_{S+t} \mid \mathcal{F}_S) = T_t f(X_S)$$

for all a.s.-finite stopping times S with respect to the given filtration, all times $t \in [0, \infty)$, and all bounded continuous functions $f \colon \Psi \to \mathbb{R}$. Fix such S, t, and f.

For $\delta > 0$, let

$$S_\delta(\omega) = \inf\{u \geq S(\omega) \colon u = k\delta \text{ for some nonnegative integer } k\}.$$

Then each S_δ is almost surely finite, and $S_\delta \to S$ a.s. as $\delta \to 0$.

By Proposition 8, for each $\delta > 0$, the random sequence

$$(X_{n\delta} \colon n = 0, 1, 2, \dots)$$

is Markov with respect to the filtration $(\mathcal{F}_{n\delta} \colon n = 0, 1, 2, \dots)$, with transition operator T_δ. By Theorem 7 of Chapter 26 (the strong Markov property of Markov sequences),

$$E(f \circ X_{S_\delta + t} \mid \mathcal{F}_{S_\delta}) = T_t f(X_{S_\delta}).$$

Since $\mathcal{F}_S \subseteq \mathcal{F}_{S_\delta}$, it follows that for any $B \in \mathcal{F}_S$,

$$E(f(X_{S_\delta + t}) I_B) = E(T_t f(X_{S_\delta}) I_B).$$

Now let $\delta \searrow 0$ and use the right-continuity of X and the continuity of $T_t f$ to obtain

$$E(f(X_{S+t}) I_B) = E(T_t f(X_S) I_B).$$

Since $T_t f(X_S)$ is clearly $\mathcal{F}_S$ measurable, (31.3) follows. $\square$

We omit the proof of the following important characterization theorem, but comment that in a proof one would only need to show 'Markov' since then 'strong Markov' would follow from Theorem 11.

Theorem 12. *Each Feller semigroup is the transition semigroup of a strong Markov family.*

31.3. Infinitesimal generators

An important aspect of the theory of Markov processes is the development of methodology for understanding global behavior in terms of an operator that describes local behavior. In the following definition, we use the term *converges boundedly* to describe a pointwise convergent sequence of $\mathbb{R}$-valued functions whose absolute values are all bounded above by a single finite constant. Similarly we speak of a limit existing *boundedly*.

Definition 13. Let $(T_t : t \in [0, \infty))$ be a transition semigroup for a Polish space Ψ. The operator G defined by

$$(31.4) \qquad Gf(x) = \lim_{t \searrow 0} \frac{T_t f(x) - f(x)}{t} \quad \text{boundedly}$$

is the *infinitesimal generator* of the transition semigroup $(T_t : t \in [0, \infty))$ and of any Markov processes and families that might correspond to this transition semigroup; the domain of G consists of those bounded measurable functions f for which the limit at (31.4) exists (boundedly).

Even though G is defined in terms of the transition semigroup (T_t), it is possible in some situations that one will know G without knowing the transition semigroup. For instance, one might only know an approximation of T_t, but this approximation may be good enough for small t so that an exact formula for G can be obtained. The following theorem gives equations 'of differential type' that one might hope to solve for the function $t \rightsquigarrow T_t$ if G is already known.

Theorem 14. *Let $(T_t : t \in [0, \infty))$ be a transition semigroup and G the corresponding infinitesimal generator for a Polish space Ψ. Then for all f in the domain of G,*

$$(31.5) \qquad T_t G f(x) = \lim_{h \to 0} \frac{T_{t+h} f(x) - T_t f(x)}{h} = G T_t f(x)$$

boundedly (with the understanding that $h > 0$ if $t = 0$).

PARTIAL PROOF. In terms of the transition distributions $\mu_{x,t}$,

$$(31.6) \qquad T_t G f(x) = \int_\Psi \lim_{h \searrow 0} \frac{T_h f(y) - f(y)}{h} \, \mu_{x,t}(dy) .$$

Since f is in the domain of G, the bounded convergence theorem applies. Thus the first equality in (31.5) holds with $h \searrow 0$ in lieu of $h \to 0$. The existence of this limit implies, by definition, that $T_t f$ is in the domain of G and that the second equality holds. The reader is asked in Problem 14 to replace the restriction $h \searrow 0$ with $h \to 0$. $\square$

Problem 14. Complete the preceding proof.

* **Problem 15.** Describe the infinitesimal generator of a compound Poisson process in terms of its Lévy measure.

Problem 16. Let G be the infinitesimal generator of a Markov family. Show that if Q_0 is an equilibrium distribution for the Markov family, then

$$\int Gf(x) \, Q_0(dx) = 0$$

for all functions f in the domain of G.

Problem 17. Let Ψ be a Polish space and G the infinitesimal generator of a transition semigroup $(T_t : t \in [0, \infty))$ for Ψ. Suppose that G is a *bounded operator* on the space of bounded measurable functions $f \colon \Psi \to \mathbb{R}$, meaning that there exists a finite constant c such that

$$\sup_{x \in \Psi} |Gf(x)| \leq c \sup_{x \in \Psi} |f(x)|$$

for all such f. Show that the transition semigroup is uniquely determined by the formula

$$T_t = e^{tG} \stackrel{\text{def}}{=} 1 + tG + \frac{t^2 G \circ G}{2!} + \frac{t^3 G \circ G \circ G}{3!} + \cdots, \quad t \in [0, \infty).$$

Hint: Under the given hypotheses, Theorem 14 implies that the function $t \rightsquigarrow GT_t f(x)$ has derivatives of all order.

31.4. The martingale problem

We adapt the main definition of Chapter 24 to the continuous-time setting.

Definition 15. A $\mathbf{D}([0, \infty), \mathbb{R})$-valued random variable Z adapted to a filtration $(\mathcal{F}_t : t \geq 0)$ is a *continuous-time martingale* with respect to that filtration if $E|Z_t| < \infty$ for all $t \in [0, \infty)$, and

$$E(Z_{t+s} \mid \mathcal{F}_s) = Z_s$$

for all $s, t \in [0, \infty)$.

Problem 18. For X a Lévy process with $E|X_1| < \infty$, show that $t \rightsquigarrow (X_t - tEX_1)$ is a continuous-time martingale with respect to the minimal filtration.

It is quite easy to adapt the main results for martingales in Chapter 24 to the continuous-time setting. Whenever we cite results from that chapter in a continuous-time context, we will assume that such an adaptation has been made.

Definition 16. Let Ψ be a Polish space, $\mathfrak{F}$ a collection of bounded continuous functions $f: \Psi \to \mathbb{R}$, and G a functional from $\mathfrak{F}$ to the space of bounded measurable functions on Ψ. A $\mathbf{D}([0,\infty), \Psi)$-valued random variable X defined on Ω is a *solution to the martingale problem* for $(G, \mathfrak{F})$ if

$$(31.7) \qquad t \rightsquigarrow f(X_t) - \int_0^t Gf(X_u)\,du$$

is a continuous-time martingale with respect to the minimal filtration of X for all $f \in \mathfrak{F}$.

The random variable defined by (31.7) is automatically $\mathbf{D}([0,\infty), \Psi)$-valued, due to the assumption that f is continuous.

For martingales and solutions to the martingale problem, we speak of *initial states* just as for Markov processes.

Denote the minimal filtration of X in Definition 16 by $(\mathcal{F}_t: t \in [0,\infty))$. Since $\int_0^s Gf(X_u)\,du$ is measurable with respect to $\mathcal{F}_s$, the statement that the expression in (31.7) is a martingale is equivalent to the following condition:

$$(31.8) \qquad E\Big(f(X_{t+s}) \,\Big|\, \mathcal{F}_s\Big) - E\Big(\int_s^{t+s} Gf(X_u)\,du \,\Big|\, \mathcal{F}_s\Big) = f(X_s),$$
$$s, t \in [0,\infty).$$

The following two theorems, which are close to being converses of each other, relate Markov families to solutions to the martingale problem.

Theorem 17. *Let G be the infinitesimal generator of a Markov process X with state space Ψ, and let $\mathfrak{F}$ denote a subset of the domain of G consisting only of continuous functions. Then X is a solution to the martingale problem for $(G, \mathfrak{F})$.*

PROOF. Let X be any member of the Markov family. To prove that it solves the martingale problem we will calculate the left side of (31.8) and show that it equals $f(X_s)$. By the Conditional Bounded Convergence Theorem, the left side of (31.8) equals

$$T_t f(X_s) - \int_s^{s+t} T_{u-s} Gf(X_s)\,du,$$

which by Theorem 14 equals

$$T_t f(X_s) - \int_0^t \frac{dT_v f(X_s)}{dv}\,dv = T_0 f(X_s) = f(X_s). \qquad \square$$

Theorem 18. *Let G and $\mathfrak{F}$ be as in Definition 16. Suppose that for each $x \in \Psi$, there is a unique cadlag solution to the martingale problem for $(G, \mathfrak{F})$ with initial state x. Then the collection of solutions obtained by varying x over Ψ is a strong Markov family whose infinitesimal generator agrees with G on $\mathfrak{F}$.*

PROOF. Let X^x be the unique solution to the martingale problem with initial state x. Now fix x, let T be a stopping time for $X \overset{\text{def}}{=} X^x$, and set $Y_t = X_{T+t}$. Let $f \in \mathfrak{F}$ and consider the random function

$$(31.9) \qquad t \rightsquigarrow f(Y_t) - \int_0^t Gf(Y_u)\, du\,,$$

which is adapted to the filtration $(\mathcal{F}_{T+t} : t \in [0, \infty))$. For $0 \le s < t$,

$$E\left(f(Y_t) - \int_0^t Gf(Y_u)\, du \;\Big|\; \mathcal{F}_{T+s} \right)$$
$$= E\left(f(X_{T+t}) - \int_0^{T+t} Gf(X_u)\, du \;\Big|\; \mathcal{F}_{T+s} \right) + E\left(\int_0^T Gf(X_u)\, du \;\Big|\; \mathcal{F}_{T+s} \right),$$

which by the Optional Sampling Theorem equals

$$f(X_{T+s}) - \int_0^{T+s} Gf(X_u)\, du + \int_0^T Gf(X_u)\, du$$
$$= f(Y_s) - \int_0^s Gf(Y_u)\, du\,.$$

Thus the conditional distribution of Y given $\mathcal{F}_T$ solves the martingale problem with initial state X_T. Since the solution to the martingale problem is unique, this conditional distribution must be that of X^{X_T}, as desired. $\square$

The key to the preceding proof is the Optional Sampling Theorem. The availability of this theorem is one reason the martingale problem is so useful for treating Markov processes. There are two other significant reasons for using the martingale problem to study Markov processes. The first of these is obvious: it provides a large collection of martingales that can be used to analyze a Markov process. The other reason for studying the martingale problem has to do with sequences of Markov distributions. It is often hard to show directly that the limit of such a sequence is itself a Markov distribution. But it turns out to be quite easy in most cases to show that such a limit is a solution to a martingale problem. Then Theorem 18 can often be used to see that the limit is strong Markov.

Problem 19. [Dynkin formula] Let X be a Markov process with infinitesimal generator G, and let $U \le V$ be a.s.-finite stopping times for the minimal filtration of X. Show that for every continuous function f in the domain of G,

$$E[f(X_V) - f(X_U)] = E\left(\int_U^V Gf(X_u)\, du \right).$$

31.5. Pure-jump Markov processes: bounded rates

To illustrate the ideas introduced in preceding sections, we now construct some Markov families with infinitesimal generators that are bounded operators on the space of bounded measurable functions.

Let Ψ be a Polish space, $\widetilde{T}$ a transition operator for Ψ with discrete generator $\widetilde{G} = \widetilde{T} - I$, x_0 a state in Ψ, and c a positive constant. We will construct a Markov process X with infinitesimal generator $G = c\widetilde{G}$ and initial state x_0. Our ingredients are a Markov sequence $\widetilde{X}$ with initial state x_0 and transition operator $\widetilde{T}$ and a random walk $S = (S_n : n = 0, 1, 2, \dots)$ having exponentially distributed steps with parameter $c \geq 0$ (the reciprocal of the mean). Assume that $\widetilde{X}$ and the sequence S are independent of each other. For $t \in [0, \infty)$, define

$$M_t = \sup\{n \geq 0 : S_n \leq t\}.$$

(Those that have read Chapter 30 will recognize M as a standard Poisson process with Lévy measure $c\delta_1$.)

The definition of X is quite simple:

$$X_t = \widetilde{X}_{M_t}, \quad t \in [0, \infty).$$

Roughly speaking, the discrete-time random sequence $\widetilde{X}$ has been converted to a continuous-time stochastic process X by using the Poisson process M to measure time.

It can be shown directly that the process X just defined is strong Markov, and as such it is called a *pure-jump Markov process* with *bounded rates*. Rather than the direct approach of showing that X is strong Markov, we will use Poisson point processes to give an alternative construction of a $\mathbf{D}([0, \infty), \Psi)$-valued random variable Y that has the same distribution as X. It will be easy to see from this construction that Y is strong Markov. An advantage of this second approach is that it is 'universal', in the sense that all pure-jump Markov processes with bounded rates are constructed on the same probability space. In Chapter 32, a similar construction is used for 'interacting particle systems', so our work here is a warm-up for that chapter.

The basis for our alternative construction is a Poisson point process Z on $(0, \infty) \times [0, 1]$, with intensity measure $\lambda = \lambda_1 \times \lambda_2$, where λ_1 is Lebesgue measure on $(0, \infty)$ and λ_2 is Lebesgue measure on $[0, 1]$. By Problem 24 of Chapter 29, we may write $Z = \{(U_n, V_n) : n = 1, 2, \dots\}$, where the sequences $U = (U_n : n = 1, 2, \dots)$ and $V = (V_n : n = 1, 2, \dots)$ are independent. That problem also implies that the sequence $(0, U_1, U_2, \dots)$ may be chosen to be a random walk with steps that are exponentially distributed with mean 1, and that V is an iid sequence of random variables that are uniformly distributed on $[0, 1]$. We will use the sequences U and V to construct a pair $(\widetilde{Y}, N)$ that has the same distribution as $(\widetilde{X}, M)$. Once we have accomplished that task, the construction is completed by defining

$$Y_t = \widetilde{Y}_{N_t}, \quad t \in [0, \infty).$$

Since $(\widetilde{Y}, N)$ and $(\widetilde{X}, M)$ have the same distribution, the $\mathbf{D}([0, \infty), \Psi)$-valued random variable $Y = (Y_t : t \in [0, \infty))$ has the same distribution as X.

The definition of N is quite simple:

$$N_t = \sup\{n \geq 0 : U_n \leq ct\}.$$

It is easy to see that N has the same distribution as M.

Before defining $\widetilde{Y}$, we temporarily restrict ourselves to the special case in which $\Psi = \mathbb{R}$. For this case, let μ_x, $x \in \mathbb{R}$, be the transition distributions associated with the transition operator $\widetilde{T}$, and let F_x, $x \in \mathbb{R}$ be the corresponding distribution functions. Define functions $f_x : [0, 1] \to \mathbb{R}$ by

$$f_x(u) = \inf\{y \in \mathbb{R} : F_x(y) \geq u\}.$$

Thus, f_x is the left-continuous inverse of F_x, and by Proposition 4 of Chapter 3, if V is any random variable that is uniformly distributed on $[0, 1]$, then $f_x \circ V$ has distribution μ_x.

We now define $\widetilde{Y}$ inductively. Let $\widetilde{Y}_0 = x_0$, and having defined $\widetilde{Y}_n$ for some $n \geq 0$, let

$$\widetilde{Y}_{n+1} = f_{\widetilde{Y}_n}(V_{n+1}).$$

It is easy to check that since V is iid, $\widetilde{Y}$ is a Markov sequence. Since the random variables in the sequence V are uniformly distributed on $[0, 1]$, it is also easy to check from the definition of the functions f_y that Y has transition operator $\widetilde{T}$. Since U and V are independent, $\widetilde{Y}$ and N are also independent. Thus, our alternative construction is complete for the case $\Psi = \mathbb{R}$.

To generalize to the case of an arbitrary Polish state space Ψ, simply use the fact that since Polish spaces are Borel (see Proposition 20 of Chapter 21), there is an isomorphism g from Ψ to $\mathbb{R}$. This isomorphism transforms the transition distributions associated with $\widetilde{T}$ to the transition distributions of a transition operator for $\mathbb{R}$. Carry out the preceding construction for these transformed transition distributions, then apply the function g^{-1} to the result to obtain the appropriate process with state space Ψ.

Let us summarize what we have done so far. For each initial state $x \in \Psi$, we have defined $\mathbf{D}([0, \infty), \Psi)$-valued random variables X^x and Y^x with the same distribution. The random variable X^x was constructed directly in terms of a Markov sequence $\widetilde{X}$ and a standard Poisson process M. The random variable Y^x was constructed by first defining a Poisson point process Z, and then using Z to construct a pair $(\widetilde{Y}, N)$ having the same distribution as $(\widetilde{X}, M)$. This construction is measurable in Z and x, in the sense that there is a measurable function $h : \{\text{discrete subsets of } \Psi\} \times \Psi$ such that $Y^x = h(Z, x)$. (See Chapter 29 for the interpretation of Z as a random discrete subset of Ψ.) The notation introduced in Problem 17 is relevant for the following result.

Theorem 19. Let $(X^x : x \in \Psi)$ and $(Y^x : x \in \Psi)$ be the two families of processes defined above in terms of a transition operator $\widetilde{T}$ with discrete generator

$\widetilde{G}$. Then for each $x \in \Psi$, X^x and Y^x have the same distribution and are strong Markov. The corresponding Markov family of distributions has generator $G = c\widetilde{G}$, with transition semigroup $(T_t \colon t \in [0, \infty))$ determined from G via

$$T_t = e^{tG} = e^{ct\widetilde{G}} = e^{-ct}e^{ct\widetilde{T}}, \quad t \in [0, \infty).$$

PROOF. We have already shown that X^x and Y^x have the same distribution for each $x \in \Psi$. We denote this distribution by Q^x. It is easy to check from the construction that the collection $(Q^x \colon x \in \Psi)$ satisfies the first two conditions of Proposition 7.

We now check the third condition of Proposition 7. Thinking of Z as a random discrete subset of $(0, \infty) \times [0, 1]$, let

$$\mathcal{F}_t = \sigma(Z \cap ((0, t] \times [0, 1])), \quad t \in [0, \infty).$$

It follows from the construction that each process Y^x is adapted to the filtration $(\mathcal{F}_t)$. Thus, it is sufficient to show for each $s \in [0, \infty)$ and $x \in \Psi$, that the conditional distribution of $t \rightsquigarrow Y^x_{s+t}$ given $\mathcal{F}_s$ is $Q^{Y^x_s}$.

Let Z^s be the point process obtained from Z by subtracting s from the first coordinate of each of the points in $Z \cap ((s, \infty) \times [0, 1])$. The construction implies that the process $t \rightsquigarrow Y^x_{s+t}$ is given by $h(Z^s, Y^x_s)$, where h is the function defined prior to the statement of the theorem. We want to show that the conditional distribution of $h(Z^s, Y^x_s)$ given $\mathcal{F}_s$ is $Q^{Y^x_s}$.

Since the basic properties of Poisson point processes imply that Z^s is independent of $\mathcal{F}_s$ and since Y^x_s is $\mathcal{F}_s$-measurable, it follows from Problem 21, Proposition 11, and Proposition 12, all of Chapter 21, that the conditional distribution of $h(Z^s, Y^x_s)$ given $\mathcal{F}_s$ is $R^{Y^x_s}$, where for each $y \in \Psi$, R^y is the distribution of $h(Z^s, y)$. It is easy to see that Z^s and Z have the same distribution, so $Q^y = R^y$ for all $y \in \Psi$, as desired.

We have shown that $(Q^x \colon x \in \Psi)$ is a Markov family. The proof that it is strong Markov is very similar to the proof of Theorem 11. Following the notation in that proof, the only difference is that we do not rely on the continuity of $T_t f$ near the end of the proof to show that $T_t f(X_{S_\delta}) \to T_t f(X_S)$ as $\delta \searrow 0$. Instead, our construction shows that the event $[X_{S_\delta} = X_S]$ contains the event $[M_{S_\delta} = M_S]$, which increases to all of Ω as $\delta \searrow 0$. The rest of the proof of Theorem 11 can be followed without change.

To calculate the infinitesimal generator and transition semigroup, we return to the construction at the beginning of this section. Let $\widetilde{X}$ and M be the Markov sequence and standard Poisson process used there to construct X. By this construction,

$$E(f \circ X_t) = \sum_{k=0}^{\infty} E(f(\widetilde{X}_k))P[M_t = k]$$

for any bounded measurable $f \colon \Psi \to \mathbb{R}$. Since M is a standard Poisson process and $EM_1 = c$, the random variable M_t is Poisson distributed with mean ct.

Thus,

$$P[M_t = k] = e^{-ct}\frac{c^k t^k}{k!}.$$

Since $\widetilde{X}$ is a Markov sequence with transition operator $\widetilde{T}$ and initial state x,

$$E(f(\widetilde{X}_k)) = \widetilde{T}^k f(x).$$

Since f and x are arbitrary, the formula $T_t = e^{-ct}e^{ct\widetilde{T}}$ follows immediately, and the formula $T_t = e^{tG}$ then follows from standard manipulations involving power series. That G is the corresponding infinitesimal generator is now an easy consequence of Definition 13. (See also Problem 17.) □

To see the relevance of the modifying phrase 'with bounded rates' for the Markov processes we have been discussing, we write the corresponding infinitesimal generators G in a different form.

By Theorem 19

$$(31.10) \qquad Gf(x) = c\widetilde{G} = c\int [f(y) - f(x)]\,\mu_x(dy), \quad x \in \Psi,$$

where μ_x, $x \in \Psi$, are the transition distributions of the transition operator $\widetilde{T}$. For each $x \in \Psi$, let

$$q(x) = c\mu_x(\Psi \setminus \{x\}).$$

If $q(x) > 0$, denote by ρ_x the probability measure on Ψ defined by

$$\rho_x(A) = \frac{\mu_x(A \setminus \{x\})}{q(x)}$$

for Borel sets $A \subseteq \Psi$. If $q(x) = 0$, let $\rho_x = \delta_x$. Now (31.10) can be rewritten as

$$(31.11) \qquad Gf(x) = q(x)\int [f(y) - f(x)]\,\rho_x(dy)$$

for all $x \in \Psi$ and bounded measurable $f : \Psi \to \mathbb{R}$.

The number $q(x)$ in the formula for G is the *jump rate at x* of the corresponding Markov family and ρ_x is the *jump distribution from x*. The function $x \rightsquigarrow q(x)$, denoted by q, is the *jump-rate function*. For an explanation of this terminology, see Problem 20. Note that our construction ensures that the jump-rate function is bounded above by c, thus explaining why these processes are said to have bounded rates. In the next section, we will discuss pure-jump Markov processes with unbounded rates.

Problem 20. For X as in Theorem 19 with initial state x, let $J = \inf\{t : X_t \neq x\}$. Suppose that $q(x) > 0$. Prove that the distribution of J is exponential with mean $1/q(x)$, that J and X_J are independent, and that the distribution of X_J is ρ_x.

* **Problem 21.** Show that Q_0 is an equilibrium distribution for the transition semigroup defined in Theorem 19 if and only if it is an equilibrium distribution for the corresponding transition operator $\widetilde{T}$.

Problem 22. Let $\widetilde{T}$ be a transition operator for Ψ. Show that if $\widetilde{T}$ takes continuous functions vanishing at ∞ to continuous functions vanishing at ∞, then the transition semigroup defined in Theorem 19 is Feller.

* **Problem 23.** For a pure-jump Markov family with bounded rates and countable state space Ψ, let $p_{xy}(t)$ denote the probability that the process with initial state x is at state y at time t, and for $y \neq x$, let $q_{xy} = q(x)\rho_x\{y\}$. [The numbers $p_{xy}(t)$ are the *transition probabilities from x to y* and q_{xy} is the *transition rate from x to y*.] Prove that

$$- p_{xy}(t)q(y) + \sum_{z \neq y} p_{xz}(t)q_{zy}$$

(31.12)

$$= \frac{dp_{xy}(t)}{dt} = -p_{xy}(t)q(x) + \sum_{z \neq x} q_{xz}p_{zy}(t)$$

for all x, y, and t.

Problem 24. Use the result in Problem 16 to find all equilibrium distributions for an arbitrary pure-jump Markov family with state space $\{0, 1\}$. Express your answer in terms of the transition rates.

In the case of a finite state space, Problem 23 gives us two systems of differential equations for the transition probabilities. When the state space is small enough, elementary methods can be used to solve either system of equations explicitly for these functions in terms of the transition rates (see Problem 25). Even in the case of a countably infinite state space, it may be possible to solve one or both of the two systems.

* **Problem 25.** Use Problem 23 to find an explicit formula for the transition semigroup of an arbitrary pure-jump Markov process with the two-point state space $\{0, 1\}$. Express your answer in terms of the transition rates. Then make an appropriate calculation to check for consistency between your answer to this problem and your answer to Problem 24.

31.6. Pure-jump Markov processes: unbounded rates

We now relax the assumption that the jump rate function q be bounded and replace it by the assumption that q is bounded on each compact subset of the state space Ψ, which we assume to be locally compact. In particular Lemma 1 of Chapter 29 implies that there exists an increasing sequence of compact sets A_n, $n = 1, 2, \ldots$, such that $A_n \nearrow \Psi$.

Under the assumptions just made, it should be clear that for each A_n, we can use either of the two constructions of the preceding section to define a process X that 'behaves like' a pure-jump Markov process with jump rate function q until the first time U_n that it jumps to a state in A_n^c. The increasing sequence (U_n) may or may not approach ∞ as $n \to \infty$. Whether it does or not, let U_∞ denote the (random) limit. If $U_\infty < \infty$, we say that an *explosion* occurs. A standard method for dealing with explosions is to adjoin a special state Δ to the state space Ψ, and then let

$$X_t = \Delta \quad \text{for} \quad t \geq U_\infty$$

if the initial state is different from Δ and $X_t = \Delta$ for all t if the initial state is Δ. Of course, one does not need to know that there is an explosion in order to adjoin Δ to Ψ. If there is no explosion, Δ is merely an 'extra state', with jump rate $q(\Delta) = 0$.

By definition the neighborhoods of Δ are taken to be those sets whose complements in $\Psi \cup \{\Delta\}$ are compact. What we have described is the one-point compactification of Ψ (see Appendix C). Our assumption that q is bounded on compact subsets of Ψ can be shown to lead to the conclusion that the construction described above leads to a $\mathbf{D}([0, \infty), \Psi \cup \{\Delta\})$-valued strong Markov family. The main difference between the bounded and unbounded case is that the domain of the infinitesimal generator no longer contains all bounded measurable functions, and e^{tG} cannot be used as a formula for the transition semigroup. Indeed, the issue of explicitly describing the domain of G is quite complex.

A process constructed as described above is called a *pure-jump Markov process*. Some may use the term to describe the process only up to the explosion time, thereby making the time domain of the random function into a random variable.

Example 1. [Pure-birth processes] Let $\Psi = \{0, 1, 2, \ldots\}$ and let $\rho_x = \delta_{x+1}$ for $x \in \Psi$. This choice of jump distributions means that when the process is in state x, its next jump is necessarily to state $x + 1$. For this reason, jumps are called 'births', and any pure-jump Markov process with such jump distributions is called a *pure-birth process*. The jump rates $(q(x), x \in \Psi)$, are called *birth rates*.

If the initial state is x for a pure-birth process, then it is easy to see from the construction that T_∞ is a sum of independent exponentially distributed random variables with parameters $q(x), q(x + 1), q(x + 2), \ldots$. Thus

$$(31.13) \qquad E(U_\infty) = \frac{1}{q(x)} + \frac{1}{q(x + 1)} + \frac{1}{q(x + 2)} + \cdots .$$

Calculation of the moment generating function of the $\overline{\mathbb{R}}^+$-valued random variable U_∞ (or alternatively the Three-Series Theorem) shows that $U_\infty < \infty$ with probability 1 or 0 according as $E(U_\infty)$ is finite or infinite. Therefore there is an explosion with probability 1 or 0 according as

$$\sum_{y=x}^{\infty} \frac{1}{q(y)} < \infty \quad \text{or} \quad = \infty .$$

Problem 26. [Birth-death processes] Let $\{q(x)\colon x \in \mathbb{Z}^+\}$ be jump rates, and let $(\rho_x \colon x \in \mathbb{Z}^+)$ be jump distributions on $\mathbb{Z}^+$ such that $\rho_0 = \delta_1$, and for each $x > 0$, ρ_x is supported by $\{x - 1, x + 1\}$. Any pure-jump Markov process constructed with such jump rates and jump distributions is called a *birth-death process*. The quantities

$$\beta_x = q(x)\rho_x\{x + 1\} \quad \text{and} \quad \delta_x = q(x)\rho_x\{x - 1\}$$

are called, respectively, the *birth rate* and *death rate* at x. Show that the probability of an explosion in a birth-death process is 0 for any initial state if the birth rates satisfy

$$\sum_{x=0}^{\infty} \frac{1}{\beta_x} = \infty.$$

Hint: Construct the birth-death process X jointly with a pure-birth process Y with the same birth rates in such a way that $X_t \le Y_t$ for all $t \in [0, \infty)$. That is, use a coupling argument.

Problem 27. Consider a birth-death process X with birth rates $\beta_x = x\beta$ and death rates $\delta_x = x\delta$, $x = 0, 1, 2, \ldots$, where β, δ are arbitrary nonnegative parameters. Show that

$$E(X_t \mid X_0) = X_0 e^{(\beta - \delta)t}.$$

Hint: Find a differential equation for the expected value on the left, as a function of t.

* **Problem 28.** Discuss how the issue of the domain of an infinitesimal generator is related to the issue of using Theorem 14 to obtain (31.12) for a pure-jump Markov process with rates that are not necessarily bounded.

* **Problem 29.** [Branching processes] For $x \in \mathbb{Z}^+$, define

$$q(x) = \gamma x$$

for some constant $\gamma > 0$. Let ρ be a probability measure on $\mathbb{Z}^+ \setminus \{1\}$, and for $x \in \mathbb{Z}^+$, define ρ_x by the formula

$$\rho_x(A) = \rho(A - x + 1).$$

Processes constructed with such jump rates and jump distributions are called *branching processes*. Show that if ρ has finite support, then the probability of an explosion is 0 for any initial state.

Problem 30. Show that a branching process with initial state x has the same distribution as the sum of x independent branching processes with initial state 1.

Problem 31. Let X be a branching process with initial state x, defined in terms of a measure ρ with finite support, as in Problem 29. Let V be the *extinction time*, defined by

$$V = \inf\{t \ge 0 \colon X_t = 0\}.$$

Find a formula for $P[V < \infty]$ in terms of ρ. *Hint:* One approach is to focus on the auxiliary Markov sequence Y used in the construction of X.

Problem 32. Suppose that the assumption that ρ has finite support in the preceding problem is replaced by the assumption that ρ is such that the probability of an explosion is 0. Does the solution of that problem remain valid? Does it remain valid if all assumptions on ρ are dropped?

31.7. ‡ Renewal theory for pure-jump Markov processes

For pure-jump Markov processes having bounded rates and countable state spaces, the concepts of 'irreducible' and 'accessible' carry over naturally from the Markov sequence setting. As we will see, 'periodicity' plays no role for the same reason that it plays no role for renewal processes. (See Theorem 8 of Chapter 30 for example.)

Proposition 20. *Let X be a pure-jump Markov process with bounded rates and initial state x. Then*

$$t \rightsquigarrow I_{[X_t=x]}$$

is a renewal process that corresponds to a subordinator with drift 1 and finite Lévy measure.

Problem 33. Prove the preceding proposition.

The next corollary now follows immediately from Theorem 8 of Chapter 30.

Corollary 21. *Let ν be the finite Lévy measure of the subordinator corresponding to a pure-jump Markov process X, as in* Proposition 20. *The set $\{t: X_t = x\}$ is bounded with probability 1 or 0 according as $\nu\{\infty\} > 0$ or $\nu\{\infty\} = 0$. Also*

$$\lim_{t\to\infty} P[X_t = x] = \frac{1}{1 + \int_{(0,\infty]} y\,\nu(dv)}.$$

In view of Proposition 20 and Corollary 21, we carry over the terms 'positive recurrent', 'null recurrent', and 'transient' from their use for renewal processes as in Chapter 30 to states and irreducible classes of pure-jump Markov processes with bounded rates. In the case of countable state spaces equilibrium distributions are related to the limits in Corollary 21 for various x in the same manner as for Markov sequences.

For an arbitrary specific pure-jump Markov process with bounded rates and nonrandom initial state x, one would like to identify the drift and Lévy measure of the subordinator corresponding to the renewal process of Proposition 20. Doing so would, in particular, enable us to decide whether x is positive recurrent, null recurrent, or transitive. The following result accomplishes this goal.

Proposition 22. *Fix $c > 0$ and let $\widetilde{X}$ be a Markov sequence with initial state x. Let X be the pure-jump Markov process with bounded rates constructed from c and $\widetilde{X}$ as in Theorem 19. Let $\widetilde{R}$ denote the waiting time distribution for the renewal sequence $n \rightsquigarrow I_{[\widetilde{X}_n = x]}$. Then the renewal process $t \rightsquigarrow I_{[X_t = x]}$ corresponds to a subordinator with drift 1 and Lévy measure ν given by*

$$\frac{d\nu}{d\lambda}(y) = ce^{-cy} \sum_{n=2}^{\infty} \frac{(cy)^{n-2}}{(n-2)!} \widetilde{R}\{n\} \quad \text{for } y \in (0, \infty)$$

and

$$\nu\{\infty\} = \widetilde{R}\{\infty\},$$

where λ denotes Lebesgue measure.

Problem 34. Prove the preceding proposition.

Problem 35. Use Proposition 22 to give a partial check of the calculations in Example 3 of Chapter 30.

* **Problem 36.** Apply Proposition 22 in case the state space is $\mathbb{Z}^+$ and the transition operator $\widetilde{T}$ for the Markov sequence $\widetilde{X}$ is given by

$$Tf(x) = \begin{cases} (1-b)\sum_{j=0}^{\infty} b^j f(j) & \text{if } x = 0 \\ f(x-1) & \text{if } x > 0 \end{cases}$$

for some $b \in [0, 1)$. Also find all equilibrium distributions, the jump-rate function, all transition probabilities, and all transition rates for the Markov process X.

Problem 37. The Markov process of the preceding problem has the property that whenever it leaves the state 0 it makes one visit to the state 1 before returning to 0 and whenever it leaves the state 1 it makes one visit to the state 0 before returning to 1. Reconcile this symmetry with the fact that the equilibrium distribution assigns different values to these two states.

Problem 38. For the setting of Proposition 22 prove that if $R\{\infty\} = 0$, then

$$1 + \int_{(0,\infty)} y\,\nu(dy) = \sum_{n=1}^{\infty} nR\{n\},$$

whether finite or infinite. Then deduce that the renewal sequence and the renewal process are both positive recurrent, both null recurrent, or both transient.

CHAPTER 32
Interacting Particle Systems

An 'interacting particle system' can be informally described as a Markov process consisting of countably many pure-jump processes that interact by modifying each other's transition rates. Each individual pure-jump process in such a system is located at a 'site' and has state space $\{0, 1, 2, \ldots, n\}$. The state of the pure-jump process at a given site is the number of 'particles' at that site, with n being the maximum particle number.

These systems have been used as models in a variety of practical applications, in such fields as physics, biology, and computer science. From a mathematical point of view, they form a rich class of Markov processes, capable of a wide variety of behaviors. They are the focus of much current research, and many fundamental questions about them remain to be answered. In this chapter, we will introduce them in a way that gives some idea about how they are constructed and how they behave, while avoiding many of the technicalities associated with the general theory.

32.1. Configuration spaces and infinitesimal generators

The state space of an interacting particle system is

$$\Xi = \{0, 1, \ldots, n\}^{\mathbb{Z}^d},$$

where d and n are positive integers. Thus, an element ξ in this space can be regarded as a collection $\xi = (\xi(x) \colon x \in \mathbb{Z}^d)$ of nonnegative integers, indexed by the d-dimensional integer lattice $\mathbb{Z}^d$. We have some special terminology for describing this state space and its members. A member of Ξ is a *configuration* of particles, and Ξ itself is *configuration space*. Configurations are typically denoted by the Greek letters ξ, η, ζ. The points in the integer lattice $\mathbb{Z}^d$ are called *sites*. Given a configuration $\xi = (\xi(x) \colon x \in \mathbb{Z}^d)$, the quantity $\xi(x)$ is the *particle number* at the site x. If $\xi(x) = 0$, we sometimes say that the site x is *vacant*, and if $\xi(x) = k \neq 0$, we say that x is *occupied* by k particles. The parameter n is the *maximum particle number*.

Since Ξ is a countable product of finite sets, it is a compact Polish space with the product topology. See Problem 2 to help gain an understanding of the notion of convergence in such a space. Interacting particle systems are random cadlag functions with values in configuration space. In other words, an interacting particle system is a $\mathbf{D}([0, \infty), \Xi)$-valued random variable X, whose state X_t at each time $t \in [0, \infty)$ is a configuration with particle numbers $X_t(x), x \in \mathbb{Z}^d$.

The interacting particle systems that we will study are allowed to make three types of transitions. These are: (i) *births*, in which the particle number at a site x increases by 1 (mod $n + 1$); (ii) *deaths*, in which the particle number at a site x decreases by 1 (mod $n + 1$); and (iii) *particle jumps*, in which a particle is transferred from one site x to another site y. When these transitions occur, we will say, respectively, that a *birth occurs* at x, a *death occurs* at x, and a *particle jumps from x to y*. One may view the occurrence of a particle jump from x to y as the simultaneous occurrence of a death at x and a birth at y.

Note that according to our definitions, if a birth occurs at a site occupied by n particles, then that site becomes vacant. This is an example of *wrap-around*. Similarly, wrap-around occurs if there is a death at a vacant site, thereby producing a site occupied by n particles. There are some interesting models for which this type of behavior is natural. In models where such transitions are not desirable, we will exclude them by setting certain transition rates equal to 0.

It will be useful to have some notation for describing the types of transitions defined above. Given a configuration ξ and a site x, let

$$\xi^x \quad \text{and} \quad {}_x\xi$$

denote the configurations obtained from ξ by, respectively, increasing or decreasing the particle number at x by 1 (mod $n + 1$). Given a configuration ξ and two sites x, y, we define

$$_x\xi^y \overset{\text{def}}{=} {}_x(\xi^y) = ({}_x\xi)^y .$$

Then ξ^x, ${}_x\xi$, and ${}_x\xi^y$ are the respective results of a birth at x, a death at x, and a particle jump from x to y, in the configuration ξ.

We will define interacting particle systems in terms of transition rates associated with the three types of transitions just introduced. The *birth* and *death* *rates* at a site x are respectively denoted by

$$b_x : \Xi \to \mathbb{R}^+ \quad \text{and} \quad d_x : \Xi \to \mathbb{R}^+ ,$$

and the particle jump rate from x to y is denoted by

$$j_{xy} : \Xi \to \mathbb{R}^+ .$$

Each of these rates is an $\mathbb{R}^+$-valued function on configuration space. Thus the birth and death rates at a site x and the particle jump rates for sites x and y may be influenced by the particle numbers at sites other than x and y. This is the 'interaction' in interacting particle systems.

Recall from Chapter 31 that an infinitesimal generator can be used to give an efficient description of a pure-jump Markov process. It should be apparent from the discussion up to this point that interacting particle systems are similar to pure-jump Markov processes, so it is natural to try to specify an infinitesimal generator for an interacting particle system. We will not in general be able to make its domain be the set of all bounded measurable functions, so our first task is to introduce a somewhat smaller set of functions on which to specify the infinitesimal generator.

For a finite set $A = \{x_1, \ldots, x_k\}$ in $\mathbb{Z}^d$, let $\mathfrak{F}_A$ denote the collection of all functions of the form

$$\xi \rightsquigarrow \varphi(\xi(x_1), \ldots, \xi(x_k)),$$

where φ is a function from $\{0, \ldots, n\}^k$ to $\mathbb{R}$. Because of the way in which they are defined, we say that the functions in $\mathfrak{F}_A$ *depend only on the sites in* A. Let

$$\mathfrak{F} = \bigcup_{\text{finite } A} \mathfrak{F}_A.$$

It is easy to check that the functions in $\mathfrak{F}$ are all bounded and measurable. In fact, they are continuous, and every continuous function from Ξ to $\mathbb{R}$ can be uniformly approximated by a member of $\mathfrak{F}$ (see Problem 3).

The infinitesimal generators of interacting particle systems will be defined in terms of birth, death, and particle jump rates. In order to avoid some of the technicalities involved in such a definition, we will restrict these rates in the following way. For $r \geq 0$, set

$$N(r) = \{z \in \mathbb{Z}^d : |z| \leq r\},$$

where $|z|$ denotes the usual Euclidean norm of z. We say that a collection of rates $b_x, d_x, j_{xy}, x, y \in \mathbb{Z}^d$ has *range* r if for all $x \in \mathbb{Z}^d$,

$$b_x, d_x, j_{xy} \in \mathfrak{F}_{x+N(r)} \text{ for } y \in \mathbb{Z}^d,$$

and

$$j_{xy} = 0 \text{ if } y \notin (x + N(r)) \setminus \{x\}.$$

A collection of rates with range r for some r has *finite range*.

We now use rates with finite range to define a class of operators that will be used as infinitesimal generators. We set

$$
\begin{aligned}
Gf(\xi) = &\sum_{x \in \mathbb{Z}^d} b_x(\xi)[f(\xi^x) - f(\xi)] \\
&+ \sum_{x \in \mathbb{Z}^d} d_x(\xi)[f(_x\xi) - f(\xi)] \\
&+ \sum_{\substack{x, y \in \mathbb{Z}^d \\ x \neq y}} j_{xy}(\xi)[f(_x\xi^y) - f(\xi)], \quad f \in \mathfrak{F}.
\end{aligned}
$$

(32.1)

Note that the sums in this expression have only finitely many nonzero terms for any given $f \in \mathfrak{F}$. When discussing G we will focus on how it acts on members of $\mathfrak{F}$, although we will eventually show under an additional assumption that it represents an infinitesimal generator of a strong Markov family, so that its domain may include functions that do not belong to $\mathfrak{F}$.

Here is the definition of 'interacting particle system' that is used in this book.

Definition 1. Let G be defined in terms of rates that have finite range, as in (32.1). A Markov process is an *interacting particle system* if its infinitesimal generator has the form (32.1) on $\mathfrak{F}$. A Markov family of such processes is a *family of interacting particle systems*.

* **Problem 1.** Suppose that there is a finite set A such that $b_x = d_x = j_{xy} = 0$ for $x \notin A$ and $j_{xy} = 0$ for $y \notin A$. Show that the formula for G in (32.1) is meaningful for all bounded measurable functions f and that G with this larger domain is the infinitesimal generator of the transition semigroup of a pure-jump Markov family with bounded rates, thereby proving the existence of an interacting particle system with infinitesimal generator G. Find the corresponding transition rates $q_{\xi\eta}$, $\xi, \eta \in \Xi$, in terms of b_x, d_x, and j_{xy}.

Problem 2. Show that a sequence $(\xi_k : k \geq 0)$ of configurations converges in Ξ to a configuration ξ if and only if for each site x, there exists an integer l such that $\xi_k(x) = \xi(x)$ for all $k \geq l$.

Problem 3. Show that the functions in $\mathfrak{F}$ are continuous, and that every continuous function from Ξ to $\mathbb{R}$ is the uniform limit of a sequence of functions in $\mathfrak{F}$.

Problem 4. Let X be an interacting particle system with infinitesimal generator G. Show that the random cadlag function Y defined by $Y_t = X_{ct}$, $t \in [0, \infty)$, is an interacting particle system with infinitesimal generator cG.

32.2. The universal coupling

In this section, we make the further assumption that all rates are bounded above by some finite constant c. In this case we say that G has *bounded rates*.

We will describe a procedure for constructing all interacting particle systems on $\mathbb{Z}^d$ having bounded finite-range rates on a common probability space, thereby 'coupling' all these systems. Briefly, this *universal coupling* involves three steps. In the first step, we define an independent collection of Poisson point processes and a corresponding probability space; this step is 'universal', in the sense that it does not depend on the rates of the interacting particle system being constructed. At the end of the step, an existence and uniqueness result is stated that specifies the precise connection between the Poisson point processes and any interacting particle system with a given set of rates. The second and third steps of the construction constitute the proof of this result.

In the second step we define certain random objects that, in some sense, represent 'regions of interaction' between various sites. This step depends on the rates of the system being constructed, but not on its initial state. In the third step, we show how to use the random objects from the second step together with the initial state to calculate the state of the system at any given time.

First step of the construction. Let

$$\{B^x, D^x : x \in \mathbb{Z}^d\} \cup \{J^{xy} : x, y \in \mathbb{Z}^d, x \neq y\}$$

be an iid collection of Poisson point processes on $(0, \infty) \times (0, \infty)$, with intensity measure λ_2, where λ_2 is two-dimensional Lebesgue measure, and denote by $(\Omega, \mathcal{F}, P)$ the probability space on which these point processes are defined. The points in these point processes will be used to indicate times at which births, deaths, and particle jumps might possibly occur.

Problem 5. Prove that with probability 1, no first coordinates of any of the random points in any of these point processes are equal.

As we continue with the first step of the construction, we note that we will always assume that we are working in the event of probability 1 identified in the preceding problem. (As we will see, it is the first coordinates that will actually be times at which things happen, so there will never be two simultaneous happenings.)

We next define some sub-σ-fields of $\mathcal{F}$: for $t \in [0, \infty)$, let

$$\mathcal{F}_t =$$

$$\sigma(B^x \cap (0, t] \times (0, \infty), D^x \cap (0, t] \times (0, \infty), J^{xy} \cap (0, t] \times (0, \infty) : x, y \in \mathbb{Z}^d, x \neq y).$$

We will construct processes that are Markov with respect to the filtration $(\mathcal{F}_t : t \in [0, \infty))$.

Remark 1. It will be clear in the construction that if the rates are bounded above by c, then the points of the Poisson processes whose second coordinates are larger than c will not play a role in the construction of the corresponding family of interacting particle systems.

Theorem 2. *Let $(\Omega, \mathcal{F}, P)$ be the probability space for the Poisson point processes described above. For each set of rates b_x, d_x, j_{xy}, $x, y \in \mathbb{Z}^d, x \neq y$, and each initial state $\xi \in \Xi$, there exists an a.s.-unique $D([0, \infty), \Xi)$-valued random variable X satisfying the following four rules for all $x, y \in \mathbb{Z}^d$ and $t \in (0, \infty)$:*

(i) Initial state rule. $X_0 = \xi$;

(ii) Birth rule. X has a birth at x at time t if and only if there exists $u \in (0, \infty)$ such that $(t, u) \in B^x$ and $b_x(X_{t-}) \geq u$;

(iii) Death rule. X has a death at x at time t if and only if there exists $u \in (0, \infty)$ such that $(t, u) \in D^x$ and $d_x(X_{t-}) \geq u$;

(iv) Particle jump rule. *X has a particle jump from x to y at time t if and only if there exists $u \in (0, \infty)$ such that $(t, u) \in J^{xy}$ and $j_{xy}(X_{t-}) \geq u$;*

Moreover, *X is a strong Markov process with respect to the filtration $(\mathcal{F}_t : t \in [0, \infty))$ defined above, whose infinitesimal generator is given by 32.1 on $\mathfrak{F}$.*

PROOF. As stated above, the bulk of the proof consists of the second and third steps in the construction.

Second step of the construction. This step depends on the Poisson point processes introduced in the preceding step, the bound c, and the range r. For each x we create a point process $\mathfrak{T}_x = \{T_{x,1} < T_{x,2} < \ldots\}$ on $\mathbb{R}^+ \setminus \{0\}$ via

$$T_{x,n} = \inf\{t > T_{x,n-1} : (t, v) \in B^x \cup D^x \cup \bigcup_{\substack{y \in x + N(r) \\ y \neq x}} (J^{xy} \cup J^{yx}) \text{ for some } v \leq c\},$$

where $T_{x,0}$ is defined to equal 0. Since projections and independent unions of Poisson point processes are themselves Poisson point processes, provided the projections of the intensity measures are Radon measures, it is clear that the point processes $\mathfrak{T}_x$ are identically distributed Poisson point processes with common intensity measure equal to $2c\sharp N(r)$ times 1-dimensional Lebesgue measure. The rules given in the statement of the theorem indicate that we want X to have the property that for each x, $t \rightsquigarrow X_t(x)$ is constant on intervals of the form $[T_{x,i-1}, T_{x,i})$. Thus, the random times in $\mathfrak{T}_x$ are the only times at which we allow possible births, deaths, or particle jumps that involve the site x. We will need to wait until the third step of the construction to see just how it is determined whether such transitions actually occur at these times.

Now fix x and a positive integer i. We will inductively define a sequence of random times $U_1 \geq U_2 \geq \ldots$ and a sequence of random finite sets $A_1 \subseteq A_2 \subseteq \ldots$ associated with the pair (x, i). Let $U_1 = T_{x,i}$. If $(U_1, v) \in J_{yx}$ for some v and y, let $A_1 = y + N(r)$. Otherwise let $A_1 = x + N(r)$. Proceeding recursively and using the convention $\sup \emptyset = 0$, we set

$$U_{j+1} = \sup\{t < U_j : t \in \mathfrak{T}_z \text{ for some } z \in A_j\}$$

if $U_j > 0$ and $= 0$ if $U_j = 0$, and we set

$$A_{j+1} = A_j \cup \bigcup_{z : U_{j+1} \in \mathfrak{T}_z} (z + N(r))$$

if $U_{j+1} > 0$ and $= A_j$ if $U_{j+1} = 0$. (Notice that $U_{j+1} \in \mathfrak{T}_z$ for either one or two values of z in case $U_{j+1} > 0$.) Having defined the random times $U_j, j = 1, 2, \ldots$, let

$$K(x, i) = \sup\{j : U_j > 0\}.$$

We will show in this step that $K(x, i)$ is a.s.-finite.

The times U_j and sets A_j are illustrated in Figure 32.1. The dots connected by curved arrows represent the locations and times of possible particle jumps; an arrow from (w, s) to (z, s) indicates that $s \in J^{wz}$. The other dots represent the locations and times of possible births or deaths. The value $K = 7$ refers to the fact that $U_7 > 0$ but $U_8 = 0$.

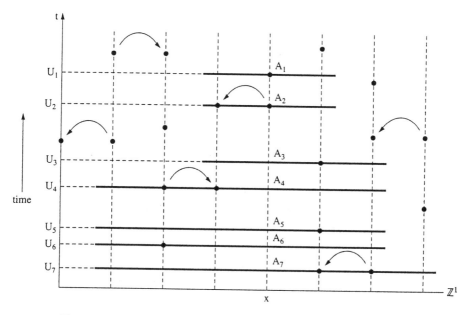

FIGURE 32.1. Second step of construction for $r = d = 1$, with $K = 7$

We want to prove that with probability one, $U_j = 0$ for some j, thereby showing that $K(x, i)$ is a.s.-finite. Let us calculate a conditional moment generating function:

$$E\big(e^{-w(U_1 - U_j)} \mid U_j \neq 0; A_1, \ldots, A_j\big)$$
$$= \prod_{m=2}^{j} E\big(e^{-w(U_{m-1} - U_m)} \mid U_j \neq 0; A_1, \ldots, A_j\big)$$
$$\leq \prod_{m=2}^{j} \frac{2cm[\sharp N(r)]^2}{w + 2cm[\sharp N(r)]^2},$$

where the last step uses $\sharp A_j < j \sharp N(r)$ and the fact that each Poisson point process $\mathcal{T}_x$ has intensity $2c \sharp N(r)$ times Lebesgue measure. For $w > 0$, this product approaches 0 as $j \to \infty$. It follows that, conditioned on $U_j \neq 0$, $U_1 - U_j \to \infty$ in distribution. By the monotonicity of $j \rightsquigarrow U_j$, we see that for almost every $\omega \in \bigcap_{j=2}^{\infty}[U_j \neq 0]$, $U_1(\omega) - U_j(\omega) \to \infty$ as $j \to \infty$. Since $U_1(\omega) - U_j(\omega) \leq U_1(\omega) < \infty$, we see that $\bigcap_{j=2}^{\infty}\{\omega : U_j(\omega) \neq 0\}$ is a null event, as desired.

We have shown that $K(x, i)$ is a.s.-finite. Since there are only countably many ordered pairs (x, i), the random function K, defined by $(x, i) \rightsquigarrow K(x, i)$, is almost surely $\mathbb{Z}^+ \setminus \{0\}$-valued.

Third step of the construction. This step will depend on the initial state ξ. For each site x and each time $T_{x,i} \in \mathfrak{T}_x$, we need to decide whether or not a transition involving x occurs at that time, and if so, whether it is a birth, death, or particle jump.

We will define $X_t(x)$ for $t \in [T_{x,i-1}, T_{x,i})$ by induction on the value of $K(x,i)$. To get the induction started, we let $K(x,0) = 0$ and define $X_t(x) = \xi(x)$ for all sites x and all times $t \in [0, T_{x,1})$, consistent with the four rules in the statement of the theorem.

Now assume inductively that $X_t(x)$ has been defined for $t \in [T_{x,i-1}, T_{x,i})$ for all pairs (x, i) such that $K(x, i) < k$, where k is some positive integer. Now fix x and i such that $K(x, i) = k$. Let A_1 be the random set corresponding to x and i defined in the second step of the construction. It follows from that step of the construction that $K(z, j) < k$ for each $z \in A_1$ and each integer $j \geq 0$ such that $T_{z,j} < T_{x,i}$. By the inductive hypothesis, the quantities $X_{T_{x,i}-}(z)$ are defined for $z \in A_1$. Now define $X_{T_{x,i}}(x)$ in terms of these quantities as follows:

- If $(T_{x,i}, v) \in B_x$ for some (no more than one) v, then we introduce a birth at time $T_{x,i}$ and site x if $v \leq b_x(X_{T_{x,i}-})$, and make no change at that site and time otherwise; that is, $X_{T_{x,i}}(x) = 1 + X_{T_{x,i}-}(x) \pmod{n+1}$ if $v \leq b_x(X_{T_{x,i}-})$, and $X_{T_{x,i}}(x) = X_{T_{x,i}-}(x)$ otherwise.
- If $(T_{x,i}, v) \in D_x$ for some v, then we introduce a death at time $T_{x,i}$ and site x if $v \leq d_x(X_{T_{x,i}-})$, and make no change at that site and time otherwise.
- If $(T_{x,i}, v) \in J_{xy}$ for some v and y, then we subtract 1 $\pmod{n+1}$ from the particle number at site x at time $T_{x,i}$ if $v \leq j_{xy}(X_{T_{x,i}-})$, and make no change at that site and time otherwise.
- If $(T_{x,i}, v) \in J_{yx}$ for some v, then we add 1 $\pmod{n+1}$ to the particle number at site x at time $T_{x,i}$ if $v \leq j_{yx}(X_{T_{x,i}-})$, and make no change at that site and time otherwise.

For $t \in [T_{x,i}, T_{x,i+1})$, let $X_t(x) = X_{T_{x,i}}(x)$. It is easy to see from the definitions that, for all sites x, $T_{x,i} \to \infty$ a.s. as $i \to \infty$, so we have defined $X_t(x)$ for all sites x and times $t \in [0, \infty)$.

Our construction of X is complete. Note that this construction has the property that if X_s is known for some $s \in [0, \infty)$, then for any time $t \geq s$, X_t can be determined by treating s as if it were time 0 and X_s as if it were an initial state, and then following the construction procedure using the portions of the Poisson point processes B^x, D^x, J^{xy} that lie in $(s, \infty) \times (0, \infty)$, suitably shifted in the negative first-coordinate direction by s units. By the independence properties of Poisson point processes, these shifted point processes are independent of the σ-field $\mathcal{F}_s$. It follows that the conditional distribution given $\mathcal{F}_s$ of the stochastic process $(X_t : t \geq s)$ depends only on X_s. Thus, X is Markov. A similar argument, in which the time s is replaced by a stopping time, can be used to show that X is strong Markov. Alternatively, one can show that X is strong Markov by showing that it is Feller, as the reader is requested to do in Problem 7. $\quad\square$

An important feature of the construction just given is that for a fixed site x and time t, the value of $X_t(x)$ can be determined from finitely many of the values $\xi(z)$ in the initial state ξ and the intersections of finitely many of the Poisson point processes with some compact subset of $(0, \infty) \times (0, \infty)$. The random sets $A_1, A_2, \ldots$, are used to indicate exactly which of the values $\xi(z)$ and what parts of the Poisson point processes are needed. One consequence of this observation is that for any site x and time t, there exists a random set A such that the processes with initial states ξ and η agree at x at time t if $\xi(y)$ and $\eta(y)$ agree at sites $y \in A$. This consequence is relevant for proving that the processes we have constructed are Feller (Problem 7).

For the purposes of understanding the behavior of interacting particle systems, it can be often useful to consider the rules in Theorem 2 to constitute the definition of X. The following example illustrates both the construction and Theorem 2.

Example 1. Let the dimension d and maximum particle number n both be 1. Define rates with range 1 as follows:

$$b_x(\xi) = \frac{1}{2}(1 - \xi(x))[\xi(x-1) + \xi(x+1)], \quad d_x(\xi) = \frac{1}{5}\xi(x), \quad j_{xy} = 0.$$

Thus, the particle jump rates are all 0, and the death rate at an occupied site is $1/5$. There are no deaths at vacant sites or births at occupied sites. The only real 'interaction' between different sites involves the birth rates at vacant sites, which can be 0, .5, or 1, depending on how many of the neighbors are occupied. We take as the initial state the configuration ξ in which only the sites -1 and 1 are occupied; that is, $\xi(1) = \xi(-1) = 1$ and $\xi(x) = 0$ if $x \neq \pm 1$.

Figure 32.2 illustrates the time evolution of X constructed as described in this section. The axes are oriented the same as in Figure 32.1. The open circles indicate possible birth times, and the black dots indicate possible death times, with the upper bound 1. We have not bothered to show the possible particle jump times, since they are irrelevant when the particle jump rates are 0. The numbers next to the possible birth and death times are the values of the second coordinate of the corresponding points in the relevant Poisson point processes. The vertical axes are thickened to indicate occupied sites. By following the rules in Theorem 2, the reader should be able to verify the correctness of the figure.

Problem 6. Using Figure 32.2, calculate the time evolution of the interacting particle system with the same rates as in Example 1, but with the initial state in which only 0 is occupied.

* **Problem 7.** Prove that the transition semigroups of the processes constructed in this section with bounded finite-range transition rates are Feller, and hence that the collection of processes constructed from given rates and various initial states is a strong Markov family.

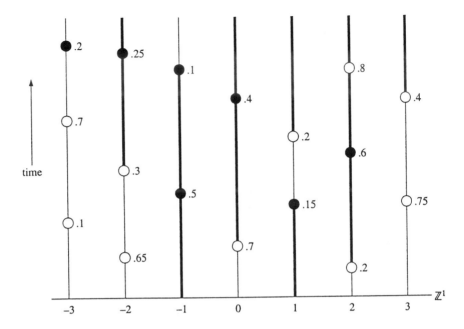

FIGURE 32.2. Time evolution of an interacting particle system

Problem 8. Show that the infinitesimal generator of a Markov family constructed as in this section with bounded finite-range transition rates agrees with G as defined in (32.1) for $f \in \mathfrak{F}$. *Hint:* One approach is to use the definition of the infinitesimal generator in terms of the transition semigroup.

Remark 2. It can be shown that 'two' Markov families of distributions each having an infinitesimal generator that agrees with some common G of the form (32.1) for $f \in \mathfrak{F}$ are identical.

Problem 9. Fix the dimension d, the range r, the maximum particle number n, and the upper bound on the rates c, and within this context, let G and $(G^{(k)} : k = 1, 2, \ldots)$ be infinitesimal generators of the form (32.1). Let X and $X^{(k)}$, $k = 1, 2, \ldots$, be corresponding interacting particle systems, all constructed by way of the universal construction, with respective initial states ξ and ξ_k, $k = 1, 2, \ldots$. Show that if $\xi_k \to \xi$ as $k \to \infty$ and for each $f \in \mathfrak{F}$, there is a $j(f)$ such that $G^{(k)}f = Gf$ for all $k > j(f)$, then with probability 1, $X_t^{(k)} \to X_t$ as $k \to \infty$, uniformly for t in bounded subsets of $[0, \infty)$.

* **Problem 10.** Weaken the hypothesis in the last sentence of Problem 9 so an 'if and only if' result is obtained.

32.3. Examples

In each of the following examples, we will give the dimension d, the maximum particle number n, the range r, and the rates, along with a brief heuristic description of the behavior of the corresponding interacting particle systems. The reader may find Theorem 2 useful in verifying that, in each case, the description matches the rates.

Example 2. [Contact process with linear birth rate] We take d and r to be arbitrary positive integers, and $n = 1$. The particle jump rates are 0, the death rates are given by

$$d_x(\xi) = \delta\xi(x)$$

for some parameter $\delta > 0$, and the birth rates are given by

$$b_x(\xi) = \frac{(1 - \xi(x))}{(\sharp N(r) - 1)} \Big(\sum_{\substack{y \in x + N(r) \\ y \neq x}} \xi(y) \Big).$$

This example generalizes Example 1. The birth rate at a vacant site is proportional to the number of occupied neighbors, with the maximum birth rate being 1. The death rate at an occupied site is the constant δ. There is no 'wrap-around' (that is, births do not occur at sites containing $n = 1$ particles, and deaths do not occur at vacant sites). The *state of extinction*, which is the configuration in which all sites are vacant, denoted by $\bar{0}$, is absorbing, since all of the rates are 0 in that state.

One imagines a population of simple organisms that do not move about, but which have offspring at neighboring vacant sites. The organisms die at a constant rate. If all of the neighbors of an occupied site are vacant, then the organism at that site has a total propagation rate of 1, and in general, the total propagation rate of an organism is proportional to the number of vacant neighboring sites. Thus, the average birth rate per capita is smaller in 'crowded' populations than in 'sparse' ones. Since it only takes a single occupied site to cause births at other sites, we sometimes say that this process exhibits 'asexual reproduction'. See Problem 12 below for a variation with 'sexual reproduction'.

An interesting question is whether a contact process starting with a single occupied site can *survive*, or in other words, avoid the state of extinction, with positive probability. It is not hard to show (see Problem 13 below) that the answer is 'no' if δ is large enough. It has been shown, using an argument that we omit, that the answer is 'yes' for sufficiently small $\delta > 0$.

Example 3. [Particle jump process with exclusion] The quantities d and r are arbitrary positive integers, and $n = 1$. The birth and death rates are all 0. The particle jump rates are defined in terms of a probability measure ρ on $N(r)$:

$$j_{xy}(\xi) = \xi(x)(1 - \xi(y))\rho\{y - x\}.$$

Since j_{xy} can only be positive if x is occupied and y is vacant, there is no wrap-around.

One imagines a system of particles, each of which attempts at rate 1 to take a step with step distribution ρ. If the step would take the particle to a vacant site, the particle jumps to that site. If it would take the particle to an occupied site, the particle stays where it is. The word 'exclusion' in the name 'particle jump process with exclusion' refers to this last part of the description. We will see in the next section that exclusion processes have very simple equilibrium distributions.

Example 4. [Particle jump process with wrap-around] We make two changes from the preceding example: we let n be an arbitrary positive integer, and define the particle jump rates as

$$j_{xy}(\xi) = \xi(x)\rho\{y - x\}.$$

Thus there is no exclusion, and when a particle jumps to a site containing n particles, that site becomes vacant, due to wrap-around. Since j_{xy} equals the number of particles at x, one may take the point of view that each particle at x is independently jumping at rate 1, with step-size distribution ρ.

It can be shown that the limit as $n \to \infty$ of this model exists (in the same sense as in Problem 9), provided the initial state is not too 'wild'. We call this limit a system of *independent random walks*, since it can be described by saying that each particle performs an independent random walk with step-size distribution ρ, the steps being taken at rate 1.

Example 5. [Cyclic threshold process] The dimension d, the maximal particle number n, and the range r are all arbitrary positive integers. The death and particle jump rates equal 0. The birth rates are defined in terms of a parameter $\theta \in [0, \sharp N(r) - 1]$:

$$b_x(\xi) = \begin{cases} 1 & \text{if } \sharp\{y \in x + N(r) \colon \xi(y) = \xi(x) + 1 \ (\text{mod } n+1)\} \geq \theta \\ 0 & \text{otherwise}. \end{cases}$$

The word 'cyclic' refers to the way in which the number of particles at a site cycle through the integers $0, \ldots, n$, with wrap-around. The parameter θ is the 'threshold'. Sometimes these models are called 'food-chain' models, particularly when $\theta = 1$; one imagines that the quantity $\xi(x)$ represents a species type at x rather than a particle number, and that individuals of species k are 'eaten' by individuals of species $k + 1$ (mod $n + 1$) in their neighborhood. There are many interesting open questions about the behavior of such models, particularly concerning the dependence of that behavior on n, r, θ, and also on the 'shape' of the neighborhood $N(r)$. (It is only for simplicity that we have made $N(r)$ spherical.)

Example 6. [Majority vote process] The quantities d and r are arbitrary positive integers, and $n = 1$. The particle jump rates are all 0. Let $A \subseteq N(r)$ be a set containing an odd number of sites, and define birth rates by

$$b_x(\xi) = [1 - \xi(x)][\varepsilon \vee \mathrm{maj}_x(\xi)] \quad \text{and} \quad d_x(\xi) = \xi(x)[\varepsilon \vee (1 - \mathrm{maj}_x(\xi))],$$

where ε is a nonnegative parameter and

$$\mathrm{maj}_x(\xi) = \begin{cases} 1 & \text{if } \sum_{y \in A+x} \xi(y) > \frac{1}{2} \sharp A \\ 0 & \text{otherwise}. \end{cases}$$

Note that the rates do not allow wrap-around.

To understand this model, first consider the case $\varepsilon = 0$. Imagine that the particle number at a site x, which can be either 0 or 1, represents the 'opinion' of an individual at x. The birth and death rates only allow the individual at x to change opinion when the majority of the individuals at sites in $A + x$ hold the opposing opinion, in which case the rate of change is 1. The set A is usually chosen to contain the origin, so that the opinion of the individual at x counts in the majority vote. Often A is chosen to equal $N(r)$.

When $\varepsilon > 0$, individuals can change their opinion, even when they are currently in the majority. This 'noise' perturbs the system away from its tendency towards unanimity. Thus, even if the initial state is $\bar{0} = $ all 0's or $\bar{1} = $ all 1's, the noise will cause infinitely many individuals to change their opinion before any time $t > 0$ (see Problem 14 below).

An interesting question is whether or not a system which starts with unanimity will stay 'close' to unanimity for all times t if $\varepsilon > 0$ is sufficiently small. More precisely, if the initial state ξ is $\bar{0}$ or $\bar{1}$, is it true that for all $p > 0$ there exists an $\varepsilon > 0$ such that

$$|E(X_t^\xi(x)) - \xi(x)| < p$$

for all $t \in [0, \infty)$? It is known that the answer is 'no' for $d = r = 1$ and any choice of $A \subseteq N(r)$, and that the answer is 'yes' for $d \geq 2$, r arbitrary, and any choice of A containing the origin such that $A \setminus \{0\}$ lies strictly inside a half-space whose boundary contains the origin. For example, take $d = 2$, $r = 1$, and $A = (0,0), (0,1), (1,0)$. The answer is unknown for most other cases. In particular, the answer is unknown for $A = N(r)$ if $d \geq 2$ and $r \geq 1$.

Problem 11. [Contact process with threshold birth rates] Change the birth rates in Example 2 so that they only take two different values: the birth rate at x is 1 if x is vacant and at least one site in $x + N(r)$ is occupied; the birth rate at x is 0 otherwise. Let $(X^\xi : \xi \in \Xi)$ be the family of interacting particle systems with these rates, and let $(Y^\xi : \xi \in \Xi)$ be the family of interacting particle systems whose rates are given in Example 2. Show that for all $\xi \in \Xi$, $x \in \mathbb{Z}^d$, and $t \in [0, \infty)$,

$$P[X_t^\xi(x) = 1] \geq P[Y_t^\xi(x) = 1].$$

Hint: Use the universal coupling to directly compare the two processes.

Problem 12. [Contact process with sexual reproduction] Modify the rates in the preceding problem so that the birth rate at x is 1 if x is vacant and at least two sites in $x + N(r)$ are occupied, and 0 otherwise. Show that if $r = 1$ and the initial state contains only finitely many occupied sites, then the hitting time of the state $\bar{0}$ is finite a.s. *Hint:* First show that if B is a d-dimensional box with sides parallel to the coordinate axes and B contains all of the sites that are occupied in the initial state, then no site outside of B can ever become occupied. Then use the Borel-Cantelli Lemma to show that there must exist a time interval $[k, k+1]$ during which deaths occur at all of the occupied sites in B and no births occur at any of the vacant sites in B.

* **Problem 13.** Show that for any of the three types of contact processes defined in this section (Example 2, Problem 11, Problem 12), if $\delta > 1$, then the hitting time of the state $\bar{0}$ is finite a.s. for any initial state with only finitely many occupied sites. *Hint:* One way to do this is to use appropriately chosen martingales involving the infinitesimal generator G.

Problem 14. Let G be an infinitesimal generator with bounded finite-range rates. Show that if the birth and death rates are all bounded below by a constant $\varepsilon > 0$, then with probability 1, infinitely many births and deaths will occur during any time interval with positive length.

Problem 15. Show that any family of interacting particle systems with $r = 0$ is an independent family of pure-jump Markov processes. Also show that in this case, the method of the previous section can be used to construct interacting particle systems even in the case that the rates are not all bounded above by some constant.

* **Problem 16.** [An example due to Blackwell] Let X be the interacting particle system with $d = n = 1$, $r = 0$, particle jump rates equal to 0, initial state $\bar{0}$, and birth and death rates given by

$$b_x(\xi) = (1 - \xi(x)) \quad \text{and} \quad d_x(\xi) = \xi(x) 2^{|x|}.$$

(See Problem 15.) Show that for all $t \in [0, \infty)$,

$$\sum_{x \in \mathbb{Z}} X_t(x) < \infty \text{ a.s. },$$

but that for all $t \in (0, \infty)$,

$$P\left[\exists s \in (0, t) : \sum_{x \in \mathbb{Z}} X_s(x) = \infty\right] = 1.$$

Problem 17. [Annihilating random walks] Explain why the wrap-around particle jump model with $n = 1$ could be called a system of 'annihilating random walks'.

* **Problem 18.** [Coalescing random walks] Define the birth, death, and particle jump rates for a family of interacting particle systems with $n = 1$ that matches the following heuristic description. Let ρ be a probability measure on $N(r) \setminus \{0\}$. Particles jump at rate 1 with step-size distribution ρ. If a particle jumps to a site y that is already occupied, the two particles coalesce into a single particle that occupies y.

32.4. Equilibrium distributions

Most of the research in interacting particle systems is related in some way to the following two questions concerning a given transition semigroup $(T_t : t \in [0, \infty))$: (i) What are the equilibrium distributions of the transition semigroup? and (ii) If μ is such an equilibrium distribution, for what distributions ν on the state space is it the case that $\nu T_t \to \mu$ as $t \to \infty$? In this section, we give some basic results that can be useful in investigating the first question. In the next section, we will have something to say about the second question for a special class of systems.

First we note that under our assumption that the maximum particle number n is finite, there always exists at least one equilibrium distribution for the transition semigroup of any interacting particle system. This fact follows from the compactness of the state space Ξ, just as in Problem 48 of Chapter 26.

Next, we wish to give a criterion for determining whether a probability measure ν is an equilibrium distribution. This criterion is analogous to the one given in Problem 16 of Chapter 31.

Theorem 3. *Let G be defined as in (32.1) with bounded finite-range rates. Then a probability measure μ on the state space Ξ is an equilibrium distribution of the transition semigroup with infinitesimal generator G if and only if*

$$(32.2) \qquad \int Gf \, d\mu = 0$$

for all $f \in \mathfrak{F}$.

PROOF. The 'only if' part of the theorem is a trivial consequence of the definition of infinitesimal generator and the Bounded Convergence Theorem. We now prove the 'if' part.

For finite $A \subseteq \mathbb{Z}^d$, let G_A be the infinitesimal generator with birth, death, and jump rates $b_{x,A}, d_{x,A}, j_{xy,A}$ given by

$$b_{x,A}(\xi) = \int \left(b_x(\eta) + \sum_{z \in A^c} j_{zx}(\xi) \right) \mu[d\eta \mid \eta(x) = \xi(x), x \in A]$$

$$d_{x,A}(\xi) = \int \left(d_x(\eta) + \sum_{z \in A^c} j_{xz}(\xi) \right) \mu[d\eta \mid \eta(x) = \xi(x), x \in A]$$

$$j_{xy,A}(\xi) = \int j_{xy}(\eta) \, \mu[d\eta \mid \eta(x) = \xi(x), x \in A]$$

for $x, y \in A$. Let the rates equal 0 for x or y in A^c.

By Problem 1, G_A is the infinitesimal generator of a pure-jump Markov process. It is straightforward to check from the definition of G_A that

$$(32.3) \qquad \int G_A f(\eta) \, (\mu_A \times \nu_A)(d\eta) = 0$$

for all $f \in \mathfrak{F}_A$, where μ_A is the marginal of μ on $\{0, \ldots, n\}^A$ and ν is any probability measure on $\{0, \ldots, n\}^{A^c}$. Since every function in $\mathfrak{F}$ can be written as a finite linear combination of products of functions in $\mathfrak{F}_A$ and functions that depend only on sites in A^c, it follows that (32.3) holds for all $f \in \mathfrak{F}$. Thus, $\mu_A \times \nu_A$ is an equilibrium distribution for the transition semigroup associated with G_A.

Clearly, no matter how the measures ν_A are chosen,

$$\lim_{A \nearrow \mathbb{Z}^d, A \text{ finite}} \mu_A \times \nu_A = \mu.$$

Since G_A agrees with G on $\mathfrak{F}_A$ for each finite A, it follows from this fact and Problem 9 that the interacting particle system with infinitesimal generator G_A and initial distribution $\mu_A \times \nu_A$ converges in distribution as $A \nearrow \mathbb{Z}^d$ to the interacting particle system with infinitesimal generator G and initial distribution μ. Since $\mu_A \times \nu_A$ is an equilibrium distribution for the transition semigroup associated with G_A, it follows that μ is an equilibrium distribution for the transition semigroup associated with G. $\square$

Example 7. Let G be the infinitesimal generator of a particle jump process with exclusion, and let μ be *Bernoulli product measure* on Ξ with parameter $p \in [0, 1]$, which is determined by the following:

$$\mu\{\xi \colon \xi(x) = 1, x \in A\} = p^{\sharp A} \quad \text{for all finite } A \subseteq \mathbb{Z}^d.$$

We will check that μ satisfies the criterion of the preceding theorem.

Since $n = 1$, it can be shown with standard arguments that every function in $\mathfrak{F}$ is a finite linear combination of the functions f_A, A a finite subset of $\mathbb{Z}^d$, where

$$f_A(\xi) = \prod_{x \in A} \xi(x).$$

Thus, it is enough to check (32.2) for $f = f_A$. Straightforward calculations give

$$\int G f_A \, d\mu = p^{\sharp A} (1 - p) \left(\sum_{x \in A, y \notin A} (\rho\{x - y\} - \rho\{y - x\}) \right),$$

where ρ is the step distribution used in the definition of the particle jump process with exclusion. We leave it to the reader to check that the sum in this expression is 0.

32.5. Systems with attractive infinitesimal generators

In the study of Markov sequences with countable state spaces, we found that an irreducible transition operator T has a unique equilibrium distribution μ if and only if T is positive recurrent, in which case any Markov sequence X with transition operator T has the property that X_n converges in distribution to μ as $n \to \infty$. For interacting particle systems, the story is not so simple. However, there is one class of systems for which we can use the theory developed in this chapter to make some useful observations. One characteristic of systems in this class is a lack of wrap-around behavior. Also for simplicity, we will restrict our attention to systems for which the particle jump rates are all 0.

For $\xi, \eta \in \Xi$, we write $\xi \geq \eta$ provided that $\xi(x) \geq \eta(x)$ for all $x \in \mathbb{Z}^d$. A function $f \colon \Xi \to \mathbb{R}$ is *increasing* if $f(\xi) \geq f(\eta)$ whenever $\xi \geq \eta$. A function f is *decreasing* if $-f$ is increasing.

We now define a property of birth and death rates which will ensure that any transition semigroup defined in terms of such rates (with particle jump rates equal to 0) will consist of transition operators that take increasing functions to increasing functions.

Definition 4. An infinitesimal generator G defined as in (32.1) with particle jump rates equal to 0 is *attractive* if for all $x \in \mathbb{Z}^d$,

    (i) there is no wrap-around at x, that is, $b_x(\xi) = 0$ if $\xi(x)$ equals the maximum particle number and $d_x(\xi) = 0$ if $\xi(x) = 0$;

    (ii) if $\xi \geq \eta$ and $\xi(x) = \eta(x)$, then $b_x(\xi) \geq b_x(\eta)$;

    (iii) if $\xi \geq \eta$ and $\xi(x) = \eta(x)$, then $d_x(\xi) \leq d_x(\eta)$.

Conditions (ii) and (iii) in Definition 4 do not say that the birth rates are increasing and the death rates are decreasing. In fact, because of (i), increasing attractive birth rates and decreasing attractive death rates are necessarily equal to 0.

Theorem 5. *Let G be an attractive infinitesimal generator having bounded finite-range rates, and let $(T_t \colon t \in [0, \infty))$ denote the corresponding transition semigroup. If $f \colon \Xi \to \mathbb{R}$ is a bounded measurable increasing function, then $T_t f$ is an increasing function for all $t \in [0, \infty)$.*

PROOF. Fix $\xi, \eta \in \Xi$ such that $\xi \geq \eta$. Let X^ξ, X^η be interacting particle systems with infinitesimal generator G, constructed by way of the universal coupling, as in Section 2. By the rules in Theorem 2, $X_t^\xi \geq X_t^\eta$ for all $t \in [0, \infty)$. Since $T_t f(\xi) = E(f(X_t^\xi))$ and $T_t f(\eta) = E(f(X_t^\eta))$, the theorem follows. $\square$

Corollary 6. *Let G be an attractive infinitesimal generator with bounded finite-range rates, $(T_t \colon t \in [0, \infty))$ the corresponding transition semigroup, and $\mu_{\xi,t}$, $\xi \in \Xi, t \in [0, \infty)$, the associated transition distributions. Then the limits*

$$\mu^- = \lim_{t \to \infty} \mu_{\bar{0},t} \quad and \quad \mu^+ = \lim_{t \to \infty} \mu_{\bar{n},t}$$

exist, where $\bar{0}$ is the configuration in which all sites are vacant and $\bar{n}$ is the config-uration in which all sites are occupied by n particles (the maximum number), and μ^- and μ^+ are both equilibrium distributions for $(T_t : t \in [0, \infty))$. Furthermore, if $\mu^- = \mu^+$, then there are no other equilibrium distributions for this transition semigroup, and

$$\lim_{t \to \infty} \mu_{\xi, t} = \mu^- = \mu^+$$

for all $\xi \in \Xi$.

PROOF. By the definition of transition semigroup,

$$T_{t+s} f(\bar{0}) = \int T_t f(\eta) \, \mu_{\bar{0}, s}(d\eta)$$

for all $s, t \in [0, \infty)$, $\xi \in \Xi$, and bounded measurable functions $f : \Xi \to \mathbb{R}$. If f is increasing, Theorem 5 implies that $T_t f$ is increasing, so

$$T_{t+s} f(\bar{0}) \geq \int T_t f(\bar{0}) \, \mu_{\bar{0}, s}(d\eta) = T_t f(\bar{0}) \, .$$

Thus the function $t \rightsquigarrow T_t f(\bar{0})$ is increasing in t for all bounded measurable increasing f, and hence

(32.4) $$\lim_{t \to \infty} \int f \, d\mu_{\bar{0}, t}$$

exists. It is easy to show that every function in $\mathfrak{F}$ is a finite linear combination of bounded measurable increasing functions, so (32.4) exists for all $f \in \mathfrak{F}$. It follows from Problem 3 that (32.4) holds for all continuous f. Since Ξ is compact, the collection $(\mu_{\bar{0}, t} : t \in [0, \infty))$ is relatively sequentially compact by Proposition 14 of Chapter 18. Since (32.4) holds for all continuous f, Proposition 15 of Chapter 18 implies that the limit μ^- exists. Similarly, the limit μ^+ exists. We leave it to the reader to show that μ^- and μ^+ are equilibrium distributions (see Problem 19 below).

As we have already seen, $T_t f$ is increasing if f is. For such f it follows that the functions $\liminf_{t \to \infty} T_t f$ and $\limsup_{t \to \infty} T_t f$ are increasing. Thus the first of these is greater than or equal to $\int f \, d\mu^-$ and the latter is less than or equal to $\int f \, d\mu^+$. Therefore, if $\mu^- = \mu^+$, then

$$\lim_{t \to \infty} \int f \, d\mu_{\xi, t} = \lim_{t \to \infty} T_t f(\xi) = \int f \, d\mu^- = \int f \, d\mu^+$$

for all increasing f and all ξ. As in the argument of the preceding paragraph, we may extend this equation to all continuous functions f. The final assertion in the corollary then follows. $\square$

The measure μ^- in the preceding corollary is the *lower equilibrium distribution* and μ^+ is the *upper equilibrium distribution* of the transition semigroup corresponding to G.

Because of Theorem 5, Corollary 6, and related results, much more is known about attractive interacting particle systems than nonattractive ones. The following example is typical of the kind of result that can be relatively easily obtained in the attractive setting.

Example 8. We take the dimension d to be 1, and let the range r and maximum particle number n be arbitrary positive integers. In this setting, let G be an attractive generator with translation invariant rates, and suppose that $b_x(\bar{0}) = 0$ for all $x \in \mathbb{Z}$. Let X be the interacting particle system with generator G and initial state ξ_0, where

$$\xi_0(x) = \begin{cases} n & \text{if } x \leq 0 \\ 0 & \text{otherwise.} \end{cases}$$

For $t \in [0, \infty)$, let

$$Z_t = \sup\{x \in \mathbb{Z} : X_t(x) > 0\}.$$

Thus, Z_t is the rightmost occupied site at time t. Of course, $Z_0 = 0$. We claim that Z_t/t converges a.s. to a constant $R \in [-\infty, \infty)$ as $t \to \infty$. The limit R is called the *right edge speed*. There is an analogous definition for the *left edge speed* L, using the initial state in which the sites $x < 0$ are vacant and the sites $x \geq 0$ are maximally occupied.

The proof of the claim uses the Kingman-Liggett Subadditive Ergodic Theorem. In order to use that theorem, we define the appropriate processes $Z_{m,n}$. Assume that X has been constructed using the universal coupling, and let $Z_{0,n} = Z_n \vee (-ln)$, $n = 1, 2, \ldots$, where l is a fixed positive integer. Eventually, we will let $l \to \infty$.

Fix $m > 0$. We will define $Z_{m,n}$ in terms of an interacting particle system $X^{(m)}$ with random initial state ξ_m, where

$$\xi_m(x) = \begin{cases} n & \text{if } x \leq Z_{0,m} \\ 0 & \text{otherwise.} \end{cases}$$

In order to construct $X^{(m)}$, we first create some new Poisson point processes from the ones used to construct X. We do this by subtracting m from the first coordinates of each of the points in all of the original Poisson point processes, and then intersecting each of the shifted point processes with $(0, \infty) \times (0, \infty)$. The resulting Poisson point processes on $(0, \infty) \times (0, \infty)$ are independent of $Z_{0,m}$, and have the same distribution as the original point processes that were used to construct X. We use these shifted point processes to construct $X^{(m)}$ according to the universal construction given in Section 2, and then for $m < n$, set

$$Z_{m,n} = (\sup\{x \in \mathbb{Z} : X_{n-m}^{(m)} > 0\} - Z_{0,m}) \vee (-l(n - m)).$$

We now indicate why the hypotheses of the Kingman-Liggett Theorem are satisfied. It should be clear from the construction that the distribution of the random sequence $(Z_{k,k+n} : n = 1, 2, \ldots)$ is the same for all $k \geq 0$, and that the random sequence $(Z_{kn,k(n+1)} : n = 0, 1, 2, \ldots)$ is iid for each such k, hence stationary and ergodic. It can be shown as in the proof of Theorem 5 that because the generator is attractive, $X^{(m)}(x) \geq X(x)$ for all $x \in \mathbb{Z}$. It follows from the definition of $Z_{m,n}$ that $Z_{0,n} \leq Z_{0,m} + Z_{m,n}$ for $0 \leq m < n$. Clearly $E(Z_{0,n}) \geq -ln$. The reader is requested in Problem 23 to show that $E(Z_{0,n}) \leq \gamma n$ for some finite constant γ. Thus, the hypotheses of the Kingman-Liggett Theorem are satisfied.

Let $R_l = \lim_{n \to \infty} (Z_{0,n}/n)$. It is not hard to show that there exists a constant $R \in [-\infty, \infty)$ such that

$$R = \lim_{l \to \infty} R_l = \lim_{n \to \infty} \frac{Z_n}{n} = \lim_{t \to \infty} \frac{Z_t}{t} \quad \text{a.s.}.$$

It can be shown that when $r = 1$, $\mu^+ = \mu^-$ if and only if $R \leq L$. The complete proof of this remarkable result is quite difficult. However, readers who have mastered the material in this chapter should be able to prove the 'if direction', for arbitrary values of the range r (see Problem 24 for the case $R < 0 < L$).

Problem 19. Complete the proof of Corollary 6 by showing that μ^- and μ^+ are equilibrium distributions of the transition semigroup $(T_t : t \in [0, \infty))$.

Problem 20. Show that μ^+ and μ^- are extremal in the set of equilibrium distributions of the corresponding transition semigroup, in the sense that neither can be written as a nontrivial convex combination of two different equilibrium distributions for that semigroup.

Problem 21. Which of the following types of interacting particle systems have attractive infinitesimal generators: contact processes with linear birth rates, contact processes with threshold birth rates, contact processes with sexual reproduction, cyclic threshold processes, majority vote processes?

Problem 22. Let X and Y be two interacting particle systems with attractive generators G_X and G_Y and initial states ξ_X and ξ_Y respectively. Suppose that each birth rate of the generator G_X is greater than or equal to the corresponding birth rate of the generator G_Y, and that each death rate of the generator G_X is less than or equal to the corresponding death rate of the generator G_Y. Show that if X and Y are coupled with the universal coupling and $X_0(x) \geq Y_0(x)$ for all $x \in \mathbb{Z}^d$, then $X_t(x) \geq Y_t(x)$ for all $t \in [0, \infty)$ and $x \in \mathbb{Z}^d$.

Problem 23. Prove the three facts about the random variables $Z_{m,n}$ that were left unproved in Example 8, namely, (i) that $E(Z_{0,n}) \leq \gamma n$ for some finite γ, (ii) that $(Z_n/n) \to R$ a.s. as $n \to \infty$, and (iii) that $(Z_t/t) \to R$ a.s. as $t \to \infty$. Hint: The 'right edge' can move at most r units to the right in one transition, and the rate at which it moves to the right is at most r times the maximum birth rate.

* **Problem 24.** Let X^ξ be an interacting particle system with initial state ξ and generator G satisfying the conditions of Example 8. Show that if $R < 0 < L$, then $\lim_{t \to \infty} X_t^\xi = \bar{0}$ a.s.

CHAPTER 33
Diffusions and Stochastic Calculus

A *diffusion* is a time-homogeneous continuous-in-time strong Markov process. Most often, the state space is $\mathbb{R}^d$, although other spaces are also considered, especially in current research.

No one knows how to characterize or construct all diffusions, even in the $\mathbb{R}^d$-setting (except for the case $d = 1$). Since our intention is only to provide a brief introduction to diffusions, we will focus most of our attention on the state space $\mathbb{R}$, and even within that restricted setting, on a class of diffusions that is particularly well understood. Our main emphasis will be on those results that generalize relatively easily to the $\mathbb{R}^d$-setting with $d > 1$, and in the final section of the chapter, we will say a little about how that generalization is accomplished.

As in the previous two chapters, it will be useful to describe these Markov processes in terms of generators and solutions to the martingale problem. This approach leads to a nice connection between diffusions and certain types of partial differential equations.

However, there is another approach to diffusions that is in many ways more natural than the generator approach. We will develop a 'stochastic calculus', and construct diffusions as solutions of 'stochastic differential equations'. This approach is the one that is most closely associated with applications in physics, signal processing, and economics, among other fields. We will see that Brownian motion and other continuous-time martingales play important roles. The generator approach will be discussed later in the chapter.

33.1. Stochastic difference equations

Let W be a Wiener process, and let $a\colon \mathbb{R} \to [0, \infty)$ and $b\colon \mathbb{R} \to \mathbb{R}$ be measurable functions. Given an $\varepsilon > 0$, consider the equations

$$(33.1) \qquad Z_{(n+1)\varepsilon} = Z_{n\varepsilon} + a(Z_{n\varepsilon})(W_{(n+1)\varepsilon} - W_{n\varepsilon}) + b(Z_{n\varepsilon})\varepsilon \,,$$

$n = 0, 1, 2, \ldots.$ Given an initial value $Z_0 = z$, these equations have a unique solution, which is a random sequence that can be calculated recursively in terms of the Wiener process W.

Let us rewrite (33.1) with some notation that emphasizes the fact that it is a 'difference equation'. Given an $\mathbb{R}$-valued function f whose domain includes the set $\{0, \varepsilon, 2\varepsilon, \dots\}$, let

$$\Delta_\varepsilon f(t) = f(t + \varepsilon) - f(t), \quad t \in \{0, \varepsilon, 2\varepsilon, \dots\}.$$

Then (33.1) becomes

$$(33.2) \qquad \Delta_\varepsilon Z_t = a(Z_t)\Delta_\varepsilon W_t + b(Z_t)\Delta_\varepsilon t, \quad t \in \{0, \varepsilon, 2\varepsilon, \dots\}.$$

We call (33.2) a *stochastic difference equation with coefficients a and b*.

Let $Z^{(\varepsilon)}$ denote the solution of (33.2) with initial value $Z_0^{(\varepsilon)} = z$. It is easy to verify that $Z^{(\varepsilon)}$ is a Markov sequence (see Problem 1). Our intention is to obtain a diffusion as some sort of limit as $\varepsilon \searrow 0$ of $Z^{(\varepsilon)}$.

As a first step in this direction, we make $Z^{(\varepsilon)}$ into a $\mathbf{C}[0, \infty)$-valued random variable by defining

$$(33.3) \qquad Z_t^{(\varepsilon)} = Z_{n\varepsilon}^{(\varepsilon)} + a(Z_{n\varepsilon}^{(\varepsilon)})(W_t - W_{n\varepsilon}) + b(Z_{n\varepsilon}^{(\varepsilon)})(t - n\varepsilon)$$

for $t \in (n\varepsilon, (n+1)\varepsilon), n = 0, 1, 2, \dots$. We will eventually show (under certain conditions on the coefficients) that the collection $(Z^{(\varepsilon)}, \varepsilon > 0)$ is Cauchy in probability as $\varepsilon \searrow 0$. Since $\mathbf{C}[0, \infty)$ is a Polish space, it follows that there is a limit Z. As will be seen, this limit is a diffusion that is the unique solution of a 'stochastic differential equation' related to (33.2).

We now introduce some more notation. For $\varepsilon > 0$, let $\tau_\varepsilon: [0, \infty) \to [0, \infty)$ be the approximation of the identity function defined by

$$(33.4) \qquad \tau_\varepsilon(t) = \varepsilon \lfloor \tfrac{t}{\varepsilon} \rfloor.$$

Then it is easy to check that $Z^{(\varepsilon)}$ is the unique solution of

$$(33.5) \qquad Z_t = z + \int_0^t a(Z_{\tau_\varepsilon(s)})\, dW_s + \int_0^t b(Z_{\tau_\varepsilon(s)})\, ds, \quad t \in [0, \infty),$$

where the first integral in this equation is a Riemann-Stieltjes integral with respect to the random continuous function W. This integral exists because the function $s \rightsquigarrow a(Z_{\tau_\varepsilon(s)})$ is piecewise constant. We regard (33.5) as the 'stochastic integral' form of the stochastic difference equation (33.2). The diffusions we will construct are solutions of equations like (33.5), with $\tau_\varepsilon(s)$ replaced by s.

Problem 1. Show that the solution of (33.1), with initial value z, is a Markov sequence with respect to the filtration $(\mathcal{F}_{n\varepsilon+}: n = 0, 1, 2, \dots)$, where for each $t \in [0, \infty)$, $\mathcal{F}_t = \sigma(W_s: s \leq t)$.

* **Problem 2.** Let $Z^{(\varepsilon)}$ be the solution of (33.5), with initial value z. Assume that the absolute values of coefficients a and b are bounded above by a function of the form $x \rightsquigarrow cx + d$ for some finite constants c, d. Show that there exist finite constants c' and d' not depending on ε such that

$$E\left[(Z_u^{(\varepsilon)})^2\right] \le d' e^{c'u}$$

for all $u \in [0, \infty)$. *Hint:* First prove the inequality by induction for $u = k\varepsilon$, $k = 0, 1, 2, \ldots$.

33.2. The Itô integral

In the previous section we encountered an integral with respect to a Wiener process. Since the integrand was piecewise constant, we were able to treat it as a Riemann-Stieltjes integral. This type of integral is not sufficient for the more general integrands that we will be considering, so in this section we develop a new type of integral.

We begin by describing an appropriate class of integrands. Let W be a Wiener process defined on a probability space $(\Omega, \mathcal{F}, P)$, and $(\mathcal{F}_{t+} : t \in [0, \infty))$ the minimal right-continuous filtration of W. A *nonanticipating W-functional* is a $\mathbf{D}[0, \infty)$-valued random variable $X = (X_t : t \in [0, \infty))$ defined on $(\Omega, \mathcal{F}, P)$ that is adapted to a filtration of the form $(\sigma(\mathcal{F}_{t+}, \mathcal{G}) : t \in [0, \infty))$, where $\mathcal{G}$ is a σ-field that is independent of W. Our goal is to define the integral with respect to W of every nonanticipating W-functional X. We will accomplish this by taking an appropriate limit as $\varepsilon \searrow 0$ of Riemann-Stieltjes integrals with respect to W of certain piecewise constant nonanticipating W-functionals. We need the following result about integrals of such functionals.

Lemma 1. *Let W be a Wiener process, and X a bounded nonanticipating W-functional for which there exists a sequence of times $0 = t_0 < t_1 < t_2 < \cdots \to \infty$ such that $X_t = X_{t_n}$ for $t \in [t_n, t_{n+1})$, $n = 0, 1, 2, \ldots$. For $t \in [0, \infty)$, let $\mathfrak{I}_t$ be the Riemann-Stieltjes integral*

$$\mathfrak{I}_t = \int_0^t X \, dW \, .$$

Then the $\mathbf{C}[0, \infty)$-valued random variable $\mathfrak{I} = (\mathfrak{I}_t : t \in [0, \infty))$ is a continuous-time martingale with respect to the minimal right-continuous filtration of W, and for each t,

$$E(\mathfrak{I}_t) = 0 \quad and \quad \mathrm{Var}(\mathfrak{I}_t) = E\left(\int_0^t X_u^2 \, du\right).$$

PROOF. To show that $\mathfrak{I}$ is a continuous-time martingale with respect to the given filtration, it is enough to show that

$$\text{(33.6)} \qquad\qquad E\left(\int_s^t X \, dW \mid \mathcal{F}_{s+}\right) = 0$$

for $0 \le s < t < \infty$. Since X is constant on each interval $[t_n, t_{n+1})$, the Riemann-Stieltjes integral $\int_s^t X\, dW$ is a sum of terms of the form

$$(33.7) \qquad \int_u^v X\, dW = X_u(W_v - W_u),$$

where $v > u \ge s$ and the intervals $[u, v]$ for various summands do not overlap except possibly at endpoints. Since W is a Wiener process, $\mathcal{F}_{u+}$ and $(W_v - W_u)$ are independent. Since X is a nonanticipating W-functional, X_u is measurable with respect to $\sigma(\mathcal{F}_{u+}, \mathcal{G})$ for some σ-field $\mathcal{G}$ that is independent of W, so X_u and $(W_v - W_u)$ are conditionally independent given $\mathcal{F}_{s+}$ by Proposition 26 of Chapter 21. Thus, the conditional expectation given $\mathcal{F}_{s+}$ of each term of the form (33.7) is 0. The equality (33.6) follows immediately.

Setting $s = 0$ in (33.6) and taking expectations of both sides gives $E(\mathfrak{I}_t) = 0$ for all t. It is an easy to show that the martingale differences (33.7) are uncorrelated for various u and v satisfying the nonoverlapping condition of the preceding paragraph. Thus the variance of $\mathfrak{I}_t$ can be obtained by summing the variances of those terms. Using the independence of X_u and $(W_v - W_u)$, we see that the variance of (33.7) is

$$(v - u)E(X_u^2).$$

The desired formula for $\mathrm{Var}(\mathfrak{I}_t)$ follows. $\square$

To now define the integral of an arbitrary nonanticipating W-functional X with respect to W, we introduce the notation $X^{(\varepsilon)}$ for the piecewise constant $\mathbf{D}[0, \infty)$-valued random variable defined by

$$(33.8) \qquad X_t^{(\varepsilon)} = X_{\tau_\varepsilon(t)}, \quad t \in [0, \infty),$$

with τ_ε defined by (33.4), and $\mathfrak{I}^{(\varepsilon)}$ for the $\mathbf{C}[0, \infty)$-valued random variable defined by

$$\mathfrak{I}_t^{(\varepsilon)} = \int_0^t X^{(\varepsilon)}\, dW, \quad t \in [0, \infty),$$

where the right side is a Riemann-Stieltjes integral. Note that $X^{(\varepsilon)}$ inherits the property of being a nonanticipating W-functional from X, and that if X is bounded, then Lemma 1 applies to $\mathfrak{I}^{(\varepsilon)}$, since in that case $X^{(\varepsilon)}$ is bounded and piecewise constant.

Definition and Proposition 2. Let W be a Wiener process on $[0, \infty)$, defined on a probability space $(\Omega, \mathcal{F}, P)$, and let X be a nonanticipating W-functional. The *Itô integral* of X is the $\mathbf{C}[0, \infty)$-valued random variable $\mathfrak{I} = (\mathfrak{I}_t : t \in [0, \infty))$, where

$$\mathfrak{I} = \lim_{\varepsilon \searrow 0} \mathfrak{I}^{(\varepsilon)} \text{ i.p.}$$

For $0 \le s \le t < \infty$, the *Itô integral of X from s to t* equals $\mathfrak{I}_t - \mathfrak{I}_s$ and is denoted by

$$\int_s^t X_u \, dW_u \quad \text{or} \quad \int_s^t X \, dW .$$

PROOF. We need to prove that the collection $(\mathfrak{I}^{(\varepsilon)} \colon \varepsilon > 0)$ is Cauchy in probability as $\varepsilon \searrow 0$. Since we are working in the Polish space $\mathbf{C}[0, \infty)$, it is sufficient to prove the following for each $t \in [0, \infty)$ and $\delta > 0$:

$$\lim_{\varepsilon, \eta \searrow 0} P\big[\sup_{s \in [0,t]} |\mathfrak{I}_s^{(\varepsilon)} - \mathfrak{I}_s^{(\eta)}| > \delta \big] = 0 .$$

First consider the case in which X is bounded. By Lemma 1, $\mathfrak{I}^{(\varepsilon)}$ is a martingale for each $\varepsilon > 0$. By the Kolmogorov Inequality of Chapter 24 (which can be extended in a straightforward manner to cadlag-valued martingales),

$$P\big[\sup_{s \in [0,t]} |\mathfrak{I}_s^{(\varepsilon)} - \mathfrak{I}_s^{(\eta)}| > \delta \big] \le \frac{E\big[(\mathfrak{I}_t^{(\varepsilon)} - \mathfrak{I}_t^{(\eta)})^2 \big]}{\delta^2} .$$

By Lemma 1, the numerator of the right side of this inequality equals

$$E\bigg(\int_0^t (X_u^{(\varepsilon)} - X_u^{(\eta)})^2 \, du \bigg) .$$

Since X is assumed to be cadlag-valued, $X_u^{(\varepsilon)}$ and $X_u^{(\eta)}$ converge to X_u as $\varepsilon, \eta \searrow 0$ for λ-a.e. $u \in [0, \infty)$, where λ denotes Lebesgue measure. The desired conclusion follows from the Bounded Convergence Theorem.

For general nonanticipating W-functionals X, we use the fact that every $\mathbb{R}$-valued cadlag function is bounded on bounded intervals. Therefore

$$\lim_{n \to \infty} P\big[X_s = (-n \vee X_s) \wedge n \text{ for all } s \in [0, t] \big] = 1 .$$

For $n = 1, 2, \ldots$, let A_n be the event in this last expression. Then the argument in the bounded case implies that

$$\lim_{\varepsilon \searrow 0} P\big(A_n \cap [\sup_{s \in [0,t]} |\mathfrak{I}_s^{(\varepsilon)} - \mathfrak{I}_s^{(\eta)}| > \delta] \big) = 0$$

for all $\delta > 0$ and $n = 1, 2, \ldots$. Since $P(A_n) \to 1$ as $n \to \infty$, the desired result follows. $\square$

Problem 3. Let X be a nonanticipating W-functional and T an a.s.-finite stopping time with respect to the minimal right-continuous filtration of W. Show that $Y = (X_{T+t} - X_T \colon t \in [0, \infty))$ is a nonanticipating $\widetilde{W}$-functional, where $\widetilde{W}$ is the Wiener process $(W_{T+t} - W_T \colon t \in [0, \infty))$.

Problem 4. Let $Z^{(\varepsilon)}$ be the solution of (33.2). Show that if the absolute values of the coefficients a and b are each bounded above by some polynomial, then $Z^{(\varepsilon)}$ is a square-integrable nonanticipating W-functional.

* **Problem 5.** If X is a nonanticipating W-functional such that the Riemann-Stieltjes integral of X with respect to W on $[0, t]$ exists with positive probability, we now have two interpretations of the expression $\int_0^t X \, dW$. Are these interpretations in agreement?

Problem 6. Show that the Itô integral is linear in the sense that if X and Y are nonanticipating W-functionals, then

$$\int_s^t (\alpha X + \beta Y) \, dW = \alpha \int_s^t X \, dW + \beta \int_s^t Y \, dW$$

for all times $0 \leq s \leq t < \infty$ and real constants α and β.

Problem 7. Let $f : [0, \infty) \to \mathbb{R}$ be a (nonrandom) cadlag function. Show that the Itô integral $\Im$ of f is a *Gaussian process*, which is to say that for all times $t_1, \ldots, t_d \in [0, \infty)$, $(\Im_{t_1}, \ldots, \Im_{t_d})$ is normally distributed. Also show that $\Im$ has mean function 0 and covariance function

$$(s, t) \rightsquigarrow \int_0^{s \wedge t} f^2(u) \, du \, .$$

We have chosen to define the Itô integral $\Im$ as a limit in probability of $\mathbf{C}[0, \infty)$-valued random variables $\Im^{(\varepsilon)}$, thereby simultaneously defining the random variables $\Im_t$, $t \in [0, \infty)$ in such a way that $\Im$ is $\mathbf{C}[0, \infty)$-valued. One consequence of this definition is that $\lim_{\varepsilon \searrow 0} \Im_t^{(\varepsilon)} = \Im_t$ i.p. for each fixed t. Under an additional assumption, we strengthen this latter result with the following lemma.

Lemma 3. *Let X be a nonanticipating W-functional, and define $X^{(\varepsilon)}$ for $\varepsilon > 0$ by (33.8). If*

(33.9) $$E\left(\int_0^t X_u^2 \, du \right) < \infty \, ,$$

then

(33.10)
$$\lim_{\varepsilon \searrow 0} E\left(\left[\int_0^t (-n \vee X_u^{(\varepsilon)}) \wedge n \, dW_u - \int_0^t (-n \vee X_u) \wedge n \, dW_u \right]^2 \right) = 0$$

and

(33.11) $$\lim_{n \to \infty} E\left(\left[\int_0^t (-n \vee X_u) \wedge n \, dW_u - \int_0^t X_u \, dW_u \right]^2 \right) = 0 \, .$$

PROOF. It follows from the Bounded Convergence Theorem and Lemma 1 that

$$\lim_{\varepsilon, \eta \searrow 0} E\left(\left[\int_0^t (-n \vee X_u^{(\varepsilon)}) \wedge n \, dW_u - \int_0^t (-n \vee X_u^{(\eta)}) \wedge n \, dW_u \right]^2 \right) = 0 \, ,$$

so

$$\varepsilon \rightsquigarrow \int_0^t (-n \vee X_u^{(\varepsilon)}) \wedge n \, dW_u$$

is Cauchy in $\mathbf{L}_2(\Omega, \mathcal{F}, P)$ as $\varepsilon \searrow 0$. We know from the definition of the Itô integral that the random variables in this collection converge in probability to $\int_0^t (-n \vee X_u) \wedge n \, dW_u$ as $\varepsilon \searrow 0$. Equation (33.10) follows.

It follows from (33.10), Lemma 1, and the Bounded Convergence Theorem that

$$E\left(\left[\int_0^t (-n \vee X_u) \wedge n \, dW_u - \int_0^t (-m \vee X_u) \wedge m \, dW_u\right]^2\right)$$

$$= E\left(\int_0^t [(-n \vee X_u) \wedge n]^2 \, du\right) + E\left(\int_0^t [(-m \vee X_u) \wedge m]^2 \, du\right)$$

$$- 2E\left(\int_0^t [(-m \vee X_u) \wedge m] \, [(-n \vee X_u) \wedge n] \, du\right)$$

for $m, n = 1, 2, \ldots$. By the Monotone Convergence Theorem, each of the expectations on the right side of this equality converges to

$$E\left(\int_0^t X_u^2 \, du\right)$$

as $m, n \to \infty$, so the sequence

$$\left(\int_0^t (-n \vee X_u) \wedge n \, dW_u : n = 1, 2, \ldots\right)$$

is Cauchy in $\mathbf{L}_2(\Omega, \mathcal{F}, P)$. It was shown at the end of the proof of Proposition 2 that this sequence converges in probability to $\mathfrak{I}_t$, so (33.11) follows. $\quad\square$

A nonanticipating W-functional that satisfies (33.9) for all $t \in [0, \infty)$ is said to be *square-integrable*. Taken together, Lemma 1 and Lemma 3 immediately imply the following important result for square-integrable nonanticipating W-functionals.

Theorem 4. *Let $\mathfrak{I}$ be the Itô integral of a square-integrable nonanticipating W-functional X. Then $\mathfrak{I}$ is a martingale with respect to the minimal right-continuous filtration of W, and for all $t \in [0, \infty)$,*

$$E(\mathfrak{I}_t) = 0 \quad and \quad \mathrm{Var}(\mathfrak{I}_t) = E\left(\int_0^t X_u^2 \, dt\right).$$

Corollary 5. *Let X, $X^{(n)}$, $n = 1, 2, \ldots$, be square-integrable nonanticipating W-functionals. Then*

$$\lim_{n\to\infty} E\left(\left[\int_s^t X^{(n)} \, dW - \int_s^t X \, dW\right]^2\right) = 0$$

if and only if

$$\lim_{n\to\infty} E\left(\int_s^t (X_u^{(n)} - X_u)^2 \, du\right) = 0.$$

Problem 8. Prove Theorem 4 and its corollary.

Problem 9. Let $\mathfrak{I}$ be the Itô integral of a square-integrable nonanticipating W-functional X. Show that $\mathfrak{I}$ is a square-integrable nonanticipating W-functional.

Example 1. Since W is itself a nonanticipating W-functional, it has an Itô integral, which we now calculate. Letting $\varepsilon = t/n$ in the definition, we have

$$\int_0^t W \, dW = \lim_{k \to \infty} \sum_{k=0}^{n-1} W_{\frac{kt}{n}} (W_{\frac{(k+1)t}{n}} - W_{\frac{kt}{n}})$$

$$= \lim_{n \to \infty} \frac{1}{2} \sum_{k=0}^{n-1} [(W_{\frac{(k+1)t}{n}}^2 - W_{\frac{kt}{n}}^2) - (W_{\frac{(k+1)t}{n}} - W_{\frac{kt}{n}})^2]$$

$$= \frac{W_t^2}{2} - \frac{1}{2} \lim_{n \to \infty} \sum_{k=0}^{n-1} (W_{\frac{(k+1)t}{n}} - W_{\frac{kt}{n}})^2 \quad \text{i.p.}$$

The terms in the last sum are iid, with mean t/n and variance $2(t/n)^2$, so the sum itself has mean t and variance $2t^2/n$. It follows that the sum converges to t in $\mathbf{L}_2(\Omega, \mathcal{F}, P)$, and hence in probability, as $n \to \infty$. We have shown that

$$\int_0^t W \, dW = \frac{W_t^2}{2} - \frac{t}{2} \, .$$

The second term on the right side of the formula just obtained is perhaps unexpected. If W were Riemann-Stieltjes integrable with respect to itself, that term would not be present. On the other hand, if that term were not present, the Itô integral of W would not be a continuous-time martingale, in contradiction to Theorem 4.

Problem 10. Let X be a nonanticipating W-functional. Show that

$$\lim_{k \to \infty} \sum_{j=0}^{k-1} X_{\frac{jt}{k}} [W_{\frac{(j+1)t}{k}} - W_{\frac{jt}{k}}]^2 = \int_0^t X_u \, du \quad \text{i.p.}$$

for $t \in [0, \infty)$.

33.3. Stochastic differentials and the Itô Lemma

The example at the end of the preceding section shows that the ordinary rules of calculus do not apply to Itô integrals. In this section, we will introduce some of the new rules of the stochastic calculus.

Suppose that $Z = (Z_t : t \in [0, \infty))$ satisfies

$$(33.12) \qquad Z_t - Z_0 = \int_0^t X_u \, dW_u + \int_0^t Y_u \, du, \quad t \in [0, \infty),$$

where X and Y are nonanticipating W-functionals. The $\mathbf{C}[0, \infty)$-valued random variable Z is called a *stochastic integral*. Notice that stochastic integrals are themselves nonanticipating W-functionals.

We now introduce a more compact way of writing (33.12):

(33.13) $$dZ = X \, dW + Y \, dt \, .$$

For example, the result of our calculation in Example 1 of $\int_0^t W \, dW$ is written compactly as

$$d(W^2) = 2W \, dW + dt \, .$$

The quantities dZ and dW in this expression are called *stochastic differentials*, and the expression itself is called a *stochastic differential equation*. The quantity dt is a *differential*, although under certain circumstances we will also call it a stochastic differential, just as we sometimes use the term 'random variable' to describe a nonrandom constant.

It is important to realize that (33.13) is simply a compact way of writing (33.12). There is no difference in meaning between the two. One reason for using the more compact form is that it naturally leads to an interpretation of integration with respect to Z. If X, Y, and Z satisfy (33.13), and if U is a nonanticipating W-functional, then the stochastic integral

$$\int_0^t XU \, dW + \int_s^t YU \, du \, , \quad t \in [0, \infty) \, ,$$

is denoted by

$$\int_0^t U \, dZ \, , \quad t \in [0, \infty) \, ;$$

or more compactly, we write

$$U \, dZ = XU \, dW + YU \, dt \, .$$

Another reason for using (33.13) to denote stochastic integrals is that the rules of stochastic calculus are easier to remember in that form. The following lemma provides the basis for deriving such rules.

Lemma 6. [Itô] *Suppose $X^{(i)}, Y^{(i)}, Z^{(i)}$ are nonanticipating W-functionals satisfying*

$$dZ^{(i)} = X^{(i)} \, dW + Y^{(i)} \, dt \, , \quad i = 1, \ldots, n \, .$$

Set

$$\widetilde{Z} = (t, Z_t^{(1)}, \ldots, Z_t^{(n)}) \, , \quad t \in [0, \infty) \, ,$$

and let $f \colon [0, \infty) \times \mathbb{R}^n \to \mathbb{R}$ be a function with continuous partial derivatives $D_i f, i = 0, \ldots, n$, and continuous second-order partial derivatives $D_{ij} f, i, j = 1, \ldots, n$. Then

$$d(f \circ \widetilde{Z}) = D_0 f \circ \widetilde{Z} \, dt + \sum_{i=1}^n D_i f \circ \widetilde{Z} \, dZ^{(i)} + \frac{1}{2} \sum_{i,j=1}^n X^{(i)} X^{(j)} \, D_{ij} f \circ \widetilde{Z} \, dt \, .$$

PROOF. The first part of the proof is a series of reductions to increasingly simple cases of the theorem. The reader is requested to verify the validity of these reductions in Problem 11. An argument based on the definitions implies that it is sufficient to prove the theorem for the case in which there exists a sequence of times $0 = t_0 < t_1 < t_2 < \ldots$ increasing to ∞ such that each $X^{(i)}$ and $Y^{(i)}$ is constant on intervals of the form $[t_n, t_{n+1})$. By considering separately the integrals over each such interval, it can be shown that it is sufficient to prove the theorem for random constant functions $X^{(i)}$ and $Y^{(i)}$. Since a nonanticipating random constant W-functional is necessarily independent of W, the theorem can be reduced to the case in which $X^{(i)}$ and $Y^{(i)}$ are nonrandom constants by conditioning. A further straightforward argument shows that this latter case is equivalent to the case in which $n = 1$, $X^{(1)} \equiv 1$, and $Y^{(1)} \equiv 0$. A limiting argument shows that it is enough to consider bounded functions f such that $D_0 f, D_1 f$, and $D_{11} f$ are also bounded, and a further limiting argument allows us to assume that the second partial derivatives $D_{00} f$ and $D_{01} f$ exist and are continuous and bounded.

Thus, it remains to show that if $f : [0, \infty) \times \mathbb{R} \to \mathbb{R}$ is bounded and has bounded continuous first- and second-order partial derivatives, then

$$
(33.14) \quad
\begin{aligned}
&f(t, W_t) - f(0, 0) \\
&= \int_0^t \left(D_0 f(u, W_u) + \tfrac{1}{2} D_{11} f(u, W_u) \right) du + \int_0^t D_1 f(u, W_u) \, dW_u
\end{aligned}
$$

for all $t \in [0, \infty)$. For each positive integer k, the left side of (33.14) equals

$$
\sum_{j=0}^{k-1} [f(\tfrac{(j+1)t}{k}, W_{\frac{(j+1)t}{k}}) - f(\tfrac{jt}{k}, W_{\frac{jt}{k}})] .
$$

By the Taylor Theorem with remainder for functions of 2 variables, we may write the terms in this sum as

$$
\begin{aligned}
&D_0 f(\tfrac{jt}{k}, W_{\frac{jt}{k}}) (\tfrac{t}{k}) + D_1 f(\tfrac{jt}{k}, W_{\frac{jt}{k}}) [W_{\frac{(j+1)t}{k}} - W_{\frac{jt}{k}}] \\
&+ \tfrac{1}{2} D_{11} f(\tfrac{jt}{k}, W_{\frac{jt}{k}}) [W_{\frac{(j+1)t}{k}} - W_{\frac{jt}{k}}]^2 + E(j, k) ,
\end{aligned}
$$

where the absolute value of the error term $E(j, k)$ is bounded above by a positive random variable $C(k)$ times

$$
(33.15) \qquad \tfrac{t}{k} \left(1 + |W_{\frac{(j+1)t}{k}} - W_{\frac{jt}{k}}| \right) + \left(W_{\frac{(j+1)t}{k}} - W_{\frac{jt}{k}} \right)^2 ,
$$

with $C(k) \to 0$ a.s. as $k \to \infty$.

By the definition of the Riemann and Itô integrals and Problem 10, the first three terms in the Taylor approximation converge i.p. to the right side of (33.14). So it is enough to show that $\sum_{j=0}^{k-1} E(j, k) \to 0$ i.p. as $k \to \infty$. By the uniform continuity of W on $[0, t]$, the sum over j of the first term in (33.15) is almost surely bounded above by some constant as $k \to \infty$. In Example 1, we showed that the sum over j of the second term converges to t in probability. The desired result follows from the fact that $C(k) \to 0$ i.p. as $k \to \infty$. $\square$

Problem 11. Justify the series of reductions in the first paragraph of the proof of the Itô Lemma.

* **Problem 12.** Use the Itô Lemma to calculate $d(f \circ W)$ for $f(x) = x^p$, $p = 1, 2, \ldots$, and $f(w) = e^{\alpha w}$, $\sin \alpha w$, $\cos \alpha w$, $\alpha \in \mathbb{R}$.

Problem 13. Use your answer to the previous problem to find a solution of the stochastic differential equation $dZ = \alpha Z\, dW$, where α is a real constant. *Hint:* The solution takes the form $t \rightsquigarrow f_1(t) f_2(W_t)$.

Example 2. [Brownian local-time process] Let us calculate $d(|W|)$. The Itô Lemma does not apply directly, since the absolute value function is not differentiable at 0, so we will need a limiting argument.

For $\delta, \varepsilon > 0$, let $f_{\delta,\varepsilon} : \mathbb{R} \to \mathbb{R}$ be the unique function determined by the following conditions: (i) $f_{\delta,\varepsilon}(0) = f'_{\delta,\varepsilon}(0) = 0$; (ii) $f''_{\delta,\varepsilon}(x) = \frac{2}{2\delta+\varepsilon}$ for $|x| \leq \delta$; (iii) $f''_{\delta,\varepsilon}(x) = 0$ for $|x| \geq \delta + \varepsilon$; (iv) $f''_{\delta,\varepsilon}$ is linear on $[-\delta - \varepsilon, -\delta]$ and $[\delta, \delta + \varepsilon]$. Thus, $f_{\delta,\varepsilon}$ approximates the absolute value function from below, and has continuous first and second derivatives, with the second derivative being piecewise linear. As $\varepsilon \searrow 0$, $f_{\delta,\varepsilon}$ converges uniformly to a function f_δ with continuous piecewise linear first derivative, and as $\delta \searrow 0$, f_δ converges uniformly to the function $x \rightsquigarrow |x|$.

We now apply the Itô Lemma to $f_{\delta,\varepsilon}$. After a slight rearrangement, we have

$$f_{\delta,\varepsilon}(W_t) - \int_0^t f'_{\delta,\varepsilon}(W_u)\, dW_u = \frac{1}{2} \int_0^t f''_{\delta,\varepsilon}(W_u)\, du\,.$$

Letting $\varepsilon \searrow 0$ and applying Corollary 5 to the left side and the Bounded Convergence Theorem to the right side, we obtain

$$(33.16) \qquad f_\delta(W_t) - \int_0^t f'_\delta(W_u)\, dW_u = \frac{1}{2\delta} \int_0^t I_{[-\delta,\delta]}(W_u)\, du\,.$$

We would like to use Corollary 5 to take the limit as $\delta \searrow 0$ of the left side of (33.16). The limit of the left side would be

$$(33.17) \qquad |W_t| - \int_0^t \operatorname{sgn}(W_u)\, dW_u\,,$$

where $\operatorname{sgn}(x) = I_{(0,\infty)}(x) - I_{(-\infty,0)}(x)$. Unfortunately $\operatorname{sgn} \circ W$ does not have cadlag paths, so its Itô integral is not defined according to the definition we have given. In a more thorough treatment, we would extend the definition to accommodate such integrands. But in this example, we content ourselves with noting that it is not hard to see that the relevant arguments of the previous section (particularly the proof of Corollary 5 and the use of the Kolmogorov Inequality in the construction of the Itô integral) can be extended to this situation to show that as $\delta \searrow 0$ the Itô integral of $f'_\delta \circ W$ converges i.p. to a $\mathbf{C}[0, \infty)$-valued random variable. It is natural to denote the limit by the integral that appears in (33.17).

Thus we have

$$d(|W|) = \operatorname{sgn} \circ W \, dW + dL \,,$$

where $L = (L_t : t \in [0, \infty))$ is the $\mathbf{C}[0, \infty)$-valued random variable defined by letting L_t equal the i.p. limit as $\delta \searrow 0$ of the right side of (33.16). We call L the *Brownian local-time process at 0*.

More generally, we may imitate the argument just given to calculate $d(|W - x|)$ for each fixed $x \in \mathbb{R}$. As a result, a $\mathbf{C}[0, \infty)$-valued random variable $L(x) = (L_t(x) : t \in [0, \infty))$ is constructed, with

$$|W_t - x| - \int_0^t \operatorname{sgn}(W_u - x) \, dW_u = L_t(x) = \lim_{\delta \searrow 0} \frac{1}{2\delta} \int_0^t I_{[-\delta+x, \delta+x]}(W_u) \, du \quad \text{i.p.}$$

This random variable is known as the *Brownian local-time process at x*. It can be shown that the random function $x \rightsquigarrow L(x)$ is a.s. continuous, and that for all Borel sets B,

$$(33.18) \qquad \int_B L_t(x) \, dx = \int_0^t I_B(W_u) \, du \,.$$

Thus, the Brownian local-time process is the density with respect to Lebesgue measure of a random measure on $\mathbb{R}$ that gives the 'occupation time' of Brownian motion.

Problem 14. Calculate $d(W^+)$. Your answer should involve Brownian local-time process and an Itô integral whose integrand is not cadlag-valued, as in Example 2.

* **Problem 15.** Prove (33.18).

33.4. Autonomous stochastic differential equations

In this section we consider the problem

$$(33.19) \qquad dZ = a \circ Z \, dW + b \circ Z \, dt, \quad Z_0 = z \,,$$

where z is an arbitrary real number and a and b are functions from $\mathbb{R}$ to $\mathbb{R}$ that satisfy certain other conditions. The requirement $Z_0 = z$ in (33.19) is the *initial condition*, z itself is the *initial value*, and the equation is an *autonomous stochastic differential equation*. The adjective 'autonomous' refers to the fact that W and t appear in it only as differentials; in particular, the *coefficients* a and b do not depend on W or t.

We restrict our attention to autonomous equations since we are interested in constructing diffusions, and because of a lack of time-homogeneity, nonautonomous equations typically do not have solutions that are diffusions. When the additional conditions that we place on a and b hold, we will show that (33.19) has a unique solution Z that is a diffusion.

Example 3. If a and b are constants, then $Z = t \rightsquigarrow aW_t + bt + z$ satisfies (33.19). Note that the solution $Z^{(\varepsilon)}$ of the stochastic difference equation (33.2) is of this form on each of the intervals $[n\varepsilon, (n+1)\varepsilon)$. Our construction of solutions of (33.19) for more general coefficients will imply that such solutions are in a sense 'locally' of this form (see also Problem 18). For this reason, the function a^2 is known as the *diffusion coefficient* and the function b is known as the *drift coefficient*.

Our method of finding a solution of (33.19) is to take a limit as $\varepsilon \searrow 0$ of the solution $Z^{(\varepsilon)}$ of the stochastic difference equation (33.2). Readers familiar with numerical solutions of ordinary differential equations should see that this approach is a stochastic version of the 'Euler method'. Since we will want to work in the Polish space $\mathbf{C}[0, \infty)$, we will actually take the limit of solutions of (33.5), which is the stochastic integral form of (33.2).

In the following, we use the terminology *bounded slope* to describe a function $f \colon \mathbb{R} \to \mathbb{R}$ if there exists a finite constant c such that

$$\left| \frac{f(x) - f(y)}{x - y} \right| \le c, \quad x \ne y.$$

(Such functions are also called 'uniformly Lipschitz continuous'.) Note that under this condition, f is automatically continuous, and

$$(33.20) \qquad |f(x)| \le c|x| + d$$

for some finite constant d, so Problem 2 applies to the solutions of stochastic difference equations with coefficients that have bounded slope.

Theorem 7. *Let $a, b \colon \mathbb{R} \to \mathbb{R}$ be functions with bounded slope, z a real number, and for $\varepsilon > 0$, let $Z^{(\varepsilon)}$ be the solution of (33.5). Then the limit*

$$Z = \lim_{\varepsilon \searrow 0} Z^{(\varepsilon)} \ i.p.$$

exists, and Z is the unique solution of (33.19). Furthermore, Z is a square-integrable nonanticipating W functional, and

$$(33.21) \qquad \lim_{\varepsilon \searrow 0} E\left(\int_0^t [Z_u^{(\varepsilon)} - Z_u]^2 \, du \right) = 0, \quad t \in [0, \infty).$$

PARTIAL PROOF. We will show that the limit Z exists and satisfies (33.19), from which it also follows that Z is a nonanticipating W-functional. In the course of the proof, we will also see that Z is square-integrable and that (33.21) holds. The proof of uniqueness is requested in Problem 16.

To show the existence of the limit Z, we prove that $\varepsilon \rightsquigarrow Z^{(\varepsilon)}$ is Cauchy in probability as $\varepsilon \searrow 0$. Fix $\varepsilon, \eta > 0$ and $t \in [0, \infty)$. From (33.5) we obtain the following:

$$\sup_{0 \le s \le t} |Z_s^{(\varepsilon)} - Z_s^{(\eta)}|$$

$$\le \sup_{0 \le s \le t} \left| \int_0^s [a(Z_{\tau^{(\varepsilon)}(u)}^{(\varepsilon)}) - a(Z_{\tau^{(\eta)}(u)}^{(\eta)})]\, dW_u \right| + \int_0^t \left| b(Z_{\tau^{(\varepsilon)}(u)}^{(\varepsilon)}) - b(Z_{\tau^{(\eta)}(u)}^{(\eta)}) \right| du .$$

Denote the two terms on the right side by A_t and B_t respectively. We have

$$P[\sup_{0 \le s \le t} |Z_s^{(\varepsilon)} - Z_s^{(\eta)}| > \delta] \le P[A_t > \delta/2] + P[B_t > \delta/2]$$

for $\delta > 0$.

By Problem 4 and Theorem 4, the Itô integral in the expression for A_t is a continuous-time martingale. By the Kolmogorov Inequality of Chapter 24 and the expression for the variance in Theorem 4,

$$P[A_t > \delta/2] \le \frac{4}{\delta^2} \int_0^t E\big([a(Z_{\tau^{(\varepsilon)}(u)}^{(\varepsilon)}) - a(Z_{\tau^{(\eta)}(u)}^{(\eta)})]^2\big)\, du .$$

Since a has bounded slope, it follows that there exists a finite constant c' such that

$$P[A_t > \delta/2] \le \frac{c'}{\delta^2} \int_0^t E\big([Z_{\tau^{(\varepsilon)}(u)}^{(\varepsilon)} - Z_{\tau^{(\eta)}(u)}^{(\eta)}]^2\big)\, du .$$

Similar reasoning (with the Chebyshev Inequality replacing the Kolmogorov Inequality) gives

$$P[B_t > \delta/2] \le \frac{c''t}{\delta^2} \int_0^t E\big([Z_{\tau^{(\varepsilon)}(u)}^{(\varepsilon)} - Z_{\tau^{(\eta)}(u)}^{(\eta)}]^2\big)\, du$$

for some finite constant c''.

Thus, in order to show that $\varepsilon \rightsquigarrow Z^{(\varepsilon)}$ is Cauchy in probability, it suffices to show that

$$\lim_{\varepsilon, \eta \searrow 0} E\big[(Z_t^{(\varepsilon)} - Z_t^{(\eta)})^2\big] = 0$$

(33.22)

uniformly for t in any bounded interval.

By definition

$$(Z_t^{(\varepsilon)} - Z_t^{(\eta)})^2$$

$$= \left[\int_0^t [a(Z_{\tau^{(\varepsilon)}(u)}^{(\varepsilon)}) - a(Z_{\tau^{(\eta)}(u)}^{(\eta)})] \, dW_u + \int_0^t [b(Z_{\tau^{(\varepsilon)}(u)}^{(\varepsilon)}) - b(Z_{\tau^{(\eta)}(u)}^{(\eta)})] \, du \right]^2$$

$$= \left[\int_0^t [a(Z_u^{(\varepsilon)}) - a(Z_u^{(\eta)})] \, dW_u + \int_0^t [b(Z_u^{(\varepsilon)}) - b(Z_u^{(\eta)})] \, du \right.$$

$$+ \int_0^t [a(Z_{\tau^{(\varepsilon)}(u)}^{(\varepsilon)}) - a(Z_u^{(\varepsilon)})] \, dW_u + \int_0^t [b(Z_{\tau^{(\varepsilon)}(u)}^{(\varepsilon)}) - b(Z_u^{(\varepsilon)})] \, du$$

$$\left. + \int_0^t [a(Z_u^{(\eta)}) - a(Z_{\tau^{(\eta)}(u)}^{(\eta)})] \, dW_u + \int_0^t [b(Z_u^{(\eta)}) - b(Z_{\tau^{(\eta)}(u)}^{(\eta)})] \, du \right]^2 .$$

The right side of this last expression is the square of a sum of 6 terms. Repeated use of the elementary inequality

$$(33.23) \qquad\qquad (x + y)^2 \le 2x^2 + 2y^2$$

shows that this squared sum is bounded above by 8 times the sum of the squares of the 6 terms. Take expectations of both sides, apply Theorem 4 to those terms on the right that contain an Itô integral, apply the Cauchy-Schwarz Inequality to the remaining terms on the right, and then use the Fubini Theorem to bring the expectation inside the integrals. The resulting inequality is

$$\frac{1}{8} E\big([Z_t^{(\varepsilon)} - Z_t^{(\eta)}]^2\big)$$

$$\le \int_0^t E\big([a(Z_u^{(\varepsilon)}) - a(Z_u^{(\eta)})]^2\big) \, du + t \int_0^t E\big([b(Z_u^{(\varepsilon)}) - b(Z_u^{(\eta)})]^2\big) \, du$$

$$+ \int_0^t E\big([a(Z_{\tau^{(\varepsilon)}(u)}^{(\varepsilon)}) - a(Z_u^{(\varepsilon)})]^2\big) \, du + t \int_0^t E\big([b(Z_{\tau^{(\varepsilon)}(u)}^{(\varepsilon)}) - b(Z_u^{(\varepsilon)})]^2\big) \, du$$

$$+ \int_0^t E\big([a(Z_u^{(\eta)}) - a(Z_{\tau^{(\eta)}(u)}^{(\eta)})]^2\big) \, du + t \int_0^t E\big([b(Z_u^{(\eta)}) - b(Z_{\tau^{(\eta)}(u)}^{(\eta)})]^2\big) \, du .$$

Letting c be the upper bound on the slopes of a and b, we obtain

$$E\big([Z_t^{(\varepsilon)} - Z_t^{(\eta)}]^2\big) \le 8c(1 + t) \int_0^t E\big([Z_u^{(\varepsilon)} - Z_u^{(\eta)}]^2\big) \, du$$

$$(33.24) \qquad\qquad + 8c(1 + t) \int_0^t E\big([Z_{\tau^{(\varepsilon)}(u)}^{(\varepsilon)} - Z_u^{(\varepsilon)}]^2\big) \, du$$

$$+ 8c(1 + t) \int_0^t E\big([Z_u^{(\eta)} - Z_{\tau^{(\eta)}(u)}^{(\eta)}]^2\big) \, du .$$

A straightforward calculation using (33.3) and (Problem 2) shows that (for $\varepsilon \le 1$)

$$E\big([Z_{\tau^{(\varepsilon)}(u)}^{(\varepsilon)} - Z_u^{(\varepsilon)}]^2\big) \le d' \varepsilon e^{c' u}$$

for some finite constants c', d'. A similar bound applies to the analogous term involving η. Put these bounds into (33.24) to see that there exists a function $g: [0, \infty) \to [0, \infty)$ that is bounded on bounded intervals such that

$$(33.25) \qquad \varphi(t) \le g(t)\left(\int_0^t \varphi(u)\, du + (\varepsilon + \eta) \right),$$

where

$$\varphi(u) = E\big([Z_u^{(\varepsilon)} - Z_u^{(\eta)}]^2\big).$$

Since $\varphi(0) = 0$, it follows from the Gronwall Inequality in the theory of ordinary differential equations that if t is restricted to a bounded interval, then there exist constants c'', d'' such that

$$\varphi(t) \le d''(\varepsilon + \eta)(e^{c''t} - 1).$$

The constants c'', d'' depend only on the right endpoint of the bounded interval and the constants c', d'. The right side of this expression converges to 0 uniformly for t in bounded intervals as $\varepsilon, \eta \searrow 0$. So we have proved (33.22), and the desired convergence in probability follows.

Let Z be the limit i.p. of $Z^{(\varepsilon)}$ as $\varepsilon \searrow 0$. It follows from (33.22) that Z is square-integrable and that (33.21) holds. Since a and b have bounded slope, it follows from Corollary 5 that the Itô integral in (33.5) converges i.p. to the Itô integral of $a \circ Z$ as $\varepsilon \searrow 0$, and it follows from the Uniform Integrability Criterion and Problem 12 of Chapter 8 that the Riemann integral in (33.5) converges to the Riemann integral of $b \circ Z$. Since the left side of (33.5) converges i.p. to Z_t as $\varepsilon \searrow 0$, we may conclude that Z satisfies (33.19), thus completing the proof. $\square$

Corollary 8. *Let a and b have bounded slope. Then the solutions of (33.19) with initial values $z \in \mathbb{R}$ form a strong Markov family with respect to the minimal right-continuous filtration of W.*

PROOF. Let T be an a.s.-finite stopping time with respect to $(\mathcal{F}_{t+}: t \in [0, \infty))$, the minimal right-continuous filtration of W. It is easy to see that $\tilde{Z} = (Z_{T+t}: t \in [0, \infty))$ satisfies the stochastic differential equation

$$d\tilde{Z} = a \circ \tilde{Z}\, d\widetilde{W} + b \circ \tilde{Z}\, dt$$

with random initial value Z_T, where $\widetilde{W}$ is the Wiener process $(W_{T+t} - W_T: t \in [0, \infty))$. It follows from the convergence portion of Theorem 7 that there is a measurable function $h: \mathbb{R} \times \mathbf{C}[0, \infty) \to \mathbf{C}[0, \infty)$ such that $h(z, W)$ is the unique solution of (33.19). Thus, $h(Z_T, \widetilde{W}) = \tilde{Z}$ a.s. Since $\widetilde{W}$ is independent of $\mathcal{F}_{T+}$ and Z_T is measurable with respect to $\mathcal{F}_{T+}$, it follows from Problem 21 and Proposition 12, both of Chapter 21, that the conditional distribution of $\tilde{Z}$ given $\mathcal{F}_{T+}$ is the same as that of the solution of (33.19) with initial value $z = Z_T$. The desired conclusion now follows from Problem 12 of Chapter 31. $\square$

As in the preceding proof, it is often useful to consider solutions of stochastic differential equations with random initial conditions. It is easy to see that Theorem 7 allows us to do so. However, in order to ensure that the solution be Markovian, it is important to assume that the initial value be independent of the Wiener process W. We will always make such an assumption.

Problem 16. Prove the uniqueness assertion in Theorem 7. *Hint:* Suppose Z and Z' are both solutions of (33.19). Let $\varphi(t) = E\big([Z_t - Z'_t]^2\big)$. Mimic the part of the proof of Theorem 7 that leads up to the use of the Gronwall Inequality to show that $\varphi(t) = 0$ for all t.

* **Problem 17.** Show that the Markov family of solutions of (33.19) has a Feller transition semigroup.

Problem 18. Let Z be the solution of (33.19), the coefficients a and b being assumed to have bounded slope. Show that for any a.s.-finite stopping time T with respect to the minimal right-continuous filtration of W, the conditional distribution of

$$\frac{Z_{T+h} - Z_T - hb(Z_T)}{\sqrt{h}}$$

given $\mathcal{F}_{T+}$ converges a.s. as $h \searrow 0$ to a normal distribution with mean 0 and variance $a^2(Z_T)$. Also show that

$$\lim_{h \searrow 0} \frac{E(Z_{T+h} - Z_T \mid \mathcal{F}_{T+})}{h} = b(Z_T) \text{ a.s.}$$

and

$$\lim_{h \searrow 0} \frac{\mathrm{Var}(Z_{T+h} - Z_T \mid \mathcal{F}_{T+})}{h} = a^2(Z_T) \text{ a.s.}$$

Problem 19. Let $f : \mathbb{R} \to \mathbb{R}$ be a bounded continuous function with bounded continuous first and second derivatives f', f''. Use the result of the preceding problem and the Taylor formula with remainder to show that if Z is the solution of (33.19) with initial value z, then

$$\lim_{h \searrow 0} \frac{E(f(Z_h) - f(z))}{h} = \frac{1}{2}a^2(z)f''(z) + b(z)f'(z).$$

Example 4. [Ornstein-Uhlenbeck processes] Let α and β be positive constants. For arbitrary initial conditions independent of W, solutions of the stochastic differential equation

$$dZ = \alpha \, dW - \beta Z \, dt$$

are known as *Ornstein-Uhlenbeck processes*, with *coefficients* α and β. Such processes are sometimes used to model the velocity of a small (but not microscopic) particle suspended in a fluid. The stochastic differential $\alpha \, dW$ represents changes in the velocity due to random bombardment by the molecules of the fluid, and $-\beta Z \, dt$ represents the effect of friction. Problem 22 shows that the position of such a particle will behave like standard Brownian motion if α and β are large and approximately equal.

By the Itô Lemma,

$$d(e^{\beta t} Z) = \alpha e^{\beta t} \, dW \, .$$

Thus

$$Z_t = e^{-\beta t} Z_0 + \alpha e^{-\beta t} \int_0^t e^{\beta u} \, dW_u \, .$$

If Z_0 is normally distributed with mean 0 and variance σ^2, then it follows from Problem 7 that Z is a Gaussian process with mean function 0 and covariance function

$$(s, t) \rightsquigarrow \sigma^2 e^{-\beta(s+t)} + \frac{\alpha^2}{2\beta} \left(e^{-\beta|s-t|} - e^{-\beta(s+t)} \right) .$$

If $\sigma^2 = (\alpha^2/2\beta)$, then the covariance function is

$$(s, t) \rightsquigarrow \sigma^2 e^{-\beta|s-t|} \, ,$$

which is a function of $|s - t|$. Thus, for this choice of σ^2, Z is a 'stationary Gaussian process' (the reader may provide a formal definition of this phrase), so Z is called a *stationary Ornstein-Uhlenbeck process*. In particular, the normal distribution with mean 0 and variance $(\alpha^2/2\beta)$ is an equilibrium distribution for the transition semigroup associated with the Markov family of Ornstein-Uhlenbeck processes with coefficients α and β.

Problem 20. Determine the relationship between Ornstein-Uhlenbeck processes and Ornstein-Uhlenbeck sequences which are defined in Chapter 28.

Problem 21. Let Z be an Ornstein-Uhlenbeck process with coefficients α and β and (random or nonrandom) initial state Z_0. Find the limiting distribution of Z_t as $t \to \infty$.

Problem 22. Let Z be a stationary Ornstein-Uhlenbeck process with coefficients α and β, and define $Y = (Y_t : t \in [0, \infty))$ by

$$Y_t = \int_0^t Z_u \, du \, .$$

Show that if $\alpha \to \infty$ and $\frac{\alpha}{\beta} \to 1$, then the distribution of Y converges to that of standard Brownian motion.

33.5. Generators and the Dirichlet problem

Problem 19 constitutes a calculation of the generator (at least on a portion of its domain) of the Markov family described in Corollary 8. There is an alternative method for calculating G using the Itô Lemma.

Example 5. Let Z be the solution of (33.19) with initial value z, the coefficients a and b being assumed to have bounded slope. Inspired by Problem 19, we set $Gf = \frac{1}{2}a^2 f'' + bf'$ for functions $f\colon \mathbb{R} \to \mathbb{R}$ having continuous first and second derivatives, with the hope of proving that G represents the generator. By (33.19),

$$(33.26) \qquad dZ = a \circ Z \, dW + b \circ Z \, dt .$$

Using this relation together with the Itô Lemma, we obtain

$$(33.27) \qquad d(f \circ Z) = (f' \circ Z) \, dZ + \tfrac{1}{2}(f'' \circ Z)(a \circ Z)^2 \, dt .$$

From (33.26) and (33.27), we obtain

$$d(f \circ Z) = (f' \circ Z)(a \circ Z) \, dW + Gf \circ Z \, dt .$$

In integral form we have

(33.28)

$$f(Z_t) - \int_0^t Gf(Z_u) \, du = f(z) + \int_0^t f'(Z_u)a(Z_u) \, dW_u , \quad t \in [0, \infty) .$$

Since f' is bounded and a has bounded slope, $|f'(x)a(x)| \le c|x|+d$ for some finite constants c, d. It follows from Theorem 7 that $(f' \circ Z)(a \circ Z)$ is square-integrable. By Theorem 4, the right side of (33.28) is a continuous-time martingale.

Thus, we have shown that Z is a solution to the martingale problem for $(G, \mathfrak{F})$, where $\mathfrak{F}$ is the collection of all bounded continuous functions $f\colon \mathbb{R} \to \mathbb{R}$ with bounded and continuous first and second derivatives. It follows that the generator of the Markov family of solutions of (33.19) agrees with G on $\mathfrak{F}$. Note that this example is an improvement on Problem 19, because we have explicitly identified the continuous-time martingale in the martingale problem as an Itô integral. Also note that our proof shows that (33.28) holds without the assumption that f, f', or f'' be bounded, as long as they are continuous.

Let G be the generator of a Markov family of diffusions with coefficients a and b having bounded slope, as calculated in the preceding example, and let (x_1, x_2) be a bounded open interval in $\mathbb{R}$. Also let f_1, f_2 be two real numbers. The *Dirichlet problem* for G on (x_1, x_2), with *boundary values* f_1, f_2, is to find a continuous function $h\colon [x_1, x_2] \to \mathbb{R}$ with bounded continuous first- and second-order derivatives on (x_1, x_2) such that

$$(33.29) \quad Gh(x) = 0, \quad x \in (x_1, x_2) \qquad \text{and} \qquad h(x_i) = f_i , \quad i = 1, 2 .$$

It is known from the theory of ordinary differential equations that the Dirichlet problem has a solution if $a(x) > 0$ for $x \in [x_1, x_2]$. The differential equation $Gu = 0$ is 'linear' in u, and there are standard nonprobabilistic methods for obtaining an explicit formula for the solution of linear 'boundary-value problems' like the Dirichlet problem.

Our goal in the remainder of this section is to use the methods of this chapter to find a probabilistic formula for the solution of the Dirichlet problem. We will also find probabilistic formulas for the solutions of other related boundary-value problems. To save space, we will use the fact that solutions of linear boundary-value problems exist to derive these formulas, even though it is possible to use probabilistic arguments to prove this fact directly.

For each $x \in \mathbb{R}$ let $Z^{(x)}$ denote the unique solution of the stochastic differential equation (33.19), with coefficients a, b having bounded slope, and initial value x. Let

$$T_x = \inf\{t \geq 0 : Z_t^{(x)} = x_1 \text{ or } x_2\}.$$

Thus, for $x \in [x_1, x_2]$, T_x is the first exit time of the interval (x_1, x_2) for the process $Z^{(x)}$. Note that T_x is a stopping time with respect to the minimal filtration of the Wiener process used to construct Z^x.

Theorem 9. *Suppose $a(x) > 0$ for $x \in [x_1, x_2]$, and let $g : [x_1, x_2] \to \mathbb{R}$ be the unique continuous function satisfying $Gg(x) = -1$, $x \in (x_1, x_2)$, with boundary values $g(x_1) = g(x_2) = 0$. Then $E(T_x) = g(x)$, $x \in [x_1, x_2]$.*

PROOF. By (33.28)

$$g(Z_t^{(x)}) - \int_0^t Gg(Z_u^{(x)}) \, du = g(x) + \int_0^t g'(Z_u^{(x)}) a(Z_u^{(x)}) \, dW_u, \quad t \in [0, \infty).$$

(In order for this equation to hold for all t, we should extend g to be defined on all of $\mathbb{R}$. However, we are only interested in $t \leq T_x$, so such an extension is irrelevant.) For $u < T_x$, $Gg(Z_u^{(x)}) = -1$. Thus

(33.30)
$$g(Z_{T_x \wedge t}^{(x)}) + (T_x \wedge t) = g(x) + \int_0^{T_x \wedge t} g'(Z_u^{(x)}) a(Z_u^{(x)}) \, dW_u, \quad t \in [0, \infty).$$

The right side has expected value $g(x)$ by Theorem 4. Letting $t \to \infty$ and taking expected values on the left side, we see from the Monotone Convergence Theorem that $E(T_x) < \infty$, so $T_x < \infty$ a.s. Hence, $Z_{T_x \wedge t}^{(x)} \to Z_{T_x}^{(x)}$ a.s. as $t \to \infty$. The desired result now follows from the Bounded Convergence Theorem and the fact that the boundary values of g are 0. $\square$

Using similar methods, we can also obtain a formula for $E(T_x^2)$. The proof is requested of the reader in Problem 24.

Theorem 10. *Suppose $a(x) > 0$ for $x \in [x_1, x_2]$, let g be the function defined in Theorem 9, and $\gamma: [x_1, x_2] \to \mathbb{R}$ the unique continuous function, with bounded continuous first and second derivatives on (x_1, x_2), satisfying $G\gamma(x) = -(g(x)a(x))^2$, $x \in (x_1, x_2)$, with boundary values $\gamma(x_1) = \gamma(x_2) = 0$. Then $E(T_x^2) = \gamma(x) - g(x)^2$, $x \in [x_1, x_2]$.*

Now that we know that T_x is a.s.-finite, the following formula for the solution of the Dirichlet problem is easy to derive from the fact that $Z^{(x)}$ is a solution to the martingale problem for $(G, \mathfrak{F})$.

Theorem 11. *Suppose that $a(x) > 0$ for $x \in [x_1, x_2]$, and let $h: [x_1, x_2] \to \mathbb{R}$ be the unique solution of the Dirichlet problem on (x_1, x_2) for G with boundary values f_1, f_2. Then*

$$h(x) = f_1 P[Z_{T_x}^{(x)} = x_1] + f_2 P[Z_{T_x}^{(x)} = x_2], \quad x \in [x_1, x_2].$$

Corollary 12. *Suppose $a(x) > 0$ for $x \in [x_1, x_2]$, and let $h: [x_1, x_2] \to \mathbb{R}$ be any nonconstant continuous function that satisfies $Gh(x) = 0$ for all $x \in [x_1, x_2]$. Then for all $x \in [x_1, x_2]$,*

$$P[Z_{T_x}^{(x)} = x_1] = \frac{h(x) - h(x_2)}{h(x_1) - h(x_2)}.$$

We can obtain an explicit formula for a suitable function h in the preceding corollary. Let

(33.31)
$$s(u) = e^{-\int_{u_0}^{u} \frac{2b(v)}{a^2(v)} \, dv},$$

where u_0 is some constant. It is easily checked that

$$h(x) = \int_{x_0}^{x} s(u) \, du$$

satisfies the conditions of the corollary for arbitrary intervals $[x_1, x_2]$ on which a is positive, and for arbitrary values of $x_0, u_0 \in [x_1, x_2]$.

Note that h is strictly increasing, and hence invertible. For $y = h(x)$, $x \in [x_1, x_2]$, let $Y^{(y)} = Z^{(x)}$ and write $S_y = T_x$, $y_1 = h(x_1)$, and $y_2 = h(x_2)$. Note that S_y is the first exit time of the interval (y_1, y_2) by the process $Y^{(y)}$. Trivially, Corollary 12 implies that

(33.32)
$$P[Y_{S_y}^{(y)} = y_1] = \frac{y - y_2}{y_1 - y_2}, \quad y \in [y_1, y_2].$$

In Problem 26, the reader is asked to show that the processes $Y^{(y)}$, $y \in [y_1, y_2]$, are diffusions that all satisfy the same differential equation. We express the fact that this collection of diffusions satisfies (33.32) by saying that it is *on its natural scale*. Since h transforms the collection $(Z^{(x)}: x \in [x_1, x_2])$ into a collection that is on its natural scale, we call h the *scale function* of the diffusions $Z^{(x)}$.

Problem 23. Let $f: \mathbb{R} \to \mathbb{R}$ have bounded continuous second derivative, but do not assume that either f or f' is bounded. Let a and b be diffusion and drift coefficients with bounded slope, and assume that a is bounded. Define $Gf(x) = \frac{1}{2}a^2(x)f''(x) + b(x)f'(x)$. Show that the left side of (33.28) is a continuous-time martingale.

Problem 24. Prove Theorem 10. *Hint:* First use (33.30) to show that

$$E(T_x^2) = g(x)^2 + E\left(\int_0^{T_x} (g'(u)a(u))^2 \, du \right).$$

Then carry out an argument similar to the proof of Theorem 9.

Problem 25. Prove Theorem 11 and its corollary.

Problem 26. Prove that the collection of processes $Y^{(y)}$, $y \in [y_1, y_2]$, defined above are all solutions of the stochastic differential equation

$$dY = (h' \circ h^{-1} \circ Y)(a \circ h^{-1} \circ Y)\, dW.$$

Problem 27. Find explicit formulas for $E(T_x)$, $E(T_x^2)$, and $P[Z_{T_x}^{(x)} = x_i]$, $i = 1, 2$, for the case in which $a > 0$ and b are constants, the so-called 'scaled' Wiener process with 'drift'. (You will need to be able to solve some simple linear ordinary differential equations with constant coefficients.)

Problem 28. Find a formula in terms of a normal distribution function for $P[Z_{T_x}^{(x)} = x_i]$, $i = 1, 2$, for the Markov family of Ornstein-Uhlenbeck processes with coefficients α, β.

33.6. Diffusions in higher dimensions

Most of what we have done in this chapter generalizes easily to dimensions $d \geq 2$. The starting point of such a generalization is the *d-dimensional Wiener process*, which is the $\mathbf{C}([0, \infty), \mathbb{R}^d)$-valued random variable $\mathbf{W}$ defined by

$$\mathbf{W} = (W^{(1)}, \ldots, W^{(d)}),$$

where the $W^{(i)}$, $i = 1, \ldots, d$ are independent (1-dimensional) Wiener processes.

By mimicking the development given in this chapter, it is straightforward to give meaning to integration with respect to $\mathbf{W}$, with results like Theorem 4 and the Itô Lemma having natural generalizations. This d-dimensional 'Itô integral' gives meaning to stochastic differential equations of the form

$$d\mathbf{Z} = \mathbf{a}(\mathbf{Z})d\mathbf{W} + \mathbf{b}(\mathbf{Z})dt,$$

where for each $x \in \mathbb{R}$, $\mathbf{a}(x)$ is a $d \times d$ matrix and $\mathbf{b}(x)$ is a member of $\mathbb{R}^d$.

Using a suitable definition of 'bounded slope', the arguments we used in the proof of Theorem 7 can be used with very little modification to show the existence and uniqueness of solutions of stochastic differential equations whose coefficients have bounded slope. For a given pair of coefficients $\mathbf{a}, \mathbf{b}$, the collection of solutions with initial states $x \in \mathbb{R}^d$ forms a Markov family whose generator G takes the form

$$(33.33) \qquad Gf(x) = \frac{1}{2} \sum_{i,j=1}^d \alpha_{ij}(x) D_{ij} f(x) + \sum_{i=1}^d b_i(x) D_i f(x),$$

where $\alpha_{ij}(x)$ is the ij-entry of the symmetric $d \times d$ matrix $\mathbf{a}^T(x)\mathbf{a}(x)$ and $b_i(x)$ is the ith entry of the vector $\mathbf{b}(x)$. This formula is valid for bounded functions f having bounded continuous first- and second-order partial derivatives $D_i f, D_{ij} f$.

We have already seen that diffusions can be used to solve the Dirichlet problem in 1 dimension. The generalization of this fact to higher dimensions has great importance in the field of partial differential equations. Let U be a bounded connected open subset of $\mathbb{R}^d$, with boundary ∂U. Given an operator G of the form (33.33) and a continuous function $f \colon \partial U \to \mathbb{R}$, the *Dirichlet problem* for G on U with *boundary condition* f is to find a continuous function $h \colon U \cup \partial U \to \mathbb{R}$ such that

$$Gh(x) = 0, \quad x \in U, \quad \text{and} \quad h(x) = f(x), \quad x \in \partial U.$$

Because the set U can have a complicated shape when $d \geq 2$, this problem is considerably more difficult in higher dimensions than in 1 dimension. Nevertheless, it can be shown that when a solution exists, it takes the form

$$(33.34) \qquad\qquad h(x) = E(f(\mathbf{Z}^{(x)}_{T_x})),$$

where T_x is the hitting time of ∂U by $Z^{(x)}$, the solution of the stochastic differential equation with coefficients $\mathbf{a}, \mathbf{b}$, and initial value x. Even when no solution exists, as can happen when U or f are not sufficiently nice, it is still useful to regard (33.34) as a 'generalized solution' of the Dirichlet problem.

* **Problem 29.** What is the generator of the d-dimensional Wiener process?

Problem 30. [A special case of the Itô Lemma] Let $f \colon \mathbb{R}^d \to \mathbb{R}$ be a continuous function with continuous first- and second-order partial derivatives $D_i f, D_{ij} f$, $i, j = 1, \ldots, d$. Appropriately interpret and prove the following formula:

$$d(f \circ \mathbf{W}) = \nabla f \circ \mathbf{W} \cdot d\mathbf{W} + \tfrac{1}{2} \Delta f \circ \mathbf{W} \, dt,$$

where '$\cdot$' denotes the usual Euclidean inner product, ∇ denotes the gradient operator, and Δ denotes the Laplacian operator.

PART 7
Appendices

Appendix A includes short descriptions of symbols and usage of terms relevant for reading the text. It focuses more on notational conventions and concepts from areas of mathematics different from probability than on the mathematics introduced in this book.

Appendices B-E contain introductions to some non-probabilistic topics that are important for certain sections of the book. Portions of Appendices B and C on metric spaces and topology are relevant for the beginning chapters.

Appendix F contains a list of books which one might want to read either concurrently with or subsequently to reading this book. References relevant to some specific propositions or theorems in the text are given in Appendix G along with some additional comments. The rule we have tried to follow for the inclusion of references in this section is: include a reference if the result is not yet part of the general body of knowledge that has appeared in other textbooks and we feel it is not known by most probabilists. We apologize for any oversights in this connection.

APPENDIX A
Notation and Usage of Terms

The first section of this appendix defines a variety of symbols used in the text, focusing on those related to prerequisite mathematical knowledge. The second section specifies how certain terms will be used, especially those which are used in various ways by different authors. A few exercises connected with notational issues constitute the third section.

A.1. Symbols

$\stackrel{\text{def}}{=}$ indicates that the expression to the left is being defined as the expression on the right, and is only used when required for clarity.

$A^c = \{x : x \notin A\}$

$A \setminus B = \{x : x \in A \cap B^c\}$

$A \triangle B = (A \setminus B) \cup (B \setminus A)$

$\emptyset$ denotes the emptyset.

$A \subseteq B$ means that A is a subset of B.

$A \subset B$ means that $A \neq B$ and $A \subseteq B$.

$A \supseteq B$ means that A is a superset of B.

$A \supset B$ means that $A \neq B$ and $A \supseteq B$.

$A - y = \{x - y : x \in A\}$

$A \vee B$ denotes the *convex hull* of A and B.

$A + B = \{x + y : x \in A, y \in B\}$, the *Minkowski sum* of A and B.

$\mathbb{R} = (-\infty, \infty)$

$\overline{\mathbb{R}} = [-\infty, \infty]$

$\mathbb{R}^+ = [0, \infty)$

$\overline{\mathbb{R}}^+ = [0, \infty]$

$\mathbb{Z} = \{\ldots, -2, -1, 0, 1, 2, 3, \ldots\}$
$\overline{\mathbb{Z}} = \{-\infty, \ldots, -2, -1, 0, 1, 2, 3, \ldots, \infty\}$
$\mathbb{Z}^+ = \{0, 1, 2, 3, \ldots\}$
$\overline{\mathbb{Z}}^+ = \{0, 1, 2, 3, \ldots, \infty\}$

$\#A$ denotes the cardinality of the set A.
$|J|$ denotes the length of the interval J.

$\mathbb{C}$ denotes the set of complex numbers. See Appendix E.
$|z|$ denotes the absolute value of the complex number z, that is, the distance in the complex plane between z and 0.
$\mathcal{R}(z)$ denotes the real part of the complex number z.
$\mathcal{I}(z)$ denotes the imaginary part of the complex number z.

$\mathbb{R}^d$ denotes d-dimensional Euclidean space.
$\mathbf{e}_i$, for $i \leq d$, denotes the ith member of the standard basis for $\mathbb{R}^d$.
$\langle x, y \rangle$ denotes the inner product (that is, dot product or scalar product) of x and y in $\mathbb{R}^d$.
$|x|$ denotes the norm (that is, distance from origin) of $x \in \mathbb{R}^d$, which reduces to the absolute value of x if $d = 1$.
B^T denotes the transpose of the matrix B.
A vector $x \in \mathbb{R}^d$ can be viewed as a row matrix. Thus x^T denotes the corresponding column matrix.

$\| \cdot \|$ denotes norm in a variety of spaces.

$\arg x$ denotes the polar coordinates angle of a point $x \in \mathbb{R}^2$.

$(a)_b^\uparrow$ denotes the *rising factorial* $\prod_{k=0}^{b-1}(a + k)$, $a \in \mathbb{R}$, $b \in \mathbb{Z}^+$.
$(a)_b^\downarrow$ denotes the *falling factorial* $\prod_{k=0}^{b-1}(a - k)$, $a \in \mathbb{R}$, $b \in \mathbb{Z}^+$.
the binomial and multinomial coefficients:

$$\binom{n}{r} = \frac{n!}{r!(n-r)!} = \frac{(n)_r^\downarrow}{r!}$$

$$\binom{n}{r_1 \ r_2 \ \cdots \ r_d} = \frac{n!}{r_1! r_2! \cdots r_d!} \quad \text{for } n = r_1 + r_2 + \cdots + r_d$$

$f: A \to B$ describes f to be a function with domain A and target B.
$x \rightsquigarrow f(x)$ is another name for the function f. Thus $x \rightsquigarrow x^2$ denotes the squaring function.

$f \circ g$ denotes the composition of the functions f and g, but see Appendix E.

$f^{-1}(C) = \{x \colon f(x) \in C\}$.

$\cdot$ is used as a place holder for the domain variable in a function. Thus if f is a function of two variables, $f(\cdot, y)$ denotes the function of the first variable obtained by fixing y to be the value of the second variable.

a.s. is an abbreviation for almost surely.

a.e. is an abbreviation for almost everywhere.

i.p. is an abbreviation for in probability.

iid is an abbreviation for independent, identically distributed.

$[A]$ denotes the equivalence class to which A belongs. Typical uses: $[A]$ for the collection of events equal to A almost surely and $[X]$ for the collection of random variables equal to X almost surely.

$a \wedge b$ denotes the smaller of a and b.

$a \vee b$ denotes the larger of a and b.

$\inf A$, the *infimum of* the set $A \subseteq \overline{\mathbb{R}}$, denotes the greatest lower bound of A, with $\inf \emptyset = \infty$.

$\sup A$, the *supremum of* the set $A \subseteq \overline{\mathbb{R}}$, denotes the least upper bound of A, with $\sup \emptyset = -\infty$. Occasionally the universal set will be $\overline{\mathbb{R}}^+$, in which case $\sup \emptyset = 0$.

$\min A$, the *minimum of* the set $A \subseteq \overline{\mathbb{R}}$, denotes the smallest member of A, and thus entails the assertion that A has a smallest member.

$\max A$, the *maximum of* the set $A \subseteq \overline{\mathbb{R}}$, denotes the largest member of A, and thus entails the assertion that A has a largest member.

$a^+ = a \vee 0$

$a^- = (-a) \vee 0$

$\lfloor a \rfloor$, the *floor of* a is the largest integer no larger than a.

$\lceil a \rceil$, the *ceiling of* a is the smallest integer no smaller than a.

$f(x-) = \lim_{y \nearrow x} f(y)$

$f(x+) = \lim_{y \searrow x} f(y)$

$f(\infty) = \lim_{y \to \infty} f(y)$, unless ∞ is a member of the domain of f in which case $f(\infty)$ has a direct meaning.

$f(-\infty) = \lim_{y \to -\infty} f(y)$, unless $-\infty$ is a member of the domain of f in which case $f(-\infty)$ has a direct meaning.

$O(f(x))$ as $x \to a \in \overline{\mathbb{R}}$ means that $\limsup \frac{|O(f(x))|}{|f(x)|} < \infty$ as $x \to a$ through the domain of $O(f(x))$.

$o(f(x))$ as $x \to a \in \overline{\mathbb{R}}$ means that $\frac{|o(f(x))|}{|f(x)|} \to 0$ as $x \to a$ through the domain of $o(f(x))$.

I_A denotes the indicator function of the set A.

exp denotes the natural exponential function. Thus $\exp(x) = e^x$. Also, see Appendix E.

log denotes the natural logarithmic function, the inverse function of exp. Also, see Appendix E.

$\mathbf{C}[a, b]$ and $\mathbf{C}[a, \infty)$ denote the spaces of continuous $\mathbb{R}$-valued functions on the intervals $[a, b]$ and $[a, \infty)$, respectively.

$\mathbf{D}[a, b]$ and $\mathbf{D}[a, \infty)$ denote the spaces of right-continuous $\mathbb{R}$-valued functions having left limits (that is *cadlag* functions) on the intervals $[a, b]$ and $[a, \infty)$, respectively.

$\mathbf{D}^+[a, b]$ and $\mathbf{D}^+[a, \infty)$ denote the spaces of increasing right-continuous $\overline{\mathbb{R}}^+$-valued functions on the intervals $[a, b]$ and $[a, \infty)$, respectively.

$\mathbf{D}([0, \infty), \Psi)$ denotes the space of right-continuous Ψ-valued functions having left limits (that is *cadlag* functions) on the interval $[0, \infty)$.

E or E_P denotes the expectation operator, the subscript P emphasizing that P is the underlying probability measure.

$E(X \, ; \, B)$ denotes the expectation of the product of X and the indicator function of B.

$\mathrm{Var}(X)$, $\mathrm{Cov}(X, Y)$, and $\mathrm{Corr}(X, Y)$, respectively, denote the variance of X, the covariance of X and Y, and the correlation of X and Y (with a subscript P being used, if appropriate, to emphasize that P is the underlying probability measure).

$X_n \xrightarrow{\mathcal{D}} X$ as $n \to \infty$ means that the sequence (X_n) converges to X in distribution.

$\int_{a-}^{b} g \, dF$ denotes $\lim_{\varepsilon \searrow 0} \int_{a-\varepsilon}^{b} g \, dF$, a limit of Riemann-Stieltjes integrals.

A.2. Usage

Binary digits is used for what some call *bits*.

$x_n \to l$ as $n \to \infty$ has the same meaning as $\lim_{n \to \infty} x_n = l$, and is suitable for modification for extra meaning: $x_n \nearrow l$ as $n \nearrow \infty$ entails the assertion that the sequence $(x_1, x_2, \dots)$ is increasing for all sufficiently large subscripts. Similarly

$x_n \searrow l$ as $n \nearrow \infty$ implies that the sequence is ultimately decreasing.

The *product* of 0 and ∞ is to be understood to equal 0, unless the surrounding discussion makes it clear that some other view is appropriate.

$0^0 = 1$, $1^\infty = 1$, unless otherwise specified.

In $\overline{\mathbb{R}}^+$-setting, $\frac{b}{0} = \infty$ for $b > 0$ is sometimes used.

The 0-fold convolution of a distribution is the identity for the convolution operation, namely the delta distribution at 0.

The derivative of order 0 of a function is the function itself.

A sum of an empty collection of summands is equal 0. Example: $\sum_{k=1}^{0} k^3 = 0$.

A product of an empty collection of factors is 1. Example: $\prod_{k=1}^{0} k^3 = 1$.

$0! = (a)_0^\uparrow = (a)_0^\downarrow = 1$ for $a \in \mathbb{R}$.

The union of an empty collection of sets is the empty set.

The intersection of an empty collection of sets is the universal set, possibly not identified explicitly.

proper difference of A and B is an appropriate description of $A \setminus B$ in the case that $B \subseteq A$; it is not required that B be different from either A or $\emptyset$.

That A be a *proper subset* of B means that $A \subset B$, the possibility that $A = \emptyset$ not being excluded.

$(A_1, A_2, \dots)$ is a *decreasing* sequence of sets if $A_n \supseteq A_{n+1}$ for every n.

$(A_1, A_2, \dots)$ is an *increasing* sequence of sets if $A_n \subseteq A_{n+1}$ for every n.

$(A_1, A_2, \dots)$ is a *monotonic* sequence of sets if it is decreasing or increasing.

$(A_1, A_2, \dots)$ is a *strictly decreasing* sequence of sets if $A_n \supset A_{n+1}$ for every n.

$(A_1, A_2, \dots)$ is a *strictly increasing* sequence of sets if $A_n \subset A_{n+1}$ for every n.

$(A_1, A_2, \dots)$ is a *strictly monotonic* sequence of sets if it is strictly decreasing or strictly increasing.

The symbols $\mathcal{A}$, $\mathcal{B}$, and $\mathcal{C}$ always denote Borel σ-fields. Arbitrary σ-fields can be denoted by symbols such as $\mathcal{F}$ and $\mathcal{G}$, whether Borel or not. And letters in this style may not even denote σ-fields. For instance, $\mathcal{E}$ is often used for an arbitrary family of sets.

image of f is the set $\{y \colon f(x) = y \text{ for some } x\}$

image of x under f is $f(x)$

target of f is any set of which the image of f is asserted to be a subset.

$f^+ = f \vee 0$ is called the *positive part* of the function f.

$f^- = (-f) \vee 0$ is called the *negative part* of the function f.

f is *cadlag* if it is right-continuous and has left limits on its domain.

f is *decreasing* if $f(x) \geq f(y)$ whenever $x \leq y$.

f is *increasing* if $f(x) \leq f(y)$ whenever $x \leq y$.

f is *monotonic* if f is decreasing or is increasing.

f is *strictly decreasing* if $f(x) > f(y)$ whenever $x < y$.

f is *strictly increasing* if $f(x) < f(y)$ whenever $x < y$.

f is *strictly monotonic* if f is strictly decreasing or is strictly increasing.

An $\mathbb{R}$-valued operator is *positive* if it assigns nonnegative values to positive elements.

An $\mathbb{R}$-valued operator is *strictly positive* if it assigns positive values to positive elements.

The term *definite* is adjoined to the terms *positive* and *strictly positive* when viewing matrices as operators, this adjective making it clear that one is not speaking of the entries of the matrix.

ω sometimes means $\{\omega\}$.

partition is used in two distinct related ways; check the index.

A.3. Exercises on subtle distinctions

The purpose of these exercises is to focus attention on some of the conventions described in the preceding section.

Problem 1. Prove that a function $f \colon \mathbb{R} \to \mathbb{R}$ that is strictly increasing on an interval $[a, b]$ and on an interval $[b, c]$ is strictly increasing on the interval $[a, c]$

* **Problem 2.** Use the preceding problem and a standard calculus theorem to prove that the function $x \rightsquigarrow x - \sin x$ is strictly increasing on the interval $[-2\pi, 2\pi]$.

Problem 3. For a one-to-one function f, the notation f^{-1} has two distinct but closely related meanings. Discuss.

Problem 4. Let $a > b > 0$. Prove that the sequence $\left((a)_c^{\uparrow}/(b)_c^{\uparrow} : c = 0, 1, \ldots \right)$ is strictly increasing. Does your proof show a strict increase from $c = 0$ to $c = 1$ or is a separate argument needed? Explain.

APPENDIX B
Metric Spaces

Often, measurable sets are Borel sets, that is, members of the smallest σ-field containing all the open sets. A natural setting for Borel σ-fields is that of metric spaces, properties of which we review here. Also, some important examples will be examined.

B.1. Definition

A *metric space* consists of a set Ψ and a function $\rho\colon \Psi \times \Psi \to \mathbb{R}^+$ satisfying the following properties:

(i) $\rho(x, y) = 0$ if and only if $y = x$;

(ii) $\rho(x, y) = \rho(y, x)$;

(iii) $\rho(x, z) \le \rho(x, y) + \rho(y, z)$.

The last two properties are called *symmetry* and *triangle inequality*, respectively. The function ρ is called the *metric* on Ψ and its value at a particular pair (x, y) is the *distance* between x and y. A metric space, thus defined, is denoted by (Ψ, ρ), or more briefly by Ψ if there is no ambiguity concerning the metric.

For x a member of a metric space (Ψ, ρ) and $\varepsilon > 0$, the sets

$$\{y\colon \rho(x, y) < \varepsilon\}, \quad \{y\colon \rho(x, y) \le \varepsilon\}, \quad \text{and} \quad \{y\colon \rho(x, y) = \varepsilon\}$$

are called the *open ball, closed ball,* and *sphere* of radius ε centered at x. A subset B of a metric space with metric ρ is an *open set* if for every $x \in B$ there exists $\varepsilon > 0$ such that the open ball of radius ε centered at x is a subset of B. It is easy to use the triangle inequality to prove that every open ball is an open set. A set is a *closed set* if its complement is open. It is easy to prove that all spheres and closed balls are closed sets.

For a set $C \subseteq \Psi$, the *interior* of C is the largest open subset of C; the *closure* of C is its smallest closed superset; and the *boundary* of C, denoted by ∂C, consists of those points in its closure that are not also in its interior. It is possible to prove the existence of the interior and closure of any subset of a metric space.

A subset B of a set C in a metric space is *dense in* C if every ball centered at a point in C contains a member of B. The modifying phrase "in C" is often omitted if C is the entire metric space. A metric space is *separable* if it contains a countable dense subset.

A set in a metric space is *bounded* if it is contained in some ball. It is *totally bounded* if for every $\varepsilon > 0$, it is contained in the union of a finite collection of balls of radius less than ε. It is *compact* if every open cover of it has a finite subcovering. (A collection of sets is called a *cover* of a set C if C is a subset of their union; the cover is *open* if each of the sets in the cover is open.) A set is *relatively compact* if it has compact closure. If any of the adjectives *bounded, totally bounded,* and *compact* apply to the set of all points in a metric space, then the corresponding adjective is also used for the metric space itself.

* **Problem 1.** Prove the following facts about sets in any metric space.
   - Finite intersections and arbitrary unions of open sets are open.
   - Finite unions and arbitrary intersections of closed sets are closed.
   - Finite unions and arbitrary intersections of compact sets are compact.
   - Every compact set is closed.
   - A closed subset of a compact set is compact.
   - The intersection of a collection of compact sets is empty if and only if the intersection of some finite subcollection is empty.
   - Any set in a metric space is itself a metric space with the inherited metric.

Problem 2. Use the first two items in the preceding problem to prove the facts mentioned above: every set has an interior, possibly empty; every set has a closure.

B.2. Sequences

A sequence $(x_1, x_2, \dots)$ in a metric space (Ψ, ρ) *converges* to a point $x \in \Psi$ if, for every $\varepsilon > 0$, there exists an integer p such that $\rho(x_n, x) < \varepsilon$ whenever $n \geq p$. The sequence is *Cauchy* if, for every $\varepsilon > 0$, there exists p such $\rho(x_m, x_n) < \varepsilon$ whenever $n \geq m \geq p$.

Proposition 1. *Every convergent sequence in a metric space is Cauchy, and every Cauchy sequence which has a convergent subsequence converges.*

Problem 3. Prove the preceding proposition.

A metric space in which every Cauchy sequence is convergent is said to be *complete*.

Proposition 2. *Every sequence in a totally bounded metric space has a subsequence that is Cauchy.*

Problem 4. Prove the preceding proposition.

A set C in a metric space is *relatively sequentially compact* if every sequence in C has a subsequence that converges. In case the subsequence can always be chosen so that its limit belongs to C, C is *sequentially compact*. The proof of the following result will be omitted.

Proposition 3. *Compactness is equivalent to sequential compactness; relative compactness is equivalent to relative sequential compactness.*

In practice one often proves that a sequence converges by simultaneously proving relative sequential compactness and that every convergent subsequence has the same limit. The next proposition entails a recipe for doing this.

Proposition 4. *A sequence $(x_n : n = 1, 2, \ldots)$ converges to a limit y if and only if every subsequence of $(x_n : n = 1, 2, \ldots)$ has a further subsequence that converges to y.*

* **Problem 5.** Prove the preceding proposition.

B.3. Continuous functions

A function g from one metric space (Ψ_1, ρ_1) to another metric space (Ψ_2, ρ_2) is *continuous* at $x \in \Psi_1$ if for every $\varepsilon > 0$, there exists $\delta > 0$ such that $\rho_2(g(y), g(x)) < \varepsilon$ for every y satisfying $\rho_1(x, y) < \delta$. Equivalently, g is continuous at x if for every sequence $(x_n : n = 1, 2, \ldots)$ in Ψ_1 converging to x, it is true that $g(x_n) \to g(x)$ as $n \to \infty$. The function g is *continuous* if it is continuous at each point. It is *uniformly continuous* if δ can be chosen to depend only on ε and g, but not on x.

Problem 6. Prove that if $f : \Psi_1 \to \Psi_2$ is a function from one metric space to another, then f is continuous if and only if for any open set $A \subset \Psi_2$, $f^{-1}(A)$ is an open subset of Ψ_1.

Problem 7. Prove that if f is a continuous function from one metric space to another, then the image under f of any compact set is compact. *Hint:* Use Problem 6.

B.4. Important metric spaces

The function $(x, y) \rightsquigarrow |y - x|$ is a metric for $\mathbb{R}$. Another metric for $\mathbb{R}$ is $(x, y) \rightsquigarrow |\arctan y - \arctan x|$. These two metrics for $\mathbb{R}$ make $\mathbb{R}$ into two different metric spaces. With the first metric, $\mathbb{R}$ is complete but not bounded, and with the second it is totally bounded but not complete. However, the open sets determined by the two metrics are easily seen to be identical. Thus these two metrics turn $\mathbb{R}$ into the same measurable space.

The metric $(x, y) \rightsquigarrow |\arctan y - \arctan x|$ for $\mathbb{R}$ can be extended to a metric for $\overline{\mathbb{R}}$ by defining $\arctan \infty = \pi/2$ and $\arctan(-\infty) = -\pi/2$. With this metric large finite real numbers are 'close' to ∞ and negative finite real numbers of large absolute value are 'close' to $-\infty$. The metric space $\overline{\mathbb{R}}$ is complete and compact. The function $x \rightsquigarrow \arctan x$ from $\overline{\mathbb{R}}$ to $[-\frac{\pi}{2}, \frac{\pi}{2}]$ is a continuous bijection with a continuous inverse, where the metric for $[-\frac{\pi}{2}, \frac{\pi}{2}]$ is $(u, v) \rightsquigarrow |u - v|$.

In $\mathbb{R}^d$ we let $|x|$ denote the Euclidean distance from the point $x \in \mathbb{R}^d$ to the origin. The function $(x, y) \rightsquigarrow |x - y|$ is a metric for $\mathbb{R}^d$ which turns $\mathbb{R}^d$ into a complete metric space.

The standard way of making $\mathbf{C}[0, 1]$, the set of continuous functions on the interval $[0, 1]$, into a metric space is to define the distance between f and g to equal $\max\{|f(t) - g(t)| : 0 \le t \le 1\}$.

Problem 8. Show that the closed ball of radius 1 centered at the 0 function in $\mathbf{C}[0, 1]$ is not totally bounded. *Hint:* Construct an infinite sequence $(f_n : n = 1, 2, \dots)$ in B such that the distance between f_m and f_n equals 2 for $m \ne n$.

Theorem 5. [Arzelà-Ascoli] *A subset A of $\mathbf{C}[0, 1]$ is relatively sequentially compact if and only if $\{f(0) : f \in A\}$ is a bounded set of real numbers and, for every $\varepsilon > 0$, there exists $\delta > 0$ such that $|f(x) - f(y)| < \varepsilon$ whenever $|x - y| < \delta$ and $f \in A$.*

We omit the proof of this theorem. Notice that δ in it is not permitted to depend on f. A set A of functions is said to be *equicontinuous* at a point x if for every $\varepsilon > 0$ there exists $\delta > 0$ such that $|f(y) - f(x)| < \varepsilon$ whenever $|y - x| < \delta$ and $f \in A$. If δ can be chosen to be independent of x, then the family A is said to be *uniformly equicontinuous*. Thus, the Arzelà-Ascoli Theorem can be stated as: a subset of $\mathbf{C}[0, 1]$ is relatively sequentially compact if and only if it is uniformly equicontinuous and the set of its values at 0 is bounded. Moreover, 'uniformly' need not be mentioned because it is a consequence of equicontinuity at each point and the fact that $[0, 1]$ is compact. [In fact, some people use 'equicontinuous' to mean 'uniformly equicontinuous'.]

APPENDIX C
Topological Spaces

Metric spaces, described in the Appendix B, are examples of a more general structure that will be described in this appendix.

C.1. Concepts

A *topological space* is a pair $(\Omega, \mathcal{O})$ where Ω is a set and $\mathcal{O}$ a family of subsets of Ω satisfying the following properties:

   (i) $\mathcal{O}$ is closed under arbitrary unions;
   (ii) $\mathcal{O}$ is closed under finite intersections;
   (iii) $\emptyset \in \mathcal{O}$;
   (iv) $\Omega \in \mathcal{O}$.

The collection $\mathcal{O}$ is called a *topology* on Ω. (Properties (iii) and (iv) are redundant in view of the standard convention that the union and intersection of an empty collection of sets are the empty set and universal set, respectively. The members of $\mathcal{O}$ are said to be *open* and their complements are *closed*. (This use of 'closed' should not be confused with its use to describe an operation as in (i) and (ii) above.) Often one refers to a topological space by mentioning the universal set Ω, rather than both Ω and the topology $\mathcal{O}$. When doing this care is required, since it is possible, as illustrated by Problem 18, for two different topological spaces having the same universal set to appear in the same discussion.

It is easily shown that a set C in a topological space has a largest open subset, which is its *interior*, and a smallest closed subset, which is its *closure*. The *boundary* of C, denoted by ∂C, consists of those points in its closure that are not also in its interior.

Problem 1. The solution of some problem in Appendix B contains a proof that every metric space is a topological space. Which problem is that?

Problem 2. Why is it true that, with one exception, every topological space has at least two sets having the property of being both open and closed? What is the one exception?

Problem 3. Prove that the collection $\mathcal{C}$ of closed sets in a topological space has the following two properties:
  (i) $\mathcal{C}$ is closed under arbitrary intersections;
  (ii) $\mathcal{C}$ is closed under finite unions.

Problem 4. Prove that every set in a topological space has both an interior and a closure.

A *neighborhood* of a point in a topological space is any set that contains some open set of which the point is a member. (Some people place an additional condition on a set for it to be a neighborhood of a point—namely, that it itself be open.) A topological space is *Hausdorff* if for any two points x and w in the space there exist neighborhoods of x and w that have empty intersection. Hausdorff spaces almost always suffice for applications to probability.

* **Problem 5.** Prove that a point x belongs to the boundary of a set B if and only if every neighborhood of x contains at least one point in B and at least one point in B^c.

* **Problem 6.** Prove that the boundary of a set is also the boundary of its complement.

Problem 7. Prove that every metric space is Hausdorff.

A subset of a topological space Ω is *compact* if every open cover of it has a finite subcovering. It is *relatively compact* if its closure is compact. In case Ω itself is compact, the topological space is a *compact space*. The next two results describe connections between compactness and closedness.

Proposition 1. *A closed subset of a compact set in a topological space is a compact subset in that topological space.*

Problem 8. Prove the preceding proposition.

Proposition 2. *Every compact set in a Hausdorff space is closed.*

PROOF. For a proof by contradiction suppose that B is a compact set that is not closed. Let $x \in \partial B \setminus B$ and let w be any member of B. Since the topological space is Hausdorff, there exist neighborhoods of x and w that have empty intersection, neighborhoods that with no loss of generality we may take to be open. The complement of the open neighborhood of w is a closed neighborhood of x. Therefore, the collection

$$\{N^c : N \text{ a closed neighborhood of } x\}$$

is an open covering of B. Because B is compact this covering contains a a finite subcovering, say
$$\{N_1^c, N_2^c, \ldots, N_k^c\}.$$
Let O_i denote the interior of N_i, $1 \leq i \leq k$. No point in $O = \cap_{i=1}^k O_i$ is covered by the finite subcovering, but O, being the intersection of a finite number of open neighborhoods of x, is itself an open neighborhood of x —and, because $x \in \partial B$, O contains a member of B by Problem 5, a member that is not covered by the finite subcovering. Therefore, we have arrived at the desired contradiction. $\square$

The following problem introduces a topological space that cannot be viewed as a metric space, no matter how one chooses to specify a metric.

* **Problem 9.** Let $\mathcal{O}$ consist of all subsets O of $\mathbb{R}$ having the property that for every $x \in O$ there exists $\varepsilon > 0$ such that the interval $[x, x + \varepsilon) \subseteq O$. Prove $(\mathbb{R}, \mathcal{O})$ is a topological space. For this topology, decide which intervals are open, which are closed, and which are compact.

C.2. Compactification

Sometimes there are strong reasons for working with a compact topological space even when the topological space of interest is not compact.

Let $(\Omega, \mathcal{O})$ be a topological space and adjoin to Ω an additional point—call it ∞ —to obtain a set $\Omega^* = \Omega \cup \{\infty\}$. Let $\mathcal{O}^*$ consist of all members of $\mathcal{O}$ and all subsets of Ω^* whose complements in Ω^* are compact subsets of Ω.

* **Problem 10.** Prove that $(\Omega^*, \mathcal{O}^*)$, as defined in the preceding paragraph, is a compact space.

The topological space $(\Omega^*, \mathcal{O}^*)$ introduced above is called the *one-point compactification* of the topological space $(\Omega, \mathcal{O})$.

Example 1. The one-point compactification of the real line $\mathbb{R}$ with the usual topology has the effect of 'putting' negative numbers of large absolute value and large positive numbers 'close' to the same member, ∞, of the compactification.

A more commonly used compactification of $\mathbb{R}$ is its *two-point compactification* $\overline{\mathbb{R}} = \mathbb{R} \cup \{-\infty, \infty\}$. The open sets in $\overline{\mathbb{R}}$ are the open sets of $\mathbb{R}$, sets of the form $[-\infty, x)$ for some $x \in \mathbb{R}$, sets of the form $(x, \infty]$ for some $x \in \mathbb{R}$, and unions of sets of these types.

Problem 11. Prove that $\overline{\mathbb{R}}$ as just described is a topological space. Also, show that this topological space is the one induced by the metric $(x, y) \rightsquigarrow |\arctan x - \arctan y|$ which was introduced in the last section of Appendix B.

C.3. Product topologies

Let J be an arbitrary index set. For $j \in J$, let $(\Omega_j, \mathcal{O}_j)$ be a topological space. Set

$$\Omega = \prod_{j \in J} \Omega_j,$$

$$\mathcal{N} = \left\{ \prod_{j \in J} O_j : O_j \text{ open in } \mathcal{O}_j \text{ and } O_j = \Omega_j \text{ for all but finitely many } j \right\},$$

and $\mathcal{O}$ equal to the collection of all unions of members of $\mathcal{N}$.

Problem 12. Prove that $(\Omega, \mathcal{O})$ as just described is a topological space.

The collection $\mathcal{O}$ defined above is called the *product topology* of the topologies $\mathcal{O}_j$, and the topological space $(\Omega, \mathcal{O})$ is the *product* of the topological spaces $(\Omega_j, \mathcal{O}_j)$.

We omit the proof of the following important theorem about product spaces.

Theorem 3. [Tychonoff] *A product of compact spaces is a compact space.*

Example 2. By the preceding theorem $\overline{\mathbb{R}}^\infty = \prod_{k=1}^\infty \overline{\mathbb{R}}$ is compact; that is, the set of all sequences in $\overline{\mathbb{R}}$ is compact.

C.4. Relative topology

For $(\Omega, \mathcal{O})$ a topological space and $\Psi \subseteq \Omega$, let

$$\mathcal{P} = \{O \cap \Psi : O \in \mathcal{O}\}.$$

Problem 13. Prove that $(\Psi, \mathcal{P})$ as just defined is a topological space.

The topological space $(\Psi, \mathcal{P})$ described above is called a *topological subspace* of $(\Omega, \mathcal{O})$ and $\mathcal{P}$ is the *relative topology* on Ψ. The next exercise shows that sets can change their topological character when a topology is replaced by its relative topology, but the subsequent proposition shows that the compactness property is stable under such a replacement.

* **Problem 14.** Give an example that shows that a set that is not a member of a topology $\mathcal{O}$ may be open in a relative topology induced by $\mathcal{O}$ on a set Ψ. Prove, however, that this phenomenon cannot happen if $\Psi \in \mathcal{O}$.

Proposition 4. *Let $(\Omega, \mathcal{O})$ be a topological space and $\Psi \subseteq \Omega$. Then a subset C of Ψ is compact with respect to the topology $\mathcal{O}$ if and only if it is compact with respect to the relative topology induced by $\mathcal{O}$ on Ψ.*

Problem 15. Prove the preceding proposition.

C.5. Limits and continuous functions

In Definition 5 and Definition 8 we essentially copy appropriate versions of definitions that are standard for the topological space $\mathbb{R}$.

Definition 5. Let f be a function from a topological space Υ to a topological space Ω. The function f is said to be *continuous at a point* $y \in \Upsilon$ if $f^{-1}(N)$ is a neighborhood of y for every neighborhood N of $f(y)$. And f is *continuous* if it is continuous at each point in Υ.

Proposition 6. *A function from one topological space to another is continuous if and only if the inverse image of every open set is open.*

PROOF. Let $f \colon \Upsilon \to \Omega$. For one direction suppose that the inverse image under f every open set in Ω is open in Υ, and consider an arbitrary $y \in \Upsilon$ and an arbitrary neighborhood N of $f(y)$. There exists an open set $O \subseteq N$ for which $f(y) \in O$. Then $f^{-1}(O)$ contains y, is open, and is a subset of $f^{-1}(N)$. Therefore, $f^{-1}(N)$ is a neighborhood of y. Since y is arbitrary, f is continuous.

For the other direction, suppose that f is continuous and consider an arbitrary open set O in Ω. Since O is a neighborhood of each of its members, $f^{-1}(O)$ is a neighborhood of all of its members and thus, for each y in $f^{-1}(O)$, there exists an open set N_y such that $y \in N_y \subseteq f^{-1}(O)$. Therefore $f^{-1}(O) = \bigcup_{y \in f^{-1}(O)} N_y$, which being the union of open sets is open. $\square$

Proposition 7. *Let $f \colon \Upsilon \to \Omega$ be continuous and suppose that Υ is compact. Then the image of f is compact.*

Problem 16. Prove the preceding theorem.

Problem 17. Use relative topology to adapt the preceding discussion to the case where the domain of f is a subset of Υ.

Problem 18. Let $f \colon \mathbb{R} \to \mathbb{R}$. Prove that f is right-continuous (as usually defined) if and only if it is continuous when the domain has the topology of Problem 9 and the target has the usual topology.

Definition 8. Let f be a function with target a topological space Ω and domain a subset of a topological space Υ. Let $y \in \Upsilon$ and suppose that every neighborhood of y contains a point different from y in the domain of f. We say that

$$\lim_{z \to y} f(z) = x$$

if for every neighborhood N of x there is a neighborhood M of y such that $f(z) \in N$ whenever $z \neq y$ is in the intersection of M and the domain of f.

Problem 19. Suppose that $f: \Upsilon \to \Omega$ is continuous. Prove that

$$\lim_{z \to y} f(z) = f(y)$$

for every $y \in \Upsilon$ for which the one-point set $\{y\}$ is not open.

Example 3. Make $\mathbb{Z}^+$ into a topological space by calling every subset open, and let $\overline{\mathbb{Z}}^+$ denote the one-point compactification of $\mathbb{Z}^+$. The compact sets in $\mathbb{Z}^+$ are the finite sets, so that the neighborhoods of ∞ in $\overline{\mathbb{Z}}^+$ are those sets that contain ∞ and have finite complements. Consider an arbitrary function $f: \mathbb{Z}^+ \to \Omega$, where Ω is any topological space. From the preceding discussion it follows that

$$\lim_{z \to \infty} f(z) = x$$

if and only if for every neighborhood N of x there is a member m of $\mathbb{Z}^+$ such that $f(z) \in N$ whenever $z > m$. Hence, we see that sequential convergence is encompassed by Definition 8.

Comment: Another way to view the topology on $\overline{\mathbb{Z}}^+$ is that it is the relative topology induced by the usual topology on $\overline{\mathbb{R}}$.

Proposition 9. *Suppose that a sequence $(x_n: n = 1, 2, \ldots)$ of points in a closed set C in a topological space Ω converges to a point $x \in \Omega$. Then $x \in C$.*

Problem 20. Prove the preceding proposition.

Theorem 10. *Any sequence in a compact Hausdorff space has a convergent subsequence.*

Problem 21. Prove the preceding theorem.

Problem 22. Let $(\Omega, \mathcal{O})$ be the product of topological spaces $(\Omega_j, \mathcal{O}_j), j = 1, 2, \ldots$. For each $n = 1, 2, \ldots$, let

$$\omega_n = (\omega_{n,1}, \omega_{n,2}, \omega_{n,3}, \ldots)$$

be a point in Ω. Show that the sequence $(\omega_n: n = 1, 2, \ldots)$ converges in Ω if and only if the sequence $(\omega_{n,j}: n = 1, 2, \ldots)$ converges in Ω_j for each fixed j.

Problem 23. In the topological space of Problem 9 find an infinite sequence that does not converge even though it would converge were the topology the usual topology for $\mathbb{R}$.

APPENDIX D
Riemann-Stieltjes Integration

The Riemann integral $\int_a^b f(x)\,dx$ is, by definition, the limit of sums of the form

$$\sum_{j=1}^n f(\xi_j)\,(x_j - x_{j-1}),$$

where $a = x_0 < x_1 < \cdots < x_n = b$ and $\xi_j \in [x_{j-1}, x_j]$ for each j. In this appendix we replace the differences $x_j - x_{j-1}$ by $g(x_j) - g(x_{j-1})$ for some function g. This procedure leads to a type of integral that lies somewhere between the Riemann integral and the Lebesgue integral in generality. One advantage that this integral has over the Lebesgue integral is that it satisfies an integration by parts formula that can be quite useful for calculational purposes.

D.1. The Riemann-Stieltjes integral

The basic setting consists of two functions f and g defined on a closed bounded interval $[a, b]$ of the real line. By a *point partition* of the interval $[a, b]$ we mean a finite subset of $[a, b]$ containing both a and b. (In many books a point partition is identified by the one-word term 'partition', which we use to denote a partition of a set.) We typically write the members of a point partition in increasing order. Thus, when we say that $\{x_0, x_1, \ldots, x_n\}$ is a point partition of $[a, b]$, it is to be understood that

$$a = x_0 < x_1 < \cdots < x_n = b.$$

To emphasize this point we may use the contrived notation $\{a = x_0 < x_1, \cdots < x_n = b\}$, possibly omitting a and b from the notation if the interval on which the point partition is based is clear from context. The *mesh* of the point partition $\{a = x_0 < x_1, \cdots < x_n = b\}$ is the maximum of the numbers $x_j - x_{j-1}$, $1 \le j \le n$.

A point partition of the interval $[a, b]$ is said to be a *refinement* of a second point partition if it contains the second point partition as a subset.

A *Riemann-Stieltjes sum* of f with respect to g corresponding to a point partition $\{x_0 < x_1 < \cdots < x_n\}$ of $[a, b]$ is a sum

$$\sum_{j=1}^{n} f(\xi_j) \left[g(x_j) - g(x_{j-1}) \right],$$

where $\xi_j \in [x_{j-1}, x_j]$ for each j. Since each ξ_j is only constrained to lie in a certain interval, there are typically many Riemann-Stieltjes sums corresponding to a particular point partition.

Definition 1. The function f is *Riemann-Stieltjes integrable* with respect to the function g on the interval $[a, b]$ if there is some number γ such that for every $\varepsilon > 0$, there is a point partition P of $[a, b]$ for which the difference between γ and any Riemann-Stieltjes sum of f with respect to g corresponding to any refinement of P has absolute value less than ε. In case there is such a γ, the *Riemann-Stieltjes integral* of f with respect to g on the interval $[a, b]$ is said to exist and equal γ, and one writes

$$\gamma = \int_a^b f \, dg = \int_a^b f(x) \, dg(x),$$

either suppressing the independent variable x or writing it explicitly.

Suppose that g is an increasing function. Any Riemann-Stieltjes sum for a given point partition is bounded above by the *upper Riemann-Stieltjes sum*

$$\sum_{j=1}^{n} \sup\{f(x) \colon x_{j-1} \le x \le x_j\} \left[g(x_j) - g(x_{j-1}) \right],$$

for that point partition and below by the *lower Riemann-Stieltjes sum*, obtained by replacing 'sup' by 'inf'. It is easy to see that f is Riemann-Stieltjes integrable with respect to g if and only if for every $\varepsilon > 0$ there is a point partition for which the corresponding upper and lower Riemann-Stieltjes sums are finite and differ by less than ε.

* **Problem 1.** Calculate $\int_{1/2}^{4} x^2 \, d\lfloor x \rfloor$, where $\lfloor x \rfloor$ denotes the largest integer that is no larger than x.

* **Problem 2.** Let

$$g(x) = \begin{cases} 0 & \text{if } x \le 0 \\ 3^{-n} & \text{if } 2^{-n} \le x < 2^{-(n-1)}, \ n \in \mathbb{Z}^+ \setminus \{0\} \\ 1 & \text{if } x \ge 1. \end{cases}$$

Evaluate $\int_{-7}^{5} (1 - x^2) \, dg(x)$.

Problem 3. Verify the following equalities:

$$\int_0^1 x \, dx^2 = \tfrac{2}{3} \, ;$$

$$\int_0^1 2F(x) \, dF(x) = F^2(1) - F^2(0) \quad \text{for } F \text{ continuous and increasing.}$$

Problem 4. Let

$$F(x) = \begin{cases} x/3 & \text{if } 0 \le x < 1 \\ 1/2 & \text{if } 1 \le x < 2 \\ x/3 & \text{if } 2 \le x \le 3 \,. \end{cases}$$

Prove that $\int_0^3 F(x) \, dF(x)$ does not exist.

Problem 5. Prove that on any closed bounded interval, every continuous function is Riemann-Stieltjes integrable with respect to every function that is the difference of two monotone functions. *Hint:* Use the uniform continuity of the continuous function.

Problem 6. Let f and g be monotone functions on an interval $[a, b]$ and suppose that f is left-continuous and g is right-continuous. Prove that f is Riemann-Stieltjes integrable with respect to g.

It is important to notice that there are no differentiability assumptions in Problem 5 or Problem 6.

D.2. Relation to the Riemann integral

The following proposition shows how to change some Riemann-Stieltjes integrals into Riemann integrals which can then often be evaluated by using the Fundamental Theorem of Calculus.

Proposition 2. *Let g be a function with a continuous first derivative on an interval $[a, b]$ and f a bounded $\mathbb{R}$-valued function on $[a, b]$. Then fg' is Riemann integrable on $[a, b]$, if and only if f is Riemann-Stieltjes integrable with respect to g on $[a, b]$ in which case*

$$(D.1) \qquad \int_a^b f(x) \, dg(x) = \int_a^b f(x)g'(x) \, dx \,.$$

PROOF. Let $P = \{a = x_0 < x_1 < \cdots < x_n = b\}$ be a point partition of $[a, b]$ and let $\xi_j \in [x_{j-1}, x_j]$ for $1 \le j \le n$. The corresponding Riemann sum of fg' is

$$\sum_{j=1}^n f(\xi_j)g'(\xi_j) \, (x_j - x_{j-1}) \,,$$

and the corresponding Riemann-Stieltjes sum of f with respect to g is

$$\sum_{j=1}^{n} f(\xi_j) \left[g(x_j) - g(x_{j-1}) \right],$$

By the Mean-Value Theorem, there exist numbers $\eta_j \in [x_{j-1}, x_j]$ such that this Riemann-Stieltjes sum equals

$$\sum_{j=1}^{n} f(\xi_j) g'(\eta_j) \left(x_j - x_{j-1} \right).$$

We conclude that the absolute value of the difference between the Riemann sum of fg' and the Riemann-Stieltjes sum of f with respect to g is bounded by

(D.2) $$\sum_{j=1}^{n} |f(\xi_j)| \cdot |g'(\xi_j) - g'(\eta_j)| \left(x_j - x_{j-1} \right).$$

The quantity (D.2) is bounded by the product of three numbers: any bound s of $|f|$, $(b-a)$, and the maximum of $|g'(v) - g'(u)|$ taken over $u, v \in [x_{j-1}, x_j]$, $1 \le j \le n$. The third of these factors can be made arbitrarily small by taking the mesh of P to be sufficiently small, say less than some ε. For all refinements of such a P there is a correspondence between Riemann sums of fg' and Riemann-Stieltjes sums of f with respect to g such that corresponding sums differ by less than $s(b-a)\varepsilon$. The desired conclusion follows. $\square$

Problem 7. Discuss how Proposition 2 might be of use in treating an integral $\int_a^b f \, dg$ even if g does not satisfy all the conditions in that proposition.

Problem 8. Evaluate $\int_{-1}^{2} |x-1| \, d|x|$ by using Proposition 2, the Fundamental Theorem of Calculus, and your response for Problem 7. Do this problem by breaking the integral into no more than two pieces for the application of the Fundamental Theorem.

D.3. Change of variables

The formula for making the same change of variables in both functions of a Riemann-Stieltjes integral is easy to remember.

Proposition 3. *Let φ be a strictly increasing continuous function on an interval $[a, b]$. Then*

$$\int_a^b (f \circ \varphi) \, d(g \circ \varphi) = \int_{\varphi(a)}^{\varphi(b)} f \, dg,$$

in the sense that if either side exists then so does the other and they are equal.

Problem 9. Prove the preceding proposition.

Problem 10. Without concerning yourself with appropriate hypotheses, show how the preceding proposition is related to the usual change of variables formula for Riemann integrals.

D.4. Integration by parts

The following theorem, which gives a general integration by parts formula, is the main reason for the existence of this appendix.

Theorem 4. *Suppose that a function f is Riemann-Stieltjes integrable with respect to a function g on an interval $[a,b]$. Then g is Riemann-Stieltjes integrable with respect to f on $[a,b]$ and*

$$\int_a^b f\,dg = f(b)g(b) - f(a)g(a) - \int_a^b g\,df\,.$$

PROOF. Set $\gamma = \int_a^b f\,dg$. Let $\varepsilon > 0$ and choose a point partition P such that for every point partition $\{x_0, x_1, \ldots, x_n\}$ that is a refinement of P and every choice of $\xi_j \in [x_{j-1}, x_j]$,

$$\left| \sum_{j=1}^n f(\xi_j)\,[g(x_j) - g(x_{j-1})] - \gamma \right| < \varepsilon\,.$$

For such a point partition consider an arbitrary Riemann-Stieltjes sum for g with respect to f:

$$\sum_{j=1}^n g(\eta_j)\,[f(x_j) - f(x_{j-1})]\,.$$

We set $\eta_0 = a$ and $\eta_{n+1} = b$ in order to rewrite this Riemann-Stieltjes sum as

$$f(b)g(b) - f(a)g(a) - \sum_{i=1}^{n+1} f(x_{i-1})\,[g(\eta_i) - g(\eta_{i-1})]$$

(D.3)
$$= f(b)g(b) - f(a)g(a) - \sum_{i=2}^{n+1} f(x_{i-1})\,[g(x_{i-1}) - g(\eta_{i-1})]$$

$$- \sum_{i=1}^n f(x_{i-1})\,[g(\eta_i) - g(x_{i-1})]\,.$$

The combination of these last two summations is the negative of a Riemann-Stieltjes sum of f with respect to g for the point partition

$$P' = \{a = x_0 = \eta_0 \le \eta_1 \le x_1 \le \eta_2 \le \cdots \le \eta_n \le x_n = \eta_{n+1} = b\}\,,$$

the possibility of equality in this description of P' causing no problem, but only indicating that there may be less than $2n$ subintervals determined by P'. Since P' is a refinement of P, (D.3) differs from $f(b)g(b) - f(a)g(a) - \gamma$ by less than ε. Since ε is arbitrary the proof is complete. $\square$

It is worth noticing that the integration by parts formula is symmetric in f and g. It is also worth observing that the above proof uses a simple technique called *summation by parts* that has a slightly messy appearance in the proof because of the interlacing of two sequences. Here is the simple useful formula isolated by itself.

Proposition 5. *Let $(a_0, a_1, \ldots, a_n)$ and $(b_0, b_1, \ldots, b_n)$ be two finite sequences of real numbers. Then*

$$\sum_{j=1}^{n} a_j \left[b_j - b_{j-1} \right] = a_n b_n - a_0 b_0 - \sum_{i=1}^{n} b_{i-1} \left[a_i - a_{i-1} \right].$$

Problem 11. Convince yourself that the preceding proposition is true.

Problem 12. Redo Problem 1 by using integration by parts.

Problem 13. Show that all monotone functions on an interval are Riemann-Stieltjes integrable with respect to every continuous function on that interval (even a continuous function whose derivative exists nowhere). *Hint:* Use Problem 5 and an important theorem.

* **Problem 14.** Let f be an $\mathbb{R}$-valued function on an interval $[a, b]$, and suppose that for every $x \in [a, b]$,

$$f(x+) = \lim_{y \searrow x} f(y) \quad \text{and} \quad f(x-) = \lim_{y \nearrow x} f(y)$$

both exist as members of $\mathbb{R}$. Show that if g has a continuous derivative on $[a, b]$, then f and g are Riemann-Stieltjes integrable with respect to each other on $[a, b]$.

D.5. Improper Riemann-Stieltjes integrals

The treatment of improper Riemann-Stieltjes integrals parallels that of improper Riemann integrals. In particular, when one is using various theorems about Riemann-Stieltjes integrals, such as integration by parts, it is wise to first write a given improper Riemann-Stieltjes integral as a limit of proper Riemann-Stieltjes integrals, then use the theorems, and finally pass to the limit.

Problem 15. Replace the interval $[-7, 5]$ of integration in Problem 2 by the interval $(-\infty, \infty)$ and then do the problem created by this replacement.

Problem 16. Does the improper integral

$$\int_0^\infty \frac{(-1)^{\lceil x \rceil}}{x} \, d\lfloor x \rfloor$$

exist as a finite number? (Here $\lceil x \rceil$ and $\lfloor x \rfloor$ denote the smallest integer larger than or equal to x and the largest integer less than or equal to x, respectively.) Give attention to the issue of existence of appropriate proper Riemann-Stieltjes integrals.

Taylor Approximations, $\mathbb{C}$-Valued Logarithms

For some portions of this book it is important to have a definition of $\log \circ \beta$ for a complex-valued function β. Under some restrictions, such a definition is presented in the Section 2. A second theme appears throughout this appendix—that of approximating or bounding transcendental functions by polynomials.

E.1. Some inequalities based on the Taylor formula

From the Taylor formula with remainder one easily gets the following families of inequalities:

$$1 - \frac{v^2}{2!} \le \cos v \le 1 \,, \qquad\qquad v \in \mathbb{R},$$

$$1 - \frac{v^2}{2!} + \frac{v^4}{4!} - \frac{v^6}{6!} \le \cos v \le 1 - \frac{v^2}{2!} + \frac{v^4}{4!} \,, \quad v \in \mathbb{R},$$

$$\vdots \quad \vdots \quad \quad \vdots \;;$$

$$v - \frac{v^3}{3!} \le \sin v \le v \,, \qquad\qquad v \in \mathbb{R}^+ \,,$$

$$v - \frac{v^3}{3!} + \frac{v^5}{5!} - \frac{v^7}{7!} \le \sin v \le v - \frac{v^3}{3!} + \frac{v^5}{5!} \,, \quad v \in \mathbb{R}^+ \,,$$

$$\vdots \quad \vdots \quad \quad \vdots \;;$$

$$1 - \frac{x^1}{1!} \le e^{-x} \le 1 \,, \qquad\qquad x \in \mathbb{R}^+ \,,$$

$$1 - x + \frac{x^2}{2!} - \frac{x^3}{3!} \le e^{-x} \le 1 - x + \frac{x^2}{2!} \,, \quad x \in \mathbb{R}^+ \,,$$

$$\vdots \quad \vdots \quad \quad \vdots \;;$$

$$x - \frac{x^2}{2} \leq \log(1 + x) \leq x, \qquad\qquad x \in \mathbb{R}^+,$$

$$x - \frac{x^2}{2} + \frac{x^3}{3} - \frac{x^4}{4} \leq \log(1 + x) \leq x - \frac{x^2}{2} + \frac{x^3}{3}, \quad x \in \mathbb{R}^+,$$

$$\vdots \quad \vdots \quad \vdots;$$

(E.1) $$e^x \geq \sum_{k=0}^{n-1} \frac{x^k}{k!}, \quad x \in \mathbb{R}^+;$$

and

(E.2) $$\log(1 - x) \leq -\sum_{k=1}^{n-1} \frac{x^k}{k}, \quad x \in [0, 1).$$

The families (E.1) and (E.2) are one-sided. One technique for getting an inequality in the opposite direction is to use a geometric bound on the tail of the infinite Taylor series. For $x \in [0, n)$,

$$e^x = \sum_{k=0}^{n-1} \frac{x^k}{k!} + \sum_{k=n}^{\infty} \frac{x^k}{k!}$$

$$\leq \sum_{k=0}^{n-1} \frac{x^k}{k!} + \frac{x^n}{n!} \sum_{k=n}^{\infty} \left(\frac{x}{n}\right)^{k-n} = \sum_{k=0}^{n-1} \frac{x^k}{k!} + \frac{x^n}{(n-1)!(n-x)}.$$

For $x \in [0, 1)$,

(E.3) $$\log(1 - x) \geq -\sum_{k=1}^{n-1} \frac{x^k}{k} - \sum_{k=n}^{\infty} \frac{x^k}{n} = -\sum_{k=1}^{n-1} \frac{x^k}{k} - \frac{x^n}{n(1-x)}.$$

For small x, the inequality $\sin x \leq x - \frac{x^3}{6} + \frac{x^5}{120}$ is an improvement over $\sin x \leq x$. For large x the simpler $\sin x \leq x$ may be better, but neither will be very good. Elementary extra information can lead to inequalities that are quite good for all values of the variable. Here are two of the most important of such inequalities:

(E.4) $$|\sin x| \leq |x| \wedge 1, \quad x \in \mathbb{R};$$

(E.5) $$1 - \cos x \leq \frac{x^2}{2} \wedge 2, \quad x \in \mathbb{R}.$$

Finite and infinite products of real numbers are often treated via their logarithms if all factors are positive. If there is at least one zero factor, the product is 0. If there are some negative factors, the sign can be treated separately and a product of positive numbers then studied.

Problem 1. Let $(a_n : n = 1, 2, \dots)$ be a sequence of numbers in $[0, 1)$. Show that $\sum_{n=1}^{\infty} a_n < \infty$ if and only if $\prod_{n=1}^{\infty} (1 - a_n) > 0$.

E.2. Complex exponentials and logarithms

Let $z = x + iy$, where x and y are real numbers and i denotes a (nonreal) number whose square equals -1. Then z is called a *complex number,* and every complex number can be written in this form. The *absolute value* of z is the nonnegative number $\sqrt{x^2 + y^2}$ and is denoted by $|z|$. The real number x is the *real part* of z and is denoted by $\Re(z)$. The real number y is the *imaginary part* of z and is denoted by $\Im(z)$. We use $\mathbb{C}$ to denote the set of all complex numbers.

If $f\colon \mathbb{R} \to \mathbb{C}$, the *derivative* of f is defined in the natural way:

$$f' \overset{\text{def}}{=} (\Re \circ f)' + i(\Im \circ f)',$$

wherever the right side is defined.

The *exponential* of a complex number z is defined via

$$e^z \overset{\text{def}}{=} e^{\Re(z)}\left(\cos(\Im(z)) + i\sin(\Im(z))\right).$$

Problem 2. Prove that $e^{w+z} = e^w e^z$ for $w, z \in \mathbb{C}$.

Problem 3. Let $c \in \mathbb{C}$. Prove that the derivative of the function $t \rightsquigarrow e^{ct}$, $t \in \mathbb{R}$, is the function $t \rightsquigarrow ce^{ct}$.

The preceding problem is a special case of the following result.

Proposition 1. *Let* $\lambda\colon \mathbb{R} \to \mathbb{C}$ *be differentiable. Then*

$$(\exp \circ \lambda)' = \lambda' \cdot (\exp \circ \lambda).$$

* **Problem 4.** Prove the preceding proposition. (Comment: The Chain Rule for functions from $\mathbb{R}$ to $\mathbb{R}$ is available, but does not give the result in one easy step since exp here is a function from $\mathbb{C}$ to $\mathbb{C}$. There is a Chain Rule for functions from $\mathbb{C}$ to $\mathbb{C}$, but it does not give a proof because the domain of λ is $\mathbb{R}$ —and it may not be possible to extend the domain of λ to $\mathbb{C}$ without sacrificing its differentiability.)

Represent the point (x, y) in polar coordinates: $x = r\cos v$, $y = r\sin v$ with $r \geq 0$. Then $x + iy = re^{iv}$ because $e^{iv} = \cos v + i\sin v$. If $r > 0$, let $u = \log r$ so that

$$z \overset{\text{def}}{=} x + iy = e^u e^{iv} = e^{u+iv} = e^w,$$

where w is the complex number $u + iv$. Of course, v is determined by z only up to additive multiples of 2π. Thus for any complex $z \neq 0$, there exist infinitely many complex numbers w such that $e^w = z$; any two such w differ by an integral multiple of $2\pi i$. Any such w is called a *logarithm* of z. Thus for example, the logarithms of i are $(\frac{\pi}{2} + 2\pi q)i$, $q \in \mathbb{Z}$.

Problem 5. Find all the logarithms of -2, where, of course, you may express your answers in terms of log as representing the real logarithm of a positive real number. Also, find all complex logarithms of the real number e and of the complex number $-3 - 2i$.

Problem 6. Let z_1 and z_2 be complex numbers different from 0. Prove that the sum of any logarithm of z_1 and any logarithm of z_2 equals a logarithm of $z_1 z_2$.

Let $\beta \colon \mathbb{R} \to \mathbb{C}$ be a continuous function satisfying $\beta(0) = 1$. Let J denote the largest open interval containing 0 in which β is never 0.

Problem 7. For β and J as just described, prove that there exists a unique continuous function $\lambda \colon J \to \mathbb{C}$ such that $\lambda(0) = 0$ and $\exp \circ \lambda = \beta$. Despite the fact that log is not a well-defined function, the notation $\log \circ \beta$ will be used to denote the function λ. Prove that for each $v \in J$, $(\log \circ \beta)(v)$ is a logarithm of $\beta(v)$.

Problem 8. Let J be an interval in $\mathbb{R}$ and, for each $n = 1, 2, \ldots$, let $\beta_n \colon J \to \mathbb{C}$ be a continuous, nonzero function such that $\beta_n(0) = 1$. Show that if $\beta_n \to \beta$ uniformly on compacts subsets of J for some function $\beta \colon J \to \mathbb{C}$, then $\log \circ \beta_n \to \log \circ \beta$ uniformly on compact subsets of J.

* **Problem 9.** Let $\beta \colon \mathbb{R} \to \mathbb{C}$ and suppose that β is continuous and $\beta(0) = 1$. Is it necessarily true that $(\log \circ \beta)(u_1) = (\log \circ \beta)(u_2)$ if $\beta(u_1) = \beta(u_2)$?

Problem 10. Let β_1 and β_2 be continuous functions from $\mathbb{R}$ to $\mathbb{C}$ taking the value 1 at 0. Prove that the domain of $\log \circ (\beta_1 \beta_2)$ is the intersection of the domains of $\log \circ \beta_1$ and $\log \circ \beta_2$ and that on this intersection

$$\log \circ (\beta_1 \beta_2) = \log \circ \beta_1 + \log \circ \beta_2 .$$

Proposition 2. *Let $\beta \colon \mathbb{R} \to \mathbb{C}$ and suppose that β is differentiable and satisfies $\beta(0) = 1$. Then $(\log \circ \beta)' = \beta'/\beta$ on the domain of $\log \circ \beta$.*

Problem 11. Prove the preceding proposition. *Hint:* This problem may not be as easy as it appears.

For real y,

$$e^{iy} = \cos y + i \sin y = \sum_{j=0}^{\infty} \frac{(-1)^j y^{2j}}{(2j)!} + i \sum_{k=0}^{\infty} \frac{(-1)^k y^{2k+1}}{(2k+1)!} = \sum_{m=0}^{\infty} \frac{(iy)^m}{m!} .$$

The Binomial Theorem is valid for complex numbers when the exponent is a member of $\mathbb{Z}^+$. Using it to multiply the power series for e^x and e^{iy}, x and y real, gives

$$\text{(E.6)} \qquad\qquad e^z = e^x e^{iy} = \sum_{m=0}^{\infty} \frac{z^m}{m!} ,$$

where $x = \Re(z)$ and $y = \Im(z)$. This series is valid for all complex numbers z. We have obtained this formula by working with functions whose domain is $\mathbb{R}$. There is a more direct approach using the theory of functions of a complex variable.

For the logarithm it is not such a trivial manner to get a series formula by only using the theory of functions of a real variable. From the theory of functions of a complex variable, one gets

$$\text{(E.7)} \qquad \log(1 + z) = 2\pi q i + \sum_{m=1}^{\infty} \frac{(-1)^{m-1} z^m}{m}, \qquad |z| < 1,$$

with the various values of the integer q giving the various logarithms of z. In view of Problem 9, care must be used in applying this series to the compositions described in the preceding section.

Possibly more important than convergence of the series (E.6) and (E.7) are bounds on errors when these series are truncated.

Problem 12. Use a bounding geometric series to prove that

$$\left| \sum_{m=0}^{n-1} \frac{z^m}{m!} - e^z \right| \leq \frac{|z|^n}{(n-1)! \, (n - |z|)}$$

for $n > |z|$.

Problem 13. Use a bounding geometric series to prove for any choice of $\log(1 + z)$, there exists an integer q such that

$$\text{(E.8)} \qquad \left| 2\pi q i + \sum_{m=1}^{n-1} \frac{(-1)^{m-1} z^m}{m} - \log(1 + z) \right| \leq \frac{|z|^n}{n \, (1 - |z|)}$$

provided that $|z| < 1$ and $n > 0$.

In this book, Problem 13 is typically used when treating $(\log \circ \beta)(v)$ in a situation where β is continuous, $\beta(0) = 1$, and $|1 - \beta(u)| < 1$ for all u between 0 and v, including $u = v$. Then q in (E.8) should be taken equal to 0.

For fixed n, the bound in Problem 12 is not applicable for large $|z|$ and the bound in Problem 13 is not very good for $|z|$ close to 1. It is important that we have global bounds for the exponential function when the argument is restricted to being pure imaginary. The following inequality follows from (E.4) and (E.5):

$$\text{(E.9)} \qquad |e^{iv} - 1| \leq 2|v| \wedge 2, \qquad v \in \mathbb{R}.$$

Successive integrations of (E.9) yields inequalities that are often more useful than (E.9) itself. Integration of (E.9) on $[0, v]$ combined with careful handling of the absolute value symbols yields

$$|e^{iv} - 1 - iv| \leq v^2 \wedge 2|v|,$$

and a second integration yields

(E.10) $$\left| e^{iv} - 1 - iv + \frac{v^2}{2} \right| \leq \frac{|v|^3}{3} \wedge v^2 .$$

E.3. Approximations of general $\mathbb{C}$-valued functions

An important feature of the following proposition is that it does not contain an assumption that the $(n+1)^{\text{st}}$ derivative exists.

Proposition 3. *Let D denote the differentiation operator. Let f be a $\mathbb{C}$-valued function defined on an open interval in $\mathbb{R}$ containing 0. If $(D^n f)(0)$ exists, then*

$$f(x) = \sum_{k=0}^{n} \frac{(D^k f)(0)}{k!} x^k + o(|x|^n) \quad \text{as } x \to 0 .$$

PROOF. We must show

$$\lim_{x \to 0} \frac{f(x) - \sum_{k=0}^{n} \frac{(D^k f)(0)}{k!} x^k}{x^n} = 0 .$$

After $n-1$ applications of the l'Hospital Rule (valid for $\mathbb{C}$-valued functions in the numerator because it is valid for the real and imaginary parts separately) the limit on the left becomes

$$\lim_{x \to 0} \frac{(D^{n-1} f)(x) - (D^{n-1} f)(0) - x (D^n f)(0)}{n! \, x} ,$$

which equals 0, by the definition of $(D^n f)(0)$. $\square$

Problem 14. Why, in the preceding argument, was the l'Hospital Rule used only $n-1$ times rather than n times?

APPENDIX F
Bibliography

The first section consists of a list of general books each of which covers a broad range of topics in probability. The second section consists of more specialized books that treat deeply a few of the topics introduced in this book. There has been no attempt to make either of the two sections comprehensive, but we have tried to make each list representative of the literature. No books that are collections of articles have been included, although one particular article in one such collection is listed.

General probability books

Ash, Robert B., *Real Analysis and Probability*, Academic Press, New York, 1972.

Bauer, Heinz, *Probability Theory and Elements of Measure Theory*, Academic Press, London, 1981.

Billingsley, Patrick, *Probability and Measure*, Third Edition, John Wiley & Sons, New York, 1995.

Breiman, Leo, *Probability*, Society for Industrial and Applied Mathematics, Philadelphia, 1992.

Chow, Yuan Shih and Teicher, Henry, *Probability Theory: Independence, Interchangeability, Martingales*, 2nd Edition, Springer-Verlag, New York, 1988.

Chung, Kai Lai, *A Course in Probability Theory*, Second Edition, Academic Press, New York, 1974.

De Finetti, Bruno, *Theory of Probability Vol. 1*, John Wiley & Sons, London, 1974.

De Finetti, Bruno, *Theory of Probability Vol. 2*, John Wiley & Sons, London, 1975.

Dudley, R. M., *Real Analysis and Probability*, Wadsworth & Brooks/Cole Advanced Books & Software, Pacific Grove, California, 1989.

Durrett, Richard, *Probability: Theory and Examples*, Wadsworth & Brooks/Cole Advanced Books & Software, Pacific Grove, California, 1991.

Feller, William, *An Introduction to Probability Theory and Its Applications, Vol. I*, Third Edition, John Wiley & Sons, New York, 1968.

Feller, William, *An Introduction to Probability Theory and Its Applications, Vol. II*, Second Edition, John Wiley & Sons, New York, 1971.

Galambos, Janos, *Advanced Probability Theory*, Marcel Dekker, New York, 1988.

Gnedenko, B. V., *The Theory of Probability*, Second Edition (translated from Russian with additions), Chelsea, New York, 1962.

Grimmett, Geoffrey and Stirzaker, David, *Probability and Random Processes*, Second Edition, Clarendon, Oxford, 1992.

Itô, Kiyosi, *Introduction to Probability Theory*, Cambridge University Press, Cambridge, 1984.

Kingman, J. F. C. and Taylor, S. J., *Introduction to Measure and Probability*, Cambridge University Press, Cambridge, 1966.

Lamperti, John, *Probability*, W. A. Benjamin, New York, 1966.

Loéve, M., *Probability Theory I*, 4th Edition, Springer-Verlag, New York, 1977.

Loéve, M., *Probability Theory II*, 4th Edition, Springer-Verlag, New York, 1977.

Moran, P. A. P., *An Introduction to Probability Theory*, Clarendon Press, Oxford, 1984.

Nelson, Edward, *Radically Elementary Probability Theory*, Princeton University Press, Princeton, New Jersey, 1987.

Port, Sidney C., *Theoretical Probability for Applications*, John Wiley & Sons, New York, 1994.

Rényi, Alfred, *Foundations of Probability*, Holden Day, San Francisco, 1970.

Shiryaev, A. N., *Probability*, Second Edition (translated from Russian, orig. 1989), Springer-Verlag, New York, 1996.

Stroock, Daniel W., *Probability Theory, an Analytic View*, Cambridge University Press, Cambridge, 1993.

Tucker, Howard G., *A Graduate Course in Probability*, Academic Press, New York, 1967.

Whittle, Peter, *Probability*, Penguin Books, Middlesex, England, 1970.

Williams, David, *Probability with Martingales*, Cambridge University Press, Cambridge, 1991.

General books on stochastic processes

Bhattacharya, Rabin and Waymire, Edward C., *Stochastic Processes with Applications*, John Wiley & Sons, New York, 1990.

Cinlar, Erhan, *Introduction to Stochastic Processes*, Prentice-Hall, Englewood Cliffs, New Jersey, 1975.

Cox, D. R. and Miller, H. D., *The Theory of Stochastic Processes*, Chapman and Hall, London, 1980.

Dellacherie, Claude and Meyer, Paul-André, *Probabilities and Potential A* (translated from French, orig. 1976), North-Holland, Amsterdam, 1978.

Doob, J. L., *Stochastic Processes*, John Wiley & Sons, New York, 1953.

Gihman, I. I. and Skorohod, A. V., *Theory of Stochastic Processes I*, Springer-Verlag, New York, 1974.

Gihman, I. I. and Skorohod, A. V., *The Theory of Stochastic Processes II*, Springer-Verlag, New York, 1975.

Gihman, I. I. and Skorohod, A. V., *The Theory of Stochastic Processes III*, Springer-Verlag, Berlin, 1979.

Gray, Robert M., *Probability, Random Processes, and Ergodic Properties*, Spring-er-Verlag, New York, 1988.

Iranpour, Reza and Chacon, Paul, *Basic Stochastic Processes*, Macmillan, New York, 1988.

Jacod, J. and Shiryaev, A. N., *Limit Theorems for Stochastic Processes*, Springer-Verlag, Berlin, 1987.

Karlin, Samuel and Taylor, Howard M., *A First Course in Stochastic Processes*, Second Edition, Academic Press, New York, 1975.

Karlin, Samuel and Taylor, Howard M., *A Second Course in Stochastic Processes*, Academic Press, New York, 1981.

Lindvall, Torgny, *Lectures on the Coupling Method*, John Wiley & Sons, New York, 1992.

Rao, M. M., *Stochastic Processes: General Theory*, Kluwer Academic Publishers, Dordrecht, 1995.

Resnick, Sidney I., *Adventures in Stochastic Processes*, Birkhäuser, Boston, 1992.

Williams, David, *Diffusions, Markov Processes, and Martingales, Vol. 1: Foundations*, John Wiley & Sons, Chichester, 1979.

Books related to Chapter 12

Lukacs, Eugene, *Stochastic Convergence*, Second Edition, Academic Press, New York, 1975.

Révész, Pál, *The Laws of Large Numbers*, Academic Press, New York, 1966.

Books related to Chapter 13

Hirschman, I. I. and Widder, D. V., *The Convolution Transform*, Princeton University Press, Princeton, New Jersey, 1955.

Lukacs, Eugene, *Characteristic Functions*, Second Edition, Charles Griffin, London, 1970.

Lukacs, Eugene, *Developments in Characteristic Function Theory*, Macmillan, New York, 1983.

Widder, D. V., *An Introduction to Transform Theory*, Academic Press, New York, 1971.

Books related to Chapters 14, 16, and 17

Gnedenko, B. V. and Kolmogorov, A. N., *Limit Distributions for Sums of Independent Random Variables*, Addison-Wesley, Reading, Massachusetts, 1954.

Gumbel, E. J., *Statistics of Extremes*, Columbia University Press, New York, 1958.

Petrov, Valentin V., *Limit Theorems of Probability Theory: Sequences of Independent Random Variables*, Clarendon Press, Oxford, 1995.

Zolotarev, V. M., *One-dimensional Stable Distributions* (translation from Russian, orig. 1983), American Mathematical Society, Providence, Rhode Island, 1986.

Books related to Chapter 15

Deuschel, Jean-Dominique and Stroock, Daniel W., *Large Deviations*, Academic Press, Boston, 1989.

Ellis, Richard S., *Entropy, Large Deviations, and Statistical Mechanics*, Springer-Verlag, New York, 1985.

Varadhan, S. R. S., *Large Deviations and Applications*, Society for Industrial and Applied Mathematics, Philadelphia, 1984.

Books related to Chapters 18, 19, 24, and 33

Billingsley, Patrick, *Convergence of Probability Measures*, John Wiley & Sons, New York, 1968.

Dellacherie, Claude and Meyer, Paul-André, *Probabilities and Potential B* (translated from French, orig. 1980), North-Holland, Amsterdam, 1982.

Durrett, Richard, *Brownian Motion and Martingales in Analysis*, Wadsworth Advanced Books & Software, Belmont, California, 1984.

Durrett, Richard, *Stochastic Calculus: A Practical Introduction*, CRC Press, Boca Raton, Florida, 1996.

Einstein, Albert, *Investigations on the Theory of the Brownian Movement*, Dover, New York, 1956.

Freedman, David, *Brownian Motion and Diffusion*, Holden-Day, San Francisco, 1971.

Friedman, Avner, *Stochastic Differential Equations and Applications, Vol. 1*, Academic Press, New York, 1975.

Friedman, Avner, *Stochastic Differential Equations and Applications, Vol. 2*, Academic Press, New York, 1976.

Gard, Thomas C., *Introduction to Stochastic Differential Equations*, Marcel Dekker, New York, 1988.

Gihman, I. I. and Skorohod, A. V., *Controlled Stochastic Processes*, Springer-Verlag, New York, 1979.

He, Sheng-wu and Wang, Jia-gang and Yan, Jia-an, *Semimartingale Theory and Stochastic Calculus*, CRC Press, Boca Raton, Florida, 1992.

Hida, Takeyuki, *Brownian Motion*, Springer-Verlag, New York, 1980.

Ikeda, Nobuyuki and Watanabe, Shinzo, *Stochastic Differential Equations and Diffusion Processes*, Second Edition, Kodansha, Tokyo, 1989.

Itô, Kiyosi, *Foundations of Stochastic Differential Equations in Infinite Dimensional Spaces*, Society for Industrial and Applied Mathematics, Philadelphia, 1984.

Itô, K. and McKean Jr., H. P., *Diffusion Processes and Their Sample Paths*, Springer-Verlag, Berlin, 1974.

Krylov, N. V., *Introduction to the Theory of Diffusion Processes*, American Mathematical Society, Providence, Rhode Island, 1995.

Ledoux, Michel and Talagrand, Michel, *Probability in Banach Spaces*, Springer-Verlag, Berlin, 1991.

Lukacs, Eugene, *Stochastic Convergence*, Second Edition, Academic Press, New York, 1975.

McKean Jr., H. P., *Stochastic Integrals*, Academic Press, New York, 1969.

Metivier, Michel and Pellaumail, J., *Stochastic Integration*, Academic Press, New York, 1980.

Metivier, Michel, *Semimartingales, a Course on Stochastic Processes*, Walter de Gruyter, Berlin, 1982.

Meyer, Paul A., *Probability and Potentials*, Blaisdell, Waltham, Massachusetts, 1966.

Nualart, David, *The Malliavin Calculus and Related Topics,* Springer-Verlag, New York, 1995.

Parthasarathy, K. R., *Probability Measures on Metric Spaces,* Academic Press, New York, 1967.

Portenko, N. I., *Generalized Diffusion Processes* (translation from Russian, orig. 1982), American Mathematical Society, Providence, Rhode Island, 1990.

Revuz, Daniel and Yor, Marc, *Continuous Martingales and Brownian Motion,* Second Edition, Springer-Verlag, Berlin, 1994.

Rogers, L. C. G. and Williams, David, *Diffusions, Markov Processes, and Martingales, Vol. 2: Itô Calculus,* John Wiley & Sons, Chichester, 1987.

Skorohod, A. V., *Asymptotic Methods in the Theory of Stochastic Differential Equations* (translated from Russian, orig. 1987), American Mathematical Society, Providence, Rhode Island, 1989.

Stroock, D. W. and Varadhan, S. R. S., *Multidimensional Diffusion Processes,* Springer-Verlag, Berlin, 1979.

Yeh, J., *Stochastic Processes and the Wiener Integral,* Marcel Dekker, New York, 1973.

Yor, Marc, *Some Aspects of Brownian Motion: Part I: Some Special Functionals,* Birkhäuser Verlag, Basel, Switzerland, 1992.

Books related to Chapters 11, 25, 26, 30, and 31

Athreya, K. B. and Ney, P. E., *Branching Processes,* Springer-Verlag, New York, 1972.

Blumenthal, R. M. and Getoor, R. K., *Markov Processes and Potential Theory,* Academic Press, New York, 1968.

Chen, Mu Fa, *From Markov Chains to Non-Equilibrium Particle Systems,* World Scientific, Singapore, 1992.

Chow, Y. S. and Robbins, Herbert and Siegmund, David, *Great Expectations: The Theory of Optimal Stopping,* Houghton Mifflin, Boston, 1971.

Chung, Kai Lai, *Markov Chains with Stationary Transition Probabilities,* Second Edition, Springer-Verlag, New York, 1967.

Chung, Kai Lai, *Lectures on Boundary Theory for Markov Chains,* Princeton University Press, Princeton, New Jersey, 1970.

Dellacherie, Claude and Meyer, Paul-André, *Probabilities and Potential C* (translated from French), Elsevier Science Publishers B. V., Amsterdam, 1988.

Dellacherie, Claude and Meyer, Paul-André, *Probabilités et Potential (Théorie du potentiel associée à une résolvante, Théorie des processus de Markov),* Hermann, Paris, 1987.

Doob, J. L., *Classical Potential Theory and its Probabilistic Counterpart,* Springer-Verlag, New York, 1984.

Doyle, Peter G. and Snell, J. Laurie, *Random Walks and Electric Networks,* Mathematical Association of America, 1984.

Dynkin, E. B., *Markov Processes, Vol. I,* Academic Press, New York, 1965.

Dynkin, E. B., *Markov Processes, Vol. II,* Academic Press, New York, 1965.

Dynkin, Evgenii B. and Yushkevich, Alexsandr A., *Markov Processes: Theorems and Problems,* Plenum Press, New York, 1969.

Edgar, G. A. and Sucheston, Louis, *Stopping Times and Directed Processes (Encyclopedia of Mathematics and Its Applications, Vol. 47),* Cambridge University Press,

Cambridge, 1992.

Ethier, Stewart N. and Kurtz, Thomas G., *Markov Processes: Characterization and Convergence*, John Wiley & Sons, New York, 1986.

Freedman, David, *Approximating Countable Markov Chains*, Holden-Day, San Francisco, 1971.

Freedman, David, *Markov Chains*, Holden-Day, San Francisco, 1971.

Harris, Theodore E., *The Theory of Branching Processes*, Dover, New York, 1989.

Hughes, Barry D., *Random Walks and Random Environments, Vol. 1: Random Walks*, Clarendon Press, Oxford, 1995.

Hughes, Barry D., *Random Walks and Random Environments, Vol. 2: Random Environments*, Clarendon Press, Oxford, 1996.

Iosifescu, Marius, *Finite Markov Processes and Their Applications*, John Wiley & Sons, Chichester, 1980.

Kalashnikov, Vladimir V., *Topics on Regenerative Processes*, CRC Press, Boca Raton, Florida, 1994.

Kemeny, John G. and Snell, J. Laurie and Knapp, Anthony W., *Denumerable Markov Chains*, D. van Nostrand, Princeton, New Jersey, 1966.

Kingman, J. F. C., *Regenerative Phenomena*, John Wiley & Sons, London, 1972.

Lawler, Gregory F., *Intersections of Random Walks*, Birkhäuser, Boston, 1991.

Maisonneuve, Bernard, *Systèmes Régénératifs (Astérique, Vol. 15)*, Société Mathématique de France, Paris, 1974.

Révész, Pál, *Random Walk in Random and Non-Random Environments*, World Scientific, Singapore, 1990.

Sharpe, Michael, *General Theory of Markov Processes*, Academic Press, Boston, 1988.

Spitzer, Frank, *Principles of Random Walk*, Second Edition, Springer-Verlag, New York, 1976.

Tackás, Lajos, *Combinatorial Methods in the Theory of Stochastic Processes*, John Wiley & Sons, New York, 1967.

Yang, Xiang-qun, *The Construction Theory of Denumerable Markov Processes*, John Wiley & Sons, Chichester, 1990.

Books related to Chapter 27

Aldous, D. J., "Exchangeability and related topics", *École d'Été de Probabilités de Saint-Flour XIII*, (ed. Hennequin, P. L.) 1-198, Springer-Verlag, Berlin, 1985.

Books related to Chapter 28

Furstenberg, Harry, *Stationary Processes and Prediction Theory*, Princeton University Press, Princeton, New Jersey, 1960.

Hida, Takeyuki, *Stationary Stochastic Processes*, Princeton University Press, Princeton, New Jersey, 1970.

Kahane, Jean-Pierre, *Some Random Series of Functions*, Second Edition, Cambridge University Press, Cambridge, 1985.

Khinchin, A. I., *Mathematical Foundations of Information Theory*, Dover, New York, 1957.

Knight, Frank B., *Foundations of the Prediction Process*, Clarendon Press, Oxford, 1992.

Leadbetter, M. R. and Lindgren, Georg and Rootzén, Holger, *Extremes and Related Properties of Random Sequences and Processes*, Springer-Verlag, New York, 1983.

Lifshits, M. A., *Gaussian Random Functions*, Kluwer Academic Publishers, Dordrecht, 1995.

Smythe, Robert T. and Wierman, John C., *First-Passage Percolation on the Square Lattice*, Springer-Verlag, Berlin, 1978.

Walters, Peter, *An Introduction to Ergodic Theory*, Springer-Verlag, New York, 1982.

Yaglom, A. M., *Correlation Theory of Stationary and Related Random Functions, Vol I: Basic Results*, Springer-Verlag, New York, 1987.

Yaglom, A. M., *Correlation Theory of Stationary and Related Random Functions, Vol II: Supplementary Notes and References*, Springer-Verlag, New York, 1987.

Books related to Chapter 29

Aldous, David, *Probability Approximations via the Poisson Clumping Heuristic*, Springer-Verlag, New York, 1989.

Ambartzumian, R. V., *Factorization Calculus and Geometric Probability (Encyclopedia of Mathematics and Its Applications, Vol. 33)*, Cambridge University Press, Cambridge, 1990.

Brémaud, Pierre, *Point Processes and Queues: Martingale Dynamics*, Springer-Verlag, New York, 1981.

Cox, D. R. and Isham, Valerie, *Point Processes*, Chapman and Hall, London, 1980.

Daley, D. J. and Vere-Jones, D., *An Introduction to the Theory of Point Processes*, Springer-Verlag, New York, 1988.

Franken, Peter and König, Dieter and Arndt, Ursula and Schmidt, Volker, *Queues and Point Processes*, John Wiley & Sons, Chichester, 1980.

Hall, Peter, *Introduction to the Theory of Coverage Processes*, John Wiley & Sons, New York, 1988.

Kallenberg, Olav, *Random Measures*, Akademie-Verlag, Berlin, 1983.

Kingman, J. F. C., *Poisson Processes*, Clarendon Press, Oxford, 1993.

Matheron, G., *Random Sets and Integral Geometry*, John Wiley & Sons, New York, 1975.

Matthes, Klaus and Kerstan, Johannes and Mecke, Joseph, *Infinitely Divisible Point Processes*, John Wiley & Sons, Chichester, 1978.

Molchanov, Ilya S., *Limit Theorems for Unions of Random Closed Sets*, Springer-Verlag, Berlin, 1993.

Resnick, Sidney I., *Extreme Values, Regular Variation, and Point Processes*, Springer-Verlag, New York, 1987.

Reiss, R.-D., *A Course on Point Processes*, Springer-Verlag, New York, 1993.

Santaló, Luis A., *Integral Geometry and Geometric Probability (Encyclopedia of Mathematics and Its Applications, Vol. 1)*, Addison-Wesley, Reading, Massachusetts, 1976.

Solomon, Herbert, *Geometric Probability*, Society of Industrial and Applied Mathematics, Philadelphia, 1978.

Stoyan, Dietrich and Stoyan, Helga *Fractals, Random Shapes and Point Fields: Methods of Geometrical Statistics*, John Wiley & Sons, Chichester, 1994.

Books related to Chapter 32

Durrett, Richard, *Lecture Notes on Particle Systems and Percolation*, Wadsworth & Brooks/Cole Advanced Books & Software, Pacific Grove, California, 1988.

Griffeath, David, *Additive and Cancellative Interacting Particle Systems*, Springer-Verlag, Berlin, 1979.

Georgii, Hans-Otto, *Gibbs Measures and Phase Transitions*, Walter de Gruyter, Berlin, 1988.

Grimmett, Geoffrey, *Percolation*, Springer-Verlag, New York, 1989.

Kesten, Harry, *Percolation Theory for Mathematicians*, Birkhäuser, Boston, 1982.

Khinchin, A. I., *Mathematical Foundations of Statistical Mechanics*, Dover, New York, 1949.

Kindermann, Ross and Snell, J. Laurie, *Markov Random Fields and their Applications*, American Mathematical Society, Providence, Rhode Island, 1980.

Liggett, Thomas M., *Interacting Particle Systems*, Springer-Verlag, New York, 1985.

Other books

Adler, Robert J., *The Geometry of Random Fields*, John Wiley & Sons, Chichester, 1981.

Alon, Noga and Spencer, Joel H. and Erdós, Paul, *The Probabilistic Method*, John Wiley & Sons, New York, 1992.

Bollobás, Béla, *Random Graphs*, Academic Press, London, 1985.

David, H. A., *Order Statistics*, Second Edition, John Wiley & Sons, New York, 1981.

Dubins, Lester E. and Savage, Leonard J., *Inequalities for Stochastic Processes (How to Gamble if You Must)*, Dover, New York, 1976.

Hida, Takeyuki and Hitsuda, Masuyuki, *Gaussian Processes* (translation from Japanese, orig. 1976), American Mathematical Society, Providence, Rhode Island, 1993.

Kac, Mark, *Statistical Independence in Probability Analysis and Number Theory*, Mathematical Association of America, 1959.

Kolchin, Valentin F., *Random Mappings*, Optimization Software, New York, 1986.

Maitra, Ashok P. and Sudderth, William D., *Discrete Gambling and Stochastic Games*, Springer-Verlag, New York, 1966.

Palmer, Edgar M., *Graphical Evolution*, John Wiley & Sons, New York, 1985.

Piterbarg, Vladimir I., *Asymptotic Methods in the Theory of Gaussian Processes and Fields* (translation from Russian, orig. 1988), American Mathematical Society, Providence, Rhode Island, 1996.

Pinsky, Mark A., *Lectures on Random Evolution*, World Scientific, Singapore, 1991.

Spencer, Joel, *Ten Lectures on the Probabilistic Method*, Second Edition, Society of Industrial and Applied Mathematics, Philadelphia, 1993.

van der Vaart, Aad W. and Wellner, Jon A., *Weak Convergence and Empirical Processes*, Springer-Verlag, New York, 1996.

Vanmarcke, Erik, *Random Fields: Analysis and Synthesis*, The MIT Press, Cambridge, Massachusetts, 1983.

Appendix G
Comments and Credits

Some comments supplementing the text will be given, and sources in the literature for further specific information will be identified. However, books that are closely related to significant portions of this book will not be cited here. Rather, there is an extensive list of such books in Appendix F.

Concerning Chapter 1

The axioms for a probability space given here are essentially the same as those given by Kolmogorov in 1933 [Kolmogorov, A. N., *Foundations of Probability,* Second English Edition, Chelsea, New York, 1956].

Parts (iii) and (iv) of Problem 12 are special cases of a problem discussed by Martin Gardiner in "Mathematical games" on pages 120-125 of the October 1974 issue of *Scientific American.* For more about this problem, see Robert W. Chen, "A circular property of the occurrence of sequence patterns in the fair coin-tossing process", *Advances in Applied Probability* **21** (1989), 938-940.

Concerning Chapter 2

Proposition 7 plays an important role in probability theory, but the result may not even be mentioned in some measure theory courses. In probability theory one works with induced measures—that is, distributions—more often than with a probability measure on an underlying probability space. Sometimes the only probability measure identified in a discussion is called a 'distribution' since the communicator wants to indicate that he or she is thinking of this probability measure as possibly being induced.

Concerning Chapter 3

The adjective 'Cauchy' for distributions is somewhat ambiguous. See the comments in this appendix concerning Chapter 17 for further information about the adjective 'Cauchy'.

A commonly used term for a distribution function F for $\overline{\mathbb{R}}$ is 'improper distribution function'. Some use this term only when $F(\infty) < 1$ or $F(-\infty) > 0$. Since we do not use this term, we must always make sure that the setting $\mathbb{R}$ or $\overline{\mathbb{R}}$ is clear.

We have defined distribution functions so that they are right-continuous. One could equally well define them to be left-continuous. If one were to do that, then the distribution functions corresponding to nonnegative random variables would be those whose value at 0 is 0.

Some people work with distribution functions of $\mathbb{R}^d$-valued random variables for $d > 1$ as well as for $d = 1$. We feel that it is easier to work with distributions than with distribution functions when $d > 1$.

Concerning Chapter 4

With respect to operators, some might use the term 'nonnegative' where we use 'positive', and 'positive' where we use 'strictly positive'.

Concerning Chapter 5

The definition of the value of a probability generating function at 1 is a matter of taste. Many would take a different view from ours and define its value at 1 to be the value that the corresponding probability measure on $\overline{\mathbb{Z}}^+$ assigns to the set $\mathbb{Z}^+$. This approach produces probability generating functions that are continuous on the closed interval $[0, 1]$, but it also gives some probability generating functions that have finite derivatives at 1 even though the corresponding distributions have infinite mean.

The proof of Corollary 5 is the same proof as given on page 8 of the book by Kahane, identified in Appendix F. It also appears in the earlier (1968) edition of the same book. Kahane treats it as elementary, does not claim it is new, does not give it a name, and does not attribute it to anyone.

For matrices, some would use the adjective 'positive definite' where we use 'strictly positive definite', and 'nonnegative definite' where we use 'positive definite'.

A good reference for generating functions is: Wilf, Herbert S., *Generatingfunctionology*, Academic Press, Boston, 1990.

The book *Enumerative Combinatorics, Vol. I,* by Richard P. Stanley and published by Wadsworth & Brooks/Cole Advanced Books and Software, Monterey, California, 1986, is a good reference for Stirling numbers and other topics in combinatorics.

Concerning Chapter 6

An argument along the lines of the proof of Corollary 5 of Chapter 4 is contained within the main proof by Simon Kochen and Charles Stone in "A note on the Borel-Cantelli Lemma", *Illinois Journal of Mathematics* **8** (1964), 248-251.

The Inclusion-Exclusion Theorem is more a combinatorial theorem than a probability theorem. It is treated as a separate topic in the book by Stanley which is identified in the comments concerning Chapter 5 in this appendix.

Concerning Chapter 7

W. Sierpiński in "Un théorème général sur les familles d'ensembles", *Fundamenta Mathematicae* **12** (1928), 206-210, proved a theorem along the lines of what we have called the Sierpiński Class Theorem. For Sierpiński, the requirement of being closed under countable disjoint unions plays the role of the requirement that we stated of being closed under countable increasing limits. In our knowledge, the first occurrence of the Sierpiński Class Theorem in the form we use it is in the book by Blumenthal and Getoor identified in Appendix F. It may be that E. B. Dynkin was the first to use the theorem extensively in probability theory. In his book, *Theory of Markov Processes*, Prentice-Hall, Englewood Cliffs, New Jersey, 1961, Dynkin calls the result the π-λ Theorem, a name that is often used.

Before 1960 the Monotone Class Theorem was often used. It says that the smallest monotone class containing a field is equal to the smallest σ-field containing that field, a monotone class being a family of sets that is closed under countable increasing and countable decreasing limits. See, for example: Halmos, Paul R., *Measure Theory*, Springer-Verlag, New York, 1974. The Monotone Class Theorem can be used to prove the following result: Let μ and ν be finite measures on a common measurable space $(\Omega, \mathcal{F})$. Suppose that $\mu(A) \leq \nu(A)$ for all A in some field $\mathcal{E}$ for which $\mathcal{F} = \sigma(\mathcal{E})$. Then $\mu(B) \leq \nu(B)$ for every $B \in \mathcal{F}$. The statement obtained by replacing "field $\mathcal{E}$" by "family $\mathcal{E}$ that is closed under finite intersections" is false, as is seen by letting Ω be some two-point set and $\mathcal{E}$ consist of a single one-point set $\{\tau\}$, and letting μ and ν be probability measures for which $\mu(\{\tau\}) < \nu(\{\tau\})$. If asked to provide a situation where the Monotone Class Theorem is useful and the Sierpiński Class Theorem is not, we would be hard-pressed to give an example other than the theorem just described. On the other hand, probability theory has many places where the Sierpiński Class Theorem is a better tool than is the Monotone Class Theorem. The term 'monotone class theorem' is used by some as a generic term indicating the Sierpiński Class Theorem, the Monotone Class Theorem, or other theorem of a similar sort.

In an early manuscript version of this book, we gave a proof of the Extension Theorem that used the full recursively defined sequence of fields $(\mathcal{E}, \mathcal{E}_1, \mathcal{E}_2, \dots)$ and more; namely the recursive definition was extended transfinitely and it was shown that for some ordinal the corresponding field and its successor are identical. John Baxter showed us the alternative approach of completing $\mathcal{E}_2$.

Concerning Chapter 8

The Riemann integral has not played a role in our development of the Lebesgue integral, although after having defined the Lebesgue integral with respect to Lebesgue measure, we have examined the relationship between the two integrals. An approach that begins with the Riemann integral notices the linearity of the integral viewed as an operator, and then extends this operator to a wider class of functions than the Riemann-integrable functions is presented in Section 9.4 of: Kingman, J. F. C. and Taylor, S. J., *Introduction to Measure and Probability*, Cambridge University Press, Cambridge, 1966. An intermediate point of view is taken in: Rudin, Walter, *Real and Complex Analysis*, Second edition, McGraw-Hill, New York, 1974. Lebesgue integrals are defined there in a manner similar to that used in this book, but Lebesgue measure is constructed in a different manner. The Riesz Representation Theorem is used: it says that there is a unique measure having the property that integration with respect to it agrees with a given bounded linear operator—for example, the Riemann integral—on the space of continuous functions on a closed bounded interval.

The argument that we have given that (iii) implies (i) in the proof of the Uniform

Integrability Criterion is one we learned from Naresh Jain.

A proof of the Stirling Formula similar to that outlined in Problem 19 through Problem 23 is contained in: Patin, J. M., "A very short proof of Stirling's Formula", *The American Mathematical Monthly* **96** (1989), 41-42. In that paper the proof commences with a clever substitution that enables one to complete the proof by using the Dominated Convergence Theorem only once. More accurate approximations of the gamma function are available via asymptotic series. See, for example: Marsaglia, George and Marsaglia John C. W., "A new derivation of Stirling's Approximation", *The American Mathematical Monthly* **97** (1990), 826-829.

The term 'absolutely continuous' is often used for certain distribution functions, namely those corresponding to measures that are absolutely continuous with respect to Lebesgue measure. It can be shown that the derivative of an absolutely continuous distribution function exists on the complement of a set having zero Lebesgue measure, and is a Radon-Nikodym derivative of the corresponding distribution with respect to Lebesgue measure. (Notice that in Problem 35 it is assumed that F' is defined everywhere.) It can also be shown that every distribution function F for $\mathbb{R}$ can be written as $F = \alpha_1 F_1 + \alpha_2 F_2 + \alpha_3 F_3$, where $0 \leq \alpha_j$ for $1 \leq j \leq 3$, $\alpha_1 + \alpha_2 + \alpha_3 = 1$, F_1 is an absolutely continuous distribution function, F_2 and F_3 are distribution functions for which $F_2' = F_3' = 0$ a.e., F_2 corresponds to a distribution that assigns probability 1 to a countable set, and F_3 is continuous.

Concerning Chapter 9

Persi Diaconis brought the zeta distribution to our attention.

For information beyond that contained in Problem 55 and Problem 57 concerning independence of the radial and angular components of random vectors see: Jennrich, Robert I. and Port, Sidney C., "Radial and directional parts of a random vector", *Statistics & Probability Letters* **6** (1987-8), 155-158.

Concerning Chapter 10

In some books, especially those not focused on probability theory, the term 'convolution' has a meaning slightly different from the one given in this book, and there is a corresponding incompatibility of notations.

It is not obvious how to generalize to $\mathbb{R}^d$ the definition given in the text of the support function p of a convex set. Here is a different definition: the *support function* $q \colon \mathbb{R}^d \setminus \{0\} \to \mathbb{R}$ of a compact convex subset A of $\mathbb{R}^d$ is given by

$$q(z) = \sup\{\langle x, z \rangle \colon x \in A\}.$$

For $z \in \mathbb{R}^2$, regarded as a complex number $re^{i\varphi}$ with $r \geq 0$, it is clear that $q(z) = rp(\varphi)$, where p is the function introduced in the text as the support function.

Concerning Chapter 11

Some solutions of Problem 29 appear as solutions of E2976 (proposed by Lee Whitt) in *The American Mathematical Monthly* **93** (1986), 62-63. The solutions there are analytic in character and apply for $0 < p < 2$.

Results for the random walk having steps equal to ± 1 each with probability $\frac{1}{2}$ may be viewed as results in combinatorics. For instance, from Example 5 we see that there are $\frac{2}{m}\binom{2(m-1)}{m-1}$ length-$2m$ sequences consisting of ± 1's and having the property that the number of 1's and the number of -1's are equal in the entire sequence but not in any initial proper subsequence.

Concerning Chapter 12

The Cantor function plays an important role in courses on real analysis. Even though it is continuous, increasing, and not the constant function, its derivative equals 0 a.e.. But it is in probability theory that functions with this 'strange' combination of properties arise most naturally. See Problem 18, for example.

Many years ago Dean Isaacson and Willis Owen brought to the attention of one of us that the Kolmogorov Three-Series Theorem can be replaced by a two-series version. The definition of Y_n needs to be changed for their version: $Y_n(\omega) = b$ if $X_n(\omega) > b$, $= -b$ if $X_n(\omega) < -b$, and otherwise $= X_n(\omega)$. Then the first of the three series can be dropped from the Kolmogorov Three-Series Theorem.

Concerning Chapter 13

An approach different from that of Example 1 for calculating the characteristic function of the standard normal distribution is to make the substitution $z = x - iv$. One obtains

$$\beta(v) = e^{-\frac{v^2}{2}} \frac{1}{\sqrt{2\pi}} \int_\gamma e^{-\frac{z^2}{2}}\, dz \,,$$

where γ is the horizontal left-to-right oriented line in the complex plane that passes through the points with imaginary part $-v$. One uses residue theory (a topic in complex analysis) to replace γ by the real axis and thus finish the calculation.

Moment generating functions can be defined for some distributions that are not supported by $[0, \infty]$. The normal distribution is an example, since its density is sufficiently small far to the left on the negative portion of $\mathbb{R}$. Some people define the 'moment generating function' as $t \rightsquigarrow E(e^{tX})$, for all t for which the expectation is finite. Here t is the negative of the variable that we use in the text. We do not favor this approach because there are distributions supported by $\mathbb{R}^+$ for which this version of the moment generating function fails to exist for all positive t.

There is inconsistency in the literature and tables concerning the definitions of the sine integral si and cosine integral ci. For instance, for the sine integral some use 0 rather than ∞ as the fixed endpoint of integration.

For $\mathbb{C}$-valued functions, some would use the adjective 'positive definite' where we use 'strictly positive definite', and 'nonnegative definite' where we use 'positive definite'.

Concerning Chapter 14

Many people attach an adjective to the convergence $F_n \to F$ that we have defined. Some of these adjectives are 'vague', 'weak', 'weak*', and 'complete'. For some, two adjectives are used in order to distinguish the $\mathbb{R}$-setting from the $\overline{\mathbb{R}}$-setting. The fact that a distribution function F generates a linear functional $g \rightsquigarrow \int g\, dF$ on the space of continuous functions having finite limits at ∞ and $-\infty$ motivates some of the terminology.

Concerning Chapter 15

Of the strictly stable distributions on $\mathbb{R}^+$, only those of index $1/2$ seem to have nice closed-form formulas for their densities. (Of course those of index 1, while not having densities, do have nice distribution functions.)

Concerning Chapter 16

Other functions $\tilde{\chi}$ can be used in lieu of χ as pictured in 16.1; $\tilde{\chi}$ must be continuous and satisfy

$$|\tilde{\chi}(y) - \chi(y)| \leq c(y^2 \wedge 1)$$

for some $c \in \mathbb{R}$. The continuity assumption can even be relaxed, but then the treatment of triangular arrays becomes slightly more complicated.

There is a Lévy-Khinchin Representation Theorem for infinitely divisible distributions in $\mathbb{R}^d$. One choice for the function χ is

$$\chi(y) = \big([(y_1 \wedge 1) \vee (-1)], \ldots, [(y_d \wedge 1) \vee (-1)]\big).$$

The measure ν on $\mathbb{R}^d \setminus \{0\}$ is a Lévy measure if and only if

$$\int_{\mathbb{R}^d \setminus \{0\}} |y|^2 \, \nu(dy) < \infty.$$

The arbitrary real number η and the nonnegative number σ^2 should be replaced by an arbitrary member of $\mathbb{R}^d$ and a positive definite symmetric matrix, respectively.

Concerning Chapter 17

The term *asymmetric Cauchy* is often used to describe the distributions in Theorem 10 with $\gamma \neq 0$. When 'Cauchy' is used without modification it must be clear, implicitly or explicitly, whether all the distributions in Theorem 10 are encompassed or only those for which $\gamma = 0$. The phrase 'symmetric Cauchy centered at b' would indicate that $\gamma = 0$, $\xi = b$, and most likely $k \neq 0$.

There are conditions on R that are equivalent to (17.21) being of regular variation of a particular index, conditions which are often separated into cases: $\alpha < 1$, $\alpha = 1$, and $\alpha > 1$. The advantage of such equivalent formulations is that they involve R more directly rather than as integrals of s^2 against R. It is not a trivial task to prove the equivalence of the various equivalent forms of Theorem 12, at least not without some key theorems about regular variation.

A distribution R is said to be in the *domain of partial attraction* of a nondegenerate distribution Q if for some choice of a_n and c_n, some subsequence of (Q_n^{*n}) converges to Q, where Q_n is defined by $Q_n(B) = R(a_n B + c_n)$. It is known that only infinitely divisible distributions have nonempty domains of partial attraction and that there exist distributions that are in the domain of partial attraction of every nondegenerate infinitely divisible distribution.

The Lévy measures on $\mathbb{R}^d \setminus \{0\}$ that correspond to stable distributions on $\mathbb{R}^d$ are the zero measure and those that can be represented as the product of some probability measure on the unit sphere centered at $\mathbf{0}$ and a radial measure of the form $cr^{-(1+\alpha)}dr$ for some $\alpha \in (0,2)$ and $c > 0$. In order that the stable distribution be spherically symmetric, the probability measure on the unit sphere must be spherically symmetric. The characteristic functions of the spherically symmetric stable distributions in $\mathbb{R}^d$ are the functions of the form $u \rightsquigarrow \exp\{-k|u|^\alpha\}$ for some $\alpha \in (0,2]$ and $k \geq 0$.

Concerning Chapter 18

It is not much more difficult to prove the full Prohorov Theorem than it is to show that a family consisting of a single probability measure is uniformly tight. An outline of an early proof of this latter fact can be found in the third footnote of Oxtoby, J. C. and Ulam, S. M., "On the existence of a measure invariant under a transformation", *Annals of Mathematics* **40** (1939), 560-566.

The limit theory for triangular arrays of $\mathbb{R}^d$-valued random variables is very similar to that for triangular arrays of $\mathbb{R}$-valued random variables.

Concerning Chapter 19

Often the σ-fields in a minimal filtration or a minimal right-continuous filtration are made larger by completing them.

Let W denote standard Brownian motion on $[0, \infty)$. For $t \in [0, 1]$ and $n = 3, 4, 5, \ldots$, set

$$Z_n(t) = \frac{W_{nt}}{\sqrt{2n \log(\log n)}}.$$

Fix t. By the Law of the Iterated Logarithm at ∞ and a little more work, the set of limit points of the sequence $(Z_n(t): n = 3, 4, \ldots)$ is the interval $[-\sqrt{t}, \sqrt{t}\,]$ with probability 1. If we consider t to be a variable, it becomes natural to ask: Which functions are limit points of the sequence $(Z_n: n = 3, 4, \ldots)$? Here also, an almost sure result exists. In Strassen, V., "An invariance principle for the Law of the Iterated Logarithm", *Zeitschrift für Wahrscheinlichkeitstheorie und Verwandte Gebiete* **3** (1964), 211-226, it is shown that the limits points are the absolutely continuous functions x on $[0, 1]$ that satisfy $x(0) = 0$ and

$$\int_{[0,1]} [x'(t)]^2 \, \lambda(dt) \leq 1,$$

where λ denotes Lebesgue measure.

Even though the Law of the Iterated Logarithm at ∞ is equivalent to the existence of $\widehat{T}$ and $T_1 < T_2 < \cdots \to \infty$ for which (19.18) and (19.19) hold, it does not tell us whether there exists $T_1 < T_2 < \cdots \to \infty$ such that

$$W_{T_n}(\omega) > \sqrt{2T_n(\omega) \log(\log T_n(\omega))}.$$

It develops that the answer is 'yes'. Then the question arises: What is the answer if $\sqrt{2t \log(\log t)}$ is replaced by some $\varphi(t)$ for which $\varphi(t)/\sqrt{2t \log(\log t)} \to 1$ as $t \to \infty$? With the assumption that $t \rightsquigarrow t^{-1}\varphi(t)$ is decreasing and $t \rightsquigarrow t^{-1/2}\varphi(t)$ is increasing (but without the limiting assumption at the end of the last sentence), the probability that there is a random sequence $T_1 < T_2 < \cdots \to \infty$ such that $W_{T_n}(\omega) > \varphi(T_n(\omega))$ is 1 or 0, according as

$$\int^{\infty} t^{-3/2} \varphi(t) e^{-\frac{[\varphi(t)]^2}{2t}} \, dt$$

diverges or converges. This criterion is known as the Kolmogorov Test.

Concerning Chapter 20

The spaces $\mathbf{L}_1(\Omega, \mathcal{F}, P)$ and $\mathbf{L}_2(\Omega, \mathcal{F}, P)$ belong to a family of metric spaces. For $1 \leq p < \infty$, the metric space $\mathbf{L}_p(\Omega, \mathcal{F}, P)$ consists of equivalence classes of those random variables X for which $E(|X|^p) < \infty$, with the distance between $[X]$ and $[Y]$ equaling $[E(|Y - X|^p)]^{1/p}$. The metric space $\mathbf{L}_\infty(\Omega, \mathcal{F}, P)$ consists of equivalence classes of almost surely bounded random variables, with the distance between $[X]$ and $[Y]$ equaling the smallest almost sure bound of $|Y - X|$. The only one of these metric spaces that is a Hilbert space is $\mathbf{L}_2$.

Concerning Chapter 21

The adjective 'regular' is often attached to the phrase 'conditional distribution' in order to emphasize that it is a random distribution rather than just a collection of conditional probabilities of events determined by some random variable. The term 'regular conditional probability' is a related term that appears in the literature. It refers to a random probability measure on the underlying probability space having certain properties. If we had introduced this concept in the text, we would have used the term 'conditional probability measure'.

The proof we have given of Lemma 21 is based on ideas obtained from Ashok Maitra and William Sudderth in conversations with them.

There are different approaches for treating conditional distributions of normal random vectors given some of its coordinates. We have learned the one presented in the last section of Chapter 21 from Morris L. Eaton. The interested reader may want to read the third chapter of his book *Multivariate Statistics, a Vector Space Approach*, John Wiley & Sons, New York, 1983.

Concerning Chapter 22

Example 1 can be extended. For instance,

$$P\big([X_0 = H] \,\big|\, [X_1 = B]\big) = \frac{P[X_0 = H, X_1 = B]}{P[X_1 = B]}$$

$$= \frac{P[X_0 = H, X_1 = B]}{P[X_0 = H, X_1 = B] + P[X_0 = T, X_1 = B]} = \frac{\frac{3}{16}}{\frac{3}{16} + \frac{7}{18}} = \frac{27}{83}.$$

Situations where one is given conditional distributions in one direction—say for the future given the past—and one wants to calculate conditional distributions in the other direction are quite common. The Bayes Theorem is a formula that gives the desired conditional probabilities in terms of the given probabilities and conditional probabilities. Some books have many examples and exercises focused on the Bayes Theorem.

Concerning Chapter 23

In "Conditional expectations of random variables without expectations", *Annals of Mathematical Statistics* **36** (1965), 1556-1559, R. E. Strauch defines conditional expectations of random variables that do not necessarily have finite expectations. This definition does not use conditional probabilities but is equivalent to our definition.

Concerning Chapter 24

The Kolmogorov Inequality can be viewed as a generalization to martingales of Markov-type inequalities for random variables. Broad classes of inequalities can be generalized in this manner as shown by Gilat, P. and Sudderth, W. D., in "Generalized Kolmogorov inequalities for martingales", *Zeitschrift für Wahrscheinlichkeitstheorie und Verwante Gebiete* **36** (1976), 67-73.

The equation (24.21) appears (with the names of the parameters changed) in Problem 18 of Chapter 12 for a different reason than that for which it appears in Chapter 24.

Concerning Chapter 25

The coupling proof we have presented of the Renewal Theorem in the case of positive recurrence is similar to the proof in Lindvall, Torgny, "A probabilistic proof of Blackwell's renewal theorem", *The Annals of Probability* **5** (1977), 482-485. Although David Freedman does not mention coupling in the proof he presents in *Markov Chains* which is listed in Appendix F, the proof we have given has a strong resemblance to his for both the positive recurrent and null recurrent cases.

Denote by $\{-a, (b-a)\}$ the support of the step distribution of a random walk in $\mathbb{Z}$ whose steps are of Bernoulli type having mean 0. For the case $a = 1$ the distribution of the first return to 0 is explicitly obtained in terms of b in: Gould, H. W., "Generalizations of Vandermonde's convolution", *The American Mathematical Monthly* **63** (1956), 84-91. In "A pentagonal pot-pourri of perplexing problems, primarily probabilistic", *The American Mathematical Monthly* **91** (1984), 559-563, Richard K. Guy indicated (referring to a problem proposed by Kai-Lai Chung) that an explicit formula for general a and b might not have yet been found. (There is no loss of generality in assuming that a and b are relatively prime.)

Concerning Chapter 26

The term 'Markov chain' is used by some as a synonym for 'Markov sequence', although others use 'Markov chain' only when the state space is countable.

For results such as Corollary 15 describing convergence of probabilities as time approaches ∞, it is natural to ask about the rate of convergence. There are a large number of papers in the probability literature about this issue. For the Ehrenfest urn sequence of Problem 55 modified slightly to remove the periodicity, Persi Diaconis and Mehrdad Shahshahani indicate in Example 1 of "Time to reach stationarity in the Bernoulli-Laplace diffusion model", *SIAM Journal on Mathematical Analysis* **18** (1987), 208-218, that starting with $b/2$ balls in each urn, it is at time approximately $\frac{1}{4}b \log b$ that the probability distribution becomes close to the equilibrium distribution.

Concerning Chapter 27

When people speak of the De Finetti Theorem they usually mean a theorem for the case of infinite exchangeable sequences. The proof of Theorem 7 we have given uses a finite sequence approximation similar to that by Heath, David and Sudderth, William, in "De Finetti's Theorem on Exchangeable Variables", *The American Statistician* **30** (1976), 188-189. In place of our identification of a reverse martingale first in Proposition 4 and then in Lemma 5 leading to a proof of Proposition 6, they use a tightness argument.

The GEM distributions introduced in Problem 53 have been named after McCloskey, J. W. [*A Model for the Distribution of Individuals by Species in an Environment*, PhD thesis (1965), Michigan State University], Engen, S. ["A note on the geometric series as a species frequency model", *Biometrika* **62** (1975), 694-699], and Griffiths, R. C.

In Blackwell, David and MacQueen, James B., "Ferguson distributions via Pólya urn schemes", *The Annals of Statistics* **1** (1973), 353-355, the Blackwell-MacQueen urns are described, though not by that name. This reference might be the first to contain the term 'Ferguson distribution'. These distributions are called 'Dirichlet processes' where they are introduced in Ferguson, Thomas S., "A Bayesian analysis of some nonparametric problems", *The Annals of Statistics* **1** (1972), 209-230.

Concerning Chapter 28

Some people call the spectral measure of a second-order stationary sequence its 'spectral distribution', even though its total measure is $\text{Var}(X_0)$ which need not equal 1.

Concerning Chapter 29

In loose mathematical conversation there is a tendency to use the term 'set' when the term 'multiset' would be better. For instance, a conversation may begin with: "Consider a set $\{X_1, X_2, \ldots, X_n\}$ of n random variables." Then later one might speak of the sum of these random variables, even though one does not intend to exclude the possibility that there may be identical random variables in the list and one intends that repeated copies of the same random variable be repeated in the sum.

In the general framework of random sets, the *distribution* of a random set X is the function $K \rightsquigarrow P[K \cap X \neq \emptyset]$ for K in an appropriate class of sets. For instance, see the book by Molchanov listed in Appendix F.

Concerning Chapter 30

The martingale approach we have outlined for proving Theorem 9 is the one used by E. S. Shatland in "On local properties of processes with independent increments", *Theory of Probability and its Applications* **10** (1965), 317-322.

Our definition of 'regenerative set' is narrower than the one typically used. In particular, our definition does not encompass the set of times when a Brownian motion is at 0, even though an appropriately general definition would apply to this situation. The set of times when Brownian motion is at 0 is uncountable with probability 1 despite the fact that at any particular time, Brownian motion has probability 0 of being at 0.

The short term 'local time' is usually used for 'local-time process'.

Concerning Chapter 31

The definition of 'infinitesimal generator' varies from treatment to treatment. One place of disagreement is the sense in which the limit in Definition 13 must exist.

Problem 20 and the fact that $\rho_x\{x\} = 0$ point to an alternate construction of pure-jump Markov processes, which one might view as more natural. The measure ρ_x describes the place to which the process jumps from x at the time it does jump, and $q(x)$ is the expected value of the exponentially distributed duration of the stay at x. Thus one can construct X recursively: alternately wait an exponentially distributed amount of time with mean equal to q(current state), and jump according to $\rho_{\text{current state}}$.

Concerning Chapter 32

Continuous-time interacting particle systems of the type considered here were first introduced by Frank Spitzer in "Interaction of Markov processes", *Advances in Mathematics* **5** (1970), 246-290. Thomas Liggett was the first to give a rigorous existence proof for such systems, in "Existence theorems for infinite particle systems", *Transactions of the American Mathematical Society* **165** (1972), 471-481. Liggett's result treats more general systems than the ones we discuss; in particular, the finite-range assumption can be considerably weakened. Discrete-time versions of these systems were studied extensively in Russia during the late 1960's. For bibliographic references, see the book by Liggett listed in Appendix F.

In most treatments, when the maximum particle number is $n = 1$, the term 'birth' is reserved for the transition $0 \to 1$ at a site, and the term 'death' is reserved for the transition $0 \to 1$. In our treatment, the transition $0 \to 1$ at a site can also be described as a 'death' (with wrap-around), and the transition $1 \to 0$ can be described as a birth, so there is an ambiguity in the way the rates such transitions are divided between the birth and death rates. In most of our examples with $n = 1$, we have maintained consistency with other treatments by making the birth rates 0 at occupied sites and the death rates 0 at vacant sites.

The original subadditive ergodic theorem of J. F. C. Kingman, in "The ergodic theory of subadditive stochastic processes", *Journal of the Royal Statistical Society B* **30** (1968), 499-510, does not apply to Example 8, since one of the conditions in that theorem requires that the distribution of the doubly indexed random sequence $(Z_{m+k,n+k} : 0 \leq m < n)$ be the same for all positive integers k. Liggett's improvement, which appeared in "An improved subadditive ergodic theorem", *Annals of Probability* **13** (1985), 1279-1285, was designed with applications like Example 8 in mind. The limit theorem in Example 8 is originally due to Richard Durrett, "On the growth of one dimensional contact processes", *Annals of Probability* **8** (1980), 890-907.

Given a generator G, equation (32.2) is usually impossible to solve for μ. On the other hand, given a probability measure μ satisfying certain mild conditions, there is a way to construct an interacting particle system (not necessarily with finite-range rates) with equilibrium distribution μ. This fact was Spitzer's original motivation for introducing interacting particle systems.

Many years ago David Griffeath told one of us of the example appearing in Problem 16. It was described by David Blackwell in "Another countable Markov process with only instantaneous states", *Annals of Mathematical Statistics* **29** (1958), 313-316. A somewhat similar example was given by W. Feller and Henry McKean in "A diffusion equivalent to a countable Markov chain", *Proceedings of the National Academy of Sciences, U. S. A.* **42** (1956), 351-354, and by R. L. Dobrushin in "An example of a countable homogeneous Markov process all states of which are transient", *Theory of Probability and its Applications* **1** (1956), 436-440.

Concerning Chapter 33

Since we have restricted the integrands in Itô integrals to be cadlag functions we have not had to impose moment conditions. Some other treatments do not make the cadlag assumption but do impose moment conditions. Care should be taken when comparing results obtained under different assumptions. Example 2 shows why it is important to consider integrands that are not cadlag functions.

The original proof of existence and uniqueness of solutions of stochastic differential equations, given by Kiyosi Itô in "On a stochastic integral equation", *Proceedings of the Japan Academy* **22** (1946), 32-35, used a stochastic version of the 'Picard iteration method' from the field of ordinary differential equations.

See the remark earlier in this appendix with reference to the term 'local-time process' that appears in Chapter 30 as well as this chapter.

Concerning Appendix A

We avoid using the phrase 'range of the function', because some people take it to mean 'image of the function' and others take it to mean 'target of the function'.

Concerning Appendix B

Two books that introduce metric spaces without treating them as examples of topological spaces are: (i) Reisel, Robert B., *Elementary Theory of Metric Spaces: A Course in Constructing Mathematical Proofs,* Springer-Verlag, New York, 1982 and (ii) Copson, E. T., *Metric Spaces,* Cambridge at the University Press, Cambridge, 1968.

Concerning Appendix C

An elementary book about general topological spaces is Moore, Theral O., *Elementary General Topology,* Prentice-Hall, Englewood Cliffs, New Jersey, 1964. It uses open sets for the starting point. For an approach that begins with neighborhoods see the first three chapters of: Wallace, Andrew H., *An Introduction to Algebraic Topology,* Pergamon Press, New York, 1957. In this latter book, neighborhoods are not necessarily open; many books on topology define neighborhood in such a way that all neighborhoods are open.

Concerning Appendix D

The definition of Riemann-Stieltjes integral used in this appendix has been taken from: Apostol, Tom M., *Mathematical Analysis, A Modern Approach to Calculus,* Addison-Wesley, Reading, Massachusetts, 1964, a book which is a good reference for analysis at the pre-measure-theoretic level.

Concerning Appendix E

It is possible to regard log as a bijective function having nice properties provided one uses an appropriate domain and an appropriate target. Take the target equal to $\mathbb{C}$. To construct the domain start with a copy of $\mathbb{C} \setminus \{0\}$ corresponding to each member of $\mathbb{Z}$, using the members of $\mathbb{Z}$ as subscripts in order to distinguish the copies. Cut each of these sets along its negative real axis. For each $n \in \mathbb{Z}$, attach the second quadrant of $(\mathbb{C} \setminus \{0\})_n$ to the third quadrant of $(\mathbb{C} \setminus \{0\})_{n+1}$ along the cuts. The object thus constructed is called a *Riemann surface*.

To define log on the Riemann surface just described, consider a point (z, n) on it, where z is a complex number different from 0 and n identifies the copy of $\mathbb{C} \setminus \{0\}$ on which it lies, regarding z as a second-quadrant complex number in case it is a negative real number. Set $\log(z, n)$ equal to that version of $\log z$ whose coefficient of i lies in the interval $(2\pi n - \pi, 2\pi n + \pi]$. It can be checked that log is a continuous one-to-one function from the Riemann surface onto $\mathbb{C}$ satisfying $\frac{d}{d(z,n)} \log(z, n) = z^{-1}$.

Index